Introduction to the METHODOLOGY OF SWITCHING CIRCUITS

George J. Klir

School of Advanced Technology
State University of New York at Binghamton

D. VAN NOSTRAND COMPANY

NEW YORK CINCINNATI
TORONTO LONDON MELBOURNE

Books by same author:

Cybernetic Modelling
An Approach to General Systems Theory

D. Van Nostrand Company Regional Offices:
New York Cincinnati Millbrae

D. Van Nostrand Company International Offices:
London Toronto Melbourne

Library of Congress Catalog Card Number: 72-181095
ISBN: 0-442-24463-0

Manufactured in the United States of America

Published by D. Van Nostrand Company
450 West 33rd Street, New York, N.Y. 10001

Published simultaneously in Canada by Van Nostrand Reinhold Ltd.

10 9 8 7 6 5 4 3 2 1

To Professor Antonin Svoboda
My best teacher and friend

Preface

The theory of switching circuits, which originated in the late thirties and displayed an extremely fast development in the fifties and the early sixties, may be considered as a well established field at present. Although a great amount of research work remains to be done in this field, basic principles, tools, and methods have already been elaborated.

A course in the theory of switching circuits has recently become a part of the typical electrical engineering and computer or system science curricula. It is usually offered to juniors or seniors. A one- or two-semester graduate course in switching circuits for computer science majors sometimes supplements the basic one- or two-semester undergraduate course.

This book is intended as a textbook for a senior or a graduate one- or two-semester course in switching circuits. Its ultimate goal is to present a comprehensive and unified methodology of switching circuits that is as independent of the state of technology as possible. Different kinds of logic and pseudo-logic relations represent a tool for the unification of the methodology of switching circuits in a manner similar to the way in which differential equations have unified the methodology of electric circuits.

For a two-semester course, Chapters 1 through 6 are suitable for the first semester, and Chapters 7 through 10 for the second semester. The book is virtually self-contained although a basic knowledge of some physical principles is helpful in Chapter 5.

Almost all sections are supplemented by exercises, and a set of comprehensive exercises accompanies each chapter. Answers to selected exercises (denoted by asterisks) are given in Appendix F. To avoid interruptions in the text, References, together with Reference Notations, are placed at the end of each chapter. A guide to the Literature (Appendix B) may be helpful in a seminar work.

Almost all topics concerning the methodology of switching circuits are covered in the book. A brief survey is given in the Guide to the Literature for each of a few topics that are missing. Some of the topics I cover are not treated in contemporary textbooks on switching theory. These include the theory of logic and pseudo-logic relations and their applications (Chapter 4), the multimodel approach to structure synthesis of sequential switching circuits (Chapter 8), and the structure synthesis of probabilistic switching circuits (Chapter 10).

The book has evolved from a set of class notes which I used at the Engineering Department of the University of California at Los Angeles in 1966–1968, and at the School of Advanced Technology, State University of New York at Binghamton since 1969. My deepest gratitude goes to the students of both these universities who, although often suffering from various imperfections of the class notes, gave me great encouragement as well as many specific comments. In particular, I am indebted to Miss Lucy Gabriel and Mr. Siraj Islam for their very valuable help. Finally, I thank my wife Milena for her patience and her help with illustrations.

BINGHAMTON, NEW YORK
MAY 1972

George J. Klir

NOTE TO THE READER

Answers to exercises that are marked with an asterisk are given in Appendix F. All mathematical symbols which are used in this book are defined in Appendix C.

The following abbreviations are used for those periodicals and collections of papers which are frequently referred to:

AAT	*Applied Automata Theory*, edited by J. T. Tou, Academic Press, New York, 1968.
AFIPS	*AFIPS (American Federation of Information Processing Societies) Conference Proceedings*, Thomson Book Co., Washington, D. C.
AMM	*American Mathematical Monthly.*
ARC	*Automation and Remote Control.*
AS	*Automata Studies*, edited by C. E. Shannon and J. McCarthy, Princeton Univ. Press, Princeton, N. J., 1956.
BSTJ	*The Bell System Technical Journal.*
CACM	*Communications of the Association for Computing Machinery.*
CD	*Computer Design.*
CRSA	*IEEE (or AIEE*, until 1962) *Conference Record* (or *Proceedings*, until 1964) *on Switching and Automata Theory* (or *Switching Circuit Theory and Logical Design*, until 1965).
IBMJ	*IBM Journal of Research and Development.*
IC	*Information and Control.*
IPM	*Information Processing Machines*, Academia (Publishing House of the Czechoslovak Academy of Sciences), Prague, Czechoslovakia.
JACM	*Journal of the Association for Computing Machinery.*
JFI	*Journal of the Franklin Institute.*
KYB	*Kybernetika.*
PC	*Problems of Cybernetics.*
PHU	*Proceedings of an International Symposium on the Theory of Switching*, Harvard University Press, Cambridge, Mass., 1959.
RDST	*Recent Developments in Switching Theory*, edited by A. Mukhopadhyay, Academic Press, New York, 1971.

SM	*Sequential Machines: Selected Papers*, edited by E. F. Moore, Addison-Wesley, Reading, Mass., 1964.
ST	*Structure Theory of Switching Circuits*, edited by M. A. Gavrilov, in Russian, Izdatelstvo Akademii Nauk U.S.S.R., Moscow, 1963.
TAMS	*Transactions of the American Mathematical Society.*
TC	*IEEE*(or *IRE*, until 1962) *Transactions on Computers* (or *on Electronic Computers*, until 1967.)
TE	*IEEE* (or *IRE*, until 1962) *Transactions on Education.*
TS	*IEEE Transactions on Systems Science and Cybernetics.*

This convention is used in all references: Volume numbers are shown in boldface; issue numbers are in parentheses following the volume numbers. Page numbers follow issue numbers.

CONTENTS

Preface vii

Note to the Reader ix

PART I Concepts, Principles, Tools

Chapter 1 Introduction 3

1.1. The System 3
1.2. Classification of Problems 13
1.3. The Switching Circuit 15
1.4. Classification of Switching Circuits 18
Reference Notations 19
References 20

Chapter 2 Fundamental Concepts 21

2.1. Logic Variables 21
2.2. Logic Functions 31
2.3. Logic Expressions 37
2.4. Polish String Notation 45
2.5. Nondegenerate Logic Functions 48
2.6. Equivalence Classes of Logic Functions 51
2.7. Complete Sets of Logic Functions 53
2.8. Logic Algebras 65
2.9. Boolean Algebra 66
2.10. Forms of Expressing Logic Functions 70
2.11. Mutual Transformations between Logic Function Forms 75
2.12. Chart Operations 80
Comprehensive Exercises 81
Reference Notations 84
References 85

Chapter 3 Minimization of Boolean Expressions 87

3.1. Discussion of the Problem 87
3.2. The Svoboda Method 88
3.3. Determination of Prime Implicants 100
3.4. The Problem of Minimal Covering 106
3.5. Minimization of Groups of Boolean Expressions 113
3.6. Multiple Level and Absolute Minimization 116
Comprehensive Exercises 121
Reference Notations 121
References 122

Chapter 4 Solution of Logic Relations 126

4.1. Formulation of the Problem 126
4.2. Transformation of a Given Set of Equations to the Standard Form 128
4.3. Solution of Logic Equations in Standard form 131
4.4. Recapitulation and Examples 134
4.5. An Alternative Processing of Discriminant 139
4.6. Possible Generalizations 143
4.7. Pseudo-Logic Relations 149
4.8. Applications: Some Examples 154
Comprehensive Exercises 160
Reference Notations 162
References 163

Chapter 5 Logic Description of Physical Systems 165

5.1. Introduction 165
5.2. Relays 166
5.3. Diodes 170
5.4. Electronic Tubes 172
5.5. Transistors 179
5.6. Magnetic Cores 183
5.7. Cryotrons 186
5.8. Pneumatic Systems 189
5.9. Some General Comments 195
Comprehensive Exercises 196
Reference Notations 196
References 196

PART II Problems and Methods

Chapter 6 Deterministic Memoryless Switching Circuits: Analysis and Synthesis 199

6.1. Analysis 199
6.2. Synthesis: Discussion of the Problem 214
6.3. General Approach to Synthesis 218
6.4. Circuits Based on Standard Boolean Forms 220
6.5. Cascade Circuits 225
6.6. Symmetric Circuits 231
6.7. Circuits with Branch-Type Elements 239
6.8. Universal Logic Primitives and Modules 252
Comprehensive Exercises 257
Reference Notations 259
References 260

Chapter 7 Deterministic Sequential Switching Circuits: Abstract Theory 264

7.1. Introduction 264
7.2. Representations of Output and Transition Functions 268
7.3. State Minimization: Completely Specified Machines 274
7.4. State Minimization: Incompletely Specified Machines 283
7.5. Abstract Synthesis: A Discussion of the Problem 304
7.6. Regular Expressions 310
Comprehensive Exercises 323
Reference Notations 324
References 325

Chapter 8 Synchronous Deterministic Sequential Switching Circuits: Structure Synthesis 327

8.1. Discussion of the Problem 327
8.2. The Concept of Synchronization 329
8.3. Memory Elements 332

8.4. Models of Sequential Switching Circuits 335
8.5. Finite-State Models 340
8.6. Finite-Memory and Combined Models 345
8.7. Svoboda's Methodical Approach 362
8.8. A Rough Evaluation of Models 369
8.9. Decomposition of Sequential Machines 374
8.10. State Assignment 385
Comprehensive Exercises 401
Reference Notations 402
References 403

Chapter 9 Asynchronous Deterministic Sequential Switching Circuits: Structure Synthesis 406

9.1. Peculiarities of Asynchronous Circuits 406
9.2. Basic Concepts 409
9.3. State Assignment 413
9.4. Hazards 426
9.5. Multiple-Input Changes 436
9.6. Summary and Examples 442
Comprehensive Exercises 453
Reference Notations 455
References 455

Chapter 10 Probabilistic Switching Circuits 458

10.1. Introduction 458
10.2. Memoryless Circuits 459
10.3. Sequential Circuits 471
10.4. Svoboda's Approach 477
10.5. State Minimization 481
Comprehensive Exercises 492
Reference Notations 492
References 493

PART III Appendixes

Appendix A. Polynomial Representations of Numbers 497
Appendix B. A Guide to the Literature 501

Appendix C. Glossary of Symbols **514**
Appendix D. Catalog of Functions Represented by the WOS Module **517**
Appendix E. Representants of PN-Equivalence Classes of Logic Functions for $n \leq 4$ **519**
Appendix F. Answers to Selected Exercises **530**
Appendix G. Selected Computer Programs **543**

Index **563**

Introduction to the METHODOLOGY OF SWITCHING CIRCUITS

PART I

Concepts, Principles, Tools

1

INTRODUCTION

Before approaching problems of switching circuits proper, let us first devote our attention to a few important concepts of wider significance. Based on these concepts we shall be able to consider the problem of switching circuits from a more general point of view and with broader implications.

1.1 THE SYSTEM

When studying nature we always confine ourselves to studying that part of it which, at the time, is of interest to us. This relevant part of nature we shall call the *object* and the remaining part, the *environment* of the object.

Our facilities do not permit us to study objects in all their complexity. We therefore always limit ourselves to a distinct point of view from which to examine the given object. On the basis of this point of view we define a *system* on the given object. This is done by defining the concrete *quantities* which interest us in the object and the *spatio-temporal resolution level* at which we want to examine these quantities, the result being a determination of the *values* (or classes of values) of the defined quantities as well as those of the time intervals which we want to consider. Then we try to find time-invariant relations between these quantities and all the properties determining these relations, which are within our sphere of interest. In general, the system need not necessarily be tied to an object but may be based solely on the mentioned definition.

When discrete time instances (or time intervals) are considered instead of continuous temporal flux and only two values (or two classes of values) are distinguished for each quantity under consideration, the system is called the *switching circuit*. Discrete time is usually represented by a sequence of integers.

Now let us present a very simple example which we shall use to demonstrate what we mean by the concept of "system." We shall then use this example to illustrate the other general concepts which we are going to introduce in the course of our exposition.

Let us imagine a room lit by an incandescent lamp supplied from the local source of electric power. Let the lamp be controlled by any one of three switches mounted in different parts of the room and placed so that their positions can be visually ascertained from a single spot. The control of the lamp is governed by the following rules:

1. If the source is live and the lamp is not alight, it can be lit by changing the position of any one of the three switches.
2. If the source is live and the lamp is alight, it can be extinguished by changing the position of any one of the three switches.
3. If the source is not live, the lamp does not light.

We shall assume that each switch has two positions, i.e., closed (c) and open (o). The position of a particular switch can thus be regarded as a quantity which can acquire two values, symbolically denoted by c and o. We shall also assume that the voltage of the source is either zero, or is sufficient to light the lamp. We are therefore again concerned with a quantity capable of acquiring two values, denoted by p (voltage present) and z (zero voltage). We can be informed of the presence of the voltage, for instance, by a neon lamp connected directly across the source terminals.

Our simple system thus comprises altogether five basic quantities: 1st quantity—position of 1st switch; 2nd quantity—position of 2nd switch; 3rd quantity—position of 3rd switch; 4th quantity—state of the lamp from the viewpoint of its momentary luminosity; and 5th quantity—state of the source from the viewpoint of the momentary presence of voltage.

In our case, the spatial resolution level consists of the assumption that the individual quantities are regarded as two-valued: in the switches we are interested only in the positions o and c, in the lamp only whether it is alight (in state a) or dark (in state d), and in the source only whether voltage is present (state p) or not (state z).

The temporal resolution level can be defined as follows: we are interested only in those instants of time when the quantities under consideration attain a steady state at the given spatial resolution level, after a change in some of the quantities, assuming the time interval between two changes to be larger than a distinct constant Δt (in our case, Δt might correspond to the time required by the lamp filament to get hot).

If we want to amplify our example, we must return for a moment to our general exposition.

The ensemble of the variations in time of all the quantities under consideration within some specific time interval will be called the *activity* of the system.

In a given activity, it is possible to distinguish three different kinds of time-invariant relations based on samples of a certain pattern, i.e., containing certain instantaneous and/or past and/or future values of the quantities for every time instant:

1. A relation which is proper to the given quantities at the resolution level, i.e., which is satisfied over the entire time interval of every possible particular activity containing the quantities at the resolution level. Let us call this relation *absolute.*

2. A relation which is satisfied anywhere within a particular activity containing the quantities at the resolution level. We shall call this *relative.*

3. Relations which apply only within some shorter time intervals of a particular activity. These we shall call *local.*

A particular time-invariant relation specified for a set of quantities and a resolution level, and based on samples of a certain pattern, will be called the *behavior* of the corresponding system. It will be useful to distinguish three basic kinds of behavior:

1. *Permanent (real) behavior*—the absolute relation, or in other words, the set of all local relations.

2. *Relatively permanent (known) behavior*—the set of all local relations of a particular activity or, according to the introduced terminology, the relative relation.

3. *Temporary behavior*—the local relation corresponding to a distinct section of a particular activity.

It should be remembered that the permanent behavior is known only in some cases, e.g., when it is directly given in some engineering system. In many instances of the experimental investigations of systems, however, it is impossible to decide whether the behavior found in some observed activity is permanent or only relatively permanent or, in other words, whether the pertinent relatively permanent behavior is equal to the permanent behavior of the system under investigation at the given resolution level.

In our system, comprising the switches, the lamp, and the source, the activity might, for instance, assume the form shown in Table 1.1.

The relatively permanent behavior, which can be ascertained in the activity presented in Table 1.1, and which represents the behavior of our system within

TABLE 1.1 Example of an Activity

nstant of time	0	1	2	3	4	5	6	7	8	9	10	11	12	13	14	15	16	17	18	19	20	21	22	23	24	25	26	27	28	29	30	31
Position of 1st switch	*o*	*c*	*o*	*o*	*o*	*o*	*o*	*c*	*c*	*o*	*o*	*c*	*c*	*c*	*c*	*c*	*c*	*c*	*c*	*o*	*o*	*o*	*c*	*o*	*o*	*c*	*c*	*c*	*c*	*c*	*o*	*o*
Position of 2nd switch	*o*	*o*	*o*	*c*	*o*	*o*	*o*	*o*	*c*	*c*	*c*	*c*	*o*	*c*	*c*	*o*	*o*	*o*	*c*	*c*	*o*	*o*	*o*	*o*	*c*	*c*	*c*	*c*	*c*	*o*	*o*	*o*
Position of 3rd switch	*o*	*o*	*o*	*o*	*o*	*c*	*o*	*o*	*o*	*o*	*c*	*c*	*c*	*c*	*o*	*o*	*o*	*c*	*c*	*c*	*c*	*o*	*o*	*o*	*o*	*o*	*c*	*c*	*o*	*o*	*o*	*c*
State of the lamp	*d*	*a*	*d*	*a*	*d*	*a*	*d*	*a*	*d*	*a*	*d*	*a*	*d*	*a*	*d*	*a*	*d*	*d*	*d*	*d*	*d*	*d*	*d*	*d*	*d*	*d*	*d*	*a*	*d*	*a*	*d*	*a*
State of the mains	*p*	*p*	*p*	*p*	*p*	*p*	*p*	*p*	*p*	*p*	*p*	*p*	*p*	*p*	*p*	*p*	*z*	*z*	*z*	*z*	*z*	*z*	*z*	*z*	*z*	*z*	*z*	*p*	*p*	*p*	*p*	*p*

the range of the given activity, can be expressed, for instance, as follows: The lamp is alight (*a*) when the source voltage is present (*p*) and an odd number of switches is simultaneously in position *c*. In all other cases the lamp is dark (*d*). Temporary behavior can also be found here. For example, within the time interval 0 to 15 the state of the lamp depends only on the state of the switches, whereas within the time interval from 16 to 26 it is constant without regard to the switch positions.

So as to exhibit a distinct behavior, a system must possess, as just mentioned, certain properties. Let us call them summarily the *organization* of the system. Since the behavior of a system may change, we must suppose that its organization can also change. It will be advantageous to define in the organization of the system its constant and its variable part. The constant part of the organization of a system will be called the *structure* of the system, its variable part the *program* of the system. It will be seen that the structure of the system produces only the absolute time invariants in the given activity, whereas the program can produce the local time invariants as well.

In our example, the organization of the system is defined as follows:

1. The system comprises the following elements: the lamp, the terminals carrying the source, and the change-over contacts of the individual switches.
2. Each element can acquire two states: the lamp can be alight or dark, the source terminals are live or dead, each switch is either in position *c* or *o*.
3. The elements of the system have the following absolute behavior: the lamp lights only when a voltage is applied to its filament, the change-over contact of a given switch connects one pair of conductors when the switch is in position *o*, and the other pair of conductors when the switch is in position *c*.
4. The elements of the system are interconnected, for instance, in the manner illustrated in Figure 1.1, where all the contacts are drawn in position *o*.

A special case occurs when a classification of external quantities to *input* and *output quantities* is given. The behavior is then regarded as a time-invariant relation between the activity of the input quantities and that of the output quantities. This case is of the greatest importance in engineering. We shall, therefore, confine ourselves in this book to switching circuits with known classification of their external quantities.

In our example we are concerned with a typical system having clearly defined input and output quantities. Its input quantities are the switch positions and the state of the source; its output quantity is the state of the lamp.

The set of instantaneous values of the input quantities of the system is regarded as the *stimulus* (excitation) whereby the environment acts upon the system. Similarly, the set of instantaneous values of the output quantities is regarded as the *response* of the system to the corresponding stimulus. Let us at the same time introduce a terminological convention, according to which one partial input is assigned to every input quantity, and one partial output to

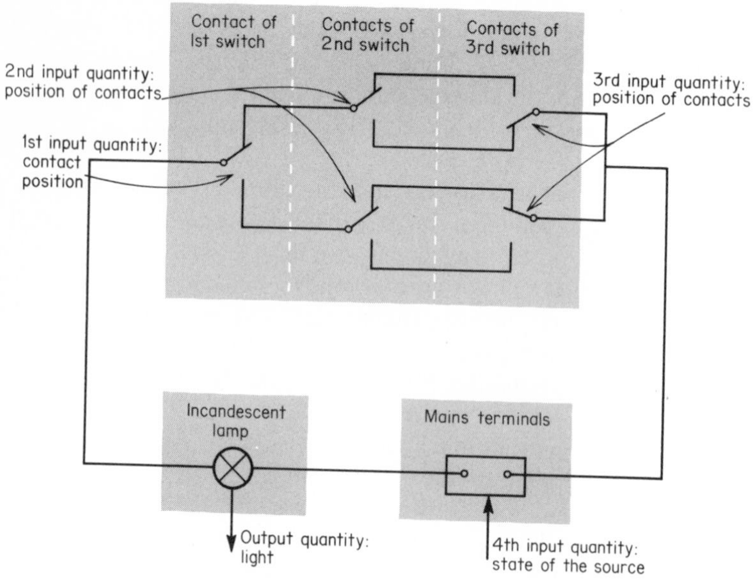

Figure 1.1 Example of a system.

every output quantity. For the sake of simplicity, the instantaneous value of a single input quantity will be called a *partial stimulus*, the instantaneous value of a single output quantity being called a *partial response*.

The basic diagram of the relation between a system and its environment is shown in Figure 1.2.

As has already been said, the behavior of a system depends on its organization, i.e., on its structure and program. When investigating the structure we shall assume that the system is made up of partial subsystems or elements of the system, each of which has a distinct behavior. The set of all elements of a system is usually called the *universe of discourse* of the system. We shall further assume that the elements of the system may be coupled in a certain manner with the environment of the system and among each other. The set of all these *couplings*, together with the set of the absolute behaviors of the individual elements, forms the structure of universe and couplings of the system (*UC-structure*).

For instance, the system illustrated in Figure 1.1 consists of seven elements and ten couplings between these elements, as shown in Figure 1.3. Comparing Figure 1.3 with Figure 1.1, we see that the input quantities act upon elements 1 through 6, whereas the output quantity originates in element 7.

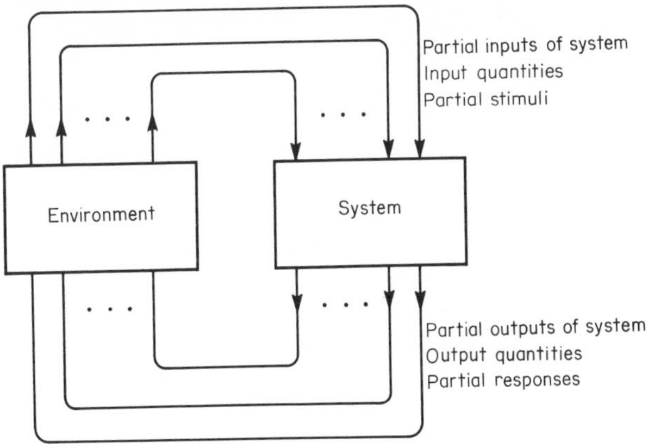

Figure 1.2 Diagram of the basic relation between a system and its environment.

Figure 1.4 shows the diagram of our system as related to its environment, i.e., according to the pattern of Figure 1.2. It can be seen that the system has four partial inputs and one partial output. Altogether there are seven couplings between the elements of the system and its environment.

Figure 1.3 also indicates the internal quantities pertaining to the individual couplings between the elements. In this case the internal quantities are the electric potentials of the pertinent conductors. These are also regarded as two-valued.

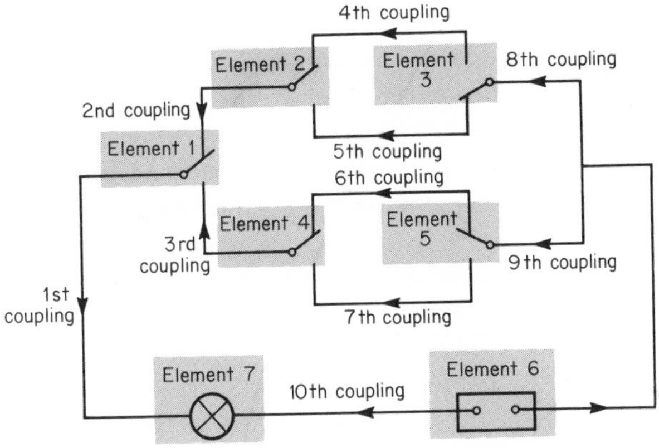

Figure 1.3 Elements and couplings in the system in Figure 1.1.

Element 1 (according to Figure 1.3) can be seen to have three partial inputs (one external and two internal ones) and one partial output (see Figure 1.5). Its absolute behavior is as follows:

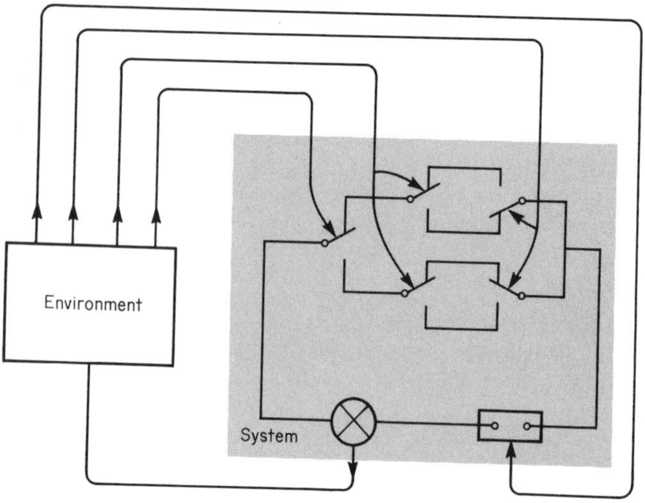

Figure 1.4 Relation between the system shown in Figure 1.1 and its environment.

(a) If the value of the quantity in the third partial input is equal to *o*, then the value of the quantity in the output (the response of the element) is the same as the value of the quantity in the first input.

(b) If the value of the quantity in the third input is equal to *c*, then the value of the quantity in the output is the same as the value of the quantity in the second input.

As we have already mentioned, the behavior of systems with known classification of external quantities is regarded as the dependence of the responses of the system on the stimuli which enter the system from the environment. This dependence need not be unique. This is because the instantaneous response of the system need not be determined by only the instantaneous stimulus, but

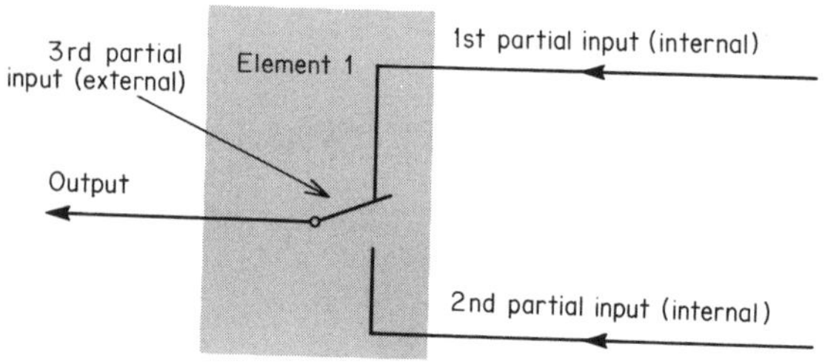

Figure 1.5 Illustration of partial inputs and outputs of element 1 of the system shown in Figure 1.3.

also by the previous input activity. This implies, however, that the required portion of the past input activity must be remembered by the system in the form of some internal quantities. Let us call the set of the instantaneous values of these internal quantities the *internal state* of the system. The internal state of the system together with the pertinent stimulus and response is then usually termed the *state* of the system.

With the aid of the internal state of the system it is possible to formulate an important property of every deterministic system: *The response of a deterministic system always depends uniquely on its internal state and on the stimulus.*

Let us denote the stimulus at time t by the symbol $\mathbf{I}^t$, the response at time t by the symbol $\mathbf{J}^t$, and the internal state of the system at time t (or $t + \Delta t$) by the symbol $\mathbf{K}^t$ (or $\mathbf{K}^{t+\Delta t}$ respectively). The behavior of the system can then be expressed in the following general form:

$$\mathbf{J}^t = \mathbf{f}(\mathbf{I}^t, \mathbf{K}^t) \tag{1.1}$$

$$\mathbf{K}^{t+\Delta t} = \mathbf{g}(\mathbf{I}^t, \mathbf{K}^t) \tag{1.2}$$

where $\mathbf{f}$ and $\mathbf{g}$ are given functions. It should be realized, however, that the variables $\mathbf{I}^t$, $\mathbf{J}^t$, $\mathbf{K}^t$ and $\mathbf{K}^{t+\Delta t}$ are usually vectors. As a rule, $\mathbf{f}$ and $\mathbf{g}$ are therefore vector functions, each of which can be replaced by the appropriate number of scalar functions $f_1, f_2, \ldots$, and $g_1, g_2, \ldots$.

If, for a given pair $\mathbf{I}^t$ and $\mathbf{K}^t$ we have

$$\mathbf{K}^{t+\Delta t} = \mathbf{K}^t \tag{1.3}$$

the system is said to be in the *stable state*. It persists in this state until the stimulus $\mathbf{I}^t$ is suitably changed. If equation (1.3) does not apply, the system is in the *unstable state*. The symbol Δt in equation (1.2) determines the time required for a change of the internal state regarded at a given resolution level. If the internal state and response of the system vary continuously with time (where $\mathbf{f}$ and $\mathbf{g}$ are continuous functions), we obtain a limiting case where $\Delta t \to 0$. Systems with this property are called *continuous*; all other systems are called *discrete*.

A direct change in the internal state of the system, i.e., a change when no intermediate state exists between the initial internal state $\mathbf{K}^t$ and the final internal state $\mathbf{K}^{t+\Delta t}$, is usually called a direct *transition*. A change from one state to another with some intermediate states is then called an indirect transition.

In the case of probabilistic systems, the relations between $(\mathbf{I}^t, \mathbf{K}^t)$ and $\mathbf{J}^t$ or $\mathbf{K}^{t+\Delta t}$, respectively, are one-to-many. In other words, they are not represented by functions (1.1) and (1.2). Elements of these relations are associated with conditional probabilities $\mathscr{P}(\mathbf{J}^t \,|\, (\mathbf{I}^t, \mathbf{K}^t))$ or $\mathscr{P}(\mathbf{K}^{t+\Delta t} \,|\, (\mathbf{I}^t, \mathbf{K}^t))$, respectively. The set of these probabilities for each of the relations constitutes a stochastic matrix.

Let us return now to the concept of the program. We defined roughly the program as being the variable part of the organization of the system. At the same time we defined a portion of the fixed part of the organization (the structure of universe of discourse and couplings or the UC-structure) as the set of elements, their permanent behaviors, and the set of couplings between the elements. In addition to the UC-structure, the organization comprises the variable contents of individual couplings, i.e., the values of the corresponding quantities (in this case both the external and internal ones), and the admissible changes of these values (transitions) determined by the structure of the system and also affected by the environment. Obviously, these properties of the organization of the system must be closely related to the program.

By the help of the concept of state, the program at any time instant can be defined as the instantaneous state of the system, a set of some other states of the system, and a set of transitions from the instantaneous state to the states under consideration in time. Thus, the program is variable from the viewpoint of the instantaneous state of the system, and different programs exist for different subsets of the set of all states of the system. It seems to be useful to distinguish three kinds of programs:

1. *Complete program* (or simply program)—instantaneous state together with the set of all other states of the system, and the set of all transitions (direct and indirect) from the instantaneous state to all states of the system in time.
2. *Subprogram*—instantaneous state together with a nonempty subset of the set of all other states of the system, and a nonempty subset of the set of all transitions (direct and indirect) from the instantaneous state to all the states under consideration in time.
3. *Instantaneous program*—instantaneous state together with the direct transitions from this state.

Being generally variable, the complete program contains a constant part, namely, the complete set of states and the set of all direct transitions between the states. This constant part should be, in accordance with our classification of the organization of the system, included in the structure. Let this part of the structure be called the state-transition structure (*ST-structure*). Thus, the ST-structure is defined as the complete set of states together with the complete set of direct transitions between the states.

Clearly, the ST-structure is represented either by the functions (1.1) and (1.2) or, in case of probabilistic systems, by the corresponding one-to-many relations supplemented by stochastic matrixes.

Exercises to Section 1.1

1.1-1 Find one system from your field of interest which can be defined by:
(a) a UC-structure.
(b) a ST-structure.

(c) a behavior.
(d) a set of quantities and a spatio-temporal resolution level.
(e) an activity.

1.1-2 Determine a UC-structure of a switching circuit similar to that discussed in Section 1.1 except that it contains four switches instead of three. If voltage is present between the source terminals, the state of the lamp can be changed at any time instant by each of the switches.

*1.1-3 Which of the following activities, each containing a single two-valued variable, x, are consistent with the behavior $(x^t - x^{t-1})^2 = 1$?
(a) 1 0 1 1 0 1 1 0 1 0 0 0 1.
(b) 0 1 0 1 0 1 0 1 0 1 0 1 0.
(c) 1 0 1 0 1 0 1 0 1 0 1 0 1.
(d) 0 0 1 1 1 0 1 1 1 0 1 1 1.

*1.1-4 Find time-invariant relations containing x^t, x^{t-1}, and x^{t-2} for the following activities, each of which contains one variable:
(a) 0 1 1 0 1 1 0 1 1 0 1 1 0 1···
(b) 1 0 1 1 0 1 1 0 1 1 0 1 1 0···
(c) 1 1 0 1 1 0 1 1 0 1 1 0 1 1···.

1.2 CLASSIFICATION OF PROBLEMS

When dealing with systems, the three fundamental classes of problems encountered most frequently are: (a) *analysis* of the system; (b) *synthesis* of the system; and (c) *investigation of* the system regarded as a "*black box.*"

The analysis of a system can be formulated as follows: The UC-structure of the system is given and we are to find the corresponding behavior. It is known that the analysis of a system with given input and output quantities has a unique solution provided that the given UC-structure does not contain any inconsistencies.

The synthesis of a system is a problem that, to a certain degree, is the reverse of analysis. It can be formulated as follows: The behavior (or sometimes only the activity) of the system and a set of types of its elements are given (the type being here understood to mean elements having the same behavior). We are to find a structure of the system which would realize the prescribed behavior and would contain only the permitted types of elements.

If an insufficient set of element types is given, the synthesis of the system cannot be accomplished. In all other cases the problem has several solutions, i.e., a given behavior can be realized by different organizations, even if their UC-structures consist of the same types of elements. It is therefore customary to prescribe further requirements concerning the organization, e.g., requirements relating to minimum costs, maximum reliability, etc. The task of synthesis is to find the structure which, out of all the possible ones, satisfies the given requirements in the best possible manner. The proposed structure is then analyzed to make sure whether the synthesis of the system was carried out correctly.

Let us consider now the problem of the "black box." The term "black box" has been adopted for every system whose organization and behavior are both unknown but which, nevertheless, can be experimented with and whose activity can be recorded. It is then possible to seek out time invariants in the recorded activity of the "black box" and thus to obtain knowledge concerning the behavior of the system. However, as long as we have no data concerning the organization of the "black box," we can never acquire full knowledge of its behavior with absolute certainty. If, on the other hand, we are in possession of some data concerning its organization (we know, for instance, the number of its states) we can sometimes and under certain assumptions (with a suitably chosen experiment) acquire complete knowledge of the behavior of the "black box." With the behavior known, we can then evolve several hypotheses concerning the probable organization of the "black box." Since the deduction of these hypotheses involves the search for possible structures fitting the given behavior, we are actually concerned with a synthesis of the system. A verification of the individual hypotheses would, of course, be possible only by "looking into" the "black box."

For clarity's sake we present in Figure 1.6 a schematic illustration of the procedures used for solving the aforesaid fundamental problems concerning the system.

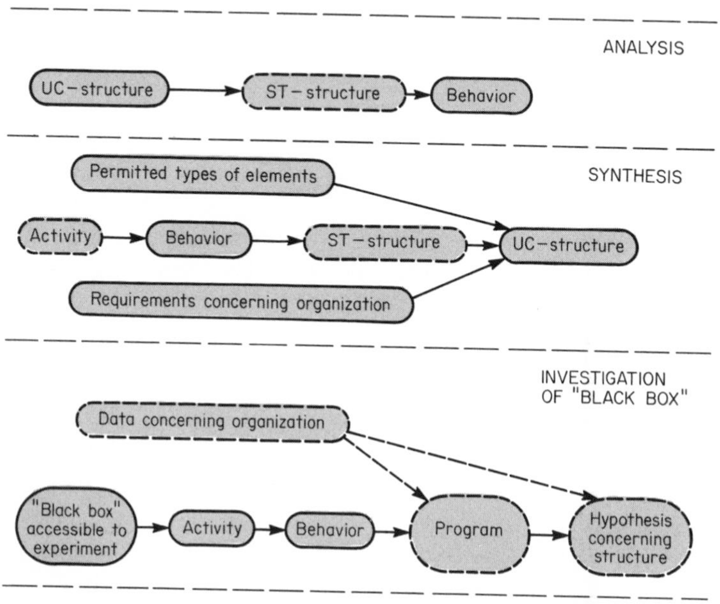

Figure 1.6 A survey of fundamental problems concerning systems.

In addition, let us note that the type of problem to be solved affects the definition of the system. This is because it is natural to define the system by the entity from which we start when solving the given problem. In the case of analysis, the system will therefore be defined by means of its UC-structure. In the case of synthesis with the aid of its behavior (or activity), and when solving "black box" problems the system will be defined by a set of quantities and by the spatio-temporal resolution level.

1.3 THE SWITCHING CIRCUIT

The *switching circuit* is a special kind of physical system, characterized by the fact that every quantity of its activity is considered only at certain discrete time instances and can acquire only values belonging to either of two disjoint sets. In principle, moreover, the actual values of a given quantity are of no importance. The question is only into which of the possible two sets the relevant values belong.

A system whose quantities can assume only two different (so-called *ideal* or *perfect*) *values* is considered as an ideal switching circuit. However, this cannot be achieved in practice with an actual (non-ideal or imperfect) switching circuit. This is because every quantity depends on many factors (e.g., the tolerances and changes in the properties of the components, the temperature, humidity, magnetic field, etc.) which cause values of the quantity to fluctuate in the neighborhood of their ideal values. Moreover, the values which are closer to the ideal value may occur more frequently, and vice versa. We are usually concerned with a normal (Gaussian) statistical distribution which, however, need not be the same for both ideal values.

Figure 1.7 illustrates an example of the probability density function $p(x)$ of a random quantity ξ which may occur in some switching circuit. The prob-

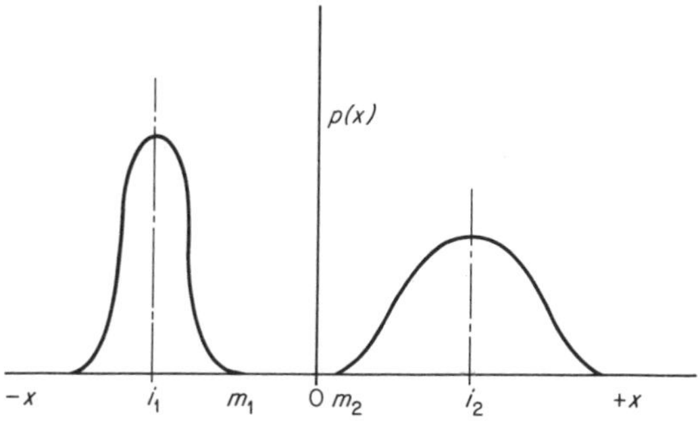

Figure 1.7 Example of the statistical distribution of the values of quantity ξ in a switching circuit.

ability that the quantity ξ has a value within the interval $\langle x, x + dx \rangle$ is approximately equal to $p(x)dx$ for small dx. The ideal values of this quantity are denoted by i_1 and i_2, respectively, and the values m_1 and m_2, where $m_1 \leq m_2$, mark the maximum admissible deviations from the ideal values i_1 and i_2. If all the values smaller than m_1 would be regarded in the corresponding switching circuit as the ideal value i_1, and all values larger than or equal to m_2 would be regarded as i_2, then the case of $m_1 = m_2$ would also be admissible. In practice, however, it is occasionally desirable that the interval $\langle m_1, m_2 \rangle$ be as large as possible.

Switching circuits are sometimes also designated as "*logic circuits*." This term has been adopted because every quantity, regarded as two valued, can be described at every instant by a simple proposition—the basic element of logic.

Any simple sentence which may be pronounced to be true or false is regarded as an *atomic proposition*. For instance, the sentence "Vienna is the capital of Austria" is a proposition, moreover a true one. Similarly the sentence "Vienna is the capital of France" is also an atomic proposition, but a false one. On the other hand, the sentence "Pass me that book" is not a proposition at all, because there is no sense in pronouncing it to be true or false.

In our case let us use, for instance, the proposition "The quantity ξ acquires the value i_1". If we know this proposition to be true, we also know that the quantity ξ acquires a value smaller than m_1. If the proposition is false, and we are aware of this, we then also know that the quantity ξ acquires values larger than or equal to m_2, because this quantity cannot acquire any value within the interval $\langle m_1, m_2 \rangle$. Of course, the opposite also applies: If the quantity ξ acquires a value smaller than m_1, the aforesaid proposition is true. If this quantity, however, acquires a value larger than or equal to m_2, our proposition is false.

Consequently, the value of any quantity of a switching circuit can present information on the truth of some proposition. We are concerned in this case with the smallest possible amount of information, which is usually called *elementary information.*

We regard as elementary information any statement concerning the truth or falsity of some proposition, in other words its *truth value*. The number of propositions whose instantaneous truth values can be determined by some statement, i.e., the number of elements of information in the statement, is then regarded as the amount of information which can be contained in this statement (provided that the probabilities of each proposition being true or false are equal).

The elementary information provides the basic quantum or unit for the objective assessment of the amount of information. The name adopted for this basic quantum or unit is the *bit*, an abbreviation of "*bi*nary digi*t*."

It is obvious that the basic quantum of information generally states which of the two possible cases has occurred. Any pair of symbols can be used to denote the two possibilities. As a rule, the symbols used are 0 and 1. This is also why the name for the unit of information (bit) was derived from the expression for the concept of "binary digit."

Every physical quantity which carries information is generally called a signal. Since every quantity of a switching circuit may carry elementary information, it will be terminologically correct to denominate the quantities of switching circuits as *elementary signals*. However, when there is no danger of confusion, the simple term "*signal*" is used to stand for elementary signal.

The elementary information says whether a given proposition is true. *Truth* is usually expressed by the symbol 1, *falsity* by the symbol 0. The elementary signal can also be characterized by two symbols, each of which expresses one ideal value of the corresponding quantity. For instance, in Figure 1.7 we use the symbols i_1 and i_2. It is clear that the signal values (i_1 and i_2) can be uniquely assigned to the information symbols (0 and 1) and vice versa. Such a one-to-one correspondence is called a *code*.

We can now enunciate a simple assertion which is of great importance to our further exposition: A quantity of a switching circuit becomes an elementary signal if and only if the values of elementary information are assigned to the ideal values of this quantity by a distinct code.

Atomic propositions can be combined in various ways to form *molecular propositions*, whose truth or falsity follows from the truth or falsity of the original atomic propositions. Each proposition is regarded as a variable which can acquire two values: 0 (falsity) or 1 (truth).

For instance, out of the atomic propositions "It rains" and "I am at home" we can make up various molecular propositions such as "I am at home and it rains," "Either I am at home or it rains," "I am at home when it rains," etc.

We see that in the given examples of molecular propositions there occur, on the one hand, the original atomic propositions, and on the other hand various *connectives* (and, either-or, when). If a connective of this kind uniquely assigns to the truth values of atomic propositions the truth values of the corresponding molecular propositions, we say that we are concerned with an extensional connection of propositions. These assignments in *extensional connections* can be written down in the form of so-called *truth tables* (see Table 1.2). It is important to realize here that two different molecular propositions may have the same values in the truth table. As an example let us quote the propositions "Either I am at home or it rains" and "I am at home and it does not rain or I am not at home and it rains."

Between the signals of switching circuits there exist relations similar to the extensional connections between propositions. The analogy of these relations is such that switching circuits can be used to model extensional relations between propositions, both with respect to behavior (modelling of the truth table, if the values of the atomic propositions are regarded as stimuli and the values

TABLE 1.2 Examples of Truth Tables

Truth Value of the Atomic Proposition		*Truth Value of the Molecular Propositions Built up by Connective*		
It Rains	I Am at Home	"And"	"Either-Or"	"When"
0	0	0	0	1
0	1	0	1	1
1	0	0	1	0
1	1	1	0	1

of the molecular propositions as responses) and with respect to structure (modelling of the atomic propositions and the appropriate connectives). Similarly, relations between signals of switching circuits can be modeled by the abstract resources of mathematical logic. We thus obtain an *abstract model* of the switching circuit.

Exercises to Section 1.3

1.3-1 Determine truth tables of all possible molecular propositions assigned to two atomic propositions. Find for each of them a proper connective.

1.3-2 Determine a truth table of the molecular proposition "Number N is even and either is divisible by 7 or contains digit 5." Identify ten smallest numbers for which the molecular proposition is true.

1.4 CLASSIFICATION OF SWITCHING CIRCUITS

From the viewpoint of their behavior, switching circuits are usually classified according to the following scheme:

Switching circuits—
- *deterministic*—
  - *memoryless (combinational)* —
    - *synchronous*
    - *asynchronous*
  - *sequential* —
    - *synchronous*
    - *asynchronous*
- *probabilistic*—
  - *memoryless* —
    - *synchronous*
    - *asynchronous*
  - *sequential* —
    - *synchronous*
    - *asynchronous*

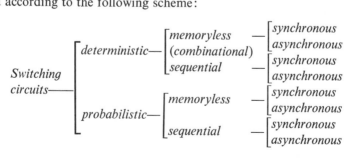

Memoryless (combinational) switching circuits are characterized by the fact that in them the values of the elementary input signals determine the values of all elementary output signals either uniquely (in case of deterministic systems) or with unique probabilities (in case of probabilistic systems). Since in these circuits the values of the output signals depend solely on the instantaneous values of the input signals, there is no need to employ elements capable of storing signals for later use.

In *sequential switching circuits* the output signals are determined not only by the instantaneous values of the elementary input signals, but also by a preceding sequence of input signals and an initial internal state. That means, of course, that essential components of the sequence must be "remembered" by the sequential switching circuit and used in the form of internal signals, together with the input signals, to produce the output signals. Thus, sequential switching circuits must contain elements capable of storing signals for later use. Such elements are usually referred to as *memory elements.*

Probabilistic switching circuits (both memoryless and sequential) must, clearly, contain some elements which generate values of some quantities with specified probabilities. Such elements are called *random generators.*

Switching circuits (both deterministic and probabilistic) are divided into synchronous and asynchronous ones according to the manner in which the signals are altered. In *synchronous circuits,* every change of both input and internal signals is controlled by *synchronizing pulses* (clock pulses) which ensure that the changes of the elementary signals occur at the same instant. In *asynchronous circuits* no precautions are taken to ensure the simultaneous change of elementary signals.

Reference Notations to Chapter 1

A comprehensive presentation of system concepts, principles, and problems, which are only outlined in this chapter, can be found in one of the author's former books [6]. This book is recommended as helpful supplementary reading, which is expected to give the student a broader insight into the methodology of switching circuits. Although this methodology has been developed as a tool for engineering design, primarily the design of digital computers, its applicability goes far beyond this area. This was demonstrated by Schoeffler et al. [10], Kempisty [4], Spiro [12], Greniewski [3], and many others [1,5,6,7,8]. Although the possibility of establishing a switching theory was recognized by M. Boda [2] as early as in the 19th century, the first important works on this subject were published by A. Nakashima [9] and C. E. Shannon [11] shortly before World War II.

References to Chapter 1

1. **Bernard, E, E.,** and **M. R. Kare,** eds., *Biological Prototypes and Synthetic Systems*, Plenum Press, New York, 1962.
2. **Boda, M.,** *Die Schaltungstheorie der Blockwerke*, Organ für die Fortschritte des Eisenbahnwesens, Vol. 35, Nos. 1–7, 1898.
3. **Greniewski, H.,** *Cybernetics Without Mathematics*, Pergamon Press, New York, 1960.
4. **Kempisty, M.,** *0–1 Cybernetic Models* (in Polish), Państwowe wydawnictwo naukowe, Warsaw, 1963.
5. **Klir, G. J.,** and **M. Valach,** *Cybernetic Modelling*, Van Nostrand, Princeton, N.J., 1967; Iliffe, London, 1967.
6. **Klir, G. J.,** *An Approach to General Systems Theory*, Van Nostrand Reinhold, New York, 1969.
7. **Klir, G. J.,** "On Organizations of Self-Organizing Systems," *Proc. Sixth Intern. Congr. on Cybernetics*, Namur, Belgium, 1970.
8. **Ledley, R. S.,** and **L. B. Lusted,** "Reasoning Foundation of Medical Diagnosis," *Science* **130** (3366) 9–21 (July 1959).
9. **Nakashima, A.,** "Theory of Relay Circuit Composition," *Nippon El. Com. Eng.*, No. 3, pp. 197–206 (1936).
10. **Schoeffler, J. D.,** et al., "Identification of Boolean Mathematical Models," in M. J. Mesarovic, ed., *Systems Theory and Biology*, Springer-Verlag, New York, 1968.
11. **Shannon, C. E.,** "A Symbolic Analysis of Relay and Switching Circuits," *Trans. AIEE* **57,** 713–723 (1938).
12. **Spiro, K.,** "A Logical Model of Differentiation and Generalization in Learning," *IPM*, No. 12, pp. 149–167, 1966.

2

FUNDAMENTAL CONCEPTS

2.1 LOGIC VARIABLES

A variable that can acquire only two distinct values will be called a *logic variable*. The values of logic variables will be denoted by the pair of digital symbols 0 and 1, in agreement with current practice. Although, for convenience, these symbols are frequently treated as numbers, zero and one, in various calculating procedures, they have essentially no quantitative meaning. To stress their formal rather than quantitative meaning, non-numerical symbols are sometimes preferred, e.g., T and F (true, false), O and I, T and ⊥. Then, however, the possibility of applying arithmetic operations to logic variables, which in some cases is a great advantage, is lost.

Let $x_1, x_2, \ldots, x_n$ be a set of n logic variables. If we substitute one of the two values 0 and 1 for each of the variables, we obtain a string of binary digits which can be interpreted as a binary number. This number will be said to express a distinct *state* of the logic variables. It is clear that with n logic variables it is possible to substitute the values 0 and 1 in altogether 2^n different ways. We thus obtain 2^n different states of n logic variables.

To get a unique representation of states of some logic variables, their values have to be written consistently in a certain order. It is a common practice that the values of $x_i (i = 1, 2, \ldots, n)$ are arranged in the decreasing order of the subscript i (sometimes called the identifier of the variables). Thus, the value of x_1 is considered as the least significant (rightmost) digit of the number and the value of x_n represents the most significant (leftmost) digit of the number. Hence, states of variables $x_1, x_2, \ldots, x_n$, are represented by binary numbers

$$\dot{x}_n \dot{x}_{n-1} \cdots \dot{x}_1$$

where $\dot{x}_i$ stands for a particular value (either 0 or 1) of variable $x_i (i = 1, 2, \ldots, n)$. Clearly, the value $\dot{x}_i$ is associated with a weight of 2^{i-1}.

It is an advantage in many cases to use the decimal equivalent for the binary number representing a state of some logic variables. This decimal equivalent is called the *state identifier* and will be denoted by s. Clearly,

$$s = \sum_{i=1}^{n} \dot{x}_i \cdot 2^{i-1} \tag{2.1}$$

is the formula for the state identifier of a state $\dot{x}_n \dot{x}_{n-1} \cdots \dot{x}_1$.

For instance, the identifier pertaining to the state 1101 is

$$s = 1 \times 2^0 + 0 \times 2^1 + 1 \times 2^2 + 1 \times 2^3 = 13$$

Notice that the digits of the binary number are applied in the order from the rightmost to the leftmost.

Two different states of n logic variables differ in at least one variable and in at most n variables. The actual number of variables in which the two states differ is called the *Hamming distance.*

For instance, the Hamming distance between the states 011010 and 111001 is 3 since they differ in three variables, namely, in the variables x_1, x_2, x_6.

Let the state identifiers of the states $\dot{x}_n \dot{x}_{n-1} \cdots \dot{x}_1$ and $\tilde{x}_n \tilde{x}_{n-1} \cdots \tilde{x}_1$ be denoted, respectively, $\dot{s}$ and $\tilde{s}$, and let the Hamming distance between these states be denoted $d(\dot{s}, \tilde{s})$. Then, the Hamming distance can be calculated by the following formula:

$$d(\dot{s}, \tilde{s}) = \sum_{i=1}^{n} |\dot{x}_i - \tilde{x}_i| \tag{2.2}$$

Other formulas can be used equally well, e.g.,

$$d(\dot{s}, \tilde{s}) = \sum_{i=1}^{n} (\dot{x}_i - \tilde{x}_i)^2$$

or

$$d(\dot{s}, \tilde{s}) = \sum_{i=1}^{n} (x_i - \tilde{x}_i)(\text{modulo } 2)$$

The set of all possible states of a collection of logic variables together with the Hamming distance defined between all pairs of states will be called the space of the logic variables or, simply, the *logic space*. The number of logic variables involved represents the dimension of the space. Thus, in the case of n logic variables we speak about the n-dimensional logic space. Such a space has all properties of an *n-dimensional unit cube* if values of the logic variables are considered as coordinates and individual states of the logic variables as vertexes of the cube.

For instance, when concerned with two or three logic variables we can easily visualize the corresponding unit cubes (Figure 2.1). It is true that for a larger number of variables our imagination fails, but the properties of the corresponding cube can be accurately defined and the vertexes and edges of the cube mapped on a plane. We know that each n-dimensional unit cube has 2^n vertexes, with one edge emerging from every vertex in the direction of each coordinate. Two vertexes interconnected by an edge, i.e., differing merely in a single coordinate, are called *adjacent*.

In the same manner, two states of logic variables will be called adjacent if they differ merely in the value of a single variable (if their Hamming distance is equal to one). It is clear that two adjacent states of logic variables correspond to two adjacent vertexes of the unit cube, and vice versa. Each state of n variables (and, similarly, each vertex of the n-dimensional unit cube) has exactly n "neighbors," i.e., adjacent states (adjacent vertexes). The concept of state adjacency plays an important role in the solution of many problems concerning switching circuits.

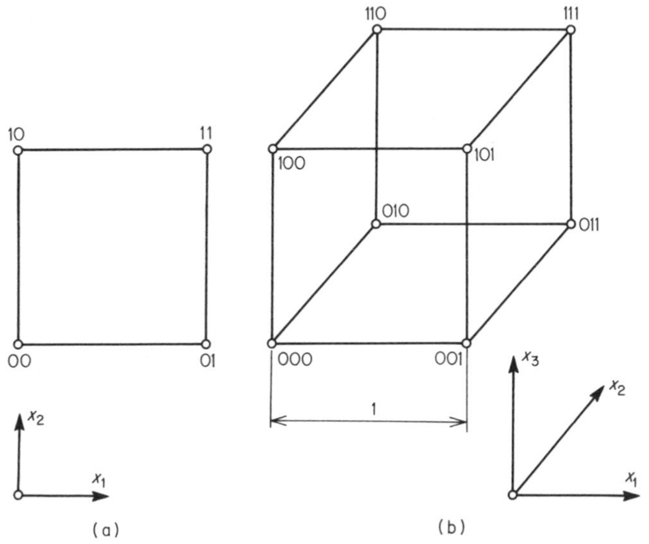

Figure 2.1 Two-dimensional and three-dimensional unit cubes.

Figure 2.2 illustrates four different ways of mapping a four-dimensional unit cube on a plane. The states of the variables which correspond to the individual vertexes are denoted by the pertinent state identifiers.

It is easy to demonstrate that the Hamming distance has the following properties:

$$(1)\ d(s_i, s_j) = 0 \quad \text{if and only if} \quad s_i = s_j$$

$$(2)\ d(s_i, s_j) = d(s_j, s_i)$$

$$(3)\ d(s_i, s_j) \leq d(s_i, s_k) + d(s_k, s_j)$$

Hence, the n-dimensional unit cube represents a metric space containing 2^n points (vertexes, states).

If one of the n logic variables is fixed, the remaining variables represent an $(n-1)$-dimensional unit cube. For instance, if $x_3 = 1$ in Figure 2.1b, we

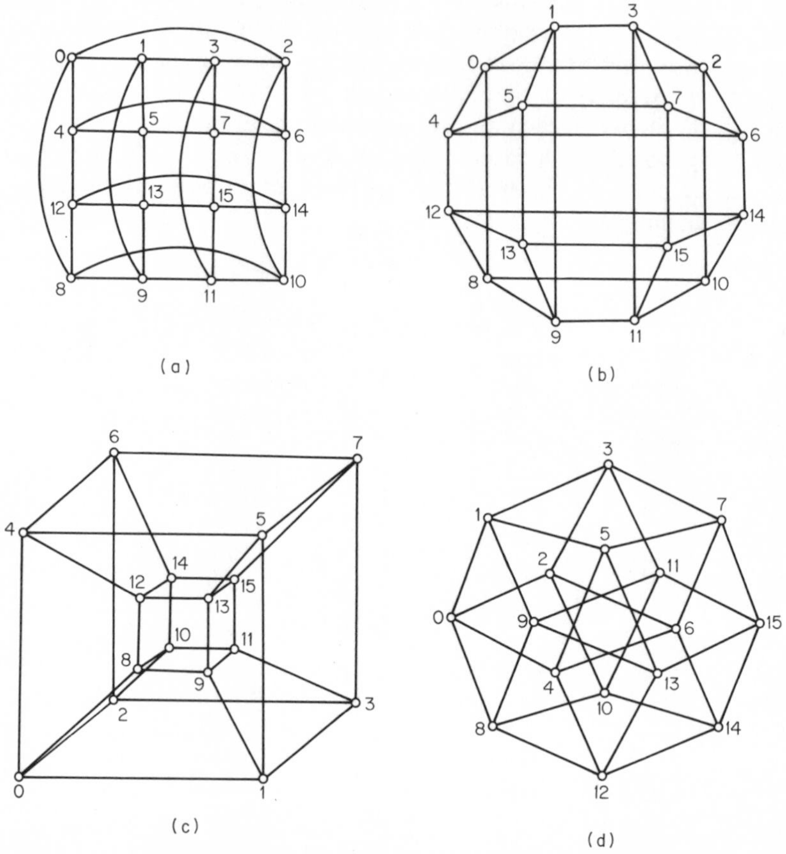

Figure 2.2 Different ways of mapping a four-dimensional unit cube on a plane.

obtain the top side of the three-dimensional unit cube, which has all the properties of a two-dimensional unit cube. If k of the n logic variables are fixed ($0 \leq k \leq n$), the remaining variables represent an $(n - k)$-dimensional unit cube. Obviously, this $(n - k)$-dimensional unit cube is completely contained in the original n-dimensional cube. We say that the former is a *subcube* or a *cell* of the latter. The number of possible choices of k out of n variables is $\binom{n}{k}$ and there are 2^k possibilities of fixing a chosen set of k logic variables. Hence, the total number C_n^k of all $(n - k)$-dimensional subcubes included in the n-dimensional unit cube is given by the formula

$$C_n^k = \binom{n}{k} \cdot 2^k$$

where $0 \leq k \leq n$.

A useful representation of the n-dimensional unit cube is a *logic chart* (*logic map*). It is a rectangle or a square regularly divided (like a chess-board) into 2^n small squares for n logic variables. A definite state of the logic variables pertains uniquely to each square in the chart, and vice versa.

Every logic chart has 2^p columns and 2^q rows where $p + q = n$ and $p, q \geq 1$. The columns of the chart are assigned to the states of p variables, its rows to the states of the remaining $q = n - p$ variables. The number of columns and rows in the chart thus depends both on the number of variables and on the manner in which all the variables are partitioned into two subsets. It is an advantage to as^ign a separate state identifier to each of these subsets. Then one of the identifiers labels columns in the chart, the other its rows. Let us denote the state identifier of the columns by s_c and that of the rows by s_r.

If variables $x_1, x_2, \ldots, x_n$ are partitioned into subsets $x_1, x_2, \ldots, x_p$ and $x_{p+1}, x_{p+2}, \ldots, x_n$, then

$$\begin{aligned} s_c &= \sum_{k=1}^{p} \dot{x}_k \cdot 2^{k-1} \\ s_r &= \sum_{l=p+1}^{n} \dot{x}_l \cdot 2^{l-p-1} \end{aligned} \tag{2.3}$$

There exist different logic charts for different assignments of s_r and s_c to rows and columns, respectively. The total number of different logic charts is, clearly, $(2^p)!(2^q)!$ for a given pair of p and q.

The *Marquand chart*, which is used in this book, is based on the natural assignment: The row r and the column c ($r,c \geq 0$) corresponding, respectively, to identifiers s_r and s_c are uniquely determined by the equations

$$\begin{aligned} r &= s_r \\ c &= s_c \end{aligned} \tag{2.4}$$

Figure 2.3 shows in what manner it would be possible to represent—by a means of the Marquand chart—the four-dimensional unit cube. Each chart

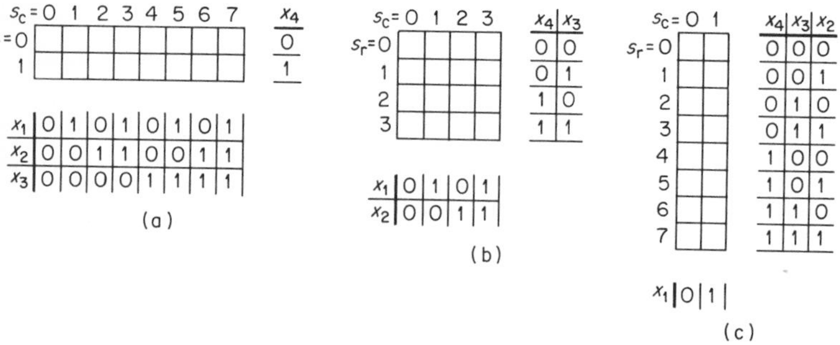

Figure 2.3 Various ways of expressing the same logic space (four-dimensional unit cube) by means of the Marquand chart.

in Figure 2.3 is marked at the left and on top by the partial identifiers s_r and s_c, respectively, and at the right and bottom by the corresponding states of the variables.

As a rule, however, we do not mark the states of the variables in the charts, and sometimes we omit even the identifiers pertaining to the individual rows and columns. It is sufficient to mark the rows and columns solely by the order of the corresponding variables in such a manner that, by substituting the values 0 and 1 for the variables, we obtain a binary number which directly defines the pertinent row or column (both rows and columns being, of course, counted from zero upward). This simplified manner of notation, which provides unique information on the assignment of the variables in the chart, is shown in Figure 2.4, which expresses the same assignment as that in Figure 2.3b.

Another logic chart, which seems to be preferred in the contemporary literature on switching circuits, is the *Karnaugh chart*. The assignment of states s_c to columns is in the Karnaugh chart such that pairs of columns placed symmetrically with respect to the vertical axes dividing the chart into two halves, the halves into four quarters, etc., are assigned to adjacent states; the same holds true for states s_r and pairs of rows placed symmetrically with regard to the corresponding horizontal axes.

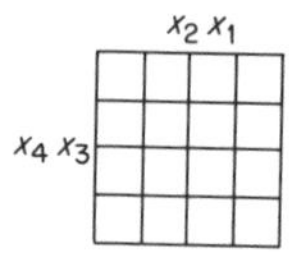

Figure 2.4 A simplified but unambiguous description of the Marquand chart shown in Figure 2.3b.

The assignment of states s_c and s_r to columns and rows, respectively, in the Karnaugh chart for four variables is shown in Figure 2.5. Each of the three alternatives in Figure 2.5 represents a particular partition of the variables into two subsets.

The rule of symmetry used in determining adjacent states in the Karnaugh is illustrated by an example in Figure 2.6. The example shows a Karnaugh chart for seven logic variables. A square corresponding to a reference state is denoted by R; squares corresponding to adjacent states of R are denoted by A. The axes of symmetry that are used at a particular stage are denoted by heavy lines. The shaded area at a particular stage denotes that portion of the chart which need not be considered at that stage since all adjacent states within it have already been determined.

It is usually claimed that, due to the rule of symmetry, it is easier to identify adjacent states of a given reference state in the Karnaugh chart than in the Marquand chart. Let us show that there is no justification for such an assertion.

Let a Marquand chart for n logic variables have 2^p columns and 2^q rows, where $p + q = n$. Let the chart be regularly divided into 2^i column strips, each containing 2^{p-i} columns ($i = 0, 1, \ldots, p-1$), and 2^j row strips, each containing 2^{q-j} rows ($j = 0, 1, \ldots, q-1$). Then squares corresponding to pairs of adjacent states lie either in the same row or in the same column (this is a rule which holds in every logic chart) and are 2^{p-i-1} columns apart or 2^{q-j-1} rows apart in a strip containing, respectively, 2^{p-i} columns or 2^{q-j} rows.

Let us note that the rule just stated is due to the natural order of identifiers s_c and s_r in the Marquand chart. Indeed, when a single logic variable x_k changes its value, the state identifier changes its value by 2^{k-1} according to equation (2.1). Since equations (2.4) hold for the Marquand chart, the change of 2^{k-1} (positive or negative) in s_c or s_r produces a shift in the chart represented by the same number of columns or rows, respectively (see Exercise 2.1-4).

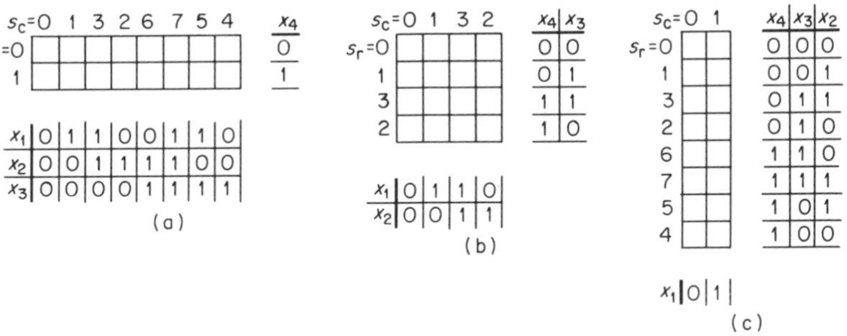

Figure 2.5 Various ways of expressing the same logic space (four-dimensional unit cube) by means of the Karnaugh chart.

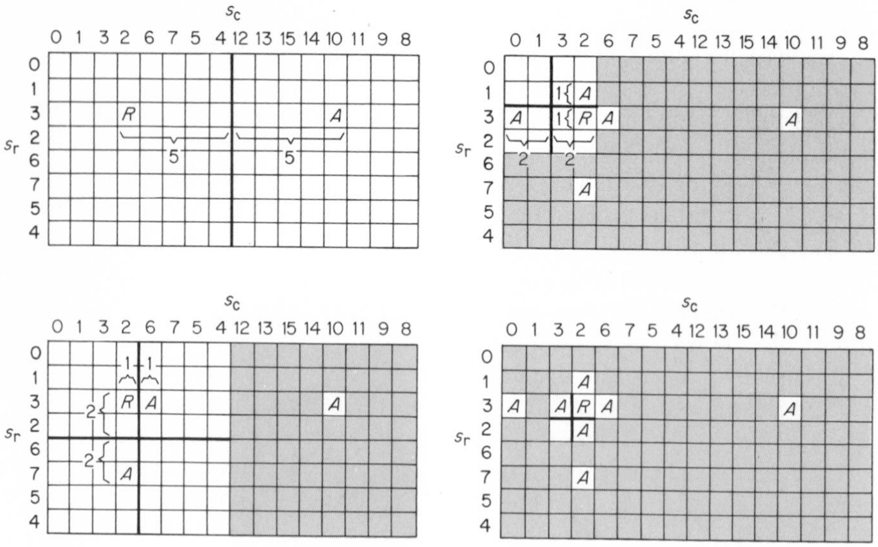

Figure 2.6 An example of application of the rule of adjacency in the Karnaugh chart (R is the reference state, A are adjacent states with regard to R).

The rule of adjacency in the Marquand chart is illustrated by an example for $n = 7$, $p = 4$, ($i = 0,1,2,3$) and $q = 3$ ($j = 0,1,2$) in Figure 2.7. R denotes the square corresponding to a reference state and A is used to denote squares corresponding to adjacent states of R. Some notes concerning the procedure in Figure 2.7 might be useful:

1. Figure 2.7a: A strip containing sixteen columns (the whole chart) is considered so that the distance of the adjacent column from the reference column within this strip is equal to eight (one-half of the total number of columns in the strip). After identifying the adjacent state in the chart, we divide the chart into two halves. The left-hand half will be ignored in the next step since it does not contain the reference square.

2. Figure 2.7b: We consider a strip containing eight columns (the right-hand half of the chart) and a strip containing eight rows. Both strips have to contain the reference square. The distance between the reference square and the adjacent squares within these strips is equal to four in both directions (one-half of the total number of columns or rows in the strips). After identifying the adjacent states, we divide each of the strips into two halves; in the next step we ignore those halves which do not contain the reference square.

3. Figure 2.7c: We consider a strip containing four columns and a strip containing four rows (one-half of the corresponding numbers used in the

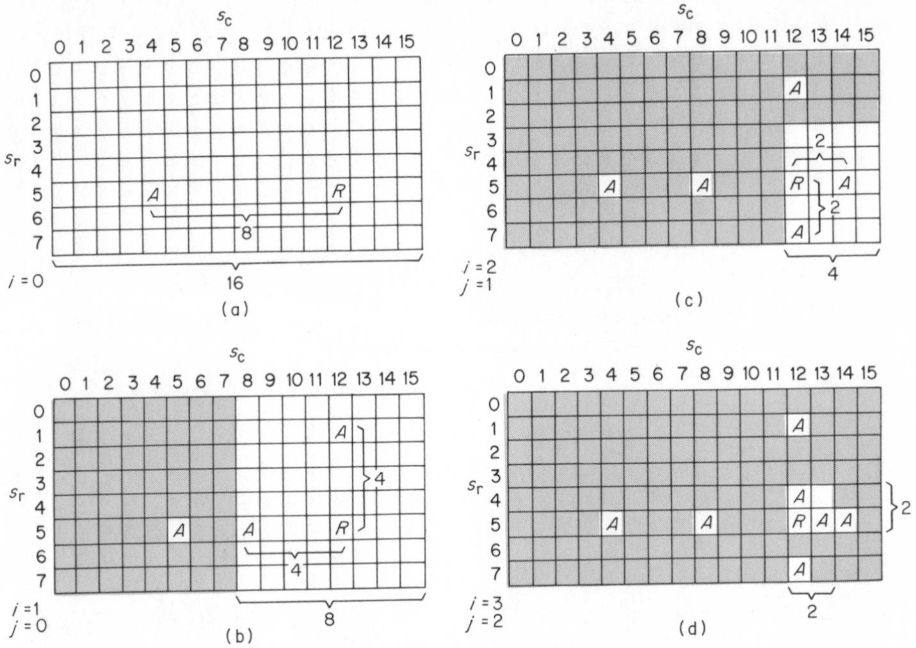

Figure 2.7 An example of application of the rule of adjacency in the Marquand chart (R is the reference state, A are the adjacent states with regard to R).

previous step). Both strips are again required to contain the reference square. Clearly, the distance between the reference square and the adjacent squares is equal to two in both directions (one-half of the total number of columns or rows in the strips). Each of the strips is again divided into two halves and in the next step only those halves which contain the reference square are considered.

4. Figure 2.7d: We consider a strip containing two columns and a strip containing two rows (one-half of the corresponding numbers used in the previous step). Both of the strips contain the reference square. The distance between the reference square and the adjacent squares is equal to one (one-half of the total number of columns or rows in the respective strip). This is the smallest possible distance. Hence, the procedure terminates at this point.

Thus there is a single and extremely simple arithmetic rule of adjacency in the Marquand chart. The application of the arithmetic rule is sometimes even more convenient than the rule of symmetry applied for the adjacency in the Karnaugh chart. This happens, for instance, if a procedure based on a logic chart is simulated by a digital computer.

A great advantage of the Marquand chart is the natural order of the state identifiers s_r and s_c associated with rows and columns, respectively. It prevents many errors which arise when logic relations are transformed from other forms to the chart form and vice versa. It is also a convenient property from the standpoint of computer simulation. There are some other advantages of the Marquand chart which will be mentioned later.

Exercises to Section 2.1

2.1-1 Determine the Hamming distance of two states of logic variables described by the following pairs of state identifiers:
*(a) $s_1 = 9$, $s_2 = 18$.
(b) $s_1 = 0$, $s_2 = 15$.
(c) $s_1 = 21$, $s_2 = 54$.

2.1-2 Prove that the Hamming distance has the following properties:
(a) $0 \leq d(s_i, s_j) \leq n$, where n is the number of logic variables.
(b) $d(s_i, s_j) = 0$ if and only if $s_i = s_j$.
(c) $d(s_i, s_j) = d(s_j, s_i)$.
(d) $d(s_i, s_j) \leq d(s_i, s_k) + (s_k, s_j)$.

2.1-3 Determine the state s_m that has the maximum distance from a reference state s for a given number n of logic variables.
*(a) $s = 11$, $n = 5$.
(b) $s = 2$, $n = 6$.
*(c) $s = a$, $n = b$ $(a < 2^b)$.

2.1-4 Prove that state identifiers s_i of all states adjacent to a given reference state $s = (\dot{x}_n\, \dot{x}_{n-1} \cdots \dot{x}_1)$ can be calculated by the formula

$$s_i = s + 2^{i-1}\, (-1)^{\dot{x}_i}$$

where $i = 1, 2, \ldots, n$.

*2.1-5 Let $s_1 = (s_{r1}, s_{c1})$ and $s_2 = (s_{r2}, s_{c2})$ be two states of n logic variables represented by partial identifiers corresponding to rows and columns in a logic chart. Calculate $d(s_1, s_2)$ if $d(s_{r1}, s_{r2})$ and $d(s_{c1}, s_{c2})$ are given.

2.1-6 Compare Figure 2.2a with Figure 2.5b. Using the analogy between these figures and properties of the Karnaugh chart, draw diagrams of n-dimensional unit cubes ($n = 5, 6, \ldots$).

2.1-7 Using the diagrams of n-dimensional unit cubes (Exercise 2.1-6), find such a sequence of states of n logic variables ($n = 4, 5\ldots$) that each state appears just once in the sequence and each pair of consecutive states is represented by adjacent states.

2.1-8 What is the number of:
(a) different Marquand charts for n logic variables?
(b) different logic charts for n logic variables?

2.2 LOGIC FUNCTIONS

A *logic function* f of n logic variables $x_1, x_2, \ldots, x_n$ is a unique assignment of values of a logic variable y to the states of the variables $x_1, x_2, \ldots, x_n$. We write $y = f(x_1, x_2, \ldots, x_n)$. If no confusion can arise, the function may be denoted by the same symbol as the corresponding variable so that we write $y = y(x_1, x_2, \ldots, x_n)$.

If the assignment is defined for every one of all the 2^n possible states of the variables $x_1, x_2, \ldots, x_n$, we are concerned with a *definite logic function.* If there is at least one state of the variables $x_1, x_2, \ldots, x_n$ to which no value of the variable y is assigned, we are concerned with an *indefinite logic function.*

The variables $x_1, x_2, \ldots, x_n$ will be called the *input variables* of the logic function f, and the variable y will be called its *output variable.* Sometimes, in the literature, "independent" and "dependent" are used instead of "input" and "output," respectively. Such a terminology may be confusing in cases where the input variables (according to our terminology) are mutually dependent or are dependent on the output variable. Existence of such cases will be documented in later chapters.

Since both the input and output variables of logic functions are associated with the same set of values, 0 and 1, we often prefer to speak about *logic operations* rather than logic functions. The terms "logic function" and "logic operation" will be considered synonymous. We may speak about unary, binary, or ternary logic operations rather than logic functions of one, two, or three variables, respectively.

The set of all states of the input variables $x_1, x_2, \ldots, x_n$ for which a logic function is defined is called the *domain* of the function. The set of all different values of the output variable y is called the *range* of the function.

Sometimes it is useful to assign to every definite logic function $f(x_1, x_2, \ldots, x_n)$ of n variables the number

$$i = \sum_{s=0}^{2^n-1} f(s) \cdot 2^s \tag{2.5}$$

where s is the state identifier of the variables $x_1, x_2, \ldots, x_n$ and $f(s)$ is the value of the output variable assigned by the function f to state s. The number i is called the identifier of definite logic functions or the *binary functional identifier.* It is easy to prove that there is a one-to-one correspondence between definite logic functions of a specified number of variables and their identifiers.

A modified version of the functional identifier, based on the formula

$$j = \sum_{s=0}^{2^n-1} f(s) \cdot 3^s \tag{2.6}$$

may be used for both definite and indefinite logic functions. Here, $f(s) = 2$ in the case that the function is not defined for state s. The number j is called the identifier of logic functions or the *ternary functional identifier*.

Now let us devote our attention to definite logic functions. We shall first present the simplest possible case, i.e., the functions of a single input variable. They may also be called unary logic operations. Let us denote the input variable by the symbol x. It is easy to find that there exist only four different ways of assigning values of the output variable y to the two possible states of the single variable x, i.e., there are only four different logic functions of a single logic input variable:

s	x	y_0	y_1	y_2	y_3
0	0	0	1	0	1
1	1	0	0	1	1

We see that for the functions y_0 and y_3 the output variables have constant values irrespective of the values of the input variable x. We may write $y_0 = 0$, $y_3 = 1$. They are called ZERO function and ONE function respectively.

For the function y_2 the output variable has the same values as the input variable x. We may therefore write $y_2 = x$. This function is usually called ASSERTION.

The most interesting function is y_1. It is called NEGATION and is characterized by the fact that the output variable assumes values opposite to those of the input variable x. This function will be denoted by a bar over the symbol of the corresponding input variable. Thus, in our case, $y_1 = \bar{x}$.

Let us note that the subscript by which we distinguished logic functions of one variable represents directly the binary functional identifier i. For example, the functional identifier i of NEGATION (function y_1) is calculated (using equation (2.5)) as follows:

$$i = 1 \times 2^0 + 0 \times 2^1 = 1$$

Let us now pass on to definite logic functions of two input variables, x_1 and x_2, for which we can write down a table similar to that used for the functions of a single input variable. They may also be called binary logic operations. Two input variables can acquire four different states. There exist sixteen different logic functions for them:

s	x_2	x_1	z_0	z_1	z_2	z_3	z_4	z_5	z_6	z_7	z_8	z_9	z_{10}	z_{11}	z_{12}	z_{13}	z_{14}	z_{15}
0	0	0	0	1	0	1	0	1	0	1	0	1	0	1	0	1	0	1
1	0	1	0	0	1	1	0	0	1	1	0	0	1	1	0	0	1	1
2	1	0	0	0	0	0	1	1	1	1	0	0	0	0	1	1	1	1
3	1	1	0	0	0	0	0	0	0	0	1	1	1	1	1	1	1	1

The subscript by which the functions of two variables are distinguished has again the meaning of the binary functional identifier.

We see immediately that some functions with properties similar to those of logic functions of one variable appear among the functions of two variables. They are: $z_0 = 0$, $z_3 = \bar{x}_2$, $z_5 = \bar{x}_1$, $z_{10} = x_1$, $z_{12} = x_2$ and $z_{15} = 1$. We have not yet encountered the remaining ten functions. Each of these functions has its own name and its symbolic designation, but there is no uniformity in the literature as to their use. For our purpose we shall adopt the names and symbols specified in Table 2.1. The symbols are often called *logic operators.*

The adopted names of the functions describe briefly their fundamental properties. For instance, the function z_8 is called AND because the output variable assigned by this function to the input variables x_1 and x_2 is equal to 1 if and only if both x_1 AND x_2 are equal to 1. Similarly, the function z_6 is called NONEQUIVALENCE because the output variable is equal to 1 if and only if x_1 and x_2 are NONEQUIVALENT, etc.

As a matter of interest let us mention that any logic function can also be represented with the aid of the ordinary arithmetic operations. For instance, the negation $\bar{x}$ can be expressed arithmetically as $1 - x$. In a similar manner it is possible to find arithmetic expressions for the other foregoing functions:

$$z_2 = x_1 - x_1x_2 \qquad z_{13} = 1 - x_1 + x_1x_2$$
$$z_6 = (x_1 - x_2)^2 \qquad z_9 = 1 - (x_1 - x_2)^2$$
$$z_8 = x_1x_2 \qquad z_7 = 1 - x_1x_2$$
$$z_{14} = x_1 + x_2 - x_1x_2 \qquad z_1 = 1 - x_1 - x_2 + x_1x_2$$

Although the arithmetic form of representation of logic functions seems to be useful at first sight, its applicability in the area of switching circuits is outmatched by other forms. We can see that AND function z_8 is represented by the arithmetic product. Therefore, we use for it the symbol of arithmetic product and, sometimes, also the name "*logic product*" or, simply, "*product.*"

In addition to the logic functions of one and two variables, names and symbols for some logic functions of more than two variables are used too. Let us mention the most important of them:

1. *Generalized* AND *function*: AND $(x_1, x_2, \ldots, x_n) = x_1x_2 \ldots x_n = 1$ if and only if $x_1 = x_2 = \cdots = x_n = 1$; otherwise AND $(x_1, x_2, \ldots, x_n) = 0$.

2. *Generalized* NAND *function*: NAND$(x_1, x_2, \ldots, x_n) = 0$ if and only if $x_1 = x_2 = \cdots = x_n = 1$; otherwise NAND$(x_1, x_2, \ldots, x_n) = 1$.

3. *Generalized* OR *function*: OR$(x_1, x_2, \ldots, x_n) = x_1 \vee x_2 \vee \cdots \vee x_n = 0$ if and only if $x_1 = x_2 = \cdots = x_n = 0$; otherwise OR$(x_1, x_2, \ldots, x_n) = 1$.

4. *Generalized* NOR *function*: NOR$(x_1, x_2, \ldots, x_n) = 1$ if and only if $x_1 = x_2 = \cdots = x_n = 0$; otherwise NOR$(x_1, x_2, \ldots, x_n) = 0$.

TABLE 2.1 Definite Logic Functions of Two Variables

x_2 x_1	1 1	1 0	0 1	0 0	Adopted Name of Function	Adopted Symbol	Other Names Used in the Literature	Other symbols Used in the Literature
z_0	0	0	0	0	ZERO function	0	FALSUM	$\emptyset, F, \perp$
z_1	0	0	0	1	NOR function	$x_1 \downarrow x_2$	Pierce function, NEITHER-NOR function	NOR(x_1, x_2), $x_1 \not\vee x_2$
z_2	0	0	1	0	PROPER INEQUALITY	$x_1 > x_2$	INHIBITION, BUT-NOT function	$x_1 \nleftarrow x_2$
z_3	0	0	1	1	NEGATION	$\bar{x}_2$	INVERTION, COMPLEMENT, NOT function	$\neg x_2, \sim x_2, x_2^0,$
z_4	0	1	0	0	PROPER INEQUALITY	$x_1 < x_2$	INHIBITION, BUT-NOT function	$x_1 \nrightarrow x_2$
z_5	0	1	0	1	NEGATION	$\bar{x}_1$	INVERTION, COMPLEMENT, NOT function	$\neg x_1, \sim x_1, x_1^0$
z_6	0	1	1	0	NONEQUIVALENCE	$x_1 \not\equiv x_2$	EXCLUSIVE OR function	$x_1 \oplus x_2, x_1 \veebar x_2$
z_7	0	1	1	1	NAND function	$x_1 \mid x_2$	Sheffer stroke, NOT-BOTH function	NAND(x_1, x_2), $x_1 \not\wedge x_2$
z_8	1	0	0	0	AND function	$x_1 \cdot x_2$ or x_1x_2	CONJUNCTION, Logic product	$x_1 \wedge x_2, x_1 \& x_2$, AND (x_1, x_2)
z_9	1	0	0	1	EQUIVALENCE	$x_1 \equiv x_2$	BICONDITIONAL	$x_1 \leftrightarrow x_2$
z_{10}	1	0	1	0	ASSERTION	x_1	IDENTITY	x_1^1
z_{11}	1	0	1	1	INEQUALITY	$x_1 \geq x_2$	IMPLICATION, CONDITIONAL	$x_1 \leftarrow x_2, x_1 \subset x_2$
z_{12}	1	1	0	0	ASSERTION	x_2	IDENTITY	x_2^1
z_{13}	1	1	0	1	INEQUALITY	$x_1 \leq x_2$	IMPLICATION, CONDITIONAL	$x_1 \rightarrow x_2, x_1 \supset x_2$
z_{14}	1	1	1	0	OR function	$x_1 \vee x_2$	DISJUNCTION, Logic Sum	OR(x_1, x_2), $x_1 + x_2$
z_{15}	1	1	1	1	ONE function	1	VERUM	I, V, T

5. *Threshold logic function* $F(x_1, x_2, \ldots, x_n)$ is defined for a set of *weights* $w_1, w_2, \ldots, w_n$ and a *threshold* T (the weights and the threshold can be positive or negative real numbers):

$$F(x_1, x_2, \ldots, x_n) = 0 \quad \text{if and only if} \quad \sum_{i=1}^{n} w_i x_i \leq T$$

$$F(x_1, x_2, \ldots, x_n) = 1 \quad \text{if and only if} \quad \sum_{i=1}^{n} w_i x_i > T$$

6. *Majority logic function* $M_n(x_1, x_2, \ldots, x_n)$:

$$M_n(x_1, x_2, \ldots, x_n) = 0 \quad \text{if and only if} \quad \sum_{i=1}^{n} x_i \leq \frac{n}{2}$$

$$M_n(x_1, x_2, \ldots, x_n) = 1 \quad \text{if and only if} \quad \sum_{i=1}^{n} x_i > \frac{n}{2}$$

The negation $\overline{M}_n$ of the majority function is called the *minority function.*

7. *Logic function "m out of n"* $(m \leq n)$: $F_n^m(x_1, x_2, \ldots, x_n) = 1$ if and only if $\sum_{i=1}^{n} x_i = m$; otherwise $F_n^m(x_1, x_2, \ldots, x_n) = 0$.

8. *Logic function "at least m out of n"* $(m \leq n)$: $F_n^{m+}(x_1, x_2, \ldots, x_n) = 1$ if and only if $\sum_{i=1}^{n} x_i \geq m$; otherwise $F_n^{m+}(x_1, x_2, \ldots, x_n) = 0$.

9. *Logic function "at most m out of n"* $(m \leq n)$: $F_n^{m-}(x_1, x_2, \ldots, x_n) = 1$ if and only if $\sum_{i=1}^{n} x_i \leq m$; otherwise $F_n^{m-}(x_1, x_2, \ldots, x_n) = 0$.

So far we have discussed definite logic functions only. As we have already mentioned, however, there also exist indefinite logic functions, in which no values of the output variable are assigned to some states of the input variables. This fact will be denoted in our tables by the symbol $\sim$.

The number of indefinite logic functions is larger than that of definite logic functions. For a single input variable x we have five different indefinite logic functions:

s	x	v_2	v_5	v_6	v_7	v_8
0	0	$\sim$	$\sim$	0	1	$\sim$
1	1	0	1	$\sim$	$\sim$	$\sim$

The subscript by which the functions are distinguished represents the ternary functional identifier. The states for which a particular logic function is not defined are usually called *don't care states.*

No names and symbols are used for indefinite logic functions. As we shall see later on, in the solution of various problems concerning switching circuits we always replace an indefinite logic function by a definite one. In this process we try to select such definite values (i.e., either 0 or 1) for the individual undefined values of the pertinent output variables, which, from the given point of view, lead to the best solution.

Exercises to Section 2.2

2.2-1 Specify in the tabular form the following logic functions:
(a) AND, OR, NAND, NOR functions of three variables.
(b) majority functions M_3 and M_5.
(c) Logic functions F_n^m, F_n^{m+}, and F_n^{m-} for $n = 3$ and $m = 2$.
(d) Threshold function $F(x_1, x_2, x_3)$ for $w_1 = 0.5$, $w_2 = 1$, $w_3 = 2$, $T = 1.5$.

2.2-2 Express each of the following functions as a threshold function:
*(a) generalized AND function.
(b) generalized OR function.
(c) majority function M_n.
*(d) logic functions F_n^m.

*2.2-3 Which of the logic functions of one or two variables can be expressed as a threshold function?

*2.2-4 NONEQUIVALENCE is identical with EXCLUSIVE OR function as far as logic functions of two variables is concerned ($n = 2$). Show that they are different for $n > 2$.

2.2-5 Determine the functional identifier of:
*(a) AND function of three variables.
(b) NAND function of three variables.
(c) functions F_3^2, F_3^{2-}, F_3^{2+}.
(d) majority function M_3.

2.2-6 Determine the definite logic functions of n variables whose binary identifier is
(a) $i = 124$, $n = 3$.
*(b) $i = 64$, $n = 3$.
(c) $i = 1$, $n = 3$.
(d) $i = 1$, $n = 4$.

2.2-7 Determine the logic functions of n variables whose ternary identifier is:
(a) $j = 64$, $n = 2$.
(b) $j = 133$, $n = 3$.
*(c) $j = 5$, $n = 2$.
(d) $j = 12$, $n = 2$.

2.2-8 Show that the functional identifier (both binary and ternary) may represent several functions if the number of input variables is not specified.

2.2-9 Find arithmetic expressions describing the following logic functions:
*(a) majority function M_3.
(b) Functions F_3^2, F_3^{2-}, F_3^{2+}.
(c) OR function of three variables.
(d) NOR function of three variables.

2.3 LOGIC EXPRESSIONS

Each of the definite logic functions that were introduced in Section 2.2 is represented by its own formal symbol. Each can also be expressed as a *composition* of some other functions. The composition of two functions is a scheme in which the output variable of one of the functions is used as an input variable of the other function. For instance, let $y_1 = x_1 \vee x_2$ and $y_2 = y_1 | x_3$. Then function y_2 is essentially a function of variables x_1, x_2, x_3 and is expressed as a composition of function y_1 (OR function) and NAND function of y_1 and x_3. We can write

$$y_2 = (x_1 \vee x_2) | x_3$$

where the *parentheses* are used to specify that the OR function is applied first. Function z_9 (EQUIVALENCE) can be expressed, for example, as

$$z_9 = (x_1 x_2) \vee ((\bar{x}_1)(\bar{x}_2))$$

$$z_9 = ((x_1 | x_1) | (x_2 | x_2)) | (x_1 | x_2)$$

$$z_9 = \overline{(x_1 \not\equiv x_2)}$$

because after any substitution of 0 or 1 for x_1 and x_2 in any of the expressions and correct applications of the formal symbols of logic functions in accordance with their meaning, introduced in Section 2.2, we always get the correct value of z_9. Parentheses specify again the order in which the functions included in an expression are applied. For instance, the expression $(x_1 x_2) \vee ((\bar{x}_1)(\bar{x}_2))$ can be evaluated step by step as shown in Table 2.2. Similarly, the evaluation of the expression $((x_1 | x_1) | (x_2 | x_2)) | (x_1 | x_2)$ is shown in Table 2.3.

Generally, every logic function can be represented by different logic expressions. The *logic expression* is defined recursively as follows:

1. The logic symbols 0 and 1 are logic expressions.
2. If x is a logic variable, then x and $\bar{x}$ are logic expressions.
3. If $A_1, A_2, \ldots, A_n$ are logic expressions, then $F(A_1, A_2, \ldots, A_n)$ is a logic expression, where F is a symbol (an operator) of a logic function (operation) of n variables $A_1, A_2, \ldots, A_n (n \geq 2)$.

TABLE 2.2. Step by Step Evaluation of the Logic Expression $(x_1 x_2) \vee ((\bar{x}_1)(\bar{x}_2))$

x_2	x_1	$\bar{x}_2$	$\bar{x}_1$	$x_1 x_2$	$\bar{x}_1 \bar{x}_2$	$(x_1 x_2) \vee ((\bar{x}_1)(\bar{x}_2))$
0	0	1	1	0	1	1
0	1	1	0	0	0	0
1	0	0	1	0	0	0
1	1	0	0	1	0	1

TABLE 2.3. Step by Step Evaluation of the Logic Expression $((x_1|x_1)|(x_2|x_2))|(x_1|x_2)$

x_2	x_1	$x_1\|x_1$	$x_2\|x_2$	$(x_1\|x_1)\|(x_2\|x_2)$	$x_1\|x_2$	$((x_1\|x_1)\|(x_2\|x_2))\|(x_1\|x_2)$
0	0	1	1	0	1	1
0	1	0	1	1	1	0
1	0	1	0	1	1	0
1	1	0	0	1	0	1

4. The only logic expressions are given by (1)–(3).

Every proper logic expression represents a single logic function. It demonstrates at the same time a certain composition of the function by some other functions (usually simpler ones) or, in other words, it demonstrates a certain decomposition of the function to some other functions. The way of composition (or decomposition) proper to a logic expression can be illustrated by a *block diagram*. Every block in the diagram corresponds to one of the logic functions used in the expression. Connections between the blocks represent the form of composition.

Let the symbols shown in Figure 2.8 be used in this text for blocks representing basic logic functions. Then block diagrams of the above mentioned expressions are shown in Figure 2.9.

A logic expression is *proper* if and only if the order in which the partial functions are to be composed is uniquely prescribed. The order of the compositions may be prescribed in any form but, usually, it is specified by parentheses. First, the partial functions enclosed by the innermost pairs of parentheses (innermost functions) are applied. Then, the obtained values of output variables are used as input variables of functions enclosed by parentheses of second order, and the same is repeated for functions enclosed in parentheses of third order, etc. The last pair of parentheses enclosing the whole expression may be omitted. Parentheses, brackets, and braces are sometimes used in order to distinguish more vividly the different orders but, principally, this is not necessary.

Let us note, that bars above logic expressions that denote negations of the expressions may substitute for parentheses. For instance, $\overline{x_1 \vee x_2}|\overline{x_3 \vee x_4}$ has the same meaning as $\overline{(x_1 \vee x_2)}|\overline{(x_3 \vee x_4)}$; it means that the functions are composed in the order: OR, NEGATION, NAND.

A *hierarchy of functions* is specified sometimes as a permanent convention in a particular algebra and the functions are applied in the order of the hierarchy unless another order is prescribed separately (e.g., by parentheses). Such a convention is called the *precedence rule*. There are other conventions concerning the order of execution of operations in algebraic expressions which

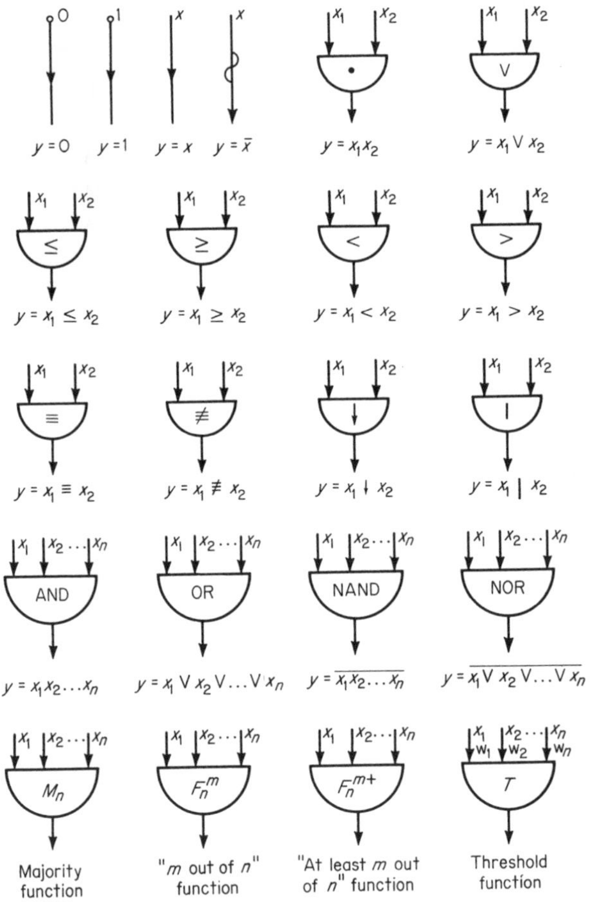

Figure 2.8 Symbols for blocks representing basic logic functions.

are used in various programming languages. For instance, if not specified otherwise by parentheses, operations are executed from right to left in APL language.

EXAMPLE 2.1 In order to illustrate that the logic function represented by a logic expression depends, generally, on the order of compositions of the partial functions included in the expression, let us consider the expression $x \leq y \leq z \leq x$ containing three INEQUALITIES. It is obviously ambiguous as far as the order of the compositions is concerned. If we denote the INEQUALITIES from left to right as a, b, c, we shall get the following possibilities:

Order of application	*Unique expressions and symbols of the corresponding functions*
abc	$((x \leq y) \leq z) \leq x = (a \leq z) \leq x = g_1$
acb or *cab*	$(x \leq y) \leq (z \leq x) = a \leq c = g_2$
bac	$(x \leq (y \leq z)) \leq x = (x \leq b) \leq x = g_3$
bca	$x \leq ((y \leq z) \leq x) = x \leq (b \leq x) = g_4$
cba	$x \leq (y \leq (z \leq x)) = x \leq (y \leq c) = g_5$

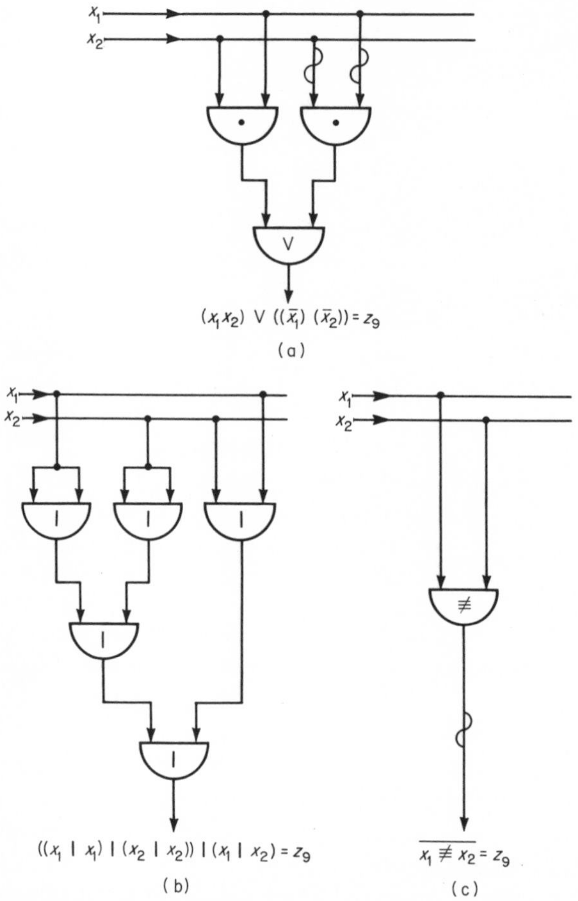

Figure 2.9 Block diagrams corresponding to the discussed logic expressions.

Both *acb* and *cab* orders represent the same function g_2 because, obviously, the single INEQUALITY (as well as any function) does not depend on the order in which the values of the input variables a and c are available. Applying the partial functions in the specified orders, we get the following functions g_1 to g_5:

$x\ y\ z$	$a\ b\ c$	$a \leq z$	$x \leq b$	$b \leq x$	$y \leq c$	g_1	g_2	g_3	g_4	g_5
0 0 0	1 1 1	0	1	0	1	1	1	0	1	1
0 0 1	1 1 0	1	1	0	1	0	0	0	1	1
0 1 0	1 0 1	0	1	1	1	1	1	0	1	1
0 1 1	1 1 0	1	1	0	0	0	0	0	1	1
1 0 0	0 1 1	1	1	1	1	1	1	1	1	1
1 0 1	0 1 1	1	1	1	1	1	1	1	1	1
1 1 0	1 0 1	0	0	1	1	1	1	1	1	1
1 1 1	1 1 1	1	1	1	1	1	1	1	1	1

We see that the functions g_1 to g_5 are, generally, different even if some of them are identical in this particular example due to special properties of INEQUALITY.

Every order of parentheses (order of execution of logic operations) participating in a logic expression specifies a *level* in the corresponding block diagram. This is illustrated in Figure 2.10.

Clearly, block diagrams of the above-mentioned type as well as the corresponding logic expressions enable us to describe uniquely structures of switching circuits. Blocks in the diagram represent elements (modules) in the corresponding switching circuit and vice versa. Interconnections between blocks in the diagram represent interconnections (wires) between terminals (pins) of the corresponding elements of the switching circuit.

We say that two logic expressions are *equal* if both of them represent the same logic function. If A and B are equal logic expressions, we write $A = B$.

If we take a closer look at Table 2.1, we find that for every function in the top half of the table, a negative of this function exists in its bottom half, and vice versa. We have thus:

$$0 = \bar{1}$$

$$x_1 \downarrow x_2 = \overline{x_1 \vee x_2}$$

$$x_1 > x_2 = \overline{x_1 \leq x_2}$$

$$x_1 < x_2 = \overline{x_1 \geq x_2}$$

$$x_1 \not\equiv x_2 = \overline{x_1 \equiv x_2}$$

$$x_1 | x_2 = \overline{x_1 x_2}$$

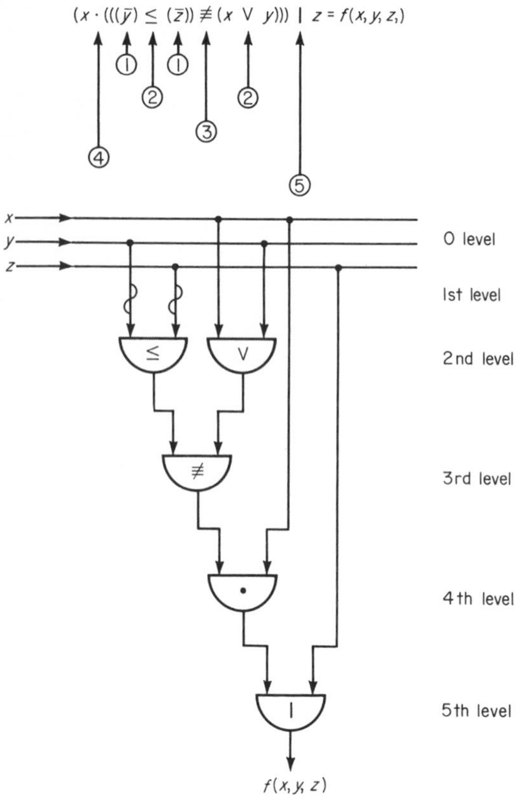

Figure 2.10 Relation between orders of parentheses in a logic expression and levels in the corresponding block diagram.

In addition to these simple equalities there exist other important equalities between logic expressions, such as:

$$z_2 = x_1 > x_2 = x_1\bar{x}_2 = \overline{\bar{x}_1 \vee x_2} = (x_1 \,|\, \bar{x}_2)|(x_1 \,|\, \bar{x}_2) = \bar{x}_1 \downarrow x_2$$

$$z_6 = x_1 \not\equiv x_2 = x_1\bar{x}_2 \vee \bar{x}_1 x_2 = (x_1 \vee x_2)(\bar{x}_1 \vee \bar{x}_2)$$

$$= (x_1 \downarrow \bar{x}_2) \vee (\bar{x}_1 \downarrow x_2) = \overline{x_1 \leq x_2} \vee \overline{\bar{x}_1 \leq \bar{x}_2}$$

$$z_8 = x_1 x_2 = \overline{\bar{x}_1 \vee \bar{x}_2} = \bar{x}_1 \downarrow \bar{x}_2 = (x_1 \,|\, x_2)|(x_1 \,|\, x_2) = x_1 > \bar{x}_2 = \overline{x_1 \leq \bar{x}_2}$$

The equality of two logic expressions which contain the same variables $x_1, x_2, \ldots, x_n$ can be verified by successively determining the values for both expressions for all 2^n states of the variables. If the determined values of the two expressions are identical for all states, the two expressions are equal. They merely represent different ways of expressing the same logic function.

Let us demonstrate this procedure by verifying the equality

$$(\bar{x}_1 \downarrow x_2) \vee (x_1 \downarrow \bar{x}_2) = \overline{x_1 \leq x_2} \vee \overline{\bar{x}_1 \leq \bar{x}_2}\,.$$

Evaluation of both these expressions is shown in Table 2.4. Since the equation is satisfied for all states of the variables involved, the expressions are equal. This procedure of verification of equality of two different logic expressions is sometimes referred to as the *proof by perfect induction.* When applying this type of proof, we verify a given theorem (e.g., an equality of two logic expressions) for every possible combination of values that the variables may assume.

So far we have considered expressions containing only variables which are mutually independent. If we admit, however, that not all the variables of two logic expressions are independent of each other, we may equate these expressions and seek the assumptions for which the resulting equality is valid. We shall treat this problem in greater detail in Chapter 4.

Now let us note that the simple logic functions introduced so far can be interpreted as the assignment of the truth values of molecular (resulting) propositions to the truth values of the relevant atomic (starting) propositions.

As an example, let us present two simple propositions: (*1*) "The number N is divisible by three." (*2*) "The number N is even," where N is a fixed number. The two propositions can be connected by different propositional connectives. For instance, when we use the conjunction "and" we obtain the molecular proposition "The number N is divisible by three and is even." In this case, the molecular proposition is true only if both the atomic propositions are true simultaneously (for instance, it is true for $N = 6$). The truth values of such a

TABLE 2.4 A way of Verifying the Equality of Two Logic Expressions

x_1	x_2	$\bar{x}_1$	$\bar{x}_2$	$x_1 \downarrow \bar{x}_2$	$\bar{x}_1 \downarrow x_2$	$(x_1 \downarrow \bar{x}_2) \vee (\bar{x}_1 \downarrow x_2)$
0	0	1	1	0	0	0
0	1	1	0	1	0	1
1	0	0	1	0	1	1
1	1	0	0	0	0	0

x_1	x_2	$\bar{x}_1$	$\bar{x}_2$	$x_1 \leq x_2$	$\overline{x_1 \leq x_2}$	$\bar{x}_1 \leq \bar{x}_2$	$\overline{\bar{x}_1 \leq \bar{x}_2}$	$\overline{(x_1 \leq x_2)} \vee \overline{(\bar{x}_1 \leq \bar{x}_2)}$
0	0	1	1	1	0	1	0	0
0	1	1	0	1	0	0	1	1
1	0	0	1	0	1	1	0	1
1	1	0	0	1	0	1	0	0

molecular proposition thus correspond to the values of the AND function, provided, of course, that we assign the symbol 1 to truth and the symbol 0 to falsity.

If we use the connective "either-or," we obtain the molecular proposition "The number N is either divisible by three or it is even." This molecular proposition is true only if one of the atomic propositions is true, but not both of them simultaneously (e.g., it is true for $N = 2$). We are thus concerned, in this case, with an interpretation of nonequivalence (exclusive OR function). In a similar manner, the use of the connective "or" corresponds to the OR function, etc.

TABLE 2.5 Assignment of Propositional Connectives to Logic Functions

Propositional Connective	Logic Function	Symbol
It is not true that X	NEGATION	$\bar{X}$
X and Y	AND function	XY
X but not Y	PROPER INEQUALITY	$X > Y$
X is not equivalent to Y	NONEQUIVALENCE	$X \not\equiv Y$
X or Y	OR function	$X \vee Y$
Neither X nor Y	NOR function	$X \downarrow Y$
X if, and only if, Y	EQUIVALENCE	$X \equiv Y$
If X then Y	INEQUALITY	$X \leq Y$
Not X or not Y	NAND function	$X \mid Y$

Some connectives are listed in Table 2.5 where the atomic propositions are denoted by the symbols X and Y. It should be noted that from the viewpoint of application of logic to switching circuits, we are interested only in the so-called *extensional propositional connectives*, i.e., in those where the truth value of the molecular proposition depends exclusively on the truth values of the atomic propositions. These are just those connectives which can be assigned to logic functions. In addition, however, there exist in logic the so-called *intensional propositional connectives* in which the value of the molecular proposition depends not only on truth values of the atomic propositions, but also on their content (e.g., "X because Y," "I believe that X," etc.).

Exercises to Section 2.3

*2.3-1 Identify which of the following expressions are proper:

(a) $x_1 \not\equiv x_2 >$

(b) $x_1 \equiv x_2 \leq x_3$

(c) $(x_1 x_2) \vee (x_1 x_3)$

(d) $(x_1 \vee x_2 \leq \equiv (x_1 \vee x_3)$
(e) $> (\bar{x}_1) < (x_2 \downarrow x_3)$
(f) $(x_1 \downarrow \bar{x}_2) \leq (x_2 x_3)$
(g) $\overline{x_1 \geq x_2}$
(h) $x_1 \mid x_2 \downarrow x_3$

2.3-2 Prove, by using the perfect induction, that the following equalities are valid:
(a) $\bar{x}_1 \vee x_2 = x_1 | (x_2 | x_2)$
(b) $x_1 \vee (\bar{x}_1 x_2) = x_1 \vee x_2$
(c) $\overline{x_1 \vee x_2} = \bar{x}_1 \bar{x}_2$
(d) $x_1 \downarrow x_2 = ((x_1 \mid x_1) | (x_2 | x_2)) | ((x_1 | x_1) | (x_2 | x_2))$
(e) $x_1 | x_2 = ((x_1 \downarrow x_1) \downarrow (x_2 \downarrow x_2)) \downarrow ((x_1 \downarrow x_1) \downarrow (x_2 \downarrow x_2))$

2.3-3 Express the function $z_6 = x_1 \not\equiv x_2$ by the following functions:
*(a) solely by the NAND function.
(b) by the NEGATION and the NAND function.
*(c) by the NEGATION and the AND and OR functions.
(d) solely by the NOR function.

2.3-4 Using results of Exercise 2.3-3, draw block diagrams for the logic expressions representing NONEQUIVALENCE.

2.3-5 A logic function $x_1 \circ x_2$ is called commutative if $x_1 \circ x_2 = x_2 \circ x_1$.
*(a) Find a necessary and sufficient condition for a logic function $x_1 \circ x_2$ to be commutative.
*(b) Determine the logic functions of two variables that are commutative.

2.3-6 A logic function $x_1 \circ x_2$ is called associative if $x_1 \circ (x_2 \circ x_3) = (x_1 \circ x_2) \circ x_3$. Determine the logic functions of two variables that are associative.

2.3-7 In addition to equalities $E_1 = E_2$ between two logic expressions E_1 and E_2, more general relations $E_1 \circ E_2$ can be used, where $\circ$ is a logic function of two variables. The relation $E_1 \circ E_2$ is valid if such pairs of values of E_1 and E_2 are obtained for all states of the independent variables for which the function $\circ$ is equal to one. Prove that the following relations are valid:
(a) $E_1 = \overline{((x_1 \leq x_2) \leq x_3) \leq x_1}$.
$E_2 = (x_1 \leq x_2) \leq (x_3 \leq x_1)$, $E_1 \not\equiv E_2$.
(b) $E_1 = \bar{x}_1 \downarrow \bar{x}_2$, $E_2 = (x_1 \downarrow \bar{x}_2) \vee (\bar{x}_1 \downarrow \bar{x}_2)$, $E_1 \leq E_2$.
(c) $E_1 = \overline{x_1 \leq \bar{x}_2}$, $E_2 = \bar{x}_1 \vee x_2$, $E_1 \geq E_2$.
(d) $E_1 = x_1 \vee (\bar{x}_1 x_2)$, $E_2 = x_1 \geq x_2$, $E_1 \vee E_2$.

2.4 POLISH STRING NOTATION

A particular logic expression represents a composition of some logic functions in a certain order. To represent a single logic function by the expression, there must not be any ambiguity in definitions of the logic functions included in the expression as well as in the order in which they are to be composed. We learned in Section 2.3 that the order of composition can always be uniquely described by a set of parentheses.

Logic expressions (as well as any algebraic expressions) with parentheses may not be suitable under certain circumstances. In such cases, so-called *parentheses-free forms* of representation are preferred. One of these forms, which was invented by a Polish logician Jan Lukasiewicz (1878–1956), is called the *Polish or Lukasiewicz notation.*

A logic expression written in the Polish notation is a string containing symbols of logic variables and operators but no parentheses (unless parentheses represent variables or operators). Such a string of symbols is referred to as the *Polish string.* There are several alternative forms of the Polish notation. The form which is described in this section is called a *reverse Polish notation.*

The reverse Polish string is scanned from left to right until the first operator is located. This operator is applied to such a number of variables immediately preceding the operator which is appropriate for it. The result is then enclosed in parentheses and considered as an input variable for some of the subsequent operators. The same procedure is repeated until all operators in the string are exhausted.

EXAMPLE 2.2 Transform the reverse Polish string

$$xy \neg z \neg \leq xy \vee \not\equiv | z \downarrow,$$

where $\neg$ is the operator of negation, to the normal notation with parentheses.

We proceed in the following steps:

1. We read the string from left to right until we come across the first operator. It is a unary operator ($\neg$, negation) so that it is applied to a single variable immediately preceding it. This variable is y; the operator modifies it to $\bar{y}$. We obtain a modified expression

$$x(\bar{y})z \neg \leq xy \vee \not\equiv | z \downarrow.$$

2. We continue our reading of the string from left to right until we come across the second operator. It is again the operator of negation. This time it is applied to variable z which produces further modification of the expression

$$x(\bar{y})(\bar{z}) \leq xy \vee \not\equiv | z \downarrow.$$

3. Now, we come across the operator of inequality. It is a binary operator so that it is applied to two symbols of variables immediately preceding it. We obtain the expression

$$x((\bar{y}) \leq (\bar{z}))xy \vee \not\equiv | z \downarrow$$

4. The next operator, which is applied to variables x and y, is the operator of the OR function. We obtain

$$x((\bar{y}) \leq (\bar{z}))(x \vee y) \not\equiv | z \downarrow$$

5. Binary operator $\not\equiv$ (NONEQUIVALENCE) is now applied to variables

$$((\bar{y}) \leq (\bar{z})) \qquad \text{and} \qquad (x \vee y),$$

so that we obtain

$$x(((\bar{y}) \leq (\bar{z})) \not\equiv (x \vee y))|z\downarrow$$

6. Binary operator | is applied to variables x and $(((\bar{y}) \leq (\bar{z})) \not\equiv (x \vee y))$ modifying thus the previous expression to the form

$$(x|(((\bar{y}) \leq (\bar{z})) \not\equiv (x \vee y)))z\downarrow$$

7. The last operator, when applied to both of the variables preceding it, produces the sought expression written in normal notation

$$(x|(((\bar{y}) \leq (\bar{z})) \not\equiv (x \vee y)))\downarrow z$$

The inverse conversion, from normal algebraic notation to reverse Polish notation, is also very simple provided that the former is fully parenthesized. We do the conversion step by step for each expression enclosed by a pair of parentheses, starting from the innermost expressions toward the outer expression.

EXAMPLE 2.3 Transform the logic expression

$$(z\downarrow((((\bar{x}) \vee y)|(x \vee (\bar{z}))) > ((\bar{y}) \not\equiv x)))$$

to the reverse Polish string. Use, as in Example 2.2, symbol $\neg$ for negation.

Let that portion of the expression which is already converted to the Polish notation at a particular step be enclosed in brackets. Then we proceed in the following steps:

1. The innermost partial expressions are the three negations included in the given expression. Each of them is represented in the Polish notation by the symbol of the variable followed by the operator representing the negation. We obtain

$$(z\downarrow((([x\neg] \vee y)|(x \vee [z\neg])) > ([y\neg] \not\equiv x)))$$

2. We consider now the innermost partial expressions included in the previous expression and convert them to the Polish notation, obtaining thus

$$(z\downarrow(([x\neg\, y \vee]|[xz\neg \vee]) > [y\neg\, x\not\equiv]))$$

3. There is only one innermost partial expression included in the previous result (the expression represented by the NAND operator |). After its conversion to the Polish notation, we obtain

$$(z\downarrow([x\neg\, y \vee xz\neg \vee |] > [y\neg\, x\not\equiv]))$$

4. Now we convert the expression associated with the operator >, obtaining thus

$$(z\downarrow[x\neg\, y \vee xz\neg \vee | y\neg\, x \not\equiv >])$$

5. The last pair of parentheses encloses an expression associated with the NOR operator $\downarrow$. After its conversion we obtain the final Polish string representing the expression given in this example:

$$zx \neg y \vee xz \neg \vee | y \neg x \not\equiv > \downarrow$$

The Polish notation is introduced in this section to show the possibility of a parentheses-free algebraic form of representation of logic functions. An understanding of the Polish notation may help the reader to understand better the principle of the multilevel composition of logic functions. It may also be of practical help, primarily in various problems associated with computer processing of logic expressions (a design of algebraic processors, translation of logic expressions to a machine language, etc.).

Exercises to Section 2.4

2.4-1 Convert the following reverse Polish strings to normal notation with parentheses:

*(a) $xyzv\downarrow w \vee | <$

(b) $x\neg yz < \geq vy \cdot \downarrow wt \neg \vee >$

(c) $xyz \vee v\downarrow wt < > |$

(d) $xy \cdot zv | \downarrow xwy \neg \vee > \downarrow \neg$

2.4-2 Convert the following logic expressions from the normal notation to the reverse Polish notation:

(a) $(x((\bar{y}) \vee z) \vee ((\bar{x})(y \vee (\bar{z})))) \vee ((\bar{x})(\bar{y}))$

*(b) $(x|y)|((x|x)|(y|y))$

(c) $\overline{x \leq y} \vee \overline{\bar{x} \leq \bar{y}}$

(d) $((xy) \vee ((\bar{x})(\bar{y})))|(((\bar{x})y) \vee (x(\bar{y})))$

2.4-3 Find which of the following expressions written in the reverse Polish notation are proper:

(a) $xyz\downarrow >$

(b) $xyz\neg \vee \downarrow$

(c) $xyz\downarrow \vee \leq$

(d) $x \vee y \neg \downarrow$

(e) $xy \vee \neg \neg$

(f) $xy\downarrow z \neg \downarrow \downarrow$

(g) $xy \vee \downarrow z \vee v$

2.5 NONDEGENERATE LOGIC FUNCTIONS

We might try to write out all logic functions of three input variables, four input variables, etc., in a manner similar to that used in writing out all the logic functions of one and two input variables. We would find very soon, however, that this is practically impossible. The number of logic functions increases far too rapidly with the number of input variables.

As we already know, the number of states of n logic variables is equal to 2^n. Any definite logic function assigns one of the two possible values, 0 or 1, to each of these states. The number of all possible ways of effecting such assignments is 2^{2^n}. This is therefore the number of different definite logic functions for n input variables. If we include also indefinite logic functions, there are 3^{2^n} possibilities (three assignments, 0, 1 or $\sim$, for each state). The number of indefinite logic functions is $3^{2^n} - 2^{2^n}$. To obtain an idea of the rapid growth of these numbers with an increasing n, the reader may consult Table 2.6, where these formulas are evaluated for $n \leq 4$.

TABLE 2.6 Numbers of Different Logic Functions

Number of Input Variables	n	0	1	2	3	4
Number of definite logic functions	2^{2^n}	2	4	16	256	65,536
Number of all logic functions	3^{2^n}	3	9	81	6,561	41,953,221
Number of indefinite logic functions	$3^{2^n} - 2^{2^n}$	1	5	65	6,305	41,887,685

We may try to reduce the number of logic functions of n input variables by considering only those functions which depend on all input variables. A logic function $f(x_1, x_2, \ldots, x_n)$ is said to be dependent on variable $x_i (1 \leq i \leq n)$ if and only if

$$f(x_1, \ldots, x_i, \ldots, x_n) \neq f(x_1, \ldots, \bar{x}_i, \ldots, x_n) \tag{2.7}$$

Otherwise, it is said to be independent of x_i. If inequality (2.7) is satisfied for all input variables, the function is called *nondegenerate*; otherwise, it is called *degenerate*.

For instance, ZERO and ONE functions are degenerate functions of one input variable. They are also degenerate functions of more than one input variable. ASSERTION and NEGATION are nondegenerate functions of one input variable but they are degenerate functions of two or more input variables.

Let $T(n) = 2^{2^n}$ denote the total number of definite logic functions of n variables and let $N(n)$ and $D(n)$ denote, respectively, the number of nondegenerate and degenerate logic functions of n variables. Then, obviously,

$$T(n) = N(n) + D(n) \tag{2.8}$$

The set of all 2^{2^n} logic functions consists of nondegenerate functions of no variables (there are $N(0) = 2$ of them, namely, ZERO and ONE functions), nondegenerate functions of one variable (there are $N(1)$ of them and each can be applied to one of $n = \binom{n}{1}$ variables), nondegenerate functions of two variables (there are $N(2)$ of them and each can be applied to one of $\binom{n}{2}$ different pairs of variables), etc., and finally, nondegenerate functions of n variables (there are $N(n)$ of them). Hence,

$$T(n) = 2^{2^n} = \sum_{k=0}^{n} \left(\binom{n}{k} \cdot N(k) \right).$$

Solving this equation for $N(n)$, we obtain

$$N(n) = 2^{2^n} - \sum_{k=0}^{n-1} \left(\binom{n}{k} \cdot N(k) \right) \tag{2.9}$$

which is a recurrent formula for $N(n)$. Using equation (2.8), we obtain a recurrent formula

$$D(n) = \sum_{k=0}^{n-1} \left(\binom{n}{k} (2^{2^k} - D(k) \right) \tag{2.10}$$

Values of $D(n)$ for $n \leq 6$ are shown in Table 2.7. The percentage of degenerate functions, which is also shown in Table 2.7, decreases rapidly with n. Although the percentage of degenerate functions is significant for $n \leq 3$, almost all logic functions are nondegenerate for $n \geq 4$. Hence, a consideration of only nondegenerate functions does not make any significant reduction of the number of logic functions under consideration.

TABLE 2.7 Numbers of Degenerate Logic Functions and their Percentage Out of the Total Numbers of Logic Functions for $n \leq 6$

n	$D(n)$	$\frac{D(n)}{T(n)} \cdot 100\%$
0	0	0
1	2	50
2	6	37.5
3	38	14.843
4	942	1.437
5	325,262	7.573×10^{-3}
6	25,768,825,638	about 1.437×10^{-7}

Exercises to Section 2.5

2.5-1 Prove, on the basis of equation (2.9), that

$$N(n) = \sum_{k=0}^{n} \left((-1)^{n-k} \binom{n}{k} \cdot 2^{2^k} \right)$$

Hint: Use the proof by induction.

2.5-2 Prove, using the equation proven in Exercise 2.5-1, that

$$\lim_{n \to \infty} \frac{N(n)}{T(n)} = 1$$

and

$$\lim_{n \to \infty} \frac{D(n)}{T(n)} = 0$$

2.5-3 Determine all degenerate functions of three input variables.

2.6 EQUIVALENCE CLASSES OF LOGIC FUNCTIONS

The number of different logic functions can be considerably reduced if we partition them on the basis of a reasonable equivalence relation. Then, all functions belonging to the same equivalence class are considered as equal and may be represented by a single function taken from this equivalence class.

From the standpoint of switching circuits, it is reasonable to establish equivalence classes of logic functions based on permutations or negations of their input variables. Obviously, a permutation of input variables does not change any block diagram representing the function, no matter what kind of composition is used and, consequently, does not change the UC-structure of the corresponding switching circuit either. It simply represents a renaming of the inputs of the block diagram or the switching circuit. Similarly, a negation of some input variables does not make any change in the structure provided negations of the input variables are available to the circuit as extra inputs. In such a case, any negation and/or permutation represents only a renaming of both the basic and the extra inputs.

Three basic equivalence relations of logic functions based on permutations or negations and applicable thus in the theory of switching circuits will be now defined.

DEFINITION 2.1 A logic function $f(x_1, x_2, \ldots, x_n)$ is called *P-equivalent* with a logic function $g(x_1, x_2, \ldots, x_n)$ if $f(x_1, x_2, \ldots, x_n)$ becomes $g(x_1, x_2, \ldots, x_n)$ by some permutation of the input variables.

For instance, the function $x < y$ is P-equivalent with the function $x > y$ since the former becomes the latter when the input variables x and y are exchanged. On the contrary, $x \not\equiv y$ is not, obviously, P-equivalent with $x \equiv y$.

DEFINITION 2.2 A logic function $f(x_1, x_2, \ldots, x_n)$ is called *N-equivalent* with a logic function $g(x_1, x_2, \ldots, x_n)$ if $f(x_1, x_2, \ldots, x_n)$ becomes $g(x_1, x_2, \ldots, x_n)$ when some input variables are negated.

For instance, $x \not\equiv y$ is N-equivalent with $x \equiv y$ since the former becomes the latter when either of the input variables is negated. Functions $x < y$ and $x > y$ are also N-equivalent since one becomes the other when both input variables are negated. On the contrary, functions $x \downarrow y$ and $x \leq y$ are not N-equivalent.

DEFINITION 2.3 A logic function $f(x_1, x_2, \ldots, x_n)$ is called *PN-equivalent* with a logic function $g(x_1, x_2, \ldots, x_n)$ if $f(x_1, x_2, \ldots, x_n)$ becomes $g(x_1, x_2, \ldots, x_n)$ by some permutation and/or negation of some input variables.

For instance, logic function $\bar{z}(\bar{y} \vee \bar{x})$ is PN-equivalent with function $\bar{y}(\bar{z} \vee x)$ though they are neither P-equivalent nor N-equivalent. Function $x \not\equiv y$ is not PN-equivalent with any logic function of two variables except $x \equiv y$.

It is easy to check that the relations introduced by Definitions 2.1 through 2.3 are reflexive, symmetric, and transitive so that they represent equivalence relations. Each partitions all logic functions of n variables into equivalence classes.

Let us note that a necessary (but not sufficient) condition for equivalence (of any of the three types) of two logic functions is that their output variables have the same number of ones (and zeros) assigned to states of input variables. This condition gives rise to another equivalence relation which, clearly, generates $2^n + 1$ equivalence classes for logic functions of n variables. Still another equivalence relation, which will be called *PNN-equivalence* relation, allows permutations and/or negations of the input variables as well as negation of the output variable.

The problem of counting numbers of equivalence classes for the P-, N-, PN-, and PNN-equivalence relations is a very difficult one. From the standpoint of switching circuits, we are not motivated to study the problem itself but rather summarize the results. This is done in Table 2.8 for $n \leq 6$. The number of N-equivalence classes for a particular n is given by the formula

$$2^{2^n - n} + 2^{2^{n-1}} - 2^{2^{n-1} - n}$$

No general formulas are known for the other three equivalence relations.

TABLE 2.8 Numbers of Equivalence Classes of Logic Functions for $n \leq 6$

Equivalence Type	1	2	3	4	5	6
P	4	12	80	3,984	37,333,248	25,626,412,338,274,304
N	3	7	46	4,336	134,281,216	288,230,380,379,570,176
PN	3	6	22	402	1,228,158	400,507,806,843,728
PNN	2	4	14	222	616,126	200,253,952,527,184

Exercises to Section 2.6

2.6-1 Determine the partition of logic functions of two variables which is based on:
(a) the P-equivalence relation.
(b) the N-equivalence relation.
(c) the PN-equivalence relation.

2.6-2 Determine all elements of an equivalence class for:
(a) the majority M_3 function and the *N*-equivalence relation.
(b) the majority M_3 function and the PN-equivalence relation.
(c) the "2 out of 3" function and the P-equivalence relation.

2.7 COMPLETE SETS OF LOGIC FUNCTIONS

If we want to avoid the difficulties arising from the large number of different logic functions for $n \geq 3$ or different equivalence classes of logic functions for $n \geq 4$, we must try to express these functions with the aid of a limited number of selected functions. Let the latter be called the *primary functions*. It can be proven that this approach is successful with some sets of primary logic functions while it fails for some other. A complete study of sets of primary functions from this point of view is the purpose of this section.

When it is possible to express any logic function by means of some primary logic functions, the latter are said to form a *complete set of logic functions*. We are interested mainly in those complete sets of logic functions which contain the least possible number of functions. We shall call them *minimal complete sets* of logic functions.

No function may be omitted from a minimal complete set of logic functions without losing the property of completeness. On the other hand, any number of functions can be added to it so that we obtain other complete sets of logic functions which, of course, are no longer minimal. It is known that the least upper bound for the number of functions in a minimal complete set is four. This means that no minimal complete set comprises more than four functions but there are minimal complete sets which contain exactly four functions.

An interesting and a very important practical question arises in connection with the complete set of logic functions: "What are the necessary and sufficient conditions which every complete set of logic functions must satisfy?" In order to be able to answer this question, we must define more precisely the complete set of logic functions and some classes of logic functions first.

DEFINITION 2.4 A set C of primary logic functions is called *complete* if and only if any logic function $f(x_1, x_2, \ldots, x_n)$ can be composed by a finite number of the trivial assertion functions $x_1, x_2, \ldots, x_n$ and the functions contained in the set C.

DEFINITION 2.5 If $f(x_1, x_2, \ldots, x_n) = 0$ for $x_1 = x_2 = \cdots = x_n = 0$, the logic function f is called the *function that preserves 0.*

Examples of functions that preserve 0: AND function, non-equivalence, OR function. Examples of functions that do not preserve 0: NOR function, equivalence, inequality.

DEFINITION 2.6 If $f(x_1, x_2, \ldots, x_n) = 1$ for $x_1 = x_2 = \cdots = x_n = 1$, the logic function f is called the *function that preserves 1.*

Examples of functions that preserve 1: AND function, equivalence, implication. Examples of functions that do not preserve 1: proper inequality, non-equivalence, NAND function.

DEFINITION 2.7 If $f(x_1, x_2, \ldots, x_n) = \bar{g}(\bar{x}_1, \bar{x}_2, \ldots, \bar{x}_n)$, the logic functions f and g are called *mutually dual.* If

$$f(x_1, x_2, \ldots, x_n) = \bar{f}(\bar{x}_1, \bar{x}_2, \ldots, \bar{x}_n),$$

the logic function f is called *self-dual.*

Examples of self-dual functions: Assertion, negation, majority function M_3. Examples of functions that are not self-dual: Equivalence, inequality, OR function.

DEFINITION 2.8 The logic function $f(x_1, x_2, \ldots, x_n)$ is called *linear* if values of its output variable can be expressed for all states of the input variables $x_1, x_2, \ldots, x_n$ by even or odd value of the linear polynomial form

$$a_0 + a_1 \cdot x_1 + \cdots + a_n \cdot x_n$$

respectively, providing that the coefficients $a_0, a_1, \ldots, a_n$ are constant for the given function and equal either to 0 or to 1. We may also write for a linear logic function:

$$f(x_1, x_2, \ldots, x_n) = (a_0 + a_1 x_1 + \cdots + a_n x_n)(\text{mod } 2).$$

Examples of linear functions: Negation, nonequivalence, equivalence. Examples of non-linear functions: AND function, OR function, proper inequality.

EXAMPLE 2.4 Determine the linear logic function $f(x_1, x_2, x_3)$ specified by the linear form $(1 + x_1 + x_3)(\text{mod } 2)$.

$f(x_1, x_2, x_3)$ is a function of three input variables so that we must evaluate the linear form for the eight states of these variables to get a complete specification of the function. This is shown in Table 2.9.

TABLE 2.9 Determination of a Linear Logic Function Given by a Linear Form Modulo 2 (Example 2.4)

x_3	x_2	x_1	$1 + x_1 + x_3$	$(1 + x_1 + x_3)(\text{mod } 2) = f(x_1, x_2, x_3)$
0	0	0	1	1
0	0	1	2	0
0	1	0	1	1
0	1	1	2	0
1	0	0	2	0
1	0	1	3	1
1	1	0	2	0
1	1	1	3	1

EXAMPLE 2.5 Find whether or not the NAND function is linear.

The NAND function is a function of two input variables so that we must consider the form $(a_0 + a_1 x_1 + a_2 x_2)(\text{mod } 2)$ with unknown coefficients a_0, a_1, a_2. Each state of the variables x_1, x_2 leads to an equation containing the unknown coefficients as illustrated in Table 2.10. The NAND function is not a linear function because the last equation is contradictory to the first three equations.

TABLE 2.10 A Test for Linearity of NAND function (Example 2.5)

x_2	x_1	NAND	Equations for the Unknown Coefficient	Results
0	0	1	$a_0(\text{mod } 2) = 1$	$a_0 = 1$
0	1	1	$(a_0 + a_1)(\text{mod } 2) = 1$	$a_1 = 0$
1	0	1	$(a_0 + a_2)(\text{mod } 2) = 1$	$a_2 = 0$
1	1	0	$(a_0 + a_1 + a_2)(\text{mod } 2) = 0$	Contradictory to the previous results

EXAMPLE 2.6 Find whether or not the NONEQUIVALENCE is a linear function.

Using the same method as in Example 2.5, we obtain the results shown in Table 2.11.

TABLE 2.11 A Test for Linearity of NONEQUIVALENCE Function (Example 2.6)

x_2	x_1	$\not\equiv$	Equations for the Unknown Coefficients	Results
0	0	0	$a_0 \pmod 2 = 0$	$a_0 = 0$
0	1	1	$a_0 + a_1 \pmod 2 = 1$	$a_1 = 1$
1	0	1	$a_0 + a_2 \pmod 2 = 1$	$a_2 = 1$
1	1	0	$(a_0 + a_1 + a_2) \pmod 2 = 0$	Consistent with the other equations

DEFINITION 2.9 Let $s_i = (\dot{x}_1, \dot{x}_2, \ldots, \dot{x}_n)$ and $s_j = (\tilde{x}_1, \tilde{x}_2, \ldots, \tilde{x}_n)$ be two states of the logic variables $x_1, x_2, \ldots, x_n (\dot{x}_k, \tilde{x}_k = 0, 1)$. Let $f(s_i)$ and $f(s_j)$ be values of the function $f(x_1, x_2, \ldots, x_n)$ for the state s_i and s_j, respectively. Let $f(s_i) \leq f(s_j)$ be satisfied for all pairs (s_i, s_j) where $\dot{x}_1 \leq \tilde{x}_1$, $\dot{x}_2 \leq \tilde{x}_2, \ldots, \dot{x}_n \leq \tilde{x}_n$ is satisfied (or in an abbreviated form $\mathbf{x}(s_i) \leq \mathbf{x}(s_j)$). Then the function f is called *monotonic*.

Examples of monotonic functions: Assertion, AND function, OR function. Examples of non-monotonic functions: NOR function, NAND function, equivalence.

EXAMPLE 2.7 Find whether or not the PROPER INEQUALITY is a monotonic function. The function is defined as follows:

State	x_2	x_1	$x_1 > x_2$
s_0	0	0	0
s_1	0	1	1
s_2	1	0	0
s_3	1	1	0

We can immediately see that:

$$
\begin{array}{lcl}
\mathbf{x}(s_0) \leq \mathbf{x}(s_1) & \text{and} & f(s_0) \leq f(s_1) \\
\mathbf{x}(s_0) \leq \mathbf{x}(s_2) & \text{and} & f(s_0) \leq f(s_2) \\
\mathbf{x}(s_0) \leq \mathbf{x}(s_3) & \text{and} & f(s_0) \leq f(s_3) \\
\mathbf{x}(s_1) \leq \mathbf{x}(s_3) & \text{and} & f(s_1) > f(s_3) \\
\mathbf{x}(s_2) \leq \mathbf{x}(s_3) & \text{and} & f(s_2) \leq f(s_3)
\end{array}
$$

The condition of a monotonic logic function is not satisfied for the pair s_1, s_3. Hence, PROPER INEQUALITY is not a monotonic function. Let us note that neither $s_1 \leq s_2$ nor $s_1 \geq s_2$ so that the pair $f(s_1), f(s_2)$ need not be tested.

In order to be able to state and prove the basic theorem specifying the necessary and sufficient conditions that must be satisfied by any complete set of logic functions, let us first prove some auxiliary theorems.

THEOREM 2.1 Every logic function that is composed only of a finite number of logic functions that preserve 0 is a logic function that preserves 0.

PROOF Let us suppose that two functions that preserve 0, e.g.,

$$f(y_1, y_2, \ldots, y_m) \quad \text{and} \quad y_1(x_1, x_2, \ldots, x_n)$$

are so composed that we get $z = f(y_1(x_1, x_2, \ldots, x_n), y_2, y_3, \ldots, y_m)$. When we substitute 0 for all variables $x_1, x_2, \ldots, x_n, y_2, y_3, \ldots, y_m$, the value of y_1 will be 0 and, consequently, the value of z will be 0 too. ▌

The composition suggested in the proof of Theorem 2.1 (and also in Theorems 2.2 through 2.5) is illustrated in Figure 2.11. Note that it suffices to consider a composition of only two functions without any loss of generality. One of these functions may be associated with y_1 since it is always possible to relabel the variables so that the variable participating in the composition will be labelled y_1.

THEOREM 2.2 Every logic function that is composed only of a finite number of logic functions that preserve 1 is a logic function that preserves 1.

PROOF Analogous to the proof of Theorem 2.1. ▌

THEOREM 2.3 Every logic function that is composed only of a finite number of self-dual logic functions is a self-dual logic function.

PROOF Let us suppose that two self-dual functions, e.g., $f(y_1, y_2, \ldots, y_m)$ and $y_1(x_1, x_2, \ldots, x_n)$ are so composed that we get $z = f(y_1(x_1, x_2, \ldots, x_n), y_2, y_3, \ldots, y_m)$. Now, let us take the function $w = \bar{f}(y_1(\bar{x}_1, \bar{x}_2, \ldots, \bar{x}_n), \bar{y}_2, \bar{y}_3, \ldots, \bar{y}_m)$. Since y_1 is a self-dual function, it must be $y_1(\bar{x}_1, \bar{x}_2, \ldots, \bar{x}_n) = \bar{y}_1(x_1, x_2, \ldots, x_n)$. Then, however, $w = \bar{f}(\bar{y}_1(x_1, x_2, \ldots, x_n), \bar{y}_2, \bar{y}_3, \ldots, \bar{y}_m)$ and, since f is a self-dual function, $w = z$. ▌

THEOREM 2.4 Every logic function that is composed only of a finite number of linear logic functions is a linear function.

PROOF Let us suppose that two linear functions, e.g., $f(y_1, y_2, \ldots, y_m)$ and $y_1(x_1, x_2, \ldots, x_n)$ are so composed that we get $z = f(y_1 (x_1, x_2, \ldots, x_n),$

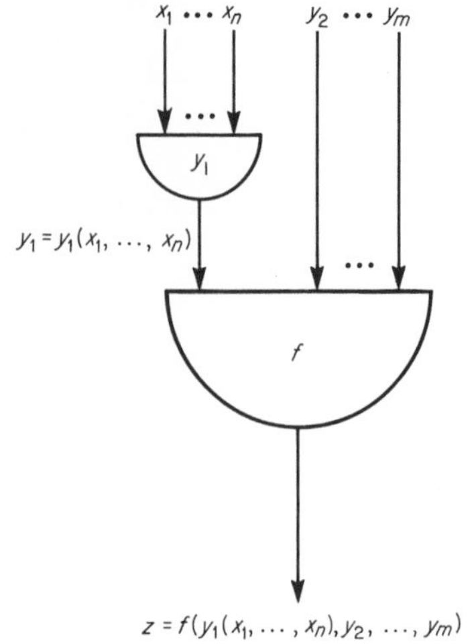

Figure 2.11 The composition considered in the proofs of Theorems 2.1 through 2.5.

$y_2, y_3, \ldots, y_m$). Let us further suppose that the polynomial forms $a_0 + a_1 \cdot y_1 + \cdots + a_m \cdot y_m$ and $b_0 + b_1 \cdot x_1 + \cdots + b_n \cdot x_n$ are assigned to the functions f and y_1, respectively. The polynomial form of the composite function can be simply obtained in the form $a_0 + a_1(b_0 + b_1 \cdot x_1 + \cdots + b_n \cdot x_n) + a_2 \cdot y_2 + \cdots + a_m \cdot y_m = A_0 + A_1 \cdot x_1 + \cdots + A_n \cdot x_n + a_2 \cdot y_2 + a_2 \cdot y_2 + \cdots + a_n \cdot y_m$, where $A_0 = a_0 + a_1 \cdot b_0$, $A_1 = a_1 \cdot b_1$, $A_2 = a_1 \cdot b_2, \ldots,$ $A_n = a_1 \cdot b_n$, which is again a linear form. ▮

THEOREM 2.5 Every logic function that is composed only of a finite number of monotonic logic functions is a monotonic function.

PROOF Let us suppose that two monotonic functions, e.g.,

$$f(y_1, y_2, \ldots, y_m) \qquad \text{and} \qquad y_1(x_1, x_2, \ldots, x_n)$$

are so composed that we get $z = f(y_1(x_1, x_2, \ldots, x_n), y_2, y_3, \cdots, y_m)$. Let s_i, s_j be states of variables $x_1, x_2, \cdots, x_n, y_1, y_2, \ldots, y_m$ such that $s_i \leq s_j$. It is obvious that each of the states s_i, s_j uniquely determines the partial states of variables $x_1, x_2, \ldots, x_n$ and $y_1, y_2, \ldots, y_m$. Let u_i, u_j be the corresponding partial states of variables $x_1, x_2, \ldots, x_n$ and v_i, v_j be the corresponding partial states of variables $y_1, y_2, \ldots, y_m$. Obviously, the inequalities $u_i \leq u_j$ and $v_i \leq v_j$ follow from $s_i \leq s_j$. Let $z(s_i)$, $z(s_j)$ be values of the function z for the states s_i, s_j, respectively, $y_1(u_i)$, $y_1(u_j)$ be values of the function y_1 for the states u_i, u_j, respectively, and $a_1, a_2, \ldots, a_m$ and $b_1, b_2, \ldots, b_m$ be values of the variables $y_1, y_2, \ldots, y_m$ for the states v_i and v_j, respectively. Then we can write

$$z(s_i) = f(y_1(u_i), a_2, a_3, \ldots, a_m)$$
$$z(s_j) = f(y_1(u_j), b_2, b_3, \ldots, b_m$$

Since the function y_1 is monotonic, $y_1(u_i) \leq y_1(u_j)$. Since $v_i \leq v_j$ and the function f is monotonic, it follows that $z(s_i) \leq z(s_j)$. ▮

THEOREM 2.6 Let $x^0 = \bar{x}$ and $x^1 = x$. Then, every logic function $f(x_1, x_2, \ldots, x_n)$ can be expressed in the form

$$f(x_1, x_2, \ldots, x_n) = \bigvee_{s=0}^{2^n-1} f(s) \prod_{i=1}^{n} x_i^{i_s} \tag{2.11}$$

where s is the state identifier, $f(s)$ is the value of the function f for the state s, i_s is the value of variable x_i for the state s, $\bigvee$ and $\prod$ represent, respectively, repeatedly applied operations OR and AND.

PROOF The function expressed by $\prod_{i=1}^{n} x_i^{i_s}$ is equal to 1 if and only if $x_i = i_s$ for all $i = 1, 2, \ldots, n$. It is equal to 0 for all other states of variables $x_1, x_2, \ldots, x_n$. The theorem follows then directly from this property and from the definition of OR function. ▮

THEOREM 2.7 The negation of any variable of a non-monotonic logic function $f(x_1, x_2, \ldots, x_n)$ can be obtained by a composition of this function, the ZERO function, and the ONE function.

PROOF Let the integers s_i, s_j identify such a pair of states of variables $x_1, x_2, \ldots, x_n$ that $s_i \leq s_j$ and $f(s_i) = 1, f(s_j) = 0$. Obviously, it is always possible to find such a pair of states because the function f is assumed to be a non-monotonic function. Some variables may have the same value in both s_i and s_j. However, if a variable has different values in s_i and s_j, it must be 0 in s_i and 1 in s_j. Thus, there exists a partition $\{\{x_a\}, \{x_b\}, \{x_c\}: a \in A, b \in B, c \in C, A \cup B \cup C = \{1, 2, \ldots, n\}, A \cap B = \varnothing, A \cap C = \varnothing, B \cap C = \varnothing\}$ of variables $x_1, x_2, \ldots, x_n$ such that $x_a = 0$ and $x_b = 1$ for both s_i and s_j whereas $x_c = 0$ for s_i and $x_c = 1$ for s_j. Let $x_a = 0$ for all $a \in A$, $x_b = 1$ for all $b \in B$, and $x_c = x$ for all $c \in C$. Then, if $x = 0$, then $f(x_a = 0, x_b = 1, x_c = x) = 1$; if $x = 1$, then $f(x_a = 0, x_b = 1, x_c = x) = 0$. Hence, $f(x_a = 0, x_b = 1, x_c = x) = \bar{x}$. ▮

THEOREM 2.8 ZERO and ONE functions can be obtained by a composition of NEGATION and a function which is not self-dual.

PROOF Let $f(x_1, x_2, \ldots, x_n)$ be a logic function that is not self-dual. Then there is at least one state $(\dot{x}_1, \dot{x}_2, \ldots, \dot{x}_n)$ of the variables $x_1, x_2, \ldots, x_n$

for which $f(\dot{x}_1, \dot{x}_2, \ldots, \dot{x}_n) = f(\bar{\dot{x}}_1, \bar{\dot{x}}_2, \ldots, \bar{\dot{x}}_n)$. Let us consider a logic function of one variable $F(y) = f(g_1(y), g_2(y), \ldots, g_n(y))$, where

$$g_i(y) = \begin{cases} y & \text{if} \quad \dot{x}_i = 0 \\ \bar{y} & \text{if} \quad \dot{x}_i = 1 \end{cases}$$

Then $F(0) = f(g_1(0), g_2(0), \ldots, g_n(0)) = f(\dot{x}_1, \dot{x}_2, \ldots, \dot{x}_n) = f(\bar{\dot{x}}_1, \bar{\dot{x}}_2, \ldots, \bar{\dot{x}}_n) = f(g_1(1), g_2(1), \ldots, g_n(1)) = F(1)$. Hence, $F(0) = F(1) = F(y) =$ CONSTANT (either 0 or 1). If $F(y) = 0$, then $F(\bar{y}) = 1$. If $F(y) = 1$, then $F(\bar{y}) = 0$. ∎

THEOREM 2.9 A set of logic functions is complete if, and only if, it contains:

1. at least one function that does not preserve 0
2. at least one function that does not preserve 1
3. at least one function that is not self-dual
4. at least one nonlinear function
5. at least one non-monotonic function

PROOF The necessity of the above conditions follows directly from Theorems 2.1 through 2.5 and from the fact that the trivial assertion functions $x_1, x_2, \ldots, x_n$, mentioned in Definition 2.4, do not satisfy any one of the conditions.

In order to prove that the conditions are sufficient, let us denote functions belonging to the respective classes by the following symbols: f_1 = a function not preserving 0; f_2 = a function not preserving 1; f_3 = a function that is not self-dual; f_4 = a nonlinear function; f_5 = a non-monotonic function. Of course, these functions need not be necessarily all different.

Now, we seek a way of composing the functions f_1, f_2, f_3, f_4, f_5 and the trivial functions $x_1, x_2, \ldots, x_n$ so that they would express any given function $f(x_1, x_2, \ldots, x_n)$.

It follows from Theorem 2.6 that NEGATION, AND function, and OR function together represent a complete set of logic functions. Since, however, it can be easily proven that $x_1 \vee x_2 = \overline{\bar{x}_1 \cdot \bar{x}_2}$, OR function can be substituted for by the AND function and NEGATION. Thus, AND function and NEGATION constitute the complete set of logic functions too. We can prove the theorem by finding a way of composing these two functions by the functions f_1, f_2, f_3, f_4, f_5 and the trivial functions $x_1, x_2, \ldots, x_n$.

Even if the output variable of a function is constant, the latter will be considered as a function of at least one input variable.

Let us compose the functions f_1 and $x_i (1 \leq i \leq n)$, i.e.,

$$g_1(x_i) = f_1(x_i, x_i, \ldots, x_i).$$

$g_1(0) = 1$ since f_1 does not preserve 0. There is either $g_1(1) = 0$ or $g_1(1) = 1$. If $g_1(1) = 0$, then $g_1(x_i) = \bar{x}_i$. If $g_1(1) = 1$, then $g_1(x_i) = 1$ (ONE function). Thus, f_1 produces either NEGATION or ONE function. Similarly, we can find that the function $g_2(x_i) = f_2(x_i, x_i, \ldots, x_i)$ produces either NEGATION or ZERO function. There are four possibilities depending on special properties of f_1 and f_2:

f_1	f_2
NEGATION	NEGATION
NEGATION	ZERO function
ONE function	NEGATION
ONE function	ZERO function

In the first three cases, NEGATION is produced by at least one of the functions. Using NEGATION and the function f_3, both ZERO and ONE functions can be produced (Theorem 2.8). In the fourth case, both ZERO and ONE functions are available. NEGATION can be produced by them and the function f_5 (Theorem 2.7). Hence, we may assume now that NEGATION and both ZERO and ONE functions are available.

It remains to obtain AND function. In order to do that we will use the function $f_4(x_1, x_2, \ldots, x_n)$. Since the function f_4 is nonlinear, obviously, its polynomial form must contain at least one product of two or more variables (a power of a single variable would not make the form nonlinear since $x^m = x$ for any logic function and $m \geq 1$). Let us suppose that such a product is formed by the variables x_1 and x_2. Then the polynomial form may be written as $x_1 x_2 h_1(x_3, x_4, \ldots, x_n) + x_1 h_2(x_3, x_4, \ldots, x_n) + x_2 h_3(x_3, x_4, \ldots, x_n) + h_4(x_3, x_4, \ldots, x_n)$.

The function h_1 is certainly not equal to zero for all states of the variables $x_3, x_4, \ldots, x_n$ (since f_4 is not linear) so that there is at least one state of the variables $x_3, x_4, \ldots, x_n$ for which $h_1(x_3, x_4, \ldots, x_n) = 1$. We can get it by a composition of the nonlinear function f_4 and the ZERO and ONE functions. The polynomial form will become $x_1 x_2 + Ax_1 + Bx_2 + C$, where A, B, C are either 0 or 1. Since we can obtain NEGATION $\bar{x}_i$, whose polynomial form is $1 + x_i$, we can obtain also the function $H(x_1, x_2)$ whose polynomial form is

$$(x_1 + B) \cdot (x_2 + A) + A(x_1 + B) + B(x_2 + A) + C$$
$$= x_1 x_2 + 2Ax_1 + 2Bx_2 + 3AB + C.$$

Since, according to Definition 2.8, we are only interested in whether the value of the polynomial expression is even or odd, it is possible to simplify it. We get $x_1 x_2 + AB + C = x_1 x_2 + D$, where $D = AB + C$. If $D = 0$, $H(x_1, x_2)$ is the AND function. If $D = 1$, $h = \overline{x_1 x_2}$. Since NEGATION is already available, the AND function can be again obtained. ▌

The construction of NEGATION, ONE, and ZERO functions is illustrated in Figure 2.12; the construction of AND function is shown in Figure 2.13. The interrupted lines express that only one of the two alternatives is used depending on specific properties of the functions involved; the interrupted boxes in Figure 2.13 mean that in each case block B is or is not used depending on the used nonlinear function f_4.

When testing logic functions of one variable we find that all of them are linear functions. It is therefore impossible to arrange a complete set by logic functions of only one variable.

There are two logic functions within the functions of two variables, namely the NOR and NAND functions, which have the property that each of them represents itself a complete set of logic functions. It can be easily verified that each of these functions satisfies all the conditions of Theorem 2.9. Obviously, each of them represents a minimal complete set of logic functions.

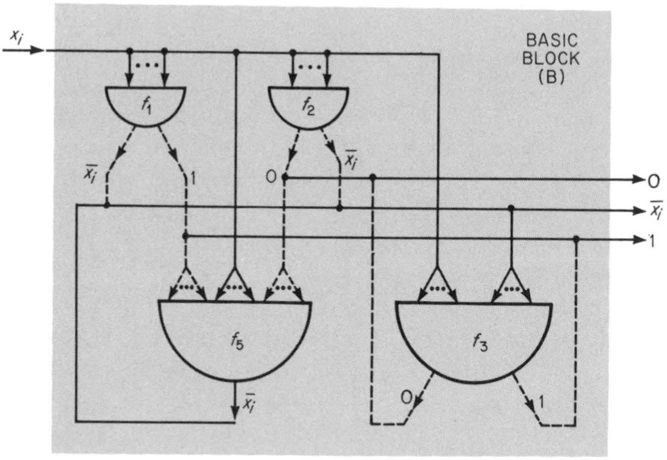

Basic block produces: $\bar{x}_i$ and either 0 or 1 or both.

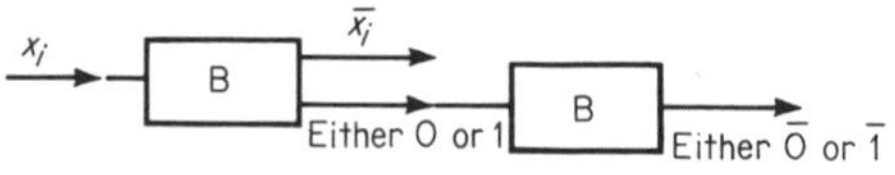

A way of producing the second constant provided that only one constant is produced by the basic block.

Figure 2.12 Proof of Theorem 2.9: a construction of NEGATION, ZERO, and ONE functions.

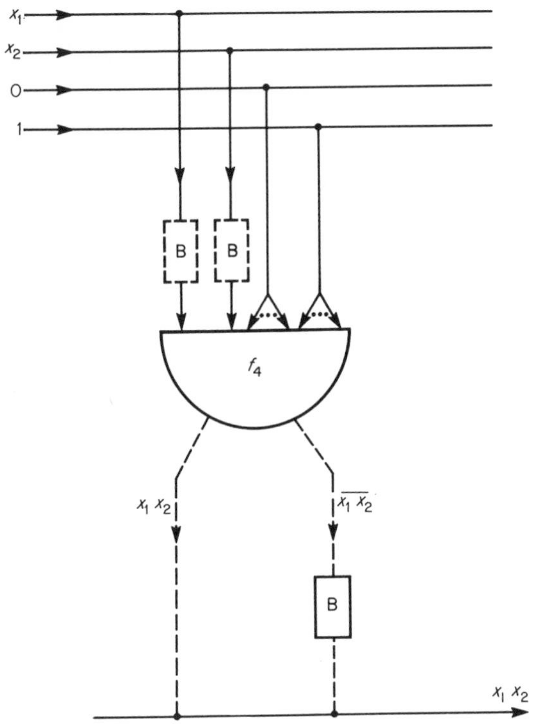

Figure 2.13 Proof of Theorem 2.9: a construction of AND function.

All other minimal sets of logic functions that can be found within logic functions of one or two variables are contained in the following list:

Inequality–negation
Inequality–ZERO function
Inequality–proper inequality
Inequality–nonequivalence
Proper inequality–equivalence
Proper inequality–negation
Proper inequality–ONE function
AND function–negation
OR function–negation
AND function–equivalence–ZERO function
AND function–equivalence–nonequivalence
AND function–nonequivalence–ONE function
OR function–equivalence–ZERO function
OR function–nonequivalence–ONE function
OR function–equivalence–nonequivalence

Exercises to Section 2.7

2.7-1 Fill out the following table. Each square in the table is identified by a logic function (columns) and a condition (rows). Write down + into a particular square if the corresponding function satisfies the assigned condition; otherwise write down – into the square.

Function E

	0	1	x	$\bar{x}$	$\cdot$	$<$ $>$	$\not\equiv$	$\vee$	$\leq$ $\geq$	$\downarrow$	$\equiv$	$\mid$	M_3	M_5	F_3^1	F_3^2	F_3^{1+}	F_3^{2-}
f preserves 0																		
f preserves 1																		
f is self-dual																		
f is linear																		
f is monotonic																		

*2.7-2 Find all logic functions of three variables each of which, by itself, forms a complete set of logic functions (analogy of NAND and NOR functions).

2.7-3 Prove the following statements applied to logic functions of n variables:
(a) There are just 2^{2^n-1} functions that preserve 0, 2^{2^n-1} functions that preserve 1, and 2^{2^n-2} functions that preserve both 0 and 1.
(b) 2^{n+1} is an upper bound for the number of linear functions.
(c) 2^{2^n-1} is the number of self-dual functions.

2.7-4 Determine all linear logic functions of two variables.

*2.7-5 Express, with the aid of a complete set of logic functions consisting of NEGATION and INEQUALITY, the remaining functions of two variables, in particular AND function, PROPER INEQUALITY, NONEQUIVALENCE, OR function, NOR function, EQUIVALENCE, and NAND function.

*2.7-6 Suppose Theorem 2.9 is not known. Let F be a complete set of logic functions $f_1, f_2, \ldots, f_p$ and let G be a given set of logic functions $g_1, g_2, \ldots, g_q$ whose completeness is to be tested. Suggest a procedure for testing of the completeness of G on the basis of F.

2.7-7 Express the following logic functions in the form (2.11):
*(a) NONEQUIVALENCE.
(b) PROPER INEQUALITY.
*(c) logic function F_3^2 ("2 out of 3").
(d) majority function M_3.
(e) threshold function of four variables with $w_1 = 1$, $w_2 = 1.5$, $w_3 = 0.5$, $w_4 = 0.5$, and $T = 0.5$.

2.7-8 In the literature universal logic functions are often called those functions which represent complete sets when supplemented by both ZERO and ONE functions. Find all universal logic functions of two input variables.

2.7-9 Give a specific meaning to functions f_1 through f_5 introduced in the proof of Theorem 2.9 and reproduce the proof of the Theorem for them.

2.8 LOGIC ALGEBRAS

A collection of a nonempty set A together with a set F of functions (operations) defined on A or on a set product (Cartesian product) of A is designated as an *abstract algebra*. Abstract algebras may be defined in either of the following two ways:

(1) The set A and the functions are specified in an explicit form.
(2) The set A and/or the functions are specified implicitly, i.e., some axioms about properties of A and/or the functions are laid down from which all the other properties of A and the functions, including their explicit specifications, can be derived by pure reasoning, provided that the axioms are consistent and complete.

Regardless of how an abstract algebra is defined, the collection of all special properties, rules, procedures, etc., concerning various compositions of the particular functions provided by the algebra constitutes the *theory of the algebra*.

Obviously, the theory of an algebra may be elaborated on the basis of either the *explicit* or the *implicit* definition of the algebra. Using the explicit definition, we speak about the *inductive* elaboration and, using the implicit definition, we speak about the *deductive* elaboration.

If $A = \{0, 1\}$ and F is a nonempty set of logic functions, then the pair (A, F) constitutes a *logic algebra*. Clearly, the set of all logic algebras is a subset of all abstract algebras.

The application of every abstract algebra is limited to the domain of A and its set products. However, if an incomplete set of functions is specified for a logic algebra, then the application of the algebra is limited even within the domain of $A = \{0, 1\}$ and its set products. This is not acceptable in the majority of applications and, therefore, logic algebras are usually required to be defined by complete sets of logic functions. Let logic algebras which are defined by complete sets of logic functions be called *proper logic algebras*.

There can be as many different proper logic algebras as there are complete sets of logic functions. We know, however, that the number of these sets is unlimited, since it is possible to add to every complete set an arbitrary number of further functions, so that we keep on obtaining new complete sets of functions.

Only a few logic algebras have so far been elaborated. They are mostly algebras based on logic functions of two independent variables.

Exercises to Section 2.8

2.8-1 Suppose that a logic algebra L_1 based on a complete set of logic functions C_1 is defined by a consistent and complete set of axioms (postulates) $P_1, P_2, \ldots, P_m$.

*(a) Using these axioms and the functions contained in C_1, suggest a procedure of introducing a set of axioms $Q_1, Q_2, \ldots, Q_n$ that define a logic algebra L_2 based on a complete set of logic functions $C_2 \neq C_1$.

*(b) Suggest a test for the completeness of $Q_1, Q_2, \ldots, Q_n$.

2.9 BOOLEAN ALGEBRA

The algebra most important to the methodology of switching circuits and to many other fields is based on the three logic functions: AND function, OR function (they will also be called, respectively, LOGIC PRODUCT and DISJUNCTION), and NEGATION. In honor of the famous Irish mathematician and logician, George Boole (1815–1864), this algebra is called *Boolean algebra.*

Let us show a possible way of deductive elaboration of Boolean algebra.

Any collection $B = \{A, \bar{x}, x \vee y, x \cdot y : x, y \in A\}$ of a set A and functions (operations) $\bar{x}$, $x \vee y$, xy that is specified by the following axioms is called a Boolean algebra:

(I) If x and y are elements of set A, then $x \vee y$ and xy are also elements of A (*laws of closure*).

(II) In set A there exists a significant element 0 such that the equality $x \vee 0 = x$ is satisfied for any element x of set A; in a similar manner, there exists a significant element 1 for which the equality $x \cdot 1 = x$ is satisfied (*laws of absorption*).

(III) The equality $x \vee y = y \vee x$ is always satisfied for the elements x and y of set A, and so is the equality $xy = yx$ (*commutative laws*).

(IV) For the elements x, y, and z of set A, the equality $x \vee (yz) = (x \vee y)(x \vee z)$ is always satisfied, and so is the equality $x(y \vee z) = xy \vee xz$ (*distributive laws*).

(V) For every element x of set A there exists an element $\bar{x}$ which always satisfies the equality $x\bar{x} = 0$ and, simultaneously, the equality $x \vee \bar{x} = 1$ (*laws of the excluded middle*).

(VI) There exist at least two elements x and y of set A such that $x \neq y$.

Let us note that the set of axioms listed above defines a special Boolean algebra in which A = {0, 1}. It is usually called the *two-valued Boolean algebra.* There exist other Boolean algebras in which the set A contains 2^k elements, where k is an integer and $k > 1$.

The two-valued Boolean algebra (called simply Boolean algebra further on) suffices for our purposes in this book. The axioms specified above were introduced by the American mathematician E. V. Huntington (1874–1952) as long ago as 1904. This is, however, only one of possible approaches to the axiomatic construction of the two-valued Boolean algebra.

Now let us demonstrate a derivation of further properties of Boolean algebra from the axioms listed; e.g., if we want to prove the validity of the equality $x \vee 1 = 1$, we can proceed as follows:

$$
\begin{aligned}
x \vee 1 &= (x \vee 1) \cdot 1 && \text{(according to II)} \\
&= (x \vee 1)(x \vee \bar{x}) && \text{(according to V)} \\
&= x \vee (1 \cdot \bar{x}) && \text{(according to IV)} \\
&= x \vee (\bar{x} \cdot 1) && \text{(according to III)} \\
&= x \vee \bar{x} && \text{(according to II)} \\
&= 1 && \text{(according to V)}
\end{aligned}
$$

In a similar manner we could prove that $x \cdot 0 = 0$. Now let us show how the relation $x \vee xy = x$ could be proved. In addition to our axioms we shall employ the property that has just been proved:

$$
\begin{aligned}
x \vee xy &= (x \cdot 1) \vee (xy) && \text{(according to II)} \\
&= x(1 \vee y) && \text{(according to IV)} \\
&= x(y \vee 1) && \text{(according to III)} \\
&= x \cdot 1 && \text{(according to the relation proved)} \\
&= x && \text{(according to II)}
\end{aligned}
$$

In the exposition to follow we abandon the deductive approach, because it would be too lengthy for our purposes. We shall content ourselves with presenting a summary of the fundamental rules of Boolean algebra, without deriving or proving them. The rules we have in mind concern the equalities which, in Boolean algebra, are always valid for certain pairs of algebraic expressions. We can ascertain the validity of any equality in Boolean algebra by transforming each of the two expressions into a tabular form in the same manner as indicated in Section 2.3.

Every logic expression of Boolean algebra consists of the symbols 0 and 1, letters denoting the logic variables $x, y, \ldots$, and symbols of the functions of Boolean algebra. To be more accurate, we should define the expressions of Boolean algebra—or *Boolean expressions*—in the following manner:

1. 0 and 1 are Boolean expressions.
2. Every logic variable x_i is a Boolean expression.

3. If E is a Boolean expression, then $\bar{E}$ is also a Boolean expression.
4. If E_1 and E_2 are Boolean expressions, then E_1E_2 is also a Boolean expression.
5. If E_1 and E_2 are Boolean expressions, then $E_1 \vee E_2$ is also a Boolean expression.
6. The only Boolean expressions are those defined by 1 to 5.

To avoid an extensive use of parentheses, a *hierarchy of the functions used in Boolean algebra* is usually established. If not specified otherwise by parentheses or by bars, the functions participating in a Boolean expression are composed in the following order: NEGATIONS, AND functions, OR functions.

If the expression includes parentheses, then the functions inside the innermost parentheses are applied first. When negating a Boolean expression, the parentheses are omitted and the symbol of negation (a bar above the expression) substitutes for them; this must be respected when determining the order of compositions.

When using the above-mentioned hierarchy of functions in Boolean algebra, the fundamental rules of the algebra (including the rules implied in the foregoing axioms) can be summed up in the following nine laws:

1. *Laws of the aggressivity and neutrality of the elements* 0 *and* 1:

$$x \vee 1 = 1 \qquad x \cdot 0 = 0$$
$$x \vee 0 = x \qquad x \cdot 1 = x$$

2. *Commutative laws:*

$$x \vee y = y \vee x \qquad xy = yx$$

3. *Associative laws:*

$$x \vee (y \vee z) = (x \vee y) \vee z \qquad x(yz) = xy(z)$$

4. *Distributive laws:*

$$x \vee yz = (x \vee y)(x \vee z) \qquad x(y \vee z) = xy \vee xz$$

5. *Laws of absorption:*

$$x \vee x = x \qquad xx = x$$
$$x \vee xy = x \qquad x(x \vee y) = x$$

6. *Laws of absorption of negation:*

$$x \vee \bar{x}y = x \vee y \qquad x(\bar{x} \vee y) = xy$$
$$\bar{x} \vee xy = \bar{x} \vee y \qquad \bar{x}(x \vee y) = \bar{x}y$$

7. *Law of double negation:*

$$\bar{\bar{x}} = x$$

8. *Laws of the excluded middle:*

$$x \vee \bar{x} = 1 \qquad x\bar{x} = 0$$

9. *Laws of the negation* (*De Morgan's laws*):

$$x \vee y = \bar{x}\bar{y} \qquad \overline{xy} = \bar{x} \vee \bar{y}$$

Let us note that the fundamental equalities of Boolean algebra listed above are, in almost all cases, arranged in pairs. The expressions in the equalities on the right-hand side can be obtained from the corresponding expressions in the equalities on the left-hand side, and vice versa, simply by replacing the element 0 by the element 1, i.e., OR function by AND function and vice versa. We are thus concerned with pairs of *dual expressions.*

The *principle of duality* has general validity in Boolean algebra. To every equality of two expressions there corresponds an equality of expressions which are dual with respect to them within the meaning of the transformation presented above, i.e.

$$0 \leftrightarrow 1$$
$$1 \leftrightarrow 0$$
$$\vee \leftrightarrow \cdot$$
$$\cdot \leftrightarrow \vee$$

The principle of duality can be used with advantage when deriving new rules of Boolean algebra. That is to say, when deducing any equality of Boolean algebra we automatically obtain an equality of the corresponding dual expressions.

Essentially, the concept of a "*Boolean function*" coincides with the concept of "*logic function.*" The difference consists only of the manner of algebraic presentation. As we know, any logic function can occur in a logic expression, whereas in Boolean expressions we are restricted to NEGATION, AND function, and OR function. We shall henceforth use occasionally the term "*Boolean function*" also in contexts where there is no difference between the properties of the Boolean functions and the logic function (e.g., in the tabular form of presentation).

Whenever speaking—for the sake of simplicity—of the values of Boolean functions, we thereby mean the values of the output variables which are assigned by the pertinent function to the values of the given input variables.

Exercises to Section 2.9

2.9-1 Prove that the equation (the De Morgan law) $\overline{x \vee y} = \bar{x}\bar{y}$ is valid.
(a) Use the tabular method of successive composition (the method of perfect induction).
(b) Use the axioms of Boolean algebra. (Hint: Prove that $x \vee y \vee \bar{x}\bar{y} = 1$ and that $(x \vee y)\bar{x}\bar{y} = 0$.)

2.9-2 Using the tabular method of successive composition (perfect induction) prove the following equalities in Boolean algebra:
(a) the distributive law.
(b) the laws of absorption.
(c) the laws of absorption of negation.

2.9-3 Using the laws of Boolean algebra, simplify the Boolean expressions on the left side of the following equations in order to obtain the expressions on the right side:

(a) $xy\bar{z} \vee \bar{x}\bar{y}z \vee \bar{x}yz \vee xyz = xy \vee \bar{x}z.$

(b) $\bar{x}\bar{y} \vee x\bar{y} \vee \bar{x}z \vee yz = \bar{y} \vee z.$

(c) $xy\bar{z} \vee \bar{x}y\bar{z} \vee x\bar{y}z \vee xyz = xz \vee y\bar{z}.$

(d) $\bar{x}\bar{z}\bar{v} \vee \bar{y}z\bar{v} \vee \bar{x}y\bar{z} \vee xzv \vee xyzv \vee \bar{x}yv = \bar{x}\bar{y}\bar{v} \vee xz\bar{v} \vee \bar{x}y\bar{z} \vee yzv.$

2.9-4 Using De Morgan's laws, determine the negations of the following Boolean expressions:

*(a) $\bar{x}y\bar{z} \vee x\bar{y}z.$

(b) $(xy \vee \bar{x}\bar{y})(z \vee \bar{x}y).$

*(c) $(\bar{x}(y \vee \bar{z}v) \vee x(\bar{v} \vee yz))w.$

(d) $\bar{x}\bar{y}\bar{z} \vee (xy(z\bar{v} \vee \bar{z}w) \vee xy)\bar{w}x.$

2.9-5 Determine the duals of the Boolean equations in Exercise 2.9-3.

2.9-6 Using the method of mathematical induction, prove De Morgan's laws for n variables:

$$\overline{\dot{x}_1\dot{x}_2\cdots\dot{x}_n} = \bar{\dot{x}}_1 \vee \bar{\dot{x}}_2 \vee \cdots \vee \bar{\dot{x}}_n$$

$$\overline{\dot{x}_1 \vee \dot{x}_2 \vee \cdots \vee \dot{x}_n} = \bar{\dot{x}}_1\,\bar{\dot{x}}_2\cdots\bar{\dot{x}}_n$$

($\dot{x}_i$ is either x_i or $\bar{x}_i$, $i = 1, 2, \cdots, n$).

*2.9-7 Prove that the Huntington axioms of Boolean algebra are complete. (Hint: Prove that $\bar{0} = 1$, $\bar{1} = 0$, $0 \cdot 0 = 0 \cdot 1 = 1 \cdot 0 = 0$, $1 \cdot 1 = 1$, $0 \vee 0 = 0$, $0 \vee 1 = 1 \vee 0 = 1 \vee 1 = 1$.)

2.9-8 Using the results of Exercise 2.8-1 and the Huntington axioms of Boolean algebra, determine a set of axioms for so-called Sheffer's algebra (based solely on NAND function) and verify that the axioms are complete. (Hint: $\bar{x} = x|x$, $x_1x_2 = (x_1|x_2)|(x_1|x_2)$, $x_1 \vee x_2 = (x_1|x_1)|(x_2|x_2)$.

2.9-9 Reduce the following Boolean expressions by taking off negations:

(a) $\overline{\overline{\overline{xy}z} \vee \bar{x}y\bar{z}}$

(b) $\overline{\overline{(\bar{x} \vee \overline{y \vee z})}\,\overline{(x \vee y \vee z)}}$

2.10 FORMS OF EXPRESSING LOGIC FUNCTIONS

There are a great number of forms of expressing logic functions. In the theory of switching circuits, however, the most frequently used fundamental means are *tables*, *charts*, and *algebraic expressions*. To them we shall therefore devote our particular attention.

The table as a means for expressing logic functions has already been introduced in Section 2.2. We therefore know that tables express logic functions without regard to the kind of algebra used. When we are concerned with a definite logic function of n input variables, the table has 2^n rows. In order to improve orientation, we sometimes write down the pertinent state identifiers in front of the individual rows. When concerned with an indefinite logic function, the undefined (" don't care ") states are usually not entered in the table. Thus, if a table has less than 2^n rows, we immediately know that it is an indefinite logic function we are concerned with.

As an example of a tabular representation let us now present a logic function $f = f(x_1, x_2, x_3, x_4)$ of four input variables:

s	x_4	x_3	x_2	x_1	$f(s)$
0	0	0	0	0	1
1	0	0	0	1	0
2	0	0	1	0	0
3	0	0	1	1	0
4	0	1	0	0	1
6	0	1	1	0	1
7	0	1	1	1	1
8	1	0	0	0	1
10	1	0	1	0	0
11	1	0	1	1	1
12	1	1	0	0	1
13	1	1	0	1	1
14	1	1	1	0	0
15	1	1	1	1	1

We see that we are concerned with an indefinite function, since the table has only fourteen rows. The states described by identifiers 5 and 9 are missing. We therefore consider them to be undefined.

The tabular representation of logic functions can be abridged by writing down only the identifiers of states of input variables and the values of the corresponding output variable assigned to them by the function. Logic functions are frequently expressed merely by the set of those identifiers for which the corresponding output variable has the value 1. If the function is indefinite, the identifiers of the undefined states are written down, for instance, in parentheses or are distinguished in some other manner.

For the function f, presented above in full tabular form, we would then obtain the following abridged notation: $f(s) = 1$ for $s = 0, 4, 6, 7, 8, 11, 12, 13, 15, (5, 9)$, where $f(s)$ is the value of f for the state s.

The logic charts which can be used to represent logic functions were introduced in Section 2.1. If the logic function to be represented in the chart assigns the value 0 to the given state, the corresponding square in the chart is left empty. If it assigns the value 1, the center of the square is marked with a plain dot. Squares corresponding to undefined states are marked with hatching.

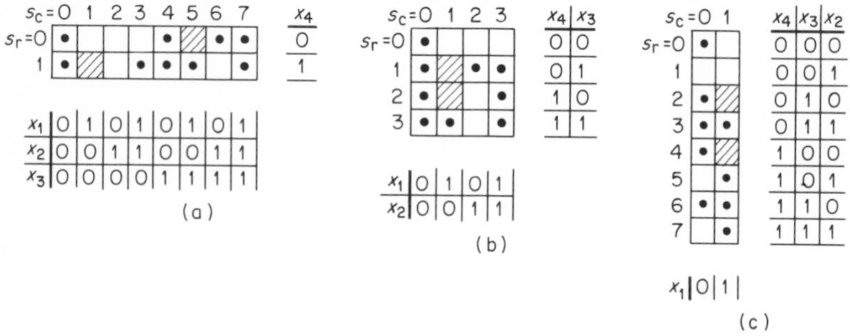

Figure 2.14 Various ways of expressing the same logic function by means of Marquand charts.

Figure 2.14 shows in what manner it would be possible to represent—by means of the Marquand charts—the logic function of four variables, expressed above in tabular form.

The possibility of using *standard Boolean expressions* to represent logic functions is implied in the following theorem.

THEOREM 2.10 Any logic function f, defined for n independent variables, can always be expressed either in its *expanded sp* (*s*um of *p*roducts)* *standard Boolean form*,

$$f = \bigvee_{s=0}^{2^n-1} f(s)M_s \tag{2.12}$$

or in its *expanded ps* (*p*roduct of *s*ums)* *standard Boolean form*,

$$f = \prod_{s=0}^{2^n-1} (f(s) \vee \overline{M}_s) \tag{2.13}$$

where s is the state identifier, $f(s)$ are values of f for the state s, and the symbol M_s denotes so-called *minterm* pertaining to the state s. The minterm is an AND operation containing all input variables of the function f, either in the asserted or negated form. Negated are only those variables which in the given state s have the value 0. All other variables are asserted. Negations of minterms are called *maxterms*. AND operations and OR operations which do not contain all input variables will be called *p-terms* and *s-terms* respectively.

The first part of Theorem 2.10 (Formula (2.12)) was proven in Section 2.7 (Theorem 2.6). The proof of the second part (Formula (2.13)) is left to the reader as an exercise.

* OR operation is sometimes referred to as a logic sum and denoted by + (see Table 2.1).

TABLE 2.12 Minterms for Four Logic Variables

s	x_4	x_3	x_2	x_1	M_s
0	0	0	0	0	$\bar{x}_4\bar{x}_3\bar{x}_2\bar{x}_1$
1	0	0	0	1	$\bar{x}_4\bar{x}_3\bar{x}_2x_1$
2	0	0	1	0	$\bar{x}_4\bar{x}_3x_2\bar{x}_1$
3	0	0	1	1	$\bar{x}_4\bar{x}_3x_2x_1$
4	0	1	0	0	$\bar{x}_4x_3\bar{x}_2\bar{x}_1$
5	0	1	0	1	$\bar{x}_4x_3\bar{x}_2x_1$
6	0	1	1	0	$\bar{x}_4x_3x_2\bar{x}_1$
7	0	1	1	1	$\bar{x}_4x_3x_2x_1$
8	1	0	0	0	$x_4\bar{x}_3\bar{x}_2\bar{x}_1$
9	1	0	0	1	$x_4\bar{x}_3\bar{x}_2x_1$
10	1	0	1	0	$x_4\bar{x}_3x_2\bar{x}_1$
11	1	0	1	1	$x_4\bar{x}_3x_2x_1$
12	1	1	0	0	$x_4x_3\bar{x}_2\bar{x}_1$
13	1	1	0	1	$x_4x_3\bar{x}_2x_1$
14	1	1	1	0	$x_4x_3x_2\bar{x}_1$
15	1	1	1	1	$x_4x_3x_2x_1$

For instance, for the four input variables x_1, x_2, x_3, x_4 we obtain the minterms shown in Table 2.12.

The function f, presented above, which we have so far expressed with the aid of a table, by a set of identifiers and by charts, has the following expanded *sp* standard form:

$$f = 1 \cdot M_0 \vee 0 \cdot M_1 \vee 0 \cdot M_2 \vee 0 \cdot M_3 \vee 1 \cdot M_4 \vee f(5) \cdot M_5 \vee 1 \cdot M_6 \vee 1 \cdot M_7$$
$$\vee 1 \cdot M_8 \vee f(9) \cdot M_9 \vee 0 \cdot M_{10} \vee 1 \cdot M_{11} \vee 1 \cdot M_{12} \vee 1 \cdot M_{13} \vee 0 \cdot M_{14}$$
$$\vee 1 \cdot M_{15}$$

The minterms M_5 and M_9 are accompanied by the general symbols $f(5)$ and $f(9)$, thus indicating that the states 5 and 9 are undefined.

Since in Boolean algebra we have $0 \cdot x = 0$ and $1 \cdot x = x$, the expanded *sp* standard form presented above can be rewritten in the simplified form

$$f = M_0 \vee M_4 \vee f(5) \cdot M_5 \vee M_6 \vee M_7 \vee M_8 \vee f(9) \cdot M_9 \vee M_{11} \vee M_{12} \vee M_{13}$$
$$\vee M_{15}$$

Substituting for the symbols M_s by the corresponding minterms, we obtain:

$$f = \bar{x}_4\bar{x}_3\bar{x}_2\bar{x}_1 \vee \bar{x}_4x_3\bar{x}_2\bar{x}_1 \vee f(5)\bar{x}_4x_3\bar{x}_2x_1 \vee \bar{x}_4x_3x_2\bar{x}_1 \vee \bar{x}_4x_3x_2x_1$$
$$\vee x_4\bar{x}_3\bar{x}_2\bar{x}_1 \vee f(9)x_4\bar{x}_3\bar{x}_2x_1 \vee x_4\bar{x}_3x_2x_1 \vee x_4x_3\bar{x}_2\bar{x}_1 \vee x_4x_3\bar{x}_2x_1$$
$$\vee x_4x_3x_2x_1$$

The representation of a logic function in its expanded *sp* form can be simplified with the aid of various rules of Boolean algebra. We may obtain different contracted *sp* standard forms depending on the used procedure. The undefined values of the Boolean function (in our case $f(5)$ and $f(9)$) can be chosen so as to obtain the simplest possible expression. In our case, if we want to obtain the simplest possible *sp* standard form, we must set $f(5) = f(9) = 1$, which leads to

$$f = \bar{x}_1\bar{x}_2 \vee x_1x_4 \vee x_3\bar{x}_4$$

Algorithmic procedures for determining the simplest Boolean forms are described in Chapter 3.

Now let us see what our function looks like in the expanded *ps* standard form. According to equation (2.13) we first obtain:

$$\begin{aligned} f = {} & (1 \vee \overline{M}_0)(0 \vee \overline{M}_1)(0 \vee \overline{M}_2)(0 \vee \overline{M}_3)(1 \vee \overline{M}_4)(f(5) \vee \overline{M}_5)(1 \vee \overline{M}_6) \\ & \cdot (1 \vee \overline{M}_7)(1 \vee \overline{M}_8)(f(9) \vee \overline{M}_9)(0 \vee \overline{M}_{10})(1 \vee \overline{M}_{11})(1 \vee \overline{M}_{12}) \\ & \cdot (1 \vee \overline{M}_{13})(0 \vee \overline{M}_{14})(1 \vee \overline{M}_{15}) \end{aligned}$$

Since in Boolean algebra we have $1 \vee x = 1$ and $0 \vee x = x$, the expression presented above can be written in the simplified form

$$f = \overline{M}_1\overline{M}_2\overline{M}_3(f(5) \vee \overline{M}_5)(f(9) \vee \overline{M}_9)\overline{M}_{10}\overline{M}_{14}$$

Substituting for the symbols $\overline{M}_s$ by the corresponding minterms we obtain:

$$\begin{aligned} f = {} & (\overline{\bar{x}_4\bar{x}_3\bar{x}_2x_1})(\overline{\bar{x}_4\bar{x}_3x_2\bar{x}_1})(\overline{\bar{x}_4\bar{x}_3x_2x_1})(f(5) \vee \overline{\bar{x}_4x_3\bar{x}_2x_1}) \\ & \cdot (f(9) \vee \overline{x_4\bar{x}_3\bar{x}_2x_1})(\overline{x_4\bar{x}_3x_2\bar{x}_1})(\overline{x_4x_3x_2\bar{x}_1}) \end{aligned}$$

Applying De Morgan's laws of negation, we get the product of sums:

$$\begin{aligned} f = {} & (x_4 \vee x_3 \vee x_2 \vee \bar{x}_1)(x_4 \vee x_3 \vee \bar{x}_2 \vee x_1)(x_4 \vee x_3 \vee \bar{x}_2 \vee \bar{x}_1) \\ & \cdot (f(5) \vee x_4 \vee \bar{x}_3 \vee x_2 \vee \bar{x}_1)(f(9) \vee \bar{x}_4 \vee x_3 \vee x_2 \vee \bar{x}_1) \\ & \cdot (\bar{x}_4 \vee x_3 \vee \bar{x}_2 \vee x_1)(\bar{x}_4 \vee \bar{x}_3 \vee \bar{x}_2 \vee x_1) \end{aligned}$$

For the following exposition it will be sufficient to confine ourselves, for instance, to the *sp* standard form. If we then want to obtain the *ps* form of some function f, we can proceed by first finding the *sp* form of the function $\bar{f}$, and by negating this form with the aid of De Morgan's laws. We thereby obtain directly the corresponding *ps* form of the function f. In the table or chart we can construct the function $\bar{f}$ very simply by negating the values $f(s)$ for all identifiers $s = 0, 1, \ldots, 2^n - 1$.

Exercises to Section 2.10

2.10-1 Determine the tabular form and all possible Marquand chart forms for the indefinite logic function $f(x_1, x_2, x_3, x_4, x_5)$ defined as follows: $f(s) = 0$ for $s = 0, 2, 4, 6, 16, 18, 24, 26, 28$, and $f(s) = 1$ for $s = 1, 3, 5, 7, 8, 9, 10, 15, 25, 30, 31$.

2.10-2 Find the expanded *sp* and *ps* standard Boolean forms of the following logic functions and, using the laws of Boolean algebra, try to reduce them to the simplest forms:

*(a) Definite logic function $f(x_1, x_2, x_3, x_4, x_5)$ defined by $f(s) = 1$ for $s = 4, 6, 8, 9, 10, 11, 12, 13, 14, 15, 16, 20, 22, 23, 24, 28, 30, 31$.

(b) Indefinite logic function $f(x_1, x_2, x_3)$ defined by $f(s) = 0$ for $s = 1, 2, 4$ and $f(s) = 1$ for $s = 0, 3, 5, 6$.

(c) Definite logic function $f(x_1, x_2, x_3, x_4)$ defined by $f(s) = 1$ for $s = 1, 2, 5, 6, 9, 10, 13, 14$.

2.10-3 Find expanded *sp* and *ps* standard forms for:

(a) All functions of two variables.

(b) Functions M_3, M_5, F_3^2, F_3^{2-}.

Try to reduce them as much as possible.

2.11 MUTUAL TRANSFORMATIONS BETWEEN LOGIC FUNCTION FORMS

The transformation of the tabular representation of a logic function into the form of a chart and vice versa is very simple, since a single square in the chart is uniquely assigned to each row of the table and vice versa.

We must first divide the input variables in the table into two groups, one of which determines the rows and the other the columns in the chart. If the manner of dividing the variables into groups does not follow directly from the specific properties of the variables (which happens, for instance, in the solution of logic equations, as will be seen in Chapter 4), we usually divide the variables so as to make both groups contain approximately the same number of variables. With an even number of input variables (i.e., an even n) we thus have $n/2$ of them in each group. If the number of variables is odd, we have $(n - 1)/2$ variables in one group and $(n + 1)/2$ in the other. For instance, seven variables are usually divided into one group of three and another group of four variables.

After having determined the two groups of variables we must decide which of them will define the columns and which the rows of the chart. We then write out in the table the partial identifiers s_c and s_r for each group of variables and mark the corresponding columns and rows of the chart with these identifiers. Now everything is ready for the transfer of the values of the pertinent function into the chart.

Conversely, when transforming a function from the chart into a table, we write out all pairs of the partial identifiers s_c and s_r (which are, as a matter of fact, the ordinal numbers of the rows and columns in the Marquand chart), and to each of them we ascribe the corresponding value of the output variable. The identifiers s_c and s_r can then be expressed in binary form, and we thus obtain the complete tabular form of the given logic function.

To learn the transformation of logic functions from charts into Boolean expressions, the so-called *Svoboda grids* are advantageous. Later, after having acquired some practice, these grids can be discarded and the reader can use their mental image instead.

A separate grid is required for each variable. The size of the grid must be adapted to that of the corresponding chart. Since it is advantageous to draw the charts on graph (square-ruled) paper, we adapt the grids to the same dimensions. The size of the chart also depends, of course, on the number of variables. We should therefore prepare a separate set of grids for each number of variables. In practice, however, we prepare separate sets of grids only for even numbers of variables, usually for $n = 4$, 6, 8, and sometimes even 10. If there is an odd number of variables (e.g., five), we employ the nearest higher set of grids (i.e., the set for six variables), one of the grids being left unused.

Figure 2.15 shows a set of grids for four variables, Figure 2.16 a set for six variables, and Figure 2.17 a set for eight variables. These sets of grids suffice

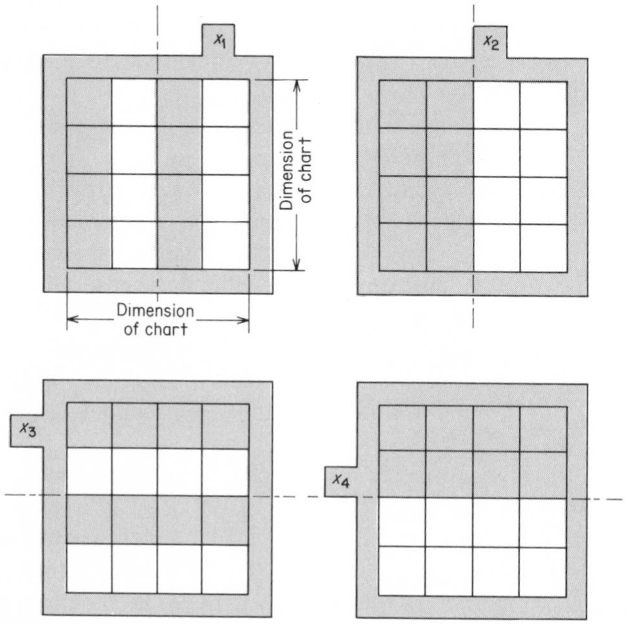

Figure 2.15 Svoboda grids for four variables.

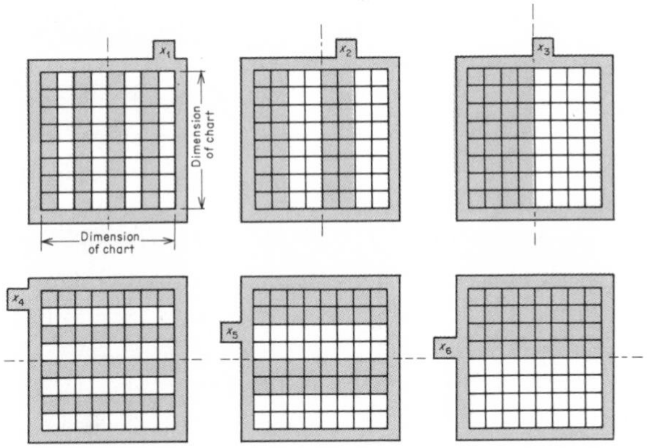

Figure 2.16 Svoboda grids for six variables.

for the majority of practical problems. If necessary, the reader will be able to construct grids for ten or twelve variables in a similar manner.

We advise the reader to cut out from cardboard the three sets of grids shown in the illustrations, adapting their dimensions to the size of the corresponding charts drawn on square-ruled paper. The chart has four rows and four columns for four variables, eight for six variables and sixteen for eight variables. The shaded areas shown in Figures 2.15, 2.16 and 2.17 should be cut out.*

Each grid is marked for one particular variable. If we place one of the grids on the chart (drawn to the same size as the grid), only those squares of the chart remain uncovered in which the corresponding variable acquires the value 1. Thus, the grid labeled by x represents the condition $x = 1$. If we turn any grid through an angle of 180° (about the axes indicated in Figures 2.15, 2.16, and 2.17), then only those squares of the chart remain uncovered in which the negation of the corresponding variable acquires the value 1. Thus, the grid labeled by $\bar{x}$ (the other side of the grid labeled by x) represents the condition $\bar{x} = 1$. A given grid can thus be used to represent on the chart either the given variable itself, or its negation. This holds, however, only if Marquand's chart is used as a basis for the grids.

* The grids together with a special graph paper can be ordered from Jewell Enterprises, 104 Jewell Street, Garfield, N.J. 07026.

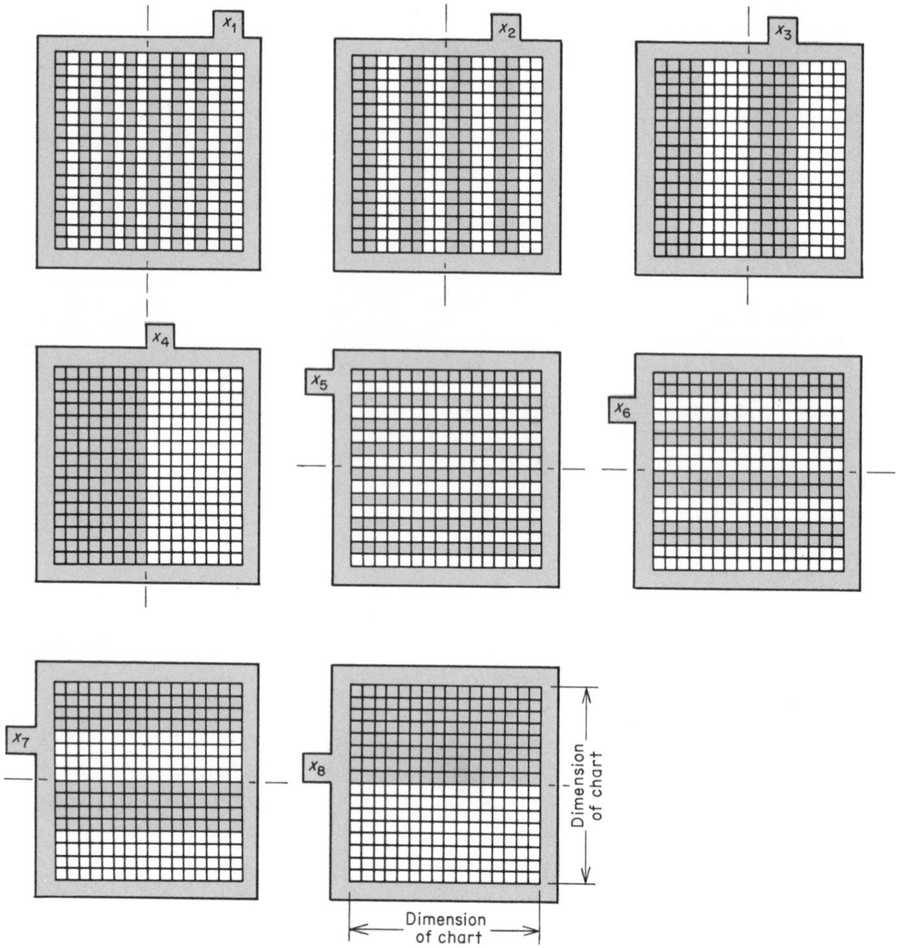

Figure 2.17 Svoboda grids for eight variables.

If we simultaneously place several grids on the chart, we thereby represent the AND operation (p-term) of the corresponding variables, or that of their negations. Thus, if we apply simultaneously grids labeled by x, $\bar{y}$, $\bar{z}$, they represent the condition $x\bar{y}\bar{z} = 1$. If we place on the chart first one group of grids and then a second group of grids, we thereby obtain the OR operation of the corresponding p-terms.

The transformation of a given logic function from the chart into a contracted sp form can be accomplished by placing suitable combinations of grids on the chart, writing out the corresponding p-terms and combining them finally by the OR operation. We start from the principle that no empty square in the chart must remain uncovered, whatever the combination of the grids placed on the chart. On the other hand, each square occupied by a dot must be successively uncovered at least once. The squares pertaining to indefinite states, which are usually hatched, may remain uncovered, but not necessarily so. In their case we may select that possibility which suits us better. Of course, we endeavor to accomplish the whole procedure with the least number of grids, since every grid used adds another symbol to the corresponding normal form.

If we want to obtain the *ps* normal form of a Boolean function f expressed by a chart, we first map the function $\bar{f}$ in the chart by replacing the empty squares by dots and vice versa, the hatched squares being left unchanged. We then seek some *sp* form of the function $\bar{f}$. When we negate the *sp* form, we obtain a *ps* form of the function f.

For instance, let us find a contracted *ps* form of the Boolean function f defined by the chart in Figure 2.18a. First we produce a chart representation of the function $\bar{f}$ (Figure 2.18b)—a trivial operation. With the aid of the Svoboda grids for four variables we then easily find:

$$\bar{f} = x_2 \bar{x}_4 \vee \bar{x}_1 \bar{x}_2 \vee x_3 x_4$$

Negating this expression, we obtain:

$$f = (\bar{x}_2 \vee x_4)(x_1 \vee x_2)(\bar{x}_3 \vee \bar{x}_4)$$

The transformation of the tabular into the algebraic form can be accomplished via the expanded standard form, which can be directly written out from the table. An abridged form can then be obtained by successive simplifications of the algebraic expression by applying the rules of Boolean algebra. This procedure is very toilsome and time-consuming, as the reader is sure to

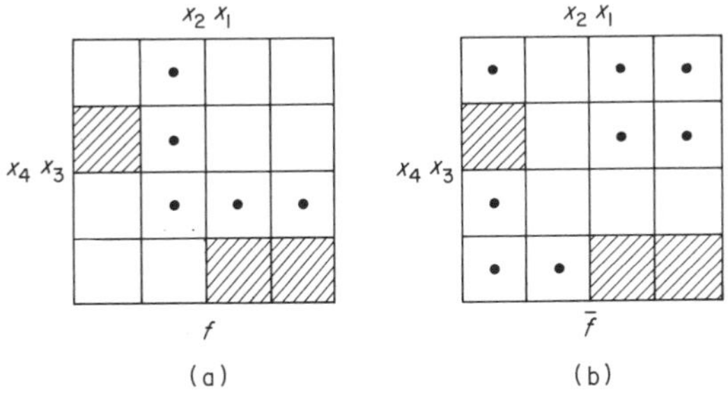

Figure 2.18 Example of chart representations of an indefinite logic function f and its negation $\bar{f}$.

have discovered when solving Exercise 2.10-2. We therefore prefer to transform the given function first from the tabular form into its chart representation, and then to find an abridged Boolean form with the aid of the Svoboda grids.

Exercises to Section 2.11

2.11-1 Using the Svoboda grids, try to find the simplest *sp* and *ps* Boolean forms for the functions specified in Exercise 2.10-2.

2.11-2 Show that a set of Svoboda grids for n variables can be used for problems concerning m variables, where $m \leq n$.

2.11-3 Let $s_c = (x_p x_{p-1} \cdots x_1)$ and $s_r = (y_q y_{q-1} \cdots y_1)$ be states of logic variables that determine columns and rows of the Marquand chart, respectively. Find a formula:

(a) Expressing the columns c_0 that are uncovered in the Svoboda grid for variable x_i.

(b) Expressing the rows r_0 that are uncovered in the Svoboda grid for variable y_i.

2.12 CHART OPERATIONS

In many problems concerning switching circuits, two Boolean expressions are composed by a logic function. For instance, let $f_1 = x \vee \bar{y}$ and $f_2 = \bar{x}z$ be composed by NONEQUIVALENCE. We have

$$\begin{aligned} f &= \bar{f}_1 f_2 \vee f_1 \bar{f}_2 \\ &= \overline{(x \vee \bar{y})}\bar{x}z \vee (x \vee \bar{y})\overline{\bar{x}z} \\ &= \bar{x}yz \vee (x \vee \bar{y})(x \vee \bar{z}) \\ &= x \vee zy \vee \bar{z}\bar{y} \end{aligned}$$

Although the operation $f = f_1 \circ f_2$ ($\circ$ stands for a logic function) can always be performed directly in the algebraic form, it is usually easier to perform it by the help of Marquand charts as follows:

(i) Using the Svoboda grids, the Boolean expressions for f_1 and f_2 are transformed to the Marquand chart form.

(ii) The function $\circ$ is applied to values in each pair of corresponding squares in the charts for f_1 and f_2. The results are registered in corresponding squares of a chart representing f.

(iii) A Boolean expression is determined for the chart form of f.

The chart operation for the previous example is shown in Figure 2.19a. Examples of other chart operations are illustrated by the same Boolean expressions in Figures 2.19b through h.

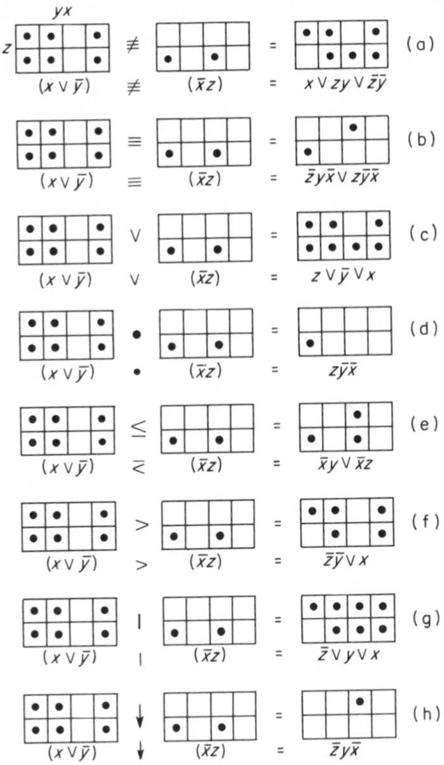

Figure 2.19 Illustration to chart operations.

Exercises to Section 2.12

2.12-1 Determine for each of the following pairs of Boolean expressions all the types of compositions that are shown in Figure 2.19.

*(a) $f_1 = x\bar{y}z \vee \bar{z}(x \vee \bar{v})$; $f_2 = x \vee \bar{y}(z \vee v)$.

(b) $f_3 = x \vee \bar{y} \vee z \vee v$; $f_4 = \bar{x} \vee \bar{y} \vee \bar{z} \vee \bar{v}$.

(c) $f_5 = (x\bar{y} \vee \bar{x}y)(\bar{z}\bar{v} \vee zv)$; $f_6 = (\bar{x} \vee \bar{y})(z \vee v)$.

2.12-2 Using the Boolean expressions given in Exercise 2.12-1, determine the following multiple compositions:

*(a) $(f_1 \leq f_2) \not\equiv (f_3 \mid f_4)$.

(b) $(f_3 \vee f_4) \leq (f_5 \vee f_1(f_2 \downarrow f_6))$.

(c) $((f_1 \downarrow f_6) > (f_2 \equiv f_4)) \mid (f_3 \leq f_4)$.

Comprehensive Exercises to Chapter 2

2.1 Show that the set of all threshold logic functions of two input variables is a complete set of logic functions.

2.2 Letting x_i take on the values 0 and 1, prove the following theorem due to Shannon:
Every logic function $f(x_1, x_2, \ldots, x_n)$ can be expressed in either of the following algebraic forms:
(a) $f(x_1, x_2, \ldots x_i, \ldots, x_n) = x_i \cdot f(x_1, x_2, \ldots, 1, \ldots, x_n) \vee \bar{x}_i \cdot f(x_1, x_2, \ldots, 0, \ldots, x_n)$.
(b) $f(x_1, x_2, \ldots, x_i, \ldots, x_n) = (\bar{x}_i \vee f(x_1, x_2, \ldots, 1, \ldots x_n)) \cdot (x_i \vee f(x_1, x_2, \ldots, 0, \ldots, x_n))$.

2.3 Applying the Shannon Theorem (Exercise 2.2), obtain both of the expressions with respect to each of the input variables for the following logic functions:
*(a) $f_1(x_1, x_2, x_3, x_4) = x_3 \bar{x}_2 (\bar{x}_1 \vee x_4) \vee x_4 \bar{x}_3 x_2$.
(b) $f_2(x_1, x_2, x_3, x_4) = x_1 \vee x_2 \bar{x}_3 \vee \bar{x}_2 x_3 x_4$.
(c) $f_3(x_1, x_2, x_3, x_4) = (\bar{x}_1 \bar{x}_2 \vee x_1 x_2)(\bar{x}_3 x_4 \vee x_3 \bar{x}_4 x_2)$.

2.4 (a) Show that Theorem 2.6 can be generalized as follows:

$$f(x_1, x_2, \ldots, x_n) = \bigvee_{s=0}^{2m-1} f(s, x_{m+1}, x_{m+2}, \ldots, x_n) \prod_{i=n}^{m} x_i^{i_s}$$

where s is the state identifier of variables $x_1, x_2, \ldots, x_m$ $(m \leq n)$, $f(s, x_{m+1}\, x_{m+1}, \ldots, x_n)$ is a function of variables $x_{m+1}, x_{m+2}, \ldots, x_n$ that is obtained by taking state of s for variables $x_1, x_2, \ldots, x_n$, and i_s is the value of variable x_i for the state s.
(b) Determine the dual formula to that specified in (a).
(c) Show that the Shannon theorems (Exercise 2.2) are special cases of the theorems mentioned in (a) and (b).

2.5 Write the following functions in the form introduced in Exercise 2.4 $(2 \leq m < n)$:
*(a) $f_1(x_1, x_2, x_3, x_4) = x_3 x_1 (x_2 \vee x_4) \vee \bar{x}_2 x_1$.
(b) $f_2(x_1, x_2, x_3, x_4) = (\bar{x}_4 x_3 \vee x_4 \bar{x}_3)(x_2 x_1 \vee \bar{x}_2 \bar{x}_1)$.

2.6 Suggest a procedure for solution of Exercise 2.5 based on the Marquand representation of the logic functions. (Hint: Assign variables $x_1, x_2, \ldots, x_m$ to columns of the chart, variables $x_{m+1}, x_{m+2}, \ldots, x_n$ to rows of the chart.)

*2.7 Using functions of Boolean algebra, write a formula for Hamming distance of the points $\dot{x} = (\dot{x}_n \dot{x}_{n-1} \ldots \dot{x}_1)$ and $\tilde{x} = (\tilde{x}_n \tilde{x}_{n-1} \ldots \tilde{x}_1)$ of logic space $\mathbf{x} = (x_n x_{n-1} \ldots x_1)$.

2.8 Show that there is a one-to-one correspondence between values of the state identifier of logic variables $x_1, x_2, \ldots, x_n$ and:
(a) Minterms of these variables.
(b) Maxterms of these variables.

2.9 Let $t = \sum_{i=1}^{n} \dot{x}_i \cdot 3^{i-1}$. Define the meaning of $\dot{x}_i$ with a view to establishing a one-to-one correspondence between the values of t and all
*(a) p-terms of logic variables $x_1, x_2, \ldots, x_n$.
(b) s-terms of logic variables $x_1, x_2, \ldots, x_n$.

2.10 Describe each of the following Boolean expressions by a set of ternary identifiers introduced in Exercise 2.9:
*(a) $x_4 \bar{x}_2 \vee \bar{x}_3 \bar{x}_2 x_1 \vee \bar{x}_4 x_2 \vee \bar{x}_1 \vee x_4 x_3 x_2 x_1$.
(b) $x_3 \bar{x}_2 \bar{x}_1 \vee \bar{x}_4 \bar{x}_3 \bar{x}_2 \bar{x}_1 \vee x_2 \vee \bar{x}_2 x_1 \vee x_2 \bar{x}_1$.
(c) $\bar{x}_4 (x_3 x_2 \vee \bar{x}_1) \vee x_4 \bar{x}_2 (x_3 x_1 \vee \bar{x}_3 \bar{x}_1)$.
(d) $(\bar{x}_2 \vee \bar{x}_1)(\bar{x}_3 \vee x_2 \vee x_1)(x_4 \vee x_3 \vee \bar{x}_2)$.

2.11 Convert the expressions specified in Exercise 2.10 to the reverse Polish notation.

2.12 Design an algorithm for converting the following expressions to the reverse Polish notation:
(a) a given minterm.
(b) a given maxterm.
(c) a given minterm form.
(d) a given maxterm form.
NOTE: Describe the algorithms in a programming language or in the form of a flow chart.

2.13 Considering n logic variables, determine the number of:
*(a) all possible minterms.
(b) all possible maxterms.
*(c) all possible p-terms.
(d) all possible s-terms.

2.14 Express the following functions by Boolean expressions:
(a) M_3 and M_5.
*(b) EXPLICIT OR for three variables.
(c) NONEQUIVALENCE for three variables.
(d) F_3^0, F_3^1, F_3^2, and F_3^3,
(e) threshold function of four variables with $w_1 = w_2 = w_3 = w_4 = 1$ and $T = 2$.

2.15 Let a term participate in a Boolean expression for a logic function of n input variables. Determine the number of states specified by the term in case of:
*(a) p-term containing m literals ($m \leq n$).
(b) s-term containing m literals ($m \leq n$).

2.16 The hierarchy of Boolean operations is: NOT, AND, OR. Explain why parentheses are needed in the ps form and are not needed in the sp form.

2.17 Prove that any logic function $f(x_1, x_2, \ldots, x_n)$ can be written in the following form: $f(x_1, x_2, \ldots, x_n) = a_0 \not\equiv a_1 x_1 \not\equiv a_2 x_2 \not\equiv \cdots \not\equiv a_n x_n \not\equiv a_{n+1} x_1 x_2 \not\equiv a_{n+2} x_1 x_3 \not\equiv \ldots \not\equiv a_{n(n+1)/2} x_{n-1} x_n \not\equiv a_{(n(n+1)/2)+1} x_1 x_2 x_3 \not\equiv \cdots \not\equiv a_{2^n-1} x_1 x_2 \cdots x_n$, where $a_i = 0$ or 1 ($i = 0, 1, \ldots, 2^n - 1$).

2.18 Express the following logic functions in the form introduced in Exercise 2.17:
(a) all logic functions of two variables.
*(b) majority function M_3.
(c) NAND (x_3, x_2, x_1) and NOR (x_3, x_2, x_1).
(d) majority function M_5.
(e) function F_4^2.

2.19 Determine both the expanded sp and ps standard Boolean forms for each of the following logic functions:
*(a) $f_1(x_1, x_2, x_3) = \bar{x}_1(\bar{x}_3 \vee x_2) \vee \bar{x}_2 x_1$.
(b) $f_2(x_1, x_2, x_3, x_4) = (\bar{x}_3 x_2 \vee x_4 x_1)(\bar{x}_2 \bar{x}_1 \vee \bar{x}_3 x_2 x_1 \vee x_4 x_2)$.
(c) $f_3(x_1, x_2, x_3, x_4) = x_1 \vee \bar{x}_2 \vee \bar{x}_1 x_2 \bar{x}_3 x_4$.

2.20 Using multiple input AND and OR elements, draw a block diagram for each of the Boolean expressions given in Exercise 2.19. Determine the number of levels in each of the block diagrams.

2.21 Determine the domain of each of the logic functions represented by the following Boolean expressions, where c_i ($i = 1, 2, \ldots$) are undetermined coefficients:
*(a) $\bar{x}_4 x_3 (x_1 \vee c_1 \bar{x}_2) \vee c_2 x_4 \vee \bar{x}_3 x_2 \vee c_3 x_3 \bar{x}_2$.

(b) $(\bar{x}_4 \vee x_2)(c_1 \vee x_3 \vee \bar{x}_1)(x_4 \vee c_2 \bar{x}_2)$.
(c) $x_4(\bar{x}_3(x_2 \vee c_1\bar{x}_1) \vee x_1) \vee \bar{x}_4 \bar{x}_3 \bar{x}_2 \bar{x}_1$.

2.22 Find whether or not each of the following equations is valid (satisfied for all states of the variables involved):
(a) $\bar{x}_1 x_2 x_3 = x_3 (\bar{x}_3 \vee x_2)(\bar{x}_2 \vee \bar{x}_1)$.
(b) $(x_1 \vee x_2)(\bar{x}_1 \vee x_2)(x_1 \vee \bar{x}_2)(\bar{x}_1 \vee \bar{x}_2) = 0$.
(c) $x_3(\bar{x}_2 \vee \bar{x}_1) \vee \bar{x}_3 x_2 = x_3 \bar{x}_2 \vee x_2 \bar{x}_1 \vee \bar{x}_3 x_2$.
(d) $\bar{x}_3 x_2 \vee x_3 x_1 = x_2 x_1 \vee x_3 x_1$.

2.23 Show that $x_2 = x_1\bar{x}_3 \vee x_1 x_3$ if $x_3 = x_1\bar{x}_2 \vee x_1 x_2$.

2.24 Find those states of logic variables x_1, x_2, x_3, x_4 for which all of the following equations are satisfied simultaneously.
$\bar{x}_1 \vee x_1 x_2 = 0$
$x_1 x_2 \vee x_1 \bar{x}_3 \vee x_3 x_4 = \bar{x}_3 x_4$
$x_1 x_2 = x_1 x_3$

2.25 Prove each of the following propositions or find a counter example:
(a) If $x\bar{y} \vee \bar{x}y = 0$, then $x = y$.
(b) If $x = y$, then $xy \vee \bar{x}\bar{y} = 1$.
(c) If $x\bar{y} \vee \bar{x}y = x\bar{z} \vee \bar{x}z$, then $y = z$.
(d) If $x \not\equiv y \not\equiv z = w$, then $x \not\equiv y = z \not\equiv w$ and $x = y \not\equiv z \not\equiv w$.

2.26 Compute the Hamming distance $d(i, j)$ for all pairs of states i and j of four logic variables. Arrange $d(i, j)$ in a square matrix (so-called distance matrix) and determine general properties of such matrixes that hold for any number of variables.

2.27 Applicants for employment at a certain company must be:
1. female, married, and over 30, or
2. female, unmarried, and over 30, or
3. male, married, and under 30, or
4. male, married, and over 30.

Establish a logic variable for each of these categories (sex, marriage, age), determine the expanded *sp* standard form describing these conditions, simplify it, and rewrite the original conditions in a simplified form.

Reference Notations to Chapter 2

The concept of distance in *n*-dimensional logic space was introduced by Hamming [7]. Lee [18] suggested a representation of logic functions by the *n*-dimensional unit cube.

The Marquand chart was originally suggested by an English archaeologist Marquand in 1881 [20]. He also designed a logic machine based on the chart [2]. The chart was rediscovered later by Veitch [35] and has been used extensively by Svoboda [31,32,34], Nadler [23], and others [15]. It is sometimes called the Veitch chart or the Svoboda chart. The Karnaugh chart, which is generally preferred in the contemporary literature concerning switching circuits, was suggested by Karnaugh in 1953 [14].

The Polish notation, suggested by a Polish logician Lukasiewicz (1878–1956), was adopted by some contemporary logicians, for instance, Prior [26]. The notation was found very useful in construction of compilers for digital computers [19].

Nondegenerate functions are well covered in Reference 10. Equivalence classes of logic functions with regard to applications in switching circuits were studied by Slepian [29], Povarov [25], Harrison [8,9] and others (further references can be found in [8] and [25]).

Properties of complete sets of logic functions were originally studied by Post [24] and Wernick [36]. All complete sets for $n \leq 3$ were determined by Kudielka and Oliva [17]. Yablonskii [37] investigated the problem of complete sets of logic functions for k-valued logics. He solved the problem for $k \leq 3$. Some other studies of the problem in this generalized form can be found in current Russian literature [3,4,38]. Menger [21] and Klir [16] investigated universal logic primitives. Among other publications on complete sets of logic functions, papers [13,22] should be mentioned.

Boolean algebra, which was originally developed by Boole [1], can be found in many books, for example, those in References 5,6,30. The set of axioms of Boolean algebra that are presented in Section 2.9 was suggested by Huntington [11] in 1904. Other sets of axioms were proposed later by Sheffer [28], Huntington [12], and others [30].

Logic grids, some of which are introduced in this chapter, were invented by Svoboda [31,32,34]. He also suggested an application of punched cards [33,34] for solution of some problems concerning switching circuits.

References to Chapter 2

1. **Boole, G.,** *The Laws of Thought*, Dover, New York, 1958. (Originally published by Macmillan in 1854.)
2. **Gardner, M.,** *Logic Machines and Diagrams*, McGraw-Hill, Book Co., New York, 1958.
3. **Gavrilov, G. P.,** "On the Functional Completeness in Logics with Countable Sets of Values," *PC*, No. 15, 5–64, 1965.
4. **Ginkin, S. G.,** and **A. A. Mucznik,** "Solution of the Problem of Completeness for a System of Logic Functions with Unreliable Realizations," *PC*, No. 15, 65–84, 1965.
5. **Halmos, P. R.,** *Lectures on Boolean Algebras*, Van Nostrand, Princeton, N.J., 1963.
6. **Hammer, P. L.,** and **S. Rudeanu,** *Boolean Methods in Operations Research*, Springer-Verlag, New York, 1968.
7. **Hamming, R. W.,** "Error Detecting and Error Correcting Codes," *BSTJ* **29**(2), 147–160 (April 1950).
8. **Harrison, M. A.,** *Introduction to Switching and Automata Theory*, McGraw-Hill, New York, 1965.
9. **Harrison, M. A.,** "Counting Theorems and Their Applications to Classification of Switching Functions," *RDST*, pp. 86–120.
10. **Hu, S.,** *Mathematical Theory of Switching Circuits and Automata*, University of California Press, Berkeley and Los Angeles, 1968.
11. **Huntington, E. V.,** "Sets of Independent Postulates for the Algebra of Logic," *TAMS* **5,** 288–309 (1904).

12. **Huntington, E. V.**, "New Sets of Independent Postulates for the Algebra of Logic," *TAMS* **35,** 274–304 (1933).
13. **Ibuki, K.**, et al., "General Theory of Complete Sets of Logical Functions," *Electronics and Communication in Japan* **46** (7) 55–65 (July 1963).
14. **Karnaugh, M.**, "The Map Method for Synthesis of Combinational Logic Circuits," *Trans. AIEE, Part I, Communication and Electronics* **72,** 593–599 (1953).
15. **Klir, G. J.**, and **L. K. Seidl**, *Synthesis of Switching Circuits*, Iliffe, London; 1968. Gordon and Breach, New York, 1969.
16. **Klir, G. J.**, "On Universal Logic Primitives," *TC* **C-20** (4) 467–469 (April 1971).
17. **Kudielka, V.**, and **P. Oliva**, "Complete Sets of Functions of Two and Three Binary Variables," *TC* **EC-15** (6) 930–931 (Dec. 1966).
18. **Lee, C. Y.**, "Switching Functions on an *n*-dimensional Cube," *Trans. AIEE, Pt. I,* **73** (14) 287-291 (1954).
19. **Lee, J. A. N.**, *The Anatomy of a Compiler*, Reinhold, New York, 1967.
20. **Marquand, A.**, "On Logical Diagrams for *n* Terms," *Philosophical Magazine* **XII,** 266–270 (1881).
21. **Menger, K. S.**, Jr., "Characterization and Cardinality of Universal Functions," *TC* **EC-14** (5) 720–721 (Oct. 1965).
22. **Mukhopadhyay, A.**, "Complete Sets of Logic Primitives," *RDST*, pp. 1–26.
23. **Nadler, M.**, *Topics in Engineering Logic*, Macmillan, New York, 1962.
24. **Post, E. L.**, "Two-Valued Iterative Systems of Mathematical Logic," *Annals of Mathematics Studies*, Vol. 5, Princeton University Press, Princeton, N. J., 1941. (The results in this paper were originally presented in 1921.)
25. **Povarov, G. N.**, "On the Group Invariance of Boolean Functions" (in Russian), *Applications of Logic in Science and Engineering*, Izd. Ak. Nauk, U.S.S.R., Moscow, 1961, pp. 263–340.
26. **Prior, A. N.**, *Formal Logic*, Oxford University Press, Oxford, 1955.
27. **Rosenbloom, P.**, *The Elements of Mathematical Logic*, Dover, New York, 1950.
28. **Sheffer, H. M.**, "A Set of Five Independent Postulates for Boolean Algebras, with Applications to Logical Constants," *TAMS* **14,** 481–488 (1913).
29. **Slepian, D.**, "On the Number of Symmetry Types of Boolean Functions of *n* Variables," *Can. J. Math.* **5** (2), 185–193 (1953).
30. **Sikorski, R.**, *Boolean Algebras*, Springer-Verlag, New York, 1964.
31. **Svoboda, A.**, "Graphico-mechanical Aids for the Analysis and Synthesis of Contact Networks" (in Czech), *IPM*, No. 4, 9–21 (1956).
32. **Svoboda, A.**, "Some Applications of Contact Grids," *PHU*, Part I, pp. 293–305.
33. **Svoboda, A.**, "Analysis of Boolean Functions by Logical Punch-Cards," *IPM*, No. 7, pp. 13–20 (1960).
34. **Svoboda, A.**, "Logical Instruments for Teaching Logical Design," *TE* **E-12** (3), 262–273 (Sept. 1969).
35. **Veitch, E. W.**, "A Chart Method for Simplifying Truth Functions," *Proc. ACM*, pp. 127–133 (1952).
36. **Wernick, W.**, "Complete Sets of Logical Functions," *TAMS* **51,** 117–132 (1941).
37. **Yablonskii, S. V.**, "Constructions of Functions in k-valued Logic" (in Russian), *Trudy Matem. Inst. V. A. Steklova*, No. 51, Izd. Akademii Nauk USSR, Moscow, 1958, pp. 6–142.
38. **Zakharova, E. Yu.**, "A Criterion of the Completeness of Functions from P_k," *PC*, No. 18, pp. 5–10 (1967).

3

MINIMIZATION OF BOOLEAN EXPRESSIONS

3.1 DISCUSSION OF THE PROBLEM

An intuitive procedure for the determination of simplified Boolean forms, which is based on Marquand charts and Svoboda grids, is outlined in Section 2.11. This procedure does not guarantee that we really obtain the simplest Boolean form. It is the purpose of this chapter to investigate the problem of simplification of Boolean expressions more thoroughly and, in particular, to present some procedures that do guarantee determination of a *minimal Boolean form* of a given logic function with respect to an *objective criterion.*

The problem of Boolean minimization may be formulated as follows: Given a logic function f and an *objective function* F defined on the set of Boolean expressions, find such a Boolean expression representing f for which F reaches its *absolute minimum.*

The objective function F can be defined in many different ways, e.g.:

(i) As the number of literals in the expression where a literal is a symbol of a variable either in the direct or in the negated form.
(ii) As the sum of the number of AND and OR operations.
(iii) As the number of negations.

(iv) As the sum of the number of literals and the number of negations.

(v) As the sum $a \cdot P + b \cdot Q + c \cdot R$, where P, Q, R denote, respectively, the number of AND, OR, and NOT functions in the expression and a, b, c are certain coefficients (weights) which should result from an engineering-cum-economical analysis of the corresponding elements (especially as far as their price and reliability are concerned).

Different definitions of the objective function are justified under different circumstances. For instance, (iii) would be used if the engineering elements representing negations were substantially more expensive and/or substantially less reliable than the elements representing AND and OR functions.

In our further exposition in this chapter, the objective function F will be defined only on *sp* Boolean forms (except Section 3.6) by the numbers of literals in these forms. It will be shown later (Section 4.8) that the most general formulation of the minimization problem, based on the above-mentioned definition (v) of objective function F, can be solved in terms of so-called pseudo-Boolean relations.

Although we consider only *sp* Boolean forms, a minimal *ps* form of f can be obtained by applying the de Morgan laws to a minimal *sp* form of $\bar{f}$. Let us note that the number of literals in a minimal *sp* form of f is, generally, different from the number of literals in a minimal *ps* form of the same function f.

3.2 THE SVOBODA METHOD

A method developed by A. Svoboda will be described here in which a Boolean *sp* form containing the least possible number of literals can be formed for a given logic function. The method consistently utilizes the chart representation of logic functions and is based at the same time on the idea of mapping logic functions of n independent variables on an n-dimensional unit cube. Some concepts, enabling us to describe the method, must be defined first.

DEFINITION 3.1 A *vertex* of a unit cube is said to be *occupied* by a logic function f if, and only if, it corresponds to a state of input variables for which $f = 1$.

DEFINITION 3.2 The *weight* W of an occupied vertex V with respect to a logic function f is the number of adjacent vertexes of V that are occupied by f.

DEFINITION 3.3 Every set of 2^k occupied vertexes that can be described by a single p-term containing $n-k$ literals, where n represents the total number of variables, is called a *k-cell*. The number k is referred to as the *order* of the cell.

DEFINITION 3.4 An occupied vertex is said to be *covered* by a k-cell if it is included among the 2^k vertexes of the cell.

DEFINITION 3.5 All occupied vertexes that are covered by a specified set of cells are called *bound vertexes*; those which are not covered by this set of cells are called *free vertexes*.

Now, we are able to describe the Svoboda method. It is based on the following two theorems.

THEOREM 3.1 A sufficient condition for including the Boolean term corresponding to a k-cell in every minimal *sp* form of the given logic function is that the k-cell contain at least one vertex with the weight $W = k$.

PROOF Let C be the k-cell under consideration and V be the vertex of this cell whose weight is equal to k. Obviously, if the weight of the vertex V is equal to k, the cell C contains necessarily all the vertexes adjacent to V. Since the vertex V has no more adjacent vertexes contained in the given logic function, dimensions of all cells except C are smaller than k. Thus vertex V can be covered either by the cell C, whose dimension is equal to k, or by some other cell, whose dimension is less than k. The p-term corresponding to C contains $n-k$ literals, where n is the number of input variables of the given logic function. p-terms corresponding to other cells incident with V contain $n-k_j$ literals, where $k_j < k$. Thus, the term corresponding to C adds the least possible number of literals to the Boolean form. ▮

THEOREM 3.2 Assume that (a) all cells satisfying Theorem 3.1 have been exhausted, (b) there exists a k-cell that contains a free vertex V_0 (the critical vertex), and/or some other free vertexes $V_1, V_2, \ldots, V_m$, and/or a number of bound vertexes.

Then, if both the following conditions are satisfied, the p-term corresponding to the k-cell may be included in the constructed minimal *sp* Boolean form:

1. There is no cell that contains the critical free vertex V_0 and a free vertex that is not among the free vertexes $V_1, V_2, \ldots, V_m$ of the considered k-cell.
2. There is $k_j \leq k$ for all other k_j-cells containing the critical vertex V_0.

PROOF Obviously, there must exist at least one cell C incident with the critical vertex V_0 and described by a p-term T belonging to at least one minimal Boolean form F of the given logic function. Assume that C is a k-cell.

Let us suppose now that the minimal form F is changed by substituting the Boolean term T for another Boolean term T_1 which corresponds to a k_1-cell C_1. Let us further suppose that all free vertexes uncovered by omitting the cell C are covered by the cell C_1 so that the cell C_1 satisfies the first condition of the theorem.

Let L, L_1 be numbers of literals in the Boolean forms F, F_1 respectively. Since F has been assumed to be a minimal form, $L \leq L_1$. The number of literals in T is equal to $n - k$ and the number of literals in T_1 is equal to $n - k_1$. Thus, $L_1 = L - (n - k) + (n - k_1) = L + k - k_1$, and $L \leq L_1$ is expressed as $L \leq L + k - k_1$. In order to satisfy the last inequality, k_1 must be smaller than or equal to k ($k_1 \leq k$). C_1 can satisfy the second condition of the theorem only for $k = k_1$. Then, $L_1 = L$.

It follows from the above mentioned facts that at least one of the two following statements is true:

(i) There exists at least one minimal *sp* Boolean form F that does not contain the term T_1; then the term T_1 belongs to the minimal form F_1 obtainable from the form F.

(ii) There is no minimal *sp* Boolean form F that does not contain the term T_1. ∎

Now, we are able to describe the minimization procedure. It contains the following steps:

1. *Determination of the weights.* Using the rules of adjacency in the Marquand chart (Section 2.1), we determine weights of all occupied vertexes. Until the reader attains sufficient practice in the use of the Marquand charts, he can determine the weights with the aid of special grids constructed in accordance with the rules mentioned. Figure 3.1 shows an example of such grids for six variables. To each column in the chart there belongs one grid denoted by a number, to each row one grid denoted by a letter, in accordance with the columns and rows in Figure 3.2. If we want to determine all adjacent vertexes of a particular vertex, we take the grids corresponding to the row and column of the reference vertex, respectively, and place them on the chart. Only those squares which belong to the vertexes adjacent to the given vertex remain uncovered in the corresponding row and column. Having found the number of occupied vertexes in the uncovered squares, we thereby immediately obtain the weight of the reference vertex. We can use each of the grids for two different rows and columns by rotating it about the axis indicated by the arrows shown in Figure 3.1.

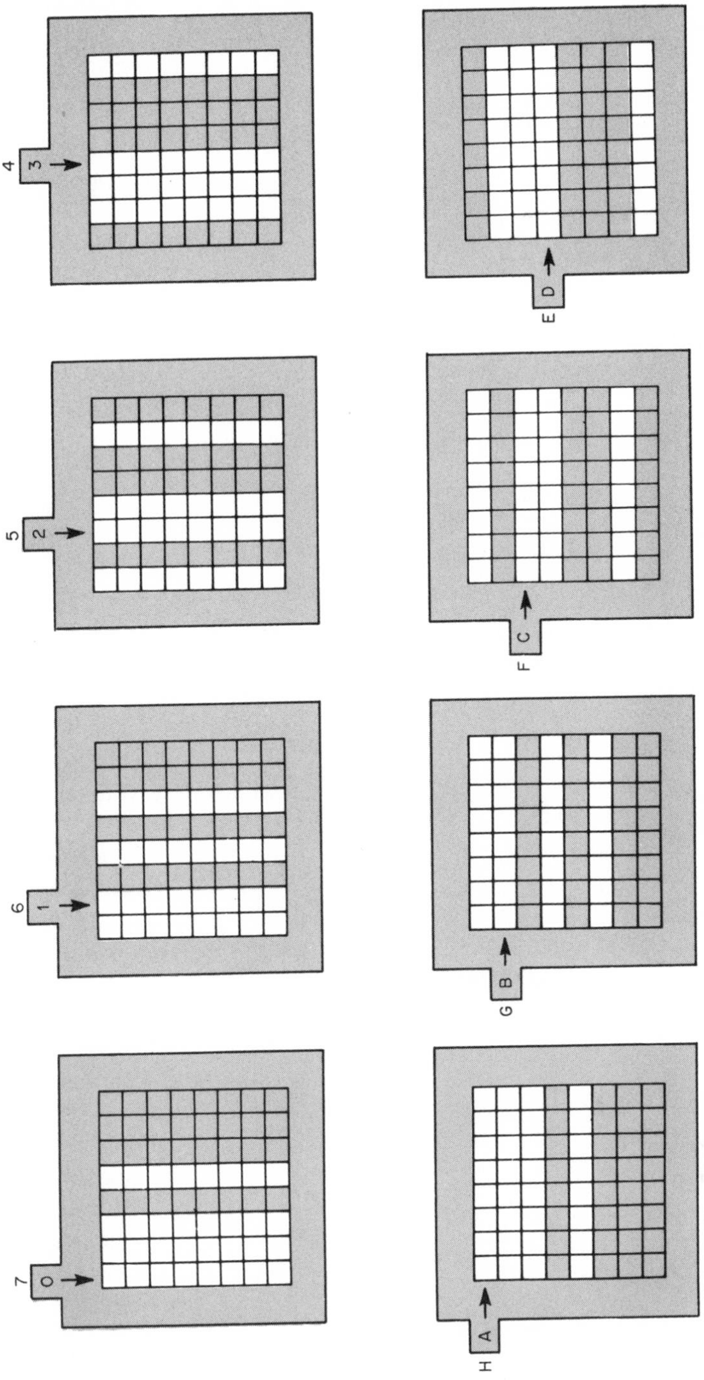

Figure 3.1 Grids for determining adjacent vertexes on a Marquand chart for six variables.

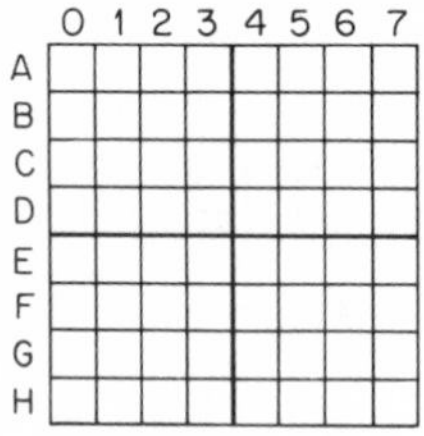

Figure 3.2   Denotation of columns and rows in a chart for six variables.

Since the weights decrease during the minimization procedure, it is useful to denote them in the chart by the symbols shown in Figure 3.3 for four through six independent variables.

In case of indefinite logic functions, the weight of a vertex must represent the number of both adjacent states occupied by the function and adjacent don't care states.

2. *Application of Theorem* 3.1.   It is reasonable to start from vertexes with smallest weights. If the weight of a vertex is equal to 0 or 1, the condition of Theorem 3.1 is always satisfied and the respective 0-cells and 1-cells must be included in the constructed minimal form. Then we proceed to consider vertexes whose weights are equal to 2, 3, etc.

Every vertex for which the condition of Theorem 3.1 is satisfied should be registered together with the selected cell and the corresponding *p*-term. Then the respective vertexes must be crossed out in the map to indicate that they are bound from now on.

The *p*-terms found by Theorem 3.1 are included in all minimal *sp* forms of the given logic function.

3. *Correction of the weights.* When the selection of cells satisfying the conditions of Theorem 3.1 is finished, we shall change (reduce) weights of the remaining free vertexes by taking into consideration only free adjacent vertexes. Both bound and don't care states account for the reduction of weight.

n = 4		*n*=5		*n*=6	
Symbol	Weight	Symbol	Weight	Symbol	Weight
				•	6
		•	5	│•	5
•	4	│•	4	│•_	4
│•	3	│•_	3	│•│	3
│•_	2	│•│	2	▫•▫	2
▫•│	1	▫•▫	1	▣	1
▫•▫	0	▣	0	▣	0

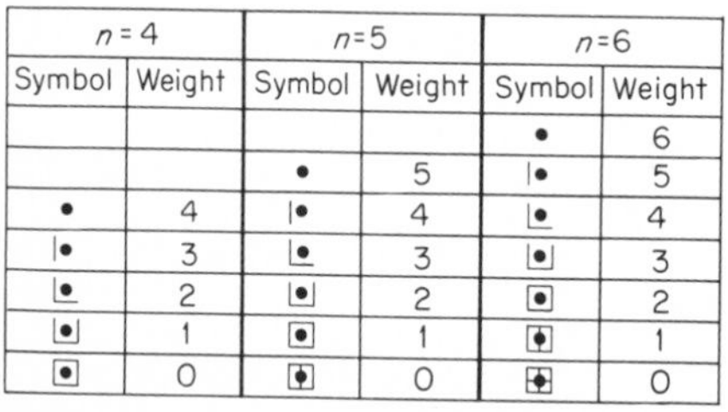

Figure 3.3   Symbols used for denoting the weights of up to six variables.

The bound vertexes may be denoted as "don't care" in the chart before we proceed to the next step.

4. *Application of Theorem* 3.2. Starting with the vertexes possessing the smallest weight we select cells that satisfy the conditions of Theorem 3.2. After a single cell is selected (and added to the list) the respective free vertexes must be crossed out and weights of the remaining free vertexes must be corrected (reduced) with regard to the vertexes which became bound.

The *p*-terms found by Theorem 3.2 are not, in general, the terms contained in all *sp* minimal Boolean forms of the given logic function.

5. *Termination of the procedure.* If Theorem 3.2 cannot be applied any more at a certain stage of minimization and some of the vertexes are still free (this sometimes happens), we must take into consideration all combinational possibilities and then select minimal forms from all the obtained forms.

The complete minimization procedure will be illustrated now by some examples.

EXAMPLE 3.1 Find a minimal *sp* Boolean form for the logic function $f(Y, X, y, x)$ given by the map in Figure 3.4.a.

The weights are denoted in Figure 3.4b. Applying Theorem 3.1 we find that the 2-cell AB23 is incident with the vertex A2 whose weight is equal to 2. Thus, the condition of Theorem 3.1 is satisfied and the term $\overline{Y}y$ corresponding to the cell AB23 is included in every *sp* minimal Boolean form of the given function *f*.

There are three other vertexes whose weights are equal to 2: B0, D0, and C1. Theorem 3.1 cannot be applied for the vertexes B0 or D0 because there are no 2-cells incident with them. It can be, however, applied for the vertex C1 since CD13 is the 2-cell incident with it. The term is Yx.

Having considered the two above-mentioned cells, the application of Theorem 3.1 terminates. Now, the weights of the remaining vertexes B0 and D0 are reduced and the bound vertexes are denoted as "don't care" (Figure 3.4c).

There is one 1-cell that satisfies the conditions of Theorem 3.2, namely, BD0. Thus, there is only one minimal *sp*-form.

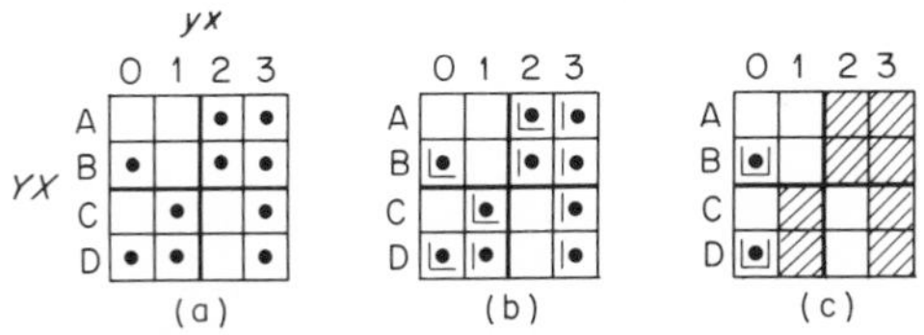

Figure 3.4 Illustration for Example 3.1.

The minimization is shortly recapitulated in the following table:

Critical Vertex	Cell	Term	A Note
A2 C1	AB23 CD13	$\overline{Y}y$ Yx	Application of Theorem 3.1
B0	BD0	$X\overline{y}\overline{x}$	Application of Theorem 3.2

Answer: There is one minimal *sp* Boolean form for the given logic function:

$$f = \overline{Y}y \vee Yx \vee X\overline{y}\overline{x}$$

EXAMPLE 3.2 Find a minimal *sp* Boolean form for the logic function $f(Z, Y, X, z, y, x)$ given in Figure 3.5a.

First, the weights must be determined. This is shown in Figure 3.5b. Then Theorem 3.1 is applied. The Theorem is always satisfied for any vertex whose weight is equal to 0 or 1. The vertex C0 is the only one whose weight is equal to 0. It itself represents the 0-cell C0. The corresponding *p*-term is $\overline{Z}YX\overline{z}\overline{y}\overline{x}$.

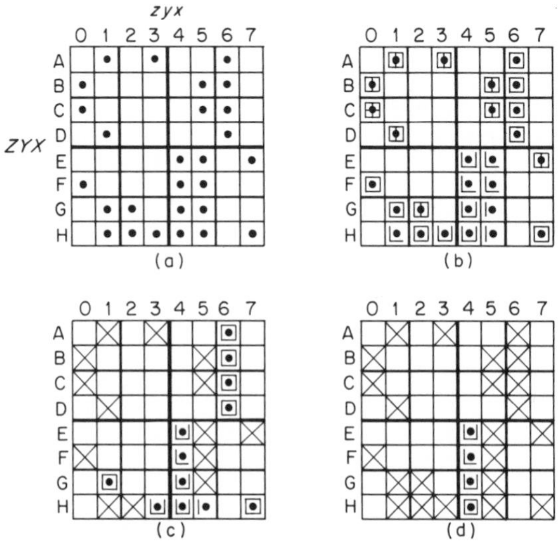

Figure 3.5 Illustration for Example 3.2.

Now, the vertexes whose weights are equal to 1 are taken into consideration and we find that the 1-cells A13, BF0, BF5, CG5, DH1, E57 and GH2 all satisfy the condition of Theorem 3.1. The corresponding *p*-terms can be easily found.

When a cell is taken, all its vertexes must be crossed out in the map. The present stage is shown in Figure 3.5c.

Now, we are looking for vertexes whose weights are equal to 2 and which satisfy the condition of Theorem 3.1. We may take any vertex of the cell ABCD6, and vertexes G1, and H7 determining 2-cells GH15, and H1357, respectively. After crossing out the respective vertexes, we get the situation shown in Figure 3.5d. Any one of the vertexes E4, G4, H4 satisfies Theorem 3.1 for 3-cell EFGH45. The procedure is completed now because no vertex occupied by the given logic function is free.

The whole procedure is again registered in a condensed form by the following table:

Critical Vertex	*Cell*	*Term*
C0	C0	$\bar{Z}Y\bar{X}\bar{z}\bar{y}\bar{x}$
A1	A13	$\bar{Z}\bar{Y}\bar{X}\bar{z}x$
B0	BF0	$\bar{Y}X\bar{z}\bar{y}\bar{x}$
B5	BF5	$\bar{Y}Xz\bar{y}x$
C5	CG5	$Y\bar{X}z\bar{y}x$
D1	DH1	$YX\bar{z}\bar{y}x$
E7	E57	$Z\bar{Y}\bar{X}zx$
G2	GH2	$ZY\bar{z}y\bar{x}$
A6	ABCD6	$\bar{Z}zy\bar{x}$
G1	GH15	$ZY\bar{y}x$
H7	H1357	$ZYXx$
E4	EFGH45	$Zz\bar{y}$

OR operation of the terms contained in the table represents the minimal *sp* Boolean form. There is only one minimal *sp*-form for the given function since the application of Theorem 3.1 has been sufficient to solve the problem.

Obviously, one map is sufficient to solve the problem. Several maps were used in Figure 3.5 only for a better understanding.

EXAMPLE 3.3 Find a minimal *sp* Boolean form for the logic function $F(Z, Y, X, z, y, x)$ given in Figure 3.6a.

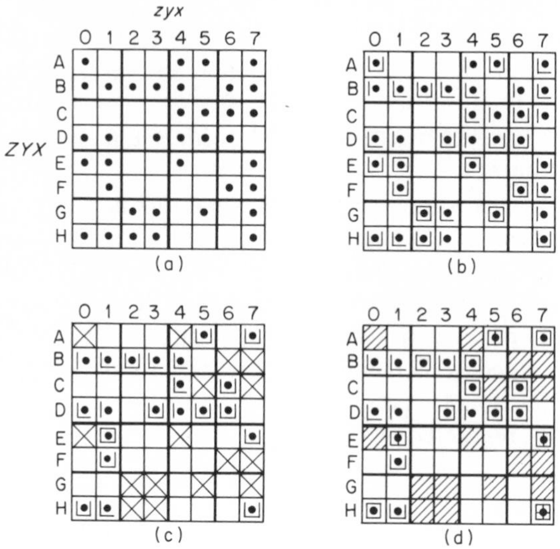

Figure 3.6 Illustration for Example 3.3.

Weights of all vertexes are denoted in Figure 3.6b. Then, applying Theorem 3.1 (for vertexes whose weights are 0, 1, 2, etc.) we get AE04, BF67, GH23, and CG57. There are no other cells for which Theorem 3.1 is applicable so that we must proceed to Theorem 3.2. The present stage is shown in Figure 3.6c.

Before applying Theorem 3.2 we change weights of all remaining free vertexes and denote the bound vertexes as "don't care." This is shown in Figure 3.6d.

Vertex H7 has the least weight so that it should be considered first. 2-Cell EFGH7, which is incident with H7, satisfies the conditions of Theorem 3.2 and, therefore, the corresponding *p*-term may be included in a minimal *sp* Boolean form. We take it and modify the weights.

Three vertexes whose weights are equal to 1 must be taken into consideration now, A5, E1, A7. We find two 2-cells incident with A5 (AC45, AC57) but neither of them satisfies the conditions of Theorem 3.2. Both the cells have the same dimension but AC45 contains free vertex C4 which is not included in AC57 and AC57 contains free vertex A7 which is not included in AC45. The cell EF1 incident with E1 satisfies the conditions of the Theorem so that the corresponding *p*-term may be included in the minimal *sp* Boolean form. Then, we consider the last vertex whose weight is equal to 1. We find that the 2-cell AC57, which has been rejected before, satisfies the conditions of Theorem 3.2 if A7 is taken as the critical vertex instead of A5.

The complete procedure is again summarized in the following table:

Critical Vertex	*Cell*	*Term*	*A Note*
E4	AE04	$\bar{Y}\bar{X}\bar{y}\bar{x}$	Theorem 3.1
F6	BF67	$\bar{Y}Xzy$	
G2	GH23	$ZY\bar{z}y$	
G5	CG57	$Y\bar{X}zx$	
H7	EFGH7	$Zzyx$	Theorem 3.2
E1	EF1	$Z\bar{Y}\bar{z}\bar{y}x$	
A7	AC57	$\bar{Z}\bar{X}zx$	
C6	CD46	$\bar{Z}Yz\bar{x}$	
D5	D0145	$\bar{Z}YX\bar{y}$	
H0	H0123	$ZYX\bar{z}$	
D3	BD13	$\bar{Z}X\bar{z}x$	
B4	B0246	$\bar{Z}\bar{Y}X\bar{x}$	

OR operation of the found Boolean terms represents one of the possible minimal *sp* Boolean forms of the given logic function. The terms obtained by Theorem 3.1 are contained in every minimal *sp* form of the given function.

EXAMPLE 3.4 Find a minimal *sp* Boolean form for the function $F(Y, X, y, x)$ given in Figure 3.7a.

First, the weights of the vertexes occupied by the given function must be determined as shown in Figure 3.7b.

Trying to apply Theorem 3.1 and then Theorem 3.2 we find that the conditions of neither of them are satisfied. For that reason we must take a vertex as a critical one and take a cell incident with it by a random choice.

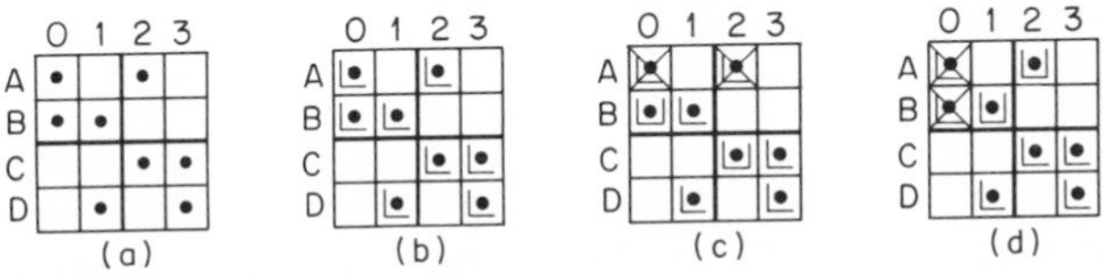

Figure 3.7 Illustration for Example 3.4.

Let us take, for instance, the vertex A0 and the cell A02. After we do it (see Figure 3.7c), Theorem 3.2 begins to be applicable and we find the following expression:

Critical Vertex	*Cell*	*Term*	*A Note*
A0	A02	$\bar{Y}\bar{X}\bar{x}$	Random choice
B0	B01	$\bar{Y}X\bar{y}$	Application of Theorem 3.2
C2	C23	$Y\bar{X}y$	
D1	D13	YXx	

Thus, $f = \bar{Y}\bar{X}\bar{x} \vee \bar{Y}X\bar{y} \vee Y\bar{X}y \vee YXx$.

Since we used a random choice we are not sure whether the found expression is really minimal. Therefore, we must return to the stage where the random choice was made and repeat the procedure for other possible choices. In our example, there is only one alternative choice for the critical vertex A0, namely, the cell AB0. After we take this cell (see Figure 3.7d), Theorem 3.2 is again applicable for the whole remaining procedure and the following expression will be found:

Critical Vertex	*Cell*	*Term*	*A Note*
A0	AB0	$\bar{Y}\bar{y}\bar{x}$	Random choice
A2	AC2	$\bar{X}y\bar{x}$	Application of Theorem 3.2
B1	BD1	$X\bar{y}x$	
C3	CD3	Yyx	

Thus, $f = \bar{Y}\bar{y}\bar{x} \vee \bar{X}y\bar{x} \vee X\bar{y}x \vee Yyx$.

We see that both the expressions contain the same number of literals so that both are minimal. This is, however, a special case. In general, different expressions obtained by random choices may contain different number of literals so that we are obliged to construct all of them and then compare the results in order to find the real minimal expression.

EXAMPLE 3.5 Find a minimal *sp* Boolean form for the indefinite logic function $F(Z, Y, X, z, y, x)$ given in Figure 3.8a.

In order to determine weights of the vertexes occupied by the given logic function we count both the adjacent vertexes where the function is equal to 1 and the adjacent vertexes where the function is not defined. The weights are denoted in Figure 3.8b.

We see that Theorem 3.1 can be successfully applied only to vertexes D0, E3, H4 which yield, respectively, cells CD02, AE37, EFGH4. When weights are modified after the application of Theorem 3.1 (Figure 3.8c), don't cares must be treated as bound vertexes. The complete procedure is summarized in the following table:

Critical Vertex	*Cell*	*Term*	*A Note*
D0	CD02	$\bar{Z}Y\bar{z}\bar{x}$	Theorem 3.1
E3	AE37	$\bar{Y}\bar{X}yx$	
H4	EFH4	$Zz\bar{y}\bar{x}$	
G7	G57	$ZY\bar{X}zx$	Theorem 3.2
C5	CG15	$Y\bar{X}\bar{y}x$	
B1	BF15	$\bar{Y}X\bar{y}x$	
H1	H13	$ZYX\bar{z}x$	
H2	CDGH2	$Y\bar{z}y\bar{x}$	
G0	EG04	$Z\bar{X}\bar{y}\bar{x}$	
F0	F0145	$Z\bar{Y}X\bar{y}$	

Exercises to Section 3.2

3.2-1 Prove that the condition of Theorem 3.1 is always satisfied for a vertex whose weight is equal to either 0 or 1.

3.2-2 Find a counter example showing that Theorem 3.1 is not always satisfied for a vertex whose weight is equal to 2 or more.

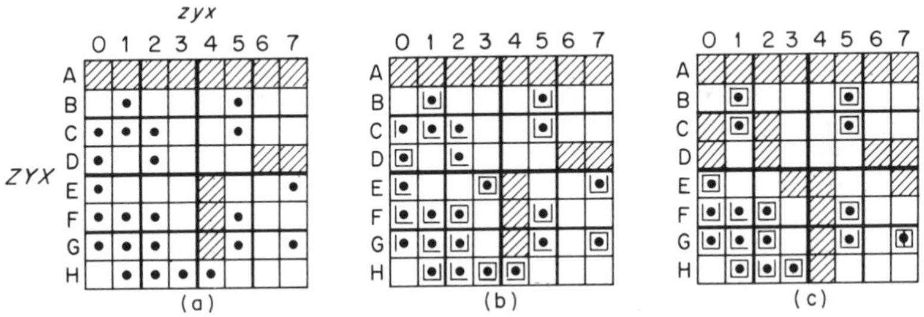

Figure 3.8 Illustration for Example 3.5.

3.2-3 Construct a situation in which:
(a) only condition 1 of Theorem 3.2 is satisfied.
(b) only condition 2 of Theorem 3.2 is satisfied.

3.2-4 Find minimal *sp* and *ps* Boolean forms of the following logic functions of four variables:
*(a) $f_1(s) = 1$ for $s = 0, 1, 3, 5, 7, 12, 13, 14, 15$
$f_1(s) = 0$ otherwise.
(b) $f_2(s) = 1$ for $s = 2, 3, 5, 7, 9, 11, 14, 15$
$f_2(s) = 0$ otherwise.
(c) $f_3(s) = 0$ for $s = 6, 15$
$f_3(s) = 1$ otherwise.
(d) $f_4(s) = 1$ for $s = 0, 2, 6, 7, 8, 9, 13, 15$
$f_4(s) = 0$ otherwise.
*(e) $f_5(s) = 1$ for $s = 1, 2, 5, 7, 14, 15$
$f_5(s)$ is don't care for $s = 3, 6, 13$
$f_5(s) = 0$ otherwise.

3.2-5 Find a minimal *sp* Boolean form of the logic function presented by:
*(a) the chart in Figure 3.9.
(b) the chart in Figure 3.10.

3.3 DETERMINATION OF PRIME IMPLICANTS

DEFINITION 3.6 An *implicant* of a logic function f is a p-term φ for which $\varphi \leq f$, i.e., $\varphi = 1$ implies $f = 1$.

DEFINITION 3.7 An implicant φ of a logic function is said to be a *prime implicant* of f if φ ceases to be an implicant of f after any one of its literals is removed.

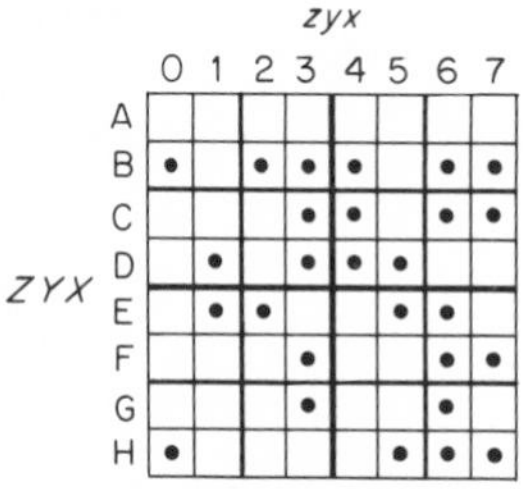

Figure 3.9 Function given in Exercise 3.2-5a.

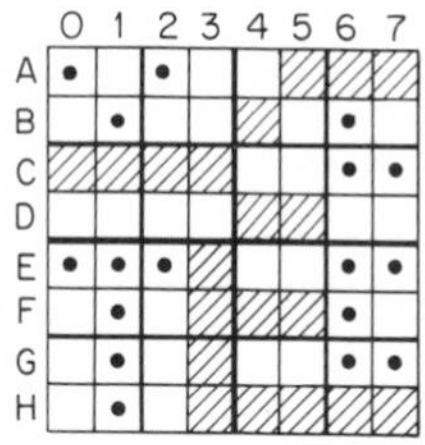

Figure 3.10 Function given in Exercise 3.2-5b.

THEOREM 3.3 Every minimal *sp* Boolean form is an OR function of prime implicants.

PROOF Let $f = \varphi_1 \vee \varphi_2 \vee \cdots \vee \varphi_m$ be a minimal *sp* Boolean form. Then, each $\varphi_i (i = 1, 2, \ldots, m)$ must be an implicant of f. If φ_i is not a prime implicant, then at least one of its literals can be dropped without changing f. This is in contradiction to the original assumption that $\varphi_1 \vee \varphi_2 \vee \cdots \vee \varphi_m$ is a minimal form. ▮

Due to the fact stated by Theorem 3.3, prime implicants play an important role in Boolean minimization theory. However, not every OR function of prime implicants represents a minimal Boolean form.

EXAMPLE 3.6 Suppose $f(s) = 0$ for $s = 2, 5$ and $f(s) = 1$ otherwise ($s = 0, 1, \ldots, 7$). We can easily prove that the following *p*-terms are all prime implicants: $\bar{x}_3 \bar{x}_2$, $\bar{x}_3 x_1$, $x_3 \bar{x}_1$, $x_3 x_2$, $\bar{x}_2 \bar{x}_1$, $x_2 x_1$. We can write $f = \bar{x}_3 \bar{x}_2 \vee \bar{x}_2 \bar{x}_1 \vee x_3 x_2 \vee x_2 x_1$. Although this form is an OR function of prime implicants and no literal can be removed from it, without destroying the equality (so called *irredundant form*), it does not represent a minimal form. Indeed, minimal *sp* forms of f contain only six literals, e.g., $f = \bar{x}_2 \bar{x}_1 \vee \bar{x}_3 x_1 \vee x_3 x_2$.

Example 3.6 shows that only a proper OR function of prime implicants creates an *sp* minimal form. To find this proper OR function, all prime implicants of a given logic function must be determined first.

Several procedures have been elaborated for the determination of all prime implicants for a given logic function. The classical procedure is based on a systematic application of the Boolean rule $a = ax \vee a\bar{x}$ to the expanded *sp* Boolean form. A method based on the *ternary representation* of Boolean *p*-terms will be presented here. The method, which is due to A. Svoboda, is exceptionally convenient for computer processing.

Let

$$t = \sum_{i=1}^{n} h_i \cdot 3^{i-1} \tag{3.1}$$

be an *identifier of Boolean p-terms* $\dot{x}_n \dot{x}_{n-1} \cdots \dot{x}_1$, where $h_i = 0$ if $\dot{x}_i = 1$ (if variable $\dot{x}_i$ does not participate in the term), $h_i = 1$ if $\dot{x}_i = x_i$ and $h_i = 2$ if $\dot{x}_i = \bar{x}_i$. All *p*-terms are ordered by the term identifier t and can be represented by a chart similar to the Marquand chart and containing 3^n little squares for n independent variables (so called *ternary chart*). An example of such a chart for $n = 4$ is shown in Figure 3.11. Symbols t_c, t_r denote partial term identifiers of t which are associated with columns and rows of the chart respectively.

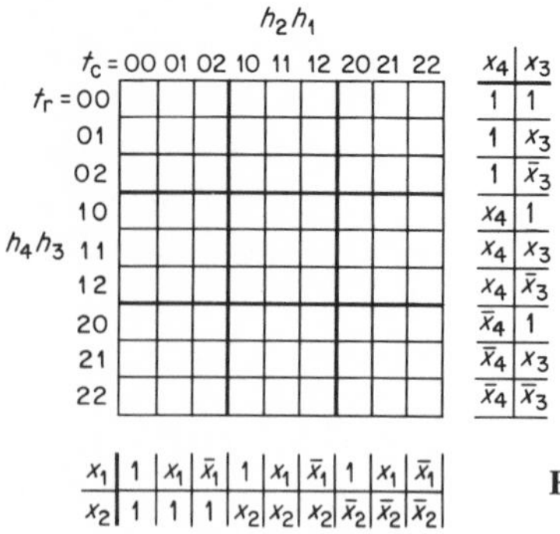

Figure 3.11 Chart representation of all Boolean p-terms.

DEFINITION 3.8 Let

$$\dot{\imath} = \sum_{i=1}^{n} \dot{h}_i \cdot 3^{i-1}$$

and

$$\tilde{\imath} = \sum_{i=1}^{n} \tilde{h}_i \cdot 3^{i-1}$$

be term identifiers of Boolean p-terms $\dot{x}_n \dot{x}_{n-1} \cdots \dot{x}_1$ and $\tilde{x}_n \tilde{x}_{n-1} \cdots \tilde{x}_1$, respectively. Then, the term $\tilde{\imath}$ is called a *formal divisor* of the term $\dot{\imath}$ if for each $\dot{h}_i (i = 1, 2, \ldots, n)$ $\tilde{h}_i$ satisfies the following relation:

$\dot{h}_i$	$\tilde{h}_i$
0	0
1	0 or 1
2	0 or 2

DEFINITION 3.9 Consider the same symbols as introduced in Definition 3.8. Then, the term $\dot{\imath}$ is *formally divisible* by the term $\tilde{\imath}$ if for each $\tilde{h}_i (i = 1, 2, \ldots, n) \dot{h}_i$ satisfies the following relation (inverse relation of that specified in Definition 3.8):

$\tilde{h}_i$	$\dot{h}_i$
0	0 or 1 or 2
1	1
2	2

Let us consider now the *space of all Boolean p-terms* of n variables. We want to exclude from the space all terms which are not prime implicants of a given logic function $f(x_1, x_2, \ldots, x_n)$.

First, all minterms for which $f = 0$ are not implicants of f and cannot be, therefore, prime implicants of f. Hence, they must be excluded from the space of terms. In addition, all of their formal divisors must be excluded too. Indeed, formal divisors of a minterm are terms which determine the same state of independent variables as the minterm and some other states.

Suppose that all minterms for which $f = 0$ and all their formal divisors have already been excluded from the space of terms. Let us denote the remaining terms in the increasing order $t_1, t_2, \ldots, t_m (t_i < t_j$ if $i < j, i \neq j)$. Some of them are necessarily prime implicants of the given logic function f. Suppose now that one of them is known. Then, we may cancel all terms which are formally divisible by this prime implicant. Indeed, these terms contain some literals in addition to all those which are included in the prime implicant. Let t_p be the prime implicant and let t_d be a term that is formally divisible by t_p. Observing the relation introduced in Definition 3.9, we find that the inequality

$$t_d > t_p$$

must be satisfied by any term t_d that is divisible by t_p. This means, however, that the term t_1 in the sequence $t_1 < t_2 < \cdots < t_m$ cannot be cancelled by any prime implicant included, eventually, in the set $t_2, t_3, \ldots, t_m$; consequently, the term t_1 itself is a prime implicant.

Thus, the first prime implicant which we discover is the term with the smallest term identifier which is not included among the cancelled minterms and their formal divisors. All terms which are formally divisible by this prime implicant can be cancelled now. Among the remaining terms, the term with the smallest term identifier is the next prime implicant, etc. The procedure terminates when each term is either cancelled or determined to be a prime implicant.

Let us summarize algorithmically the whole procedure of determination of all prime implicants for a given logic function $f(x_1, x_2, \ldots, x_n)$:

ALGORITHM 3.1 Given a logic function $f(x_1, x_2, \ldots, x_n)$, to find all of its prime implicants:

(1) Consider the space of all Boolean p-terms of variables $x_1, x_2, \ldots, x_n$, e.g., in the form of a ternary chart.
(2) Cancel each minterm for which $f = 0$ and all its formal divisors.
(3) Cancel the term with the smallest term identifier and register it as a prime implicant.
(4) Cancel all terms which are divisible by the prime implicant determined in (3).
(5) If all terms are cancelled, this algorithm terminates, otherwise go to (3).

EXAMPLE 3.7 Determine all prime implicants of the logic function f specified in Figure 3.12.

The space of all Boolean p-terms of four variables is shown in the form of a chart in Figure 3.13a. Minterms for which $f = 0$ are denoted by little crosses. All formal divisors of the minterm $x_4 x_3 \bar{x}_2 x_1$ are cancelled in Figure 3.13b. Cancellation of all prohibited minterms and their formal divisors is shown in Figure 3.13c. The first prime implicant, denoted by ①, is shown in Figure 3.13d. Among those terms which have not been cancelled yet, there are only two terms which are divisible by this prime implicant. They are denoted by 1. Further procedure is demonstrated in Figure 3.13e., where prime implicants are denoted by ②, ③, ... and the terms which they cancel by 2, 3, ..., respectively. To summarize, the prime implicants are discovered in the following order:

Order of Determination	*Term Identifier*		*Prime Implicant*
	Ternary	*Decimal*	
1	0112	14	$x_3 x_2 \bar{x}_1$
2	0201	19	$\bar{x}_3 x_1$
3	0220	24	$\bar{x}_3 \bar{x}_2$
4	1010	30	$x_4 x_2$
5	1200	45	$x_4 \bar{x}_3$
6	2020	60	$\bar{x}_4 \bar{x}_2$
7	2102	65	$\bar{x}_4 x_3 \bar{x}_1$

Obviously, the whole procedure can be done in one chart and some useful properties of the chart can be employed with advantage.

In case of an incompletely specified logic function, the functional value 1 may be selected for each "don't care" state. After all prime implicants are determined, those which contain solely "don't care" states must be excluded.

Exercises to Section 3.3

3.3-1 Prove the following propositions:

(a) If a term is not an implicant of a logic function, then all its formal divisors are not implicants of the function.

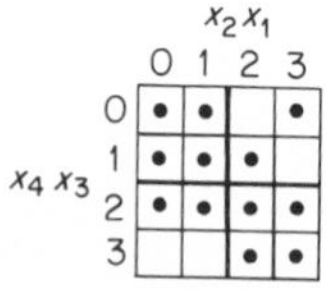

Figure 3.12 Illustration for Example 3.7.

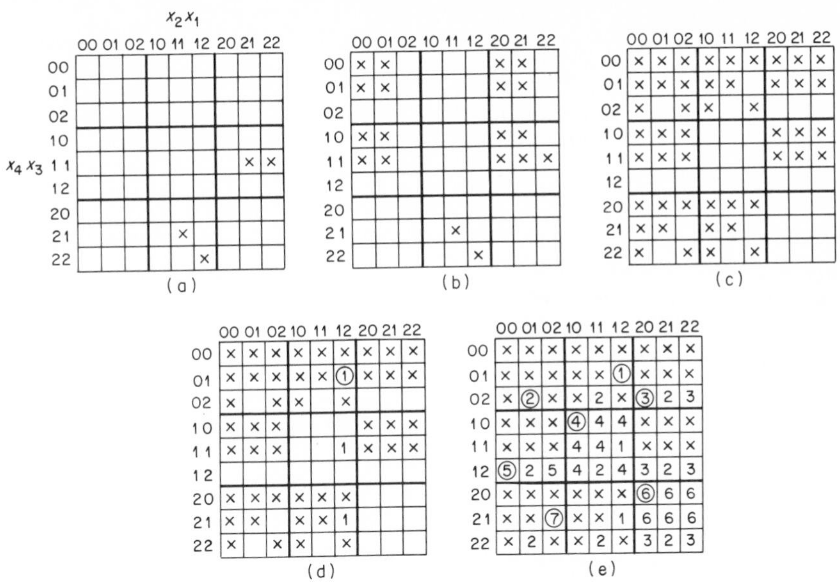

Figure 3.13 Illustration of Algorithm 3.1 (Example 3.7).

(b) If t_p is a prime implicant, then any term t divisible by t_p such that $t \neq t_p$ is not a prime implicant.

(c) Let t_c and t_r be the column identifier and the row identifier of a Boolean p-term. All formal divisors t'_c, t'_r, of this term satisfy the inequalities $t'_c \leq t_c$, $t'_r \leq t_r$ and all terms t''_c, t''_r divisible by it satisfy the inequalities $t''_c \geq t_c$, $t''_r \geq t_r$.

(d) If t'_c, t_r and t_c, t'_r are formal divisors of a term t_c, t_r, then t'_c, t'_r is its formal divisor too.

(e) If t'_c, t_r and t_c, t'_r are both divisible by t_c, t_r, then t'_c, t'_r is divisible by t_c, t_r too.

3.3-2 Suggest a procedure based on a systematic application of the Boolean rule $a = ax \vee a\bar{x}$ by which all prime implicants of a given logic function can be determined.

3.3-3 Determine all prime implicants of the following logic functions of n variables:

*(a) $f(s) = 0$ for $s = 1, 6, 7, 9, 11, 13$
$f(s) = 1$ otherwise, $n = 4$.

*(b) $f(s) = 0$ for $s = 0, 11$
$f(s) = 1$ otherwise, $n = 4$.

(c) $f(s) = 0$ for $s = 0, 2, 5, 8, 10, 15$
$f(s) = 1$ otherwise, $n = 4$.

(d) $f(s) = 1$ for $s = 0, 3, 5, 6, 9, 10, 15$
$f(s)$ is not defined for $s = 2, 12, 13$
$f(s) = 0$ otherwise, $n = 4$.

(e) $f(s) = 0$ for $s = 4, 7, 8$
$f(s)$ is not defined for $s = 0, 1, 2, 3, 12, 13$
$f(s) = 1$ otherwise, $n = 4$.

*(f) $f(s) = 1$ for $s = 1, 2, 3, 5, 9, 10, 11, 18, 19, 20, 21, 23, 25, 26, 27$
$f(s) = 0$ otherwise, $n = 5$.
*(g) $f(s) = 0$ for $s = 3, 7, 8, 12, 14, 15$
$f(s) = 1$ otherwise, $n = 4$.

3.4 THE PROBLEM OF MINIMAL COVERING

After all prime implicants of a given logic function are determined, we prepare a so-called *prime implicant table.* Each row of this table corresponds to a prime implicant and each column represents a state for which the function is equal to 1. A dot is placed at the intersection of a row and a column if the corresponding state is occupied by the respective prime implicant.

The prime implicant table for the function, whose prime implicants were determined in Example 3.7, is shown in Figure 3.14a. Rows of the table are grouped according to the number of literals in the respective prime implicants (or according to the number of nonzero digits in their ternary identifiers).

Using the prime implicant table, we want to cover all states (columns) included in the table by a set of prime implicants (rows) with the minimal sum of literals (nonzero digits in ternary identifiers). Each state (column) must be covered by at least one prime implicant (row). Hence, if a state is covered just by one prime implicant, then the prime implicant must necessarily be included in every minimal form of the function. Such a prime implicant is called an *essential prime implicant* (essential *row*).

Initial Table

	Ternary form of the prime implicants	Column and row identifiers of the states 00	01	03	10	11	12	20	21	22	23	32	33
*	0 2 0 1		•	•					•		•		
	0 2 2 0	•	•					•	•				
*	1 0 1 0									•	•	•	•
	1 2 0 0							•	•	•	•		
*	2 0 2 0	•	•		•	•							
	0 1 1 2						•					•	
	2 1 0 2				•		•						
		+	+	*	+	*			+	+	+	+	*

(a)

Reduced Table

	12	20
0 2 2 0		•
1 2 0 0		•
0 1 1 2	•	
2 1 0 2	•	

(b)

$$f = \bar{x}_3 \bar{x}_1 \vee x_4 x_2 \vee \bar{x}_4 \bar{x}_2 \begin{cases} \vee \bar{x}_3 \bar{x}_2 \vee x_3 x_2 \bar{x}_1 \\ \vee \bar{x}_3 \bar{x}_2 \vee \bar{x}_4 x_3 \bar{x}_1 \\ \vee x_4 \bar{x}_3 \vee x_3 x_2 \bar{x}_1 \\ \vee x_4 \bar{x}_3 \vee \bar{x}_4 x_3 \bar{x}_1 \end{cases}$$

(c)

Figure 3.14 Illustration of the Quine-McCluskey covering procedure (Example 3.7).

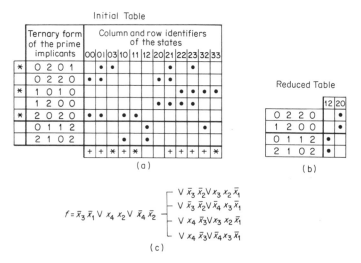

The columns that are covered by a single row are called *distinguished columns* and are denoted in Figure 3.14a by asterisks, as are the essential rows. The columns which are not distinguished but are covered by the essential rows are denoted by crosses. When we exclude from the prime implicant table all essential rows (whose prime implicants must be registered for the minimal form), all covered columns, and all rows representing only columns covered by the essential rows, we obtain a *reduced prime implicant table*. Then, we can apply the following theorem.

THEOREM 3.4 Let $S(p_i)$ and $S(p_j)$ denote the sets of states covered, respectively, by the prime implicants p_i and p_j in a reduced prime implicant table. Let $L(p_i)$ and $L(p_j)$ denote the number of literals (the number of nonzero digits in the ternary identifiers) in the prime implicants p_i and p_j, respectively. Then, if both

$$S(p_i) \subseteq S(p_j)$$

and

$$L(p_i) \geq L(p_j)$$

are satisfied, there exists a minimal *sp* Boolean form which does not include p_i.

PROOF Left to the reader as an exercise.

If both conditions of Theorem 3.4 are satisfied, we say that prime implicant p_i is *dominated by* prime implicant p_j. It follows from Theorem 3.4 that all dominated prime implicants (rows) can be excluded from the prime implicant table. After we exclude them, we can again look for columns each of which is covered by a single row. If such a column exists, then the prime implicant corresponding to the row belongs to the minimal form under construction (so-called secondary essential prime implicant).

The reduced prime implicant table derived from the prime implicant table in Figure 3.14a is shown in Figure 3.14b. An application of Theorem 3.4 is very easy in this case. Either the first or the second row may be excluded and either the third or the fourth row may be excluded. This leads to four minimal forms shown in Figure 3.14c.

To summarize, the described procedure of minimal prime implicant covering, often called the *Quine-McCluskey method*, is based on the following two rules:

Rule 1. Determination and recording of essential prime implicants and exclusion of the respective columns and rows for each of them.

Rule 2. Exclusion of dominated prime implicants (rows).

These rules are applied as specified by the flow chart in Figure 3.15.

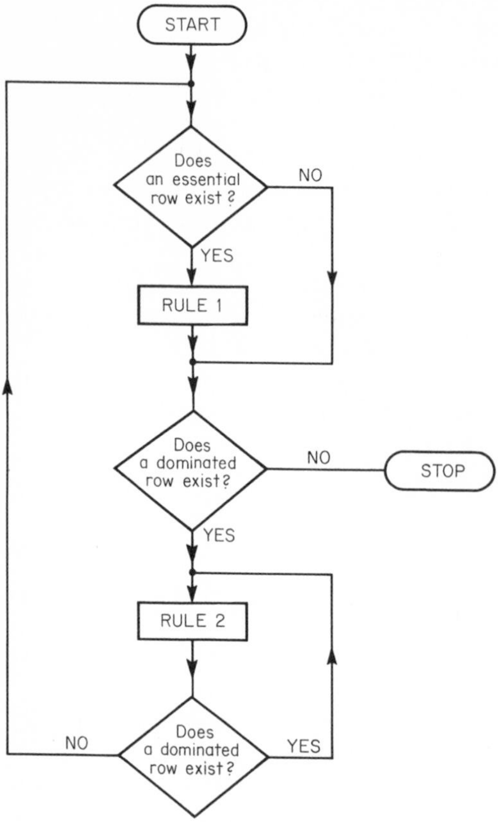

Figure 3.15 Flow chart of the Quine-McCluskey procedure for minimal prime implicant covering.

EXAMPLE 3.8 Solve the problem of minimal covering for the prime implicant table in Figure 3.16a.

Using the rule of distinguished columns, we obtain the reduced prime implicant table in Figure 3.16b. The second row is dominated by the fifth row so that the former may be excluded. Similarly, either the third or the fourth row in Figure 3.16b can be excluded. Exclusion of the latter leads to the table shown in Figure 3.16c. Here, the second row and the third row represent secondary essential prime implicants because each of them covers alone one of the columns. Hence, we have determined the following minimal form:

$$f = \bar{x}_3 x_2 \vee x_4 \bar{x}_3 x_1 \vee x_5 \bar{x}_4 x_3 \bar{x}_2 \vee x_5 \bar{x}_4 x_2 x_1 \vee \bar{x}_5 \bar{x}_4 \bar{x}_2 x_1$$

It happens sometimes that no essential (or secondary essential) prime implicant exists so that we cannot proceed. In such a case we cover a column

	Term identifier	1	2	3	5	9	10	11	18	19	20	21	23	25	26	27
*	0 0 2 1 0		•	•			•	•	•	•					•	•
*	0 1 2 0 1					•		•						•		•
	2 0 2 0 1	•		•		•		•								
	0 2 1 2 1				•							•				
	1 2 0 1 1								•				•			
	1 2 1 0 1											•	•			
*	1 2 1 2 0										•	•				
	2 2 0 2 1	•			•											
			*	+		+	*	+	+	*	*	+		*	*	+

(a)

Term identifier	1	5	23
2 0 2 0 1	•		
0 2 1 2 1		•	
1 2 0 1 1			•
1 2 1 0 1			•
2 2 0 2 1	•	•	

(b)

	Term identifier	1	5	23
	2 0 2 0 1	•		
*	1 2 0 1 1			•
*	2 2 0 2 1	•	•	
		+	*	*

(c)

Figure 3.16 Illustration for Example 3.8.

by a randomly selected row from the set of rows covering this column. Then we complete the minimization procedure employing again Theorem 3.4, the rule of distinguished columns, and, if necessary, a random choice. After the procedure is completed, we must return to the stage where a random choice was used and repeat the procedure for other possible choices.

EXAMPLE 3.9 Solve the problem of minimal covering for the prime implicant table in Figure 3.17a.

The prime implicant table is in Figure 3.17a and the reduced table in 3.17b. There is no distinguished column and no row can be excluded. This means that we have to make a random choice of a prime implicant. In this case, however, we can immediately see that each pair of the three rows covers all the columns in the secondary table. All these pairs are acceptable because all the prime implicants have the same number of literals. Thus, there are three minimal *sp* forms shown in Figure 3.17b.

EXAMPLE 3.10 Solve the problem of minimal covering for the prime implicant table in Figure 3.18a.

Only one essential prime implicant (corresponding to the first row) is selected from the prime implicant table. The reduced table is shown in Figure 3.18b. There is no distinguished column and no row can be excluded on the basis of Theorem 3.4. This means that we have to make a random choice

	Term identifier	State identifier 0	2	3	4	5	8	10	12	14	15
	0 0 2 2	•			•		•		•		
	0 2 0 2	•	•				•	•			
	1 0 0 2						•	•	•	•	
*	1 1 1 0									•	•
*	2 1 2 0				•	•					
*	2 2 1 0		•	•							
			+	*	+	*				+	*

(a)

Term identifier	State identifier 0	8	10	12
0 0 2 2	•	•		•
0 2 0 2	•	•	•	
1 0 0 2		•	•	•

$$f = x_4 x_3 x_2 \vee \bar{x}_4 x_3 \bar{x}_2 \vee \bar{x}_4 \bar{x}_3 x_2 \begin{cases} \vee \bar{x}_2 \bar{x}_1 \vee \bar{x}_3 \bar{x}_1 \\ \vee \bar{x}_2 \bar{x}_1 \vee x_4 \bar{x}_1 \\ \vee \bar{x}_3 \bar{x}_1 \vee x_4 \bar{x}_1 \end{cases}$$

(b)

Figure 3.17 Illustration for Example 3.9.

	Term identifier	State identifier 1	2	3	4	5	6	7	8	9	10	12	13	14	15
*	0 1 0 0				•	•	•	•				•	•	•	•
	0 0 1 2		•				•				•			•	
	0 0 2 1	•				•				•			•		
	1 0 0 2								•		•	•		•	
	1 0 2 0								•	•		•	•		
	2 0 0 1	•		•		•		•							
	2 0 1 0		•	•			•	•							
					*	+	+	+				+	+	+	*

(a)

	Term identifier	State identifier 1	2	3	8	9	10
	0 0 1 2		•				•
*	0 0 2 1	•				•	
	1 0 0 2				•		•
	1 0 2 0				•	•	
*	2 0 0 1	•		•			
	2 0 1 0		•	•			
		*					

(b)

Term identifier	State identifier 2	3	8	10
0 0 1 2	•			•
1 0 0 2			•	•
1 0 2 0			•	
2 0 0 1		•		
2 0 1 0	•	•		

(c)

	Term identifier	State identifier 2	3	8	10
	0 0 1 2	•			•
*	1 0 0 2			•	•
*	2 0 1 0	•	•		
		+	*	*	+

(d)

	Term identifier	State identifier 2	8	9	10
*	0 0 1 2	•			•
	1 0 0 2		•		•
*	1 0 2 0		•	•	
		*	+	*	+

(e)

Figure 3.18 Illustration for Example 3.10.

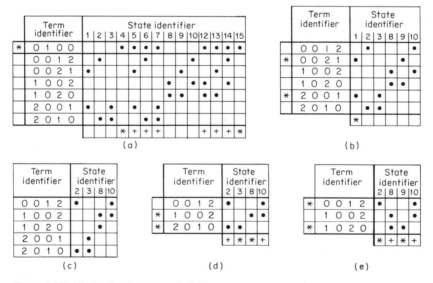

of a prime implicant. Referring, for instance, to the first column, it can be covered either by the second row or by the fifth row. One of these two rows represents necessarily a prime implicant belonging to a minimal Boolean form. We are obliged to try both these possibilities. When we choose the second row, we obtain the table in Figure 3.18c. Now, the third row and the fourth row can be excluded because they are dominated, respectively, by the second row and the fifth row. This gives us the table in Figure 3.18d, which, when the rule of distinguished columns is applied, leads immediately to the solution.

We are obliged to return to the table in Figure 3.18b to do the same procedure for the second random choice. The final table corresponding to this choice is shown in Figure 3.18e. We have:

$$f = x_3 \vee \bar{x}_2 x_1 \vee x_4 \bar{x}_1 \vee \bar{x}_4 x_2$$

for the first choice, and

$$f = x_3 \vee \bar{x}_4 x_1 \vee x_2 \bar{x}_1 \vee x_4 \bar{x}_2$$

for the second choice. Both of the alternative solutions contain the same number of literals so that both are minimal forms.

Let us describe another procedure for minimal coverings, which is based on a Boolean description of the prime implicant table. The procedure is sometimes called the *Petrick method*. It can be described by the following statements:

1. A special logic variable is assigned to each prime implicant (row).
2. When a state occupied by the given function (column) is included in several prime implicants (rows), it must be covered at least by one of them. This condition is represented by the OR function of those variables which are assigned to the respective prime implicants (rows).
3. Conditions of type (2) are satisfied simultaneously for all states occupied by the given function (columns). This is represented by the AND function of all OR functions obtained under (2).
4. When the Boolean form specified in (3) is transformed to a *sp* form, each *p*-term in the latter represents a possible covering. We select those which contain the smallest number of literals.

It is reasonable to combine the Petrick method with the Quine-McCluskey method. The former is advantageous especially in cases where the latter requires the investigation of several possibilities based on random choices of prime implicants.

EXAMPLE 3.11 Suppose that a problem of minimal covering has been solved by the Quine-McCluskey method and we obtained the reduced prime implicant table shown in Figure 3.19a. Clearly, the rule of distinguished

Number of literals	Prime implicant	States occupied by the function S_1	S_2	S_3	S_4	S_5	S_6
2	A	•	•				
2	B			•			•
3	C		•	•			
3	D					•	•
3	E				•	•	
3	F	•			•		

(a)

Covering	Number of literals
ABE	7
$ABDF$	10
$ACDE$	11
CDF	9
$BCEF$	11

(b)

Figure 3.19 Illustration for Example 3.11.

columns cannot be applied and no row can be cancelled due to equality or domination of another row. Thus, we have to make a random choice in the Quine-McCluskey method. Instead, we will apply the Petrick method.

The prime implicants are represented by logic variables $A, B, \ldots, F$. The prime implicant table is described by the Boolean expression

$$p = (A \vee F)(A \vee C)(B \vee C)(E \vee F)(D \vee E)(B \vee D).$$

This expression can be step-by-step multiplied out as follows:

$$p = (A \vee CF)(BE \vee BF \vee CE \vee CF)(D \vee BE)$$
$$p = (ABE \vee ABF \vee ACE \vee CF)(D \vee BE)$$
$$p = ABE \vee ABDF \vee ACDE \vee CDF \vee BCEF$$

Each of the p-terms in the last form represents a possible covering of all the occupied states by the prime implicants. All of these coverings are irredundant and some of them are minimal. To find the minimal coverings, we have to calculate the number of literals in each particular covering and compare these numbers. This is shown in Figure 3.19b. There is only one minimal covering represented by prime implicants A, B, E.

Exercises to Section 3.4

3.4-1 Compare essential prime implicants with cells selected on the basis of Theorem 3.1 in the Svoboda method.

3.4-2 Prove Theorem 3.4 and compare it with Theorem 3.2 applied in the Svoboda method.

*3.4-3 Solve the problem of minimal covering for the prime implicants obtained in Exercises 3.3-3c, d, e, g.

3.4-4 Construct some prime implicant tables whose covering can be solved only by applying random choices. Determine chart representations of the logic functions corresponding to the prime implicant tables.

3.5 MINIMIZATION OF GROUPS OF BOOLEAN EXPRESSIONS

So far we have considered only the minimization of Boolean expressions of a single logic function. However, in the synthesis of switching circuits, we frequently need to implement a number of logic functions of the same variables. Although we can implement each function independently, using a single function minimization, considerable savings can often be obtained if we employ a sharing of the same prime implicants by several functions.

EXAMPLE 3.12 Figure 3.20 illustrates charts of two logic functions, f_1 and f_2. We can easily ascertain that there exists only one minimal *sp* form for the function f_1, namely,

$$f_1 = \bar{x}_3 \bar{x}_1 \vee x_3 x_2 x_1 \vee \bar{x}_2 \bar{x}_1$$

whereas for the indefinite function f_2 there exists two minimal *sp* forms,

$$^1f_2 = x_3 \bar{x}_1 \vee x_3 x_2$$

$$^2f_2 = \bar{x}_2 \bar{x}_1 \vee x_3 x_2$$

The two expressions $^1f_2$ and $^2f_2$ have the same number of literals. Whereas *p*-terms contained in $^1f_2$ are all different from those included in f_1, the expressions f_1 and $^2f_2$ share the term $\bar{x}_2 \bar{x}_1$. This is an advantageous property which often results in a saving of hardware for the corresponding switching circuit. Indeed, the shared term can be implemented by a single element. Block diagrams corresponding to f_1, $^1f_2$ and f_1, $^2f_2$ are shown in Figures 3.21 a and b, respectively. One AND element (shaded in Figure 3.21) is saved when we use f_1, $^2f_2$ instead of f_1, $^1f_2$.

Let us describe a generalization of the Quine-McCluskey method for the minimization of groups of Boolean expressions. To identify the terms shared by several functions, all prime implicants are determined for each function separately, for logic products of all pairs of the functions, for logic products of all triplets of the functions, etc.

For instance, consider the group of logic functions specified in Figure 3.22a. Their prime implicants are:

$$\bar{x}_2 \bar{x}_1,\ x_3 \bar{x}_1,\ x_3 x_2 \text{ for } f_1$$

$$x_2 x_1,\ x_3 x_1,\ \bar{x}_3 \bar{x}_2 \bar{x}_1 \text{ for } f_2$$

$$\bar{x}_2 \bar{x}_1,\ x_3 \bar{x}_1,\ \bar{x}_3 x_2 x_1 \text{ for } f_3$$

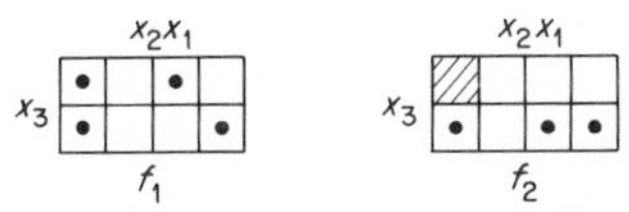

Figure 3.20 Illustration for Example 3.12.

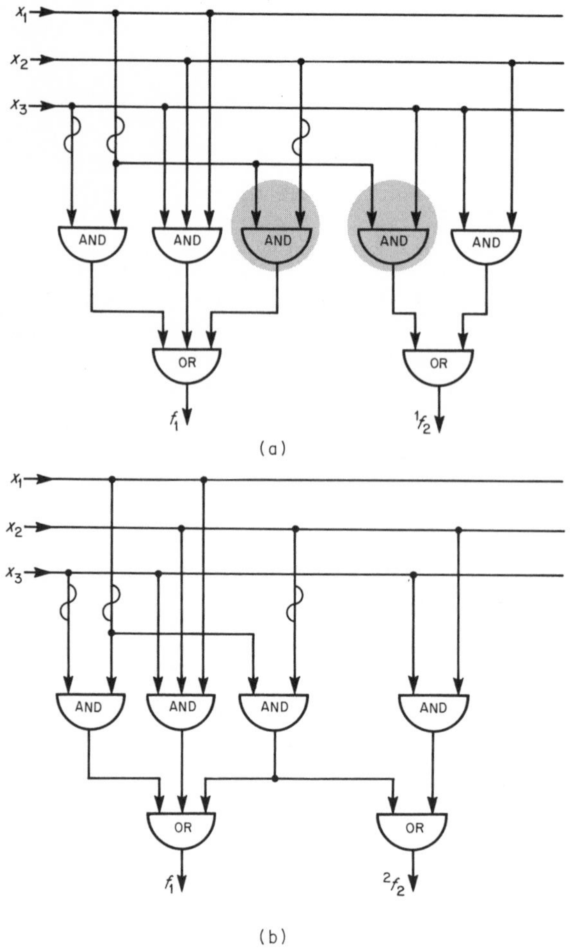

Figure 3.21 Illustration for Example 3.12.

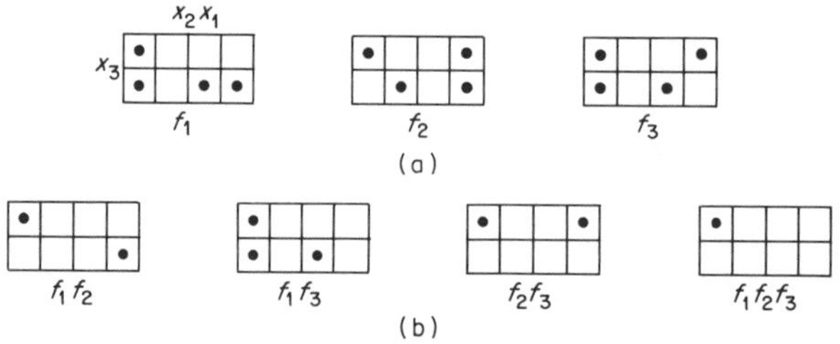

Figure 3.22 A group of logic functions and all their logic products.

Prime implicants of logic products of the functions, which are shown in Figure 3.22b, are:

$$x_3 x_2 x_1,\ \bar{x}_3 \bar{x}_2 \bar{x}_1 \text{ for } f_1 f_2$$
$$\bar{x}_2 \bar{x}_1,\ x_3 \bar{x}_1 \text{ for } f_1 f_3$$
$$\bar{x}_3 x_2 x_1,\ \bar{x}_3 \bar{x}_2 \bar{x}_1 \text{ for } f_2 f_3$$
$$\bar{x}_3 \bar{x}_2 \bar{x}_1, \text{ for } f_1 f_2 f_3$$

After prime implicants of all given functions and their logic products are determined, we prepare a *generalized version of the prime implicant table.* The columns correspond to states occupied by each function separately. The rows correspond to prime implicants grouped by the number of literals. Each row is also identified by the function or function product, into which the respective prime implicant belongs. Each prime implicant appears only once in the table and is identified by the highest order product of the functions in which it belongs. The prime implicant table for the example started above is shown in Figure 3.23.

Using the generalized prime implicant table, the problem of minimal covering can be solved by exactly the same procedure as that elaborated for a single function: application of distinguished columns, equal and dominated rows, and random choices or the Petrick function. Note that the entire table must be considered, not just a part of it corresponding to a single function.

The prime implicant table in Figure 3.23 contains four distinguished columns. They determine four essential prime implicants which cover all states occupied by the functions except state 7 occupied by f_1. This state can be covered either by $x_3 x_2$ or by $x_3 x_2 x_1$ (see Figure 3.23). Apparently, $x_3 x_2$ is preferable since it contains a fewer number of literals. Thus the group minimal forms are:

$$f_1 = x_3 x_2 \vee x_3 \bar{x}_1 \vee \bar{x}_3 \bar{x}_2 \bar{x}_1$$
$$f_2 = x_3 x_1 \vee \bar{x}_3 x_2 x_1 \vee \bar{x}_3 \bar{x}_2 \bar{x}_1$$
$$f_3 = x_3 \bar{x}_1 \vee \bar{x}_3 x_2 x_1 \vee \bar{x}_3 \bar{x}_2 \bar{x}_1$$

	Functions	Term identifier	f_1 0	4	6	7	f_2 0	3	5	7	f_3 0	3	4	6
	f_1	1 1 0			•	•								
	f_2	0 1 1						•		•				
*	f_2	1 0 1							•	•				
	$f_1 f_3$	0 2 2	•	•							•		•	
*	$f_1 f_3$	1 0 2		•	•								•	•
	$f_1 f_2$	1 1 1				•				•				
*	$f_2 f_3$	2 1 1						•				•		
*	$f_1 f_2 f_3$	2 2 2	•				•				•			
			+	+	+		*	+	*	+	+	*	+	*

Figure 3.23 Prime implicant table for the functions and their products shown in Figure 3.22.

If we minimized each of the functions separately, we would obtain uniquely the forms:

$$f_1 = \bar{x}_2\bar{x}_1 \vee x_3x_2$$
$$f_2 = x_3x_1 \vee \bar{x}_3\bar{x}_2\bar{x}_1 \vee x_2x_1$$
$$f_3 = \bar{x}_2\bar{x}_1 \vee x_3\bar{x}_1 \vee \bar{x}_3x_2x_1$$

Although these forms contain less literals than the group minimal forms, the advantage of an effective *term sharing* makes the latter preferable. If literals of repeated p-terms are counted only once (since repeated terms are implemented by a single element), we obtain twelve literals for the group minimal forms and 16 literals for the individual minimal forms. The group minimization (Figure 3.24a) saves two AND elements in this example in comparison with the individual minimization (Figure 3.24b).

Exercises to Section 3.5

3.5-1 Determine a set of group minimal forms for the logic functions given in Exercises:
- (a) 3.3-3a, b, and c.
- (b) 3. 3-3a, c, and d.
- (c) 3.3-3b, c, and d.

3.5-2 Determine a set of group minimal forms for the following groups of logic functions:
- (a) $f_1(s) = 1$ for $s = 2, 4, 5$; $f_1(s) = 0$ for $s = 0, 1, 6, 7$; $f_2(s) = 1$ for $s = 5, 6$; $f_2(s) = 0$ for $s = 0, 2, 3, 4, 7$; $f_3(s) = 1$ for $s = 2, 3, 5$; $f_3(s) = 0$ otherwise; $n = 3$.
- *(b) The functions specified in Figure 3.25a.
- *(c) The functions specified in Figure 3.25b.
- (d) The functions shown in Figure 3.25c.

3.6 MULTIPLE LEVEL AND ABSOLUTE MINIMIZATION

At the beginning of this chapter we indicated several ways of formulating the problem of minimization of the standard Boolean form of a given logic function. The reader is certain to have noticed that the standard form, which is minimal for the given objective function, need not be *the absolute minimal Boolean expression* from the viewpoint of this objective function. This is because the distributive laws of Boolean algebra, which permit factoring, do not find application in the standard form.

At first sight it might seem that the absolute minimal expression can be obtained by modifying the minimal standard form with the aid of the distributive laws. It can be shown, however, that for some logic functions there

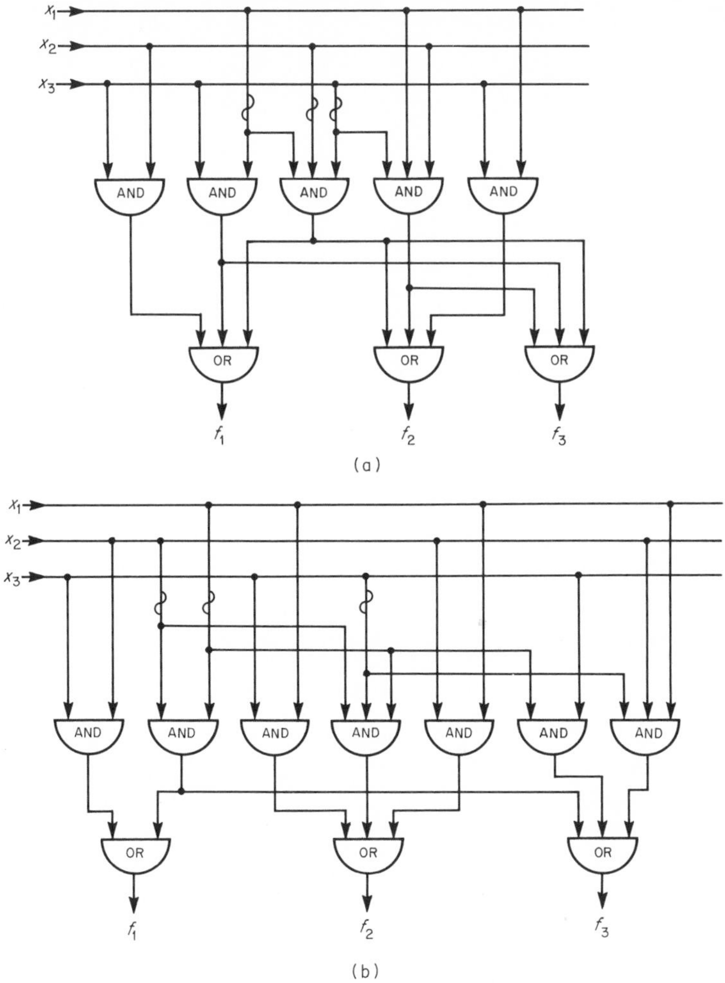

Figure 3.24 Block diagrams based on group and individual minimization for the functions shown in Figure 3.22a.

exist absolute minimal forms which cannot be derived from the minimal standard form solely by the distributive laws.

EXAMPLE 3.13 The logic function f expressed by the chart in Figure 3.26 has only one minimal *sp* Boolean form:

$$f = \bar{x}_6 \bar{x}_5 x_4 x_3 x_2 \vee \bar{x}_6 x_5 x_4 x_3 \bar{x}_1 \vee x_6 \bar{x}_5 x_4 x_2 \bar{x}_1 \vee x_6 x_5 x_4 x_3 x_2 \vee \bar{x}_4 x_3 x_2 x_1 \vee \bar{x}_4 \bar{x}_3 \bar{x}_2 x_1 \vee x_4 \bar{x}_3 \bar{x}_2 \bar{x}_1$$

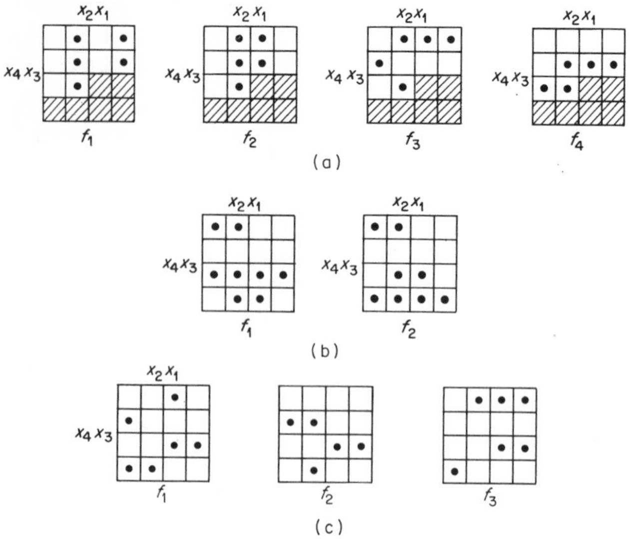

Figure 3.25 Illustration for Exercise 3.5-1.

It is easy to ascertain that the absolute minimal expression

$$f = x_4\{x_3[x_2(\bar{x}_6\bar{x}_5 \vee x_6x_5) \vee \bar{x}_6x_5\bar{x}_1] \vee \bar{x}_3x_6\bar{x}_5\bar{x}_1\} \vee (\bar{x}_4x_1 \vee x_4\bar{x}_1)(x_2x_3 \vee \bar{x}_2\bar{x}_3)$$

cannot be derived from the foregoing minimal standard form merely by applying the distributive laws.

Methods for constructing absolute minimal Boolean expressions of logic functions are already well known. It appears, however, that these methods are so complicated as to be practically unusable. A more practical problem arises when the number of levels in the Boolean expression is limited. Then we speak about Boolean forms *sps*, *psp*, *spsp*, etc.

To show that the multiple level minimization is a generalization of techniques used for minimization of standard Boolean forms (forms *sp* or *ps*), let us consider the problem of finding a minimal *sps* Boolean form for a given logic function *f*.

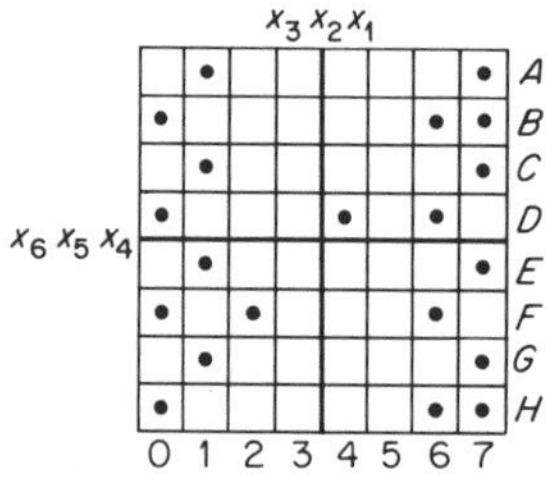

Figure 3.26 Example of function whose absolute minimal form cannot be derived from the minimal standard form.

By definition, a minimal *sps* form is an OR function of *ps* expressions. Each of these *ps* expressions must: (1) represent a function g_i which implies f ($g_i \leq f$), and (2) be a minimal *ps* form for the function g_i it represents. Clearly, if f occupies m states, there are $2^m - 1$ functions g_i that imply f provided that the ZERO function is not counted. Minimal *ps* forms of functions g_i are usually called *ps prime implicants* of f. We try to find such a covering of f by some of its *ps* prime implicants which contains the smallest possible number of literals. The method described in Section 3.4 can be used without any modification for this covering problem.

Thus, the essence of the multiple level minimization consists in determination of all *ps* minimal forms for the functions g_i. It will be shown that, as a rule, not all functions g_i must be considered.

EXAMPLE 3.14 Find a minimal *sps* Boolean form for the function f defined by the chart in Figure 3.27a.

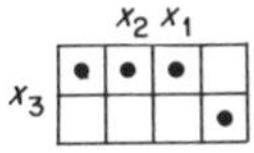

(a)

m_j	Occupied states of g_i	Minimal ps form m_j for g_i such that $g_i \rightarrow m_j \rightarrow f$	States covered by m_j
m_1	0	$\bar{x}_3\bar{x}_1$	0,2
m_2	0	$\bar{x}_3\bar{x}_2$	0,1
m_2	1	$\bar{x}_3\bar{x}_2$	0,1
m_1	2	$\bar{x}_3\bar{x}_1$	0,2
m_3	7	$x_3x_2x_1$	7
m_4	0,7	$(\bar{x}_3 \vee x_2)(x_3 \vee \bar{x}_2)(\bar{x}_2 \vee x_1)$	0,1,7
m_5	0,7	$(\bar{x}_3 \vee x_2)(x_3 \vee \bar{x}_2)(\bar{x}_3 \vee x_1)$	0,1,7
m_6	0,7	$(\bar{x}_3 \vee x_1)(x_3 \vee \bar{x}_1)(x_2 \vee \bar{x}_1)$	0,2,7
m_7	0,7	$(\bar{x}_3 \vee x_2)(\bar{x}_3 \vee x_1)(x_3 \vee \bar{x}_1)$	0,2,7
m_8	0,7	$(\bar{x}_3 \vee x_2)(x_3 \vee \bar{x}_1)(\bar{x}_2 \vee x_1)$	0,7
m_9	0,7	$(\bar{x}_3 \vee x_1)(x_3 \vee \bar{x}_2)(x_2 \vee \bar{x}_1)$	0,7
m_{10}	1,2	$\bar{x}_3(\bar{x}_2 \vee \bar{x}_1)$	0,1,2
m_{11}	1,7	$(\bar{x}_3 \vee x_2)(x_3 \vee \bar{x}_2)x_1$	1,7
m_{12}	2,7	$(\bar{x}_3 \vee x_1)(x_3 \vee \bar{x}_1)x_2$	2,7
m_{13}	1,2,7	$(\bar{x}_3 \vee x_2)(x_3 \vee \bar{x}_2 \vee \bar{x}_1)(\bar{x}_3 \vee x_1)$	0,1,2,7

(b)

m_j	States covered by				Number of literals in m_j
	0	1	2	7	
m_1	•		•		2
m_2	•	•			2
m_3				•	3
m_{10}	•	•	•		3
m_{11}		•		•	5
m_{12}			•	•	5
m_4	•	•		•	6
m_5	•	•		•	6
m_6	•		•	•	6
m_7	•		•	•	6
m_8	•			•	6
m_9	•			•	6
m_{13}	•	•	•	•	7

(c)

Figure 3.27 Example of a multiple-level minimization (Example 3.14).

First, we have to determine minimal *ps* Boolean forms for the functions g_i through *sp* forms of $\bar{g}_i$. It is convenient to consider them in order of increasing number of occupied states.

The list of all different minimal *ps* forms m_j of functions g_i (*ps* prime implicants of f) is given in Figure 3.27b. Note that not all functions g_i need be considered. For example, the function that occupies states 0 and 2 need not be considered because it is represented by m_1. Similarly, the function occupying 0 and 1 is not considered because it is represented by m_2, etc.

The prime implicant table is shown in Figure 3.27c. Applying the procedure described in Section 3.4, we find that the pair m_3 and m_{10} represents a unique minimal covering. Therefore,

$$f = m_3 \vee m_{10} = \bar{x}_3(\bar{x}_2 \vee \bar{x}_1) \vee x_3 x_2 x_1$$

Thus, minimal *sps* Boolean forms of f are those OR operations of *ps* prime implicants of f which contain the smallest number of literals. Similarly, minimal *psp* Boolean forms are logic products of *sp* prime implicants with the smallest number of literals, minimal *psps* forms are logic products of *sps* prime implicants with the smallest number of literals, etc.

Exercises to Section 3.6

3.6-1 Find *sp* and *ps* standard Boolean minimal forms of the logic function specified in Figure 3.26.

3.6-2 Show that the absolute minimal Boolean expression given in Example 3.13 cannot be derived by modifying either of the standard Boolean minimal forms, obtained in Exercise 3.6-1, with the aid of the distributive laws.

3.6-3 Prove the following proposition: If a literal is contained in a prime implicant of a logic function, then every expression representing the function must contain this literal.
Hint: Show that every expression can be converted to an OR operation of exactly all prime implicants of the logic function.

3.6-4 Find a minimal *sps* Boolean form for the function:
- *(a) $f(s) = 1$ for $s = 0, 2, 4, 7$
  $f(s) = 0$ otherwise, $n = 3$.
- *(b) $f(s) = 1$ for $s = 3, 5, 6, 7, 11, 13, 15$
  $f(s) = 0$ otherwise, $n = 4$.
- *(c) $f = \bar{x}_4 \bar{x}_3 x_1 \vee \bar{x}_4 \bar{x}_3 \bar{x}_2 \vee \bar{x}_5 \bar{x}_3 \bar{x}_2 \vee \bar{x}_5 \bar{x}_4 \bar{x}_2 \vee \bar{x}_5 \bar{x}_4 x_1$
- *(d) $f(s) = 1$ for $s = 0, 3, 5, 6$
  $f(s) = 0$ otherwise, $n = 3$.
- (e) $f = x_3 \bar{x}_2 \bar{x}_1 \vee x_3 x_2 x_1 \vee \bar{x}_3 \bar{x}_2 x_3$.
- (f) $f(s) = 0$ for $s = 1, 2, 3, 6, 10, 14$
  $f(s) = 1$ otherwise, $n = 4$.

*3.6-5 Find all minimal *spsp* Boolean forms for the function $f(s) = 1$ for $s = 3, 5, 6$; $f(s) = 0$ for $s = 1, 2, 4, 14$; and $f(s)$ is "don't care" otherwise; $n = 4$.

Comprehensive Exercises to Chapter 3

3.1 Prove the following statement: When an objective function is used such that its value does not increase when a literal is removed from an *sp* standard Boolean form, there is at least one *sp* minimal form which is an OR operation of prime implicants.

3.2 Prove that each implicant of the Petrick function defined on a logic function represents an irredundant *sp* Boolean form of the logic function.

3.3 Suppose an *sp* Boolean form that contains at least two *p*-terms and each of them contains at least two literals. Let each *p*-term of the Boolean form be replaced by a NAND function of the participating variables or their negations and let all OR operations in the Boolean form also be replaced by NAND functions. Show that the resulting NAND form represents the same function as the original *sp* Boolean form.

3.4 Extend the rules of replacement specified in Exercise 3.3 for the case where the *sp* Boolean form contains:
 (a) Only one *p*-term.
 (b) A *p*-term with a single literal.

3.5 Using the idea outlined in Exercises 3.3 and extended in Exercise 3.4, transform the following *sp* Boolean forms to the corresponding NAND forms:
 *(a) $\bar{x}_1 \vee x_2 \bar{x}_3 \vee \bar{x}_2 x_3$.
 *(b) $\bar{x}_1 x_2 x_3 \bar{x}_4$.
 (c) $x_1 \bar{x}_2 \vee \bar{x}_3 \bar{x}_4 \vee x_2 x_4 x_3$.
 (d) $\bar{x}_1 \bar{x}_2 \bar{x}_3 x_4 \vee x_2 x_3 x_4 \vee x_1 \bar{x}_2 x_3 \bar{x}_4$.

*3.6 Replace negations in the NAND forms determined in Exercise 3.5 by NAND functions.

3.7 Determine rules of replacement, analogous to those which are the subject of Exercises 3.3 and 3.4, that can be applied for the transformation from a *ps* Boolean form to the corresponding NOR form.

3.8 Transform the following *ps* Boolean forms to the corresponding NOR forms:
 *(a) $(\bar{x}_1 \vee x_2 \vee x_3)(\bar{x}_3 \vee \bar{x}_2)$.
 *(b) $(x_2 \vee \bar{x}_3)x_1$.
 *(c) $x_1 \vee x_2$.
 (d) $(x_1 \vee x_2)(x_3 \vee x_4)\bar{x}_1 \bar{x}_3$.
 (e) $(x_1 \vee \bar{x}_2 \vee \bar{x}_3 \vee x_4)(\bar{x}_1 \vee \bar{x}_2 \vee x_3)\bar{x}_4$.

3.9 A definite logic function $f(x_1, x_2, \ldots, x_n)$ is equal to one for $2^n - 1$ states.
 (a) How many *p*-terms and literals are there in its minimal *sp* Boolean form.
 (b) Show the pattern of this minimal form.

3.10 Let $f = \bar{x}_3 \bar{x}_2 \bar{x}_1 \vee \bar{x}_4 x_3 \bar{x}_1 \vee x_4 \bar{x}_3 x_2 x_1 \vee x_4 \bar{x}_3 \bar{x}_1$. Show that:
 (a) $(\bar{x}_4 \vee \bar{x}_3)(x_4 x_2 \vee \bar{x}_1(x_3 \vee \bar{x}_2))$ is an absolute minimal Boolean expression for f.
 (b) The expression specified in (a) cannot be derived from any minimal *sp* or *ps* form by the distributive laws.

Reference Notations to Chapter 3

The problem of minimization of standard Boolean forms was first studied by Quine [50, 51, 52] and the Staff of the Harvard Computational Laboratory

[59]. A modified version of the Quine method was elaborated by McCluskey [34]. A topological approach to Boolean minimization was initiated by Urbano and Mueller [68]. This approach was also used by Roth [54, 55] and Svoboda [61]. The Svoboda method is described in Section 3.2. Its description can also be found in Reference 42. An alternate version of the method is suggested in Reference 64. The method of determination of all prime implicants described in Section 3.3 and its successful hardware implementation is also due to Svoboda [62, 63]. The Petrick function was first suggested in Reference 46. Several other methods concerning minimization of normal Boolean forms have been suggested [e.g., 5, 8, 14, 15, 21, 27, 33, 38, 39, 43, 44, 45, 49, 66, 67]. Several authors pointed out that the problem can be solved in a more general form by linear programming [10, 31]. Application of the pseudo-Boolean approach [20] in combination with special hardware units [63] seems to be promising in this direction (see also Section 4.8).

The problem of minimization of groups of Boolean standard forms was discovered in the late fifties and several methods for its solution have been elaborated [3, 19, 30, 32, 35, 41, 47, 56, 58, 60, 65, 70].

Methods for multiple level and absolute minimization were elaborated by Abhyankar [1, 2], Lawler [24, 25] and others. Kazakov [22] gives estimates for a number of operations in absolute minimization. Gimpel investigated a special problem of three-level NAND form minimization [18, 53].

Boolean minimization is included in practically every book concerning switching circuits. A comprehensive presentation of this topic can be found in References 4, 38, and 48. A detailed classification of Boolean minimization methods was prepared by Gavrilov [16]. There are several references which contain surveys of methods for minimization of Boolean expressions [e.g., 13, 70].

For students interested seriously in switching circuits, some theoretical papers concerning various aspects of Boolean minimization should be mentioned [23, 26, 28, 29, 36, 37, 57, 69].

References to Chapter 3

1. **Abhyankar, S.,** "Minimal Sum of Products of Sums Expressions of Boolean Functions," *TC* **EC-7** (4) 268–276 (Dec. 1958).
2. **Abhyankar, S.,** "Absolute Minimal Expressions of Boolean Functions," *TC* **EC-8** (1) 3–8 (March 1958).
3. **Bartee, T. C.,** "Computer Design of Multiple Output Logical Networks," *TC* **EC-10** (1) 21-30 (March 1961).
4. **Bartee, T. C.,** et al., *Theory and Design of Digital Machines*, McGraw-Hill, New York, 1962.
5. **Booth, T. M.,** "The Vertex-Frame Method for Obtaining Minimal Proposition-Letter Formulae," *TC* **EC-11** (2) 144–154 (April 1962).

6. **Bowman, R. M.,** and **E. S. McVey,** "A Method for the Fast Approximate Solution of Large Prime Implicant Charts," *TC* **C-19** (2) 169–173 (Feb. 1970).
7. **Breuer, M. A.,** "Simplification of the Covering Problem with Application to Boolean Expressions," *JACM* **17** (1) 166–181 (Jan. 1970).
8. **Choudhury, A. K.,** and **M. S. Basu,** "A Mechanized Chart for Simplification of Switching Functions," *TC* **EC-11** (5) 713–714 (Oct. 1962).
9. **Chu, J. T.,** "Some Methods for Simplifying Switching Circuits Using 'Don't Care' Conditions," *JACM* **8** (4) 497–512 (Oct. 1961).
10. **Cobham, A. R.,** et al., "An Application of Linear Programming to the Minimization of Boolean Functions," *CRSA* **2,** 3–9 (Oct. 1961).
11. **Das, S. R.,** "An Approach for Simplifying Switching Functions by Utilizing the Cover Table Representation," *TC* **C-20** (3) 355–359 (March 1971).
12. **DeVries, R. C.,** "Comment on Lawler's Multilevel Boolean Minimization," *CACM* **13** (4) 265–266 (April 1970).
13. **Didenko, V. P.,** "Some Methods for Boolean Function Minimization," *ST,* pp. 148–162.
14. **Gaines, R. S.,** "Simplification Techniques for Boolean Functions," *CRSA* **5,** 174–182 (Oct. 1964).
15. **Gavrilov, M. A.,** "Boolean Function Minimization," *ARC* **20** (9) 1217–1238 (1959).
16. **Gavrilov, M. A.,** "The Present State of the Theory of Switching Circuits," *ST,* pp. 5–73.
17. **Gimpel, J. F.,** "A Reduction Technique for Prime Implicant Tables," *TC* **EC-14** (4) 535–541 (Aug. 1965).
18. **Gimpel, J. F.,** "The Minimization of TANT Networks," *TC* **EC-16** (1) 18–38 (Feb. 1967).
19. **Gordon, B. B.,** et al., "Simplifications of the Covering Problem for Multiple Output Logical Networks," *TC* **EC-15** (6) 891–897 (Dec. 1966).
20. **Hammer, P. L.,** and **S. Rudeanu,** *Boolean Methods in Operations Research,* Springer-Verlag, New York, 1968.
21. **Harris, B.,** "An Algorithm for Determining Minimal Representations of a Logic Function," *TC* **EC-6** (2) 103–108 (June 1957).
22. **Kazakov, V. D.,** "Minimization of Boolean Functions with the Aid of Factoring," *ST,* pp. 163–169.
23. **Kellerman, E.,** "A Formula for Logical Network Cost," *TC* **C-17** (9) 881–884 (Sept. 1968).
24. **Lawler, E. L.,** "Minimal Boolean Expressions with More Than Two Levels of Sums and Products," *CRSA* **3** 49–59 (Sept. 1962).
25. **Lawler, E. L.,** "An Approach to Multi-Level Boolean Minimization," *JACM* **11** (3) 283–295 (July 1964).
26. **Lin, S.,** "On Comparison of Complexities of Minimal and Shortest Disjunctive Normal Forms for Logic Functions," *PC,* No. 18, pp. 11–44 (1967).
27. **Luccio, F.,** "A Method for Selection for Prime Implicants," *TC* **EC-15** (2) 205–212 (April 1966).
28. **Lupanov, O. B.,** "On the Complexity of Realizing Boolean Functions," *PC,* No. 3, pp. 61–80 (1960).
29. **Lupanov, O. B.** "On Realization of Logic Functions by Expressions from Finite Classes (Expressions of a Limited Depth) in the Base AND, OR, NOT," *PC,* No. 6, pp. 5–14 (1961).
30. **Mage, J. M.,** "Application of Iterative Consensus to Multiple-Output Functions," *TC* **C-19** (4) 359 (April 1970).

31. **Maistrova, T. L.,** "Linear Programming and the Problem of Minimizing Normal Forms of Boolean Functions" (in Russian), in *Problems of Communication*, No. 12, 5–15, Izd. Akademii Nauk SSSR, Moscow, 1962.
32. **Marcus, M. P.,** and **W. H. Niehoff,** "Iterated Consensus for Multiple-Output Functions," *IBMJ* **14** (6) 677–679 (Nov. 1970).
33. **McCluskey, E. J.,** "Minimal Sums for Boolean Functions Having Many Unspecified Fundamental Products," *CRSA*, pp. 10–17 (Oct. 1961).
34. **McCluskey, E. J.,** "Minimization of Boolean Functions," *BSTJ* **35** (6) 1417-1444 (Nov. 1956).
35. **McNaughton, R.,** and **B. Mitchell,** "The Minimality of Rectifier Nets with Multiple Outputs Incompletely Specified," *JFI* **264** (6) 457–480 (June 1957).
36. **Mileto, F.,** and **G. Putzolu,** "Average Values of Quantities Appearing in Boolean Function Minimization," *TC* **EC-13** (2) 87–92 (April 1964).
37. **Mileto, F.,** and **G. Putzolu,** "Average Values of Quantities Appearing in Multiple Output Boolean Minimization," *TC* **EC-14** (4) 542–552 (Aug. 1965).
38. **Miller, R. E.,** *Switching Theory*, Vol. I, *Combinational Circuits*, John Wiley, New York, 1965.
39. **Morreale, E.,** "Partitioned List Algorithm for Prime Implicant Determination from Cononical Forms," *TC* **EC-16** (5) 611–620 (Oct. 1967).
40. **Morreale, E.,** "Recursive Operators for Prime Implicant and Irredundant Normal Form Determination," *TC* **C-19** (6) 504–509 (June 1970).
41. **Muller, D. E.,** "Application of Boolean Algebra to Switching Circuit Design and to Error Detection," *TC* **EC-3** (3) 6–12 (Sept. 1954).
42. **Nadler, M.,** *Topics in Engineering Logic*, Macmillan, New York, 1962.
43. **Necula, N. N.** "A Numerical Procedure for Determination of the Prime Implicants of a Boolean Function," *TC* **EC-16** (5) 687–689 (Oct. 1967).
44. **Necula, N. N.,** "An Algorithm for the Automatic Approximate Minimization of Boolean Functions," *TC* **C-17** (8) 770–782 (Aug. 1968).
45. **Nelson, R. J.,** "Simplest Normal Truth Functions," *J. Symbolic Logic* **20** (2), 105–108 (June 1955).
46. **Petrick, S. R.,** "A Direct Determination of the Irredundant Forms of a Boolean Expression from the Set of Prime Implicants," Air Force Cambridge Res. Lab., AFCRC-TR-56-100, Bedford, Mass., 1956.
47. **Polansky, R. B.,** "Minimization of Multiple-Output Switching Circuits," *AIEE Trans., Part I, Comm. and El.*, **80**, 67–73 (1961).
48. **Prather, R. E.,** *Introduction to Switching Theory: A Mathematical Approach*, Allyn and Bacon, Boston, Mass., 1967.
49. **Pyne, I. B.,** and **E. J. McCluskey,** "An Essay on Prime Implicant Tables," *J. Soc. Indust. Appl. Math.* **9** (4) 604–631 (Dec. 1961).
50. **Quine, W. V.,** "The Problem of Simplifying Truth Functions," *AMM* **59** (8) 521–531 (1952).
51. **Quine, W. V.,** "A Way to Simplify Truth Functions," *AMM* **62** (9) 627–631 (Nov. 1955).
52. **Quine, W. V.,** "On Cores and Prime Implicants of Truth Functions," *AMM* **66** (9) 755–760 (Nov. 1959).
53. **Robinson, S. U., II** and **R. W. Hoose,** "Gimpel's Reduction Technique Extended to the Covering Problem with Cost," *TC* **EC-16** (4) 509–514 (Aug. 1967).
54. **Roth, J. P.,** "Algebraic Topological Methods for the Synthesis of Switching Systems," *TAMS* **88** (2) 301–326 (July 1958).
55. **Roth, J. P.,** "Algebraic Topological Methods in Synthesis," *PHU*, Part I, pp. 57–73.

56. **Seidl, L. K.,** " Multiple Output Two-Stage Switching Circuits with a Minimum Number of Diodes" (in Russian), *IPM*, No. 9, pp. 229–248 (1963).
57. **Sholomov, L. A.,** "Complexity Criteria of Boolean Functions," *PC*, No. 17, pp. 91–127 (1966).
58. **Slagle, J. R.,** et al., "A New Algorithm for Generating Prime Implicants," *TC* **C-19** (4) 304–310 (April 1970).
59. Staff of the Harvard Computational Laboratory, *Synthesis of Electronic Computing and Control Circuits*, Harvard University Press, Cambridge, Mass., 1951.
60. **Su, Y.,** and **D. L. Dietmeyer,** "Computer Reduction of Two-Level, Multiple Output Switching Circuits," *TC* **C-18** (1) 58–63 (Jan. 1969).
61. **Svoboda, A.,** "Some Applications of Contact Grids," *PHA*, Part I, pp. 293–305 (1959).
62. **Svoboda, A.,** "Ordering of Implicants," *TC* **EC-16** (1) 100–105 (Feb. 1967).
63. **Svoboda, A.,** "Boolean Analyzer," Proc. IFIP Intern. Congress in Edinburgh, Scotland, August 1968, Booklet D, pp. 97–102; North-Holland Publ. Co., Amsterdam, 1968.
64. **Svoboda, A.,** "Logical Instruments for Teaching Logical Design," *TE* **E-12** (3) 262–273 (Dec. 1969).
65. **Svoboda, A.,** "Mosaics of Boolean Functions," Proc. 3rd Intern. Hawaii Conf. on System Science, Western Periodicals, Honolulu, Jan. 1970.
66. **Tison, P.,** "Generalization of Consensus Theory and Application to the Minimization of Boolean Functions," *TC* **EC-16** (4) 446–456 (Aug. 1967).
67. **Troye, N. C.,** "Classification and Minimization of Switching Functions," *Philips Res. Rep.* **14** (3) 151–193 (April 1959).
68. **Urbano, R. H.,** and **R. K. Mueller,** "A Topological Method for the Determination of the Minimal Forms of a Boolean Function," *TC* **EC-5** (3) 126–132 (Sept. 1956).
69. **Vasilev, YU. L.,** "On a Comparison of Complexities of Irredundant and Minimal Disjunctive Normal Forms," *PC* (10) 5–61 (1963).
70. **Weiner, P.,** and **T. F. Dwyer,** "Discussion of Some Flows in the Classical Theory of Two-Level Minimization of Multiple-Output Switching Networks," *TC* **C-17** (2) 184–186 (Feb. 1968).

4

SOLUTION OF LOGIC RELATIONS

When solving various problems concerning switching circuits, it is sometimes necessary to express explicitly some logic variables contained implicitly in a set of logic equations or other relations between logic expressions. In such cases we say that we solve the given set of logic equations (or, more generally, the set of logic relations) with respect to the pertinent variables.

In this chapter we are going to describe a highly effective method for the solution of simultaneous logic equations. The author of this method is A. Svoboda. The method is based on the chart representation of logic functions. We will also investigate possible generalizations and applications of the method.

4.1 FORMULATION OF THE PROBLEM

Let us assume that $x_1, x_2, \ldots, x_n, y_1, y_2, \ldots, y_m$ are logic variables. Let us further assume that we have a *set of simultaneous logic equations*:

$$\begin{aligned}
f_1(x_1, x_2, \ldots, x_n, y_1, y_2, \ldots, y_m) &= g_1(x_1, x_2, \ldots, x_n, y_1, y_2, \ldots, y_m)\\
f_2(x_1, x_2, \ldots, x_n, y_1, y_2, \ldots, y_m) &= g_2(x_1, x_2, \ldots, x_n, y_1, y_2, \ldots, y_m)\\
&\vdots\\
f_p(x_1, x_2, \ldots, x_n, y_1, y_2, \ldots, y_m) &= g_p(x_1, x_2, \ldots, x_n, y_1, y_2, \ldots, y_m)
\end{aligned} \tag{4.1}$$

where $f_1, f_2, \ldots, f_p, g_1, g_2, \ldots, g_p$ are algebraically expressed logic functions in the variables $x_1, x_2, \ldots, x_n, y_1, y_2, \ldots, y_m$. Some of these variables can be regarded as independent, the remaining variables are assumed to depend upon them.

The division into *independent* and *dependent* variables does not directly follow from the given set of equations. It depends on our choice which, of course, results from the application of the set of equations.

Let us assume that $x_1, x_2, \ldots, x_n$ are the independent variables in the set (4.1). The transformation of this set of equations into the equations of the form

$$\begin{aligned} y_1 &= {}^kF_1(x_1, x_2, \ldots, x_n) \\ y_2 &= {}^kF_2(x_1, x_2, \ldots, x_n) \\ &\vdots \\ y_m &= {}^kF_m(x_1, x_2, \ldots, x_n) \end{aligned} \tag{4.2}$$

is then regarded as the *solution of the set of equations* (4.1). The superscript $k = 1, 2, \ldots, r$ is used to mark different solutions, i.e., solutions that differ by their chart (or table) representation. Solutions which differ by the form of their algebraic expression but have the same chart representation are regarded as equal.

Hence, in general, the set (4.1) has r different solutions (4.2). Moreover, only those m functions (for a fixed k) are regarded as solutions of (4.1) for which, after substituting the corresponding expressions for $y_1, y_2, \ldots, y_m$ in (4.1), all the equalities of (4.1) are satisfied in the variables $x_1, x_2, \ldots, x_n$.

The variables $y_1, y_2, \ldots, y_m$, which are considered in the equations (4.1) as dependent, can also be regarded as *unknowns* which we have to determine from (4.1). This is why we speak of the solution of logic equations.

The procedure to be described consists of two basic parts. In the first part, the given set of equations (4.1) is transformed into a single standard equation with unity on its right side, which is satisfied if (4.1) is satisfied. The second part consists in the solution of this standard form of equation.

Exercises to Section 4.1

*4.1-1 A dictator decided to give one of his prisoners a chance to free himself. The prisoner had to decide which of two possible doors from his cell he would use. He knew that one of them led to an exit from the prison whereas the other led to a place where he would be executed. The prisoner was allowed to ask a question of one of two personal servants of the dictator. He knew that one of them had been instructed to tell a lie and the other to tell the truth. But he did not know which of them was supposed to tell the truth. After several minutes of thinking he asked one of the servants: "Would the other servant say that this door leads to an exit from the prison?"

Construct a Boolean equation that describes this problem and show that the prisoner can uniquely find the correct door.
(Hint: Logic variables are represented by the following statements: $x =$ "The answer of the servant is positive," $y =$ "The asked servant is a liar," $z =$ "The door leads to an exit from the prison.")

4.1-2 (a) Suggest a trivial trial-and-error method for solution of logic equations (4.1) consisting of successive substitution of all possible sets of m functions $y_i (i = 1, 2, \dots, m)$ of n variables $x_j (j = 1, 2, \dots, n)$ into (4.1) and checking the equalities.

*(b) Determine the number of trials necessary to obtain all possible solutions.

(c) Using this method, solve the equation obtained in Exercise 4.1-1.

*(d) Using this method, solve the equation

$$\bar{y}\bar{x} \vee \bar{z}(\bar{y} \vee x) = \bar{z}(\bar{y} \vee \bar{x})$$

with respect to variable z.

4.2 TRANSFORMATIONS OF GIVEN SET OF EQUATIONS TO THE STANDARD FORM

The given form (4.1) expresses equalities between the functions f_j and $g_j (j = 1, 2, \dots, p)$. If $f_j = 0$, then necessarily $g_j = 0$, and vice versa. If $f_j = 1$, then necessarily $g_j = 1$, and vice versa. For each state of the variables $x_1, x_2, \dots, x_n, y_1, y_2, \dots, y_m$ we thus have either $f_j = g_j = 0$ and $\bar{f}_j \bar{g}_j = 1$ or $f_j = g_j = 1$ and $f_j g_j = 1$. Hence, $\bar{f}_j \bar{g}_j \vee f_j g_j = 1$. Equations (4.1) can, therefore, be transformed into another set of equations,

$$\begin{aligned} \bar{f}_1 \bar{g}_1 \vee f_1 g_1 &= 1 \\ \bar{f}_2 \bar{g}_2 \vee f_2 g_2 &= 1 \\ &\vdots \\ \bar{f}_p \bar{g}_p \vee f_p g_p &= 1 \end{aligned} \tag{4.3}$$

The transformation of (4.1) into (4.3) can be performed very easily with the aid of logic charts.

An example will help to illustrate this point. Let (4.1) be represented by the following two equations:

$$\begin{aligned} \bar{x}_2(\bar{y}_1 \vee \bar{x}_1 y_2) &= x_2 \bar{y}_2 \vee \bar{x}_2 y_1 y_2 \\ \bar{x}_1 x_2 \vee y_2(y_1 \vee x_2) &= \bar{x}_2 \bar{y}_1 \vee x_2(y_2 \vee x_1 y_1) \end{aligned} \tag{4.4}$$

These are equations containing two independent variables, x_1 and x_2, and two unknowns, y_1 and y_2.

The algebraic expressions on the left-hand and right-hand sides of the given equations are represented by the charts by assigning the independent variables to the columns and the unknowns to the rows of the charts (Figure 4.1). This assignment is essential to the method described and, in the explanation to follow, it will therefore be tacitly assumed to have been performed.

Figure 4.2a shows the chart representation of the algebraic expressions on the left-hand sides of the given equations, Figure 4.2b those on their right-hand sides. From the chart representations of the functions f_j and g_j ($j = 1, 2$ in this example) we can now very easily determine the chart representation of the function $\bar{f}_j\bar{g}_j \vee f_j g_j$. This function has the value 1 for those states of variables $x_1, x_2, \ldots, x_n, y_1, y_2, \ldots, y_m$, for which the functions f_j and g_j have equal values. We are thus concerned with equivalence such as we already came across when dealing with functions of two variables in Chapter 2.

When constructing a chart representation of this equivalence we proceed by successively going through the equally positioned squares in the chart representation of the functions f_j and g_j. Wherever we find equal values, we place a dot in the corresponding square of the chart intended to represent the function $\bar{f}_j\bar{g}_j \vee f_j g_j$ (Section 2.12). In our case we obtain the functions shown in Figure 4.2c. We have thus transformed the chart representation of the form (4.1) into that corresponding to the form (4.3).

All the equations in (4.1) must be simultaneously satisfied. The equations

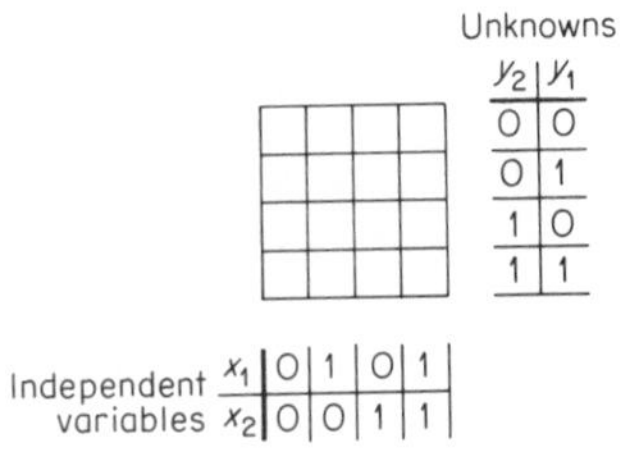

Figure 4.1 Assignment of independent logic variables and unknowns in the chart when solving logic equations.

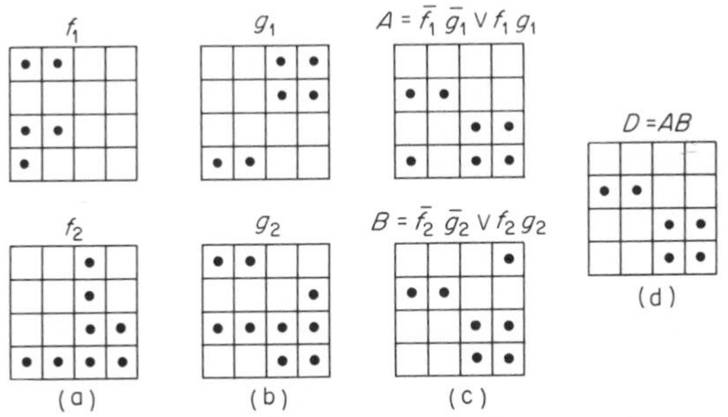

Figure 4.2 Determination of the discriminant for the Boolean equations (4.4).

of (4.3) must therefore also be valid simultaneously. This condition can be expressed by performing the logic product of all the equations of (4.3), i.e.,

$$(\bar{f}_1\bar{g}_1 \vee f_1 g_1) \cdot (\bar{f}_2\bar{g}_2 \vee f_2 g_2) \cdots (\bar{f}_p\bar{g}_p \vee f_p g_p) = 1 \tag{4.5}$$

The logic product of the expressions on the left-hand sides of equations (4.3) is, of course, also performed in the charts. The resulting chart, which represents the expression on the left-hand side of equation (4.5), is called the *discriminant* of the set of equations (4.1) and denoted by D.

For our example, the discriminant of the pair of equations (4.4) is shown in Figure 4.2d.

The discriminant of the simultaneous equations (4.1) embodies those states of the variables $x_1, x_2, \ldots, x_n, y_1, y_2, \ldots, y_m$ for which all the equations are satisfied. As will be seen later, the discriminant of the equations (4.1) forms a basis from which we can directly determine the solution proper in the form (4.2) as well as the number of different solutions and the set of states of independent variables in which these solutions exist (the domain of the solution).

Instead of the form (4.5), we may use the form

$$(\bar{f}_1 g_1 \vee f_1\bar{g}_1) \vee (\bar{f}_2 g_2 \vee f_2\bar{g}_2) \vee \cdots \vee (\bar{f}_p g_p \vee f_p\bar{g}_p) = 0 \tag{4.6}$$

which represents the negation $\bar{D}$ of the discriminant D. It will be shown later that the application of the form (4.6) is sometimes more advantageous than that of the form (4.5).

Two sets of logic equations will be called equivalent if they have the same discriminant. It can be easily verified that this equivalence obeys the reflexive law, the symmetric law, and the transitive law. Consequently, it represents an ordinary equivalence relation. All equations containing the same variables are thus partitioned into equivalence classes. Each of these classes is represented by a single logic function corresponding to the left-hand side of equation (4.5) or by a single logic function corresponding to the left-hand side of equation (4.6). Thus, either the form (4.5) or the form (4.6) may be used as a *standard representation* of the respective equivalence class.

Let us note that either of the standard forms corresponding to a class of logic equations can be uniquely determined if a set of equations (4.1) is given. On the other hand, the way from either of the standard forms to the original form (4.1) is ambiguous.

EXERCISES TO SECTION 4.2

4.2-1 Determine the discriminant of equations (4.4) solely by algebraic manipulations and compare the result with Figure 4.2d.

*4.2-2 Using equations (4.3), we have

$$\bigvee_{i=1}^{p}(\bar{f}_i\bar{g}_i \vee f_i g_i) = 1$$

$$\prod_{i=1}^{p}(\bar{f}_i\bar{g}_i \vee f_i g_i) = 1$$

Do both these forms represent the discriminant?

4.2-3 Show algebraically that $\bar{f}_i g_i \vee f_i \bar{g}_i$ is the negation of $\bar{f}_i\bar{g}_i \vee f_i g_i$.

*4.2-4 (a) Can equations (4.3) be determined from the discriminant?
(b) Can equations (4.1) be determined from equations (4.3)?

4.2-5 Suppose that all variables participating in equations (4.1) are mutually independent. Can this property be found from the discriminant corresponding to (4.1)?

4.2-6 Show that the above defined relation of equivalence of logic equations is reflexive, symmetric, and transitive.

4.3 SOLUTION OF LOGIC EQUATIONS IN STANDARD FORM

In our further exposition we shall assume that the given set of logic equations (4.1) has already been transformed into a single standard equation (4.5) that is implied by (4.1). For our further procedure, however, we shall start from the chart representation of the equation (4.5) which we have called the discriminant of the equations to be solved.

Each column of the discriminant pertains to a single state of the independent variables and each of its rows corresponds to a single state of the unknowns. A solution of the equations exists only if the value 1 (marked by a dot) is represented at least once in every column. If there are any columns not containing the value 1, this means that the corresponding states of independent variables never satisfy the given set of equations. Such a state would therefore have to be eliminated. We would thus no longer be concerned with independent variables but with mutually related variables, the relation being expressed, for instance, by the equation

$$h(x_1, x_2, \ldots, x_n) = 0 \tag{4.7}$$

A solution of the given equations then exists only if the secondary condition (4.7) is satisfied. Equation (4.7) actually specifies those states of independent variables that are prohibited.

Now let us assume that each column in the discriminant contains the value 1 at least once. The variables $x_1, x_2, \ldots, x_n$ are then independent and there exists a solution of (4.1) without any secondary condition (4.7), i.e., in the entire domain of the independent variables.

If the value 1 occurs in some columns more than once, this means that several states of unknowns can be assigned to the corresponding states of the independent variables. However, since we want the solution (4.2) to exist in the form of functions, we require for a particular solution that the states of the unknowns be uniquely assigned to the states of the independent variables. Therefore, if there are several values 1 in some column of the discriminant, then each of them pertains to a different solution.

We can now easily determine the number of different solutions. Let a_c be the number showing how often the value 1 occurs in the c-th column of the discriminant. For the number of different solutions A we then have

$$A = \prod_{c=0}^{2^n-1} a_c \tag{4.8}$$

We see that equation (4.8) includes even the case when there is some column devoid of the value 1 and when no solution exists unless the secondary condition (4.7) is satisfied.

If there is more than one solution, the discriminant falls into *partial discriminants*, ${}^1D, {}^2D, \ldots, {}^rD$, each of which corresponds to a single solution. In addition, the original discriminant is equal to the disjunction of all the partial discriminants, i.e.,

$$D = {}^1D \vee {}^2D \vee \cdots \vee {}^rD \tag{4.9}$$

Thus, for instance, for the discriminant of our example (Fig. 4.2d) we obtain four partial discriminants as shown in Figure 4.3.

Each partial discriminant pertains to a single solution, since it expresses the unique assignment of the states of unknowns to the states of the independent variables. The functional dependence of a particular unknown on the states of the independent variables is obtained by selecting from the corresponding partial discriminant only those states for which the pertinent unknown acquires the value 1, i.e., we produce, as a matter of fact, the logic product of the partial discriminant and the corresponding unknown. As a result, the value 1 is left in some columns only. These are the columns corresponding to those states of the independent variables which determine the pertinent unknown. We can thus express a given unknown immediately in the expanded

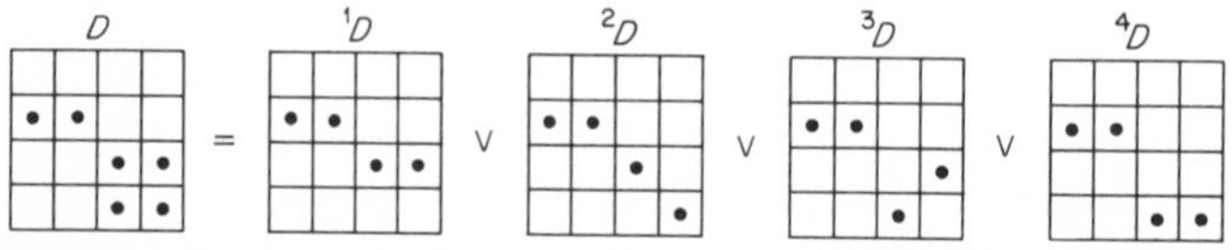

Figure 4.3 Partition of the discriminant shown in Figure 4.2d into partial discriminants.

sp standard form. The procedure further consists of the contraction or, if necessary, the minimization of this form.

On performing the logic product of the partial discriminants (see Fig. 4.3) and the unknowns y_1 and y_2 we obtain, in our example, the charts shown in Figure 4.4. In addition, each chart is accompanied by the minimal Boolean expression for the given unknown.

The validity of the solution can be checked by substituting for the unknowns in the original equations (4.1) the expressions found. If the solution is correct, all the equations will be identically satisfied within the domain of the independent variables.

EXERCISES TO SECTION 4.3

*4.3-1 Using the logic function $h(x_1, x_2, \ldots, x_n)$, which is defined by equation (4.7), express the domain of the solution of (4.1).

4.3-2 Solve the following sets of simultaneous Boolean equations for the unknowns $y_i (i = 1, 2, 3)$:

*(a) $y_2 \bar{y}_1 x_2 \vee x_1 y_2 \vee x_1(x_2 \vee \bar{y}_2) = y_2(\bar{x}_1 \vee x_2 \vee \bar{y}_1)$
$x_1 \bar{y}_1 \vee \bar{x}_1 y_1 = \bar{x}_1 x_2 \bar{y}_1 \bar{y}_2 \vee x_1(\bar{y}_1 y_2 \vee y_1 \bar{y}_2) \vee y_1(\bar{x}_1 \bar{x}_2 \vee x_1 x_2)$

(b) $x_1 \vee \bar{y}_2 = y_1 \bar{y}_2 \vee \bar{x}_1(y_1 \vee \bar{y}_2)$
$\bar{x}_1 \bar{x}_2 \bar{y}_2 \vee y_1 = y_1 y_2 \vee y_1 \bar{y}_2$

(c) $y_1 y_2 = x_3 \vee x_1 x_2$
$y_2 y_3 = x_1 \vee x_2 x_3$
$y_1 y_3 = x_2 \vee x_1 x_3$

*(d) $\bar{x}_2 x_1 \vee \bar{y}_2 \bar{y}_1 \bar{x}_2 = \bar{y}_2 \bar{x}_2 \vee y_2 \bar{x}_2 \bar{x}_1 \vee \bar{y}_1 \bar{x}_2 \bar{x}_1$
$\bar{x}_2 \vee \bar{y}_2 x_1 = y_2 \bar{x}_1 \vee \bar{y}_2(\bar{x}_2 \vee x_1)$

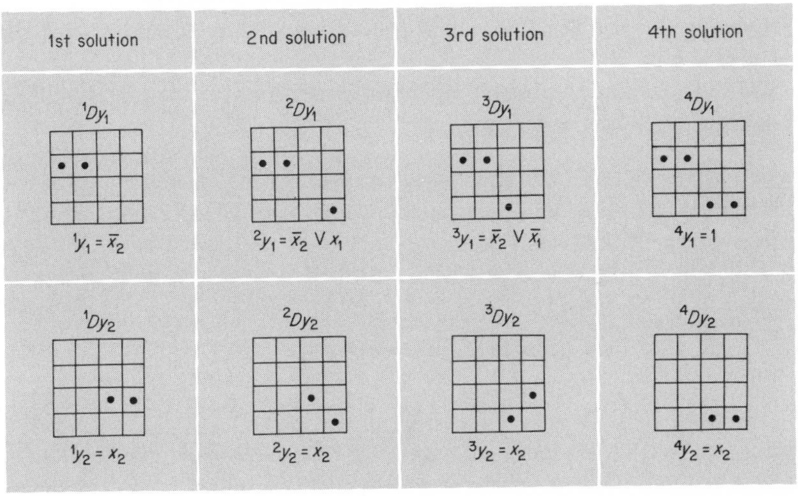

Figure 4.4 Solutions corresponding to the partial discriminants presented in Figure 4.3.

4.4 RECAPITULATION AND EXAMPLES

For the convenience of the reader we shall now sum up the entire procedure suggested for the solution of simultaneous logic equations and illustrate it by means of some more examples.

The problem of solving simultaneous logic equations consists of finding, for a given set of equations (4.1), the implied form (4.2) for $k = 1, 2, \ldots, r$. We proceed as follows:

1. We express the left-hand and right-hand sides of all equations of (4.1) in the form of logic charts.
2. For each equation of (4.1) we express in the chart the function

$$\bar{f}_j \bar{g}_j \vee f_j g_j (j = 1, 2, \ldots, p)$$

We thereby obtain a chart representation of (4.3).

3. We express the logic product which forms the left-hand side of (4.5) by a chart, thus obtaining the discriminant of the set of equations (4.1).
4. We ascertain the number of solutions by means of the formula (4.8). If no solution exists over the entire domain of the independent variables, we determine the secondary condition (4.7) which restricts the domain of the independent variables in which a solution of (4.1) exists.
5. We resolve the discriminant D into partial discriminants ${}^1D, {}^2D, \ldots, {}^rD$ so as to satisfy the relation (4.9).
6. We produce the logic product of the partial discriminants and the individual unknowns. The columns retaining the value 1 determine the states of the independent variables, for which the corresponding unknown has the value 1.
7. We find the simplest possible algebraic expressions for the corresponding unknowns in some convenient manner.

EXAMPLE 4.1 Let us solve the following set of simultaneous Boolean equations for the unknowns y_1 and y_2:

$$x_1(y_2 \vee \bar{y}_1) \vee \bar{y}_2 = \bar{x}_1\bar{x}_2\bar{y}_2 \vee y_1y_2 \tag{I}$$

$$y_2(x_1x_2 \vee x_3\bar{x}_2) = x_2y_2 \vee x_3y_1 \tag{II}$$

$$x_3(x_1\bar{y}_2 \vee x_2y_1) \vee y_1y_2 = \bar{x}_2x_1y_1 \vee x_1x_2 \tag{III}$$

First we enter the left-hand sides (Figure 4.5a) and the right-hand sides (Figures 4.5b) into charts so that the independent variables x_1, x_2, x_3 determine the columns and the unknowns y_1, y_2 the rows in the corresponding

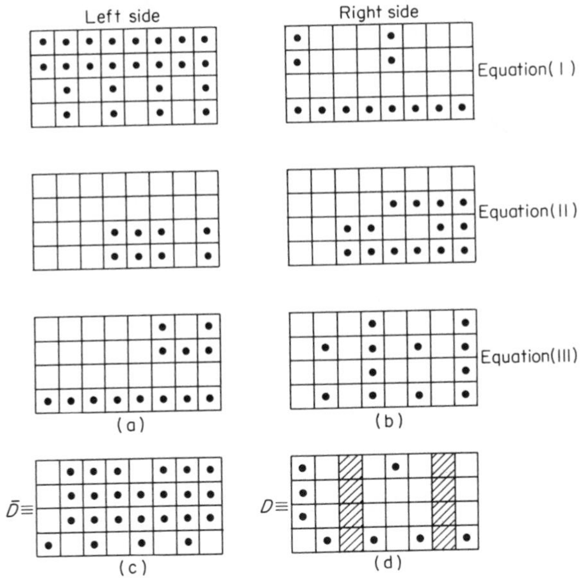

Figure 4.5 Determination of the discriminant in Example 4.1.

charts (see Figure 4.6). For each equation we now construct a chart expressing the equivalence of its left and right-hand sides. We then obtain the discriminant by forming the logic product of all these equivalences. However, we can use a faster method based on the form (4.6). This consists in producing—in place of the equivalences—nonequivalences for the individual equations, and then forming their OR function. We thereby obtain the negation of the discriminant, shown for our example in Figure 4.5c. The actual discriminant (Figure 4.5d) is obtained from its negation by negating the values in the individual squares of the chart. The advantage of this procedure consists in that the OR function can be entered into a single chart without having to delete the dots already marked in it.

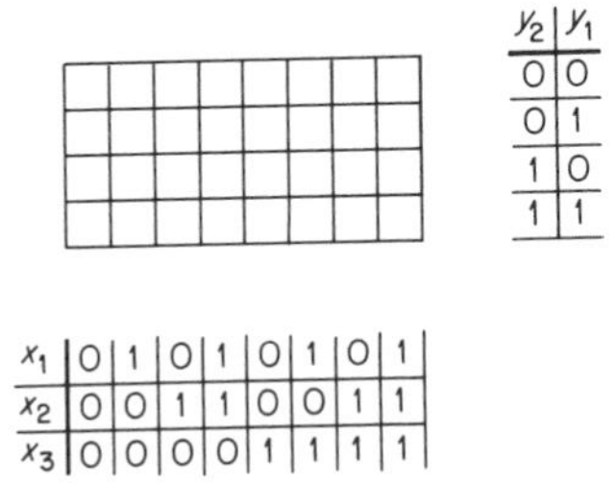

Figure 4.6 Assignment of independent variables and unknowns for Example 4.1.

From the discriminant shown in Figure 4.5d we see that, in our case, there exists no solution over the entire domain of the independent variables, since the states of the independent variables corresponding to the third and seventh column of the discriminant are not related to the dependent variables. These states must therefore be eliminated, in our case by means of the equation $\bar{x}_1 x_2 = 0$. This relation represents the secondary condition which must be satisfied if the solution of the given equations in the form (4.2) is to have a meaning.

It is clear that, when choosing a solution, we can assign any state of the unknowns to the eliminated states of the independent variables. We can thus regard them as undefined states (don't care states), and we have therefore marked them in Figure 4.5d by hatching.

The discriminant of our problem can now be resolved into three partial discriminants ${}^1D$, ${}^2D$, and ${}^3D$, as shown in Figure 4.7. We now form the logic products of the partial discriminants and the individual unknowns (Figure 4.8) and from the resulting chart mappings we can readily determine the algebraic expressions for the unknowns. We must not forget, however, that the solutions shown in Figure 4.8 are valid only under the assumption that the aforementioned secondary condition, which restricts the domain of the independent variables, is satisfied.

EXAMPLE 4.2 Let us solve the following Boolean equations:

$$x_1\bar{x}_2 \vee x_2(\bar{y}_1\bar{y}_2 \vee y_2y_3) \vee \bar{x}_2\bar{y}_2y_3 = \bar{y}_3(x_2 \vee y_1 \vee \bar{x}_1y_2) \vee \bar{x}_2y_3 \qquad \text{(i)}$$

$$x_1(x_2 \vee \bar{y}_2\bar{y}_3) \vee \bar{x}_1y_3 = \bar{x}_1x_2 \vee x_1[y_3(y_1 \vee y_2) \vee \bar{x}_2\bar{y}_1\bar{y}_2] \qquad \text{(ii)}$$

Figure 4.9a shows the chart mappings of the left and right-hand sides of the two equations; Figure 4.9b illustrates the discriminant of the equations. We see that a solution exists over the whole domain of the independent variables, moreover a single one:

$$y_1 = x_1 \vee x_2$$
$$y_2 = x_1\bar{x}_2$$
$$y_3 = x_2$$

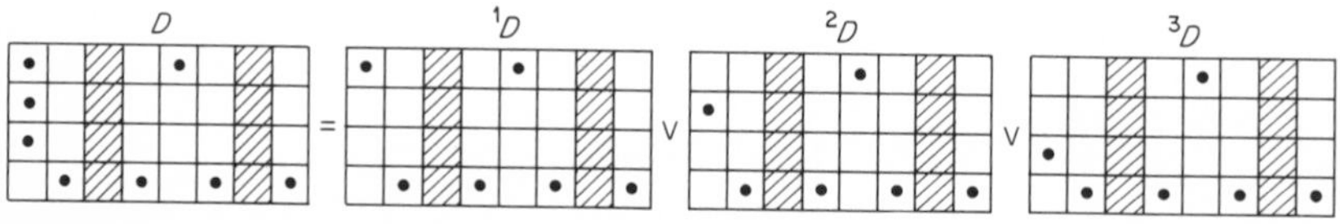

Figure 4.7 Partial discriminants in Example 4.1.

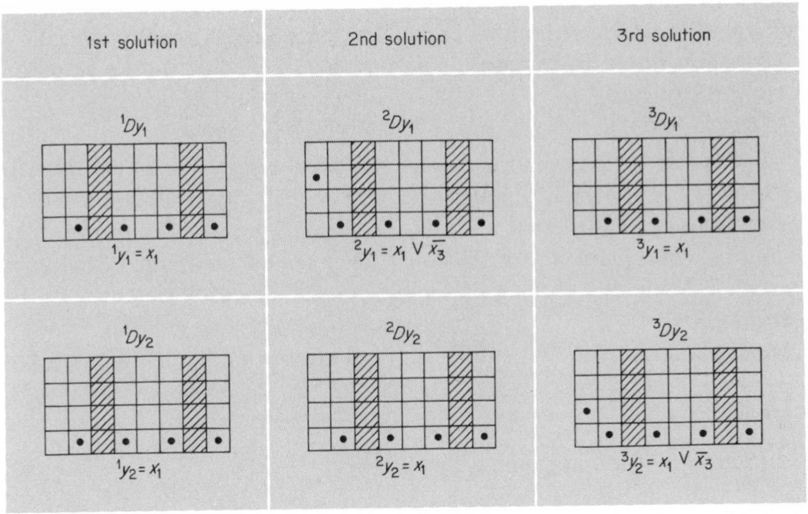

Figure 4.8 Solutions of Example 4.1.

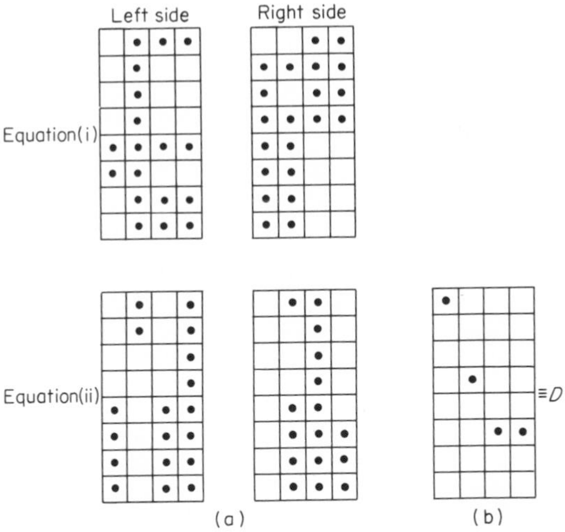

Figure 4.9 Determination of the discriminant in Example 4.2.

EXAMPLE 4.3 To show that the logic expressions participating in the set of equations (4.1) need not be Boolean expressions, let us solve the following logic (non Boolean) equation:

$$(x_1 \,|\, \bar{x}_2) \leq y_1 = (((y_2 \not\equiv y_1) \downarrow x_2)\bar{y}_2) \vee y_2\bar{y}_1 x_2$$

Chart representations for the left-hand side, the right-hand side, and the discriminant are shown in Figure 4.10. The easiest way to obtain the chart representations consists in transforming the logic expressions on the left-hand side and the right-hand side of the equation to the corresponding Boolean *sp* forms first. Then, the chart representations can be easily obtained by the Svoboda grids.

The given logic equation is equivalent to the following Boolean equation:

$$\overline{\overline{x_1\bar{x}_2}} \vee y_1 = \overline{y_2\,\bar{y}_1 \vee \bar{y}_2 y_1 \vee x_2} \cdot \bar{y}_2 \vee y_2\,\bar{y}_1 x_2$$

After exclusion of the negations, we obtain

$$x_1\bar{x}_2 \vee y_1 = (\bar{y}_2 \vee y_1)(y_2 \vee \bar{y}_1)\bar{x}_2\,\bar{y}_2 \vee y_2\,\bar{y}_1 x_2$$

Executing some of the AND operations on the right side of the equation, we obtain *sp* Boolean forms on both sides of the equation:

$$x_1\bar{x}_2 \vee y_1 = \bar{y}_2\,\bar{y}_1\bar{x}_2 \vee y_2\,\bar{y}_1 x_2$$

Now, we are able to obtain the chart representation in Figure 4.10. We see from the discriminant D that there exists only one solution of the given equation:

$$y_1 = 0,\ y_2 = \bar{x}_2\,\bar{x}_1$$

Exercises to Section 4.4

*4.4-1 Specify those cases of (4.1) for which it is more advantageous to determine $\bar{D}$ than D.

*4.4-2 Formula (4.8) gives the number of different solutions of (4.1) in the entire domain of the independent variables. Modify the formula for the case that the solution exists only in a restricted domain of the independent variables.

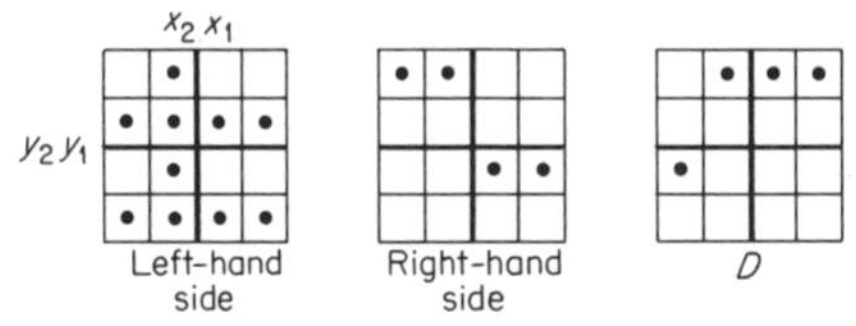

Figure 4.10 Determination of the discriminant in Example 4.3.

4.4-3 Solve the following logic equations for the unknowns y_i $(i = 1, 2, 3)$:

*(a) $((y_1 \downarrow x_1) \vee (x_2 \downarrow x_1)) \vee y_1 x_2 x_1 = y_1((x_1 \not\equiv x_2) \vee x_1) \vee x_2 \bar{x}_1$

(b) $\overline{(y_1 \leq y_2) > y_3} = (\bar{x}_1 \leq x_2) > \bar{x}_3$

(c) $x_1 | \bar{x}_2 = y_1 \downarrow y_2$
$\bar{x}_2 \leq x_3 = y_1 > y_2$

(d) $x_1 = y_1 \geq y_2$
$x_2 = y_1 \not\equiv y_2$
$x_3 = y_1 \downarrow y_2$
$x_4 = y_1 \vee y_2$

4.5 AN ALTERNATIVE PROCESSING OF DISCRIMINANT

An alternative way of processing the discriminant is based on the fact that the number of different solutions for a particular unknown may be considerably smaller than the total number of different solutions of the whole set of unknowns as given by formula (4.8). The procedure described in Section 4.3 can be simplified if we determine for each of the unknowns only the different solutions and, then, combine them properly.

An example will help to illustrate the outlined procedure. Let us solve the Boolean equation

$$\bar{x}_2 \vee y_2 \bar{y}_1 x_1 \vee y_2 y_1 \bar{x}_1 = y_2 \bar{y}_1 \vee \bar{y}_1 \bar{x}_2 \bar{x}_1 \vee y_2 x_1$$

where x_1, x_2 are independent logic variables and y_1, y_2 are logic unknowns.

Determination of the discriminant is shown in Figure 4.11a. Applying formula (4.8) to the discriminant, we find $A = 24$ (the number of different solutions of the given equation).

Let us consider solutions for each unknown separately. Considering $y_i (i = 1, 2$ in our example), we determine for each state of independent variables (each column of the chart representation of the discriminant) whether the discriminant relates that particular state to: (i) both $\bar{y}_i$ and y_i, or (ii) only $\bar{y}_i$, or (iii) only y_i, or (iv) neither $\bar{y}_i$ nor y_i. Results can be summarized in a single chart representing the logic space of independent variables x_1, x_2. The above mentioned cases have, respectively, the following meaning:

(i) y_i can be equal either to 0 or to 1; let us denote this condition in the respective square of the chart for y_i by a cross (×).
(ii) $y_i = 0$ so that the respective square of the chart for y_i is left blank.
(iii) $y_i = 1$ so that we place a dot in the respective square.
(iv) y_i is not defined (don't care state of independent variables) so that the respective square is hatched.

The charts summarizing all solutions for y_1 and y_2 separately are shown in Figure 4.11b. Let us call them *contracted solutions*. All particular solutions for

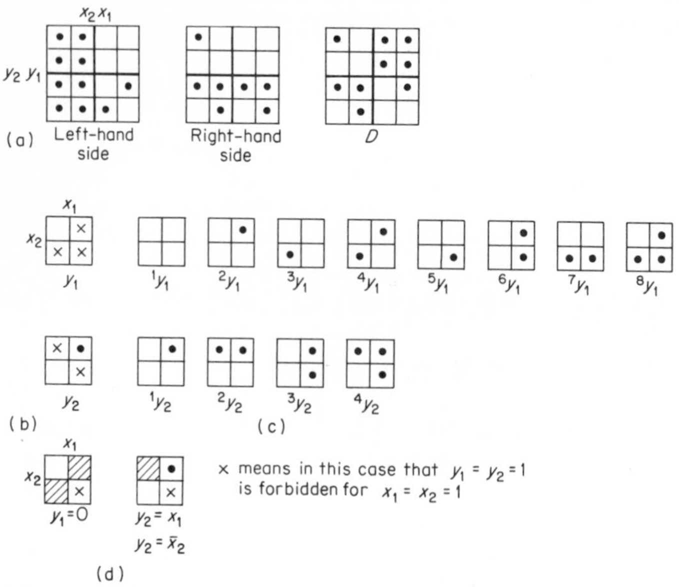

Figure 4.11 Illustration of the alternative method of solution of logic equations.

y_1 and y_2 are shown in Figure 4.11c. Although there are 24 different solutions for the given equation, we obtained only 8 different solutions for y_1 and 4 different solutions for y_2. Now, the solutions for y_1 must be properly paired with those for y_2. They cannot be paired arbitrarily. The reader himself should be able to prove the following general rules of combining the particular solutions:

1. If either $y_i = 0$ or $y_i = 1$ for a particular state s of independent variables, then s does not impose any restriction for combining particular solutions of y_i with particular solutions of other unknowns $y_j (j \neq i)$.

2. If $y_i = \times$ for a state s of independent variables, then s may impose a restriction for combining particular solutions of y_i with particular solutions of other unknowns $y_j (j \neq i)$ providing that also $y_j = \times$. Blank squares in column s of the discriminant specify the states of unknowns which are prohibited for s. Therefore, such combinations of particular solutions of $y_i (i = 1, 2, \ldots, m)$ have to be excluded which represent the prohibited states of unknowns for s.

Applying the rules of combining particular solutions to our example, we find that only the state $s = 3(x_1 = x_2 = 1)$ of independent variables may impose a restriction. The last column in the discriminant (Figure 4.11a) which represents state $s = 3$ has one blank square, for $y_1 = y_2 = 1$. Hence, such pairs of the particular solutions for y_1 and y_2 are prohibited where both $y_1 = 1$ and $y_2 = 1$ for $x_1 = x_2 = 1$ $(s = 3)$. This means that ${}^5y_1$ through ${}^8y_1$ may not be combined with ${}^3y_2$ and ${}^4y_2$ (see Figure 4.11c). All other pairs are possible.

We can generalize the above suggested rules for contracted solutions of several logic unknowns $y_i (i = 1, 2, \ldots, m)$ as follows:

1. If $y_i = 0$, $y_i = 1$ or y_i is don't care for a state s of independent variables, then s does not impose any restriction for combining particular solutions of y_i with particular solutions of other unknowns $y_j (j \neq i)$.

2. If both $y_i = 0$ and $y_i = 1$ for s $(y_i = \times)$, then the following two cases must be distinguished:

(i) If $y_j = \times$ for k unknowns $y_j (0 \leq k \leq m - 1)$ and unknown $y_i (i \neq j)$, and all 2^{k+1} states of variables y_i and y_j are related by the discriminant to s, then the value of y_i for s in the contracted solution may be considered as don't care. In this case we denote the corresponding square in the chart for the contracted solution of y_i by hatching.

(ii) If $y_j = \times$ for k unknowns y_j $(0 \leq k \leq m - 1)$ and unknown $y_i (i \neq j)$, and some of the 2^{k+1} states of variables y_i and y_j are not related by the discriminant to s, then these states are forbidden in a set of particular solutions of y_i and y_j. In this case we denote the corresponding square in the chart of the contracted solution for y_i by a cross or by the state identifiers of either the prohibited or the permitted states of variables y_i, y_j.

When these rules are applied to the discriminant in Figure 4.11a, we obtain the contracted solution shown in Figure 4.11d. Without considering all particular solutions for y_1 and y_2, we can determine pairs of minimal solutions directly from the contracted forms if we consider states denoted by crosses as conditional don't care states. We obtain $y_1 = 0$ and $y_2 = x_1$ or $y_2 = \bar{x}_2$. Both pairs of minimal solutions, $y_1 = 0$, $y_2 = x_1$, and $y_1 = 0$, $y_2 = \bar{x}_2$ are possible in this case because neither of them is based on the forbidden combination $y_1 = y_2 = 1$ for $x_1 = x_2 = 1$. An alternative solution is $y_2 = \bar{x}_2$ and $y_1 = x_1$ or $y_1 = x_2$.

EXAMPLE 4.4 Solve the Boolean equation

$$\bar{y}_1 x_1 \vee \bar{y}_3 \bar{x}_3 (\bar{y}_2 \bar{x}_1 \vee x_2 \bar{x}_1) \vee \bar{y}_2 \bar{y}_1 \bar{x}_3 x_2 = \bar{y}_3 \bar{y}_1 \bar{x}_3 x_1 \vee y_3 (y_1 x_1 \vee \bar{y}_2 \bar{x}_3 \bar{x}_2 \bar{x}_1) \vee x_3 x_2 x_1 (\bar{y}_2 \vee y_1)$$

Determination of the discriminant is shown in Figure 4.12a. Using formula (4.8), we find that the total number of solutions is 49,152 so that the method of decomposition of the discriminant to partial discriminants (Section 4.3) is

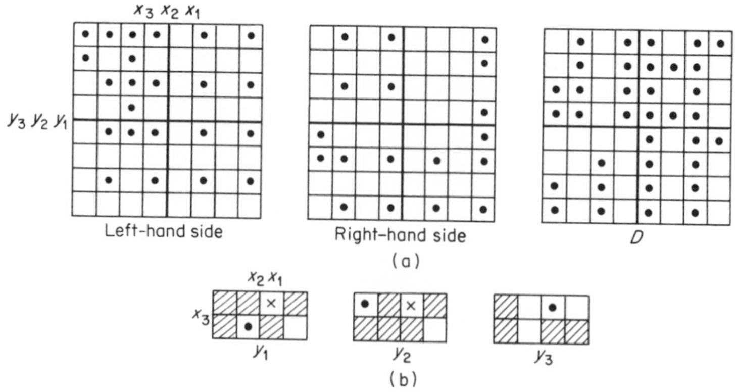

Figure 4.12 Illustration to Example 4.4.

impractical in this case. Nevertheless, we can use the method of contracted solutions.

The contracted solutions are shown in Figure 4.12b. We can immediately find the following minimal Boolean forms:

$$y_1 = \bar{x}_2$$
$$^1y_2 = \bar{x}_2 \text{ or } {}^2y_2 = \bar{x}_1 \text{ or } {}^3y_2 = \bar{x}_3$$
$$y_3 = \bar{x}_1$$

Observing column 2 in the discriminant we find that $y_1 = y_2 = 0$ is forbidden for $s = 2(x_3 = x_1 = 0, x_2 = 1)$. Thus, $y_1 = \bar{x}_2$ may not be combined with ${}^1y_2 = \bar{x}_2$ but may be combined with ${}^2y_2 = \bar{x}_1$ or ${}^3y_2 = \bar{x}_3$. Hence, we obtain the following set of minimal solutions for the given Boolean equation:

$$y_1 = \bar{x}_2,\ y_2 = \bar{x}_1 \text{ or } y_2 = \bar{x}_3,\ y_3 = \bar{x}_1$$

Exercises to Section 4.5

4.5-1 Prove the general rules for combining particular solutions of individual unknowns.

4.5-2 Using the method of contracted solutions, find minimal Boolean solutions for y_i $(i = 1, 2, 3)$ of the following sets of simultaneous logic equations:

*(a) $\bar{y}_3 y_1 \vee y_2(\bar{y}_1\bar{x}_1 \vee x_2 x_1) = y_3 x_3 \vee y_1 x_1 \vee \bar{x}_3 x_2 x_1$.

*(b) $\bar{x}_1(\bar{y}_1 \vee y_2) \vee \bar{x}_2 y_1 \vee \bar{y}_1 y_2 = (x_1\bar{y}_2 \vee \bar{x}_1 y_2)y_1 \vee (\bar{x}_1\bar{y}_2 \vee x_1 y_2)\bar{x}_2$
$x_1(x_2 \vee y_1 y_2) \vee x_2 y_1 = (x_2\bar{y}_1 \vee \bar{x}_2 y_1)x_1 \vee (\bar{x}_2\bar{y}_1 \vee x_2 y_1)\bar{y}_2$.

(c) $\bar{y}_1 \vee y_2 x_2 x_1 = \bar{x}_2(y_1 \not\equiv y_2) \vee \bar{y}_1(\bar{y}_2(x_1 \not\equiv x_2) \vee x_2 x_1)$.

*4.5-3 How many particular solutions exists for y_i if $y_i = \times$ for p states and y_i is don't care for q states of independent variables?

4.6 POSSIBLE GENERALIZATIONS

We are going to consider the following three directions of possible generalization of simultaneous logic equations:

(i) f_i can be related to g_i by any logic function, not necessarily only by equivalence. Different functions can be used for different i.
(ii) The concept of simultaneity, which is represented by the logic function AND, can be replaced by any logic function of p variables, e.g., OR, EXCLUSIVE OR, NAND, NOR, "q out of p" ($q \leq p$), M_p, etc.
(iii) f_i and g_i can be *pseudo-logic functions.* A pseudo-logic function is a function with integer values (not necessarily only 0 and 1) defined on a logic space (if desirable, the range of pseudo-logic functions can be extended to a set of real numbers).

Generalizations (i) and (ii) are discussed in this section. Generalization (iii), concerning pseudo-logic functions, is the subject of Section 4.7.

There are ten possible nontrivial logic functions (binary relations) by which f_i can be related to g_i, namely: $f_i \equiv g_i$ (equivalence, which has the same meaning here as the equality $f_i = g_i$); $f_i \not\equiv g_i$ (nonequivalence); $f_i > g_i$ or $f_i < g_i$ (proper inequality); $f_i \geq g_i$ or $f_i \leq g_i$ (inequality); $f_i \cdot g_i$ (AND function); $f_i \vee g_i$ (OR function); $f_i | g_i$ (NAND function); and $f_i \downarrow g_i$ (NOR function).

For every binary logic relation $f_i R_i g_i$, we can find a Boolean expression $B_i(f_i, g_i)$ such that

$$B_i(f_i, g_i) = 1$$

and

$$\bar{B}_i(f_i, g_i) = 0$$

when f_i is related to g_i by R_i. Boolean expressions B_i for all possible relations $f_i R_i g_i$ are summarized in Table 4.1.

Using Table 4.1, every set of binary logic relations $f_i R_i g_i$ ($i = 1, 2, \ldots, p$) can be transformed either to the equations

$$\begin{aligned} B_1(f_1, g_1) &= 1 \\ B_2(f_2, g_2) &= 1 \\ &\vdots \\ B_p(f_p, g_p) &= 1 \end{aligned} \tag{4.10}$$

or to the equations

$$\begin{aligned} \bar{B}_1(f_1, g_1) &= 0 \\ \bar{B}_2(f_2, g_2) &= 0 \\ &\vdots \\ \bar{B}_p(f_p, g_p) &= 0 \end{aligned} \tag{4.11}$$

TABLE 4.1 Boolean Expressions Transforming R_i to B_i or $\bar{B}_i$

$f_i R_i g_i$	$B_i(f_i, g_i) = 1$	$\bar{B}_i(f_i, g_i) = 0$
$f_i = g_i$	$\bar{f}_i \bar{g}_i \vee f_i g_i = 1$	$\bar{f}_i g_i \vee f_i \bar{g}_i = 0$
$f_i \neq g_i$	$\bar{f}_i g_i \vee f_i \bar{g}_i = 1$	$\bar{f}_i \bar{g}_i \vee f_i g_i = 0$
$f_i < g_i$	$\bar{f}_i g_i = 1$	$f_i \vee \bar{g}_i = 0$
$f_i > g_i$	$f_i \bar{g}_i = 1$	$\bar{f}_i \vee g_i = 0$
$f_i \leq g_i$	$\bar{f}_i \vee g_i = 1$	$f_i \bar{g}_i = 0$
$f_i \geq g_i$	$f_i \vee \bar{g}_i = 1$	$\bar{f}_i g_i = 0$
$f_i \cdot g_i$	$f_i \cdot g_i = 1$	$\bar{f}_i \vee \bar{g}_i = 0$
$f_i \vee g_i$	$f_i \vee g_i = 1$	$\bar{f}_i \cdot \bar{g}_i = 0$
$f_i \mid g_i$	$\bar{f}_i \vee \bar{g}_i = 1$	$f_i \cdot g_i = 0$
$f_i \downarrow g_i$	$\bar{f}_i \cdot \bar{g}_i = 1$	$f_i \vee g_i = 0$

TABLE 4.2 Standard Equations for Some Typical Functions Relating a Set of Binary Relations

Function F	Meaning	Standard equation in the Form (4.12)	Standard equation in the Form (4.13)
AND	Simultaneous validity of all relations	$\prod_{j=1}^{p} B_j = 1$	$\bigvee_{j=1}^{p} \bar{B}_j = 0$
OR	At least one relation is valid	$\bigvee_{j=1}^{p} B_j = 1$	$\prod_{j=1}^{p} \bar{B}_j = 0$
NAND	All relations are not valid simultaneously	$\bigvee_{j=1}^{p} \bar{B}_j = 1$	$\prod_{j=1}^{p} B_j = 0$
NOR	None of the relations is valid	$\prod_{j=1}^{p} \bar{B}_j = 1$	$\bigvee_{j=1}^{p} B_j = 0$
EXCLUSIVE OR	Only one relation is valid	$\bigvee_{i=1}^{p} B_i \cdot \bigvee_{j \neq 1}^{p} \bar{B}_j = 1$	$\prod_{i=1}^{p} \bar{B}_i \vee \prod_{j \neq 1} B_i = 0$
EQUIVALENCE	Either all relations are valid or none is valid	$\prod_{i=1}^{p} B_i \vee \prod_{i=1}^{p} \bar{B}_i = 1$	$\bigvee_{i=1}^{p} \bar{B}_i \cdot \bigvee_{i=1}^{p} B_i = 0$

Partial relations $f_i R_i g_i (i = 1, 2, \ldots, p)$ are related together by a given logic function F of p variables. Clearly, any of 2^{2^p} possible logic functions F can be used. In case of simultaneous relations, F is represented by the generalized AND function.

Function F can be directly applied to functions $B_1, B_2, \ldots, B_p$ in order to obtain the discriminant $D = F(B_1, B_2, \ldots, B_p)$. Thus, the discriminant is represented by the standard equation

$$F(B_1, B_2, \ldots, B_p) = 1 \tag{4.12}$$

Similarly, the negation of the discriminant is represented by the standard equation.

$$\bar{F}(B_1, B_2, \ldots, B_p) = 0 \tag{4.13}$$

Standard equations for some typical functions relating a set of binary relations $f_i R_i g_i (i = 1, 2, \ldots, p)$ are shown in Table 4.2.

Let us note that both the functions in Table 4.1 and those in Table 4.2 (as well as other possible functions F) can again be easily accomplished in terms of chart representations of functions $f_i, g_i, B_i (i = 1, 2, \ldots, p)$.

EXAMPLE 4.5 Solve the following three inequalities provided they are related by EXCLUSIVE OR function:

$$[\bar{y}_2 y_1 \vee y_2 \bar{y}_1 \vee x_2] \leq [\bar{x}_2 \bar{x}_1 \vee \bar{y}_2 \bar{y}_1 x_2]$$
$$[\bar{y}_2 \bar{x}_2 \vee y_2 x_2] \leq [y_2(\bar{y}_1 \bar{x}_1 \vee y_1 \bar{x}_2)]$$
$$[x_2 \vee \bar{y}_2 \bar{y}_1 \vee y_2 x_1] \leq [\bar{y}_2 x_2 \bar{x}_1 \vee \bar{y}_1 \bar{x}_2 x_1]$$

Chart forms of the functions corresponding to the left-hand sides and the right-hand sides of the inequalities are shown in Figure 4.13a and b respectively. Applying the chart operation $\bar{f}_i \vee g_i$ (see Table 4.1), we obtain the functions $B_i (i = 1, 2, 3)$ in Figure 4.13c. The inequalities are related by EXCLUSIVE OR so that the function F has the form

$$F = B_1 \bar{B}_2 \bar{B}_3 \vee B_2 \bar{B}_1 \bar{B}_3 \vee B_3 \bar{B}_1 \bar{B}_2 \text{ (see Table 4.2)}$$

When this operation is accomplished with the charts in Figure 4.13c, we obtain the discriminant D as shown in Figure 4.13d. We see that there is only one solution of the set of inequalities:

$$y_1 = x_1, y_2 = \bar{x}_1 x_2$$

The obtained result must be interpreted as follows: When we replace in the inequalities each y_1 by x_1 and each y_2 by $\bar{x}_1 x_2$, we obtain inequalities between Boolean expressions containing variables x_1, x_2:

$$[x_2 \vee x_1] \leq [\bar{x}_2 \bar{x}_1]$$
$$[\bar{x}_2 \vee \bar{x}_1] \leq [x_2 \bar{x}_1]$$
$$[x_2 \vee \bar{x}_1] \leq 0$$

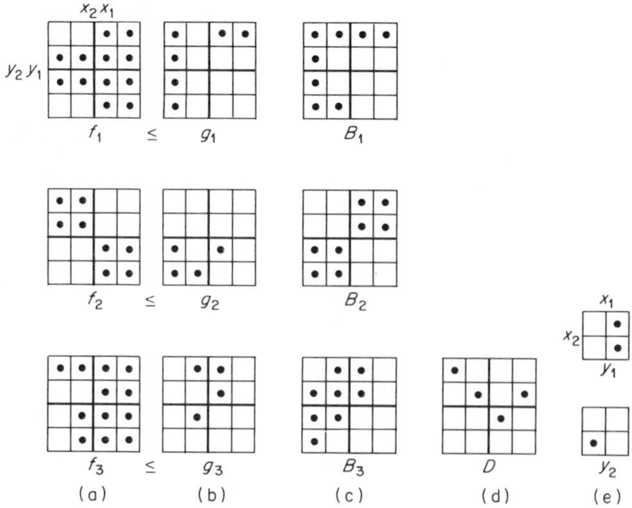

Figure 4.13 Solution of EXCLUSIVE OR inequalities given in Example 4.5.

For each particular state of these variables, just one of the inequalities is satisfied. We can easily find that $x_1 = x_2 = 0$ satisfies the first inequality, $x_1 = 1$ and $x_2 = 0$ satisfies the third inequality, and $x_1 = 0$ and $x_2 = 1$ or $x_1 = x_2 = 1$ satisfy the second inequality.

EXAMPLE 4.6 Solve the following proper inequalities provided they are related by OR function:

$$[y_2 y_1 \bar{x}_2] < [x_1(\bar{x}_2 \vee y_1)]$$
$$[\bar{y}_2(\bar{y}_1 \vee \bar{x}_2)] < [\bar{y}_1 x_1 \vee y_1 x_2 \bar{x}_1]$$
$$[\bar{y}_2(\bar{y}_1 \vee \bar{x}_2) \vee x_2(x_1 \vee y_2)] < [\bar{x}_2 \vee \bar{x}_1 \vee y_2]$$
$$[\bar{y}_2 \bar{y}_1 \vee \bar{x}_2(\bar{x}_1 \vee \bar{y}_1)] < [x_1 \vee y_2 y_1]$$

The whole procedure of solution is shown in Figure 4.14. Let us note that $B_i = \bar{f}_i g_i (i = 1, 2, 3, 4)$ in this case (see Table 4.1) and $F = B_1 \vee B_2 \vee B_3 \vee B_4$ (see Table 4.2). Pairs of minimal Boolean solutions follow directly from the contracted form in Figure 4.14:

$${}^1y_1 = 1,\ {}^1y_2 = 1 \qquad {}^3y_1 = x_2,\ {}^1y_2 = 1$$
$${}^1y_1 = 1,\ {}^2y_2 = \bar{x}_1 \qquad {}^3y_1 = x_2,\ {}^2y_2 = \bar{x}_1$$
$${}^1y_1 = 1,\ {}^3y_2 = \bar{x}_2 \qquad {}^3y_1 = x_2,\ {}^3y_2 = \bar{x}_2$$
$${}^2y_1 = \bar{x}_1,\ {}^1y_2 = 1$$

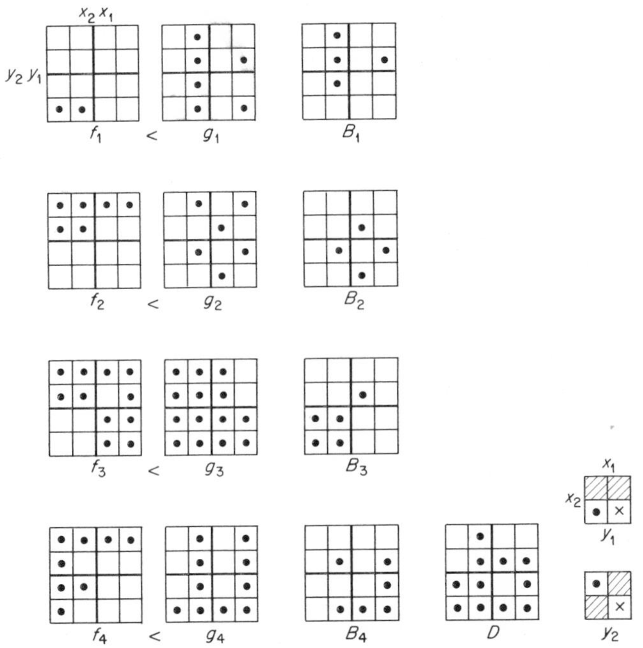

Figure 4.14 Solution of OR proper inequalities given in Example 4.6.

EXAMPLE 4.7 Solve the following set of binary relations which are mutually related by the majority function M_3 (the logic expressions involved are enclosed in brackets):

$$[(y_2 < (y_1 | x_2)) \vee (x_2((x_1 > y_1) \vee (\bar{x}_1 y_1 y_2)))] \not\equiv [y_1(\bar{y}_2 \vee x_2)]$$
$$[(y_2 \geq y_1) | (x_2 < x_1)] < [(y_1 \geq x_1) \vee (y_2 > x_2)]$$
$$[\bar{x}_1((x_2 \vee y_2) \leq (y_1 x_2))] \geq [(x_2 < x_1) | (y_2 > y_1)]$$

First, we transform all logic expressions to the corresponding *sp* Boolean forms:

$$[\bar{y}_2 \bar{y}_1 \vee \bar{y}_2 \bar{x}_2 \vee \bar{y}_1 x_2 x_1 \vee y_2 y_1 x_2 \bar{x}_1] \not\equiv [\bar{y}_2 y_1 \vee y_1 x_2]$$
$$[\bar{y}_2 y_1 \vee x_2 \bar{x}_1] < [y_1 \vee \bar{x}_1 \vee y_2 \bar{x}_2]$$
$$[\bar{y}_2 \bar{x}_2 \bar{x}_1 \vee y_1 x_2 \bar{x}_1] \geq [\bar{y}_2 \vee y_1 \vee x_2 \vee \bar{x}_1]$$

The Boolean expressions can be easily transformed to the chart forms (see Figure 4.15a). Applying now the formulas corresponding to binary relations $\not\equiv$, $<$, $\geq$, which are presented in Table 4.1, we obtain chart representations

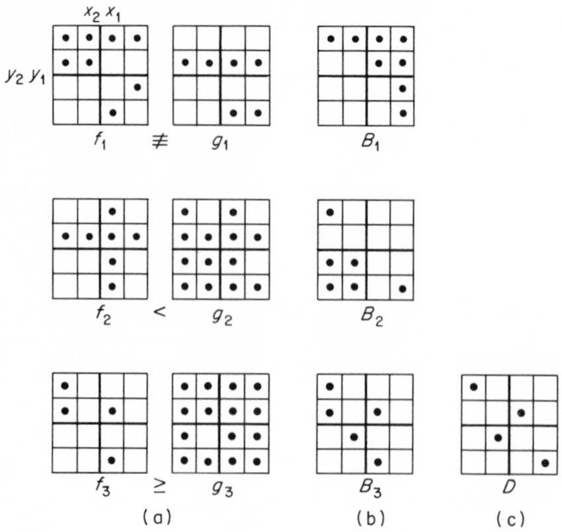

Figure 4.15 Solution of mixed binary relations related by the majority function M_3 (Example 4.7).

of the functions B_1, B_2, B_3 (Figure 4.15b). These functions are related by the majority function M_3 whose Boolean expression is

$$B_1 B_2 \vee B_1 B_3 \vee B_2 B_3$$

This expression specifies the discriminant as shown in Figure 4.15c. The discriminant contains only one solution

$$y_1 = x_2, y_2 = x_1$$

Exercises to Section 4.6

4.6-1 Suppose a set of binary logic relations $f_i R_i g_i$ $(i = 1, 2, 3)$. Find a Boolean form describing the discriminant of the relations in case that:
- *(a) $f_i = g_i$, first two equations are simultaneous and are related to the third one by EXCLUSIVE OR.
- *(b) $f_i < g_i$ related by the equivalence relation.
- (c) $f_1 \leq g_1$, $f_2 = g_2$, $f_3 \not\equiv g_3$ related by the "2 out of 3" function.
- (d) $f_1 \not\equiv g_1$, $f_2 \geq g_2$, $f_3 = g_3$ related by OR function.

4.6-2 Show that:
- (a) A set of simultaneous equations $f_i = g_i$ is represented by the same discriminant as the set of nonequivalences $f_i \not\equiv g_i$ $(i = 1, 2, \ldots, p)$ related by NOR function.
- (b) A set of OR inequalities $f_i < g_i$ is represented by the same discriminant as the set of NAND inequalities $f_i \geq g_i$ $(i = 1, 2, \ldots, p)$.

4.6-3 Following the scheme of Exercise 4.6-2, find some other pairs of sets of binary logic relations that are represented by identical discriminants.

4.6-4 Solve the following sets of binary logic relations with respect to variables y_i provided they are related by EXCLUSIVE OR function:

*(a) $[\bar{x}_2 \vee x_1 \vee y_1] \leq [\bar{x}_2 x_1 \vee y_1\bar{x}_2]$
$[x_2 x_1 \vee \bar{y}_1 x_1] < [\bar{x}_2 \vee \bar{x}_1]$
$[\bar{y}_1\bar{x}_1 \vee \bar{y}_1(\bar{x}_2 \vee x_1)] = [x_1 \vee y_1\bar{x}_2]$.

(b) $[x_1 \vee x_2] \leq [y_1 \vee y_2]$
$[x_1 \vee y_1] \leq [x_2 \vee y_2]$.

(c) $[x_1\bar{x}_2 y_1 \vee x_2 y_2] = [y_2 x_1 \bar{x}_2 \vee y_1]$
$[y_2(x_1 \vee y_1)] \not\equiv [\bar{y}_2 \bar{y}_1 x_2 \bar{x}_1 \vee \bar{x}_2]$
$[\bar{y}_2 y_1(x_2 \vee \bar{x}_1)] < [y_2 y_1(\bar{x}_2 \vee x_1)]$.

(d) $[x_1 \leq ((x_2 \leq y_1) \leq \bar{x}_2)] = [\bar{x}_1 \leq (x_2 \leq \bar{y}_2)]$
$[(\bar{x}_1 \leq x_2) \leq (x_1 \leq y_1)] = [(\bar{y}_2 \leq y_1) \leq \bar{x}_2]$.

(e) $[y_2 \vee y_1 y_3] = [x_2 \vee x_3 x_1]$
$[y_3 \vee y_1 y_2] = [x_3 \vee x_1 x_2]$
$[x_1 \vee x_2 x_3] \leq [y_1 \vee y_2 y_3]$
$[y_1] \leq [x_1 \vee x_2 x_3]$.

4.6-5 Solve all the partial problems of Exercise 4.6-4 in case that the relations are:

(a) simultaneous.
(b) related by OR function.
(c) related by equivalence.

4.7 PSEUDO-LOGIC RELATIONS

A *pseudo-logic relation* is a relation between *pseudo-logic functions* (functions with integer values defined on a logic space). An example of a pseudo-logic function of three logic variables is shown in Figure 4.16.

Every pseudo-logic function can be represented algebraically by the form

$$f = \sum_{s=0}^{2^n - 1} f(s) \cdot M_s \tag{4.14}$$

where s is the state identifier, $f(s)$ is the value of f for the state s, and M_s is the minterm corresponding to the state s. Although the form (4.14) is not a Boolean expression, it shows certain similarities with the expanded *sp* standard Boolean form (compare (2.12), p. 72). Therefore, (4.14) is called the *expanded sp standard pseudo-Boolean form.*

s	$x_3\ x_2\ x_1$	$f(s)$
0	0 0 0	3
1	0 0 1	0
2	0 1 0	−2
3	0 1 1	−5
4	1 0 0	5
5	1 0 1	2
6	1 1 0	3
7	1 1 1	1

x_3 \ $x_2 x_1$				
	3	0	−2	−5
	5	2	3	1

f

Figure 4.16 An example of pseudo-logic function.

The pseudo-logic function f in Figure 4.16 has the following expanded *sp* pseudo-Boolean form:

$$f = 3\bar{x}_3\bar{x}_2\bar{x}_1 - 2\bar{x}_3x_2\bar{x}_1 - 5\bar{x}_3x_2x_1 + 5x_3\bar{x}_2\bar{x}_1 + 2x_3\bar{x}_2x_1 + 3x_3x_2\bar{x}_1 + x_3x_2x_1$$

The same pseudo-logic function can also be represented by simplified pseudo-Boolean forms. For instance,

$$f = 2x_3\bar{x}_2 + 3\bar{x}_1 - 5\bar{x}_3x_2 + x_3x_2x_1 = 3\bar{x}_1 + x_3(2\bar{x}_2 + x_2x_1) - 5\bar{x}_3x_2$$

Using Svoboda grids, a pseudo-logic function represented by a simplified pseudo-Boolean *sp* form can be transformed to the chart form as follows:

1. We take grids corresponding to the first Boolean p-term in the *sp* form and apply them on a proper chart. We write down the integer associated with the term into each of uncovered squares of the chart (see Figure 4.17a).
2. We repeat Step (1) for each term of the given pseudo-Boolean *sp* form (see Figures 4.17b through d).
3. We add all integers in each square of the chart thus obtaining the chart representation of the pseudo-logic function (see Fig. 4.17e).

Let us consider now a set of formulas

$$f_i(x_1, x_2, \ldots, x_n, y_1, y_2, \ldots, y_m)R_i g_i(x_1, x_2, \ldots, x_n, y_1, y_2, \ldots, y_m) \tag{4.15}$$

where f_i, g_i are pseudo-logic functions and R_i are binary relations between these functions ($i = 1, 2, \ldots, p$). Some typical binary relations R_i are $=, \neq, <, \leq$, each of which is represented by pairs of integers. When we write $f_i R_i g_i$, we actually classify states of the logic variables $x_1, x_2, \ldots, x_n, y_1, y_2, \ldots, y_m$ to those for which the relation is satisfied and those for which it is not satisfied. This classification can be described by a logic equation

$$B_i(x_1, x_2, \ldots, x_n, y_1, y_2, \ldots, y_m) = 1 \tag{4.16}$$

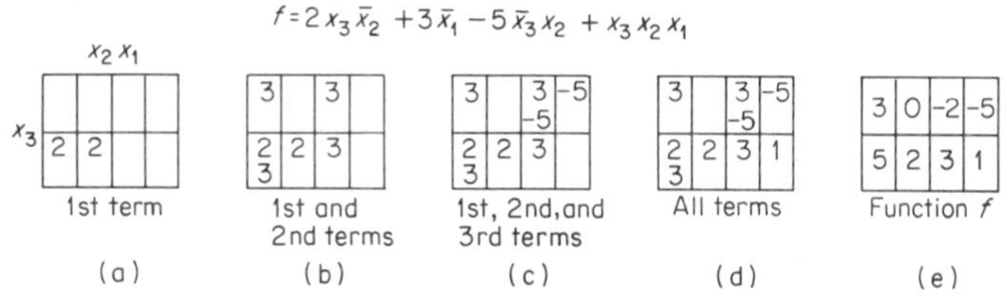

Figure 4.17 Transformation of a simplified pseudo-Boolean *sp* form into the chart form.

whose validity reflects validity of the pseudo-logic relation $f_i R_i g_i (i = 1, 2, \ldots, p)$. The set of pseudo-logic relations (4.15) implies thus the set of logic equations (4.16).

The relations (4.15) are mutually related by a given logic functions F of p variables. This function can be directly applied to functions B_i in order to obtain the discriminant of (4.15). The processing of the discriminant is the same as for logic relations.

EXAMPLE 4.8 Solve the following pair of simultaneous pseudo-Boolean relations:

$$[3\bar{y}_2 y_1 - 2y_2\bar{y}_1 + \bar{y}_2 x_2 + 5x_2 x_1 - y_2\bar{x}_2 x_1] \leq [-2x_2 + 7y_2 y_1 - 3\bar{y}_2\bar{y}_1\bar{x}_2\bar{x}_1]$$

$$[3\bar{y}_2\bar{x}_2\bar{x}_1 - 5\bar{y}_2 y_1 + 2y_2\bar{x}_2] = [2\bar{x}_1\bar{y}_1 + y_2\bar{x}_2 x_1]$$

The functions on the left-hand sides (f_1, f_2) and the right-hand sides (g_1, g_2) of the relations are plotted in logic charts shown in Figure 4.18a and b respectively. Values of f_1 and g_1 are compared for all states of logic variables x_1, x_2, y_1, y_2, and the validity of the inequality $f_1 \leq g_1$ is expressed by the function B_1 (Fig. 4.18c). If $f_1(s) \leq g_1(s)$, where s is a state of the logic variables x_1, x_2, y_1, y_2, then $B_1(s) = 1$; otherwise $B_1(s) = 0$. Similarly, the validity of the equality $f_2 = g_2$ is expressed by the function B_2. Since the relations are to hold simultaneously, their discriminant is represented by the logic function

$$F = B_1 \cdot B_2 \text{ (Figure 4.18d)}$$

The discriminant yields only one solution:

$$y_1 = x_2 \text{ and } y_2 = \bar{x}_1 + x_2$$

We can see now that the procedure for solution of pseudo-logic relations is a direct generalization of the procedure for solution of logic relations. The

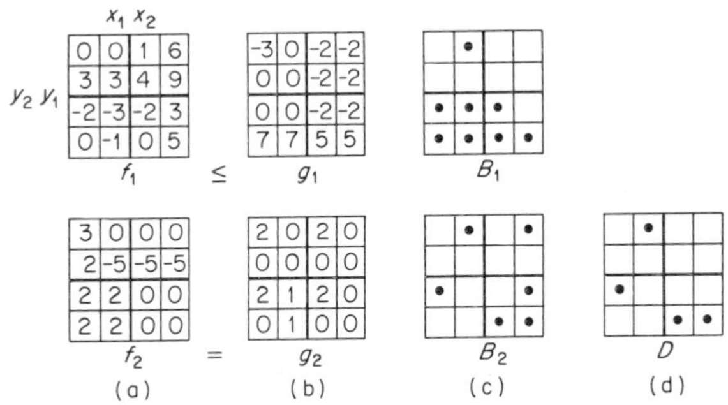

Figure 4.18 Solution of simultaneous pseudo-Boolean inequality and equation (Example 4.8).

generalization consists in the range of f_i and g_i, which is extended from 0 and 1 to integers, and in the relations R_i between f_i and g_i, which are also extended to integers. Otherwise, both the procedures are identical.

Let us consider now the problem of the determination of minima and/or maxima of a pseudo-logic function $f(x_1, x_2, \ldots, x_n)$ subject to some constraints expressed by a set of binary relations between some pseudo-logic functions of logic variables x_1, x_2, ..., x_n. The following procedure is suggested:

1. The function f is transformed to a corresponding *sp* pseudo-Boolean form (if represented by another form) and, then, to the chart form. The latter can be done very easily by the Svoboda grids.
2. The constraints, which are represented by a set of binary pseudo-logic relations, are transformed to the discriminant. No classification to independent and dependent variables exists in this case so that the assignment of variables in charts is arbitrary.
3. The discriminant identifies the domain of the function f. The minimum (or maximum) of the function f can be obtained from its chart representation by inspection.

EXAMPLE 4.9 Find both the minima and maxima of the function

$$f = \bar{x}_4\bar{x}_3(2\bar{x}_2 + 5x_1) + 3\bar{x}_2x_3 - x_2(6x_4 - 7\bar{x}_1 + 4\bar{x}_4x_3) + x_4\bar{x}_3(\bar{x}_2(x_1 - 7\bar{x}_1))$$

subject to the following constraints:

$$[x_3(3\bar{x}_4 + x_4) + 2x_4\bar{x}_3x_2x_1] \leq [\bar{x}_4(2\bar{x}_3 + 5x_3) + x_4(3\bar{x}_3\bar{x}_2\bar{x}_1 + x_3x_2x_1)]$$
$$[x_4\bar{x}_3(3\bar{x}_2\bar{x}_1 + 2x_2x_1) + \bar{x}_4\bar{x}_3(6x_1 - 5x_2)] \leq 0$$
$$[\bar{x}_4x_3(3\bar{x}_2\bar{x}_1 + 2x_2x_1) + \bar{x}_4\bar{x}_3x_2\bar{x}_1] \leq [\bar{x}_2x_1]$$

These inequalities hold simultaneously. Their transformation to the form of discriminant is shown in Figure 4.19a. The chart form of the function f is in Figure 4.19b. The effect of the constraints on f is expressed by the discriminant D, which specifies the domain of f as shown in Figure 4.19c. Clearly, there is one minimum for $x_4 = x_3 = x_2 = x_1 = 1$, where $f = -6$, and two maxima for $x_4 = 0$, $x_3 = 1$, $x_2 = 0$ or 1 and $x_1 = 1$ or 0, respectively, where $f = 3$.

Exercises to Section 4.7

4.7-1 Prove that the form (4.14) represents the pseudo-logic function whose values for states s are $f(s)$.

4.7-2 Solve the following sets of binary pseudo-logic relations with respect to variables y_i provided they are related by EXCLUSIVE OR function:

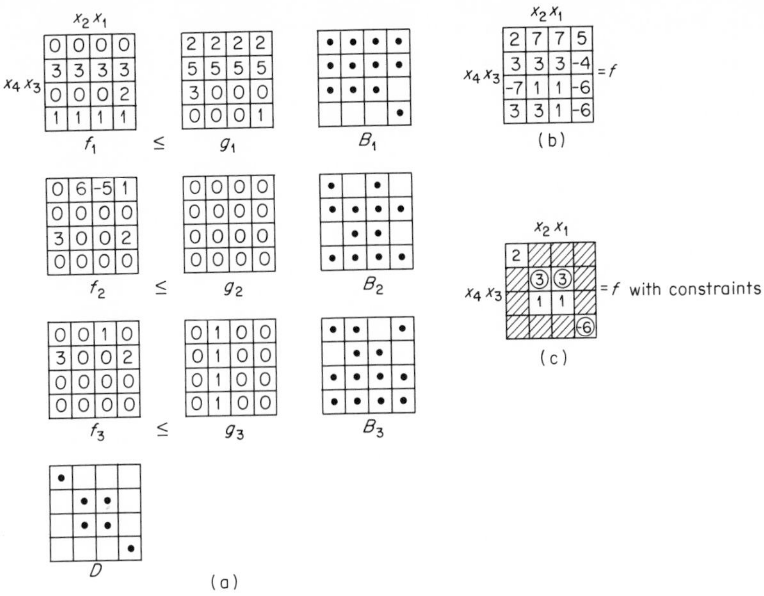

Figure 4.19 Determination of minima and maxima of a pseudo-logic function with constraints (Example 4.9).

*(a) $[2y_1\bar{x}_1 - 5\bar{x}_2\bar{x}_1] \leq [y_1x_2 - 3\bar{y}_1\bar{x}_1]$
$[4x_1x_2 + 3\bar{x}_1\bar{x}_2 - 2\bar{y}_1x_1x_2] = [3y_1\bar{x}_1 - 2\bar{y}_1x_1]$.
(b) $[3\bar{y}_2y_1 - 2y_2\bar{y}_1 + \bar{y}_2x_2] \geq 0$
$[2\bar{y}_2\bar{x}_1 + 3y_1x_1 + 2] \geq 0$.
(c) $[x_1 + y_1] > 0$
$[x_2 + y_2] < 0$
$[\bar{x}_1 + \bar{x}_2] = 1$
$[y_1 + y_2] = 1$.

4.7-3 Solve all the partial problems of Exercise 4.7-2 in case that the relations are:
(a) simultaneous.
(b) related by OR function.
(c) related by equivalance.
(d) related by NAND function.

4.7-4 Find minima and maxima of the pseudo-logic function

$$f = 2\bar{x}_1x_2 + 5x_5\bar{x}_3 - 7x_4x_2 + x_3x_1$$

subject to the following constraints:
*(a) Inequalities related by EXCLUSIVE OR function:
$[x_1 + x_2 + x_3 + x_4 + x_5] > 0$
$[\bar{x}_1 + \bar{x}_2 + \bar{x}_3 + \bar{x}_4 + \bar{x}_5] \leq 0$.
*(b) Inequalities related by function F_3^2:
$[\bar{x}_4x_2 - 7x_3x_1] > 0$
$[2x_2\bar{x}_1 + 3\bar{x}_5x_4] > 0$
$[x_4 + x_3] > [2x_2 - 3x_1]$.

(c) Simultaneous inequalities
$[x_1\bar{x}_2 - 7x_3\bar{x}_4] \leq [3x_4\bar{x}_5 - 2\bar{x}_1]$
$[4x_3x_2x_1 + 2\bar{x}_1] \leq [5x_3x_2 + \bar{x}_5\bar{x}_4\bar{x}_3]$.

4.7-5 Applying the constraints given in Exercise 4.7-4, find maxima and minima of the pseudo-logic function $f = x_6\bar{x}_3\bar{x}_2 - 9x_2\bar{x}_1 + 7\bar{x}_6x_4x_1 - 8x_3 + 5\bar{x}_4\bar{x}_3\bar{x}_1 + 5\bar{x}_2x_1$.

4.8 APPLICATIONS: SOME EXAMPLES

Suppose that we are to design a memoryless switching circuit S with n inputs and m outputs (Figure 4.20a) whose behavior is represented by given logic functions

$$y_i = f_i(x_1, x_2, \ldots, x_n)$$

where $i = 1, 2, \ldots, m$. Let an element E (available building block, module), whose behavior is represented by logic functions

$$v_j = g_j(z_1, z_2, \ldots, z_p)$$

where $j = 1, 2, \ldots, q$, be given (Figure 4.20b). By connecting outputs of E to first q outputs of S, we can decompose q functions f_i with respect to functions $g_j(i, j = 1, 2, \ldots, q)$. This situation, which is shown in Figure 4.20c, leads to the simultaneous logic equations

$$f_i(x_1, x_2, \ldots, x_n) = g_j(z_1, z_2, \ldots, z_p)$$

where $i, j = 1, 2, \ldots, q$. Unknowns are the intermediate variables $z_1, z_2, \ldots, z_p$ (see Figure 4.20c), which represent the *decomposition of functions* $f_i(x_1, x_2, \ldots, x_n)$ *with respect to functions* $g_j(z_1, z_2, \ldots, z_p)$. After solving the set of logic equations, we obtain all possible decompositions in the form

$$\begin{aligned} z_1 &= {}^kF_1(x_1, x_2, \ldots, x_n) \\ z_2 &= {}^kF_2(x_1, x_2, \ldots, x_n) \\ &\vdots \\ z_p &= {}^kF_p(x_1, x_2, \ldots, x_n) \end{aligned}$$

where k distinguishes different solutions. We may select such a solution which satisfies best certain criteria of goodness.

A general approach to synthesis of memoryless switching circuits that is based on the principle just outlined is described in Chapter 6.

EXAMPLE 4.10 Determine all possible decompositions of the majority function M_3 with respect to nonequivalence (see Fig. 4.21).

The problem leads to the logic equation

$$M_3(x_1, x_2, x_3) = z_1 \not\equiv z_2$$

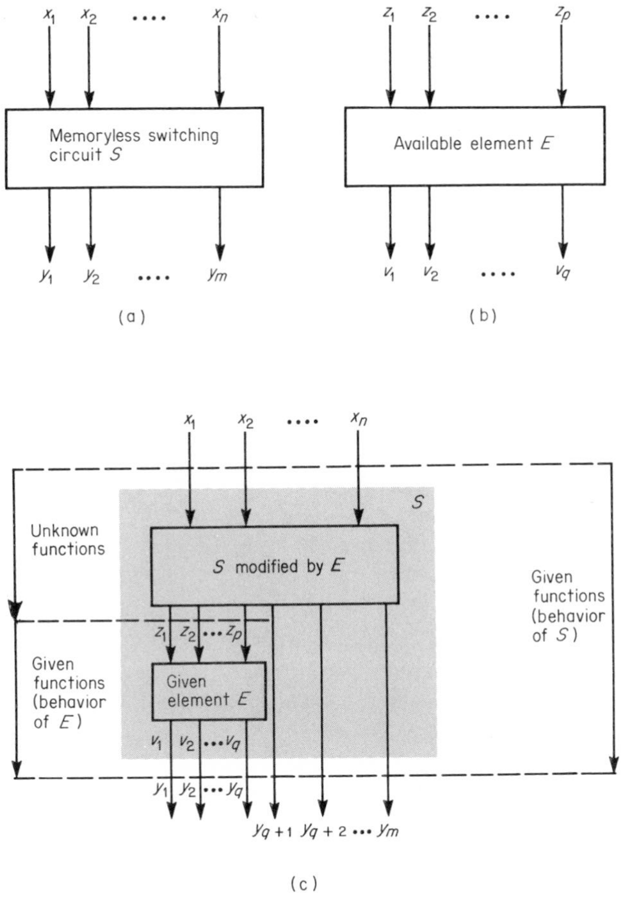

Figure 4.20 Application of simultaneous Boolean equations to decomposition of switching circuits.

whose Boolean form is

$$x_1x_2 \vee x_1x_3 \vee x_2x_3 = \bar{z}_1z_2 \vee z_1\bar{z}_2$$

The solution of the equation for z_1 and z_2 is left to the reader as an exercise.

Let us show now that the problem of *minimum covering of a logic function by its prime implicants*, which is discussed in Section 3.4, can be solved in terms of pseudo-logic relations.

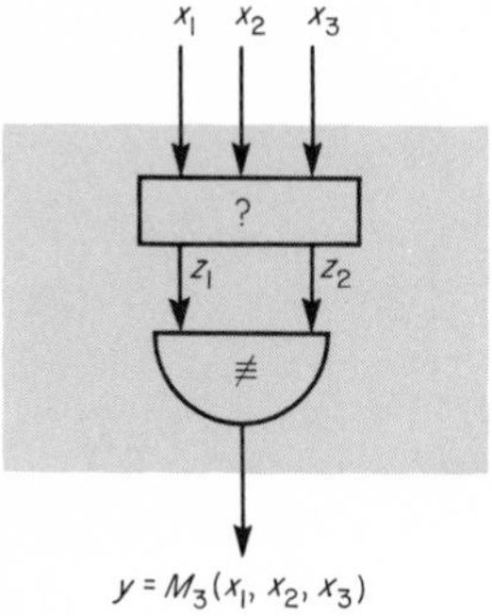

Figure 4.21 Decomposition of M_3 with respect to NONEQUIVALENCE (Example 4.10).

Let $s_i (i = 1, 2, \ldots, m)$ be states of independent variables $x_1, x_2, \ldots, x_n$ covered by a logic function $f(x_1, x_2, \ldots, x_n)$, let $P_j (j = 1, 2, \ldots, q)$ be prime implicants of f and let

$$p_j = \begin{cases} 1 \text{ if } P_j \text{ is included in the Boolean form} \\ 0 \text{ otherwise} \end{cases}$$

Let v_j be the number of literals which occur in P_j and let

$$a_{ij} = \begin{cases} 1 \text{ if } s_i \text{ is covered by } P_j \\ 0 \text{ otherwise} \end{cases}$$

Then the covering problem can be stated as follows:
Find minima of the pseudo-logic function

$$\sum_{j=1}^{q} v_j p_j \tag{4.17}$$

under the restrictions

$$\sum_{j=1}^{q} a_{ij} \cdot p_j \geq 1 \qquad (i = 1, 2, \ldots, m) \tag{4.18}$$

EXAMPLE 4.11 Find all minimal *sp* Boolean forms of the logic function f whose chart form and prime implicant table are in Figure 4.22a and b respectively.

The problem can be solved by finding all minima of the pseudo-logic function

$$2p_1 + 2p_2 + 2p_3 + 2p_4 + 2p_5 + 2p_6$$

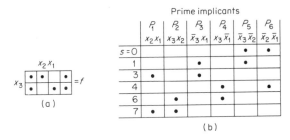

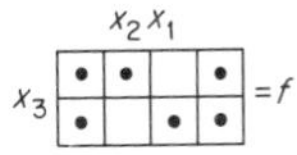

(a)

Prime implicants

	P_1 $x_2 x_1$	P_2 $x_3 x_2$	P_3 $\bar{x}_3 x_1$	P_4 $x_3 \bar{x}_1$	P_5 $\bar{x}_3 \bar{x}_2$	P_6 $\bar{x}_2 \bar{x}_1$
$s=0$					•	•
1			•		•	
3	•		•			
4				•		•
6		•		•		
7	•	•				

(b)

Figure 4.22 Illustration for Example 4.11.

subject to pseudo-Boolean constraints

$$p_1 + p_2 \geq 1 \qquad p_3 + p_5 \geq 1$$
$$p_1 + p_3 \geq 1 \qquad p_4 + p_6 \geq 1$$
$$p_2 + p_4 \geq 1 \qquad p_5 + p_6 \geq 1$$

These constraints can also be expressed in the Boolean form:

$$p_1 \vee p_2 = 1 \qquad p_3 \vee p_5 = 1$$
$$p_1 \vee p_3 = 1 \qquad p_4 \vee p_6 = 1$$
$$p_2 \vee p_4 = 1 \qquad p_5 \vee p_6 = 1$$

There are two minima:

(i) $p_2 = p_3 = p_6 = 0,\ p_1 = p_4 = p_5 = 1$
(ii) $p_1 = p_4 = p_5 = 0,\ p_2 = p_3 = p_6 = 1$

Hence there are two minimal *sp* forms:

(i) $f = x_1 x_2 \vee \bar{x}_1 x_3 \vee \bar{x}_2 \bar{x}_3$
(ii) $f = x_2 x_3 \vee x_1 \bar{x}_3 \vee \bar{x}_1 \bar{x}_2$

The *objective pseudo-logic function* (4.17) can be modified. For example, we may include number of negations in individual prime implicants. Also the number of OR operations can be included because it is uniquely determined for each state of variables p_j.

The objective function may include costs of elements representing the individual logic functions. Let the number of AND, OR, and NOT functions be denoted respectively, $N_a(p_j)$, $N_o(p_1, p_2, \ldots, p_q)$ and $N_n(p_j)$. Notice that N_a and N_n are considered as functions of single variables p_j while N_o is a function of all variables $p_1, p_2, \ldots, p_q$. Let C_a, C_o, C_n denote respectively, the cost of elements AND, OR, NOT. Then, the objective pseudo-Boolean function expressing the *total cost* of the circuit has the form

$$C_o \cdot N_o(p_1, p_2, \ldots, p_q) + \sum_{j=1}^{q} (C_a \cdot N_a(p_j) + C_n \cdot N_n(p_j)) \tag{4.19}$$

Sometimes it is required, from the standpoint of reliability, that each state covered by a logic function be represented by r prime implicants (or implicants) of the function (so-called *realization with the reliability level r*). In such a case, the inequalities (4.18) become

$$\sum_{j=1}^{q} a_{ij} \cdot p_j \geq r(i = 1, 2, \ldots, m) \tag{4.20}$$

The method just described can be applied to other covering problems concerning switching circuits too. For instance, minimization of groups of Boolean expressions, minimization of TANT (*T*hree-level *AN*D-*NOT*) switching circuits, state minimization of incompletely specified sequential switching circuits are typical applications.

Suppose again that we want to design a memoryless switching circuit S with n inputs and m outputs (Figure 4.20a) whose behavior is represented by logic functions

$$y_i = f_i(x_1, x_2, \ldots, x_n) \tag{4.21}$$

where $i = 1, 2, \ldots, m$. Suppose further that some functions

$$z_j = g_j(x_1, x_2, \ldots, x_n) \tag{4.22}$$

where $j = 1, 2, \ldots, r$, are available in addition to the trivial assertion functions $x_1, x_2, \ldots, x_n$. Find minimal Boolean forms of functions

$$y_i = F_i(x_1, x_2, \ldots, x_n, z_1, z_2, \ldots, z_p) \tag{4.23}$$

such that

$$F_i(x_1, x_2, \ldots, x_n, g_1(x_1, \ldots, x_n), \ldots, g_r(x_1, \ldots, x_n)) = f_i(x_1, x_2, \ldots, x_n)$$

for all $i = 1, 2, \ldots, m$.

This situation is illustrated in Figure 4.23.

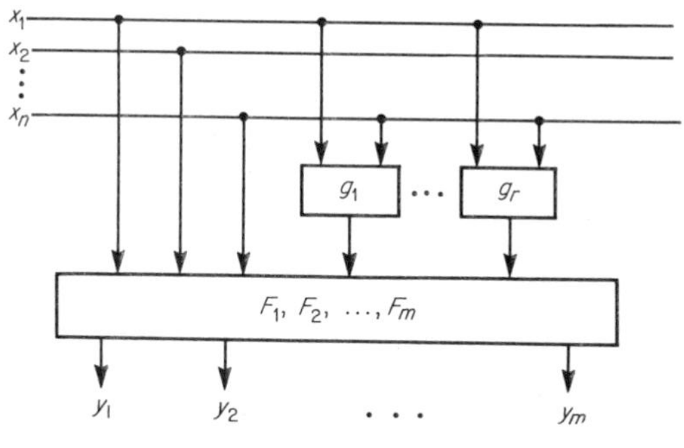

Figure 4.23 Minimization with available functions.

Functions F_i can be determined by solving simultaneous equations (4.21) and (4.22) with respect to $y_i (i = 1, 2, \ldots, m)$. Clearly, $x_1, x_2, \ldots, x_n, z_1, z_2, \ldots, z_r$ are not independent because of (4.22).

EXAMPLE 4.12 Find a minimal Boolean form of the function

$$y = F(x_1, x_2, z)$$

provided that

$$y = (x_1 \not\equiv x_2)$$

and (a)

$$z = (x_1 < x_2)$$

as shown in Figure 4.24a.

The solution of (a) is illustrated in Figure 4.24b. The restriction of the domain of the solution is due to the relation between variables x_1, x_2, z represented by the second equation in (a).

There are other known applications of logic and pseudo-logic equations and other binary relations, some of which are described in later chapters. Since a systematic study of these applications has been started only recently, many new applications are expected in the future, the ultimate goal being a unification of a large portion of the methodology of switching circuits in terms of logic and pseudo-logic relations.

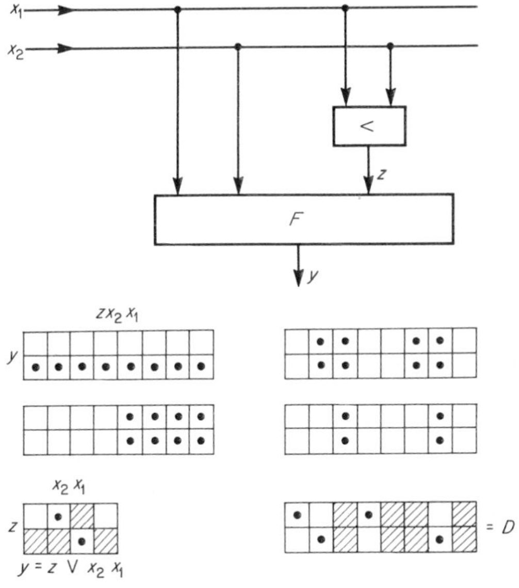

Figure 4.24 Illustration for Example 4.12.

Exercises to Section 4.8

4.8-1 Using the pseudo-logic function approach, find all minimal *sp* Boolean forms of a logic function f of four variables such that:
  (a) $f(s)=1$ for $s=0, 2, 6, 7, 10, 11, 14, 15$, $f(s)=0$ otherwise.
  (b) $f(s)=0$ for $s=0, 11$, $f(s)=1$ otherwise.

4.8-2 Find the number of possible decompositions of the following definite logic functions of three variables with respect to nonequivalence:
  *(a) $f(s)=1$ for $s=1, 3, 5, 6$, $f(s)=0$ otherwise.
  (b) $f(s)=1$ for $s=0, 3, 5, 6$, $f(s)=0$ otherwise.
  (c) $f(s)=0$ for $s=1, 6, 7$, $f(s)=1$ otherwise.
  (d) $f(s)=0$ for $s=3, 7$, $f(s)=1$ otherwise.
  Determine one particular decomposition for each of the functions and draw a block diagram of the decomposition.

4.8-3 Repeat Exercise 4.8-2 with respect to:
  (a) NAND function.
  (b) implication.
  (c) OR function.

4.8-4 Use Boolean equations to realize (if possible) the following functions by the block diagram in Figure 4.25:
  *(a) $f=x_1x_2 \vee \bar{x}_2 x_3$.
  *(b) $f=(x_1 \not\equiv \bar{x}_2) \not\equiv (\bar{x}_1 \not\equiv x_3)$.
  (c) $f=(\bar{x}_1 | x_2) | (x_2 | \bar{x}_3)$.
  Hint: Solve $f(x_1, x_2, x_3)=y_1y_2$ with respect to y_1 and y_2.

Comprehensive Exercises to Chapter 4

4.1 Consider a single relation $f \circ g$ between two logic functions f, g of n independent and m dependent variables whose discriminant contains u zeros and v ones. How many different pairs of f, g lead to the same discriminant if the relation $\circ$ is:
  (a) equality (equivalence) $f=g$.
  (b) inequality (implication) $f \leq g$.
  (c) nonequivalence $f \not\equiv g$.
  (d) proper inequality (inhibition) $f < g$.
  (e) OR function $f \vee g$.

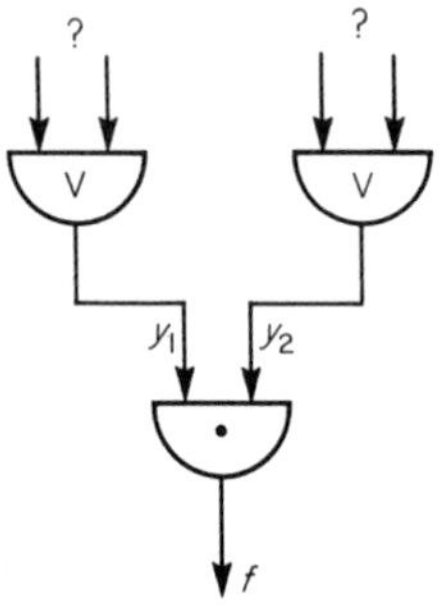

Figure 4.25 Illustration for Exercise 4.8-4.

(e) OR function $f \vee g$.
(f) NAND function $f | g$.

4.2 Generalize results of Exercise 4.1 for p relations of the same type provided that they are related by:
(a) the generalized AND function (simultaneous relations).
(b) the generalized OR function.
(c) the generalized EXCLUSIVE OR function ("one out of p").
(d) the function "q out of p."
(e) the majority function M_p.

4.3 Let $f\,R\,g$ be a binary relation between pseudo-logic functions f and g whose range is represented by all integers from 0 to N. What is the number of all possible mutually different relations (including trivial relations)?

*4.4 Solve the pseudo-Boolean relation

$$(2x_2\bar{x}_1y_1 + 7x_1\bar{y}_2\bar{y}_1 - 3y_2)R(4x_2 - 5x_1y_1 - 4\bar{x}_2\bar{y}_2)$$

where the relation R is specified by the following set of pairs: $R = \{(-3, 0), (-2, 0), (-1, 0), (7, 4), (-1, 4), (-1, 3), (-1, 2)\}$.

4.5 A pseudo-logic function $f(x_1, x_2, \ldots, x_n)$ is said to have a local minimum at state $(\dot{x}_1, \dot{x}_2, \ldots, \dot{x}_n)$ if the inequality

$$f(\dot{x}_1, \dot{x}_2, \ldots, \dot{x}_n) \leq f(\dot{x}_1, \dot{x}_2, \ldots, \dot{x}_{i-1}, x_i, \dot{x}_{i+1}, \ldots, \dot{x}_n)$$

is satisfied for all $i = 1, 2, \ldots, n$. Show that the states with local minima of f coincide with the discriminant of the set of inequalities

$$(2x_i - 1)g_i(x_1, x_2, \ldots, x_{i-1}, x_{i+1}, \ldots, x_n) \leq 0$$

where $i = 1, 2, \ldots, n$, and where g_i is defined by the equation

$$f(x_1, x_2 \ldots, x_n) = x_i g_i(x_1, x_2, \ldots, x_{i-1}, x_{i+1}, \ldots, x_n) + h_i(x_1, x_2, \ldots, x_{i-1}, x_{i+1}, \ldots, x_n).$$

4.6 (a) Let $f_1 = x_2\bar{x}_1$ and $f = x_3x_2x_1 \vee \bar{x}_3\bar{x}_2 \vee x_3x_1$ for the circuit in Figure 4.26a. Find $f_2(x_1, x_2, x_3)$ and $f_3(x_1, x_2, x_3)$.

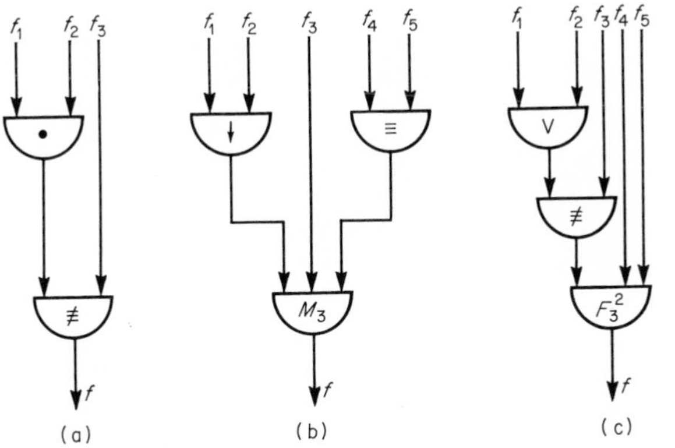

Figure 4.26 Illustration for Exercise 4.6.

(b) Let $f_1 = x_1 \vee \bar{x}_2 \bar{x}_3$, $f_3 = x_2 \bar{x}_3 \vee x_1 \vee \bar{x}_3 x_2$ and $f = x_1 \vee \bar{x}_2 \vee x_3$ for the circuit in Figure 4.26b. Find $f_2(x_1, x_2, x_3)$, $f_4(x_1, x_2, x_3)$, and $f_5(x_1, x_2, x_3)$.

(c) Let $f_1 = x_2 x_1 \vee x_3 \bar{x}_1$, $f_5 = x_3 x_1 \vee \bar{x}_3 x_2$ and $f = \bar{x}_3 \bar{x}_2 \bar{x}_1 \vee x_3 x_2 x_1$ for the circuit in Figure 4.26c. Find $f_2(x_1, x_2, x_3)$, $f_3(x_1, x_2, x_3)$ and $f_4(x_1, x_2, x_3)$.

4.7 Explain the meaning of

(a) an empty discriminant (all squares in the chart are blank).

(b) a full discriminant (all squares in the chart contain dots).

Reference Notations to Chapter 4

The first study of simultaneous logic equations was presented by Boole in 1854 [3]. Although many works on logic equations have been published since, applications of logic equations in problems concerning switching circuits and methods capable of their use in these applications have been suggested only recently [1, 6, 9, 12, 15, 16, 17, 18, 20, 21, 23, 25, 26]. Applicability of logic equations for a study of Boolean models in biology was demonstrated by Schoeffler and others [22]. A comprehensive presentation of algebraic methods for solution of simultaneous Boolean and pseudo-Boolean equations, inequalities, and their applications was presented by Hammer and Rudeanu [10].

The method used in this book for solution of logic and pseudo-logic relations is a direct generalization of the method proposed by Svoboda [23] for solution of simultaneous Boolean equations. Ledley [15] and Maitra [16] published similar methods.

Marin [17, 18] shows that a general method for the synthesis of memoryless switching circuits, which is independent of the types of elements used and the objective criteria, can be based on the solution of Boolean equations. The method is expected to be extremely efficient in the case that a general-purpose digital computer be supplemented by the Boolean analyzer—a special hardware operation unit designed by Svoboda [24]. These expectations are based on computer simulation of the Boolean analyzer done by Marin [17].

Minimization subject to reliability specifications was initiated by Gill [7] and further discussed by Lawler [14]. The problem of minimization with available functions was raised by Bell [2]. Givone and others [8, 13] discussed logic equations involving many-valued variables.

Applications of pseudo-Boolean equations and inequalities in linear and nonlinear integer mathematical programming have recently attracted considerable attention [10, 11, 19, 27].

Further references concerning Boolean and pseudo-Boolean equations and inequalities can be found in 10.

References to Chapter 4

1. **Bazilevskii, Yu. Ya.**, "Transformation and Solution of Logic Equations" (in Russian), *Voprosy Teorii Matem. Machines*, Vol. 2, Gos. Izd. Fiziko-matem. Lit, Moscow, 1962, pp. 107–121.
2. **Bell, N. R.**, "A Map Method for the Teaching of the Fundamental Concepts of Compound-Input Logic Circuits," *TE*, **E-11** (3) 173–177 (Sept. 1968).
3. **Boole, G.**, *An Investigation of the Laws of Thought*, Dover, New York, 1958. (Originally published by Macmillan in 1854.)
4. **Brown, F. M.**, "Reduced Solutions of Boolean Equations," *TC* **C-19** (10) 976–981 (Oct. 1970).
5. **Davio, M.**, and **J. P. Deschamps**, "Classes of Solutions of Boolean Equations," *Philips Research Report* **24** (5) 373–378 (Oct. 1969).
6. **Even, S.**, and **A. Meyer**, "Sequential Boolean Equations," *TC* **C-18** (3) 230–240 (March 1969).
7. **Gill, A.**, "Minimization of Contact Networks Subject To Reliability Specifications," *TC* **EC-9** (1) 122–123 (March 1960).
8. **Givone, D. D.**, et al., "A Method of Solution for Multiple-Valued Logic Expressions," *TC* **C-20** (4) 464–467 (April 1971).
9. **Grasselli, A.**, and **U. Montanari**, "On the Minimization of READ-ONLY Memories in Microprogrammed Digital Computers," *TC* **C-19** (11) 1111–1114 (Nov. 1970).
10. **Hammer, P. L.**, and **S. Rudeanu**, *Boolean Methods in Operations Research*, Springer-Verlag, New York, 1968.
11. **Inagaki, Y.**, and **T. Fukumura**, "Pseudo-Boolean Programming With Constraints," *Electronics and Communication in Japan* **50** (6) 26–34 (July 1967).
12. **Klir, G. J.**, "Solution of Systems of Boolean Equations" (in Czech), *Aplikace matematiky* **7** (4) 265–271 (1962).
13. **Klir, G. J.**, and **L. K. Seidl**, *Synthesis of Switching Circuits*, Iliffe, London, 1968, Gordon and Breach, New York, 1969.
14. **Lawler, E. L.**, "Minimization of Switching Circuits Subject to Reliability Conditions," *TC* **EC-10** (4) 781–782 (Dec. 1961).
15. **Ledley, R. S.**, *Digital Computer and Control Engineering*, McGraw-Hill, New York, 1960.
16. **Maitra, K. K.**, "A Map Approach to the Solution of a Class of Boolean Equations," *Communication and Electronics*, No. 59, pp. 34–36 (1962).
17. **Marin, M. A.**, "Investigation of the Field of Problems for the Boolean Analyzer," Ph. D. Dissertation, University of California at Los Angeles, Report No. 68-28, 1968.
18. **Marin, M. A.**, "On a General Synthesis Algorithm of Logical Circuits Using a Restricted Inventory of Integrated Circuits," Proc. of the Share ACM and IEEE Design Automation Workshop, Washington, D.C., July 1968.
19. **Mine, H.**, and **H. Narihisa**, "An Algorithm for Solving the Linear Programming Problem with Bivalent Variables by Boolean Algebra," Proc. Second Hawaii Int. Conf. on System Science, Western Periodicals Co., Honolulu, Jan. 1969, pp. 759–762.
20. **Phister, M.**, *Logical Design of Digital Computers*, John Wiley, New York, 1958.
21. **Rudeanu, S.**, "Boolean Equations and Their Applications to the Study of Bridge Circuits," *Bull. Math. de la Soc. Sci. Math. Phys. de la R. P. Roumaine* **3** (51), No. 4, pp. 445–473 (1959).

22. **Schoeffler, J. D.,** et al., "Identification of Boolean Mathematical Models," in M. D. Mesarovic, ed., *Systems Theory and Biology*, Springer-Verlag, New York; 1968.
23. **Svoboda, A.,** "An Algorithm for Solving Boolean Equations," *TC* **EC-12** (5) 557–558 (Oct. 1963).
24. **Svoboda, A.,** "Boolean Analyzer," Proc. of the IFIP Congress, Edinburgh, Scotland, August 1968; Booklet D, North-Holland Publ. Co., Amsterdam, 1968, pp. 97–102.
25. **Tomashpol'skiy, A. M.,** "Algorithm for Solution of a System of Logical Equations for Majorital Decomposition of Two-Valued Logic Functions," *Engineering Cybernetics*, No. 6, pp. 117–122 (1967).
26. **Wang, H.,** "Circuit Synthesis by Solving Sequential Boolean Equations," *Zeitschrift für Math. Logik und Grundlagen die Mathematik*, **5**, pp. 291–322 (1959).
27. **Yoshida, Y.,** et al., "Algorithms of Pseudo-Boolean Programming Based on the Branch and Bound Method," *Electronics and Communication in Japan* **50** (10) 277–285 (Oct. 1967).

5

LOGIC DESCRIPTION OF PHYSICAL SYSTEMS

5.1 INTRODUCTION

In Chapter 2, we presented the properties of a complete set of logic functions. We demonstrated that each logic function contained in some particular complete set of functions is a *primary function* of the corresponding logic algebra.

Conforming to current convention, we shall call the physical model of any primary logic function of some complete set a *logic element*.

In the same way in which we obtained some particular logic expression by combining logic variables with the aid of primary functions, we obtain a switching circuit—which is a model of the pertinent logic expression—by correspondingly combining suitable logic elements.

Theoretically, we can use any physical principle for the design of logic elements. However, since it is usually desirable (especially in the case of computers) that the logic elements be as small as possible and have fast response, only a few principles have found wide application. To date, these are primarily electrical, magnetic, and pneumatic principles. Optical, chemical, and possibly even biological principles appear promising.

In this chapter, we concentrate our attention upon the introduction of basic principles associated with the interface between *physical behavior* and *logic behavior*. These principles are illustrated by some classes of logic elements that have been frequently used in various fields. These are relays and elements involving the use of diodes, electronic tubes, transistors, ferrite cores, superconductive devices, and pneumatic components. We discuss chiefly some of their properties which are directly associated with their applications in switching circuits. Their physical properties are described only briefly.

On every logic element we distinguish *input signals*, *output signals*, and *internal signals*. The following conditions must be satisfied: (a) the input signals must be independent of both the internal and output signals; (b) the internal signals must be independent of the output signals.

The signal is formed in such a manner that information is assigned to a certain measurable physical quantity (voltage, contact position, current, air pressure, etc.), in our case by means of a proposition which divides the values of the physical quantity into two groups. To each proposition formed in this manner we assign a logic variable. The value of the assigned logic variable is given by the truth value of the proposition concerning the signal. The truth values of the propositions relating to the internal and external signals are determined by the state of the logic element.

5.2 RELAYS

First, let us present the common *polarized relay*. We shall assume the relay to carry *two windings*, W_1 and W_2, two rest positions of the *armature*, left and right, and two contacts (one *make* and one *break contact*). The elementary input signals are represented by the voltages across the windings, the internal signal by the position of the armature, and the elementary output signals by the contact positions.

The *physical process* proceeding in the polarized relay is as follows: When a sufficiently large voltage is applied to winding W_1 of the polarized relay, the current flowing through the winding produces a magnetic flux which causes the armature to move to the right-hand rest position without regard to its prior position. It reaches the right-hand rest position not later than after a period Δt provided that the voltage at W_1 is sufficiently large during the whole period of Δt. Similarly, a voltage applied to winding W_2 causes the armature to move to its left-hand rest position. If the voltages are applied to both windings simultaneously, the position of the armature remains unchanged. When the armature is in its right-hand position, the break contact is closed and the make contact is open.

The *signals* in a polarized relay can be defined, for instance, by the following set of propositions:

$x_1(t)$: Within the time interval $t_0 \le t < t_0 + \Delta t$, the voltage across winding W_1 is sufficient to cause the relay armature to move to the right-hand rest position.

$x_2(t)$: Within the time interval $t_0 \le t < t_0 + \Delta t$, the voltage across winding W_2 is sufficient to cause the relay armature to move to its left-hand rest position.

$z(t_0)$: At the instant t_0, the armature is in its left-hand rest position.

$y_1(t_0)$: At the instant t_0, the break contact is closed.

$y_2(t_0)$: At the instant t_0, the make contact is closed.

Δt must be such an interval of time to guarantee that the effect of the input signals becomes manifested.

The relations between the logic variables that correspond to the aforesaid propositions are expressed in Table 5.1, where the truth of a given proposi-

TABLE 5.1. Relations Between the Logic Variables Corresponding to the Signals in a Polarized Relay

$z(t_0)$	$x_2(t)$	$x_1(t)$	$y_1(t_0)$	$y_2(t_0)$	$z(t_0+\Delta t)$
0	0	0	1	0	0
0	0	1	1	0	0
0	1	0	1	0	1
0	1	1	1	0	0
1	0	0	0	1	1
1	0	1	0	1	0
1	1	0	0	1	1
1	1	1	0	1	1

tion is denoted by the symbol 1 and its falsity by the symbol 0. On the basis of this table we can easily construct the following algebraic expressions:

$$\begin{aligned} y_1(t_0) &= \bar{z}(t_0) \\ y_2(t_0) &= z(t_0) \\ z(t_0 + \Delta t) &= z(t_0)[\bar{x}_1(t) \vee x_2(t)] \vee \bar{x}_1(t)x_2(t) \end{aligned} \tag{5.1}$$

The future value of the internal signal is thus a function of the present value of the internal signal and of the present values of the input signals. Moreover, it is evident that the value of the output signals is a function of the value of the internal signal. In the general case, we have

$$\begin{aligned} \mathbf{z}(t_0 + \Delta t) &= \mathbf{f}[\mathbf{x}(t), \mathbf{z}(t_0)] \\ \mathbf{y}(t_0) &= \mathbf{g}[\mathbf{x}(t), \mathbf{z}(t_0)] \end{aligned} \tag{5.2}$$

where **x**, **y**, and **z** are vectors. This pair of functions (5.2) will be called the *operator* of the corresponding logic element.

It is often possible to simplify (5.2) by introducing additional assumptions concerning the structure of the switching circuit. For instance, assume that at the beginning of each operating time specified by an external source of synchronizing pulses, the armature of the polarized relay is always in its right-hand rest position. Then, $z(t_0) = 0$ and the last equation in (5.1) assumes the form

$$z(t + \Delta t) = \bar{x}_1(t)x_2(t)$$

The internal variable z can thus be eliminated from equations (5.1). We obtain

$$\begin{aligned} y_1(t_0 + \Delta t) &= \overline{\bar{x}_1(t)x_2(t)} = x_1(t) \vee \bar{x}_2(t) \\ y_2(t_0 + \Delta t) &= \bar{x}_1(t)x_2(t) \end{aligned} \tag{5.3}$$

If the time delay can be neglected, (5.3) becomes

$$\begin{aligned} y_1 &= x_1 \vee \bar{x}_2 \\ y_2 &= \bar{x}_1 x_2 \end{aligned} \tag{5.4}$$

Operator (5.4) is the one usually employed to describe the polarized relay.

If the aforesaid propositions concerning the signals of a polarized relay were subjected to a more rigorous examination, it would appear that concepts such as "sufficient voltage," "closed contact" etc., should be defined. That is to say, we are concerned here with a very coarse classification of the values of continuous quantities which need not change monotonically with time (e.g., the armature or contacts may vibrate). Thus, it might appear that the propositions should be improved upon, supplemented, or replaced by others (e.g., substituting the proposition "the contact is closed" by the proposition "the contact has a contact resistance lower than 0.1 Ω"). However, we do not intend to treat this aspect. The task of the designer of switching circuits is to conceive the logic element proper as well as the auxiliary circuits (e.g., sources of synchronizing pulses, power supplies, etc.) so that the corresponding signals can be regarded as two-valued.

The signals need not all be expressed by the same physical quantities. However, it should be possible to transform the output signals of one logic element into the input signals of another element in a simple manner. In a relay this is done by applying to one side of the contact an electric potential which, when the contact is closed, is brought to its other side and hence to other contacts until it is finally conducted to one end of the relay winding which is connected to the output of the contact circuit.

When the input quantities are changed, a *transient physical process* takes place in the logic element during the period Δt and terminates in a new steady state. As we shall see later, the time interval Δt may be functionally utilized. It can be determined by external synchronizing pulses, in which case it is substantially longer than the reaction time proper. We then speak of the *synchronous* connection of logic elements. If the time interval Δt is determined solely by the proper physical action of the logic element, which is generally different for different logic elements (owing to tolerances in the technical parameters), we speak of the *asynchronous* connection of logic elements.

Equations (5.2) are steady-state equations, i.e., they do not tell us anything about transients. As a rule, this is sufficient for the design of switching circuits, since these are mostly arranged so as to prevent transient phenomena from taking effect. The problems that crop up when transient states must also be considered are discussed in Chapter 9.

A *neutral relay* consists of an electromagnet* which carries several windings, all connected, as a rule, in the same sense. The following physical events proceed in the neutral relay:

The electromagnet acts upon the armature which, if we disregard the transient process, can take up two positions: the rest position, when it is not attracted to the core of the electromagnet, and the working position, when it is attracted. The armature moves to its rest position (or remains in it) only if no sufficient voltage is applied to any of the relay windings. It moves to the working position whenever a sufficiently large voltage is applied to at least one of the windings. The magnitude of the voltage required to attract the armature depends, of course, on the construction of the particular relay. The armature acts mechanically upon a set of contacts which it either closes or opens. Its position thus characterizes the state of the relay.

The signals in a neutral relay can be defined, for instance, by the following set of propositions:

$x_i(t)$: Within the time interval $t_0 \leq t < t_0 + \Delta t$, the voltage across winding $W_i (i = 1, 2, \ldots, n)$ is sufficient to cause the relay armature to be attracted (i.e., to move to its working position).

$z(t_0)$: At the instant t_0 the relay armature is attracted to the core (it is in its working position).

$y_i(t_0)$: At the instant t_0, the contact K_j is closed (it is conducting).

* Besides the electromagnetic type, relays based on other principles, e.g., piezoelectric, magnetostrictive, electrostatic, electrodynamic, etc., are also known. However, they are not widely used and therefore we shall not discuss them here.

It would be easy to prove that the operator of the neutral relay has the form

$$\begin{aligned} z(t_0 + \Delta t) &= x_1(t) \vee x_2(t) \vee \cdots \vee x_n(t) \\ y_j(t_0) &= \bar{z}(t_0) \text{ for break contacts} \\ y_j(t_0) &= z(t_0) \text{ for make contacts} \end{aligned} \tag{5.5}$$

Eliminating the internal variable and neglecting the time relations we obtain

$$\begin{aligned} y_j &= \bar{x}_1\bar{x}_2 \cdots \bar{x}_n \quad \text{for the break contact} \\ y_j &= x_1 \vee x_2 \vee \cdots \vee x_n \quad \text{for the make contact} \end{aligned} \tag{5.6}$$

It would even be possible to consider a more general example, in which we would assume each winding to have a different effect upon the operation of the armature, each effect being expressed by a separate coefficient. In such a case we would express the total effect of all the windings as the sum of these coefficients multiplied by the corresponding values of the voltage across the individual windings. The operator of a relay considered in this manner is a threshold function.

Exercises to Section 5.2

*5.2-1 Determine the operator of a neutral relay with two windings, where the winding W_2 is connected via a make contact of the same relay to a signal source to which the logic variable v is assigned. The relay has two make and one break contacts.

5.2-2 Determine the operator of a logic element consisting of two neutral relays, each carrying one winding. A signal (with the variable v_1 assigned to it) is applied to the winding of relay 1 via the break contact of relay 2; in a similar manner, another signal (with the variable v_2 assigned to it) is applied to the winding of relay 2 via the make contact of relay 1. Each relay has one make and one break contact.

5.3 DIODES

In these elements we utilize the nonlinear relationship between voltage and current in semiconductor diodes. With the voltage applied with a particular polarity, the semiconductor diode passes a large current since its resistance is only a few tens of ohms. In this case we speak of the forward direction. With the opposite polarity of the voltage, only a small current is passed in the opposite direction (the reverse direction), since in this direction the resistance of the diode is several hundred kiloohms.

The two basic circuit arrangements of logic elements made up of diodes and resistors are shown in Figure 5.1. In the first case (Figure 5.1a) the

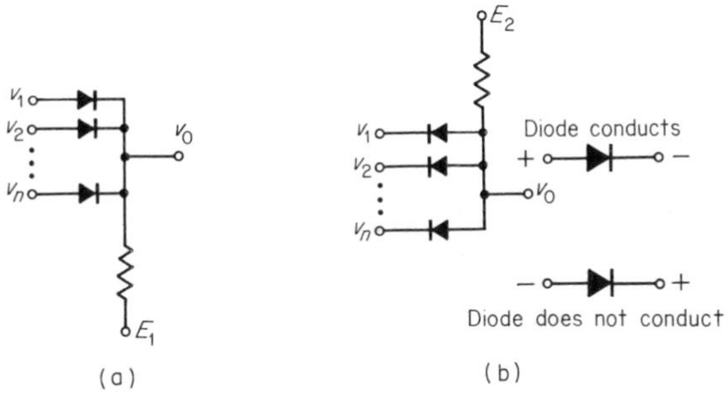

value of the voltage E_1 must be smaller than, or at the most equal to, any of the input voltages; the output voltage is approximately equal to the highest of all the input voltages. Conversely, in the second case (Figure 5.1b) the value of the voltage E_2 must be larger than any of the input voltages and the output voltage V is approximately equal to the lowest input voltage.

We assume that in the steady state the input voltages applied to the circuit are two-valued, i.e., each of them is either at a low level (L) or at a high level (H). Similarly, the current in resistor R is either low (L) or high (H).

In the diode logic element we can define the signals, for instance, by the following set of propositions:

$x_i(t)$: Within the time interval $t_0 \leq t < t_0 + \Delta t$, the input voltage V_i ($i = 1, 2, \ldots, n$) has a high level (H).
$z(t_0)$: At the time t_0, the current in resistor R is high (H).
$y(t_0)$: At the time t_0, the output voltage V_0 is at a high level (H).

In each of the propositions we can replace the word "high" by the word "low" changing thus the type of the proposition. If propositions for z and y are of the same type, we obtain four possible sets of propositions defining the signals in diode logic elements. We shall denote them by HH, HL, LH, and LL, the first letter referring to the internal and output signals, the second letter to the input signals.

For the given set of propositions (i.e., HH) and for the circuit arrangement of Figure 5.1a we obtain an operator of the form

$$\begin{aligned} y(t_0) &= z(t_0) \\ z(t_0 + \Delta t) &= x_1(t) \vee x_2(t) \vee \cdots \vee x_n(t) \end{aligned} \tag{5.7}$$

For the same set of propositions, but with the circuit arrangement of Figure 5.1b, we obtain an operator of the form

$$\begin{aligned} y(t_0) &= z(t) \\ z(t_0 + \Delta t) &= x_1(t)x_2(t) \cdots x_n(t) \end{aligned} \tag{5.8}$$

In a manner similar to relay elements we can here also often neglect the time relations and, in eliminating the internal variable, we obtain the operator

$$y = x_1 \vee x_2 \vee \cdots \vee x_n \tag{5.9}$$

or the operator

$$y = x_1 x_2 \cdots x_n \tag{5.10}$$

The simplified operators for both types of circuit arrangement and for all sets of propositions are summarized in Table 5.2.

TABLE 5.2. Operators of Diode Logic Elements

	Circuit Diagram of Figure 5.1a	Circuit Diagram of Figure 5.1b
HH	$y = x_1 \vee x_2 \vee \cdots \vee x_n$	$y = x_1 x_2 \cdots x_n$
HL	$y = \bar{x}_1 \vee \bar{x}_2 \vee \cdots \vee \bar{x}_n$	$y = \bar{x}_1 \bar{x}_2 \cdots \bar{x}_n$
LH	$y = \bar{x}_1 \bar{x}_2 \cdots \bar{x}_n$	$y = \bar{x}_1 \vee \bar{x}_2 \vee \cdots \vee \bar{x}_n$
LL	$y = x_1 x_2 \cdots x_n$	$y = x_1 \vee x_2 \vee \cdots \vee x_n$

Exercises to Section 5.3

*5.3-1 Using the set of propositions HL (or LH), the element arranged according to Figure 5.1a (or 5.1b) can be described by an operator which by itself forms a complete set of logic functions. Is it possible to model any Boolean function by means of such elements? If not, explain the contradiction.

*5.3-2 Derive all possible operators for the two-circuit arrangements of Figure 5.1, assuming that the voltage E_1 (or E_2) is regarded as a two-valued input signal.

5.3-3 Determine all logic functions which can be represented by either of the connections of diode elements shown in Figure 5.2.

5.4 ELECTRONIC TUBES

Logic description of systems containing electronic tubes will be illustrated with examples of circuits comprising vacuum triodes.

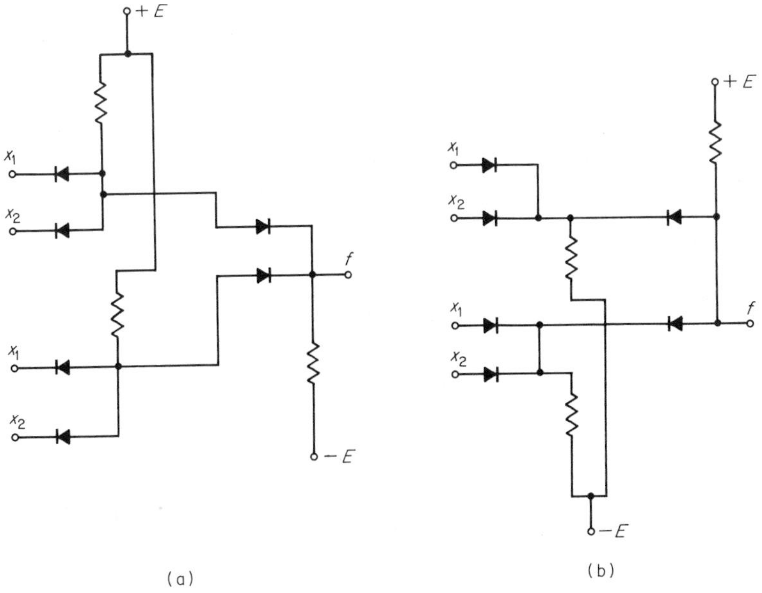

Figure 5.2 Diode logic circuits modeling the logic function given in Example 5.2.

Let us define signals associated with triodes by the following set of propositions:

- $x_i(t)$: During the time interval $t_0 \leq t < t_0 + \Delta t$ there is a positive voltage of sufficient amplitude across the input i.
- $y_j(t_0)$: At time instant t_0, there is a positive voltage of sufficient amplitude across the output j.
- $z_k(t_0)$: At time instant t_0, a sufficiently large current flows through the triode k.
- $\delta x_i(t_0)$: At time instant t_0, there is a positive change of voltage of sufficient amplitude and slope across the input i.
- $\delta y_j(t_0)$: At time instant t_0, there is a positive change of voltage of sufficient amplitude and slope across the output j.

One of the basic circuit arrangements of a triode logic element is shown in Figure 5.3. The choice of the circuit parameters of the element $(R_1, R_2, R_3, E_a, E_g)$ will not be discussed here and we shall proceed directly to derive its operator.

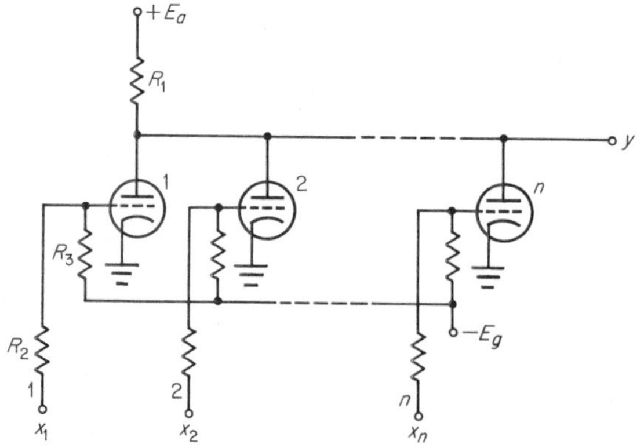

Figure 5.3 Example of a triode logic element.

In the case where a positive voltage is applied to the grid of triode i, a current of sufficient intensity flows through the triode. The output voltage is positive only in the case where none of the triodes passes current.

The operator of the element under consideration has the form

$$\begin{aligned} z_1(t_0 + \Delta t) &= x_1(t) \\ z_2(t_0 + \Delta t) &= x_2(t) \\ &\vdots \\ z_n(t_0 + \Delta t) &= x_n(t) \\ y(t_0) &= \bar{z}_1(t_0)\,\bar{z}_2(t_0) \cdots \bar{z}_n(t_0) \end{aligned} \tag{5.11}$$

Neglecting the time relations and eliminating the internal variables we obtain

$$y = \bar{x}_1 \bar{x}_2 \bar{x}_3 \cdots \bar{x}_n \tag{5.12}$$

Figure 5.4 illustrates an electronic triode logic element with a common resistor in the cathode circuit. A positive voltage appears across the output only if at least one of the triodes conducts. This element is described by the operator

$$\begin{aligned} z_i(t_0 + \Delta t) &= x_i \quad (i = 1, 2, \ldots, n) \\ y(t_0) &= z_1(t_0) \vee z_2(t_0) \vee \cdots \vee z_n(t_0) \end{aligned} \tag{5.13}$$

Neglecting the time relations and eliminating the internal variables, we obtain the operator

$$y = x_1 \vee x_2 \vee \cdots \vee x_n \tag{5.14}$$

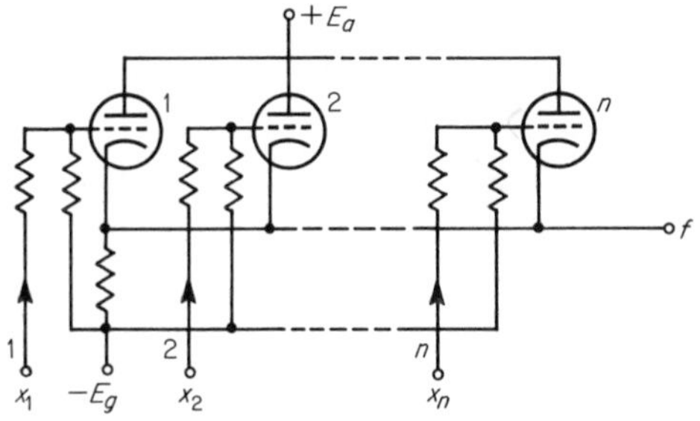

Figure 5.4 Electronic tube logic element with common resistor in the cathode circuit.

Figure 5.5 shows the circuit diagram which represents a threshold logic function. Resistances $R_1, R_2, \ldots, R_n$ are proportional to the weights, the type of the tube and resistances R_g, R_a affect the threshold.

An important instance is the *bi-stable logic element*. One of its possible circuit arrangements is presented in Figure 5.6.

The bi-stable logic element has two stable states, for which the relation $z_1(t_0) = \bar{z}_2(t_0)$ holds true. In the explanation to follow we shall assume that in the first stable state a current of sufficient intensity flows through triode 1, whereas insufficient current passes through triode 2, i.e., $z_1 = 1$ and $z_2 = 0$. In the second stable state, conditions are reversed, i.e., $z_1 = 0$, $z_2 = 1$. We shall further assume that signals of constant amplitude (i.e., those corresponding to the variables x_1 and x_2) are never applied to the input of the element simultaneously with step-function variables (δx_1, δx_2, and δx_3).

A positive voltage of sufficient amplitude, applied to input 1 (or 2), always switches the bi-stable element to the first (or, respectively, second) state. These voltages must not be applied to both inputs simultaneously.

In a similar manner, a positive voltage jump of sufficient amplitude and slope, when applied to input 1′ (or 2′), brings the element to the first (or, respectively, second) state. The simultaneous application of a positive

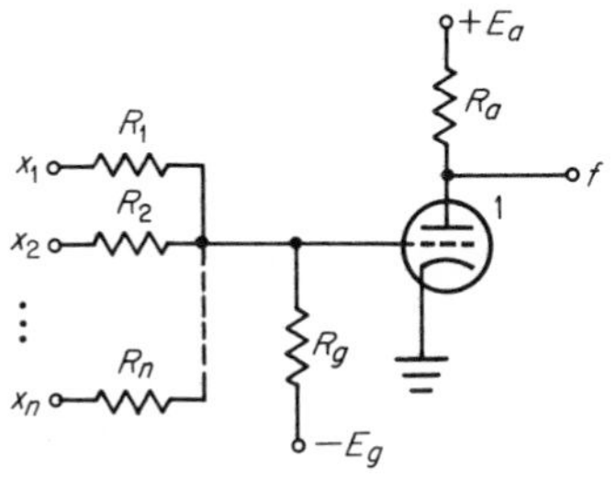

Figure 5.5 Threshold logic element using electronic tubes.

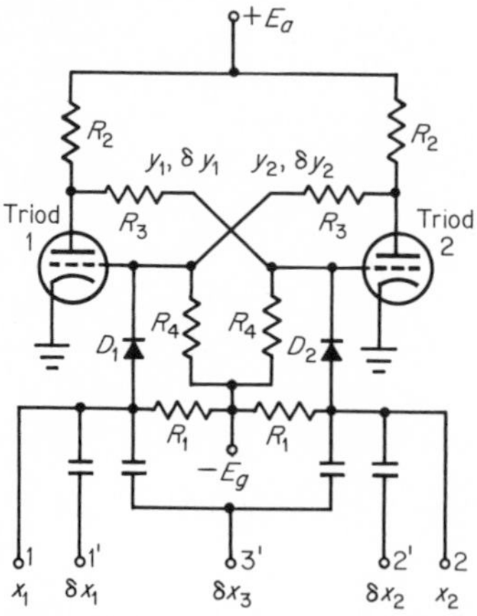

Figure 5.6 Examples of bi-stable logic element using electronic tubes.

voltage jump of sufficient amplitude and slope to both triode inputs 1′ and 2′, or an application of the same signal to the input 3′, will bring the element to the state opposite to that which it was in before the arrival of the signal.

For the input variables x_1 and x_2, the operator of the bi-stable logic element has the form

$$\begin{aligned} z_1(t_0+\Delta t) &= x_1(t) \vee \bar{x}_2(t) z_1(t_0) \\ z_2(t_0+\Delta t) &= x_2(t) \vee \bar{x}_1(t) z_2(t_0) = \bar{z}_1(t_0+\Delta t) \\ y_1(t_0) &= z_2(t_0) = \bar{z}_1(t_0) \\ y_2(t_0) &= z_1(t_0) = \bar{z}_2(t_0) \end{aligned} \tag{5.15}$$

together with the following condition imposed upon the input variables,

$$x_1(t)x_2(t) = 0 \tag{5.16}$$

If the variables δx_1, δx_2, δx_3 are taken into consideration instead of the variables x_1 and x_2, we obtain an operator of the form

$$\begin{aligned} z_1(t_0+\Delta t) &= \delta x_3(t_0)\bar{z}_1(t_0) \vee \overline{\delta x_3}(t_0)[z_1(t_0)\,\overline{\delta x_2}(t_0) \vee \bar{z}_1(t_0)\,\delta x_1(t_0)] \\ z_2(t_0+\Delta t) &= \delta x_3(t_0)\bar{z}_2(t_0) \vee \overline{\delta x_3}(t_0)[\bar{z}_2(t_0)\,\delta x_2(t_0) \vee z_2(t_0)\,\overline{\delta x_1}(t_0)] \\ \delta y_1(t_0+\Delta t) &= z_1(t_0)[\delta x_2(t_0) \vee \delta x_3(t_0)] \\ \delta y_2(t_0+\Delta t) &= z_2(t_0)[\delta x_1(t_0) \vee \delta x_3(t_0)] \end{aligned} \tag{5.17}$$

In synchronous switching circuits a synchronous *delay element* is frequently used besides bi-stable elements. This delay element is characterized by having only one input and one output, and its output signal, which is derived from the input signal, is in phase with the synchronizing pulses. Moreover, the output signal is required to be delayed and to have a standard waveform even if the input signal is deformed in shape and phase within admissible limits.

The time diagram in Figure 5.7 illustrates the action of a delay element which produces a delay of one time interval. Figure 5.7a shows synchronizing pulses, Figure 5.7b standard input signals, and Figure 5.7c an example of actual (deformed) input signals. We see that the output signals of the delay element, illustrated in Figure 5.7d, are of standard waveform and that they are delayed by one time interval relative to the undeformed input signals.

The operator of a delay element which produces a delay of k time intervals contains k internal variables z_1 to z_k, one input variable and, as a rule one output variable. It has the following form:

$$
\begin{aligned}
z_1(t_0 + \Delta t) &= x(t) \\
z_2(t_0 + \Delta t) &= z_1(t_0) \\
&\vdots \\
z_k(t_0 + \Delta t) &= z_{k-1}(t_0) \\
y(t_0) &= z_k(t_0)
\end{aligned}
\tag{6.18}
$$

In this case, the time relations are essential so that they cannot be neglected.

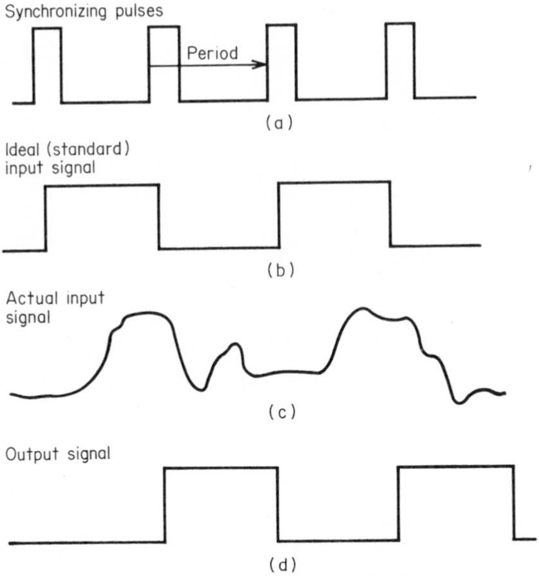

Figure 5.7 Diagram illustrating the action of a delay element.

However, if we eliminate all the internal variables, we obtain the operator

$$y(t_0 + k\,\Delta t) = x(t) \tag{5.19}$$

Exercises to Section 5.4

*5.4-1 Compare the operator (5.1) of a polarized relay with the operator (5.15) of a bi-stable circuit using signals of constant amplitude, and point out their similarities and differences.

5.4-2 A standard Boolean form is given, and we are to model it by means of the logic element shown in Figure 5.3. Formulate a simple rule by means of which it would be possible to derive directly from the given expression and in a simple manner: (a) the number of logic elements, (b) the number of electronic tubes required.

5.4-3 Use operator (5.11) to determine the operator of the elements shown in Figure 5.8.

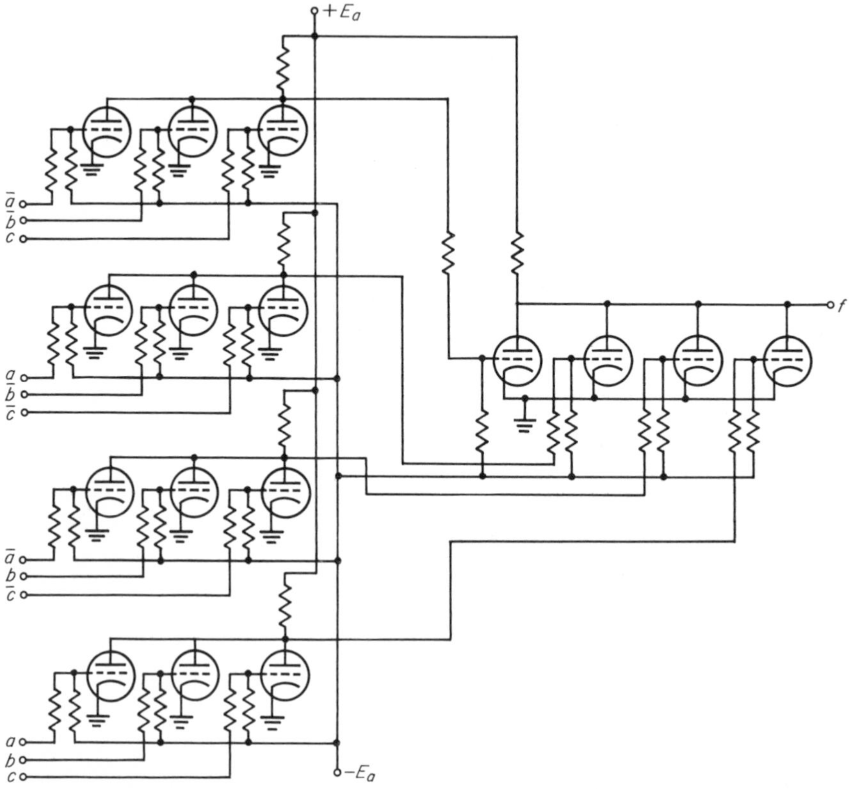

Figure 5.8 Electronic tube circuit modeling the logic function given in Example 5.3.

5.5 TRANSISTORS

In our exposition in this section we assume the reader to be acquainted with the fundamental physical properties of the transistor.

The basic circuits of transistor logic elements are analogous to those employing vacuum triodes. However, by making use of the special properties of transistors (utilizing the complementary properties of *p-n-p* and *n-p-n* transistors, the small differences between the potentials of the individual electrodes, etc.) we obtain additional logic elements which cannot be realized with the aid of electronic tubes.

Our further discussion is based on the following four sets of propositions defining the signals.

Set A

$x_i(t)$: During the interval $t_0 \leq t < t_0 + \Delta t$ a positive voltage of sufficient amplitude exists at input i.

$y_j(t_0)$: At time instant t_0, there is a positive voltage of sufficient amplitude at the output j.

$z_k(t_0)$: At time instant t_0 a current of sufficient intensity flows through the collector junction of the transistor k.

Set B

This is identical with set A except that, in the proposition $x_i(t)$ and $y_j(t)$, the concept of "positive voltage" is replaced by that of "negative voltage."

Set A'

$\delta x_i(t_0)$: At time instant t_0 there occurs a positive change of voltage of sufficient amplitude and slope at the input i.

$\delta y_i(t_0)$: At time instant t_0 there occurs a positive change of voltage of sufficient amplitude and slope at the output j.

$z_k(t_0)$: At time instant t_0, a current of sufficient intensity flows through the collector junction of the transistor.

Set B'

This is identical with set A′ except that, in the propositions $\delta x_i(t_0)$ and $\delta y_j(t_0)$, the concept of "positive change" is replaced by that of "negative change."

Set A (and A', respectively) applies to elements employing *n-p-n* transistors, set B (and B', respectively) to elements employing *p-n-p* transistors. When dealing with logic elements in which both types of transistors are utilized, it must be explicitly stated which set of propositions is to be used.

Figure 5.9 presents a survey of transistor elements, some of which correspond to the electronic tube elements described in Section 5.4. The operators of these elements, together with an indication of the type of proposition employed, are summarized in Table 5.3. Figure 5.9 shows only elements

TABLE 5.3. Summary of Operators for Transistor Logic Elements

Circuit Diagram of Figure 5.9	Operator	Set of Propositions
(a)	(5.11)	A
(b)	(5.13)	A
(c)	$z(t+\Delta t)=\{m, m+1, \cdots, n\}(x_1(t), x_2(t), \ldots, x_n(t))$ $y(t)=\bar{z}(t)$	A
(d)	$z_i(t+\Delta t)=x_i(t), \quad i=1, 2, \ldots, n$ $y(t)=z_1(t)z_2(t)\cdots z_n(t)$	A
(e)	(5.11)	A
	(5.15)	A
(f)	(5.17)	A'

with *n-p-n* transistors. Elements with *p-n-p* transistors have the same circuit diagram, but differ by the polarity of the voltages in all the nodes of the circuit and by the set of propositions which define the signals (either the propositions B or B' being employed).

Let us note that, for instance, the transistors in the elements shown in Figure 5.9d and 5.9e are connected in the same manner as the relay contacts in a contact circuit. This is made possible by special types of transistors, for which the voltage between collector and emitter is smaller than the base voltage required for a current of sufficient intensity to flow through the transistor. An example of such an element is the circuit shown in Figure 5.10. There is a positive voltage of sufficient amplitude at the output 7 only if no current of sufficient intensity flows through transistor T_6, or through both the transistors T_4 and T_5, or through any of the transistors T_1, T_2, and T_3. This condition can be expressed by the relation

$$y = \bar{x}_6 \vee \bar{x}_4 \bar{x}_5 \vee \bar{x}_3 \bar{x}_2 \bar{x}_1 \tag{5.20}$$

which is the simplified operator of the circuit (with the internal variables eliminated and the time relations neglected).

Figure 5.11 represents a transistor logic element employing current control. The collector voltage swing is very much smaller than in the foregoing elements. The diodes D_1 serve to limit this amplitude, and the diodes D_2 and D_3 serve to shift the dc levels so as to enable such elements to be directly

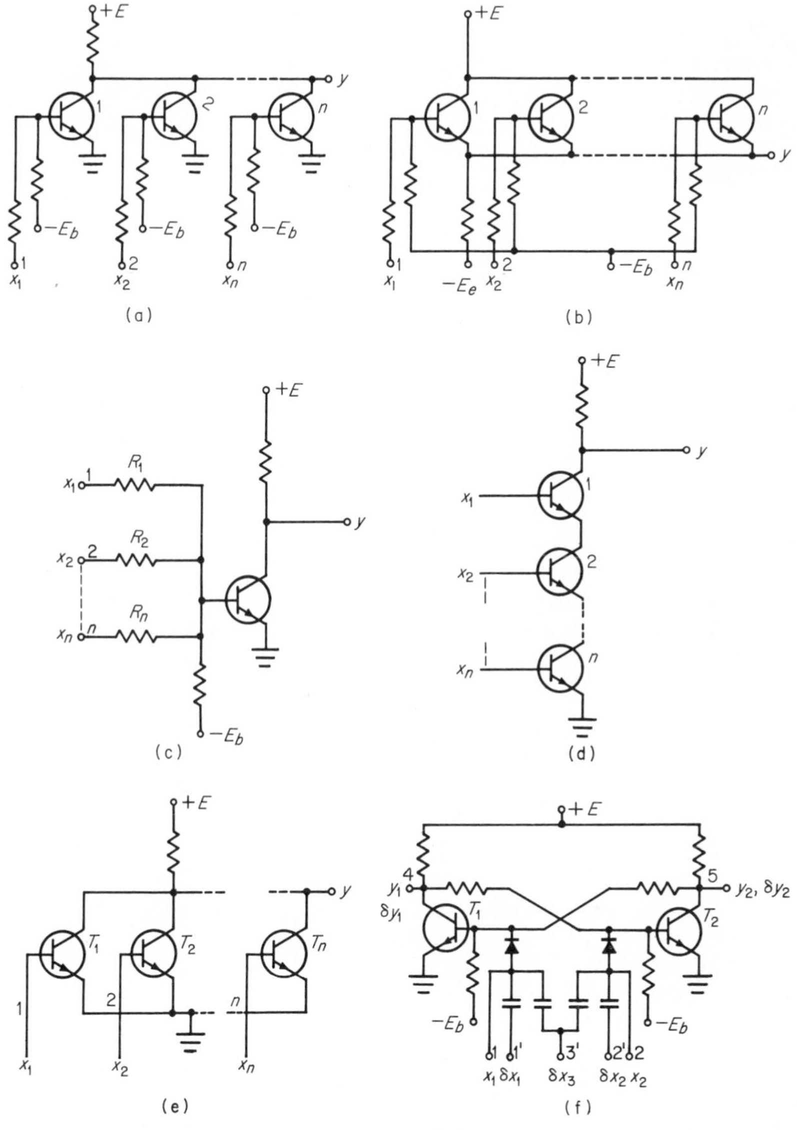

Figure 5.9 Survey of basic transistor logic elements.

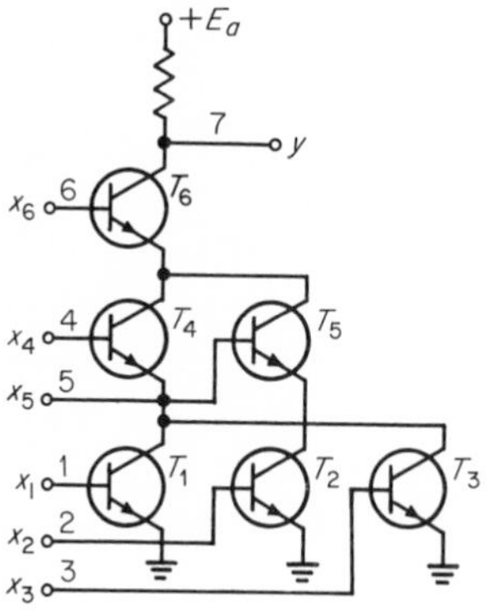

Figure 5.10 Example of circuit using transistor logic elements.

interconnected. These elements have two outputs, and their simplified operator (with the time relations neglected and the internal variables eliminated) has the form

$$\begin{aligned} y_1 &= \bar{x}_1 \bar{x}_2 \bar{x}_3 \cdots \bar{x}_n \\ y_2 &= \bar{y}_1 = x_1 \vee x_2 \vee \cdots \vee x_n \end{aligned} \tag{5.21}$$

Exercises to Section 5.5

*5.5-1 What change do the operators of the transistor elements presented in Table 5.3 for *n-p-n* transistors undergo if we employ the set of propositions B or B'?

*5.5-2 What change does the operator of the transistor element shown in Figure 5.9b undergo if we regard the voltages $+E$ and $-E_b$ as further input signals?

5.5-3 How will the transistor logic element shown in Figure 5.9f operate if we connect

(a) output 5 with input 2′.

(b) output 5 with input 2′ and output 4 with input 1′.

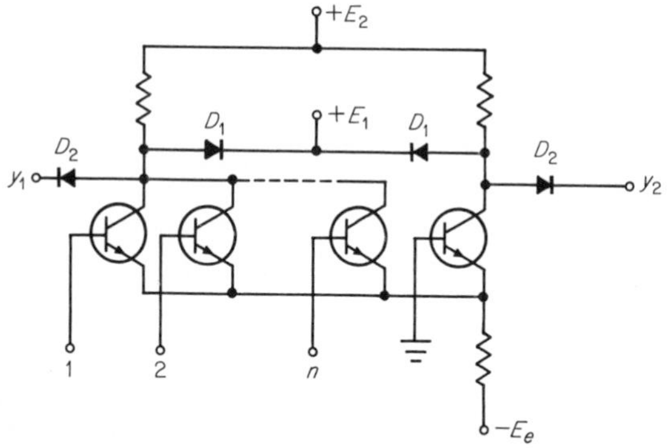

Figure 5.11 Current-controlled transistor logic element.

5.6 MAGNETIC CORES

An essential part of magnetic logic elements is a toroidal core of magnetically soft material having an approximately square hysteresis loop and carrying several windings. Each winding can be wound in either of two senses, which we shall denote by N and P. The steady-state magnetic flux through the core is always non-zero and has either of two directions corresponding to the magnetic states of the core, $+B$ and $-B$. The transition from one magnetic state of the core to the other is called core switching.

In magnetic logic elements the signals are defined by the following set of propositions:

$^0x_i(t)$: During the time interval $t_0 \leq t < t_0 + \Delta t$, the positive voltage applied to the input i that is connected to a winding of sense N has a magnitude sufficient to switch the core to the magnetic state $-B$.

$^1x_j(t)$: During the time interval $t_0 \leq t < t_0 + \Delta t$, the positive voltage applied to the input j that is connected to a winding of sense P has a magnitude sufficient to switch the core to the magnetic state $+B$.

$y_k(t_0)$: At time instant t_0, the positive voltage across the output k is of sufficient magnitude.

$z(t_0)$: At time instant t_0, the core is in the state $+B$.

A voltage pulse can be obtained across the output in two ways:

(a) By switching the core to the state $-B$ with the aid of an auxiliary source of pulse voltage, connected in series with the output winding. Provided that the core was in the $+B$ state, the output winding has an impedance so large that the output terminals do not carry a sufficient positive voltage. If the core was in the $-B$ state, the output winding has, on the contrary, an impedance so small that a sufficiently high voltage (essentially the voltage of the source) appears across the output terminals. Thus, a signal appears at the output only if the core was not switched over to the $+B$ state.

(b) By switching the core to the $-B$ state with the aid of a current pulse supplied by a source of current. However, for the sake of uniformity we shall replace the current source by a voltage source with a resistance connected in series. Then it will be possible in this case also to use the aforementioned set of propositions defining the signal. A positive voltage of sufficient amplitude is then induced across the ouput terminals only when the core is being switched from state $+B$ to $-B$.

In principle, the core may carry any number of input windings, the effects of which add up algebraically (the effect of a winding of sense N being considered with a minus sign). However, in order to ensure reliable operation, only the following combinations of input windings are used in the majority of cases: (a) any number of windings of sense P; (b) one winding of sense N and one winding of sense P.

Two of the most frequently employed circuit arrangements of magnetic core elements are shown in Figure 5.12 where N-type windings are marked by a dot. The diode (or transitor, respectively) acts here as a component which prevents the cores connected to the output from reacting upon the winding. The RC (or RCL) network serves to delay the output signal. Without this network the output signal would prematurely affect the next core.

Magnetic logic elements are chiefly employed in synchronous switching circuits. Their operating cycle is divided into two successive time intervals—the preparatory and the working interval. At the beginning of the preparatory interval ($t_0 \leq t < t_0 + \Delta t$) the core is in the $-B$ state and during this interval it is either switched over or remains in its original state according to the kind of signal applied. During the working interval ($t_0 + \Delta t \leq t + \Delta t < t_0 + 2\Delta t$), the core is brought to the $-B$ state again by the auxiliary voltage

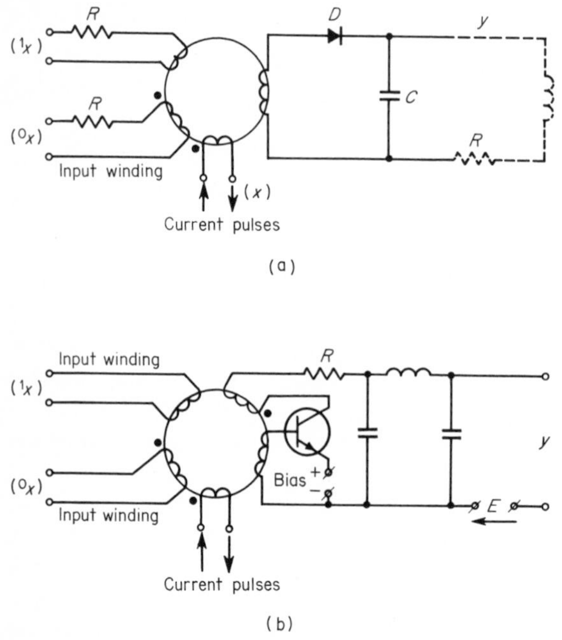

Figure 5.12 Most frequently used magnetic core logic elements: (a) with diode, (b) with transistor.

source (the variable which corresponds to this voltage being denoted by x), and across the output terminals of the element there appears a signal corresponding to the variable y. The operator of magnetic elements depends on the sense of the windings and whether current or voltage sources are used:

$$
\begin{aligned}
&PN\text{, current operated: } && z(t_0 + \Delta t) = z(t_0)\overline{{}^0x(t)} \vee \overline{{}^0x(t)} \cdot {}^1x(t) \\
& && y(t_0) = x(t) \cdot z(t_0) \\
&PN\text{, voltage operated: } && z(t_0 + \Delta t) = z(t_0)\overline{{}^0x(t)} \vee \overline{{}^0x(t)} \cdot {}^1x(t) \\
& && y(t_0) = \overline{z(t_0)} \cdot x(t) \\
&PP\text{, current operated: } && z(t_0 + \Delta t) = z(t_0) \vee {}^1x_1(t) \vee {}^1x_2(t) \\
& && y(t_0) = x(t) \cdot z(t_0) \\
&PP\text{, voltage operated: } && z(t_0 + \Delta t) = z(t_0) \vee {}^1x_1(t) \vee {}^1x_2(t) \\
& && y(t_0) = \overline{z(t_0)} \cdot x(t)
\end{aligned}
\tag{5.22}
$$

On introducing the conditions $z(t_0) = 0$, $x(t + \Delta t) = 1$, eliminating the internal variable z and neglecting the time relations, these operators acquire the form

$$
\begin{aligned}
y &= {}^1x_1 \cdot {}^0\bar{x}_1 && (PN\text{, current switching}) \\
y &= {}^1\bar{x}_1 + {}^0x_1 && (PN\text{, voltage switching}) \\
y &= {}^1x_1 + {}^1x_2 && (PP\text{, current switching}) \\
y &= {}^1\bar{x}_1 \cdot {}^1\bar{x}_2 && (PP\text{, voltage switching})
\end{aligned}
\tag{5.23}
$$

Interesting properties are exhibited by a magnetic core which has several holes with windings. This element is called a *transfluxor*. An example of a transfluxor with three holes is depicted in Figure 5.13a. This transfluxor has five windings: 2 and 3 are input windings, 1 is an auxiliary winding (which serves to reset the transfluxor to its basic magnetic state), 4 and 5 are working windings (which mediate the signal transfer).

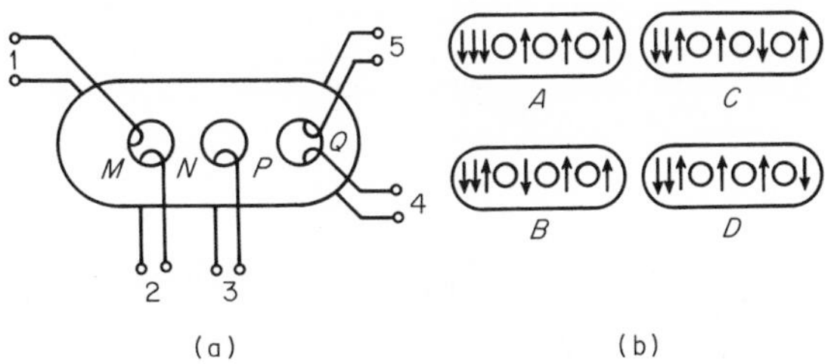

Figure 5.13 Example of a transfluxor.

The transfluxor of Figure 5.13a can assume either of four stable magnetic states. These are indicated in Figure 5.13b. In the magnetic states A and B, the magnetic coupling between windings 4 and 5 is weak because it is established by the change of the magnetic flux in the material in the immediate vicinity of the hole, the material being, moreover, saturated by the magnetic flux traversing the branches P and Q in the same direction. Windings 4 and 5 cannot produce any additional magnetic flux since, whatever its direction, it would always increase, in one of the branches P or Q, the flux already existing there. However, in the states A and B this is impossible, since the material is already saturated. In the magnetic states C and D it is possible to produce a magnetic coupling between the windings 4 and 5 by reducing the flux and alternating its direction in both the branches P and Q simultaneously.

The conditions for a transition of the transfluxor into its individual magnetic states are presented, together with the necessary explanations, in Table 5.4. It is left to the reader to define the signals and construct the operator of this logic element (Exercise 5.6-2).

TABLE 5.4. Conditions for Transition in Transfluxors

State	State into which the Transfluxor Can Turn	Conditions under which the Transition is Realized
A	B	$x_2 = 1, \quad x_1 = 0$
A	A	$x_2 = 0, \quad x_1 = 1$
B	A	$x_1 = 1$
B	B	$\bar{x}_1 \bar{x}_3 = 1$
B	C	$x_1 = 0, \quad x_3 = 1$
C	A	$x_1 = 1$
C	D	$x_4 = 1, \quad x_5 = 0$
D	C	$x_4 = 0, \quad x_5 = 1$

Exercises to Section 5.6

*5.6-1 Derive the operators of current and voltage switched magnetic elements illustrated in Figure 5.14.

*5.6-2 Set up the operators for a transfluxor with three holes, using Table 5.4.

5.7 CRYOTRONS

Superconductive logic elements—*cryotrons*—utilize the fact that some metals become superconductive (their resistance disappears completely) at temperatures near absolute zero. The temperature at which the superconductive state sets in is called the *critical temperature*, T_c.

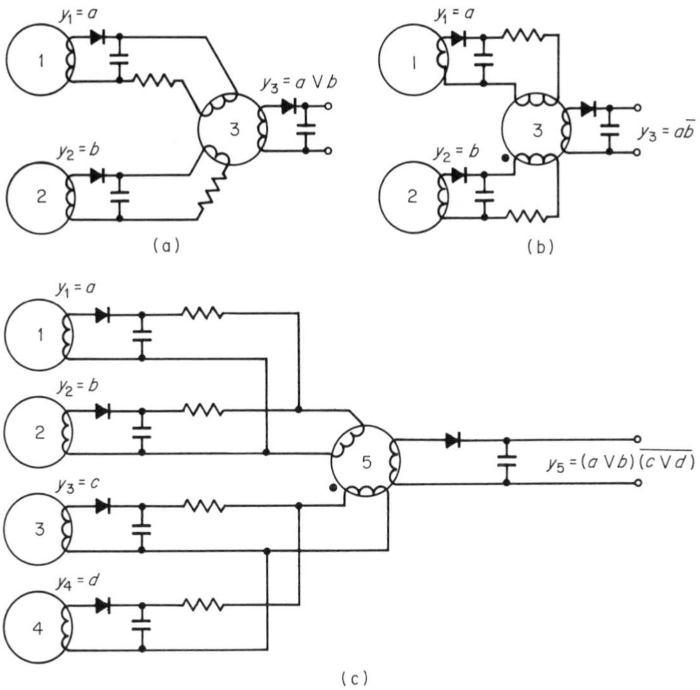

Figure 5.14 Fundamental circuit arrangements of magnetic core logic elements with diodes.

The critical temperature is a function of the magnetic field. As the magnetic field increases in intensity, the critical temperature is reduced. Examples of the dependence of the critical temperature on the magnetic field are shown for some metals in Figure 5.15.

The principle of the cryotron is illustrated in Figure 5.16. This device consists of two metal strips, vacuum-deposited on a glass base, insulated from each other and arranged crosswise. One of the strips forms the input

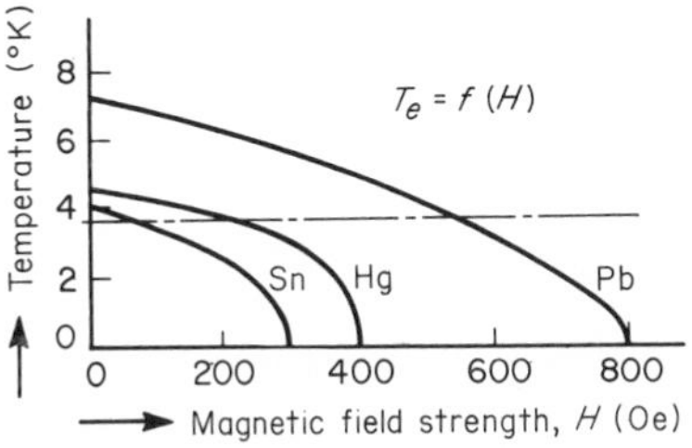

Figure 5.15 Dependence of critical temperature on the magnetic field.

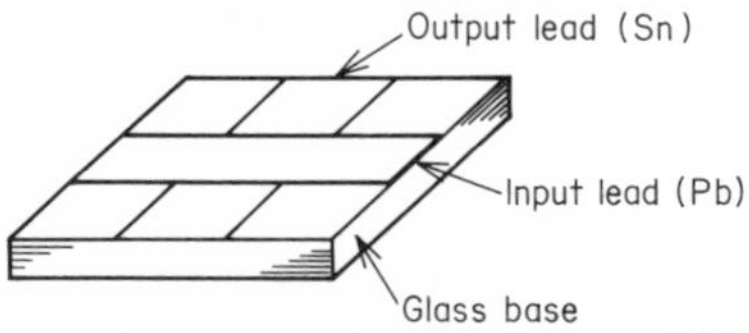

Figure 5.16 Principle of cryotron.

lead, and this is always in the superconductive state. When current, supplied by an external source, is passed through it, a magnetic field is produced which acts on the other strip—the output lead—and switches it from superconductive state to the state in which it possesses non-zero resistance.

The signals of the cryotron are defined by the following set of propositions:

$x_i(t)$: During the time interval $t_0 \leq t < t_0 + \Delta t$, a current of sufficient intensity flows through the input lead i.

$y(t_0)$: At time instant t_0, the output resistance across the terminals of the output lead is zero.

$z(t_0)$: At time instant t_0, the output lead is in the superconductive state.

These propositions and the description of the function of the cryotron yield an operator of the following form:

$$\begin{aligned} z(t_0 + \Delta t) &= \bar{x}(t) \\ y(t_0) &= z(t_0) \end{aligned} \tag{5.24}$$

Eliminating the internal variable and neglecting the time relations we get

$$y = \bar{x}$$

The cryotron resembles to a considerable degree a relay with a single break contact. The input lead corresponds to the relay winding, the output lead to the break contact. Another property the cryotron has in common with the neutral relay is that its behavior is independent of the direction of the current, in the input as well as in the output lead. Of course, a number of other physical conditions must be fulfilled to ensure the correct operation of cryotron switching circuits. For instance, a constant temperature and pressure must be maintained, the leading edges of the current supplied by the current source must not be too steep, etc.

Similarly to contact switching circuits, the parallel connection of the output leads of a cryotron models the OR function, their series connection models the AND function.

Exercises to Section 5.7

*5.7-1 Derive the operator of a cryotron fitted with two input leads.

5.7-2 Compare cryotrons with relays. Specify their similarities and differences.

5.8 PNEUMATIC SYSTEMS

The signal carrier in pneumatic systems is the pressure of gases conducted through pipes. Two types of pneumatic systems are used: (a) Systems with moving members consisting, for instance, of a diaphragm connected with a valve cone, a valve, ball, etc.; (b) Systems without any moving member—so-called *fluidic elements.*

We define the signals of a pneumatic logic element with a moving member by the following set of propositions:

$x_i(t)$: During the time interval $t_0 \leq t < t_0 + \Delta t$, there is a sufficiently high pressure in the inlet pipe.

$y_{jk}(t_0)$: At time instant t_0, the pipes j and k are interconnected by a sufficiently low pneumatic resistance.

$z(t_0)$: At time instant t_0, the piston of the pneumatic element is in its working position.

The signals of the fluidic element are defined by another set of propositions:

$x_i(t)$: During the time interval $t_0 \leq t < t_0 + \Delta t$, there is a sufficiently high pressure in the inlet pipe i.

$y_j(t_0)$: At time instant t_0, there is a sufficiently high pressure in the outlet pipe j.

$z(t_0)$: At time instant t_0, the jet of gas is in its working position.

The basic pneumatic logic element with piston is the *pneumatic relay*, one of whose many possible versions is illustrated in Figure 5.17. It has three inlet pipes 1, 2, and 3, which are connected with the corresponding chambers I, II, and III. The chambers are separated from each other by mechanically linked diaphragms. The movement of any diaphragm causes a movement of the piston which is fitted with two valve cones. The piston can take up two positions, the top (or rest) position and the bottom (or working) position. In the rest position, the piston interconnects the chambers IV and V and interrupts the connection between chambers V and VI. With the piston in its working position, conditions are reversed. A sufficient pressure in chambers I and II causes the piston to move to the working position. However, a sufficiently high pressure in chamber III can prevent the piston from doing so. If there is no pressure in any of the inlets, the piston a returns to its rest position owing to the action of the spring b. If there is no spring in the pneumatic relay, the piston remains in the position taken up last.

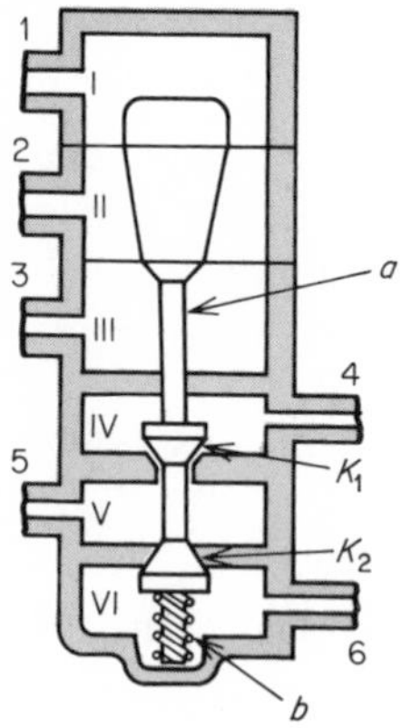

Figure 5.17 Example of pneumatic relay.

The description of this pneumatic element leads to the following operators:

(a) *for the pneumatic relay with spring*:

$$\begin{aligned} z(t_0 + \Delta t) &= \bar{x}_3(t)[x_1(t) \vee x_2(t)] \\ y_{4,5}(t_0) &= \bar{z}(t_0) \\ y_{5,6}(t_0) &= z(t_0) \end{aligned} \tag{5.25}$$

(b) *for the pneumatic relay without spring*:

$$\begin{aligned} z(t_0 + \Delta t) &= \bar{x}_3(t)[x_1(t) \vee x_2(t) \vee z(t_0)] \\ y_{4,5}(t_0) &= \bar{z}(t_0) \\ y_{5,6}(t_0) &= z(t_0) \end{aligned} \tag{5.26}$$

An isomorphic relation will be found to exist between the pneumatic element described and the electromagnetic relay, assuming that the assignment presented in Table 5.5 is respected. An electromagnetic relay modeling a pneumatic relay is illustrated in Figure 5.18. This assignment permits the properties of pneumatic elements and circuits to be studied on isomorphic relay elements and circuits which are easier to follow. It is also possible fully to utilize the analogy between pneumatic and electromagnetic relays when designing pneumatic switching circuits. However, when realizing the circuits, the specific physical properties of pneumatic relays must be taken into account, especially the limitation imposed upon the number of contacts that can be connected in series or parallel.

A miniature pneumatic logic element with a ball replacing the piston is illustrated in Figure 5.19. It consists of a small cylinder with a ball moving

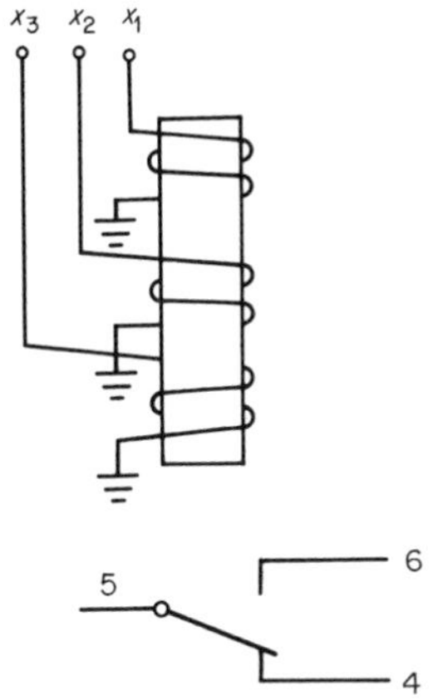

Figure 5.18 Electromagnetic relay modeling a pneumatic relay.

TABLE 5.5. Assignment Between Pneumatic and Electromagnetic Relay, Used for Mutual Modeling

Pneumatic Relay	Electromagnetic Relay
Pneumatic relay without spring	Polarized electromagnetic relay
Pneumatic relay with spring	Neutral electromagnetic relay
Upper pressure level	Potential of one terminal of source
Lower pressure level	Potential of second terminal
Pressure difference	Voltage
Piston	Armature
Pipe, chamber	Conductor
Pneumatic resistance	Electrical resistance
Jet of gas	Electric current
Gas leakage	Leakage current
Cone k_1 with chamber IV and V	Break contact
Cone k_2 with chamber V and VI	Make contact

inside. The ball can take up two positions—the rest and the working position. In the rest position the ball increases the pneumatic resistance between the pipes 1 and 3 and interconnects the pipes 2 and 4 by a pneumatic resistance that is practically zero. In the working position, conditions are reversed. The movement of the ball from one position to the other is effected by an increase in pressure acting on one side of the ball. One pair of pipes (1, 3 or 2, 4) always functions as the input, the other pair (2, 4 or 1, 3) as the output. A peculiar property of this element is that the output can temporarily function as the input, and vice versa.

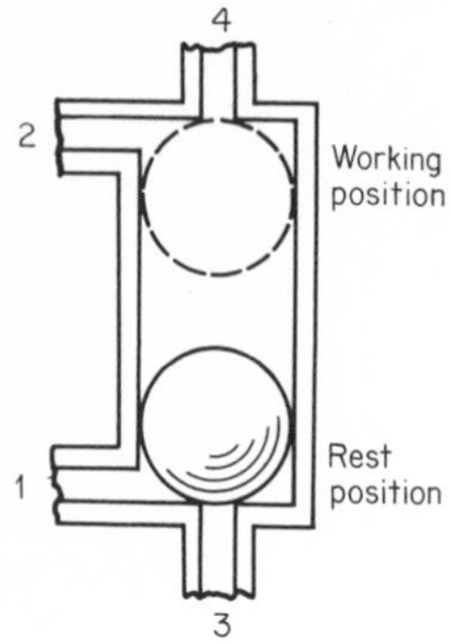

Figure 5.19 Pneumatic logic element with ball.

The operator of this pneumatic logic element has the form

$$
\begin{aligned}
z(t_0 + \Delta t) &= x_1(t)x_3(t) \vee z(t_0)[\bar{x}_2(t) \vee \bar{x}_4(t)] \\
y_{1,3}(t_0) &= z(t_0) \\
y_{2,4}(t_0) &= \bar{z}(t_0)
\end{aligned}
\tag{5.27}
$$

Figure 5.20 presents various possibilities of interconnecting this pneumatic element when modeling more complicated logic functions, together with the isomorphic contact switching circuits.

Fluidic elements make use of the mutual effect of flowing gases. Hydraulic flow elements which utilize the mutual effect of flowing liquids are also known, but are used less frequently. Both pneumatic and hydraulic fluidic elements have the same operators. It will therefore suffice to discuss, for instance, pneumatic elements.

The function of pneumatic fluidic elements is based on the pneumatic amplifier, whose principle is illustrated in Figure 5.21. A jet of gas at a sufficient pressure p_1 (the power jet) issues from pipe 1. As long as it remains unaffected by a transverse gas jet of sufficient pressure p_c (the control jet), it enters pipe 2 where it goes on at nearly the same pressure ($p_2 = p_1$). However, when a gas jet of sufficient pressure p_c emerges from the transverse pipe (or nozzle), the power jet is deflected and enters pipe 3 at nearly unchanged pressure ($p_3 = p_1$). The undeflected jet will be said to be in the rest position, the deflected jet in the working position.

A section through the fundamental type of fluidic element is shown in Figure 5.22 (in general, it may have several inlet pipes). The signals of this element were already defined. Its operator has the form

$$
\begin{aligned}
z(t_0 + \Delta t) &= x_1(t) \vee x_2(t) \\
y_1(t_0) &= \bar{z}(t_0) \\
y_2(t_0) &= z(t_0)
\end{aligned}
\tag{5.28}
$$

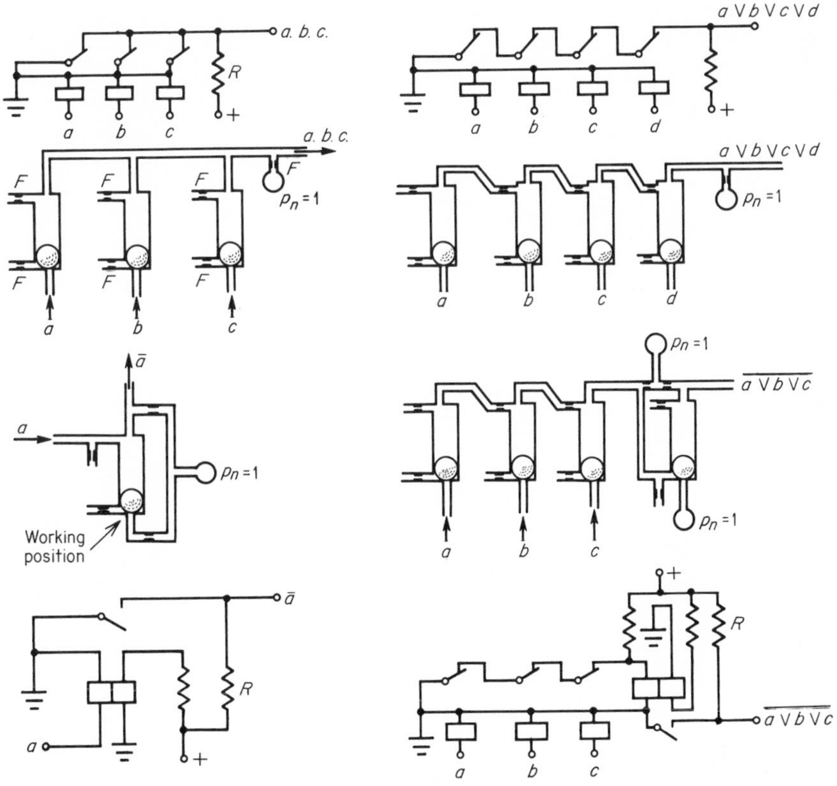

Figure 5.20 Examples of the interconnection of pneumatic elements which employ balls, and the isomorphic contact switching circuits.

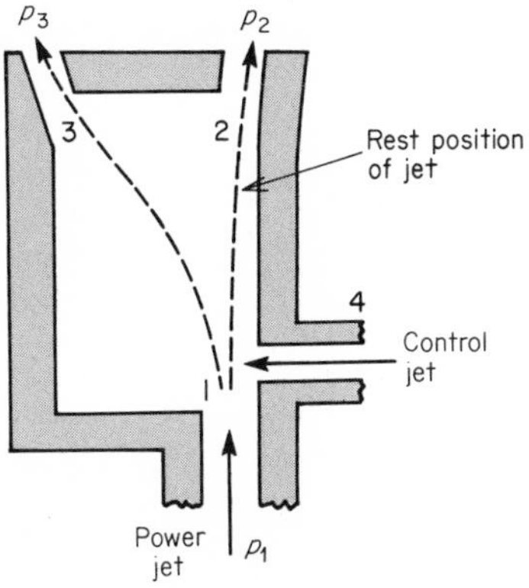

Figure 5.21 Principle of pneumatic amplifier.

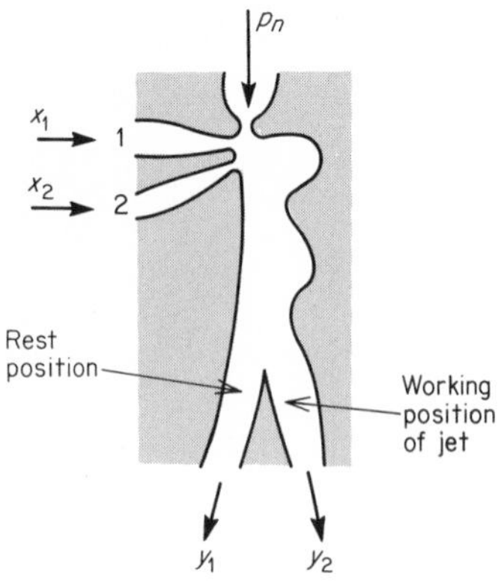

Figure 5.22 Schematic section through pneumatic fluidic logic element.

Neglecting the time relations and eliminating the internal variable, we obtain

$$\begin{aligned} y_1 &= \bar{x}_1\bar{x}_2 \\ y_2 &= x_1 \vee x_2 \end{aligned} \tag{5.29}$$

Feedback couplings which affect the function of the fluidic element can be introduced by suitably shaping its internal cavities. For example, Figure 5.23 shows a section through a bi-stable fluidic element. In this case, the jet remains in the position into which it was brought last by one of the two control jets. The operator of this element has the form

$$\begin{aligned} z(t_0 + \Delta t) &= x_1(t) \vee z(t_0)\bar{x}_2(t) \\ y_1(t_0) &= \bar{z}(t_0) \\ y_2(t_0) &= z(t_0) \\ x_1(t)x_2(t) &= 0 \text{ (a limiting condition for the input variables)} \end{aligned} \tag{5.30}$$

Examples of switching circuits using pneumatic fluidic elements are presented in Figure 5.24.

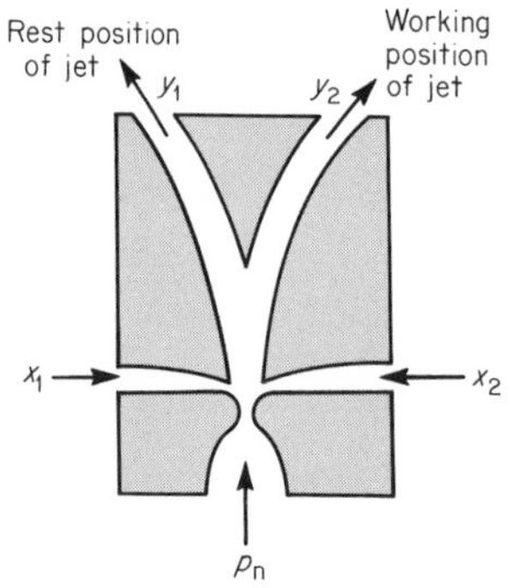

Figure 5.23 Section through bi-stable pneumatic flow element.

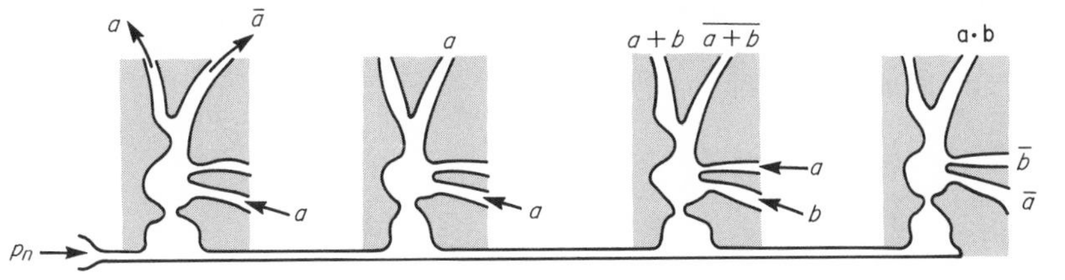

Figure 5.24 Examples of circuits using pneumatic fluidic logic elements.

Exercises to Section 5.8

5.8-1 Suggest a way of connecting a pneumatic relay with spring, described by the operator (5.25), so as to obtain a bi-stable pneumatic logic element with two inputs.

5.8-2 For the element shown in Figure 5.19 design a working cycle analogous to that of magnetic core elements (see Section 5.6).

5.9 SOME GENERAL COMMENTS

A considerable study concerning logic properties of physical systems has been presented by L. Löfgren. He has shown that every memoryless physical system with more than one input quantity can be used as a switching circuit if and only if its behavior is represented by a nonlinear function with respect to the ordinary algebra.

Löfgren established also an exhaustive classification of physical elements to gate and bıanch elements. He based his considerations on the classification of physical quantities into those which are measured by attaching the instrument to two points without cutting any coupling (so called across-quantities, e.g., electric voltage), those which are measured by cutting a coupling and connecting it again by the instrument (so called through-quantities, e.g., electric current), and those which require both the across and the through measurements (across- and through-quantities, e.g., electric resistance). The conclusion is that there are essentially two types of physical elements:

(1) Elements, whose quantities all have the same physical dimension and the same resolution level. They are called *gate elements.* Examples of gate elements are: diode elements, transistor elements, fluidics.

(2) Elements for which (1) does not hold. They are called *branch elements.* Examples of branch elements: Contacts, cryotrons.

Löfgren has shown that there are only continuous or binary branch elements whereas the gate elements exist for any resolution level.

Special techniques have been elaborated for the analysis and synthesis of switching circuits built by either type of elements. Some of them are presented in subsequent chapters of this book.

Comprehensive Exercises to Chapter 5

5.1 Find a possible realization of the operator $y = \bar{x}_1 x_2 + x_1 x_3$ by:
(a) triods.
(b) transistors.
(c) magnetic cores.
(d) a transfluxor.
(e) cryotrons.
(f) pneumatic elements with balls.
(g) fluidics.

5.2 Repeat Exercise 5.1 for the following operators:
(a) $y = x_1 \bar{x}_2 \vee \bar{x}_1 x_2$.
(b) $y = x_1 y \vee x_2$.
(c) $y = \bar{x} y \vee x \bar{y}$.

5.3 Show all possible modifications of the following operators arising from changes of the respective propositions:
(a) operators (5.15).
(b) operators (5.28).

Reference Notations to Chapter 5

The first comprehensive study of physical elements employed in switching circuits was presented by R. K. Richards [4]. Later he published another book devoted to this subject [5]. This book contains an excellent bibliography. Good surveys of physical principles employed in switching circuits are presented by A. W. Lo [2] and J. K. Hawkins [1]. The results mentioned in Section 5.9 are due to L. Löfgren [3].

References to Chapter 5

1. **Hawkins, J. K.,** *Circuit Design of Digital Computers*, John Wiley, New York, 1968.
2. **Lo, A. W.,** *Introduction to Digital Electronics*, Addison-Wesley, Reading, Mass., 1967.
3. **Löfgren, L.,** *Structure of Switching Nets*, Electrical Engineering Research Laboratory, University of Illinois, Technical Report No. 7, Urbana, Ill., 1961.
4. **Richards, R. K.,** *Digital Computer Components and Circuits*, Van Nostrand Reinhold, New York, 1958.
5. **Richards, R. K.,** *Electronic Digital Components and Circuits*, Van Nostrand Reinhold, New York, 1967.

PART II

Problems and Methods

6

DETERMINISTIC MEMORYLESS SWITCHING CIRCUITS: ANALYSIS AND SYNTHESIS

6.1 ANALYSIS

As outlined in Chapter 1, the *analysis* of a switching circuit is the following problem: The switching circuit is given by its UC-structure. Find its behavior or ST-structure.

The UC-structure was defined in Chapter 1 as a set of elements with given behaviors and a set of couplings between these elements. However, not every UC-structure satisfying the definition is a proper one. It may happen that the UC-structure of a memoryless switching circuit contains some contradictions which make the problem of analysis meaningless. Essentially, this may be due to the following reasons:

(i) A single variable is simultaneously controlled from two or more elements whose outputs are inconsistent. An example is the variable y in Figure 6.1a.

(ii) The UC-structure contains closed loops of inconsistent dependencies in which a variable y depends on another variable x in such a way that two

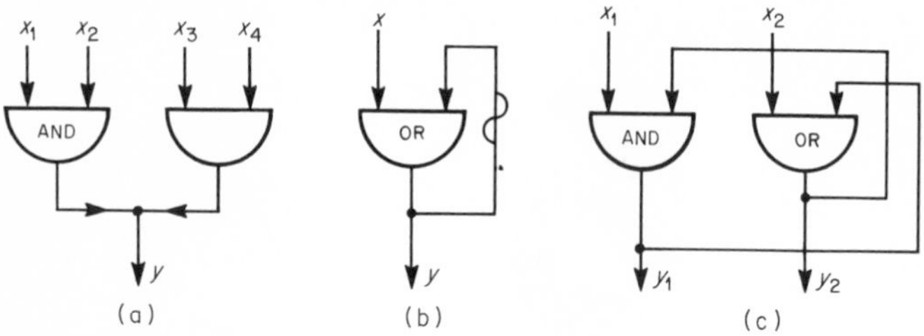

Figure 6.1 Examples of improper UC-structures.

different values of y are required at a single time instant. An example is the variable y in Figure 6.1b. The inconsistency happens only if $x = 0$. If $x = 1$, then $y = 1$.

(iii) There are several different behaviors that can be satisfied by a single UC-structure. An example is the UC-structure shown in Figure 6.1c. It is represented by the following pair of simultaneous Boolean equations:

$$y_1 = x_1 y_2$$
$$y_2 = x_2 \vee y_1$$

When solving these equations for y_1 and y_2 we obtain either

$$\begin{aligned} y_1 &= x_2 x_1 \\ y_2 &= x_2 \end{aligned} \qquad \text{(a)}$$

or

$$\begin{aligned} y_1 &= x_1 \\ y_2 &= x_2 \vee x_1 \end{aligned} \qquad \text{(b)}$$

Clearly, if $x_2 = 0$ and $x_1 = 1$, (a) gives $y_1 = y_2 = 0$ but (b) gives $y_1 = y_2 = 1$. Thus, unless the state $x_2 = 0$, $x_1 = 1$ is don't care, the response of the circuit is ambiguous.

Although it is easy to check that each variable is controlled from a single element, it is more difficult to find contingent closed loops of inconsistent dependencies or to demonstrate that the UC-structure leads to an ambiguous behavior. Nevertheless, memoryless switching circuits that do not contain any feedback loops (sometimes called feed-forward circuits) cannot certainly contain either closed loops of inconsistent dependencies or an ambiguous behavior.

To ascertain that a circuit does not contain feedback loops, we can proceed in the following manner: All nodes in the circuit are numbered by natural numbers so that (a) different nodes get different numbers, and (b) the output node of each element gets a number higher than any of its input nodes. If it proves possible to number the whole block diagram according to the rules presented, then the circuit does not contain any feedback loop. In the opposite case, the circuit incorporates feedback loops which pass through those logic elements for whose nodes it proved impossible to satisfy the conditions which relate to numbering.

If a memoryless switching circuit contains some feedback loops, it may contain inconsistent dependencies. It turns out that in many cases these inconsistent dependencies do not effect output variables at all. In any event, the procedure of analysis itself should give the correct answer in this respect.

Two possibilities arise as far as the analysis of memoryless switching circuits is concerned: (*1*) response times of all elements are completely neglected; (*2*) real time relations of the elements are considered, including possible imperfections.

Although there are cases where real response times are critical, the idealization specified in (1) is acceptable in many cases. In fact, it is always acceptable in feed-forward circuits if we are not interested in transient processes.

It is the purpose of this section to describe basic principles of analysis of memoryless switching circuits with idealized time relations. Problems arising due to real time relations are discussed in Section 9.4.

The analysis of switching circuits built up by branch-type elements differs from the analysis of the gate-type switching circuits. The latter will be considered first.

The analysis of memoryless switching circuits built up by gate-type elements consists of an elimination of internal variables from a set of logic equations describing behaviors of the used elements. We may proceed from the output terminals toward the input terminals. Every output variable of an element is substituted by the respective function of its input variables and this is repeated until only the external input variables are reached.

EXAMPLE 6.1 Analyze the circuit whose UC-structure is shown in Figure 6.2. All nodes (variables) can be numbered according to the rules specified above so that the circuit does not contain any feedback loops.

The UC-structure is described by the following set of logic equations:

$$y_1 = x_1 \vee x_3 \qquad z_1 = y_1 x_2$$
$$y_2 = \bar{x}_1 \qquad z_2 = y_2 y_3$$
$$y_3 = \bar{x}_2 \qquad v = z_1 \vee z_2$$

The behavior of the circuit is represented by a single function $v = f(x_1, x_2, x_3)$. Since the switching circuit does not contain any feedbacks, v can be determined

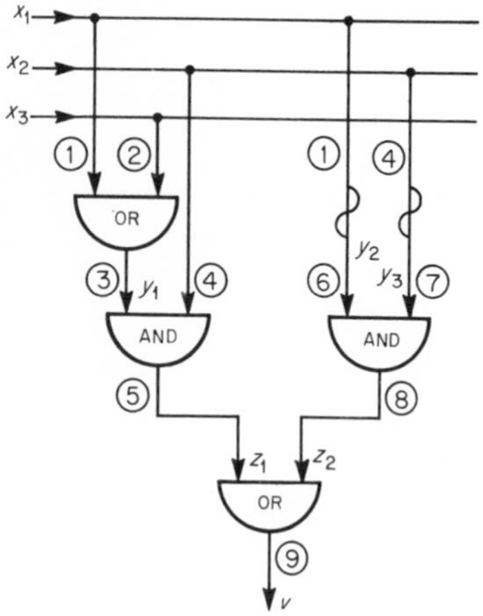

Figure 6.2 Illustration of the analysis of a memoryless switching circuit (Example 6.1).

by the following step-by-step elimination of the internal variables y_1, y_2, y_3, z_1, z_2:

$$v = z_1 \vee z_2$$
$$v = y_1 x_2 \vee y_2 y_3$$
$$v = (x_1 \vee x_3)x_2 \vee \bar{x}_1 \bar{x}_2$$

The last equation represents the behavior of the circuit.

EXAMPLE 6.2 Analyze the circuit whose UC-structure is in Figure 6.3a.

In this case, nodes (variables) of the circuit cannot be numbered according to the rules specified above so that the circuit contains feedback loops.

The UC-structure is described by six Boolean equations:

$$z_1 = \bar{x}_1 \bar{y}_3 \quad \text{(a)} \qquad y_1 = \bar{z}_1 \bar{y}_3 \quad \text{(a')}$$
$$z_2 = \bar{x}_2 \bar{y}_1 \quad \text{(b)} \qquad y_2 = \bar{z}_2 \bar{y}_1 \quad \text{(b')}$$
$$z_3 = \bar{x}_3 \bar{y}_2 \quad \text{(c)} \qquad y_3 = \bar{z}_3 \bar{y}_2 \quad \text{(c')}$$

The behavior of the circuit is represented by three functions:

$$z_1 = f_1(x_1, x_2, x_3)$$
$$z_2 = f_2(x_1, x_2, x_3)$$
$$z_3 = f_3(x_1, x_2, x_3)$$

To find these functions, we must determine functions:

$$y_1 = g_1(x_1, x_2, x_3)$$
$$y_2 = g_2(x_1, x_2, x_3)$$
$$y_3 = g_3(x_1, x_2, x_3)$$

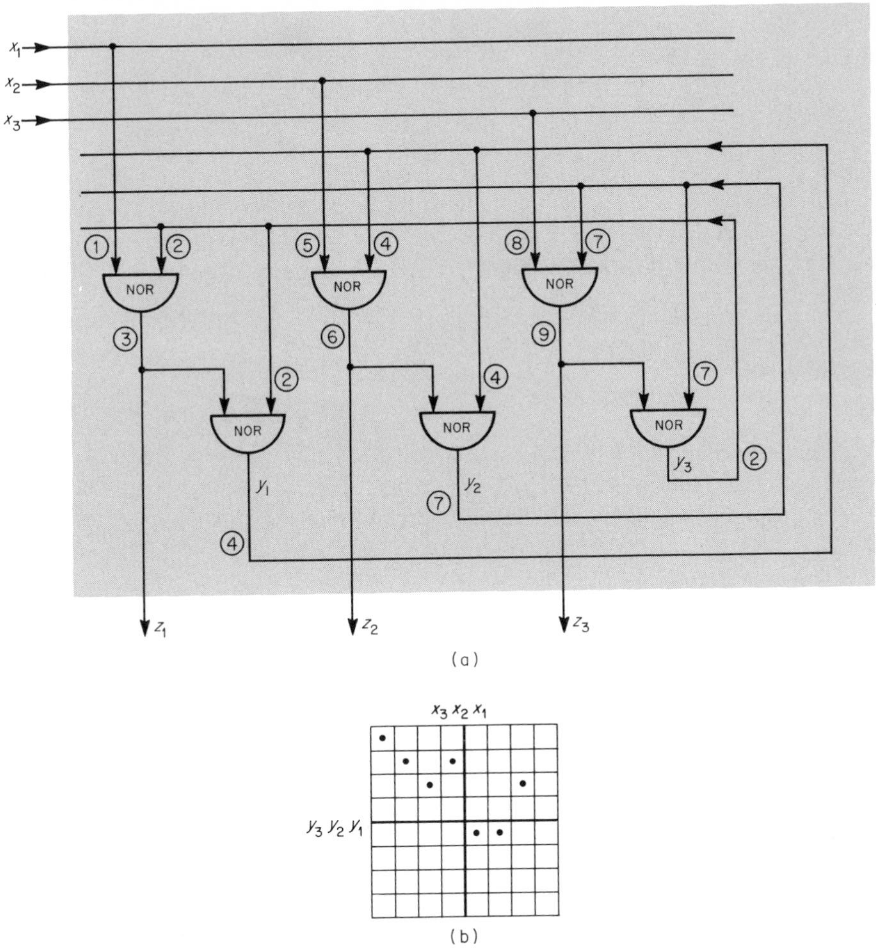

Figure 6.3 A proper memoryless switching circuit with feedbacks (Example 6.2).

First, when substituting for $\bar{z}_1$ in (a′) from (a), for $\bar{z}_2$ in (b′) from (b), and for $\bar{z}_3$ in (c′) from (c), we obtain

$$y_1 = (x_1 \vee y_3)\bar{y}_3$$
$$y_2 = (x_2 \vee y_1)\bar{y}_1$$
$$y_3 = (x_3 \vee y_2)\bar{y}_2$$

or in a simplified form

$$y_1 = x_1\bar{y}_3$$
$$y_2 = x_2\,\bar{y}_1 \qquad \text{(i)}$$
$$y_3 = x_3\,\bar{y}_2$$

When solving these simultaneous equations, we obtain

$$
\begin{aligned}
y_1 &= \bar{x}_3 x_1 \\
y_2 &= \bar{x}_1 x_2 \\
y_3 &= \bar{x}_2 x_3
\end{aligned}
\tag{ii}
$$

as a unique solution of (i) which is satisfied for all states of variables x_1, x_2, x_3 except the state $x_1 = x_2 = x_3 = 1$. (The discriminant of (i) is shown in Figure 6.3b). No solution exists for the latter state. When substituting from (ii) into (a), (b) and (c), we obtain

$$
\begin{aligned}
z_1 &= \bar{x}_1(x_2 \vee \bar{x}_3) \\
z_2 &= \bar{x}_2(x_3 \vee \bar{x}_1) \\
z_3 &= \bar{x}_3(x_1 \vee \bar{x}_2)
\end{aligned}
\tag{iii}
$$

Although solution (ii), which is a part of (iii), does not exist for $x_1 = x_2 = x_3 = 1$, this fact does not have any effect on (iii). Clearly, if and only if $x_1 = x_2 = x_3 = 1$, then all the output variables are independent of variables y_1, y_2, y_3. Thus, functions (iii) represent the behavior of the circuit. It is interesting that while variables y_1, y_2, y_3 are inconsistent (oscillate for $x_1 = x_2 = x_3 = 1$) the output variables z_1, z_2, z_3 (which depend on y_1, y_2, y_3 except when $x_1 = x_2 = x_3 = 1$) are perfectly consistent.

EXAMPLE 6.3 The circuit shown in Figure 6.4a can be described by the following equations:

$$
\begin{aligned}
y_1 &= \bar{y}_3 \bar{y}_2 \vee x_3 \bar{y}_2 \\
y_2 &= x_1 \vee \bar{y}_1 \\
y_3 &= x_3 \bar{x}_2
\end{aligned}
$$

When solving these equations, we obtain the discriminant shown in Figure 6.4b. Clearly, if $x_1 = 0$, then $y_2 \neq y_1$ (either $y_2 = 0$, $y_1 = 1$ or $y_2 = 1$, $y_1 = 0$). Thus, the behavior is ambiguous so that the given UC-structure is not proper.

In case of switching circuits built up by *branch-type elements*, UC-structures are represented by graphs rather than block diagrams. A basic component of these graphs is a single branch controlled by a logic variable which may be a function of some other logic variables (Figure 6.5a). The following properties of these graphs can be seen immediately:

(1) If two single branches controlled by y and z, respectively, are connected *in series*, the final branch is controlled by y and z, i.e., by the AND function yz (Figure 6.5b).

(2) If two single branches controlled by x and y, respectively, are connected *in parallel*, the final branch is controlled by x or y, i.e., by the OR function $x \vee y$ (Figure 6.5c).

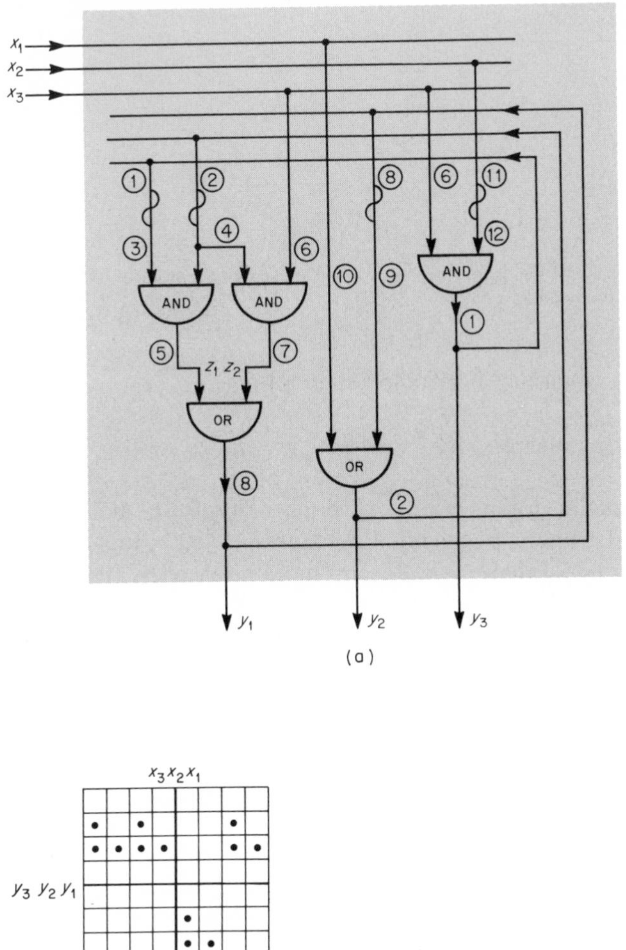

Figure 6.4 An improper memoryless switching circuit with feedbacks (Example 6.3).

It is convenient to divide switching circuits with branch elements into two classes:

(a) *Series-parallel circuits*, in which it is possible to decide uniquely on each branch, with respect to which branches of the same circuit it is connected in series, and with respect to which it is connected in parallel.

(b) *Bridge circuits*, in which it is impossible to decide uniquely on at least one branch, with respect to which branches it is connected in series and with

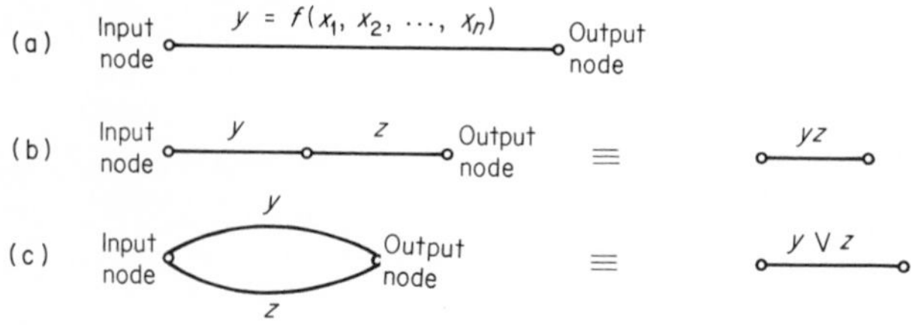

Figure 6.5 Basic properties of the graph representation of switching circuits built up by branch-type elements.

respect to which it is connected in parallel; such a branch is said to form a *bridge*.

Examples of series-parallel and bridge circuits are shown, respectively, in Figures 6.6a and b.

Using the AND function for a series connection of branches and the OR function for a parallel connection, the Boolean expression can be immediately written for a given series-parallel graph. For instance, the graph in Figure 6.6a leads to the expression

$$(a \vee b)c \vee de(f \vee g)$$

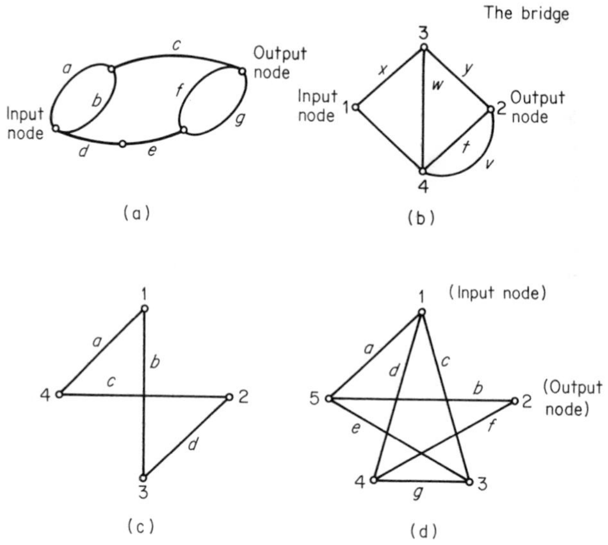

Figure 6.6 Examples of series-parallel and bridge circuits built up by branch-type elements.

which represents the behavior of the circuit and, at the same time, reflects its UC-structure.

The UC-structure of a bridge circuit cannot be described by a Boolean expression. To describe the behavior of a bridge circuit, we must find all possible paths between the input node and the output node. Each of these paths is represented by a Boolean p-term. The OR function of all these terms describes the behavior of the circuit. For instance, the bridge circuit whose UC-structure is given by the graph in Figure 6.6b has the behavior

$$xy \vee xwt \vee xwv \vee zt \vee zv \vee zwy$$

A more sophisticated approach to analysis of switching circuits built up by branch elements (including bridge circuits) has been developed in terms of Boolean matrices. A brief summary is presented as follows.

DEFINITION 6.1 Every square matrix whose elements b_{ij} are Boolean expressions is a *Boolean matrix* $[b_{ij}]$.

As a rule, we shall operate with Boolean matrices for which

$$b_{ij} = b_{ji} \text{ for all } i \text{ and } j \tag{6.1}$$

$$b_{ij} = 1 \text{ for } i = j \tag{6.2}$$

If (6.1) is satisfied, the matrix is said to be *symmetric*; if (6.2) is satisfied, it is said to be *normal.*

For instance, the matrix

$$\mathbf{F} = \begin{bmatrix} 1 & x_1x_2 \vee \bar{x}_3 & x_3 \vee \bar{x}_4 \\ x_1x_2 \vee \bar{x}_3 & 1 & \bar{x}_3\bar{x}_4 \\ x_3 \vee \bar{x}_4 & \bar{x}_3\bar{x}_4 & 1 \end{bmatrix}$$

is a Boolean matrix which is both symmetric and normal.

The following relations will be introduced for Boolean matrices of order m:

1. *Equality*

   $\mathbf{F} = \mathbf{G}$ if and only if $f_{ij} = g_{ij}$ for every i and j.

2. *Inequality*

   $\mathbf{F} \leq \mathbf{G}$ if and only if $f_{ij} \leq g_{ij}$ for every i and j.

3. *Negation*

$$\bar{\mathbf{F}} = [\bar{f}_{ij}]$$

   For a normal matrix, $\bar{\mathbf{F}} = [a_{ij}]$, where $a_{ij} = \bar{f}_{ij}$ for $i \neq j$, but where $a_{ij} = f_{ij} = 1$ for $i = j$. For instance, if

$$\mathbf{F} = \begin{bmatrix} 1 & x & y \\ x & 1 & z \\ y & z & 1 \end{bmatrix}$$

then

$$\overline{\mathbf{F}} = \begin{bmatrix} 1 & \bar{x} & \bar{y} \\ \bar{x} & 1 & \bar{z} \\ \bar{y} & \bar{z} & 1 \end{bmatrix}$$

4. *OR operation*

$$\mathbf{F} \vee \mathbf{G} = [f_{ij} \vee g_{ij}]$$

5. *Product*

$$\mathbf{F} \times \mathbf{G} = [(\bigvee_{k=1}^{m} f_{ik}\, g_{kj})]$$

For instance, for

$$\mathbf{F} = \begin{bmatrix} 1 & x & y \\ x & 1 & z \\ y & z & 1 \end{bmatrix} \qquad \mathbf{G} = \begin{bmatrix} 1 & a & b \\ a & 1 & c \\ b & c & 1 \end{bmatrix}$$

we have

$$\mathbf{F} \times \mathbf{G} = \begin{bmatrix} 1 & x \vee a \vee yc & y \vee b \vee xc \\ x \vee a \vee zb & 1 & xb \vee c \vee z \\ y \vee b \vee za & z \vee c \vee ya & 1 \end{bmatrix}$$

We see that the product preserves the normality of the matrices, but spoils their symmetry. Let us also note that the commutative law does not apply to the product, since

$$\mathbf{G} \times \mathbf{F} = \begin{bmatrix} 1 & a \vee x \vee bz & b \vee y \vee az \\ a \vee x \vee cy & 1 & c \vee z \vee ay \\ b \vee cx \vee y & bz \vee c \vee z & 1 \end{bmatrix}$$

6. *AND operation*

$$\mathbf{FG} = [f_{ij}\, g_{ij}]$$

7. *Power*

$\mathbf{F}^p$, where p is a positive integer, is defined as the product of p equal matrices, i.e., $\mathbf{F} \times \mathbf{F} \times \cdots \times \mathbf{F}$. For example, if

$$\mathbf{F} = \begin{bmatrix} 1 & x & y \\ x & 1 & z \\ y & z & 1 \end{bmatrix}$$

then

$$\mathbf{F}^2 = \begin{bmatrix} 1 & x \vee yz & y \vee xz \\ x \vee yz & 1 & z \vee xy \\ y \vee xz & z \vee xy & 1 \end{bmatrix}$$

We see that in the power of a matrix both its normality and symmetry are preserved.

8. *AND operation of a matrix and a Boolean expression w*

$$w\mathbf{F} = [wf_{ij}]$$

For a normal matrix we define $w\mathbf{F} = [c_{ij}]$, where

$$\begin{aligned} c_{ij} &= wf_{ij} && \text{for} \quad i \neq j \\ c_{ij} &= 1 && \text{for} \quad i = j \end{aligned}$$

For a Boolean matrix of the mth order we define the *determinant* as the logical sum of $m!$ components (products of the elements of the matrix) obtained by methods known from the theory of determinants. *Minors* are defined in a similar manner.

Let us consider a switching circuit built up by branch-type elements with p external nodes $1, 2, \ldots, p$. Between these nodes there exist certain *connectivities.* Each connectivity between two nodes i and $j (i, j = 1, 2, \ldots, p)$ is given by a logic function f_{ij}. For the given circuit we obtain altogether p^2 functions f_{ij}, which can be arranged in the form of a square Boolean matrix $\mathbf{F} = [f_{ij}]$ of the pth order. We shall call it the *output matrix* of the circuit.

Since each node is permanently connected to itself, we have

$$f_{ij} = 1 \text{ for all } i = j = 1, 2, \ldots, p$$

Thus $\mathbf{F}$ is a normal matrix.

As long as we do not use one-way elements (e.g., diodes) in the circuit, the connectivity between two nodes must be the same in both directions, i.e.,

$$f_{ij} = f_{ji} \text{ for all } i, j = 1, 2, \ldots, p$$

Thus, in such cases the matrix is *symmetric.*

Now let us consider, in addition to the external nodes, all the internal nodes of the circuit, which we denote by $p + 1, p + 2, \ldots, p + k$. The connectivity of the branches between the nodes i and j $(i, j = 1, 2, \ldots, p + k)$ is denoted by $e_{ij}(i, j = 1, 2, \ldots, p + k)$. A matrix, whose elements are the connectivities e_{ij}, will be called a *primitive connection matrix* and denoted by $\mathbf{E} = [e_{ij}]$. This matrix is of the $(p + k)$th order; it is normal, as a rule symmetric, and each of its elements is made up of a single variable or one of the constants 0, 1.

Let us introduce, for later use, still another type of matrix which we shall call the *connection matrix* and denote it by $\mathbf{V} = [v_{ij}]$. The elements of this matrix are again the connectivities of the branches between individual nodes, but in contrast to the matrix $\mathbf{E}$ we do not consider the internal nodes at all, or consider only some of them. If we denote the order of the matrix **V** by the symbol v, the inequalities

$$p \leq v < p + k$$

hold true.

For the limiting case, $v = p$, we obtain the so-called *reduced connection matrix*, which we denote by $\mathbf{V}_0$.

We shall now present some fundamental theorems, which form the basis for the matrix method of analysis of switching circuits with branch type elements.

THEOREM 6.1 A necessary and sufficient condition for the equality of two branch-type switching circuits from the viewpoint of behavior is that their **F**-matrices be equal.

PROOF Follows directly from the definition of the **F**-matrix. ∎

THEOREM 6.2 If $\mathbf{A} = [a_{ij}]$ is some normal Boolean matrix of the mth order, then there exists a natural number $q \leq m - 1$ such that the relations

$$\mathbf{A} \leq \mathbf{A}^2 \leq \mathbf{A}^3 \leq \cdots \leq \mathbf{A}^q = \mathbf{A}^{q+1} = \mathbf{A}^{q+2} = \cdots$$

hold true.

PROOF Let $\mathbf{A}^b = [b_{ij}]$. Then

$$b_{ij} \leq \bigvee_{k=1}^{m} b_{ik} a_{kj}$$

since **A** is normal ($a_{ii} = 1$). This means that $\mathbf{A}^b \leq \mathbf{A}^{b+1}$ and, consequently, $\mathbf{A} \leq \mathbf{A}^2 \leq \cdots$. It remains to prove that there exists such a natural number $q \leq m - 1$ that $\mathbf{A}^q = \mathbf{A}^{q+1} = \mathbf{A}^{q+2} = \cdots$. If we prove $\mathbf{A}^q = \mathbf{A}^{q+1}$ then the equalities $\mathbf{A}^{q+1} = \mathbf{A}^{q+2} = \cdots$ follow as a trivial consequence. It is sufficient to show that $\mathbf{A}^{m-1} = \mathbf{A}^m$. This can be done by showing that $\mathbf{A}^{m-1} \geq \mathbf{A}^m$ because $\mathbf{A}^{m-1} \leq \mathbf{A}^m$ has been already demonstrated.

It follows immediately from the definition of matrix product that a product of two normal Boolean matrices is again a normal Boolean matrix. This means that the diagonal entries of $\mathbf{A}^{m-1}$ and $\mathbf{A}^m$ are all equal, i.e., all are equal to 1. Consider the i, j off-diagonal entry of $\mathbf{A}^m$. It is of the form

$$\bigvee_{k_{m-1}=1}^{m} \bigvee_{k_{m-2}=1}^{m} \cdots \bigvee_{k_1=1}^{m} a_{ik_1} a_{k_1k_2} \cdots a_{k_{m-2}k_{m-1}} a_{k_{m-1}j}$$

There are $m + 1$ subscripts in each of these products, namely, $i, k_1, k_2, \ldots, k_{m-1}, j$. Hence, at least one of them is repeated. Three possibilities must be considered:

(1) $j = k_r$ for some integer r. The term takes the form

$$a_{ik_1} a_{k_1k_2} \ldots a_{k_{r-1}j} a_{jk_{r+1}} \ldots a_{k_{m-1}j}$$

This term is contained in the term

$$a_{ik_1}a_{k_1k_2}\dots a_{k_{r-1}j}$$

since $xy \leq x$ holds in Boolean algebra. The last term is the i, j entry of $\mathbf{A}^r$. Hence, the i, j entry of $\mathbf{A}^m$ is contained in the i, j entry of $\mathbf{A}^r$ and thus also in the i, j entry of $\mathbf{A}^{m-1}$, by what has already been demonstrated.

(2) $i = k_r$ for some integer r. Using a similar consideration as in (1), we obtain the same conclusions. This is left to the reader as an exercise.

(3) $k_s = k_r$ for some integers r and s. The term under consideration takes the form

$$a_{ik_1}a_{k_1k_s}\dots a_{k_{s-1}k_r}a_{k_rk_{s+1}}\dots a_{k_{r-1}k_r}a_{k_rk_{r+1}}\dots a_{k_{m-1}j}$$

which is included in the term

$$a_{ik_1}a_{k_1k_2}\dots a_{k_{s-1}k_r}a_{k_rk_{r+1}}\dots a_{k_{m-1}j}$$

But the latter term is included in the i, j entry of $\mathbf{A}^{m-1}$.

Since every term of the i, j entry of $\mathbf{A}^m$ is included in the i, j entry of $\mathbf{A}^{m-1}$ we may conclude that $\mathbf{A}^m \leq \mathbf{A}^{m-1}$. ▮

THEOREM 6.3 If $\mathbf{F}$ is the output matrix of a switching circuit with branch-type elements and p external nodes, and $\mathbf{V}_0$ is the reduced connection matrix of the circuit, then there exists an integer $k(1 \leq k < p)$ such that $\mathbf{V}_0^{p-k} = \mathbf{F}$.

PROOF Follows directly from the definitions of $\mathbf{V}_0$ and $\mathbf{F}$, and from Theorem 6.1. ▮

COROLLARY 6.1 The necessary and sufficient condition for the matrix $\mathbf{V}_0$ to be identical with the matrix $\mathbf{F}$ is that $\mathbf{V}_0^2 = \mathbf{V}_0$.

COROLLARY 6.2 The matrix $\mathbf{V}_0$ of a two-terminal switching circuit is identical with the matrix $\mathbf{F}$.

COROLLARY 6.3 Let $\mathbf{V}$ be a connection matrix with q rows and columns. Then there exists an integer $k(1 \leq k < q)$ such that elements of $\mathbf{V}^{q-k}$ which are associated with pairs of external nodes of the circuit described by $\mathbf{V}$ are elements of $\mathbf{F}$.

The analysis of a switching circuit with branch-type elements consists in the determination of matrix $\mathbf{F}$ for the given matrix $\mathbf{V}_0$. This can be accomplished by the following algorithm, which is based on Theorems 6.2 and 6.3.

ALGORITHM 6.1 Given a connection matrix **V** of a switching circuit with branch-type elements, to find the output matrix **F** of the circuit:

(1) Let $i = 1$.

(2) Construct $\mathbf{A}^i \times \mathbf{A} = \mathbf{B}$.

(3) If $\mathbf{B} \neq \mathbf{A}^i$, replace $\mathbf{A}^i$ by **B** and go to (2).

(4) If $\mathbf{B} = \mathbf{A}^i$, the algorithm ends and elements of **B** associated with pairs of external nodes are elements of **F** (if the given matrix is $\mathbf{V}_0$, then $\mathbf{B} = \mathbf{F}$).

EXAMPLE 6.4 Make the analysis of the switching circuit whose UC-structure is expressed by the graph in Figure 6.6c. In this case, matrix $\mathbf{V}_0$ is identical with the matrix **E**:

$$\mathbf{V}_0 = \mathbf{E} = \begin{bmatrix} 1 & 0 & b & a \\ 0 & 1 & d & c \\ b & d & 1 & 0 \\ a & c & 0 & 1 \end{bmatrix}$$

$$\mathbf{V}_0^2 = \begin{bmatrix} 1 & ac \vee bd & b & a \\ ac \vee bd & 1 & d & c \\ b & d & 1 & ab \vee cd \\ a & c & ab \vee cd & 1 \end{bmatrix}$$

$$\mathbf{V}_0^3 = \mathbf{F} = \begin{bmatrix} 1 & bd \vee ac & b \vee acd & a \vee bcd \\ bd \vee ac & 1 & d \vee abc & c \vee abd \\ b \vee acd & d \vee abc & 1 & ab \vee cd \\ a \vee bcd & c \vee abd & ab \vee cd & 1 \end{bmatrix}$$

EXAMPLE 6.5 Analyze the circuit in Figure 6.6b by the following connection matrix **V** which incorporates the connectivities between nodes 1, 2, 3:

$$\mathbf{V} = \begin{bmatrix} 1 & zt \vee zv & x \vee zw \\ zt \vee zv & 1 & y \vee tw \vee vw \\ x \vee zw & y \vee tw \vee vw & 1 \end{bmatrix}$$

By Corollary 6.3, elements in first two rows and columns of $\mathbf{V}^2$ are elements of **F**. We have

$$\mathbf{V}^2 = \begin{bmatrix} 1 & zt \vee zv \vee xy \vee xwt \vee xwv \vee yzw & x \vee yzt \vee yzv \vee zw \\ zt \vee zv \vee xy \vee xwt \vee xwv \vee yzw & 1 & xzt \vee xzv \vee y \vee wt \vee wv \\ x \vee yzt \vee yzv \vee zw & xzt \vee xzv \vee y \vee wt \vee wv & 1 \end{bmatrix}$$

Hence,

$$\mathbf{F} = \begin{bmatrix} 1 & zt \vee zv \vee xy \vee xwt \vee xwv \vee yzw \\ zt \vee zv \vee xy \vee xwt \vee xwv \vee yzw & 1 \end{bmatrix}$$

For the sake of completeness, let us quote a theorem, without its proof, which offers an alternative matrix method for the analysis of branch-type switching circuits.

THEOREM 6.4 Each element f_{ij} of the matrix $\mathbf{F}$ is equal to the Boolean expression representing the minor $\mathbf{V}_{ij}$ of the matrix $\mathbf{V}_0$.

EXAMPLE 6.6 Make the analysis of the switching circuit whose UC-structure is shown in Figure 6.6d.

We have

$$\mathbf{V}_0 = \begin{bmatrix} 1 & 0 & c & d & a \\ 0 & 1 & 0 & f & b \\ c & 0 & 1 & g & e \\ d & f & g & 1 & 0 \\ a & b & e & 0 & 1 \end{bmatrix}$$

We want to determine f_{12}. It follows from Theorem 6.4 that

$$f_{12} = f_{21} = V_{21} = \begin{vmatrix} 0 & c & d & a \\ 0 & 1 & g & e \\ f & g & 1 & 0 \\ b & e & 0 & 1 \end{vmatrix}$$

Using Laplace's expansion, we obtain the following minors for the first two rows of V_{21}:

$$m_1 = \begin{vmatrix} 0 & c \\ 0 & 1 \end{vmatrix} = 0 \qquad m_2 = \begin{vmatrix} 0 & d \\ 0 & g \end{vmatrix} = 0 \qquad m_3 = \begin{vmatrix} 0 & a \\ 0 & e \end{vmatrix} = 0$$

$$m_4 = \begin{vmatrix} c & d \\ 1 & g \end{vmatrix} = cg \vee d \qquad m_5 = \begin{vmatrix} c & a \\ 1 & e \end{vmatrix} = ce \vee a \qquad m_6 = \begin{vmatrix} d & a \\ g & e \end{vmatrix} = de \vee ag$$

We then determine the minors complementary to the non-zero minors m_4, m_5 and m_6:

$$m_4' = \begin{vmatrix} f & 0 \\ b & 1 \end{vmatrix} = f \qquad m_5' = \begin{vmatrix} f & 1 \\ b & 0 \end{vmatrix} = b \qquad m_6' = \begin{vmatrix} f & g \\ b & e \end{vmatrix} = fe \vee bg$$

$$\begin{aligned} f_{12} = f_{21} = \bigvee_i m_i m_i' &= f(cg \vee d) \vee b(ce \vee a) \vee (fe \vee bg)(de \vee ag) \\ &= ab \vee aefg \vee bce \vee dbeg \vee cfg \vee df \end{aligned}$$

By referring to Figure 6.6d we can ascertain that the expression found really represents the connectivity of the circuit between the input node and the output node.

Exercises to Section 6.1

6.1-1 Analyze gate-type switching circuits whose UC-structures are shown in Figures 3.21 and 3.24.

6.1-2 Analyze the branch-type switching circuit whose UC-structure is shown in Figure 6.6b by Theorem 6.4.

6.1-3 Prove Theorem 6.4.

6.1-4 Let **B** be a symmetric and normal Boolean matrix of order n. Prove that

$$\bigvee_{i=1}^{n} \mathbf{B}^i = \bigvee_{i=1}^{n+r} \mathbf{B}^i$$

for any $r \geq 0$.

6.1-5 Using the procedure explained in Chapter 4, solve the equations:
(a) describing the UC-structure in Figure 6.1c.
(b) denoted (i) in Example 6.2.
(c) specified in Example 6.3.

6.1-6 Let x_1, x_2 be input and y_1, y_2 output variables of a switching circuit and let the UC-structure of the circuit be represented by the following Boolean expressions:

$$y_1 = \bar{y}_1 x_2(y_2 \vee \bar{x}_1) \vee y_1(y_2 \bar{x}_1 \vee \bar{y}_2 \bar{x}_2)$$
$$y_2 = \bar{x}_2(\bar{y}_2 \vee x_1) \vee y_2(y_1 \vee x_2).$$

Determine:
(a) a block diagram of the UC-structure.
*(b) the behavior of the circuit.
(c) a feed-forward UC-structure representing the behavior.

*6.1-7 Show that the UC-structure represented by Boolean expressions

$$y_1 = \bar{x}_4 \bar{x}_2 \vee x_4 \bar{x}_3 x_2 \bar{x}_1$$
$$y_2 = \bar{x}_4 \bar{x}_2 \bar{x}_1 \vee x_4 \bar{x}_3 x_2$$

could be simplified if feedbacks were used.

6.1-8 Let the behavior of a memoryless switching circuit be specified by the following two functions:

$$y_1 = \bar{x}_4 \bar{x}_2 \vee x_4 x_2(\bar{x}_3 x_1 \vee x_3 \bar{x}_1)$$
$$y_2 = \bar{x}_4 \bar{x}_2 \bar{x}_1 \vee x_4 x_2(\bar{x}_3 \vee \bar{x}_1)$$

Analyze the following suggested ways of implementing the behavior by circuits with feedbacks:
(a) $y_1 = \bar{x}_4 \bar{x}_2 \vee y_2 y_1$
$y_2 = y_1 \bar{x}_1 \vee x_4 x_2(\bar{x}_3 \vee \bar{x}_1).$
(b) $y_1 = \bar{x}_4 \bar{x}_2 \vee y_2(x_1 \vee x_3)$
$y_2 = y_1 \bar{x}_1 \vee x_4 x_2(\bar{x}_3 \vee \bar{x}_1).$

6.2 SYNTHESIS: DISCUSSION OF THE PROBLEM

As already mentioned in Chapter 1, the synthesis of a switching circuit is the following problem: The switching circuit is given by its behavior. Find a UC-structure which: (a) produces the given behavior; (b) contains only elements from a specified set of elements; (c) has a required form (number of levels, fan-in, fan-out, etc.); and (d) represents the minimum or a value reasonably near to the minimum of a specified objective function defined on a set of UC-structures.

As a rule, the synthesis of complex memoryless switching circuits requires that the following two basic steps be carried out in succession:

(1) A formulation of the concrete problem, on the basis of general requirements. This involves, primarily, the specification of the number of input and output logic variables, of the set of logic elements, of the general conception of the whole system under consideration (distribution in time of the variables, selection of the code, division into partial blocks, etc.), specification of some requirements imposed on the final UC-structure, and of the objective criteria (cost, reliability, diagnostics, etc.). As a result of this step we obtain a complete specification of a set of logic functions for each individual block, a list of the primary logic functions (operators of the given elements), a list of requirements imposed on the UC-structure, and a concrete expression of some objective viewpoints (definition of a particular objective function).

(2) Such an implementation of logic functions pertaining to individual blocks which uses only available types of elements, satisfies the requirements imposed on the UC-structure, and for which the objective function reaches its minimum or its value is reasonably near to the minimum.

Let us note that the formulation of the concrete problem influences strongly the actual implementation. If the problem is not properly formulated, it may happen that the implementation cannot be accomplished at all. This may be due to an incomplete set of logic elements or to some conflicting requirements. It turns out that in many problems of synthesis of memoryless switching circuits it is impossible to precisely formulate our real objectives (a combination of the real cost, reliability, etc.) or our formulations represent very complex problems which are not practically manageable. In such cases, we try to find a reasonably good approximation of our objectives in terms of a relatively simple objective function. Thus, for example, we define the objective function by the numbers of AND and OR elements rather than by the real cost. Such an approximation may considerably simplify the procedure of implementation of the given logic functions, thus decreasing the expenses of the design (cost of man time or computer time). This may prove to be an advantage from the viewpoint of the overall cost evaluation.

Some aspects of the first step of synthesis (a problem formulation) belong to the so-called *macro-design*. This includes the partition of the whole system into individual blocks, the specification of a behavior for each block (an assignment of input and output states), and specification of objective criteria. Other aspects of the first step, like the selection of logic elements and various requirements imposed on the UC-structures of individual blocks (fan-in, fan-out, number of levels, etc.), belong, together with the second step (the implementation), to the so-called *micro-design*.

Although the methodology of the micro-design of memoryless switching circuits has been well elaborated for some types of logic elements and some objective functions, the macro-design still depends considerably on the intuition and experience of the designer. Various aspects of the micro-design are usually considered as a subject matter of the methodology of switching circuits. The macro-design, on the other hand, is considered as a part of the so-called systems architecture.

The boundary between the macro-design and the micro-design is somewhat fuzzy and changes, to a certain extent, with the development of technology. Logic elements become more and more complex. Algebraic expressions (Boolean or other), which are able to describe uniquely UC-structures built up by simple gates (AND, OR, NOT, NAND, etc.), are unable to describe UC-structures based on elements with considerably more complex behaviors (integrated circuits, large-scale integration). Therefore, new approaches are needed to solve synthesis problems on the new micro-level.

Let us discuss some aspects of the formulation of the problem of synthesis of a complex memoryless switching circuit. First, a set of types of logic elements, usually called *modules* or *building blocks*, has to be selected. The whole circuit is then supposed to be built up only by these modules. Each module represents one or more logic functions of a certain number of variables. Clearly, the set of all functions available within the given modules has to be a complete set of logic functions (see Section 2.7).

The completeness of logic functions represented by the modules is not the only aspect applicable in our selection of the modules. Usually we want to use a small number of modules but, at the same time, we require that they are efficient as far as their ability to implement logic functions is concerned. This efficiency can be expressed, for instance, by the number of logic functions that can be implemented by single modules taken from a particular set. As an example, let us consider a module representing NAND function of three variables,

$$y = \bar{a} \vee \bar{b} \vee \bar{c}$$

and a module representing function

$$y = b\bar{c} \vee a\bar{b}c \vee \bar{a}\bar{c}$$

which is usually called in the literature the *WOS-module* (well-organized sequence-module). It can be easily proven that each of these functions represents in itself a complete set of logic functions. By applying the constants 0 and 1, variables x_1, x_2, x_3, and their negations to various inputs (a, b, c) of a single

NAND module, we can implement twenty eight different logic functions out of the total number of 256 logic functions of three variables. By doing the same with the WOS module, we can implement sixty two functions (Appendix D). Hence, the WOS module can be considered as more efficient than the NAND module.

Occasionally, the behavior of a memoryless switching circuit is given as a mapping of input states into output states without any restriction as far as coding of input or output states by two-valued variables is concerned. In such cases we want to find such coding within which the absolute minimum of the objective function can be reached. This problem is usually referred to as the *input/output state assignment.* This problem is closely related to the state assignment problem in sequential switching circuits which has received considerable attention in the literature. Whereas the former problem appears only rarely in the practical design, the latter constitutes always an essential step in the synthesis of sequential switching circuits. Since both of the problems have the same essence and the latter is more important, discussion of the problem and a presentation of some methods for its solution is included among the topics of structure synthesis of sequential circuits (Chapters 8 and 9).

It often happens that the given behavior is incomplete in the sense that no values of output variables are prescribed for some states of the input variables. This case occurs, as a rule, when the required circuit is included in a system in which some states of the logic variables involved have no meaning whatsoever (e.g., as a result of the code used). This fact is expressed by indefinite logic functions which have undefined values for the meaningless states of their arguments. From the viewpoint of synthesis, this is an advantage since such a circumstance provides greater freedom in the design of the circuit and often permits better results to be obtained from the viewpoint of the objective criteria.

Various methods of synthesis of memoryless switching circuits have been designed for specific sets of elements and certain *objective criteria.* Some of these methods are described later in this chapter.

At the present time, there is a growing need for more general methods, less dependent on specific features of individual problems of synthesis. Whereas types of elements were considerably limited in the past, contemporary technology offers more and more types of elements. At the same time, it offers highly complex elements. Such a development reduces, obviously, the significance of specialized methods. Their number grows rapidly and each of them is less and less applicable. Under such circumstances, a general approach to synthesis, independent of the used inventory becomes very important. It is shown in Section 6.3 that logic equations can be employed to formalize such a general approach to synthesis of memoryless switching circuits.

6.3 GENERAL APPROACH TO SYNTHESIS

Let us suppose that the behavior of a memoryless switching circuit with n inputs and m outputs (Figure 6.7a) is given by the tabular form of logic functions

$$y_i = f_i(x_1, x_2, \ldots, x_n) \tag{6.3}$$

where $i = 1, 2, \ldots, m$. Let us further suppose, for the sake of simplicity, that the functions are to be implemented with a single element type (integrated

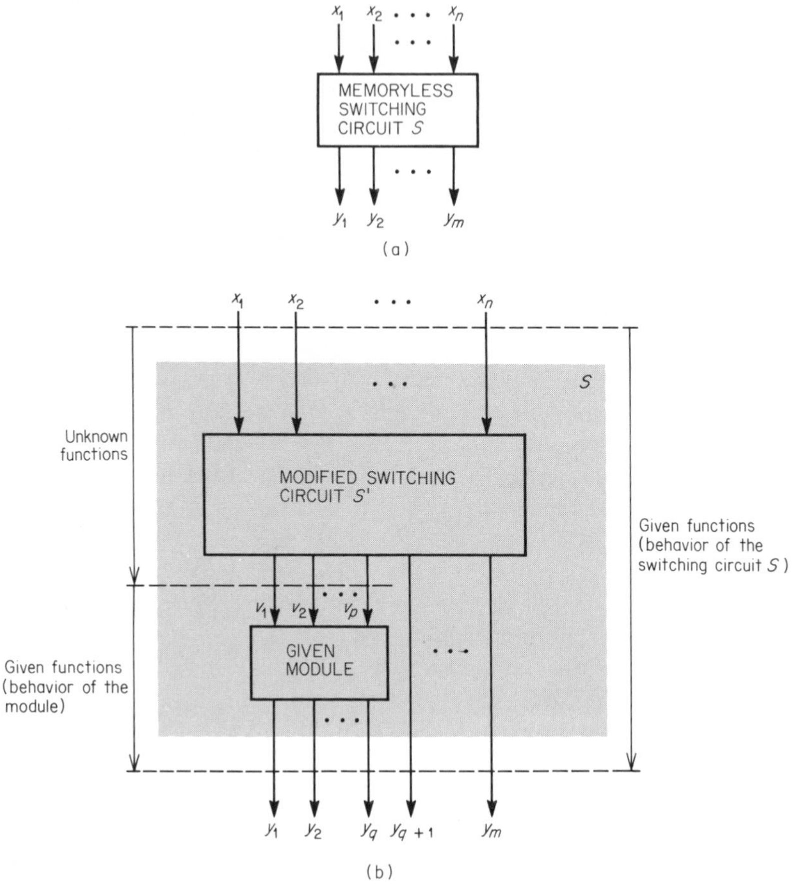

Figure 6.7 An illustration of the general approach to synthesis of memoryless switching circuits.

circuit module) with p inputs and q outputs whose behavior is specified by logic functions

$$w_j = F_j(v_1, v_2, \ldots, v_p) \tag{6.4}$$

where $j = 1, 2, \ldots, q$. Let $q \leq m$.

Suppose now that outputs of an element are connected to q outputs of the switching circuit under design. Assume, for the sake of simplicity, that

$$y_i = w_j$$

for $i, j = 1, 2, \ldots, q$ and $i = j$. Then

$$f_i(x_1, x_2, \ldots, x_n) = F_j(v_1, v_2, \ldots, v_p) \tag{6.5}$$

for $i, j = 1, 2, \ldots, q$ and $i = j$ (generally, $j = 1, 2, \ldots, q$ and $i \in I \subset \{1, 2, \ldots, m\}$). The situation expressed by (6.5) is shown in Figure 6.7b. Original switching circuit is decomposed to the element and a modified switching circuit. A behavior of the latter is represented by logic functions

$$y_k = f_k(x_1, x_2, \ldots, x_n) \tag{6.6a}$$

$$v_l = g_l(x_1, x_2, \ldots, x_n) \tag{6.6b}$$

where $k = q + 1, q + 2, \ldots, m$ and $l = 1, 2, \ldots, p$.

Functions (6.6a) are given, function (6.6b) can be obtained by solution of simultaneous equations (6.5). As shown in Chapter 4, the solution of equations (6.5) is not, generally, unique. In other words, there are, generally, several functions g_l (for each particular l) such that their composition with functions (6.4) produces functions (6.3). The designer may select such a set of functions g_l which satisfies the objective criteria best. If there are too many solutions, the designer may add some constraints, expressed in the form of additional logic or pseudo-logic equations, that are somehow related to the objective criteria.

Suppose that the objective is a memoryless switching circuit implementing a given behavior by the minimum number of modules of a chosen type (equation (6.4)). Suppose further that the module has just one output ($q = 1$). Let the chosen module be denoted by M. Then the following procedure may serve as an example of a possible approach:

(1) We prepare a catalog of all functions that can be implemented by a single module M. Each function in the catalog can be uniquely identified by its binary identifier (Appendix D).

(2) We apply successively the module M to individual outputs of the designed circuit and for each application we solve the corresponding logic equation (equation (6.5) for $q = 1$). If M represents a complete set of logic functions, the solution exists in the whole domain of the input variables.

(3) If there is more than one solution, we assign a weighting factor W_l to each function g_l (associated with a dependent variable v_l, see equation (6.6b)) of individual solutions according to the following rules:

(i) If $v_l = 0, 1, x_i, \bar{x}_i$ (assume that $\bar{x}_i$, 0 and 1 are available), then $W_l = 0$.
(ii) If g_l is a function included in the catalog of functions implemented by a single module M, then $W_l = 1$.
(iii) If g_l satisfies neither (i) nor (ii), we determine variables x_i on which g_l essentially depends (we exclude dummy variables). Let their number be N_l. Then, $W_l = N_l (2 \leq N_l \leq n, l = 1, 2, \ldots, p)$.

(4) Each solution is evaluated on the basis of the maximum value of W_l, max W_l, and the sum $W_1 + W_2 + \cdots + W_p$. Those solutions are considered as successful which have the smallest value of max W_l. If there are several of them, we select those with the smallest value of the sum $\sum W_l$.

After applying steps (1) through (4) to all output functions of the original switching circuit, we obtain a set of new functions. If some of these functions are not constants (0, 1), assertions or negations of the input variables $(x_1, \bar{x}_i)$, we again apply steps (1) through (4) to them. This process is repeated until all functions become constants, assertions, or negations.

Let us note that the approach outlined in this section, although considerably general and flexible as far as used modules and objective criteria is concerned, has little value for manual processing. On the other hand, it is well suited for computer processing.

6.4 CIRCUITS BASED ON STANDARD BOOLEAN FORMS

A set of standard Boolean forms (either *sp* or *ps*) represents a UC-structure of the so-called *two stage switching circuit*. This structure is shown in Figure 6.8.

In case of *sp* form, $G_i (i = 1, 2, \ldots, p)$ are AND elements, $F_j (j = 1, 2, \ldots, q)$ are OR elements, $\dot{x}_k = x_k$ or $\bar{x}_k$ or $1 (k = 1, 2, \ldots, n)$, and $v_{ij} = y_i$ or $0 (i = 1, 2, \ldots, p$ and $j = 1, 2, \ldots, q)$. Couplings between elements G_i and F_j can be described by a matrix $[d_{ij}]$ such that

$$d_{ij} = 1 \quad \text{if} \quad v_{ij} = y_i$$
$$d_{ij} = 0 \quad \text{if} \quad v_{ij} = 0$$

In case of *ps* form, $G_i (i = 1, 2, \ldots, p)$ are OR elements, $F_j (j = 1, 2, \ldots, q)$ are AND elements, $\dot{x}_k = x_k$ or $\bar{x}_k$ or $0 (k = 1, 2, \ldots, n)$, and $v_{ij} = y_i$ or

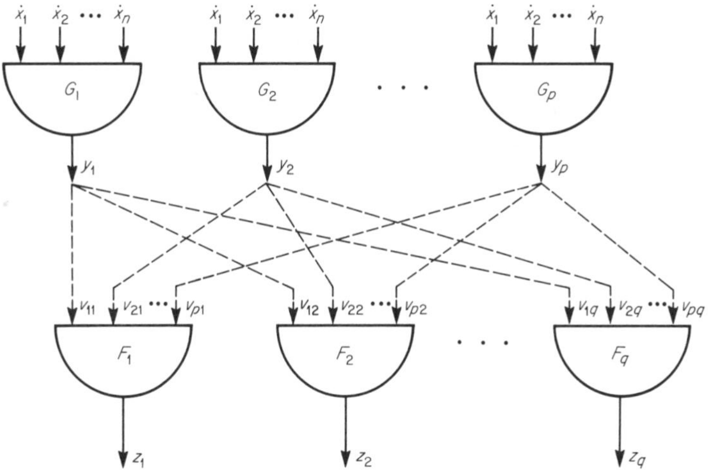

Figure 6.8 UC-structure of two-stage switching circuits based on standard Boolean forms.

$1 (i = 1, 2, \ldots, p$ and $j = 1, 2, \ldots, q)$. Couplings between elements G_i and F_j can be described by a matrix $[c_{ij}]$ such that

$$\begin{aligned} c_{ij} &= 1 \quad \text{if} \quad v_{ij} = y_i \\ c_{ij} &= 0 \quad \text{if} \quad v_{ij} = 1 \end{aligned}$$

An example of the two stage switching circuit is shown in Figure 6.9, where a and b are constants (0 or 1). Couplings between elements G_i and F_j are described by the matrix

$$d_{ij} = c_{ij} = \begin{bmatrix} 1 & 0 & 0 \\ 1 & 0 & 1 \\ 1 & 1 & 1 \\ 0 & 1 & 1 \\ 0 & 1 & 0 \end{bmatrix}$$

If G_i are AND elements, F_j are OR elements, $a = 1$, and $b = 0$, we obtain:

$$\begin{aligned} z_1 &= y_1 \vee y_2 \vee y_3 = x_2 x_3 \vee \bar{x}_1 x_3 \vee \bar{x}_1 \bar{x}_2 \bar{x}_3 \\ z_2 &= y_3 \vee y_4 \vee y_5 = \bar{x}_1 \bar{x}_2 \bar{x}_3 \vee x_1 x_2 \bar{x}_3 \vee x_1 x_3 \\ z_3 &= y_2 \vee y_3 \vee y_4 = \bar{x}_1 x_3 \vee \bar{x}_1 \bar{x}_2 \bar{x}_3 \vee x_1 x_2 \bar{x}_3 \end{aligned}$$

If G_i are OR elements, F_j AND elements $a = 0$, and $b = 1$, we obtain:

$$\begin{aligned} z_1 &= y_1 y_2 y_3 = (x_2 \vee x_3)(\bar{x}_1 \vee x_3)(\bar{x}_1 \vee \bar{x}_2 \vee \bar{x}_3) \\ z_2 &= y_3 y_4 y_5 = (\bar{x}_1 \vee \bar{x}_2 \vee \bar{x}_3)(x_1 \vee x_2 \vee \bar{x}_3)(x_1 \vee x_3) \\ z_3 &= y_2 y_3 y_4 = (\bar{x}_1 \vee x_3)(\bar{x}_1 \vee \bar{x}_2 \vee \bar{x}_3)(x_1 \vee x_2 \vee \bar{x}_3) \end{aligned}$$

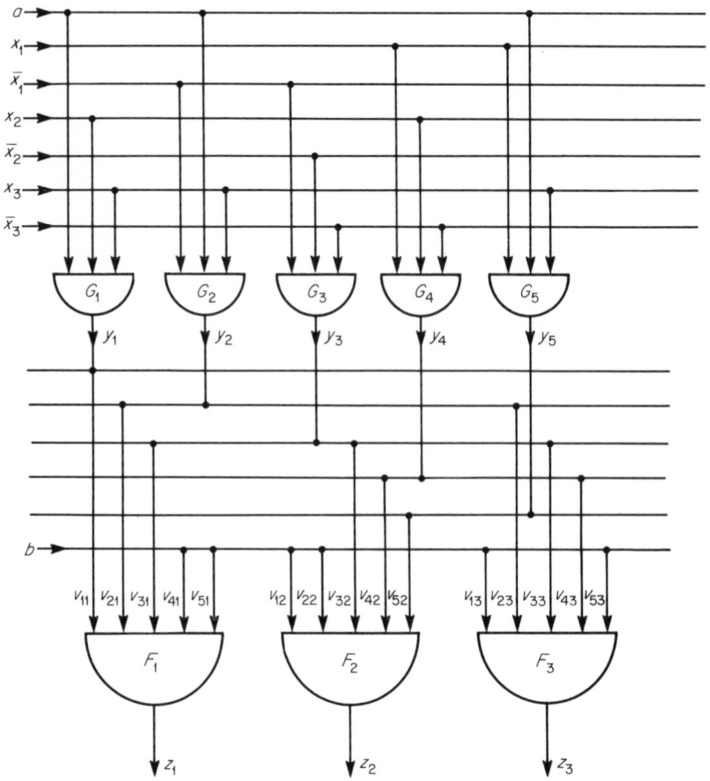

Figure 6.9 An example of the two-stage circuit.

Let us show some other algebraic forms that satisfy the structure of the two stage circuit (Figure 6.8) and can be derived either from the *sp* or *ps* Boolean forms. Consider, for the sake of simplicity, that the circuit has just one output associated with logic variable z. A generalization for more outputs is trivial.

Using the De Morgan laws, the *sp* Boolean form

$$z_1 = y_1 \vee y_2 \vee \cdots \vee y_p = (\dot{x}_1\dot{x}_2 \cdots \dot{x}_n)_1 \vee (\dot{x}_1\dot{x}_2 \cdots \dot{x}_n)_2 \vee \cdots \vee (\dot{x}_1\dot{x}_2 \cdots \dot{x}_n)_p = \text{OR}\,[\text{AND}(\dot{x}_1, \dot{x}_2, \cdots, \dot{x}_n)_1, \text{AND}(\dot{x}_1, \dot{x}_2, \cdots, \dot{x})_2, \cdots \text{AND}(\dot{x}_1, \dot{x}_2, \ldots, \dot{x}_n)_p] \quad (6.7)$$

where $\dot{x}_k = x_k$ or $\bar{x}_k$ or 1, can be rewritten as

$$z_1 = \overline{\bar{y}_1\bar{y}_2 \cdots \bar{y}_p} = \overline{\overline{(\dot{x}_1\dot{x}_2 \cdots \dot{x}_n)_1}\,\overline{(\dot{x}_1\dot{x}_2 \cdots \dot{x}_n)_2}\,\overline{(\dot{x}_1\dot{x}_2 \cdots \dot{x}_n)_p}} = \text{NAND}[\text{NAND}(\dot{x}_1, \dot{x}_2, \ldots, \dot{x}_n)_1, \text{NAND}(\dot{x}_1, \dot{x}_2, \ldots, \dot{x}_n)_2, \ldots, \text{NAND}(\dot{x}_1, \dot{x}_2, \ldots, \dot{x}_n)_p] \quad (6.8)$$

where $\dot{x}_k = x_k$ or $\bar{x}_k$ or 1. Hence, if both G_i and F_j in the structure shown in Figure 6.8 represent NAND functions, we obtain the same output functions as in the case where the G_i and F_j represent AND and OR functions.

Similarly, the *ps* Boolean form

$$\begin{aligned} z_2 &= y_1 y_2 \cdots y_p = (\dot{x}_1 \vee \dot{x}_2 \vee \cdots \vee \dot{x}_n)_1 (\dot{x}_1 \vee \dot{x}_2 \vee \cdots \vee \dot{x}_n)_2 (\dot{x}_1 \vee \dot{x}_2 \vee \cdots \vee \dot{x}_n)_p \\ &= \text{AND}[\text{OR}(\dot{x}_1, \dot{x}_2, \ldots, \dot{x}_n)_1, \text{OR}(\dot{x}_1, \dot{x}_2, \ldots, \dot{x}_n)_2, \ldots, \text{OR}(\dot{x}_1, \dot{x}_2, \ldots, \dot{x}_n)_p] \end{aligned} \tag{6.9}$$

where $\dot{x}_k = x_k$ or $\bar{x}_k$ or 0, can be rewritten as

$$\begin{aligned} z_2 &= \overline{\bar{y}_1 \vee \bar{y}_2 \vee \cdots \vee \bar{y}_p} \\ &= \overline{\overline{(\dot{x}_1 \vee \dot{x}_2 \vee \cdots \vee \dot{x}_n)_1} \vee \overline{(\dot{x}_1 \vee \dot{x}_2 \vee \cdots \vee \dot{x}_n)_2} \vee \overline{(\dot{x}_1 \vee \dot{x}_2 \vee \cdots \vee \dot{x}_n)_p}} \\ &= \text{NOR}[\text{NOR}(\dot{x}_1, \dot{x}_2, \ldots, \dot{x}_n)_1, \text{NOR}(\dot{x}_1 \dot{x}_2, \ldots, \dot{x}_n)_2, \ldots, \\ &\qquad \text{NOR}(\dot{x}_1, \dot{x}_2, \ldots, \dot{x}_n)_p] \end{aligned} \tag{6.10}$$

where $\dot{x}_k = x_k$ or $\bar{x}_k$ or 0. Hence, if both G_i and F_j in the structure shown in Figure 6.8 represent NOR functions, we obtain the same output functions as in the case where G_i and F_j represent OR and AND functions. Notice that $z_1 \neq z_2$ if the same structure of the circuit is used.

Each of the forms (6.7) through (6.10) can be further modified by exchanging x_k and $\bar{x}_k$ for all $k = 1, 2, \ldots, n$ and modifying the operations correspondingly. For example, the form (6.7) can be rewritten as

$$\begin{aligned} z_1 &= \overline{(\bar{x}_1 \vee \bar{x}_2 \vee \cdots \vee \bar{x}_n)_1} \vee \overline{(\bar{x}_1 \vee \bar{x}_2 \vee \cdots \vee \bar{x}_n)_2} \vee \cdots \vee \overline{(\bar{x}_1 \vee \bar{x}_2 \vee \cdots \vee \bar{x}_n)_p} \\ &= \text{OR}[\text{NOR}(\bar{x}_1, \bar{x}_2, \ldots, \bar{x}_n)_1, \text{NOR}(\bar{x}_1, \bar{x}_2, \ldots, \bar{x}_n)_2, \ldots, \\ &\qquad \text{NOR}(\bar{x}_1, \bar{x}_2, \ldots, \bar{x}_n)_p] \end{aligned} \tag{6.11}$$

Similarly, the forms (6.8), (6.9), (6.10) can be rewritten, respectively, as

$$\begin{aligned} z_1 &= \overline{(\bar{x}_1 \vee \bar{x}_2 \vee \cdots \vee \bar{x}_n)_1 (\bar{x}_1 \vee \bar{x}_2 \vee \cdots \vee \bar{x}_n)_2 \cdots (\bar{x}_1 \vee \bar{x}_2 \vee \cdots \vee \bar{x}_n)_p} \\ &= \text{NAND}[\text{OR}(\bar{x}_1, \bar{x}_2, \ldots, \bar{x}_n)_1, \text{OR}(\bar{x}_1 \bar{x}_2, \ldots, \bar{x}_n)_2, \\ &\qquad \text{OR}(\bar{x}_1, \bar{x}_2, \ldots, \bar{x}_n)_p] \end{aligned} \tag{6.12}$$

$$\begin{aligned} z_2 &= \overline{(\bar{x}_1 \bar{x}_2 \cdots \bar{x}_n)_1}\, \overline{(\bar{x}_1 \bar{x}_2 \cdots \bar{x}_n)_2} \cdots \overline{(\bar{x}_1 \bar{x}_2 \cdots \bar{x}_n)_p} \\ &= \text{AND}[\text{NAND}(\bar{x}_1, \bar{x}_2, \ldots, \bar{x}_n)_1, \text{NAND}(\bar{x}_1, \bar{x}_2, \ldots, \bar{x}_n)_2, \ldots, \\ &\qquad \text{NAND}(\bar{x}_1, \bar{x}_2, \ldots, \bar{x}_n)_p] \end{aligned} \tag{6.13}$$

$$\begin{aligned} z_2 &= \overline{(\bar{x}_1 \bar{x}_2 \cdots \bar{x}_n)_1 \vee (\bar{x}_1 \bar{x}_2 \cdots \bar{x}_n)_2 \vee \cdots \vee (\bar{x}_1 \bar{x}_2 \cdots \bar{x}_n)_p} \\ &= \text{NOR}[\text{AND}(\bar{x}_1, \bar{x}_2, \ldots, \bar{x}_n)_1, \text{AND}(\bar{x}_1, \bar{x}_2, \ldots, \bar{x}_n), \ldots, \\ &\qquad \text{AND}(\bar{x}_1, \bar{x}_2, \ldots, \bar{x}_n)_p] \end{aligned} \tag{6.14}$$

All possible derivatives of standard Boolean forms are summarized and illustrated by examples in Table 6.1. The meaning of *a* and *b* is the same as introduced in Figure 6.9.

TABLE 6.1. Derivatives of *sp* and *ps* Boolean Forms

Example	Operators of Elements Used: G_i	F_j	x_k or $\bar{x}_k$	a	b	Example
						sp Form
(6.7)	AND	OR	x_k	1	0	$z_1 = \bar{x}_1\bar{x}_2 \vee x_1x_2 = \text{OR}[\text{AND}(\bar{x}_1, \bar{x}_2), \text{AND}(x_1, x_2)]$
(6.11)	NOR	OR	$\bar{x}_k$	0	0	$z_1 = \overline{x_1 \vee x_2} \vee \overline{\bar{x}_1 \vee \bar{x}_2} = \text{OR}[\text{NOR}(x_1, x_2), \text{NOR}(\bar{x}_1, \bar{x}_2)]$
(6.8)	NAND	NAND	x_k	1	1	$z_1 = \overline{\overline{\bar{x}_1\bar{x}_2} \cdot \overline{x_1x_2}} = \text{NAND}[\text{NAND}(\bar{x}_1, \bar{x}_2), \text{NAND}(x_1, x_2)]$
(6.12)	OR	NAND	$\bar{x}_k$	0	1	$z_1 = \overline{(x_1 \vee x_2)(\bar{x}_1 \vee \bar{x}_2)} = \text{NAND}[\text{OR}(x_1, x_2), \text{OR}(\bar{x}_1, \bar{x}_2)]$
						ps Form
(6.9)	OR	AND	x_k	0	1	$z_2 = (\bar{x}_1 \vee \bar{x}_2)(x_1 \vee x_2) = \text{AND}[\text{OR}(\bar{x}_1, \bar{x}_2), \text{OR}(x_1, x_2)]$
(6.13)	NAND	AND	$\bar{x}_k$	1	1	$z_2 = \overline{x_1x_2} \cdot \overline{\bar{x}_1\bar{x}_2} = \text{AND}[\text{NAND}(x_1, x_2), \text{NAND}(\bar{x}_1, \bar{x}_2)]$
(6.10)	NOR	NOR	x_k	0	0	$z_2 = \overline{\overline{\bar{x}_1 \vee \bar{x}_2} \vee \overline{x_1 \vee x_2}} = \text{NOR}[\text{NOR}(\bar{x}_1, \bar{x}_2), \text{NOR}(x_1, x_2)]$
(6.14)	AND	NOR	$\bar{x}_k$	1	0	$z_2 = \overline{x_1x_2 \vee \bar{x}_1\bar{x}_2} = \text{NOR}[\text{AND}(x_1, x_2), \text{AND}(\bar{x}_1, \bar{x}_2)]$

Exercises to Section 6·4

*6.4-1 Let $z = \bar{x}_4(\bar{x}_2 \vee x_3) \vee x_3 x_2 \bar{x}_1$. Determine:
(a) A two stage NAND-NAND circuit implementing this function with minimal number of elements.
(b) A two stage NOR-NOR circuit implementing this function with minimal number of elements.

6.4-2 Determine the NOR-OR and OR-NAND derivatives of the circuit from Exercise 6.4-1a.

6.4-3 Determine the NAND-AND and AND-OR derivatives of the circuit from Exercise 6.4-1b.

6.4-4 Discuss the differences between the two-stage circuits derived from the *sp* Boolean form and those derived from the *ps* Boolean form in the preceding exercises.

6.5 CASCADE CIRCUITS

Cascade switching circuits are those which are built up solely by elements modeling the logic function

$$y = \bar{v}_1 v_2 \vee v_1 v_3 \tag{6.15}$$

This element will be called a *cascade element*; it will be represented by the symbol shown in Figure 6.10.

Function (6.15) is neither linear nor self-dual nor monotonic but preserves both 0 and 1. Hence, it does not itself represent a complete set of logic functions. However, if both assertions and negations of input variables are available, we shall obtain a complete set of logic functions since the negation preserves neither 0 nor 1. In the method for synthesis of cascade circuits that is described in this section (let us call it the *cascade method*) it is assumed that the constants 0 and 1 are available at the input of the circuit too. Thus a circuit implementing some logic function of n variables $x_1, x_2, \ldots, x_n$ has, in the general case, $2n + 2$ input nodes, namely, nodes for the variables x_i and $\bar{x}_i$ $(i = 1, 2, \ldots, n)$ and the constants 0 and 1.

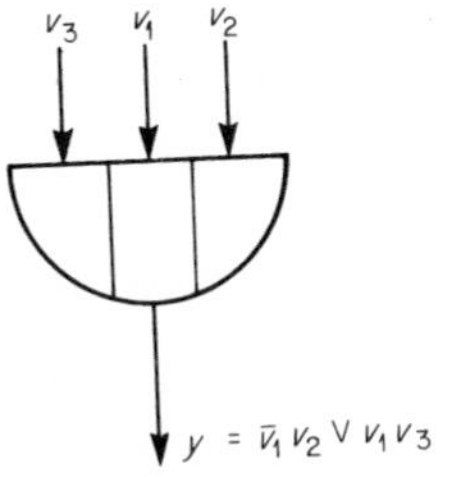

Figure 6.10 Symbol of the element used in cascade switching circuits.

It is well known (see Exercise 2.2) that every logic function $f(x_1, x_2, \ldots, x_n)$ can be expressed in the form:

$$f(x_1, x_2, \ldots, x_n) = \bar{x}_k f(x_1, x_2, \ldots, x_{k-1}, 0, x_{k+1}, \ldots, x_n) \\ \vee x_k f(x_1, x_2, \ldots, x_{k-1}, 1, x_{k+1}, \ldots, x_n)$$

This decomposition can be performed by the operator (6.15) under the assumption that $v_1 = x_k$, $v_2 = f(x_1, x_2, \ldots, x_{k-1}, 0, x_{k+1}, \ldots, x_n)$, and $v_3 = f(x_1, x_2, \ldots, x_{k-1}, 1, x_{k+1}, \ldots, x_n)$.

Let a switching circuit contain m output nodes that pertain to the nondegenerate functions $f_1, f_2, \ldots, f_m$ of n variables $x_1, x_2, \ldots, x_n$. Assume that each of these functions is decomposed with the aid of the cascade element. The result is the creation of $2m$ new nodes corresponding to the functions $f_i(x_1, x_2, \ldots, x_{k-1}, 0, x_{k+1}, \ldots, x_n)$ and $f_i(x_1, x_2, \ldots, x_{k-1}, 1, x_{k+1}, \ldots, x_n)$, where $i = 1, 2, \ldots, m$. These are functions of $n - 1$ variables.

The aforementioned decomposition (with respect to the variable x_k) is illustrated in Figure 6.11. That part of the resulting switching circuit which corresponds to the decomposition with respect to a single variable (e.g., as shown in Figure 6.11) will be called a *stage*.

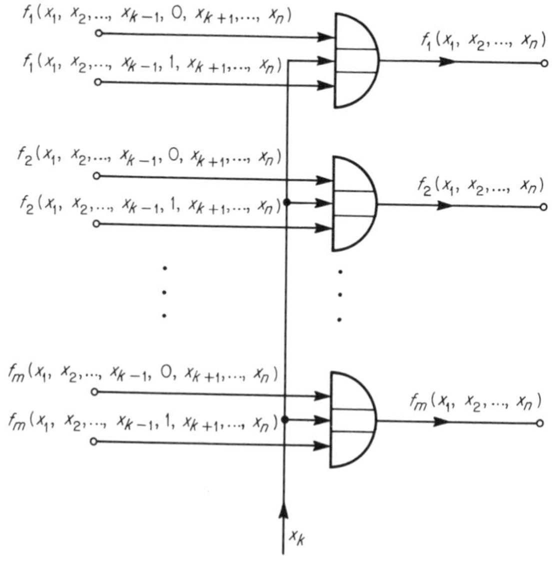

Figure 6.11 Decomposition of functions $f_1, f_2, \ldots, f_m$ with respect to x_k made by the elements described by the operator (6.15).

The cascade method prescribes that at the input of each stage we

(i) identify all nodes corresponding to equal functions (if $v_2 = v_3 = v$, then $\bar{v}_1 v_2 \vee v_1 v_3 = v$),
(ii) connect those nodes that pertain to trivial functions directly to the corresponding input nodes of the switching circuit.

Let us assume that there are r_j nodes corresponding to the nontrivial and mutually differing functions at the input of the jth stage. Clearly, $r_j \leq m \cdot 2^j$. If $r_j = 0$, the synthesis is concluded; if $r_j > 0$, we have to construct another stage. Evidently, the switching circuit is made up of at most $n - 1$ stages since at the input of the jth stage there occur functions of $n - j$ variables and when $n - j = 1$ (only the trivial functions occur), the synthesis terminates.

It is clear from this explanation that two mutually opposed processes find application in the cascade method: (1) the creation of new nodes; (2) the identification and merging of nodes pertaining to equal logic functions which may sometimes lead to the omission of some elements from the stage.

When synthesizing cascade switching circuits, we perform the decomposition successively in some distinct order of the input variables or their negations. Different orders lead to circuits of different complexity. The cascade method does not say how to determine the most advantageous order. However, practical experience shows that it produces results comparable with those of other methods even if an unsuitable order is chosen.

EXAMPLE 6.7 Design a cascade switching circuit (a decoder) which transforms a decimal number represented by the weight code 5121 into the code $3N + 2$ (N is the decimal digit). The transformation is specified by Table 6.2.

TABLE 6.2. Conversion of Decimal Numbers from Code 5121 to the Code $3n + 2$

N	5	1	2	1	$3n + 2$				
0	0	0	0	0	0	0	0	1	0
1	0	0	0	1	0	0	1	0	1
2	0	0	1	0	0	1	0	0	0
3	0	0	1	1	0	1	0	1	1
4	0	1	1	1	0	1	1	1	0
5	1	0	0	0	1	0	0	0	1
6	1	1	0	0	1	0	1	0	0
7	1	1	0	1	1	0	1	1	1
8	1	1	1	0	1	1	0	1	0
9	1	1	1	1	1	1	1	0	1

The sought-for switching circuit will have altogether ten input and five output nodes. Let us randomly select some fixed sequence of $n-1$ of the input variables, for instance, x_2, x_3, x_1 and let us successively decompose the output functions of individual stages with respect to the variables taken in this order.

We shall mark the nodes of the logic net by natural numbers in the order in which they arise. The logic function in node i will be denoted by f_i. Let $f_1 = 1$, $f_2 = 0$, $f_3 = x_1$, $f_4 = \bar{x}_1$, $f_5 = x_2$, $f_6 = \bar{x}_2$, $f_7 = x_3$, $f_8 = \bar{x}_3$, $f_9 = x_4$, $f_{10} = \bar{x}_4$. The output functions, which are specified in Table 6.2, will be denoted by $f_{11}, f_{12}, f_{13}, f_{14}, f_{15}$.

Let us describe now, with the aid of Figure 6.12, the procedure of synthesis. First we draw, next to the output nodes, the charts of the corresponding functions f_{11} to f_{15}. Functions $f_{11} = x_4$ and $f_{12} = x_2$ are trivial so that the output nodes 11 and 12 are directly connected to the respective input nodes. The remaining output functions, which are nontrivial, are then decomposed with respect to the first variable in the sequence chosen, i.e., in our case with respect to the variable x_2. We obtain the indefinite functions f_{16} to f_{21} of variables x_1, x_3, x_4. Inspecting the chart representations of these functions, we easily find that the functional values for the undefined states can be chosen so that $f_{20} = f_{21} = f_{22}$. One of the elements in the first stage becomes thereby superfluous and we can cross its symbol out from the block diagram.

The nontrivial functions f_{16} through f_{19} and f_{22} at the input of the first stage have to be decomposed now with respect to the second variable of the sequence chosen, i.e., with respect to the variable x_3. We obtain the indefinite functions f_{23} through f_{32}. By a suitable specification of the don't care conditions we can again make some of them identical obtaining thus:

$$f_{23} = f_{26} = f_3 = x_1,\ f_{24} = f_1 = 1,\ f_{25} = f_2 = 0,$$
$$f_{27} = f_{28} = f_{32} = f_{33},\ f_{29} = f_{30} = f_{31} = f_{34}$$

Two elements become superfluous at this stage and we cross out their symbols.

There are only two nontrivial functions of two variables at the input of the second stage. They have to be decomposed with respect to the third variable in the sequence, in our case x_1. This leads to four new functions of one variable: $f_{35} = f_{38} = f_{10} = \bar{x}_4$, $f_{36} = f_{37} = f_9 = x_4$. The synthesis is completed now. A scheme of the final circuit is shown in Figure 6.13.

The identification and merging of nodes is not a unique procedure due to don't care conditions. A proper identification leads to an economy in logic elements of the stage just produced (some of them can be eliminated) as well as of the next stage (a lower number of nodes means that a smaller number of elements is required).

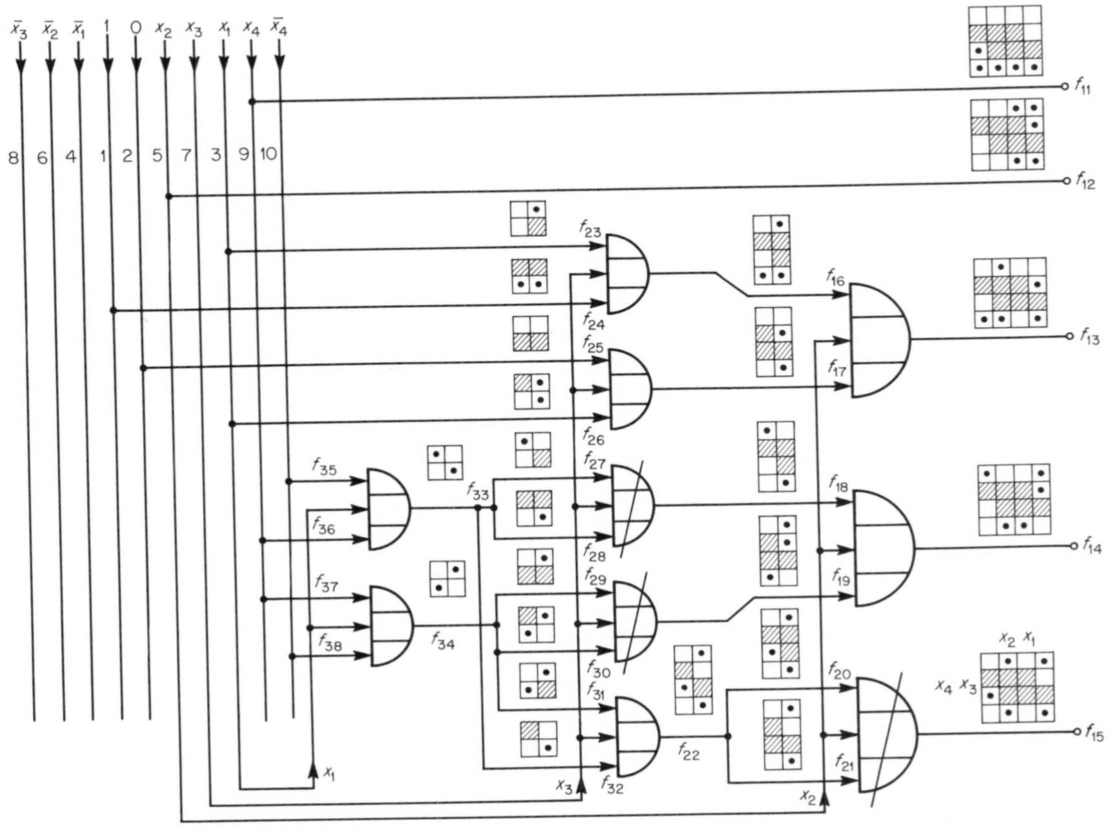

Figure 6.12 Illustration of the procedure of synthesis of a cascade switching circuit (Example 6.7).

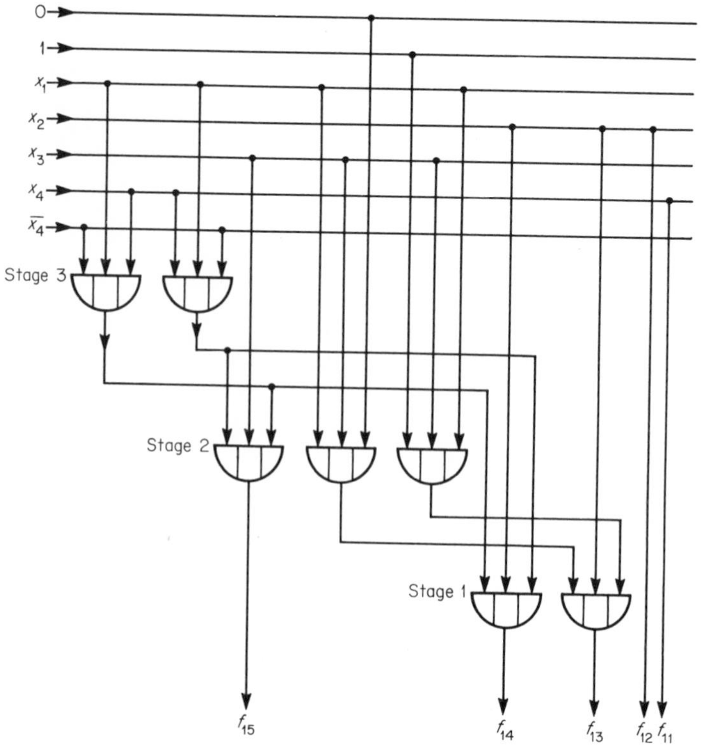

Figure 6.13 Final scheme of the cascade circuit whose synthesis is shown in Figure 6.12 (Example 6.7).

The cascade element described by operator (6.15) can be considered as the simplest member of a *family of cascade elements* whose operators have the form

$$y = \bigvee_{i=0}^{2^n-1} M_i \cdot x_{n+i}$$

where M_i are minterms of variables $x_1, x_2, \ldots, x_n$. Number n will be called the *order* of the cascade element. When $n = 1$, we obtain the operator (6.15) representing the first order cascade element. When $n = 2$, we obtain the operator

$$y = \bar{x}_1\bar{x}_2 x_3 \vee x_1\bar{x}_2 x_4 \vee \bar{x}_1 x_2 x_5 \vee x_1 x_2 x_6$$

The element described by this operator (the second order cascade element) decomposes a given logic function with respect to two variables. Generally, the nth order cascade element decomposes a given function with regard to n variables.

Exercises to Section 6.5

6.5-1 Test the function (6.15) and the function given in Exercise 6.5-3 for the conditions required by the complete set of logic functions.

*6.5-2 Compare the cascade circuit in Figure 6.13 with a minimal two-stage NAND implementation (see Section 6.4). How many NAND gates are needed?

6.5-3 In what manner can the cascade method be modified so as to be applicable to circuits built up solely by elements modeling the logic function

$$y = (\bar{v}_1 \vee v_2)(v_1 \vee v_3)$$

A hint: The second Shannon theorem can be applied (Exercise 2.2).

*6.5-4 Find the upper bound for the number of the first-order cascade elements in a circuit implementing a logic function of n variables.

6.5-5 Find a logic function whose cascade circuit implementations for two different sequences of variables contain different number of elements.

6.5-6 Design a cascade circuit implementing:
(a) the majority function M_5.
(b) WOS module.
(c) NAND function of five variables.
(d) each of the logic functions of two variables.

6.5-7 Solve Example 6.7 with the second-order cascade elements at the first stage and the first-order cascade elements at the second stage.

6.5-8 Design the third-order cascade element by the first-order elements.

6.5-9 Design the fourth-order cascade element by the first- and second-order cascade elements.

6.6 SYMMETRIC CIRCUITS

It is true that any memoryless switching circuit can be designed with the aid of the cascade method (Section 6.5) but it is of special advantage to circuits whose output nodes correspond to the so-called *symmetric logic functions*. We shall call them *symmetric circuits*.

DEFINITION 6.2 A logic function $f(x_1, x_2, \ldots, x_n)$ is called *symmetric in the pair of variables* x_i, x_j if it does not change when x_i is replaced by x_j and vice versa.

For example, the function

$$f(x_1, x_2, x_3) = x_1\bar{x}_2 x_3 \vee x_1 x_2 \bar{x}_3$$

is symmetric in the pair x_2, x_3 because

$$f(x_1, x_3, x_2) = x_1\bar{x}_3 x_2 \vee x_1 x_3 \bar{x}_2 = f(x_1, x_2, x_3).$$

The function is not symmetric in the pairs x_1, x_2 and x_1, x_3 because

$$f(x_2, x_1, x_3) = x_2\bar{x}_1 x_3 \vee x_2 x_1 \bar{x}_3 \neq f(x_1, x_2, x_3)$$
$$f(x_3, x_2, x_1) = x_3\bar{x}_2 x_1 \vee x_3 x_2 \bar{x}_1 \neq f(x_1, x_2, x_3)$$

DEFINITION 6.3 A logic function $f(x_1, x_2, \ldots, x_n)$ is called *totally symmetric* if it is symmetric in all pairs of variables $x_i, x_j (1 \le i, j \le n)$.

It is well known that any permutation of some elements can be obtained by successive interchanges of pairs of elements. Hence, a totally symmetric function $f(x_1, x_2, \ldots, x_n)$ does not change under any permutation of the variables $x_1, x_2, \ldots, x_n$.

Totally symmetric functions are often simply referred to as symmetric functions and those functions which are symmetric in some but not all pairs are referred to as *partially symmetric functions.*

For example, the function

$$f(x_1, x_2, x_3) = \bar{x}_1 \bar{x}_2 \bar{x}_3 \vee x_1 x_2 x_3$$

is totally symmetric because

$$f(x_1, x_2, x_3) = f(x_2, x_1, x_3) = f(x_3, x_2, x_1) = f(x_1, x_3, x_2)$$

The function

$$f(x_1, x_2, x_3) = \bar{x}_1 \bar{x}_2 \bar{x}_3 \vee x_1 \bar{x}_2 x_3 \vee x_1 x_2 \bar{x}_3$$

although symmetric only in the pair x_2, x_3, can be made a totally symmetric function by negating variable x_1. Clearly,

$$f(\bar{x}_1, x_2, x_3) = x_1 \bar{x}_2 \bar{x}_3 \vee \bar{x}_1 \bar{x}_2 x_3 \vee \bar{x}_1 x_2 \bar{x}_3$$

is a totally symmetric function.

We just saw that even if a function $f(x_1, x_2, \ldots, x_n)$ is not totally symmetric in variables $x_1, x_2, \ldots, x_n$ it may be totally symmetric in variables $\dot{x}_i, \dot{x}_2, \ldots, \dot{x}_n$, where $\dot{x}_i$ is either x_i or $\bar{x}_i (i = 1, 2, \ldots, n)$. The variables $\dot{x}_1, \dot{x}_2, \ldots, \dot{x}_n$ in which a function is totally symmetric will be called the *variables of symmetry.*

THEOREM 6.5 A logic function $f(x_1, x_2, \ldots, x_n)$ is totally symmetric in the variables $(\dot{x}_1, \dot{x}_2, \ldots, \dot{x}_n)$ if and only if it may be specified by a set of integers $\{a_1, a_2, \ldots, a_m\}$ such that $f(\dot{x}_1, \dot{x}_2, \ldots, \dot{x}_n) = 1$ if exactly $a_i (i = 1, 2, \ldots, m)$ of the variables of symmetry $\dot{x}_1, \dot{x}_2, \ldots, x_n$ are equal to 1; otherwise $f(\dot{x}_1, \dot{x}_2, \ldots, \dot{x}_n) = 0$.

PROOF (i) Let $f(\dot{x}_1, \dot{x}_2, \ldots, \dot{x}_n)$ be a symmetric function and let s_1 and s_2 be two different states of variables $\dot{x}_1, \dot{x}_2, \ldots, \dot{x}_n$ with the same number of variables equal to 1. Then s_2 can be obtained from s_1 (and vice versa) by a permutation of variables $\dot{x}_1, \dot{x}_2, \ldots, \dot{x}_n$. Hence, $f(s_1) = f(s_2)$ is necessary to satisfy the assumption that f is a symmetric function.

(ii) Suppose that $f(\dot{x}_1, \dot{x}_2, \ldots, \dot{x}_n) = 1$ if and only if exactly a_i of the variables $\dot{x}_1, \dot{x}_2, \ldots, \dot{x}_n$ are equal to 1. Applying any permutation of the variables, the function will still be equal to 1 only when any a_i of the variables are equal to 1. This holds for any $a_i (i = 1, 2, \ldots, m)$ so that the function is symmetric. ∎

If we want to test a given logic function $f(x_1, x_2, \ldots, x_n)$ for symmetry in the given variables $\dot{x}_1, \dot{x}_2, \ldots, \dot{x}_n$, where $\dot{x}_i$ is either x_i or $\bar{x}_i (i = 1, 2, \ldots, n)$, we can apply the following theorem due to Povarov.

THEOREM 6.6 A logic function $f(x_1, x_2, \ldots, x_n)$ is totally symmetric in the variables $\dot{x}_1, \dot{x}_2, \ldots, \dot{x}_n$ if and only if

$$f(\dot{x}_1, \dot{x}_2, \ldots, \dot{x}_n) = f(\dot{x}_2, \dot{x}_1, \dot{x}_3, \ldots, \dot{x}_n) \tag{6.16}$$

and

$$f(\dot{x}_1, \dot{x}_2, \ldots, \dot{x}_n) = f(\dot{x}_2, \dot{x}_3, \dot{x}_4, \ldots, \dot{x}_n, \dot{x}_1) \tag{6.17}$$

PROOF (i) Necessity of both (6.16) and (6.17) follows directly from the definition of totally symmetric logic function.

(ii) To prove that (6.16) and (6.17) are sufficient we have to show that interchanges of all pairs of variables can be obtained by applying only the two permutations involved in (6.16) and (6.17). Suppose that we want to interchange a general pair of variables, x_i and x_j, where $i < j$. We can repeat (6.17) until we obtain

$$f(x_i, x_{i+1}, \ldots, x_j, \ldots, x_n, x_1, x_2, \ldots, x_{i-1})$$

One application of (6.16) leads now to

$$f(x_{i+1}, x_i, x_{i+2}, \ldots, x_j, \ldots, x_n, x_1, x_2, \ldots, x_{i-1})$$

When (6.17) is applied once and followed by one application of (6.16), we obtain

$$f(x_{i+2}, x_i, x_{i+3}, \ldots, x_j, \ldots, x_n, x_1, x_2, \ldots, x_{i-1}, x_{i+1})$$

This pair of permutations can be applied repeatedly until we obtain

$$f(x_j, x_i, x_{j+1}, \ldots, x_n, x_1, x_2, \ldots, x_{i-1}, x_{i+1}, \ldots, x_{j-1})$$

When (6.17) is repeated exactly $(n - 1)$ times, we obtain

$$f(x_{j-1}, x_j, x_i, x_{j+1}, \ldots, x_n, x_1, x_2, \ldots, x_{i-1}, x_{i+1}, \ldots, x_{j-2})$$

A single application of (6.16) produces now

$$f(x_j, x_{j-1}, x_i, x_{j+1}, \ldots, x_n, x_1, x_2, \ldots, x_{i-1}, x_{i+1}, \ldots, x_{j-2})$$

Next we apply repeatedly (6.16) followed by (6.17) until we obtain

$$f(x_{i-1}, x_j, x_{i+1}, \ldots, x_{j-2}, x_{j-1}, x_i, x_{j+1}, \ldots, x_n, x_1, x_2, \ldots, x_{i-2})$$

Applying now repeatedly (6.17) we obtain the form

$$f(x_1, x_2, \ldots, x_{i-1}, x_j, x_{i+1}, \ldots, x_{j-1}, x_i, x_{j+1}, \ldots, x_n)$$

Clearly, x_i and x_j are interchanged in this form so that the proof is completed. ∎

We shall introduce now some other well known theorems concerning symmetric logic functions. Proofs of the theorems are very simple and, therefore, are left to the reader. Symmetric functions will be expressed in the symbolic form

$$\{a_1, a_2, \ldots, a_m\}(\dot{x}_1, \dot{x}_2, \ldots, \dot{x}_n)$$

where $\dot{x}_i (i = 1, 2, \ldots, n)$ are variables of symmetry ($\dot{x}_i$ is either x_i or $\bar{x}_i$) and $a_j (j = 1, 2, \ldots, m)$ denote numbers of ones in states of the variables of symmetry for which the function is equal to one (see Theorem 6.5).

THEOREM 6.7 A totally symmetric function of the form

$$\{a_1, a_2, \ldots, a_m\}(\dot{x}_1, \dot{x}_2, \ldots, \dot{x}_n)$$

can also be represented as the following totally symmetric function

$$\{n - a_1, n - a_2, \ldots, n - a_m\}(\bar{\dot{x}}_1, \bar{\dot{x}}_2, \ldots, \bar{\dot{x}}_n)$$

THEOREM 6.8 The negation

$$\overline{\{a_1, a_2, \ldots, a_m\}(\dot{x}_1, \dot{x}_2, \ldots, \dot{x}_n)}$$

of the totally symmetric function

$$\{a_1, a_2, \ldots, a_m\}(\dot{x}_1, \dot{x}_2, \ldots, \dot{x}_n)$$

is also a totally symmetric function, i.e.,

$$\{b_1, b_2, \ldots, b_r\}(\dot{x}_1, \dot{x}_2, \ldots, \dot{x}_n)$$

in the same variables of symmetry $\dot{x}_1, \dot{x}_2, \ldots, \dot{x}_n$, where

$$\{b_1, b_2, \ldots, b_r\} = \{0, 1, \ldots, n\} - \{a_1, a_2, \ldots, a_m\}$$

THEOREM 6.9 The AND operation

$$[\{a_1, a_2, \ldots, a_p\}(\dot{x}_1, \dot{x}_2, \ldots, \dot{x}_n)] \cdot [\{b_1, b_2, \ldots, b_q\}(\dot{x}_1, \dot{x}_2, \ldots, \dot{x}_n)]$$

of two symmetric functions which have the same variables of symmetry is also a symmetric function

$$\{c_1, c_2, \ldots, c_r\}(\dot{x}_1, \dot{x}_2, \ldots, \dot{x}_n)$$

in the same variables of symmetry, where

$$\{c_1, c_2, \ldots, c_r\} = \{a_1, a_2, \ldots, a_p\} \cap \{b_1, b_2, \ldots, b_q\}$$

THEOREM 6.10 The OR operation

$$[\{a_1, a_2, \ldots, a_p\}(\dot{x}_1, \dot{x}_2, \ldots, \dot{x}_n)] \vee [\{b_1, b_2, \ldots, b_q\}(\dot{x}_1, \dot{x}_2, \ldots, \dot{x}_n)]$$

of two symmetric functions which have the same variables of symmetry is also a symmetric function

$$\{c_1, c_2, \ldots, c_r\}(\dot{x}_1, \dot{x}_2, \ldots, \dot{x}_n)$$

in the same variables of symmetry, where

$$\{c_1, c_2, \ldots, c_r\} = \{a_1, a_2, \ldots, a_p\} \cup \{b_1, b_2, \ldots, b_q\}$$

THEOREM 6.11 Every symmetric function of the form

$$\{a_1, a_2, \ldots, a_m\}(\dot{x}_1, \dot{x}_2, \ldots, \dot{x}_n)$$

can be uniquely represented by the expansion

$$\bigvee_{i=1}^{m} \{a_i\}(\dot{x}_1, \dot{x}_2, \ldots, \dot{x}_n)$$

THEOREM 6.12 A totally symmetric function

$$\{a_1, a_2, \ldots, a_m\}(\dot{x}_1, \dot{x}_2, \ldots, \dot{x}_n)$$

can be expressed in the form

$$\bar{x}_i \cdot [\{a_1, a_2, \ldots, a_m\}(\dot{x}_1, \dot{x}_2, \ldots, \dot{x}_{i-1}, \dot{x}_{i+1}, \ldots, \dot{x}_n)]$$
$$\vee\, \dot{x}_i \cdot [\{a_1 - 1, a_2 - 1, \ldots, a_m - 1\}(\dot{x}_1, \dot{x}_2, \ldots, \dot{x}_{i-1}, x_{i+1}, \ldots, \dot{x}_n)]$$

for any $i = 1, 2, \ldots, n$, where a_j is eliminated from the set $\{a_1, a_2, \ldots, a_m\}$ if $a_j = n$ and $a_k - 1$ is eliminated from the set $\{a_1 - 1, a_2 - 1, \ldots, a_m - 1\}$ if $a_k = 0$.

EXAMPLE 6.8 To illustrate Theorems 6.7 through 6.12, let us consider two symmetric functions:

$$f(\bar{x}_1, x_2, \bar{x}_3) = \{1, 2\}(\bar{x}_1, x_2, \bar{x}_3)$$
$$g(x_1. \bar{x}_2, x_3) = \{0, 1, 3\}(x_1, \bar{x}_2, x_3)$$

Then:

$$f(x_1, \bar{x}_2, x_3) = \{1, 2\}(x_1, \bar{x}_2, x_3) \quad \text{(Th.6.7)}$$

$$g(\bar{x}_1, x_2, \bar{x}_3) = \{0, 2, 3\}(\bar{x}_1, x_2, \bar{x}_3) \quad \text{(Th.6.7)}$$

$$\bar{f}(\bar{x}_1, x_2, \bar{x}_3) = \{0, 3\}(\bar{x}_1, x_2, \bar{x}_3) \quad \text{(Th.6.8)}$$

$$g(x_1, \bar{x}_2, x_3) = \{2\}(x_1, \bar{x}_2, x_3) \quad \text{(Th.6.8)}$$

$$f \cdot g = \{2\}(\bar{x}_1, x_2, \bar{x}_3) \quad \text{(Th.6.7 and 6.9)}$$

$$f \cdot g = \{1\}(x_1, \bar{x}_2, x_3) \quad \text{(Th.6.7 and 6.9)}$$

$$f \vee g = \{0, 1, 2, 3\}(\bar{x}_1, x_2, \bar{x}_3) = \{0, 1, 2, 3\}(x_1, \bar{x}_2, x_3) \quad \text{(Th.6.7 and 6.10)}$$

$$f = \{1\}(\bar{x}_1, x_2, \bar{x}_3) \vee \{2\}(\bar{x}_1, x_2, \bar{x}_3) \qquad \text{(Th.6.11)}$$
$$g = \{0\}(x_1, \bar{x}_2, x_3) \vee \{1\}(x_1, \bar{x}_2, x_3) \vee \{3\}(x_1, \bar{x}_2, x_3) \qquad \text{(Th.6.11)}$$
$$f = x_1 \cdot [\{1, 2\}(x_2, \bar{x}_3)] \vee \bar{x}_1 \cdot [\{0, 1\}(x_2, \bar{x}_3)] \qquad \text{(Th.6.12)}$$
$$f = \bar{x}_2 \cdot [\{1, 2\}(\bar{x}_1, \bar{x}_3)] \vee x_2 \cdot [\{0, 1\}(\bar{x}_1, \bar{x}_3)] \qquad \text{(Th.6.12)}$$
$$g = x_2 \cdot [\{0, 1\}(x_1, x_3)] \vee \bar{x}_2[\{0, 2\}(x_1, x_3)] \qquad \text{(Th.6.12)}$$
$$g = \bar{x}_3 \cdot [\{0, 1\}(x_1, \bar{x}_2)] \vee x_3[\{0, 2\}(x_1, \bar{x}_2)] \qquad \text{(Th.6.12)}$$

It follows from Theorem 6.6 that two equations involving three permutations are sufficient to test a function for symmetry in the given variables $\dot{x}_1, \dot{x}_2, \ldots, \dot{x}_n$. There are 2^n possible choices of suitable variables $\dot{x}_1, \dot{x}_2, \ldots, \dot{x}_n$. However, because of the property stated in Theorem 6.7, we need to consider only half the possible choices, i.e., 2^{n-1} possibilities.

Hence, to find whether a logic function is symmetric for given variables of symmetry, we have to consider two possibilities; to find the variables of symmetry we have to consider at most 2^{n-1} possibilities, i.e., at most a total number of 2^n possibilities need to be tried.

EXAMPLE 6.9 Determine if the function

$$f(x_1, x_2, x_3) = x_1\bar{x}_2 \vee x_1x_3 \vee \bar{x}_1x_2\bar{x}_3 \vee \bar{x}_2x_3$$

is symmetric with respect to suitable variables of symmetry $\dot{x}_1, \dot{x}_2, \dot{x}_3$.

Clearly, we have to try eight conditions at most. Applying Theorem 6.6, we find that

$$f(x_1, x_2, x_3) \neq x_2\bar{x}_1 \vee x_2x_3 \vee \bar{x}_2x_1\bar{x}_3 \vee \bar{x}_1x_3 = f(x_2, x_1, x_3)$$

so that x_1, x_2, x_3 are not variables of symmetry.

We select another set of variables now and apply Theorem 6.6 again. Let $\bar{x}_1, \bar{x}_2, x_3$ be the chosen set of variables. Then,

$$\begin{aligned} f(\bar{x}_1, x_2, x_3) &= \bar{x}_1\bar{x}_2 \vee \bar{x}_1x_3 \vee x_1x_2\bar{x}_3 \vee \bar{x}_2x_3 \\ &= \bar{x}_2\bar{x}_1 \vee \bar{x}_2x_3 \vee x_2x_1\bar{x}_3 \vee \bar{x}_1x_3 = f(x_2, \bar{x}_1, x_3) \end{aligned}$$

but

$$f(x_2, x_3, \bar{x}_1) = \bar{x}_2\bar{x}_3 \vee \bar{x}_2x_1 \vee x_2x_3\bar{x}_1 \vee \bar{x}_3x_1 \neq f(\bar{x}_1, x_2, x_3)$$

so that the theorem is not satisfied.

Let $x_1, \bar{x}_2, x_3$ be the next set of variables under investigation. We get

$$\begin{aligned} f(x_1, \bar{x}_2, x_3) &= x_1x_2 \vee x_1x_3 \vee \bar{x}_1\bar{x}_2\bar{x}_3 \vee x_2x_3 = f(\bar{x}_2, x_1, x_3) \\ &= x_2x_1 \vee x_2x_3 \vee \bar{x}_2\bar{x}_1\bar{x}_3 \vee x_1x_3 = f(\bar{x}_2, x_3, x_1) \\ &= x_2x_3 \vee x_2x_1 \vee \bar{x}_2\bar{x}_3\bar{x}_1 \vee x_3x_1 \end{aligned}$$

This proves that f is a symmetric function in the variables $x_1, \bar{x}_2, x_3$, namely,

$$\{0, 2, 3\}(x_1, \bar{x}_2, x_3)$$

As already mentioned, switching circuits which realize symmetric functions by means of elements modeling the logic function (6.15) can be designed very advantageously with the aid of the cascade method. In this case, it is sufficient to denote the functions which pertain to the individual nodes merely by sets $\{a_1, a_2, \ldots, a_m\}$. The decomposition by the function (6.15) at a node is performed by a simple transformation of the respective set $\{a_1, a_2, \ldots, a_m\}$. The transformation rules, which are summarized in Figure 6.14, are based on Theorem 6.12.

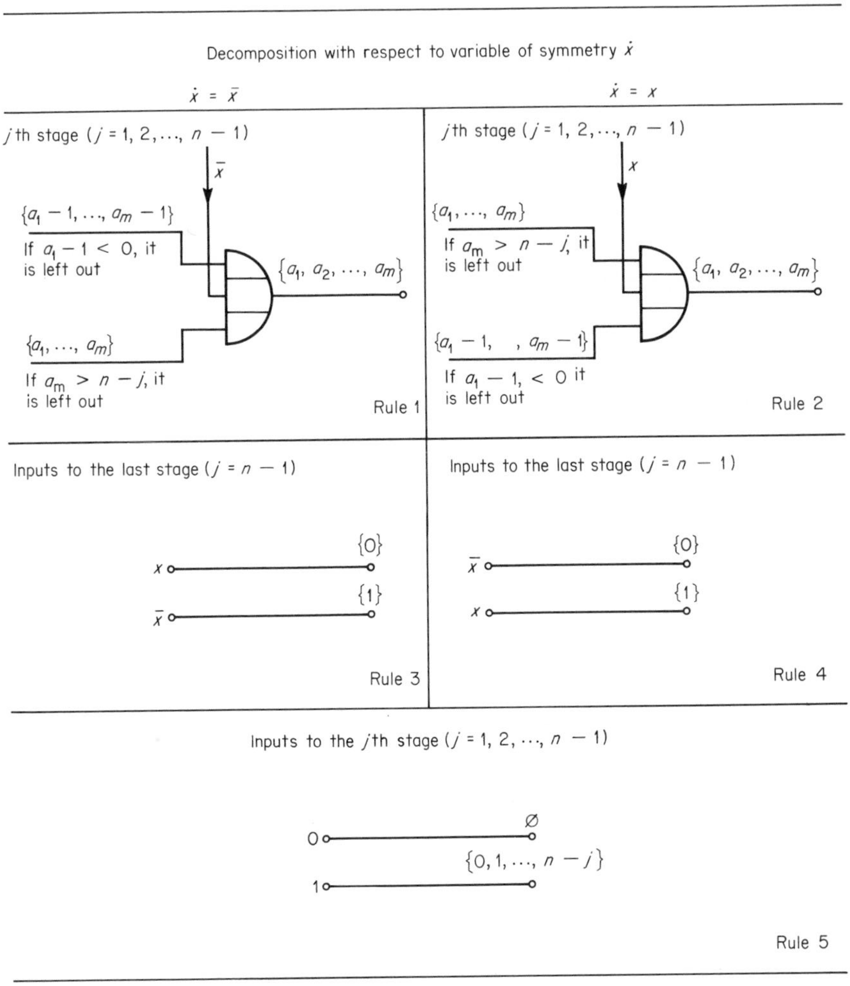

Figure 6.14 Decomposition rules in cascade circuits representing symmetric functions.

EXAMPLE 6.10 Design a cascade circuit implementing the symmetric function $\{1\}(\bar{x}_1, x_2, \bar{x}_3)$.

A possible implementation is shown in Figure 6.15. The output function is decomposed at the first stage ($j = 1$) with respect to variable $\bar{x}_3$. We use Rule 1 because the variable of symmetry is $\bar{x}_3$. This gives us functions $\{0\}(\bar{x}_1, x_2)$ and $\{1\}(\bar{x}_1, x_2)$. Variable x_2 is used for the decomposition at the second stage ($j = 2$). Since x_2 is the variable of symmetry now, we use Rule 2. We obtain functions $\{0\}(x_1)$, $\{1\}(x_1)$, and $\varnothing(x_1)$. We apply Rule 3 for the first two of them because the second stage is the last one and $\bar{x}_1$ is the variable of symmetry in this case. We apply Rule 5 for the function $\varnothing(x_1)$.

The designed circuit is described by equations

$$z_1 = \bar{x}_2 x_1$$

$$z_2 = x_2 x_1 \vee \bar{x}_2 \bar{x}_1$$

$$y = x_3 z_2 \vee \bar{x}_3 z_1$$

where z_1, z_2, y have the meaning introduced in Figure 6.15. When substituting for z_1 and z_2 in the third equation from the other equations, we obtain the given function; hence the design is correct.

Exercises to Section 6.6

6.6-1 Design a cascade circuit implementing symmetric function
- (a) $\{2\}(x_1, x_2, \bar{x}_3)$.
- (b) $\{3\}(\bar{x}_1, \bar{x}_2, \bar{x}_3)$.
- (c) $\{0, 1\}(x_1, x_2, x_3)$.
- (d) $\{1, 2, 3\}(x_1, \bar{x}_2, x_3)$.
- (e) $\{0, 3\}(\bar{x}_1, x_2, \bar{x}_3)$.

6.6-2 Design a cascade circuit whose outputs represent symmetric functions $\{1, 3\}(\bar{x}_1, x_2, \bar{x}_3, x_4)$, $\{3, 4\}(\bar{x}_1, x_2, \bar{x}_3, x_4)$, and $\{1, 2, 3\}(\bar{x}_1, x_2, \bar{x}_3, x_4)$. Use several different sequences of variables for the decomposition.

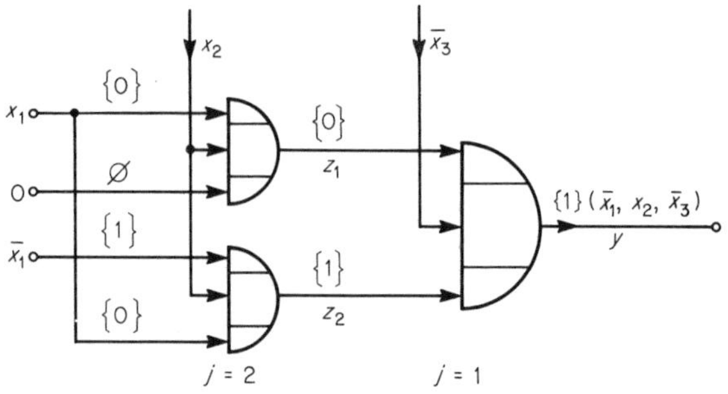

Figure 6.15 A cascade circuit implementing a symmetric function (Example 6.9).

6.6-3 Find which of the following definite logic functions are symmetric:
(a) $y = \bar{x}_1\bar{x}_2\bar{x}_3 \vee x_1\bar{x}_2 x_3 \vee \bar{x}_1 x_2 x_3$.
(b) $y = \bar{x}_4(x_1 \vee x_2) \vee x_4 x_3$.
(c) $y = \bar{x}_2\bar{x}_1(\bar{x}_4 \vee \bar{x}_3) \vee \bar{x}_4\bar{x}_3(\bar{x}_2 \vee \bar{x}_1)$.
(d) $f(s) = 1$ for $s = 3, 5, 6, 9, 10, 12$; $n = 4$.
(e) $f(s) = 1$ for $s = 7, 11, 13, 14, 15$; $n = 4$.
(f) $f(s) = 1$ for $s = 1, 2, 4, 7, 8, 11, 13, 14$; $n = 4$.

6.6-4 Prove Theorems 6.7 through 6.12.

6.7 BRANCH TYPE CIRCUITS

In this section we discuss some features pertaining to switching circuits built up by branch type elements. Relay contacts and cryotrons are the most common examples of branch type elements (Chapter 5).

Contact networks are used in this section to illustrate ways of analyzing and synthesizing memoryless switching circuits built up by branch-type elements. Whatever will be said about contact networks can be easily adapted for circuits containing other branch-type elements.

Every contact has two positions—the *rest* and the *working* position. In one position it is closed, in the other open. A contact closed in the rest position is called a *break contact*, that closed in the working position is called a *make contact*. By connecting a break contact with a make contact we obtain a *transfer contact*.

Contacts are always drawn in their rest position, in the manner shown in Figure 6.16a. A simplified way of drawing contacts is shown in Figure 6.16b. In this case a simple make (or break) contact is depicted by a line segment designated by the logic variable (or its negation, respectively) assigned to the physical variable (voltage, current) which controls the position of the pertinent contact. If we depict a contact network in this manner, we obtain an undirected graph, each link of which is marked by some variable or its negation. Such

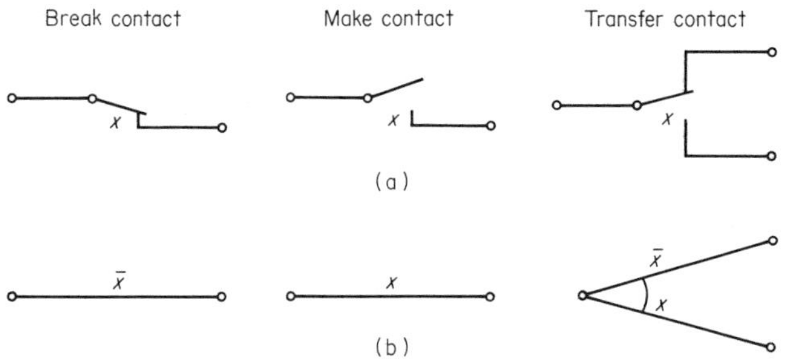

Figure 6.16 Ways of drawing contacts.

a graph is a universal tool for describing switching circuits built up by branch-type elements no matter what the specific physical principle involved is.

Assume that a switching circuit is built up of neutral relays. Then its basic block diagram is shown in Figure 6.17. *Input signals* (variables) $x_1, x_2, \ldots, x_p$ control relays $R_1, R_2, \ldots, R_p$ whose contacts form the contact network. *Output signals* (variables) $y_1, y_2, \ldots, y_q$ are determined from a node with a constant signal value corresponding to the logic value 1. This node, which will be called the *source* and denoted by S, is connected to the nodes representing the output signals through an appropriate contact network. An output variable is equal to one for a particular state of the input variables involved if and only if the contact network connects the source S to the corresponding output node for this state. Otherwise, the output variable is equal to zero.

One approach to synthesis of the contact network in Figure 6.17 consists of determining minimal Boolean forms representing the output variables as functions of the input variables. Each of these forms represents a series-parallel contact configuration which connects the source S with the respective output node and can be represented by a graph based on the conventions expressed by Figures 6.5 and 6.16. After the graph of the whole contact network (representing all the Boolean forms) is drawn, further simplifications can usually be made. For instance, we can merge partial branches pertaining to the same variable or introduce advantageous bridges which enable us to eliminate some branches.

EXAMPLE 6.11 Design a relay decoder that transforms—according to Table 6.3—decimal numbers binary coded in a given manner to a different binary-coded decimal representation.

First, we find minimal Boolean forms for the given logic functions:

$$y_1 = \bar{x}_1$$

$$y_2 = \bar{x}_2 x_1 \vee x_2 \bar{x}_1$$

$$y_3 = \bar{x}_3 x_1 \vee \bar{x}_3 x_2 \vee x_3 \bar{x}_2 \bar{x}_1$$

$$y_4 = x_4 x_3 \vee x_4 \bar{x}_2 \bar{x}_1$$

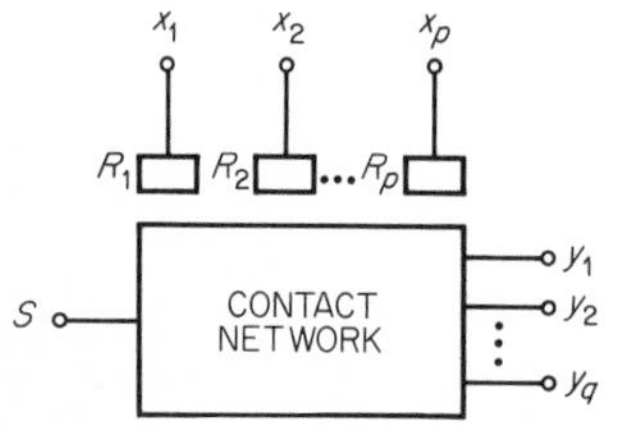

Figure 6.17 Basic block diagram of relay switching circuits.

The corresponding graph is in Figure 6.18a. Next, we try to merge some branches of this graph which have the same label. After each modification we must verify that the logic functions represented by the connections between the node S and the output nodes have not changed. In our example, the graph can be transformed to a simpler form that is shown in Figure 6.18b; the corresponding contact network is in Figure 6.18c. A still more simplified graph is shown in Figure 6.18d. This simplification has been achieved by introducing the bridge $\bar{x}_3$ between the nodes A and B. This permits the branch $\bar{x}_1\bar{x}_2$ between the node S and node B to be eliminated and leads to the saving of two contacts in the relay contact assembly. The corresponding contact network is in Figure 6.18e.

TABLE 6.3. Conversion of Codes Used in Example 6.11

Decimal Numbers Coded	x_4	x_3	x_2	x_1	y_4	y_3	y_2	y_1
0	0	0	0	0	0	0	0	1
1	0	1	1	1	0	0	0	0
2	0	1	1	0	0	0	1	1
3	0	1	0	1	0	0	1	0
4	0	1	0	0	0	1	0	1
5	1	0	1	1	0	1	0	0
6	1	0	1	0	0	1	1	1
7	1	0	0	1	0	1	1	0
8	1	0	0	0	1	0	0	1
9	1	1	1	1	1	0	0	0

In relay circuits it is often necessary to ascertain the instantaneous state of a particular group of relays. As a rule, the resource used for this purpose is a contact circuit called a "*tree*," in which two links, namely x_i and $\bar{x}_i$ for a particular i, emerge from each node. Hence, the tree contains solely transfer contacts.

Figure 6.19 shows all possible *complete trees* for three variables (relays). We see that the distribution of the number of contacts for the individual relays is 1,2,4 for the first tree and 1,3,3 for the second one. The number of possibilities rises sharply with an increasing n. For $n = 4$, there are already eight possibilities:

1248	1257	1266	1338
1347	1356	1446	1455

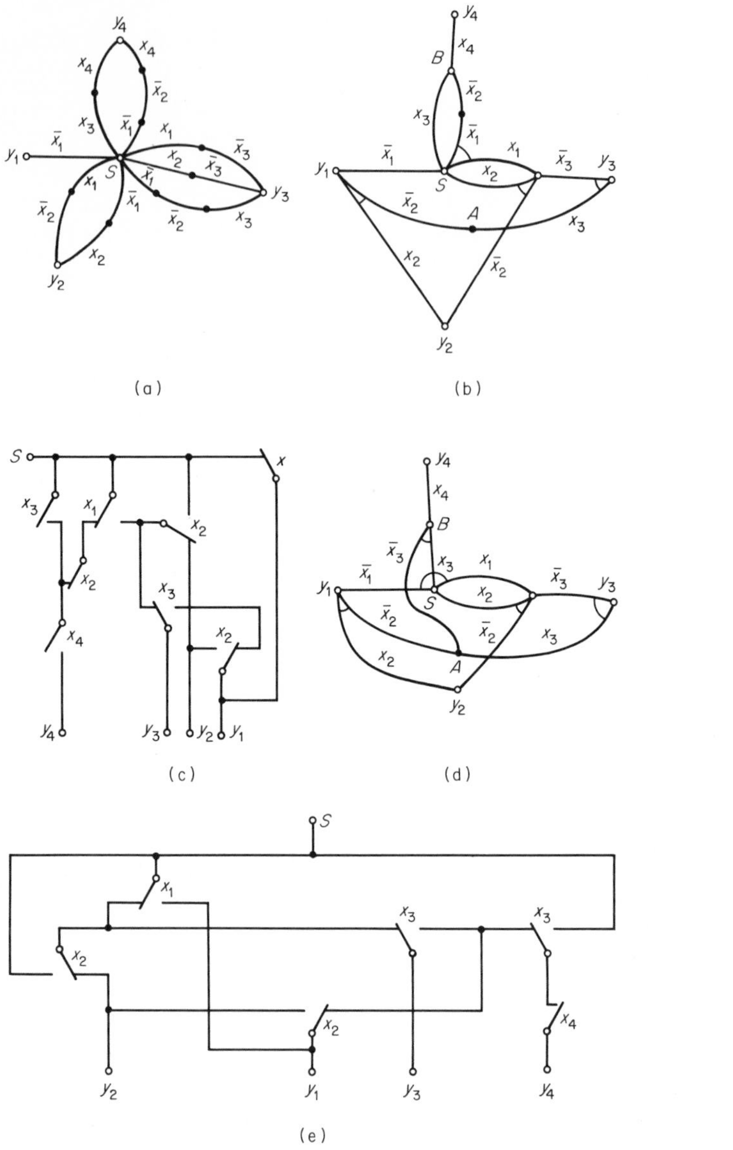

Figure 6.18 Modifications of the contact network discussed in Example 6.11.

For $n = 5$ there are as many as 105 possibilities, which are all listed in Table 6.4.

TABLE 6.4. Distribution of the Number of Contacts in a Complete Tree for Five Variables

1	2	2	10	16	1	3	3	11	13	1	4	5	9	12
1	2	2	11	15	1	3	3	12	12	1	4	5	10	11
1	2	2	12	14	1	3	4	7	16	1	4	6	6	14
1	2	2	13	13	1	3	4	8	15	1	4	6	7	13
1	2	3	9	16	1	3	4	9	14	1	4	6	8	12
1	2	3	10	15	1	3	4	10	13	1	4	6	9	11
1	2	3	11	14	1	3	4	11	12	1	4	6	10	10
1	2	3	12	13	1	3	5	6	16	1	4	7	7	12
1	2	4	8	16	1	3	5	7	15	1	4	7	8	11
1	2	4	9	15	1	3	5	8	14	1	4	7	9	10
1	2	4	10	14	1	3	5	9	13	1	4	8	8	10
1	2	4	11	13	1	3	5	10	12	1	4	8	9	9
1	2	4	12	12	1	3	5	11	11	1	5	5	5	15
1	2	5	7	16	1	3	6	6	15	1	5	5	6	14
1	2	5	8	15	1	3	6	7	14	1	5	5	7	13
1	2	5	9	14	1	3	6	8	13	1	5	5	8	12
1	2	5	10	13	1	3	6	9	12	1	5	5	9	11
1	2	5	11	12	1	3	6	10	11	1	5	5	10	10
1	2	6	6	16	1	3	7	7	13	1	5	6	6	13
1	2	6	7	15	1	3	7	8	12	1	5	6	7	12
1	2	6	8	14	1	3	7	9	11	1	5	6	8	11
1	2	6	9	13	1	3	7	10	10	1	5	6	9	10
1	2	6	9	12	1	3	8	8	11	1	5	7	7	11
1	2	6	10	11	1	3	8	9	10	1	5	7	8	10
1	2	7	7	14	1	3	9	9	9	1	5	7	9	9
1	2	7	8	13	1	4	4	6	16	1	5	8	8	9
1	2	7	9	12	1	4	4	7	15	1	6	6	6	12
1	2	7	10	11	1	4	4	8	14	1	6	6	7	11
1	2	8	8	12	1	4	4	9	13	1	6	6	8	10
1	2	8	9	11	1	4	4	10	12	1	6	6	9	9
1	2	8	10	10	1	4	4	11	11	1	6	7	7	10
1	2	9	9	10	1	4	5	5	16	1	6	7	8	9
1	3	3	8	16	1	4	5	6	15	1	6	8	8	8
1	3	3	9	15	1	4	5	7	14	1	7	7	7	9
1	3	3	10	14	1	4	5	8	13	1	7	7	8	8

In a complete tree it is thus possible to choose different distributions of the number of contacts for individual relays. This is sometimes very advantageous since the relays which form the tree are utilized as a rule, simultaneously in other circuits, where different numbers of contacts are used up for various relays.

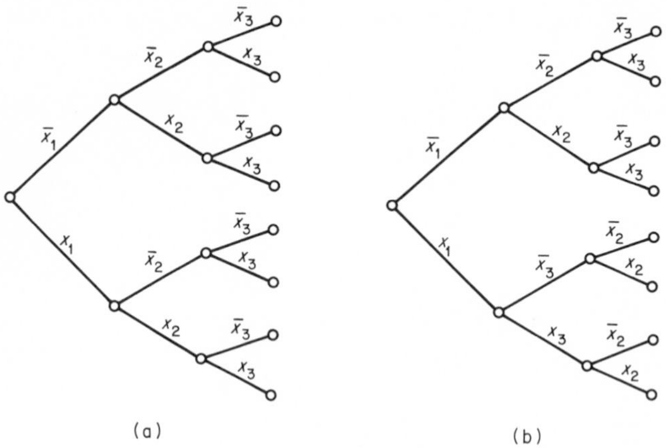

Figure 6.19 Complete trees for three variables.

EXAMPLE 6.12 Consider a complete tree which is to be made up of contacts of relays A,B,C,D, assuming that each relay has eight transfer contacts. If all the relay contacts are to be used up solely in the tree, then we shall employ the distribution 1,6,8,8,8, or the distribution 1,7,7,8,8. If, however, some of the contacts are used in other circuits, we choose such a distribution which makes the number of contacts in the individual relays, as far as possible, equal to an integer multiple of eight. The number of relays used will then be minimal. An example is presented in Table 6.5.

If the tree does not ascertain all states, we call it *partial*. The problem of constructing a partial tree may be formulated as follows. Let us have v different logic functions of n independent variables (v being the number of states

TABLE 6.5. Example Illustrating the Design of Trees

Relay	*A*	*B*	*C*	*D*	*E*		
Number of contacts already used	5	7	6	2	5	Number of relays	Number of transfer contacts
Suitable distributions	11	1	2	6	11	7	56
	3	1	10	14	3	7	56
Unsuitable distributions	10	1	2	16	2	8	64
	7	1	7	7	9	9	72
	8	1	3	15	4	10	80
	4	1	6	16	4	12	96

ascertained by the tree), each of which has the value 1 prescribed for only a single state of the independent variables. The states for which none of the functions has the value 1 are undefined for all functions. We are to design a contact switching circuit with one input node (the source) and v output nodes, where the connectivities between the source and the individual output nodes correspond to the given logic functions.

If we want to construct a partial tree with the minimal number of contacts, we can proceed as follows:

1. The states to be ascertained are consecutively numbered in a single chart.
2. We divide the chart into halves by placing on it some of the fundamental logic grids and then divide these halves into two halves again by means of another grid, etc., so that at least one ascertained state will always be left in each of the two halves. We thus obtain sub-charts of diminishing size.
3. To each of the newly obtained sub-charts we ascribe that variable or its negation, with the aid of whose grid it was produced.
4. If the sub-chart contains only one ascertained state, we do not divide it any further.
5. The procedure is completed when each sub-chart contains only one ascertained state. The corresponding tree is then determined directly from the variables or their negations ascribed to the individual sub-charts.

An example of the foregoing procedure, as applied to the functions illustrated in Figure 6.20 is shown in Figure 6.21.

It is easy to prove that the number of transfer contacts required to ascertain v states in a correctly designed tree is $v - 1$.

Special forms of contact networks have been suggested for the implementation of symmetric functions. These forms are usually referred to as *symmetric networks*. Except for some trivial cases, symmetric networks require significantly fewer contacts than networks based on the standard series-parallel implementation.

Any symmetric function can be implemented by the so-called *basic symmetric network*. For a particular number of variables, the basic symmetric network has a fixed structure. These structures are shown for two through six variables in Figure 6.22. For n variables, the basic symmetric network has one input node (source), labeled S, and $n + 1$ output nodes, labeled 0 through n.

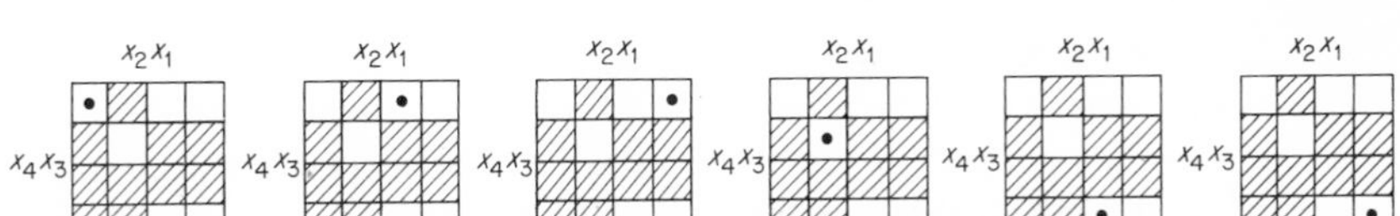

Figure 6.20 Example of specification for the construction of a partial tree.

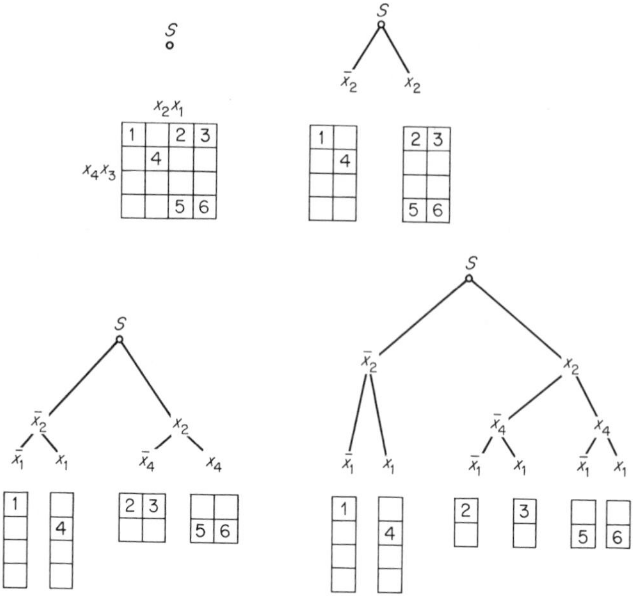

Figure 6.21 Design procedure for the partial tree specified in Figure 6.20.

If the input node is associated with logic value 1, then the output node labeled k represents the symmetric function $\{k\}(x_1, x_2, \ldots, x_n)$. By connecting two output nodes labeled k_1 and k_2, we obtain a new node which represents the symmetric function $\{k_1, k_2\}(x_1, x_2, \ldots, x_n)$. This rule can be clearly extended to three or more output nodes (Theorem 6.11). Thus, any symmetric function of n variables can be implemented by simply connecting together appropriate output nodes of the basic symmetric network. For instance, the node F in Figure 6.23a represents the symmetric function $\{1, 2, 4\}(x_1, x_2, x_3, x_4)$.

Symmetric networks formed by connecting together certain output nodes can be simplified. The first type of simplification consists in *excluding superfluous branches* in the graph (contacts) which lead only to the unused output nodes. An example is the elimination of branches leading to the output nodes 0 and 3 in Figure 6.23a, which produces Figure 6.23b.

The second type of simplification consists in *merging* two or more branches of the graph that pertain to the same variable. For instance, the branches labeled x_4 and connected to node 4 and node 1 can be combined in Figure 6.23b; the final result of the symmetric network representing function $\{1, 2, 4\}(x_1, x_2, x_3, x_4)$ is shown in Figure 6.23c.

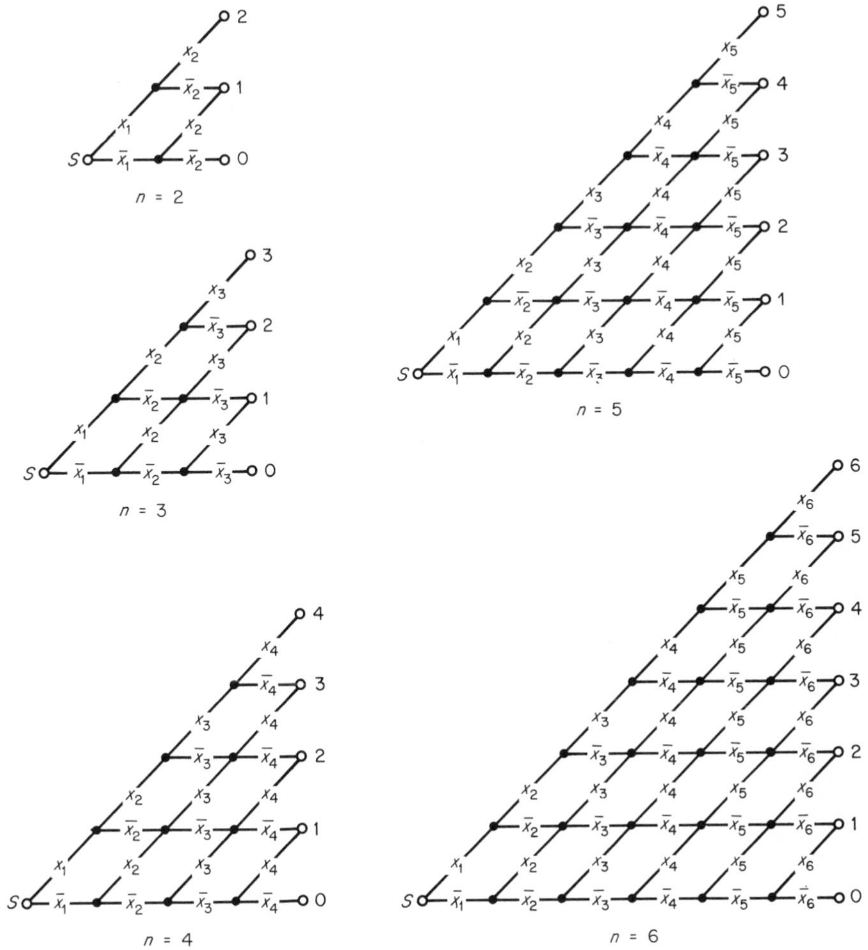

Figure 6.22 Basic symmetric networks.

The third type of simplification, usually called "*folding*," can be used for such symmetric functions $\{a_1, a_2, \ldots, a_m\}(x_1, x_2, \ldots, x_n)$ whose numbers $a_1, a_2, \ldots, a_m$ form an arithmetic progression containing all possible numbers between the starting number and n. For instance, functions $\{0, 2, 4\}(x_1, x_2, x_3, x_4)$ and $\{2, 3, 4\}(x_1, x_2, x_3, x_4)$ satisfy this condition. On the other hand, functions $\{0, 2\}(x_1, x_2, x_3, x_4)$ or $\{1, 2, 4\}(x_1, x_2, x_3, x_4\}$ do not satisfy it.

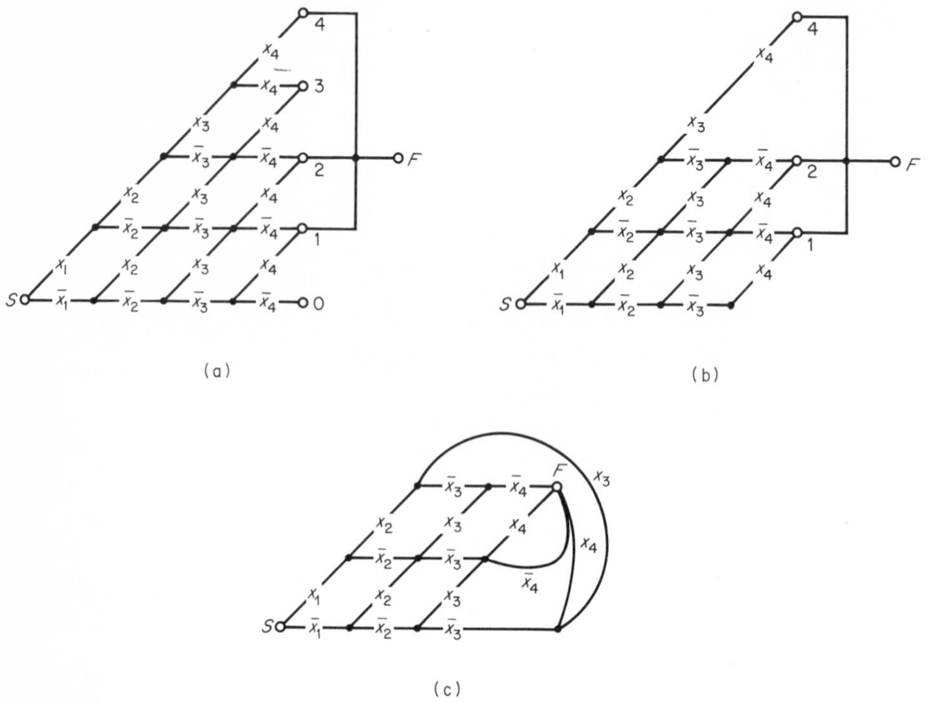

Figure 6.23 Symmetric network implementing symmetric function {1, 2, 4} (x_1, x_2, x_3, x_4).

The folding of the basic symmetric network means that diagonal branches on a certain level are changed to go down one or more levels rather than up. It is important that each of them ends in the same column of nodes as before.

The folding technique can be used to fold a level down d levels if the arithmetic progression $a_1, a_2, \ldots, a_m$ has a difference $d > 1$. Some examples are shown in Figure 6.24. If the folding technique is used, care must be taken to make sure that no sneak paths are introduced and the network really represents the given symmetric function.

A single basic symmetric network of n variables can be used for implementing several symmetric functions provided that the sets of numbers characterizing the individual functions are mutually disjoint. An example is shown in Figure 6.25.

So far we have considered only symmetric networks representing functions

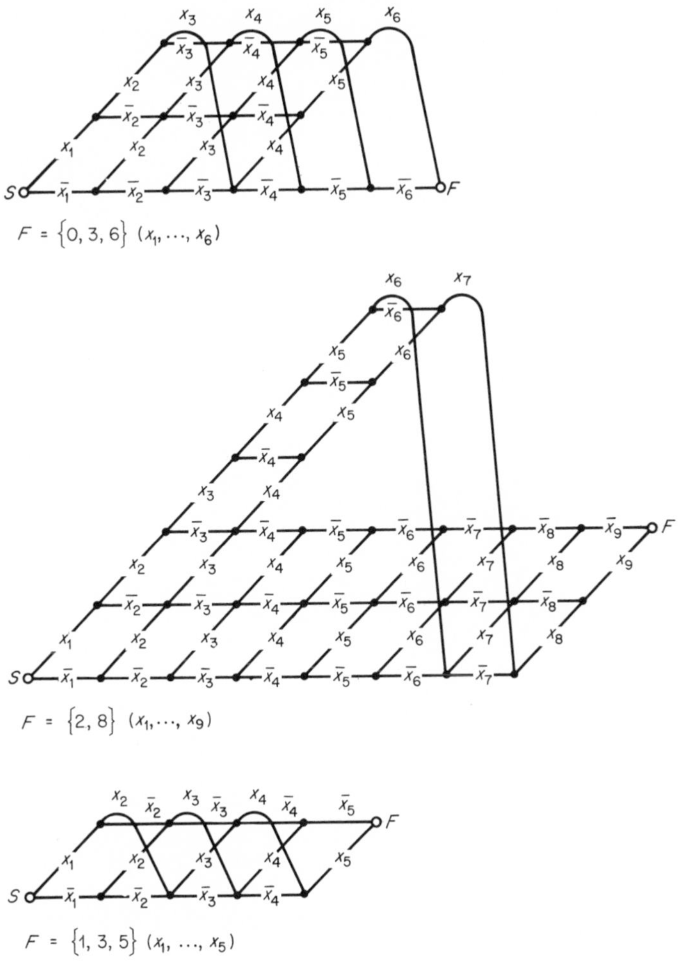

Figure 6.24 Illustration of the folding technique.

with asserted variables of symmetry. If negations are used as variables of symmetry, then we simply negate all labels in the basic symmetric network which represent the negated variables of symmetry. For instance, the basic network in Figure 6.26 is prepared for the variables of symmetry $\bar{x}_1, x_2, \bar{x}_3, \bar{x}_4, x_5$.

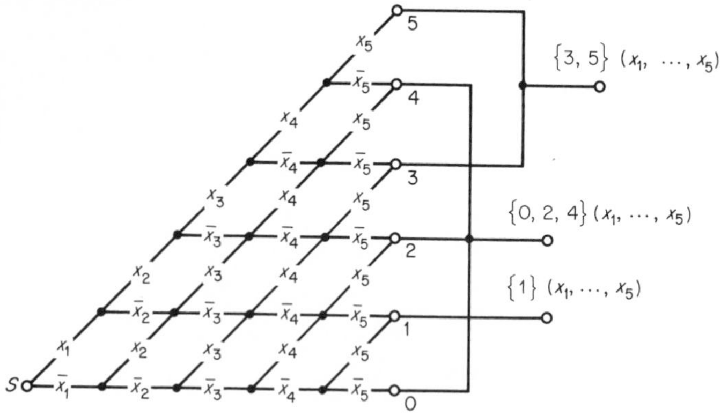

Figure 6.25 Basic symmetric circuit implementing three symmetric functions.

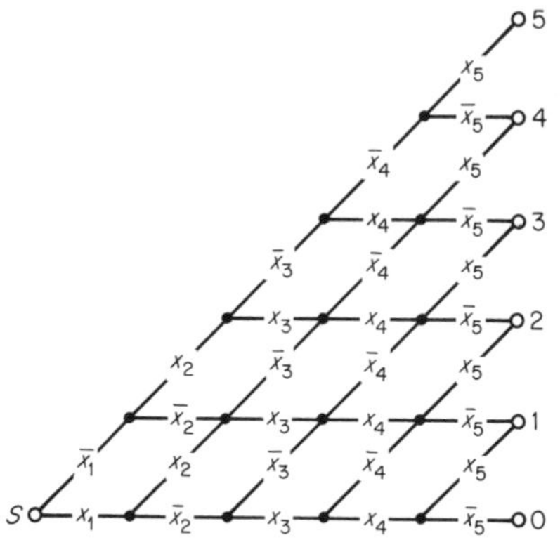

Figure 6.26 Basic symmetric network for variables of symmetry $\bar{x}_1$, x_2, $\bar{x}_3$, $\bar{x}_4$, x_5.

Exercises to Section 6.7

6.7-1 What is the minimum number of relays required to ascertain the states 0, 2, 7, 16, 21, 22, 32, 35, 37, 38, 49, 50, 61, 63 by means of a contact tree provided that each relay has only four transfer contacts?

6.7-2 Find contact models of various equalities of Boolean Algebra (distributive laws, laws of absorption, etc.).

6.7-3 Assume that the contact networks shown in Figure 6.27 have equal connectivities between the corresponding pairs of terminals (*A-B*, *A-C*, *B-C*). Assume further that f_A, f_B, f_C, f_{AB}, f_{AC}, f_{BC} are generally Boolean forms describing certain contact configurations. Using simultaneous Boolean equations find:

*(a) f_A, f_B, f_C if f_{AB}, f_{AC}, f_{BC} are given ($\Delta - Y$ transformation).

(b) f_{AB}, f_{AC}, f_{BC} if f_A, f_B, f_C are given ($Y - \Delta$ transformation).

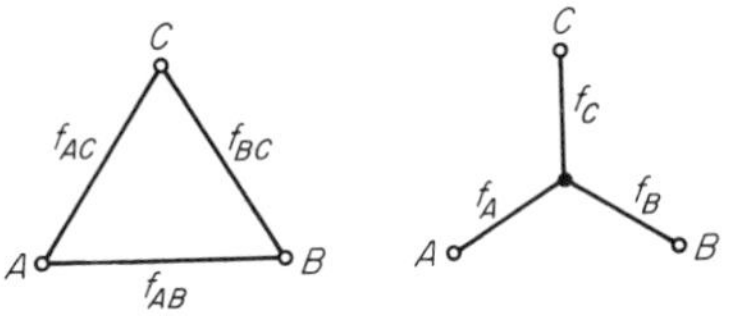

Figure 6.27 $\Delta - Y$ and $Y - \Delta$ transformations.

In each case, determine the number of solutions and the simplest Boolean solution.

6.7-4 Describe properties of the basic symmetric network for n variables.

6.7-5 Draw the basic symmetric network for $n = 7$.

6.7-6 Design symmetric networks implementing the following symmetric functions:

(a) $\{1, 2, 4, 6\}(x_1, \ldots, x_6)$.
(b) $\{4, 8\}(x_1, \ldots, x_9)$.
(c) $\{0, 2, 4\}(x_1, \ldots, x_5)$.
(d) $\{3, 4, 5, 6\}(x_1, \ldots, x_8)$.
(e) $\{3, 10\}(x_1, \ldots, x_{12})$.
(f) $\{1, 2, 4, 5\}(x_1, \ldots, x_6)$.
(g) $\{0, 3\}(x_1, x_2, x_3)$.

In each case, apply all three types of simplification, if possible.

6.7-7 Design such a contact network containing contacts of three switches that a lamp can be turned on or off by each switch independently of the other two.

6.7-8 Prove that the basic symmetric network of n variables contains $(n + 1)n/2$ transfer contacts.
Hint: Use mathematical induction.

6.7-9 Suppose that the contact networks shown in Figure 6.28 are equivalent as far as the connectivity between nodes A and B is concerned. Find a Boolean equation which describes the transformation of one of them to the other and vice versa.

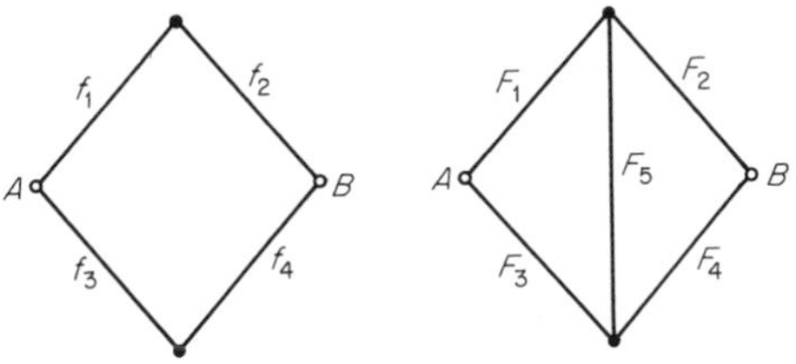

Figure 6.28 A transformation of a series-parallel contact network to an equivalent bridge network (Exercise 6.7-9).

6.8 UNIVERSAL LOGIC PRIMITIVES AND MODULES

A logic function that satisfies all the five conditions of a complete set (Section 2.7) will be called a *universal logic primitive*. Clearly, if an integrated circuit module represents a universal logic primitive, then it becomes a *universal building block module* in the sense that any logic function can be implemented by properly interconnected copies of this module. It is well known that the cost of switching circuits would be reduced if: (1) all modules within the circuits were identical, and (2) the function realized by this (universal) building-block module were selected so as to minimize on the average the number of copies of the module needed to implement any logic function.

Thus, universal logic primitives are candidates for being represented by integrated circuit modules. Each of the universal logic primitives with a fixed number n of logic variables which generates the maximum number of logic functions of m variables ($m \leq n$) by *biasing and duplicating its input variables* is considered as a successful candidate.

TABLE 6.6. Numbers of Functions Generated by Universal Logic Primitives of Three Variables

I. Negations are not available
II. Negations are available
A. Constants 0 and 1 are not available
B. Constants 0 and 1 are available
1. Total number of generated functions
2. Number of generated P-equivalence classes
3. Number of PN-equivalence classes generated

	Representants of P-Equivalence Classes of Universal Logic Primitives of Three variables													
	1	7	9	11	25	27	31	41	45	47	61	107	111	127
I, A, 1	7	9	9	12	12	15	9	9	15	12	12	9	9	7
I, A, 2	3	3	3	3	4	4	3	3	4	3	4	3	3	3
I, A, 3	3	3	3	3	4	4	3	3	4	3	4	3	3	3
I, B, 1	9	14	23	29	26	32	14	26	38	29	26	26	23	9
I, B, 2	5	6	8	8	9	9	6	9	11	8	9	9	8	5
I, B, 3	5	6	6	6	7	6	6	6	7	6	7	6	6	5
II, A, 1	27	43	31	43	62	62	43	27	62	43	62	27	31	27
II, A, 2	10	12	10	12	18	16	12	10	18	12	18	10	10	10
II, A, 3	4	4	4	4	7	7	4	4	7	4	7	4	4	4
II, B, 1	28	56	38	56	62	62	56	34	62	56	62	34	38	28
II, B, 2	11	16	13	16	18	16	16	13	18	16	18	13	13	11
II, B, 3	5	6	6	6	7	7	6	6	7	7	7	6	6	5

NAND and NOR functions are the only universal logic primitives of two variables. Either of them generates five functions if negated variables are not available and ten functions if they are available. Both of these functions are thus successful candidates for modules with two inputs.

For three variables, there are fifty nondegenerate universal logic primitives which can be classified into fourteen P-equivalence classes (Section 2.6). A list of representatives of these equivalence classes together with numbers of functions generated under various circumstances is given in Table 6.6, where functions are specified by their identifiers.

Next we shall derive a general formula for the number of all nondegenerate universal logic primitives of n variables and a procedure for their construction.

Universal logic primitives can be found only among the logic functions that are both non-zero and non-unity preserving. However, all of these functions, the number of which is 2^{2^n-2}, are nonmonotonic since $f(0, 0, \ldots, 0) > f(1, 1, \ldots, 1)$. To determine which of them are universal logic primitives, we have to test their self-duality and linearity. In this respect, we can take full advantage of the following theorem.

THEOREM 6.13 There does not exist any nonuniversal logic primitive which (i) does not preserve zero, (ii) does not preserve unity, (iii) is not monotonic, (iv) is not self-dual, (v) is linear.

PROOF Let us assume that a linear logic function $f(x_1, x_2, \ldots, x_n)$ satisfying conditions (i) through (iv) exists. Then this function is expressed by a linear form, say $(a_0 + a_1 x_1 + a_2 x_2 + \cdots + a_n x_n)$ (modulo 2).
We get $a_0 = 1$ from (i) and

$$1 + \sum_{i=1}^{n} a_i = 0 \text{ (modulo 2)}$$

from (ii). This equation is satisfied only if

$$\sum_{i=1}^{n} a_i = 1 \text{ (modulo 2)} \tag{6.18}$$

Notice that (iii) is always satisfied when both (i) and (ii) are satisfied. Now, owing to (iv), there exists a partition $\{\{x_j\}, \{x_k\} : j \in J, k \in K, J \cup K = \{1, 2, \ldots, n\}, J \cap K = \varnothing\}$ of variables $x_1, x_2, \ldots, x_n$ such that

$$a_0 + \sum_{j \in J} a_j = A \text{ (modulo 2)}$$

and

$$a_0 + \sum_{j \in K} a_k = A \text{ (modulo 2)}$$

where $A \in \{0, 1\}$. Hence,

$$2a_0 + \sum_{j \in J} a_j + \sum_{k \in K} a_k = 2A \text{ (modulo 2)}$$

This can be written as

$$\sum_{j \in J} a_j + \sum_{k \in K} a_k = 0 \text{ (modulo 2)}$$

or

$$\sum_{i=1}^{n} a_i = 0 \text{ (modulo 2)} \tag{6.19}$$

Equation (6.19) is contradictory to equation (6.18). Hence, the assumption of a linear logic function satisfying conditions (i) through (iv) is false. ▮

COROLLARY 6.4 All nonuniversal logic primitives that satisfy conditions (i) through (iii) of Theorem 6.13 are self-dual logic functions.

COROLLARY 6.5 A function is a universal logic primitive if and only if it satisfies conditions (i), (ii), and (iv) of Theorem 6.13.

THEOREM 6.14 The total number of all nonuniversal logic primitives of n variables that satisfy conditions (i) through (iii) of Theorem 6.13, $R(n)$, is given by

$$R(n) = 2^{2^{n-1}-1}$$

PROOF From Corollary 6.4, $R(n)$ must be the number of all self-dual functions that satisfy conditions (i) and (ii) (and, consequently, also condition (iii)). The condition of self-duality partitions states of n logic variables into 2^{n-1} pairs $(\dot{x}_1, \dot{x}_2, \ldots, \dot{x}_n)$ and $(\bar{\dot{x}}_1, \bar{\dot{x}}_2, \ldots, \bar{\dot{x}}_n)$ for which

$$f(\dot{x}_1, \dot{x}_2, \ldots, \dot{x}_n) = A$$

and

$$f(\bar{\dot{x}}_1, \bar{\dot{x}}_2, \ldots, \bar{\dot{x}}_n) = \bar{A}$$

where $A \in \{0, 1\}$. One of the pairs is fixed due to (i) and (ii). The remaining $2^{n-1} - 1$ pairs yield $2^{2^{n-1}-1}$ functions. ▮

COROLLARY 6.6 The total number $T(n)$ of all universal logic primitives of n variables is given by the formula

$$T(n) = 2^{2^n-2} - 2^{2^{n-1}-1} \tag{6.20}$$

It is desirable to find also the total number $N(n)$ of all nondegenerate universal logic primitives of n variables. Since $N(n) = 0$ for $n < 2$, clearly, the set of $T(n)$ universal logic primitives consists of nondegenerate universal logic primitives of two variables (there are $N(2) = 2$ of them and each can be applied to $\binom{n}{2}$ different pairs of variables), nondegenerate universal logic primitives of three variables (there are $N(3)$ of them and each can be applied to $\binom{n}{3}$ different triads of variables), etc., and finally, nondegenerate functions of n variables (there are $N(n)$ of them). Hence,

$$T(n) = \sum_{k=2}^{n} \binom{n}{k} N(k)$$

Solving this equation for $N(n)$ and substituting for $T(n)$ from (6.20), we obtain

$$N(n) = 2^{2^n-2} - 2^{2^{n-1}-1} - \sum_{k=2}^{n-1} \binom{n}{k} N(k)$$

By a rather tedious derivation, which is not included here, we can find that

$$N(n) = \sum_{k=2}^{n} (-1)^{n-k} \binom{n}{k} (2^{2^k-2} - 2^{2^{k-1}-1})$$

Some values of $T(n)$, $N(n)$, and $T(n)/2^{2^n}$ are in Table 6.7. It is trivial to prove

TABLE 6.7. Numbers of Universal Logic Primitives

n	1	2	3	4	5
$T(n)$	0	2	56	16,256	1,073,709,056
$N(n)$	0	2	50	16,044	1,073,628,316
$T(n)/2^{2^n}$	0	0.125	0.219	0.248	0.25

that

$$\lim_{n\to\infty} \frac{T(n)}{2^{2^n}} = 0.25$$

Hence, for a large number of variables, about 25% of all logic functions are universal logic primitives.

A procedure for constructing all universal logic primitives for a given number, n, of variables, which follows directly from Corollary 6.4, is suggested as follows:

(a) Determine all logic functions $f_i(x_1, x_2, \ldots, x_n)$ such that $f_i(0, 0, \ldots, 0) = 1$ and $f_i(1, 1, \ldots, 1) = 0$. There are 2^{2^n-2} such functions.

(b) Eliminate from the functions obtained under (a) all self-dual functions. There are $2^{2^{n-1}-1}$ of them.

Step (a) of the procedure is rather trivial. The construction of all self-dual functions in Step (b) can be considered, for instance, as the following construction of a matrix with 2^n rows representing naturally ordered states of the variables (and counted from zero) and $2^{2^{n-1}-1}$ columns representing the self-dual functions that preserve neither zero nor unity:

(α) Row 0 contains only ones.
(β) Row i, where $1 \leq i < 2^{n-1}$, repeats a sequence containing 2^{i-1} ones followed by 2^{i-1} zeroes.
(δ) Row $2^n - 1 - j$, where $1 \leq j < 2^{n-1}$, repeats a sequence of 2^{j-1} zeroes followed by 2^{j-1} ones.
(σ) Row $2^n - 1$ contains only zeroes.

The term "*universal logic module*" has been recently used in the literature for a memoryless logic module with m inputs that is capable of realizing every logic function of n variables ($n < m$) by an assignment of these variables, their negations, and constants 0 and 1 to the inputs of the module. Since the number of inputs of a module strongly effects its cost in the medium and large-scale integrated electronic circuits, major efforts have been devoted to the design of universal logic modules with minimum number of inputs.

So far, universal logic modules with the minimum number of inputs have been determined for $n \leq 3$. Examples of logic functions representing these modules are in Figure 6.29a. The function specified in Figure 6.29b represents a universal logic module for $n = 4$. It is not known whether this module has the minimum number of inputs since the theoretical lower bound of inputs for $n = 4$ is 6. However, the existence of the universal logic module with six inputs for $n = 4$ has been neither demonstrated nor disproved.

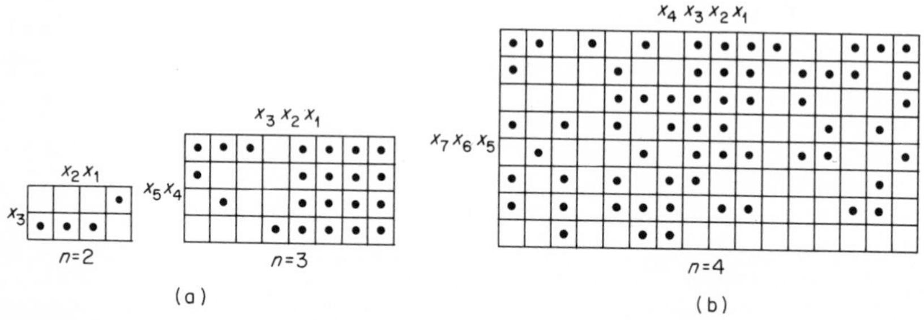

Figure 6.29 Examples of functions representing universal logic modules for $n = 2, 3, 4$.

The lower and upper bounds for the minimum number of inputs of the universal logic modules for $n \leq 10$, as known at this time, are given in the following table:

n	1	2	3	4	5	6	7	8	9	10
Lower bound	1	3	5	6	10	17	32	62	119	230
Upper bound	1	3	5	7	16	28	39	72	137	266

Exercises to Section 6.8

6.8-1 Determine the catalog of all functions generated by:
(a) NAND function of three variables.
(b) NOR function of three variables.

6.8-2 Show that the WOS module is a universal logic module for $n = 2$.

6.8-3 Determine all universal logic primitives for three variables.

6.8-4 Order functions M_3, F_3^2, F_3^1, AND (of three variables) on the basis of their logic power measured by the number of functions in their catalogs. Consider that:
(a) Only assertions of the variables are available.
(b) Assertions and negations of the variables are available.
(c) Assertions of the variables and constants 0 and 1 are available.
(d) Assertions and negations of the variables, and constants 0 and 1 are available.

Comprehensive Exercises to Chapter 6

6.1 Formulate the problem of synthesis of contact networks in terms of Boolean matrices.

6.2 Find a rule by means of which we can ascertain whether a particular element v_{ij} of the branch connection matrix is redundant.

6.3 Describe the change in the branch connection matrix if an internal node is
(a) eliminated.
(b) introduced.

6.4 Let s_x be the state identifier of logic variables x_4, x_3, x_2, x_1, let $s_x \leq 9$, and let s_y be the state identifier of logic variables y_4, y_3, y_2, y_1. Design a cascade circuit realizing the behavior:
(a) $s_y = s_x + 1$ (modulo 10).
(b) $s_y = s_x + 2$ (modulo 10).
(c) $s_y = s_x^2$ (modulo 10).
(d) $s_y = s_x^3 - 3s_x^2 + 2s_x - 5$ (modulo 10).
(e) $s_y = \lceil \log_2 s_x^3$ (modulo 10).

6.5 Perform Exercise 6.4 in terms of a circuit with:
(a) NAND elements.
(b) WOS elements.
(c) relay contacts.

6.6 Modify the cascade method to switching circuits with branch-type elements.

6.7 The binary full adder is a block capable of performing binary addition of three binary digits: Augent x, addend y, and carry-in c_i. It generates the sum s and the carry-out c_o given by the formulas:

$$s = x + y + c_i \text{ (modulo 2)}$$
$$c = (x + y + c_i - s)/2$$

Design the binary full adder by:

(a) AND, OR, NONEQUIVALENCE elements.
(b) relay contacts.
(c) WOS elements.
(d) the cascade method.

6.8 Assume that the augent x, the addend y, and the sum s are binary coded decimal digits. Let c_i and c_o denote, respectively, the carry-in and the carry-out. Then the behavior of the decimal binary coded full adder is described by the formulas

$$s = x + y + c_i \text{ (modulo 10)}$$
$$c = (x + y + c_i - s)/10$$

Design this adder in case that the so-called Rubinoff code (Table 6.8) is used for x, y, s and

(a) only NAND elements are used.
(b) the cascade method is used.

TABLE 6.8. Rubinoff Code

Decimal Digit	Binary Digits
0	0 0 0 0
1	0 1 1 1
2	0 1 1 0
3	0 1 0 1
4	0 1 0 0
5	1 0 1 1
6	1 0 1 0
7	1 0 0 1
8	1 0 0 0
9	1 1 1 1

A note: Denote binary digits of x, y, and s, respectively, by (x_4, x_3, x_2, x_1), (y_4, y_3, y_2, y_1), and (s_4, s_3, s_2, s_1).

6.9 Modify the concept of the binary full adder (Exercise 6.7) to define the binary full subtractor. Derive formulas for difference and borrow and implement them by:

(a) AND, OR, NONEQUIVALENCE elements.
(b) a cascade circuit.

*6.10 Suppose that only one NOT element and an unlimited number of AND and OR elements are available. Design, under these constraints, a switching circuit that realizes the logic function $f = \bar{x}_4 x_1 \vee x_2 \bar{x}_1 \vee \bar{x}_3 x_1$.
Hint: Consider nonstandard Boolean forms.

6.11 Let $(x_2 x_1)$ and $(y_2 y_1)$ be two two-digit binary numbers and let $(z_4 z_3 z_2 z_1)$ be the binary number representing their product. Design a switching circuit capable of performing this form of multiplication.

6.12 Let an element which implements the logic function $f = \bar{v}_1 \vee v_3 \bar{v}_2 \vee \bar{v}_3 v_2$ be called SPECIAL.
(a) Using only SPECIAL elements, design a switching circuit which realizes function $F = x_1 \vee \bar{x}_3 x_2 \vee x_3 \bar{x}_2$.
*(b) Can any logic function be realized solely by SPECIAL elements?

6.13 Show that the first order cascade element is a universal logic module for $n = 2$.

6.14 Design the fourth-order cascade element by:
(a) NAND elements.
(b) relay contacts.
(c) WOS module.

Reference Notations to Chapter 6

The idea of applying Boolean algebra to the analysis of branch-type switching circuits is due to Shannon [61]. The algebra of Boolean matrices and its application to the analysis of branch-type switching circuits was proposed by Aranovich as early as 1949 [1]. Later, other applications of Boolean matrices to switching circuits were investigated by Lunts [30, 31], Tsetlin [69], Hohn and Schissler [23], Semon [60], and others.

A precise formulation of structure properties of deterministic switching circuits was given by Burks and Wright [7]. The possibility of feedbacks in memoryless switching circuits, though suggested by Short in 1960 [65], has only recently received some publicity [24, 25].

The general approach to synthesis of memoryless switching circuits described in Section 6.3 was suggested by Marin [35, 36]. According to his observations, the method would be very efficient if the Boolean analyzer proposed by Svoboda [67] were a part of a general-purpose computer. Marin also presents a simulation of the Boolean analyzer [35]. Other general approaches to the synthesis of memoryless switching circuits employ various techniques of decomposition of logic functions. First extensive study of the functional decomposition was done by Ashenhurst [3]. Other significant publications dealing with this subject are References 58, 59, and 63. An excellent survey of decomposition techniques is in the book by Curtis [9].

Many papers are devoted to various problems associated with the synthesis of two-stage switching circuits based on the standard Boolean forms, e.g., References 34 and 68. The synthesis of three level NAND (sometimes referred to as TAND) or NOR circuits, which is meaningful when negations of input variables are not available, is also well represented in the literature [18, 21, 37, 40]. Other problems of circuits with NAND and NOR elements are discussed in References 10, 13. Hellerman [22] presents a catalog of the simplest NAND and NOR circuits implementing all logic functions of three variables. Mukhopadhyay and Schmitz [42] study the problem of minimizing the total number of equivalence and nonequivalence elements.

The development of the cascade method, which is due to Klir and Seidl [27], was stimulated by Povarov [50] and by the graphical method of Roginskij [56, 57]. A similar method was suggested by Mazer [38]. Lowenschuss [29] discusses some of the cascade elements.

The study of symmetric functions, which is well represented in the literature, was started by Pankajam [47, 48]. The basic symmetric circuits described in Section 6.7 were introduced by Shannon [62]. Several authors have investigated the problem of recognition of symmetric functions and identification of variables of symmetry [2, 5, 8, 12, 19, 33, 39, 44, 64]. Theorem 6.6 is due to Povarov [50, 51]. A transformation of a given logic function to a completely symmetric function was suggested by Born and Scidmore [6]. Algebraic properties of symmetric functions were studied by Arnold and Harrison [2].

The study of tree circuits was initiated by Shannon [62] and followed by Marcus [32], Meo [41], Prather [52, 53], and others. A good survey is given in Reference 43.

The material concerning universal logic primitives in Section 6.8 is taken from Reference 28. The idea of universal logic modules was suggested by Dunham [16, 17]. Various aspects of universal logic modules were investigated by Forslund and Waxman [20], King [26], Patt [49], Muller and Preparata [45, 54, 55], and others. A good survey is given in Reference 66.

From other problems associated with memoryless switching circuits, we should mention the problem of optimal coding of input and output states [4, 11, 14, 15, 46].

References to Chapter 6

1. **Aranovich, B. I.**, "An Application of Matrix Methods in Problems of Structure Analysis of Relay-Contact Circuits" (in Russian), *Avtomatika i telemekhanika* **10** (6) 437–451 (1949).
2. **Arnold, R. F.**, and **M. A. Harrison**, "Algebraic Properties of Symmetric and Partially Symmetric Functions," *TC* **EC-12** (3) 244–251 (June 1963).
3. **Ashenhurst, R. L.**, "The Decomposition of Switching Functions," *PHU*, 74–116.
4. **Bernstein, A. J.**, "Reducing Variable Dependency in Combinational Circuits," *CRSA* **5**, 156–164 (Oct. 1964).
5. **Biswas, N. N.**, "On Identification of Totally Symmetric Boolean Functions," *TC* **19** (7) 645–648 (July 1970).

6. **Born, R. C.,** and **A. K. Scidmore,** "Transformation of Switching Functions to Completely Symmetric Switching Functions," *TC* **C-17** (6) 596–599 (June 1968).
7. **Burks, A. W.,** and **J. B. Wright,** "Theory of Logical Nets," *Proc. IRE* **41** (10) 1357–1365 (Oct. 1953); reprinted in SM.
8. **Caldwell, S. H.,** "The Recognition and Identification of Symmetric Switching Functions," *TC* **EC-12** (3) 244–251 (June 1963).
9. **Curtis, H. A.,** *A New Approach to the Design of Switching Circuits.* Van Nostrand Reinhold, New York, 1962.
10. **Davidson, E. S.,** "An Algorithm for NAND Decomposition Under Network Constraints," *TC* **C-18** (12) 1098–1109 (Dec. 1969).
11. **Davis, W. A.,** "An Approach to the Assignment of Input Codes," *TC* **EC-16** (4) 435–442 (Aug. 1967).
12. **Dietmeyer, D. L.,** and **P. R. Schneider,** "Identification of Symmetry Redundancy and Equivalence of Boolean Functions," *TC* **EC-16** (6) 804–817 (Dec. 1967).
13. **Dietmeyer, D. L.,** and **Y. Su,** "Logic Design Automation of Fan-In Limited NAND Networks, *TC* **C-18** (1) 11–22 (Jan. 1969).
14. **Dolotta, T. A.,** and **E. J. McCluskey,** "Encoding of Incompletely Specified Boolean Matrices," *Western Joint Computer Conference, AFIPS Proc.* **17,** 231–238 (1960).
15. **Dolotta, T. A.,** "The Coding Problem in the Design of Switching Circuits," Ph.D. Dissertation, Princeton University, Princeton, N.J., May 1961.
16. **Dunham, B.,** and **J. H. North,** "The Use of Multipurpose Logical Device," *PHU*, 192–200.
17. **Dunham, B.,** et al., "The Multipurpose Bias Device," *IBMJ*, Part I, **1** (2) 117–129 (April 1957); Part II, **3** (1) 46–53 (Jan. 1959).
18. **Ellis, D. T.,** "A Synthesis of Combinational Logic with NAND or NOR Elements," *TC* **EC-14** (5) 701–705 (Oct. 1965).
19. **Epstein, G.,** "Synthesis of Electronic Circuits for Symmetric Functions," *TC* **EC-7** (1) 57–60 (March 1960).
20. **Forslund, D. C.,** and **R. Waxman,** "The Universal Logic Block (ULB) and Its Application to Logic Design," *CRSA* **7,** 236–250 (Oct. 1966).
21. **Gimpel, J. F.,** "The Minimization of TANT Networks," *TC* **EC-16** (1) 18–38 (Feb. 1967).
22. **Hellerman, L.,** "A Catalogue of Three-Variable OR-Invert and AND-Invert Logical Circuits," *TC* **EC-12** (3) 198–223 (June 1963).
23. **Hohn, F. E.,** and **L. R. Schissler,** "Boolean Matrices and the Design of Combinational Relay Switching Circuits," *BSTJ* **34** (1) 177–202 (Jan. 1955).
24. **Huffman, D. A.,** "Combinational Circuits with Feedback," *RDST*, 28–55.
25. **Kautz, W. H.,** "The Necessity of Closed Circuit Loops in Minimal Combinational Circuits," *TC* **C-19** (2) 162–164 (Feb. 1970).
26. **King, W. F., III,** "The Synthesis of Multipurpose Logic Devices," *CRSA* **7,** 227–235 (Oct. 1966).
27. **Klir, G. J.,** and **L. K. Seidl,** *Synthesis of Switching Circuits*, Iliffe, London, 1968; Gordon and Breach, New York, 1969.
28. **Klir, G. J.,** "On Universal Logic Primitives," *TC* **C-20** (4) 467–469 (April 1971).
29. **Lowenschuss, O.,** "Universal LSI Package for Implementing Control Logic Functions," *CD* **9** (9) 67–70 (Sept. 1970).
30. **Lunts, A. G.,** "The Application of Boolean Matrix Algebra to the Analysis and Synthesis of Relay Networks," *Dokl. Akad. Nauk U.S.S.R.* **70** (3) 421–423 (1950).

31. **Lunts, A. G.,** "Algebraic Methods of the Analysis and Synthesis of Contact Networks" (in Russian), *Izv. Akad. Nauk. U.S.S.R.* **16,** 405–426 (1952).
32. **Marcus, M. P.,** "Minimization of the Partially Developed Transfer Tree," *TC* **EC-6** (2) 92–95 (June 1957).
33. **Marcus, M.P.,** "The Detection and Identification of Symmetric Switching Functions with the Use of Tables of Combinations," *TC* **EC-5** (4) 237–239 (Dec. 1956).
34. **Marin, M. A.,** and **M. A. Melkanoff,** "Canonical and Minimal Forms for NAND and NOR Logic," Report No. 65-57, Dept. of Engineering, University of California at Los Angeles, 1965.
35. **Marin, M.A.,** "Investigation of the Field of Problems for the Boolean Analyzer," Ph.D. Dissertation, UCLA Report No. 68-28, Los Angeles, 1968.
36. **Marin, M. A.,** "On a General Synthesis Algorithm of Logical Circuits Using a Restricted Inventory of Integrated Circuits," Proc. of the Share ACM and IEEE Design Automation Workshop, Washington, D.C., 1968.
37. **Marin, M. A.,** "Synthesis of TANT Networks Using a Boolean Analyzer," *The Computer Journal* **12** (3) 259–267 (Aug. 1969).
38. **Mazer, L.,** "Topological Solution of Bilateral Switching Networks," *TC* **C-20** (2) 234–238 (Feb. 1971).
39. **McCluskey, E. J.,** "Detection of Group Invariance or Total Symmetry of a Boolean Function," *BSTJ* **35** (6) 1445–1453 (Nov. 1956).
40. **McCluskey, E. J.,** "Logical Design Theory of NOR Gate Networks with No Complemented Inputs," *CRSA* **4,** 137–148 (1963).
41. **Meo, A. R.,** "Modules Tree Structures," *TC* **C-17** (5) 432–442 (May 1968).
42. **Mukhopadhyay, A.,** and **G. Schmitz,** "Minimization of Exclusive OR and Logical Equivalence Switching Circuits," *TC* **C-19** (2) 132–140 (Feb. 1970).
43. **Mukhopadhyay, A.,** "Lupanov Decoding Networks," *RDST,* 57–83.
44. **Mukhopadhyay, A.,** "Detection of Total or Partial Symmetry of a Switching Function with the Use of Decomposition Charts," *TC* **EC-12** (5) 553–557 (Oct. 1963).
45. **Muller, D. E.,** "Universal Boolean Functions," *ACM SICACT News,* No. 1, 8-11 (April 1969).
46. **Nichols, A. J.,** and **T. H. Mott, Jr.,** "State Assignment in Combinational Networks," *TC* **EC-14** (3) 343–349 (June 1965).
47. **Pankajam, S.,** "On Symmetric Functions of n Elements in a Boolean Algebra," *J. Indian Math. Soc.* **2** (5) 198–210 (1936/37).
48. **Pankajam, S.,** "On Symmetric Functions of n Symmetric Functions in a Boolean Algebra," *Proc. Indian Acad. Sci. Sect. A* **9** (2) 95–102 (1939).
49. **Patt, Y. N.,** "A Complex Logic Module for the Synthesis of Combinational Switching Circuits," Proc. AFIPS Spring Joint Computer Conf., 699–705, 1967.
50. **Povarov, G. N.,** "Networks with One Input and k Outputs," *PHU,* Part II, 74–94.
51. **Povarov, G. N.,** "On the Group Invariance of Boolean Functions" (in Russian), in *Application of Logic in Science and Engineering,* Izd. Akad. Nauk U.S.S.R., Moscow, 1960, 263–340.
52. **Prather, R. E.,** "On Tree Circuits," *TC* **EC-14** (6) 841–851 (Dec. 1965).
53. **Prather, R. E.,** "Three Variable Multiple Output Tree Circuits," *TC* **EC-15** (1) 3–13 (Feb. 1966).
54. **Preparata, F. P.,** and **D. E. Muller,** "Generation of Near-Optimal Universal Boolean Functions," *J. Comp. and System Sci.* **4,** 93–102 (1970).

55. **Preparata, E. P.,** "On the Design of Universal Boolean Functions," *TC* **C-20** (4) 418–423 (April 1971).
56. **Roginskij, V. M.,** "A Graphical Method for the Synthesis of Multiterminal Contact Networks," *PHU*, Part II, 302–315.
57. **Roginskij, V. N.,** *The Synthesis of Relay Switching Circuits*, Van Nostrand Reinhold, New York, 1963.
58. **Roth, J. P.,** and **R. M. Karp,** "Minimization over Boolean Graphs," *IBMJ* **6,** (2) 227–238 (April 1962).
59. **Schneider, P. R.,** and **D. L. Dietmeyer,** "An Algorithm for Synthesis of Multiple-Output Combinational Logic," *TC* **C-17** (2) 117–128 (Feb. 1968).
60. **Semon, W.,** "Matrix Methods in the Theory of Switching," *PHU*, Part II, 13–50.
61. **Shannon, C. E.,** "A Symbolic Analysis of Relay and Switching Circuits," *AIEE Trans.* **57,** 713–723 (1938).
62. **Shannon, C. E.,** "The Synthesis of Two-Terminal Switching Circuits," *BSTJ* **28** (1) 59–98 (Jan. 1949).
63. **Shen, V. Y.,** and **A. C. McKellar,** "An Algorithm for the Disjunctive Decomposition of Switching Functions," *TC* **C-19** (3) 239–248 (March 1970).
64. **Sheng, C. L.,** "Detection of Totally Symmetric Boolean Functions," *TC* **EC-14** (6) 924–926 (Dec. 1965)
65. **Short, R. A.,** "A Theory of Relations between Sequential and Combinational Realizations of Switching Functions," Stanford Electronics Lab., Techn. Dept. 098-1, Menlo Park, Calif., 1960.
66. **Stone, H. S.,** "Universal Logic Modules," *RDST*, 230–254.
67. **Svoboda, A.,** "Boolean Analyzer," Proc. of the IFIP Congress, Edinburgh, Scotland, Aug. 1968. Booklet D, 97–102, North-Holland Publ. Co., Amsterdam, 1968.
68. **Todd, C. D.,** "An Annotated Bibliography on NOR and NAND Logic," *TC* **EC-12** (5) 462–464 (Oct. 1963).
69. **Tsetlin, M. L.,** "The Application of Matrix Calculus to the Synthesis of Relay-Contact Networks," *Dokl. Akad. Nauk. U.S.S.R.*, **86** (3) 525–528 (1952).

7

DETERMINISTIC SEQUENTIAL SWITCHING CIRCUITS: ABSTRACT THEORY

7.1 INTRODUCTION

We encountered the concept of the *sequential switching circuit* in Chapter 1. We ascribed this designation to those switching circuits in which the values of the output signals are determined not only by the instantaneous values of the input signals but also depend on some sequences of past input signals. These sequences must somehow be stored within the system. Although there exist various forms (direct or indirect) of storing the input sequences, we often do not care to consider them all. We are satisfied with a single paradigm (or an abstract model) provided that it is sufficiently general (can substitute for all other forms at the abstract level).

An important paradigm of deterministic switching circuits, which has been widely used, is based on the assumption that, at any discrete time, all information from the past pertinent to a determination of the present response of the system is represented by the so-called *internal state*. It follows from this assumption that at any particular discrete time: (i) only the internal state is stored in the system; (ii) the response of the system is determined by the stimulus and the internal state stored in the system; (iii) a new internal state is generated on the basis of the present internal state and the present stimulus.

The paradigm outlined above is usually called the *Mealy finite-state machine* (or *Mealy sequential machine*). Let us define it more precisely now.

DEFINITION 7.1 The Mealy finite-state machine is a quintuple

$$M = (X, Y, Z, \mathbf{f}, \mathbf{g})$$

where X is a finite set of input states (stimuli), Y is a finite set of output states (responses), Z is a finite set of internal states, **f** is an *output function*.

$$\mathbf{y}^t = \mathbf{f}(\mathbf{x}^t, \mathbf{z}^t) \tag{7.1}$$

and **g** is a *transition* (or *next state*) *function*

$$\mathbf{z}^{t+1} = \mathbf{g}(\mathbf{x}^t, \mathbf{z}^t) \tag{7.2}$$

where $\mathbf{x}^t \in X$, $\mathbf{y}^t \in Y$, $\mathbf{z}^t \in Z$ are, respectively, the *stimulus*, the *response*, and the *internal state* at time $t (t = 0, 1, \ldots)$, and $\mathbf{z}^{t+1} \in Z$ is the internal state at time $t + 1$ (*next state*).

An alternative finite-state machine, which is usually referred to as the *Moore machine*, differs from the Mealy machine only in the output function. Responses of the Moore machine are completely determined by the internal states. Thus,

$$\mathbf{y}^t = \mathbf{f}(\mathbf{z}^t) \tag{7.3}$$

is the output function for the Moore machine.

The Mealy machine can be considered as a coupling of two elements shown in Figure 7.1. The element called the *functional generator* is a deterministic memoryless system which implements the output and the transition (next state) functions. The element called the *memory* represents a delay of one discrete time. Any signal which enters the memory at time t appears at its output without any change at time $t + 1$. In the case of the finite-state machine, the memory is used only for the internal state. At any time t, the memory stimulates the functional generator with the present state $\mathbf{z}^t$. At the same time, the functional generator produces the next state $\mathbf{z}^{t+1}$. The next state enters the memory and becomes the present state one unit of time later.

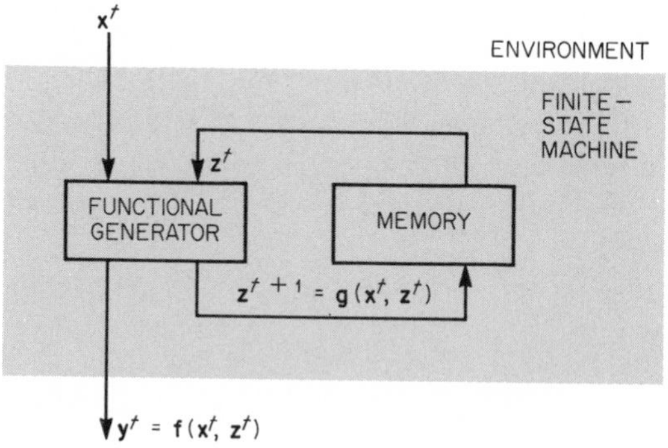

Figure 7.1 A block diagram of the Mealy machine.

An example of the Mealy machine is shown in Figure 7.2a. It has one input variable x, one output variable y, and one variable z representing the internal state. All these variables are two-valued. The memory contains one delay element, which is denoted by the triangle. The functional generator contains one element of equivalence and one of nonequivalence. The output and transition functions are shown in Figure 7.2b.

A block diagram of the Moore machine is shown in Figure 7.3. Two functional generators are shown here stressing thus that the domain of the output function differs from the domain of the transition function in the Moore machine.

The Mealy machine is used consistently in this text. The Moore machine is referred to only occasionally. However, the transformation from a particular Mealy machine to the equivalent Moore machine and vice versa is very simple. The equivalence is considered here from the standpoint of the external performance.

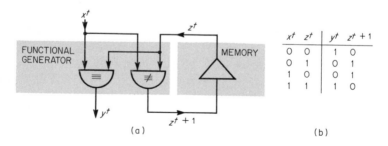

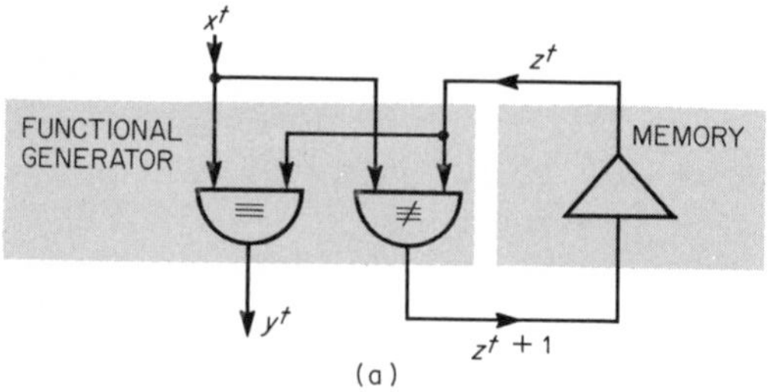

(a)

x^t	z^t	y^t	z^{t+1}
0	0	1	0
0	1	0	1
1	0	0	1
1	1	1	0

(b)

Figure 7.2 An example of the Mealy machine.

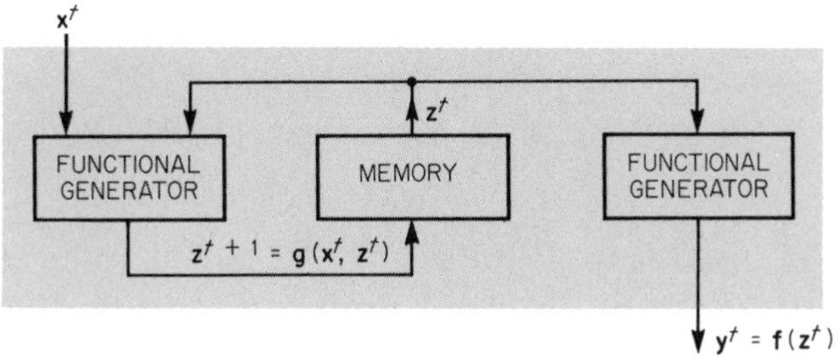

Figure 7.3 A block diagram of the Moore machine.

Let us note that the finite-state machine (no matter whether Mealy or Moore) represents a *deterministic discrete system* defined by its state-transition structure (or ST-structure) according to the terminology introduced in Chapter 1. Let us further note that the definition of the finite-state machine does not assign any meaning to the stimuli, responses, and internal states involved. When concerned with switching circuits, each of these is represented by states of some two-valued variables. The finite-state machine includes this case but, at the same time, covers many other cases. Thus, the finite-state machine is an abstract model of ST-structure of deterministic sequential switching circuits. It reflects some but not all properties of this class of switching circuits.

Some (but by no means all) problems concerning deterministic sequential switching circuits can be completely solved in terms of the finite-state model. This means that a methodology elaborated for finite-state machines constitutes a portion of the methodology of deterministic switching circuits. This portion is usually referred to as the *abstract theory of deterministic switching circuits*. Problems belonging to the abstract theory can be classified as follows:

(i) A transformation from an input/output description of an external performance of a system to the standard form of an equivalent finite-state machine. This problem, which often is called the *abstract synthesis*, is remarkably resistant to a fully general and algorithmic solution. Some approaches to its solution are outlined in this chapter.

(ii) Various modifications of a given finite-state machine. We are mainly interested in modifications converging to the minimum of the number of internal or input states but preserving, at the same time, all properties concerning the external performance of the system. For such cases we speak about the *state minimization*. It has been completely solved and many methods for its solution are available now. Some of these methods are described in this chapter.

(iii) Determination of various properties of the system from its finite-state model. Some of these problems are relevant to structure theory of switching circuits and, consequently are included in Chapter 8.

(iv) Various black-box problems.

Exercises to Section 7.1

7.1-1 Find at least three systems from the field of your interest which can be described as finite-state machines. Determine for each of them the output and transition functions for both the Mealy and the Moore alternatives.

7.1-2 Show that $\mathbf{y}^t$ is the response to $\mathbf{x}^t$ in the Mealy machines but y^{t+1} is the response to $\mathbf{x}^t$ in the Moore machine.

7.2 REPRESENTATIONS OF OUTPUT AND TRANSITION FUNCTIONS

There are three principal, widely used forms of representing the output and transition functions of the finite-state machine. These are tables, diagrams, and matrices. All of them represent, essentially, the ST-structure. Therefore, they will be called ST-*tables*, ST-*diagrams*, and ST-*matrices*.

Let $X = \{\mathbf{x}_1, \mathbf{x}_2, \ldots, \mathbf{x}_p\}$, $Y = \{\mathbf{y}_1, \mathbf{y}_2, \ldots, \mathbf{y}_q\}$, and $Z = \{\mathbf{z}_1, \mathbf{z}_2, \ldots, \mathbf{z}_r\}$ be, respectively, the set of stimuli, responses, and internal states. Then the ST-*table* has the format shown in Figure 7.4. A specific example of a Mealy machine with three stimuli, two responses, and five internal states is shown in Figure 7.5. Elements of all the three sets are represented by integers here.

Another type of the *ST-table* was used in Figure 7.2b. Its format is

$\mathbf{x}^t$ $\mathbf{z}^t$	$\mathbf{y}^t$ $\mathbf{z}^{t+1}$

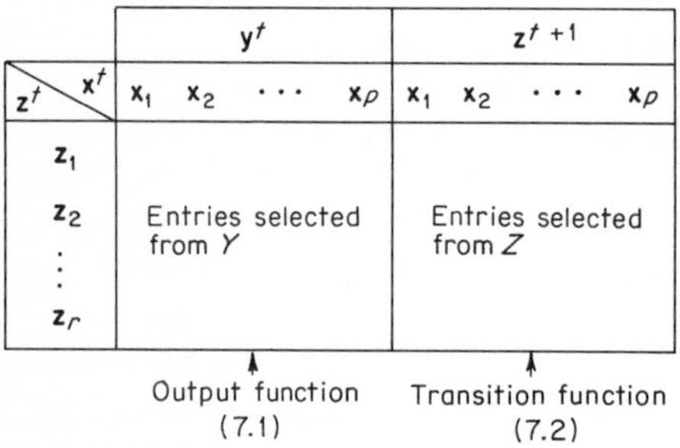

Figure 7.4 A format of the ST-table.

	y^t			z^{t+1}		
z^t \ x^t	0	1	2	0	1	2
1	1	0	0	2	2	3
2	0	1	1	1	2	2
3	1	0	0	4	2	1
4	0	1	1	1	5	4
5	0	1	1	3	5	3

	y^t		z^{t+1}	
z^t \ x^t	0	1	0	1
1	–	–	–	2
2	–	–	4	3
3	0	–	1	–
4	1	–	1	–

Figure 7.5 An example of the ST-table.

In the left-hand side of this table, all pairs $\mathbf{x}_i$, $\mathbf{z}_j$ are included ($i = 1, 2, \ldots, p$ and $j = 1, 2, \ldots, r$). In its right-hand side, the assigned values of $\mathbf{y}^t$ and $\mathbf{z}^{t+1}$ are placed.

The ST-*diagram* is a directed graph whose nodes (drawn as small circles) represent internal states and whose arcs (drawn as directed lines between pairs of the circles) represent the output and transition functions. Each directed line (arc) is described by at least one pair of a stimulus $\mathbf{x}_i$ and a response $\mathbf{y}_j$ ($i = 1, 2, \ldots, p$ and $j = 1, 2, \ldots, q$). Pairs of stimuli and responses are usually written in the form $\mathbf{x}_i/\mathbf{y}_j$ and are called *input-output pairs*.

An element of the ST-diagram is shown in Figure 7.6. It has this meaning: If the present internal state of the Mealy machine is $\mathbf{z}_k$ and the present stimulus is $\mathbf{x}_i$, then the response is $\mathbf{y}_j$ and the next state is $\mathbf{z}_l$. Obviously, a line in the diagram may be described by several input-output pairs, one for each stimulus.

An example of the ST-diagram is shown in Figure 7.7. It represents the Mealy machine whose ST-table is in Figure 7.5. The advantage of the ST-diagram is that it enables the investigator to visualize various properties of the machine enhancing his intuition.

The ST-*matrix* $\mathbf{M} = [m_{ij}]$ of a Mealy machine with r internal states has r rows and r columns. The rows correspond to the present state and the columns correspond to the next state of the machine. The entry m_{ij} at the intersection of the ith row and jth column is the set union of all input-output pairs associated with the transition from state $\mathbf{z}_i$ to state $\mathbf{z}_j$.

The Mealy machine, whose ST-table is in Figure 7.5 and whose ST-diagram is in Figure 7.7 is represented by the following ST-matrix $\mathbf{M}$, where rows and

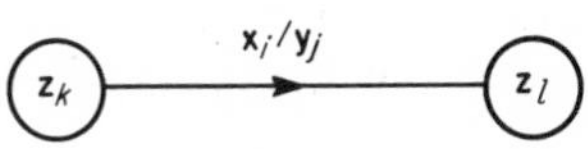

Figure 7.6 An element of the ST-diagram.

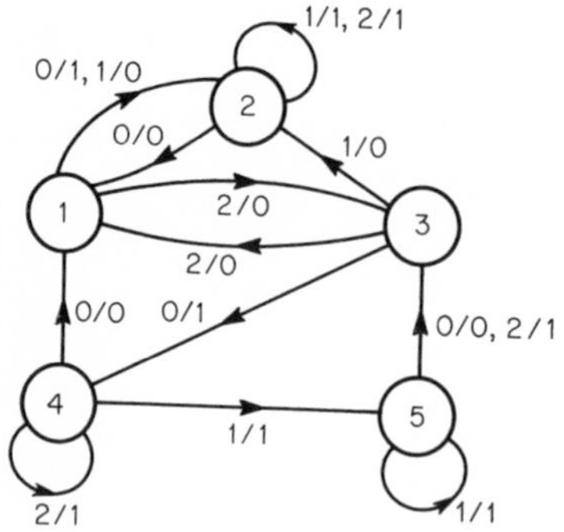

Figure 7.7 An example of the ST-diagram.

columns are labeled for clarity by symbols of the corresponding states:

$$\begin{array}{c} \\ 1\\2\\3\\4\\5 \end{array}\begin{array}{c} \begin{array}{ccccc} \quad 1 \quad & \quad 2 \quad & \quad 3 \quad & 4 & 5 \end{array} \\ \begin{bmatrix} \varnothing & 0/1 \cup 1/0 & 2/0 & \varnothing & \varnothing \\ 0/0 & 1/1 \cup 2/1 & \varnothing & \varnothing & \varnothing \\ 2/0 & 1/0 & \varnothing & 0/1 & \varnothing \\ 0/0 & \varnothing & \varnothing & 2/1 & 1/1 \\ \varnothing & \varnothing & 0/0 \cup 2/1 & \varnothing & 1/1 \end{bmatrix} = \mathbf{M} \end{array}$$

Symbol $\cup$ denotes the set union, $\varnothing$ represents the empty set.

Let $[a_{ij}]$ and $[b_{ij}]$ be two ST-matrices with r rows and r columns each. Then, the *matrix product* of these matrices, which is a ST-matrix $[c_{ij}]$, is defined by the formula

$$c_{ij} = \bigcup_{k=1}^{r} a_{ik} b_{kj} \tag{7.4}$$

The formula resembles the formula for the ordinary matrix multiplication. However, there are two differences: (1) Formula (7.4) uses the operation of set union instead of addition. (2) Formula (7.4) uses the operation of concatenation instead of multiplication.

The *concatenation* (or linking together) is an operation by which sequences of certain elements are constructed. In our case, the elements will be input-output pairs. For instance, the concatenation of 0/1 and 2/0 is the sequence (0/1, 2/0), the concatenation of the latter with the sequence (1/1, 0/0, 1/0) produces the sequence (0/1, 2/0, 1/1, 0/0, 1/0), etc.

To perform correctly the matrix operations (7.4), we have to treat properly the set union and the concatenation. Let a, b, c, d be some input-output pairs. Then the following list is a summary of basic properties of the operations involved:

1. $a \cup b = b \cup a$ but $ab \neq ba$ (the concatenation is not commutative).

2. $(a \cup b) \cup c = a \cup (b \cup c)$ and $(ab)c = a(bc)$ (both of the operations are associative).

3. $a(b \cup c) = (ab) \cup (ac)$ but $a \cup (bc) \neq (a \cup b)(a \cup c)$ (the distributive law holds only with respect to the concatentation); $(a \cup b)\,(c \cup d) = ac \cup ad \cup bc \cup bd$.

4. $a \cup a = a$ but $aa \neq a$ (the concatenation is not idempotent).

5. $a \cup \varnothing = a$ and $a\varnothing = \varnothing a = \varnothing$ (identity laws for the set union and concatenation).

Application of formula (7.4) in switching circuits consists primarily of calculations of higher-order ST-matrices. It follows directly from the matrix product (7.4) and from the properties of the set union and the concatenation that the kth order matrix $\mathbf{M}^k$ of a ST-matrix $\mathbf{M}$ contains all sequences containing k input-output pairs which are meaningful for the corresponding finite-state machine. More specifically, its entry at the intersection of the ith row and the jth column is a union of all possible sequences containing k input-output pairs which start in state $\mathbf{z}_i$ and terminate in state $\mathbf{z}_j$. Such sequences are usually said to have *length k*.

As an example, let us calculate square $\mathbf{M}^2$ of the above given ST-matrix $\mathbf{M}$. To simplify the procedure, let the symbol m_{ij}, be substituted for that entry of $\mathbf{M}$ which is different from $\varnothing$ and lies in the ith row and jth column, i.e.,

$$\mathbf{M} = \begin{bmatrix} \varnothing & m_{12} & m_{13} & \varnothing & \varnothing \\ m_{21} & m_{22} & \varnothing & \varnothing & \varnothing \\ m_{31} & m_{32} & \varnothing & m_{34} & \varnothing \\ m_{41} & \varnothing & \varnothing & m_{44} & m_{45} \\ \varnothing & \varnothing & m_{53} & \varnothing & m_{55} \end{bmatrix}$$

Then,

$$\mathbf{M}^2 = \begin{bmatrix} m_{12}m_{21} \cup m_{13}m_{31} & m_{12}m_{22} \cup m_{13}m_{32} & \varnothing & m_{13}m_{34} & \varnothing \\ m_{22}m_{21} & m_{21}m_{12} \cup m_{22}m_{22} & m_{21}m_{13} & \varnothing & \varnothing \\ m_{32}m_{21} \cup m_{34}m_{41} & m_{31}m_{12} \cup m_{32}m_{22} & m_{31}m_{13} & m_{34}m_{44} & m_{34}m_{45} \\ m_{44}m_{41} & m_{41}m_{12} & m_{41}m_{13} \cup m_{45}m_{53} & m_{44}m_{44} & m_{44}m_{45} \cup m_{44}m_{55} \\ m_{53}m_{31} & m_{53}m_{32} & m_{55}m_{53} & m_{53}m_{34} & m_{55}m_{55} \end{bmatrix}$$

For instance, $m_{11}^2 = m_{12}m_{21} \cup m_{13}m_{31} = (0/1 \cup 1/0)0/0 \cup (2/0, 2/0) = (0/1, 0/0) \cup (1/0, 0/0) \cup (2/0, 2/0)$. Indeed, these are all possible input-output sequences of length 2 which start in state 1 and terminate in the same state. Figure 7.7 is suitable for a verification of this result. Similarly, $m_{32}^2 = m_{31} m_{12} \cup m_{32}m_{22} = 2/0(0/1 \cup 1/0) \cup 1/0(1/1 \cup 2/1) = (2/0,\ 0/1) \cup (2/0,\ 1/0) \cup (1/0,\ 1/1) \cup (1/0,\ 2/1)$. This, again, is the union of all possible input-output sequences of length 2 which start in state 3 and terminate in state 2.

So far, we have considered that the output and transition functions are completely specified. In many practical cases, however, these functions are incompletely specified. This may be due to the following reasons:

1. The system under consideration is an element of a higher-order system and we know the constraints of the latter. More specifically, we know that only some input sequences are meaningful. Since internal states represent certain input sequences, we may conclude that certain stimuli cannot be applied at some states.

2. Sometimes, we are not interested in responses associated with certain internal states.

If the output function or the transition function is not completely specified, the corresponding machine is called *incompletely specified.* Otherwise, it is called *completely specified.*

An example of the incompletely specified Mealy machine is given in Figure 7.8 in the form of the ST-table and diagram. We can see that the response is of no interest when the stimulus is 1 or when it is 0 and the machine is in state 1 or 2. The stimulus cannot be 0 if the machine is in state 1, and it cannot be 1 if the machine is in state 3 or 4.

It is often desirable to define some other functions on the basis of the output and transition functions. Three functions of input sequences $\mathbf{I} = (\mathbf{x}_{i_1}, \mathbf{x}_{i_2}, \ldots, \mathbf{x}_{i_k})$ and *initial (reference) internal states* $\mathbf{z}_R$ at which the input sequences are applied are defined as follows.

DEFINITION 7.2 *The output-sequence function* $\mathbf{F}(\mathbf{I}, \mathbf{z}_R)$ is an assignment of output sequences to input sequences $\mathbf{I}$ and initial internal states $\mathbf{z}_R$ such that

$$\begin{aligned}\mathbf{F}(\mathbf{I}, \mathbf{z}_R) &= (\mathbf{f}(\mathbf{x}_{i_1}, \mathbf{z}_R), \mathbf{f}(\mathbf{x}_{i_2}, \mathbf{g}(\mathbf{x}_{i_1}, \mathbf{g}(\mathbf{x}_{i_1}, \mathbf{z}_R)), \ldots,\\ &\qquad \mathbf{f}(\mathbf{x}_{i_k}, \mathbf{g}(\mathbf{x}_{i_{k-1}}, \mathbf{g}(\mathbf{x}_{i_{k-2}}, \ldots, \mathbf{g}(\mathbf{x}_{i_1}, \mathbf{z}_R)) \ldots)\\ &= (\mathbf{f}(\mathbf{x}_{i_1}, \mathbf{z}_R), \mathbf{f}(\mathbf{x}_{i_2}, \mathbf{z}_{i_2}), \ldots, \mathbf{f}(\mathbf{x}_{i_k}, \mathbf{z}_{i_k}))\end{aligned}$$

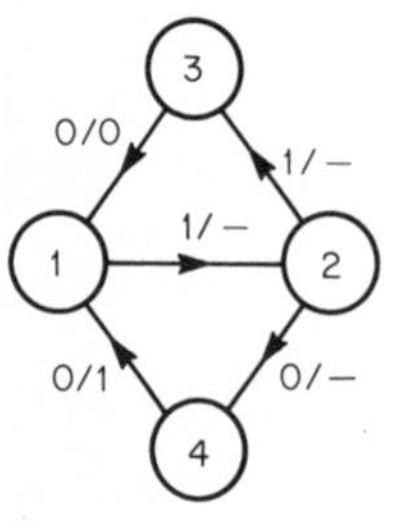

Figure 7.8 An example of an incompletely specified Mealy machine.

For instance, $\mathbf{F}(012202, 5) = 001100$ for the machine whose ST-diagram is shown in Figure 7.7. Similarly $\mathbf{F}(012202, 3) = 111011$, $\mathbf{F}(212001, 2) = 111011$, and $\mathbf{F}(210, 4) = 110$ for the same machine.

DEFINITION 7.3 *The terminal-state function* $\mathbf{G}(\mathbf{I}, \mathbf{z}_R)$ is an assignment of terminal internal states $\mathbf{z}_T$ reached by input sequences $\mathbf{I}$ applied at initial states $\mathbf{z}_R$. Formally,

$$\mathbf{z}_T = \mathbf{G}(\mathbf{I}, \mathbf{z}_R) = \mathbf{g}(\mathbf{x}_{i_k}, \mathbf{g}(\mathbf{x}_{i_{k-1}}, \ldots, \mathbf{g}(\mathbf{x}_{i_1}, \mathbf{z}_R)) \ldots)$$

For instance, $\mathbf{G}(012202, 5) = 3$, $\mathbf{G}(1212, 2) = 2$ and $\mathbf{G}(21020, 5) = 4$ for the machine in Figure 7.7.

DEFINITION 7.4 *The last-output function* $\mathbf{L}(\mathbf{I}, \mathbf{z}_R)$ is an assignment of last responses produced by input sequences $\mathbf{I}$ applied at initial states $\mathbf{z}_R$. Formally,

$$\mathbf{L}(\mathbf{I}, \mathbf{z}_R) = \mathbf{f}(\mathbf{x}_{i_k}, \mathbf{G}(\mathbf{x}_{i_1}, \mathbf{x}_{i_2}, \ldots, \mathbf{x}_{i_{k-1}}, \mathbf{z}_R))$$

For instance, $\mathbf{L}(012202, 5) = 0$, $\mathbf{L}(2210, 2) = 0$, $\mathbf{L}(012, 3) = 1$ for the machine in Figure 7.7.

Exercises to Section 7.2

7.2-1 Determine the ST-table, diagram, and matrix for the switching circuit shown in Figure 7.2

7.2-2 Determine the ST-diagram and the ST-matrix of a Mealy machine with the following ST-table:

	$\mathbf{y}^t$		$\mathbf{z}^{t+1}$	
$\mathbf{z}^t$ \ $\mathbf{x}^t$	0	1	0	1
1	0	0	1	1
2	0	0	2	1
3	0	1	3	2
4	0	1	4	2
5	1	0	5	3
6	1	0	6	3

7.2-3 Suggest a procedure for the transformation
(a) from a Moore machine to a corresponding Mealy machine.
(b) from a Mealy machine to a corresponding Moore machine.

*7.2-4 Compute the total number of all Mealy machines with p stimuli, q responses, and r internal states (so called (p, q, r) machines).

7.2-5 Show that $r \leq L \leq p \cdot r$, where L is the number of directed lines in a ST-diagram of a (p, q, r) Mealy machine.

7.2-6 Find for the machine whose ST-diagram is in Figure 7.7
- *(a) F(012012012, 4).
- (b) F(221100000, 3).
- *(c) G(2222222, 5).
- (d) G(122010021, 1).
- (e) L(220122102, 2).
- *(f) L(12120001212, 5).

7.2-7 Define the output-sequence, terminal-state, and last-output functions for the Moore machine.

7.3 STATE MINIMIZATION: COMPLETELY SPECIFIED MACHINES

A finite-state machine may contain two or more internal states which have equal output-sequence functions. These states can be combined together and represented by a single state. If a machine does not contain any pair of internal states with equal output-sequence functions, it is said to be represented by its minimal form. A transformation of a given finite-state machine to its minimal form is a problem of a great importance for the methodology of switching circuits. This transformation is usually referred to as the *state minimization*. For completely specified machines, the state minimization is based on the concepts of state equivalence and machine equivalence defined as follows:

DEFINITION 7.5 State $\mathbf{z}_i$ of machine M_1 and state $\mathbf{z}_j$ of machine M_2 are said to be *equivalent*, written as $\mathbf{z}_i = \mathbf{z}_j$, if and only if

$$\mathbf{F}_1(\mathbf{I}, \mathbf{z}_i) = \mathbf{F}_2(\mathbf{I}, \mathbf{z}_j) \tag{7.5}$$

for all possible input sequences $\mathbf{I}$ (of any length). If equation (7.5) is not satisfied for at least one input sequence, then the states $\mathbf{z}_i$ and $\mathbf{z}_j$ are said to be *distinguishable* and we write $\mathbf{z}_i \neq \mathbf{z}_j$. M_1 and M_2 may refer to the same machine.

DEFINITION 7.6 Two finite-state machines M_1 and M_2 are *equivalent*, written as $M_1 = M_2$, if and only if for each state $\mathbf{u}_i$ of M_1 there exists a state $\mathbf{v}_j$ of M_2 such that $\mathbf{u}_i = \mathbf{v}_j$ and, conversely, for each state $\mathbf{v}_j$ of M_2 there exists a state $\mathbf{u}_i$ of M_1 such that $\mathbf{v}_j = \mathbf{u}_i$.

It can be readily verified that both the state equivalence and the machine equivalence obey the *reflexive*, *symmetric*, and *transitive* laws. Consequently, they can be treated as ordinary equivalence relations. They partition a set of states or a set of machines into equivalence classes of states or machines, respectively.

We can see now that the problem of the state minimization is identical with the problem of partitioning the set of states of a given machine into equivalence classes of states. Further discussion of this problem can be considerably simplified if we introduce the concept of k-equivalence first.

DEFINITION 7.7 State $\mathbf{z}_i$ of machine M_1 and state $\mathbf{z}_j$ of machine M_2 are said to be *k-equivalent*, written as $\mathbf{z}_i \overset{k}{=} \mathbf{z}_j$, if and only if

$$\mathbf{F}_1(\mathbf{I}_k, \mathbf{z}_i) = \mathbf{F}_2(\mathbf{I}_k, \mathbf{z}_j) \tag{7.6}$$

for all possible input sequences $\mathbf{I}_k$ of length k. If equation (7.6) is not satisfied for at least one input sequence of length k, then the states $\mathbf{z}_i$ and $\mathbf{z}_j$ are said to be *k-distinguishable*. This is written as $\mathbf{z}_i \overset{k}{\neq} \mathbf{z}_j$. M_1 and M_2 may refer to the same machine.

Obviously, the k-equivalence is a relation, defined on a set of states, which is reflexive, symmetric, and transitive. Thus, it is an equivalence relation which partitions the set of states into equivalence classes. Let the partition induced by the equivalence of states (Definition 7.5) and the k-equivalence of states (Definition 7.7) be denoted, respectively, P and $P_k (k = 1, 2, \ldots)$. Let $P_0 = Z$ (P_0 is equal to the original set of internal states).

A Theorem will be presented now on the basis of which various procedures for the state minimization of completely specified machines can be elaborated.

THEOREM 7.1 If $P_k = P_{k+1} (k = 0, 1, 2, \ldots)$, then $P_k = P$.

PROOF Let $\mathbf{z}_i$ and $\mathbf{z}_j$ be two internal states of a Mealy machine. Assume that $P_k = P_{k+1}$ for this machine. Then $\mathbf{z}_i \overset{k}{=} \mathbf{z}_j$ implies $\mathbf{z}_i \overset{k+1}{=} \mathbf{z}_j$. Then, by Definition 7.7,

$$\mathbf{F}(\mathbf{x}\mathbf{I}_k, \mathbf{z}_i) = \mathbf{F}(\mathbf{x}\mathbf{I}_k, \mathbf{z}_j) \tag{7.7}$$

where $\mathbf{x}$ and $\mathbf{I}_k$ stand for, respectively, an arbitrary stimulus and an arbitrary input sequence of length k. The last equation can be written in the form

$$\mathbf{f}(\mathbf{x}, \mathbf{z}_i)\mathbf{F}(\mathbf{I}_k, \mathbf{g}(\mathbf{x}, \mathbf{z}_i)) = \mathbf{f}(\mathbf{x}, \mathbf{z}_j)\mathbf{F}(\mathbf{I}_k, \mathbf{g}(\mathbf{x}, \mathbf{z}_j)) \tag{7.8}$$

It follows immediately that

$$\mathbf{f}(\mathbf{x}, \mathbf{z}_i) = \mathbf{f}(\mathbf{x}, \mathbf{z}_j) \tag{7.9}$$

and

$$\mathbf{F}(\mathbf{I}_k, \mathbf{g}(\mathbf{x}, \mathbf{z}_i)) = \mathbf{F}(\mathbf{I}_k, \mathbf{g}(\mathbf{x}, \mathbf{z}_j)) \tag{7.10}$$

The last equation says, according to Definition 7.7, that

$$\mathbf{g}(\mathbf{x}, \mathbf{z}_i) \overset{k}{=} \mathbf{g}(\mathbf{x}, \mathbf{z}_j) \tag{7.11}$$

Thus, the necessary condition that two states are $(k + 1)$-equivalent is that they are k-equivalent and their successors, with respect to any stimulus, are k-equivalent (equation (7.11)).

Suppose now that the necessary condition is satisfied. Then equations (7.9) and (7.10) hold and, consequently, equation (7.8) holds. Since equation (7.8) can be rewritten as equation (7.7), we have $\mathbf{z}_i \overset{k+1}{=} \mathbf{z}_j$.

Hence, $\mathbf{z}_i \overset{k+1}{=} \mathbf{z}_j$ if and only if $\mathbf{z}_i \overset{k}{=} \mathbf{z}_j$ and $\mathbf{g}(\mathbf{x}, \mathbf{z}_i) \overset{k}{=} \mathbf{g}(\mathbf{x}, \mathbf{z}_j)$ for every stimulus $\mathbf{x}$.

Mathematical induction will be used now to prove that if $P_k = P_{k+1}$, then $P_k = P_{k+m}$ for all integers $m \geq 1$.

Assume that $P_{k+l} = P_k$ for all positive integers $l \leq m$ and let $m \geq 1$. To prove that, under this assumption, $P_{k+m+1} = P_k$ it suffices to prove that

$$\mathbf{z}_i \overset{k+m}{=} \mathbf{z}_j \quad \text{and} \quad \mathbf{g}(\mathbf{x}, \mathbf{z}_i) \overset{k+m}{=} \mathbf{g}(\mathbf{x}, \mathbf{z}_j)$$

whenever $\mathbf{z}_i \overset{k}{=} \mathbf{z}_j$. Since, according to our assumption, $P_k = P_{k+m}$, it must be both

$$\mathbf{z}_i \overset{k+m-1}{=} \mathbf{z}_j \tag{7.12}$$

and

$$\mathbf{g}(\mathbf{x}, \mathbf{z}_i) \overset{k+m-1}{=} \mathbf{g}(\mathbf{x}, \mathbf{z}_j) \tag{7.13}$$

whenever $\mathbf{z}_i \overset{k}{=} \mathbf{z}_j$. If, on the other hand, $\mathbf{z}_i \overset{k}{\neq} \mathbf{z}_j$, then neither (7.12) nor (7.13) is satisfied. Furthermore, since $P_{k+m-1} = P_{k+m}$, it must be

$$\mathbf{z}_i \overset{k+m}{=} \mathbf{z}_j \tag{7.14}$$

whenever (7.12) is satisfied. But (7.12) is, under our assumptions, always accompanied by (7.13). Therefore,

$$\mathbf{g}(\mathbf{x}, \mathbf{z}_i) \overset{k+m}{=} \mathbf{g}(\mathbf{x}, \mathbf{z}_j) \tag{7.15}$$

provided that $P_{k+m-1} = P_{k+m}$. Equations (7.14) and (7.15) represent necessary and sufficient conditions for

$$\mathbf{z}_i \overset{k+m+1}{=} \mathbf{z}_j$$

Hence, $P_{k+m} = P_{k+m+1}$. ▮

COROLLARY 7.1 $\mathbf{z}_i \overset{k+1}{=} \mathbf{z}_j$ if and only if $\mathbf{z}_i \overset{k}{=} \mathbf{z}_j$ and $\mathbf{g}(\mathbf{x}, \mathbf{z}_i) \overset{k}{=} \mathbf{g}(\mathbf{x}, \mathbf{z}_j)$ for every $\mathbf{x}$.

Theorem 7.1 directly suggests the following procedure of state minimization: Determine partitions P_1, P_2, ... until a value of k is found such that $P_k = P_{k+1}$. The procedure terminates at this point and $P_k = P$.

Various techniques of determination of the partitions $P_1, P_2, \ldots, P_k$ can be applied. Before going into details, let us first show that the outlined procedure of state minimization always converges after a finite number of steps.

THEOREM 7.2 Let r be the number of internal states of a Mealy machine ($r \geq 2$). Then, $P_k = P$, where $k \leq r - 1$.

PROOF Let $|P_i|$ denote the number of equivalent classes in P_i. Then, clearly, $1 \leq |P_i| \leq r$. Let $|P_1| = 1$. Then the output function is independent of internal states. Hence, $P_1 = P_2$ and the procedure terminates for $k = 1$. Since $r \geq 2$ and $k = 1$, the theorem is satisfied. Let $|P_1| \geq 2$. If $P_1 = P_2$, the theorem is satisfied for the same reason. If $P_1 \neq P_2$, then $|P_2| \geq |P_1| + 1$. Similarly, if $P_2 \neq P_3$, then $|P_3| \geq |P_2| + 1 \geq |P_1| + 2$, etc. Finally, if $P_1 \neq P_2 \neq \cdots \neq P_k$, then $|P_k| \geq |P_1| + k - 1$. But $|P_1| + k - 1 \leq r$ leads to the inequality $k \leq r + 1 - |P_1|$. Since $|P_1| \geq 2$ we get $k \leq r - 1$. ∎

Theorem 7.2 shows that the above outlined procedure of state minimization terminates after at most $r - 1$ steps each of which represents a determination of a partition $P_k (k = 1, 2, \ldots)$ of the internal states. It remains now to describe some techniques for the determination of the state partitions P_k. To simplify this description, let states belonging to the same equivalence class be called *adjoint states* and those belonging to different classes be called *disjoint states*. Furthermore, let equivalence classes containing a single state be called *singletons*.

Partitions $P_k (k = 1, 2, \ldots)$ can be determined on the basis of the following rules:

1. Determination of P_1. States $\mathbf{z}_i$ and $\mathbf{z}_j$ are adjoint in P_1 if and only if $\mathbf{f}(\mathbf{x}, \mathbf{z}_i) = \mathbf{f}(\mathbf{x}, \mathbf{z}_j)$ for every stimulus $\mathbf{x}$ (Definition 7.7).
2. Determination of P_{k+1} from $P_k (i \geq 1)$.
   (a) States that are disjoint in P_k are disjoint in P_{k+1} (Corollary 7.1).
   (b) A pair of adjoint states in P_k that, for every stimulus, pass into adjoint states in P_k is a pair of adjoint states in P_{k+1} (Corollary 7.1).

Two techniques implementing these rules will be described now. The first technique consists in constructing of so-called P_k-tables ($k = 1, 2, \ldots$). The second technique uses a table of 1-equivalent states and their successors.

The P_k-table of a given Mealy machine contains the transition function portion ($\mathbf{z}^{t+1}$ portion) of its ST-table modified as follows:

1. Rows corresponding to states belonging to the same k-equivalence class are grouped together and separated from rows belonging to other equivalence classes.

2. Each k-equivalence class (an element of P_k) is provided with a label.

3. A subscript is attached to every next state $\mathbf{z}^{t+1}$ in the table, which identifies the k-equivalence class to which the state belongs.

A determination of the P_1-table from a given ST-table consists, essentially, in grouping together rows which are not distinguished by the output function (have the same values of $\mathbf{y}^t$ for all stimuli). Then, the resulting elements of P_1 are labeled and entries of $\mathbf{z}^{t+1}$ identified with these elements.

A determination of the P_{k+1}-table from a given P_k-table consists in regrouping the rows and modifying the labels. Rows representing the same element of P_k must be split in P_{k+1} if the subscripts attached to the values of $\mathbf{z}^{t+1}$ in these rows of the P_k-table are not equal for all stimuli. Thus, the elements of P_{k+1} in the P_{k+1}-table can be determined by inspection of the subscripts attached to the values of $\mathbf{z}^{t+1}$ in the P_k-table. If no splitting occurs in the P_k-table, the procedure terminates and $P_k = P$.

The technique of state minimization based on P_k-tables is illustrated by an example in Figure 7.9. In this case, $P_3 = P$.

When P is determined, the minimal ST-table is constructed as follows:

1. It has one row for each element of P. The same labels as in P-table are used for elements of P.

2. Its values of $\mathbf{z}^{t+1}$ are identical with the subscripts identifying values of $\mathbf{z}^{t+1}$ in the P-table.

3. Its values of $\mathbf{y}^t$ are determined for the elements of P from the original ST-table (all states belonging to the same element of P must have the same values of $\mathbf{y}^t$ for all stimuli). Formally: Let $\mathbf{z}_a \in a \in P$, $\mathbf{z}_b \in b \in P$, $\mathbf{x}_i \in X$, $\mathbf{y}_j \in Y$. Furthermore, let $\mathbf{f}$, $\mathbf{g}$ and $\mathbf{f}_m$, $\mathbf{g}_m$ be the output and transition functions for the original and the minimal machine respectively. Then, if

$$\mathbf{f}(\mathbf{x}_i, \mathbf{z}_a) = \mathbf{y}_j \quad \text{and} \quad \mathbf{g}(\mathbf{x}_i, \mathbf{z}_a) = \mathbf{z}_b$$

then

$$\mathbf{f}_m(\mathbf{x}_i, \mathbf{a}) = \mathbf{y}_j \quad \text{and} \quad \mathbf{g}_m(\mathbf{x}_i, \mathbf{a}) = \mathbf{b}$$

The last table in Figure 7.9 illustrates the construction of the minimal ST-table from the P_3-table and the original ST-table.

A technique of state minimization using a table of pairs of 1-equivalent states and their successors will be described now. The advantage of this technique is that only one table is used during the whole procedure. Let the table be called the *pairs table*.

The pairs table derived from the original ST-table in Figure 7.9 is shown in Figure 7.10.

The first column of the pairs table contains all distinct pairs of states which are 1-equivalent. They can be easily determined by inspection of values of $\mathbf{y}^t$ in the original ST-table. Only if $\mathbf{y}^t$ is equal in two rows for all stimuli, the

z^t \ x^t	y^t: 0	y^t: 1	z^{t+1}: 0	z^{t+1}: 1
1	0	1	4	8
2	1	1	6	3
3	0	1	4	6
4	0	1	3	5
5	1	1	3	4
6	1	1	4	4
7	1	1	4	3
8	1	1	2	1

Original ST-table
(P_0 table)

P_1	z^t \ x^t	z^{t+1}: 0	z^{t+1}: 1
a	1	4*a*	8*b*
	3	4*a*	6*b*
	4	3*a*	5*b*
b	2	6*b*	3*a*
	5	3*a*	4*a*
	6	4*a*	4*a*
	7	4*a*	3*a*
	8	2*b*	1*a*

P_1-table

P_2	z^t \ x^t	z^{t+1}: 0	z^{t+1}: 1
a	1	4*a*	8*b*
	3	4*a*	6*c*
	4	3*a*	5*c*
b	2	6*c*	3*a*
	8	2*b*	1*a*
c	5	3*a*	4*a*
	6	4*a*	4*a*
	7	4*a*	3*a*

P_2-table

P_3	z^t \ x^t	z^{t+1}: 0	z^{t+1}: 1
a	1	4*b*	8*d*
b	3	4*b*	6*e*
	4	3*b*	5*e*
c	2	6*e*	3*b*
d	8	2*c*	1*a*
e	5	3*b*	4*b*
	6	4*b*	4*b*
	7	4*b*	3*b*

P_3-table

z^t \ x^t	y^t: 0	y^t: 1	z^{t+1}: 0	z^{t+1}: 1
a	0	1	b	d
b	0	1	b	e
c	1	1	e	b
d	1	1	c	a
e	1	1	b	b

Minimal ST-table

Figure 7.9 State minimization by P_k-tables.

corresponding pair of states is included in the first column of the pairs table. Other columns of the pairs table represent successors of the reference pair of state for individual stimuli.

It follows immediately from Corollary 7.1 that if a row in the pairs table contains at least one pair of distinct successors which is not included among the pairs in the first column, than the states in the reference pair of the row are not 2-equivalent. Such a row may be crossed out from the table because it cannot be k-equivalent ($k \geq 2$). It should be stressed that the pairs of states must be treated as unordered, i.e., $(\mathbf{z}_i, \mathbf{z}_j)$ and $(\mathbf{z}_j, \mathbf{z}_i)$ have the same meaning.

After eliminating from the pairs table all rows corresponding to pairs of states which are not 2-equivalent, we scan the pairs of successors again.

	Successors	
Pairs	0	1
1,3	4,4	8,6
1,4	4,3	8,5
2,5	6,3	3,4
2,6	6,4	3,4
2,7	6,4	3,3
2,8	6,2	3,1
3,4	4,3	6,5
5,6	3,4	4,4
5,7	3,4	4,3
5,8	3,2	4,1
6,7	4,4	4,3
6,8	4,2	4,1
7,8	4,2	3,1

1-equivalent states

	Successors	
Pairs	0	1
1,3	4,4	8,6
1,4	4,3	8,5
~~2,5~~	~~6,3~~	~~3,4~~
~~2,6~~	~~6,4~~	~~3,4~~
~~2,7~~	~~6,4~~	~~3,3~~
2,8	6,2	3,1
3,4	4,3	6,5
5,6	3,4	4,4
5,7	3,4	4,3
~~5,8~~	~~3,2~~	~~4,1~~
6,7	4,4	4,3
~~6,8~~	~~4,2~~	~~4,1~~
~~7,8~~	~~4,2~~	~~3,1~~

2-equivalent states

	Successors	
Pairs	0	1
~~1,3~~	~~4,4~~	~~8,6~~
~~1,4~~	~~4,3~~	~~8,5~~
~~2,5~~	~~6,3~~	~~3,4~~
~~2,6~~	~~6,4~~	~~3,4~~
~~2,7~~	~~6,4~~	~~3,3~~
~~2,8~~	~~6,2~~	~~3,1~~
3,4	4,3	6,5
5,6	3,4	4,4
5,7	3,4	4,3
~~5,8~~	~~3,2~~	~~4,1~~
6,7	4,4	4,3
~~6,8~~	~~4,2~~	~~4,1~~
~~7,8~~	~~4,2~~	~~3,1~~

3-equivalent states

Figure 7.10 State minimization by the pairs table.

Whenever we find a distinct pair of states which is not among the remaining pairs in the first column, we cross out the corresponding row. This is repeated again and again until no row is crossed out during the last scanning of remaining rows. Then, the pairs of states remaining in the first column are equivalent states. States which are not included in the first column of the pairs table at this point are singletons.

The technique of state minimization by pairs tables is illustrated in Figure 7.10. The same example of Mealy machine is used as in Figure 7.9. Obviously, the whole procedure shown in Figure 7.10 can be carried out in a single pair table; the pair table is repeated in Figure 7.10 only to illustrate the successive steps involved in the procedure.

To show that the minimal form of a Mealy machine is unique, we need to define *isomorphic machines* first.

DEFINITION 7.8 Let $M_1 = (X, Y, Z_1, \mathbf{f}_1, \mathbf{g}_1)$ and $M_2 = (X, Y, Z_2, \mathbf{f}_2, \mathbf{g}_2)$ be two equivalent Mealy machines that have the same number of internal states. Then, M_1 is said to be isomorphic to M_2 if there exists a one-to-one mapping $\mathbf{h}$ of Z_1 onto Z_2 such that

$$\mathbf{f}_1(\mathbf{x}, \mathbf{z}_1) = \mathbf{f}_2(\mathbf{x}, \mathbf{h}(\mathbf{z}_1))$$

and

$$\mathbf{h}(\mathbf{g}_1(\mathbf{x}, \mathbf{z}_1)) = \mathbf{g}_2(\mathbf{x}, \mathbf{h}(\mathbf{z}_1))$$

for all $\mathbf{x} \in X$ and all $\mathbf{z}_1 \in Z_1$. At the same time, for the inverse mapping $\mathbf{h}^{-1}$ of $\mathbf{h}$,

$$\mathbf{f}_1(\mathbf{x}, \mathbf{h}^{-1}(\mathbf{z}_2)) = \mathbf{f}_2(\mathbf{x}, \mathbf{z}_2)$$

and

$$\mathbf{g}_1(\mathbf{x}, \mathbf{h}^{-1}(\mathbf{z}_2)) = \mathbf{h}^{-1}(\mathbf{g}_2(\mathbf{x}, \mathbf{z}_2))$$

for all $\mathbf{x} \in X$ and all $\mathbf{z}_2 \in Z_2$.

Note that isomorphic machines are identical except for state labeling. Note also that if M_1 is isomorphic to M_2, then $M_1 = M_2$. On the contrary, if M_1 and M_2 are equivalent machines with the same number of states, this does not necessarily mean that they are isomorphic. However, as can be immediately concluded from the following theorem, they are isomorphic, if both M_1 and M_2 are minimal forms.

THEOREM 7.3 If M_m is the minimal form of a Mealy machine M, then M_m is unique up to isomorphism.

PROOF Suppose that the k-equivalence partition $P_k = \{a_1, a_2, \ldots, a_m\}$ is not unique. Then, there exists another k-equivalence partition, say $P'_k = \{b_1, b_2, \ldots, b_n\}$, for the same machine. Let $a_i = \{\mathbf{z}_{i_1}, \mathbf{z}_{i_2}, \ldots, \mathbf{z}_{i_l}\}$. States of a_i are pairwise equivalent. Furthermore, there is no state outside a_i which is equivalent to any state in a_i. Consequently, there must be an element of P'_k, say b_j, which contains all of the states $\mathbf{z}_{i_1}, \mathbf{z}_{i_2}, \ldots, \mathbf{z}_{i_l}$ and no other states. The same argument can be applied to all $i = 1, 2, \ldots, m$. Hence, P'_k contains all elements of P_k. Since P_k is a state partition and P'_k is required to be a state partition too, P'_k cannot contain any additional elements. P_k and P'_k are thus identical which means that P_k is unique. The same argument can be applied for any $k \geq 1$. Since $P = P_{r-1}$ (Theorem 7.2), P is unique. If P is specified, then the construction of M_m from M is unique except for labeling the internal states. Hence, all minimal forms of a machine M are pairwise isomorphic. ▮

Exercises to Section 7.3

*7.3-1 Using both the technique of the P_k tables and the technique of the pairs table, determine the minimal form of each Mealy machine specified in Figure 7.11.

7.3-2 Show that distinguishability and k-distinguishability are not equivalence relations.

7.3-3 Show that if all rows are identical in the output function portion of a ST-table, the corresponding machine represents a memoryless system.

7.3-4 Show that if $\mathbf{z}_i = \mathbf{z}_j$ and $\mathbf{z}_j \neq \mathbf{z}_k$, then $\mathbf{z}_i \neq \mathbf{z}_k$.

	y^t			z^{t+1}		
z^t \ x^t	0	1	2	0	1	2
1	1	0	0	2	2	5
2	0	1	1	1	4	4
3	1	0	0	2	2	5
4	0	1	1	3	2	2
5	1	0	0	6	4	3
6	0	1	1	8	9	6
7	1	0	0	6	2	8
8	1	0	0	4	4	7
9	0	1	1	7	9	7

M_1

	y^t		z^{t+1}	
z^t \ x^t	0	1	0	1
1	0	0	5	4
2	0	1	3	3
3	0	0	7	9
4	1	1	3	3
5	0	0	3	3
6	0	0	2	5
7	0	0	8	6
8	0	0	5	5
9	0	0	1	8

M_2

$$M_3 = \begin{bmatrix} \varnothing & 0/0 & \varnothing & \varnothing & 1/0 \\ \varnothing & \varnothing & \varnothing & 1/0 & 0/0 \\ 1/0 & \varnothing & \varnothing & 0/1 & \varnothing \\ \varnothing & \varnothing & 0/1 & \varnothing & 1/0 \\ \varnothing & 0/0 & \varnothing & 1/0 & \varnothing \end{bmatrix}$$

	y^t		z^{t+1}	
z^t \ x^t	0	1	0	1
1	0	1	6	2
2	0	1	7	1
3	0	1	2	3
4	0	1	3	2
5	0	1	4	1
6	1	1	5	6
7	1	1	5	7

M_4

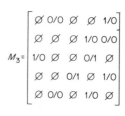

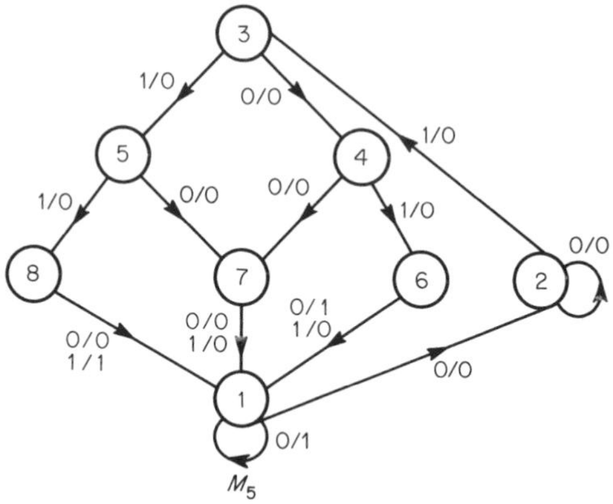

M_5

Figure 7.11 Mealy machines involved in Exercise 7.3-1.

7.3-5 Show that if $M_i = M_j$ and $M_j \neq M_k$, where M_i, M_j, M_k are Mealy machines, then $M_i \neq M_k$.

7.3-6 Prove that if $\mathbf{z}_i = \mathbf{z}_j$ then $\mathbf{G}(I, \mathbf{z}_i) = \mathbf{G}(\mathbf{I}, \mathbf{z}_j)$ for all input sequences $\mathbf{I}$.

7.3-7 Prove that $\mathbf{z}_i = \mathbf{z}_j$ if and only if

$$\mathbf{f}(\mathbf{x}, \mathbf{z}_i) = \mathbf{f}(\mathbf{x}, \mathbf{z}_j)$$

and

$$\mathbf{g}(\mathbf{x}, \mathbf{z}_i) = \mathbf{g}(\mathbf{x}, \mathbf{z}_j)$$

are satisfied for every stimulus $\mathbf{x}$.

7.3-8 Show that if two rows in a ST-table have identical values of $\mathbf{y}^t$ and $\mathbf{z}^{t+1}$ for all stimuli, then the corresponding states are equivalent.

7.3-9 Prove the following statements:

(a) If two states are k-equivalent, then they are k_1-equivalent for every $k_1 \leq k$.

(b) If two states are k-distinguishable, then they are k_1-distinguishable for every $k_1 \geq k$.

7.3-10 Prove that if $\mathbf{z}_i \overset{k}{=} \mathbf{z}_j$ and $\mathbf{G}(\mathbf{I}_k, \mathbf{z}_i) = \mathbf{G}(\mathbf{I}_k, \mathbf{z}_j)$, with respect to every input sequence $\mathbf{I}_k$ of length k, then $\mathbf{z}_i = \mathbf{z}_j$.

7.3-11 Prove that if $\mathbf{z}_i = \mathbf{z}_j$ and $\mathbf{G}(\mathbf{I}_k, \mathbf{z}_i) \overset{l}{=} \mathbf{G}(\mathbf{I}_k, \mathbf{z}_j)$ with respect to every input sequence $\mathbf{I}_k$ of length k, then $\mathbf{z}_i = \mathbf{z}_j$ provided that $k + l \geq r - 1$.

*7.3-12 Find a formula for calculating the numbers of rows in the pairs table provided that the ith 1-equivalence class ($i = 1, 2, \ldots, n$) contains r_i states.

7.3-13 Let a Mealy machine have r states and let $|P_1| = m$ for this machine.

*(a) Find the minimum and the maximum of $|P_k|$ provided that $P_k \neq P_{k-1} (k \geq 2)$.

(b) Find the smallest value of k for which P_k is guaranteed to be identical to P_{k+1}.

7.3-14 Derive the following lower and upper bounds of the number $N(p, q, r)$ of minimal Mealy machines with p stimuli, q responses and r internal states such that no two machines are equivalent:

$$\frac{1}{r!} r^{pr} \prod_{r=0}^{r-1} (q^p - i) \leq N(p, q, r) \leq \frac{1}{r!} \prod_{i=0}^{r-1} ((qr)^p - i)$$

7.4 STATE MINIMIZATION: INCOMPLETELY SPECIFIED MACHINES

The *state minimization of incompletely specified finite-state machines* is a more general problem than that of the state minimization of completely specified machines. This is essentially due to the fact that some input sequences are not applicable in some internal states of incompletely specified machines. At the same time, input sequences which are not applicable in a particular state may be applicable in some other states of the same machine. Consequently, the output sequence functions $\mathbf{F}(\mathbf{I}, \mathbf{z}_i)$ and $\mathbf{F}(\mathbf{I}, \mathbf{z}_j)$ have, generally, different domains for different states $\mathbf{z}_i$ and $\mathbf{z}_j$. This prohibits us from applying the concept of state equivalence (Definition 7.5). Instead, we have to use a more general concept of *state compatibility* which includes cases where the output sequence function is not defined. However, the state compatibility is not an equivalence relation and, therefore, it does not partition the states into equivalence classes. On the contrary, it produces a collection of compatible subsets of states which are not disjoint (some of their intersections are not empty). This collection constitutes a basis for the construction of a minimal form of the machine.

Before being able to speak about the state minimization of incompletely specified machines, we have to define some new concepts.

DEFINITION 7.9 An input sequence $\mathbf{I}_k = \mathbf{x}_1, \mathbf{x}_2, \ldots, \mathbf{x}_k$ is called *applicable* to a Mealy machine in state $\mathbf{z}_i$ if (i) all states $\mathbf{z}_1 = \mathbf{g}(\mathbf{x}_1, \mathbf{z}_i)$, $\mathbf{z}_2 = \mathbf{g}(\mathbf{x}_2, \mathbf{z}_1), \ldots, \mathbf{z}_{k-1} = \mathbf{g}(\mathbf{x}_{k-1}, \mathbf{z}_{k-2})$ are defined, and (ii) $\mathbf{L}(\mathbf{I}_k, \mathbf{z}_i) = \mathbf{f}(\mathbf{x}_k, \mathbf{z}_{k-1})$ is defined.

Note that only the last output function is required to be defined for the sequence $\mathbf{I}_k$ applied in state $\mathbf{z}_i$. Neither the output sequence function nor the last output function is required to be defined for any input sequence $\mathbf{I}_j$ whose length is smaller than that of $\mathbf{I}_k$. On the contrary, the transition function is required to be defined for every sequence $\mathbf{I}_j$ whose length is smaller than (but not equal to) the sequence $\mathbf{I}_k$.

For instance, for the machine specified in Figure 7.8, the input sequence 0, 1, 0, 0 is applicable in state 3 though the corresponding output sequence 0, -, -, 1 is not defined due to its second and third members. On the other hand, the same input sequence is not applicable in state 2 since the first condition of Definition 7.9 is not satisfied. Input sequence 0, 1 is not applicable in state 4 since the second condition of Definition 7.9 is not satisfied.

On the basis of Definition 7.9 all meaningful input sequences of a machine can be partitioned with respect to an internal state $\mathbf{z}_i$ into sequences which are applicable in z_i and those which are not. Let A_i denote the set of all input sequences applicable in state $\mathbf{z}_i$. The notion of the set of all applicable input sequences in a state enables us to establish the concepts of compatible and k-compatible states.

DEFINITION 7.10 State $\mathbf{z}_i$ of machine M_1 and state $\mathbf{z}_j$ of machine M_2 are said to be *compatible* (or *k-compatible*), written as $\mathbf{z}_i \simeq \mathbf{z}_j$ (or $\mathbf{z}_i \overset{k}{\simeq} \mathbf{z}_j$, respectively), if and only if

$$\mathbf{L}_1(\mathbf{I}, \mathbf{z}_i) = \mathbf{L}_2(\mathbf{I}, \mathbf{z}_j) \tag{7.16}$$

for all input sequences (or all input sequences of length not greater than k, respectively) such that

$$\mathbf{I} \in A_i \cap A_j \tag{7.17}$$

If (7.16) is not satisfied for at least one input sequence (or one input sequence of length not greater than k) that satisfies (7.17), then the states are said to be *incompatible* (or *k-incompatible*, respectively). We write $\mathbf{z}_i \not\simeq \mathbf{z}_j$ (or $\mathbf{z}_i \overset{k}{\not\simeq} \mathbf{z}_j$, respectively). M_1 and M_2 may refer to the same machine.

It is easy to verify that the compatibility relation is reflexive ($\mathbf{z}_i \simeq \mathbf{z}_i$) and symmetric (if $\mathbf{z}_i \simeq \mathbf{z}_j$ then $\mathbf{z}_j \simeq \mathbf{z}_i$). However, it is not transitive (both $\mathbf{z}_i \simeq \mathbf{z}_j$ and $\mathbf{z}_j \simeq \mathbf{z}_k$ but $\mathbf{z}_i \not\simeq \mathbf{z}_k$ for some states $\mathbf{z}_i, \mathbf{z}_j, \mathbf{z}_k$) since $A_i \cap A_k = A_j \cap A_k$ is not guaranteed by Definition 7.10. Hence, the compatibility is not an equivalence relation. Although internal states of a machine can be classified by the compatibility relation, as suggested by the following definition, this relation does not partition them.

DEFINITION 7.11 A *compatible* (or *k-compatible*) *class* is a set of internal states such that all states in the set are pairwise compatible (or pairwise k-compatible, respectively).

Note that every subset of a compatible class is a compatible class too. Thus, the collection of all compatible classes is ordered according to the relation of set inclusion. For a given machine, there are some compatible classes each of which is not a proper subset of any other compatible class. These are usually referred to as maximal compatible classes. The collection of all maximal compatible classes is called the *final class* of the machine.

Similarly, k-compatible sets can be ordered by the set inclusion. The collection of all maximal k-compatible sets is called the *k-class*.

DEFINITION 7.12 Let c_i be a compatible class of states and let c_j be the set of states $\mathbf{g}(\mathbf{x}, \mathbf{z}_i)$ for all $\mathbf{z}_i \in c_i$ and a particular stimulus $\mathbf{x}$. Then, c_j is said to be implied by c_i.

A collection $C = \{c_1, c_2, \ldots, c_m\}$ of compatible classes of a Mealy machine can be used instead of the original set of states in an alternative representation of the machine provided that the used compatible classes satisfy the following two conditions:

1. $\bigcup_{i=1}^{m} c_i = Z$, where Z is the original state set.
2. There exists at least one j such that $\mathbf{g}(\mathbf{x}, c_i) \subseteq c_j$ for every stimulus $\mathbf{x}$.

The first condition says that the used compatible classes must cover all the internal states of the original representation of the machine (*covering condition*). The second condition expresses the requirement that the used set of compatible classes must be closed with respect to the transition function (*closure condition*). Note that images of the transition function are the compatible classes $c_1, c_2, \ldots, c_m$ or subsets of them.

Every collection of compatible classes of states of a Mealy machine that satisfies the two above mentioned conditions and for which $c_i \nsubseteq c_j$ if $i \neq j$ is called a *preserved cover* of the machine. The condition $c_i \nsubseteq c_j$ for $i \neq j$ says that none of the used compatible classes is either repeated in the collection or included in another compatible class.

Suppose that $C = \{c_1, c_2, \ldots, c_m\}$ is the final class of a Mealy machine. Then every compatible class is either an element of C or a subset of at least one element of C. Thus, the final class is a form of representation of all compatible classes of the machine. Since any form of the machine can be based only on compatible classes and all of them are embraced in the final class, a determination of the final class represents the first step in the state minimization of incompletely specified finite-state machines.

Let C_k and C denote, respectively, the k-class and the final class of a Mealy machine. Then it can be shown that if $C_k = C_{k+1}$ for a particular $k \geq 1$, then $C_k = C$. This statement is a modification of Theorem 7.1 for incompletely specified machines. Its proof, which is similar to that of Theorem 7.1, is left to the reader as an exercise.

The property "if $C_k = C_{k+1}$, then $C_k = C$" is directly employed in the following procedure of determination of the final class: Determine k-classes $C_1, C_2, \ldots$ until a value of k is found such that $C_k = C_{k+1}$. The procedure terminates at this point and $C_k = C$.

The outlined procedure always terminates after a finite number of steps due to the following simple facts: Elements of C_k are subsets of the state set Z. Since Z is finite, C_k is finite too. If $C_{k+1} \neq C_k$, then $|C_{k+1}| > |C_k|$. If $C_{k+1} = C_k$, the procedure terminates.

A technique of C_k-tables can be used to implement the procedure. This technique is, essentially, a modification of the technique of P_k-tables (Section 7.3). Table C_1 is determined from a given ST-table by grouping together rows which are 1-compatible. Two states $\mathbf{z}_i$ and $\mathbf{z}_j$ are in the same set if $\mathbf{f}(\mathbf{x}, \mathbf{z}_i) = \mathbf{f}(\mathbf{x}, \mathbf{z}_j)$ for all $\mathbf{x} \in A_i \cap A_j$. Elements of C_1 (1-compatible sets) are labeled and entries of $\mathbf{z}^{t+1}$ identified with these elements. Since every state may belong to several 1-compatible sets, a single entry of $\mathbf{z}^{t+1}$ may exhibit several identifiers.

When determining the C_{k+1}-table from a given C_k table, we inspect the identifiers associated with entries of $\mathbf{z}^{t+1}$ in the C_k-table. Rows representing the same element of C_k must be split in C_{k+1} if defined entries $\mathbf{z}^{t+1}$ in these rows have no common identifier for at least one stimulus.

An example of the application of C_k-tables for a determination of the final class for a given Mealy machine is shown in Figure 7.12. Clearly, no elements of C_3 have to be split. Hence, $C_3 = C$.

Another procedure for the determination of the final class, which is based on the following theorem, will be described. The theorem is a modification of the theorem suggested in Exercise 7.3-7.

THEOREM 7.4 Two states $\mathbf{z}_i$ and $\mathbf{z}_j$ are compatible if and only if, for every stimulus $\mathbf{x} \in A_i \cap A_j$,

1. $\mathbf{f}(\mathbf{x}, \mathbf{z}_i) = \mathbf{f}(\mathbf{x}, \mathbf{z}_j)$

and

2. $\mathbf{g}(\mathbf{x}, \mathbf{z}_i) \simeq \mathbf{g}(\mathbf{x}, \mathbf{z}_j)$.

PROOF Necessity of the first condition follows immediately from Definition 7.10. If $\mathbf{z}_i \simeq \mathbf{z}_j$, then by Definition 7.10,

$$\mathbf{L}(\mathbf{x}\mathbf{I}, \mathbf{z}_i) = \mathbf{L}(\mathbf{x}\mathbf{I}, \mathbf{z}_j) \tag{7.18}$$

	y^t		z^{t+1}	
z^t \ x^t	0	1	0	1
1	0	–	2	–
2	0	0	1	6
3	1	–	4	–
4	0	0	5	3
5	–	1	–	6
6	0	–	4	–

Given ST-table

		z^{t+1}	
C_1	z^t \ x^t	0	1
a	1	$2a$	–
	2	$1ab$	$6ab$
	4	$5bc$	$3c$
	6	$4a$	–
b	1	$2a$	–
	5	–	$6ab$
	6	$4a$	–
c	3	$4a$	–
	5	–	$6ab$

C_1-table

		z^{t+1}	
C_2	z^t \ x^t	0	1
a	1	$2a$	–
	2	$1ac$	$6ac$
	6	$4b$	–
b	4	$5cd$	$3d$
c	1	$2a$	–
	5	–	$6ac$
	6	$4b$	–
d	3	$4b$	–
	5	–	$6ac$

C_2-table

Final class =
= {(1,2), (6), (4), (1,5), (3,5)}

		z^{t+1}	
C_3	z^t \ x^t	0	1
a	1	$2a$	–
	2	$1ad$	$6b$
b	6	$4c$	–
c	4	$5de$	$3e$
d	1	$2a$	–
	5	–	$6b$
e	3	$4c$	–
	5	–	$6b$

C_3-table = C-table

Figure 7.12 State minimization by C_k-tables.

for every $\mathbf{xI} \in A_i \cap A_j$, where $\mathbf{x}$ and $\mathbf{I}$ stand for single stimuli and sequences of stimuli, respectively. Equation (7.18) can be written as

$$\mathbf{L}(\mathbf{I}, \mathbf{g}(\mathbf{x}, \mathbf{z}_i)) = \mathbf{L}(\mathbf{I}, \mathbf{g}(\mathbf{x}, \mathbf{z}_j)) \tag{7.19}$$

Hence,

$$\mathbf{g}(\mathbf{x}, \mathbf{z}_i) \simeq \mathbf{g}(\mathbf{x}, \mathbf{z}_j) \tag{7.20}$$

is a necessary condition for $\mathbf{z}_i \simeq \mathbf{z}_j$.

If (7.20) holds, then, by Definition 7.10, (7.19) holds for every $\mathbf{I} \in A_{\mathbf{g}(\mathbf{x}, \mathbf{z}_i)} \cap A_{\mathbf{g}(\mathbf{x}, \mathbf{z}_j)}$. If the length of $\mathbf{I}$ is at least one, then (7.19) can be written as (7.18), so that $\mathbf{z}_i \simeq \mathbf{z}_j$. When $\mathbf{I}$ is an empty sequence (its length is zero), (7.19) has no meaning and (7.18) cannot be obtained from it. In such a case, however, (7.18) becomes

$$\mathbf{f}(\mathbf{x}, \mathbf{z}_i) = \mathbf{f}(\mathbf{x}, \mathbf{z}_j)$$

If this equality is satisfied, as required by the theorem, then $\mathbf{z}_i \simeq \mathbf{z}_j$. ∎

The procedure to be described consists in a determination of all pairs of compatible states. The final class is then constructed from these pairs. Various techniques based on Theorem 7.4 have been elaborated to implement this procedure. A technique employing the so-called *compatibility table* will be described.

A standard form of the compatibility table is illustrated in Figure 7.13b. Each cell of the table corresponds to a pair of states defined by corresponding row and column headings. If two states are shown to be incompatible, the corresponding cell is crossed out. If they are shown to be compatible, a check mark (√) is placed into the cell. If we do not know at a particular stage whether they are compatible or not, pairs of states whose compatibility is implied by the compatibility of the reference pair are listed in the cell.

At the beginning, the compatibility table is filled in directly from the ST-table in the following way:

1. All cells corresponding to pairs of states for which the first condition of Theorem 7.4 is not satisfied are crossed out. Indeed, these are incompatible pairs of states since the condition is necessary for compatibility of two states. Practically, we can determine them in the ST-tables as states for which the outputs differ and are both specified for at least one stimulus.

2. A check mark is placed in cells corresponding to pairs of states whose compatibility can be directly recognized. This is the case of states for which the

	y^t			z^{t+1}		
z^t \ x^t	0	1	2	0	1	2
1	–	0	–	–	–	3
2	0	–	–	4	6	–
3	1	1	–	5	2	1
4	1	–	0	6	–	2
5	–	1	–	3	5	–
6	–	1	1	–	5	2

(a)

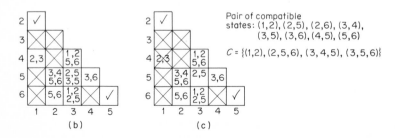

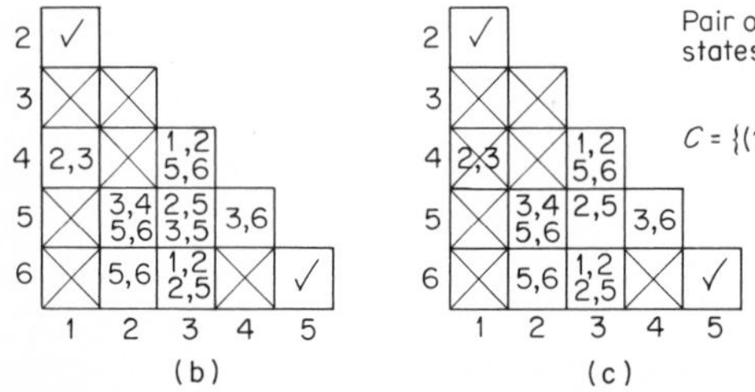

Pair of compatible states: (1,2), (2,5), (2,6), (3,4), (3,5), (3,6), (4,5), (5,6)

$C = \{(1,2), (2,5,6), (3,4,5), (3,5,6)\}$

Figure 7.13 Determination of compatible pairs of states by the compatibility table.

outputs do not differ for any stimulus when both are specified and next state pair, whenever both next states are specified, is the same as the reference pair or consists of identical states.

3. If a pair of states cannot be directly recognized as either compatible or incompatible, the set of all specified pairs of next states that are different from the reference pair are written in the corresponding cell.

Once the initial form of the compatibility table has been prepared, further procedure consists in the elimination of additional incompatible pairs of states on the basis of the second condition of Theorem 7.4. All pairs of states listed in a cell have to be compatible to make the pair corresponding to the cell compatible. If at least one of these pairs is already identified as being incompatible, the cell has to be crossed out. We update the table step by step by successive passes through the table during which some cells are crossed out. The procedure terminates when the table cannot be further updated. Pairs of states whose cells in the final form of the table are not crossed out are compatible.

The above described procedure is illustrated by an example in Figure 7.13. Table (b) is the initial form of the compatibility table for the *ST*-table (a) and table (c) is its final form. The compatible pairs are: (1, 2), (2, 5), (2, 6), (3, 4), (3, 5), (3, 6), (4, 5), (5, 6).

After all pairs of compatible states are determined, the next step consists in incorporating them into the maximal compatible classes. This is a rather trivial problem. It can be easily solved on the basis of the following rule which follows directly from Definition 7.11:

Let A be the set of all pairs of compatible states and let c be a compatible class containing m states ($m \geq 2$). Then a state $\mathbf{z}_i \notin c$ can be included into c if $(\mathbf{z}_i, \mathbf{z}_j) \in A$ for every $\mathbf{z}_j \in c$. When $\mathbf{z}_i$ is included in c, a new class of $m + 1$ compatible states is created.

The reader himself should be able to incorporate this rule into a systematic procedure for determining the final class from the set of all compatible pairs.

It is easy to find that for the example in Figure 7.13 we obtain the final class

$$C = \{(1, 2), (2, 5, 6), (3, 4, 5), (3, 5, 6)\}$$

It could be easily proven that the final class is always a preserved cover. Although, due to this fact, it can be used to represent the machine, it is not guaranteed that this representation is minimal. The final class may even contain more elements than the original state set. In any event, the smaller number of these two may serve as an upper bound of the number of states in a minimal form. The minimum number of maximum compatible sets that must be selected from the final class so that each state is included in at least one of these sets represents a lower bound.

For instance, the machine specified in Figure 7.13 has six states and four elements of the final class. Hence, the upper bound for the number of states in a minimal form of the machine is 4. Clearly, at least three elements must be taken from the final class {(1, 2), (2, 5, 6), (3, 4, 5), (3, 5, 6)} to cover all states. Hence, 3 is the lower bound. Two selections from the final class correspond to this lower bound: {(1, 2), (2, 5, 6), (3, 4, 5)} and {(1, 2), (3, 4, 5), (3, 5, 6)}. Neither of them, however, is closed with respect to the transition function. For the first selection $\mathbf{g}(0, (3, 4, 5)) = (3, 5, 6)$ but (3, 5, 6) is not included in this selection. For the second selection $\mathbf{g}(1,(3, 5, 6)) = (2, 5)$ but (2, 5) is neither an element nor a subset of any element of this selection.

A question arises at this point: Are the elements of the final class the only potential candidates for inclusion in a minimal form? They certainly cannot be less economical than any other compatible classes in covering all of the original states. On the other hand, they may be less economical than some other compatible classes in satisfying the requirement of closure. Hence, the elements of the final class are not the only potential candidates for a minimal form. To identify properties of the potential candidates, we have to introduce some new concepts.

DEFINITION 7.13 The *implied class set* $I(c_i) = \{c_j\}$ of a compatible class c_i is the collection of all sets c_j implied by c_i such that:

1. c_j contains more than one state
2. $c_j \nsubseteq c_i$
3. $c_j \nsubseteq c_k$ if $c_k \in I(c_i)$

Note that the implied class set $I(c_i)$ expresses the closure requirements imposed by the class c_i. Implied classes which do not satisfy all the three conditions of Definition 7.13 are not included in the implied class set because of the following reasons:

1. Since all states of a given machine must be included in every preserved cover regardless of the closure conditions, c_j containing a single state does not impose any additional requirement on the collection of classes.
2. If $c_j \subseteq c_i$, the closure requirement is satisfied directly by the class c_i.
3. If $c_j \subseteq c_k \in I(c_i)$, then the closure requirements imposed by c_j are satisfied, as a by-product, by the closure requirements imposed by c_k.

DEFINITION 7.14 A *prime compatible class* c_i of a given Mealy machine is a compatible class which satisfies, for every compatible class $c_j \neq c_i$ of the machine, at least one of the following conditions:

1. $c_i \nsubseteq c_j$
2. $I(c_i) \nsupseteq I(c_j)$

We can see now that every prime compatible set is a potential candidate for a minimal class set. The following theorem states this property formally.

THEOREM 7.5 There exists at least one minimal preserved cover for a given Mealy machine which contains only prime compatible classes.

PROOF Assume that $A = \{c_1, c_2, \ldots, c_k, \ldots, c_m\}$ is a minimal preserved cover. Furthermore, assume that c_k is not a prime compatible class. Then, $c_k \subset c_j$ and $I(c_k) \supseteq I(c_j)$ for at least one class c_j. When c_k is replaced by c_j, the new set of classes is again a minimal preserved cover since neither the cover nor the closure requirements were violated by this replacement and the new set contains the same number of classes as A. Either c_j is a prime compatible class or the described replacement can be repeated until a prime compatible class is obtained and substituted for c_k. If any other elements in the new minimal preserved cover are not prime compatible classes, the above argument may be repeated for each of them so that the last minimal preserved cover contains only prime compatible classes. ∎

Note that all maximal compatible classes are also prime compatible classes because all of them satisfy the first condition of Definition 7.14. Furthermore, they are the only prime compatible classes that satisfy this condition. Other prime compatible classes may exist which satisfy the second condition of Definition 7.14 but not its first condition. Clearly, all of these prime classes are subsets of some maximal compatible classes and, thus, can be generated from the latter. The following procedure for generating all prime compatible sets from the final class is suggested:

1. Given the final class, select the maximal compatible classes with the greatest number of states, say $m (m \geq 2)$. They constitute the initial list of prime compatible classes representing all prime compatible classes with m states. Determine for each of them the implied class set.
2. Let $i = 0$.
3. Suppose that all prime compatible classes containing $m - i$ states are in the list of prime compatible classes. Then, from the maximal classes with m, $m - 1, \ldots, m - i$ states, form all subsets containing $m - i - 1$ states and determine their implied class sets. Add those of them which satisfy either conditions of Definition 7.14 to the list of prime compatible classes together with their implied compatible classes.
4. Add all maximal compatible classes containing $m - i - 1$ states to the list of prime compatible classes and determine their implied class sets.
5. (a) If $m - i - 1 = 1$, the procedure terminates. (b) If $m - i - 1 > 1$, increase i by one and return to step 3.

The procedure is illustrated in Figure 7.14 for the final class corresponding to the machine specified in Figure 7.13a.

Final class	Implied class-sets
(1,2)	∅
(2,5,6)	(3,4)
(3,4,5)	(1,2), (2,5), (3,5,6)
(3,5,6)	(1,2), (2,5)

v_i	Prime classes	Implied class-sets	Steps of the procedure
v_1	(2,5,6)	(3,4)	Step 1
v_2	(3,4,5)	(1,2), (2,5), (3,5,6)	
v_3	(3,5,6)	(1,2), (2,5)	
v_4	(2,6)	(5,6)	Steps 2 and 3 for i = 0
v_5	(5,6)	∅	
v_6	(3,4)	(1,2), (5,6)	
v_7	(3,5)	(2,5)	
v_8	(4,5)	(3,6)	
v_9	(1,2)	∅	Step 4 for i = 0
v_{10}	(3)	∅	Step 5(b), steps 3 and 4 for i = 1, step 5(a)
v_{11}	(4)	∅	

Figure 7.14 Determination of all prime classes.

The last stage of the state minimization consists in finding a selection of prime compatible classes which represents a minimal preserved cover. The selection must satisfy the following conditions:

1. The selected prime classes must cover all the internal states.
2. The selected set of prime classes must be closed.
3. The number of selected prime classes must be minimal.

Let a logic variable v_i be assigned to every prime compatible class c_i. Let $v_i = 1$ if the corresponding prime class c_i is selected. Otherwise, $v_i = 0$.

Suppose now that a state z_i is included in prime classes associated with logic variables $v_{i_1}, v_{i_2}, \ldots, v_{i_m}$. Then, at least one of these prime classes must be included in the minimal preserved cover. This condition can be described by the following Boolean equation:

$$v_{i_1} \vee v_{i_2} \vee \cdots \vee v_{i_m} = 1 \tag{7.21}$$

We obtain one equation of this form for every internal state.

Similarly, the closure conditions can be expressed in the form of Boolean equations. Suppose that the implied class set implied by a prime class c_i is $\{i_a, i_b, \ldots, i_m\}$. Furthermore, suppose that i_a is a subset of prime classes c_{a_1}, $c_{a_2}, \ldots, c_{a_p}$, i_b is a subset of prime classes $c_{b_1}, c_{b_2}, \ldots, c_{b_q}$, etc., and i_m is a subset of prime classes $c_{m_1}, c_{m_2}, \ldots, c_{m_r}$. Then the closure constraints imposed by the prime class c_i can be described by the proposition: c_i is not selected or at least one prime class must be selected from the subset $c_{a_1}, c_{a_2}, \ldots, c_{a_p}$, one from the subset $c_{b_1}, c_{b_2}, \ldots, c_{b_q}$, etc., and one from the subset $c_{m_1}, c_{m_2}, \ldots,$ c_{m_r}. This proposition can be written either as a single Boolean equation

$$\bar{v}_i \vee (v_{a_1} \vee v_{a_2} \vee \cdots \vee v_{a_p})(v_{b_1} \vee v_{b_2} \vee \cdots \vee v_{b_q}) \cdots (v_{m_1} \vee v_{m_2} \vee \cdots \vee v_{m_r}) = 1 \tag{7.22}$$

or a set of simultaneous Boolean equations

$$
\begin{aligned}
\bar{v}_i \vee v_{a_1} \vee v_{a_2} \vee \cdots \vee v_{a_p} &= 1 \\
\bar{v}_i \vee v_{b_1} \vee v_{b_2} \vee \cdots \vee v_{b_q} &= 1 \\
\vdots \qquad\qquad & \ \ \vdots \\
\bar{v}_i \vee v_{m_1} \vee v_{m_2} \vee \cdots \vee v_{m_r} &= 1
\end{aligned}
\tag{7.23}
$$

Given a set of all prime compatible classes, the problem of selection of a minimal preserved cover can be stated as follows: Find a minimum of

$$\sum_i v_i$$

subject to constraints represented by the simultaneous Boolean equations (7.21) and (7.22).

Before the Boolean equations are written, it is an advantage to describe the constraints in the form of a table similar to the covering table used for prime implicants when minimizing Boolean expressions (Chapter 3). This table will be called the *covering and closure table.*

For instance, for the prime classes and class sets in Figure 7.14, we obtain the covering and closure table in Figure 7.15. It is divided into two sections, the covering and closure conditions respectively. In the covering section, the column corresponding to state z_i has a cross in every row corresponding to a prime class that contains z_i. The closure section contains two kinds of entries, crosses and circles. Each column represents one closure condition which can be described by a single equation included in (7.23). Circles correspond to the negated variables and crosses represent the other variables involved in the equations.

Prime classes	Covering conditions						Closure conditions											
	1	2	3	4	5	6	*a*	*b*	*c*	*d*	*e*	*f*	*g*	*h*	*i*	*j*	*k*	v_i
(2,5,6)		×			×	×	○		×			×	×		×	×		v_1
(3,4,5)			×	×	×		×	○	○	○								v_2
(3,5,6)			×		×	×				×	○	○	×		×		×	v_3
(2,6)		×				×							○					v_4
(5,6)					×	×							×		×			v_5
(3,4)			×	×			×							○	○			v_6
(3,5)			×		×											○		v_7
(4,5)				×	×												○	v_8
(1,2)	×	×						×			×			×				v_9
(3)			×															v_{10}
(4)				×														v_{11}

Figure 7.15 Covering and closure table for the prime classes and class-sets in Figure 7.14.

Although the Boolean equations describing the covering and closure constraints can be written directly from the table of prime classes with their implied class sets, the application of the covering and closure table proves to be useful. Some simple *rules of simplification* of this table can be used in many cases. These rules enable us to eliminate certain rows and columns from the table, diminishing thus the required number of Boolean variables and equations. An obvious rule says: If a column has only a single cross in row r and blank entries elsewhere, the prime class corresponding to row r must be selected (the corresponding logic variable is equal to 1). Row r together with all columns in which it has crosses, can be then eliminated from the table. Applying this rule to the table in Figure 7.15, we find that there is only one cross in the first column. Thus, the prime class (1, 2) must be selected so that $v_9 = 1$. Then, one row and five columns can be eliminated as shown by interrupted lines.

Other rules of simplification could be applied, some of which are formulated in Exercise 7.4-4. At any stage of simplification, we can write the respective Boolean equations (one equation for each column). The discriminant of these equations provides us with all preserved covers of the given machine built by prime classes. Preserved covers with the minimum number of prime classes are the sought minimal forms of the given machine.

Starting from the covering and closure table in Figure 7.15, the problem of determining minimal preserved covers can be formulated as follows:

Find minima of the objective function

$$\sum_{i=1}^{11} v_i$$

for $v_9 = 1$ (determined by the simplification rule) and subject to covering constraints

$$\begin{aligned} v_2 \vee v_3 \vee v_6 \vee v_7 \vee v_{10} &= 1 \\ v_2 \vee v_6 \vee v_8 \vee v_{11} &= 1 \\ v_1 \vee v_2 \vee v_3 \vee v_5 \vee v_7 \vee v_8 &= 1 \\ v_1 \vee v_3 \vee v_4 \vee v_5 &= 1 \end{aligned}$$

and closure constraints

$$\begin{aligned} \bar{v}_1 \vee v_2 \vee v_6 &= 1 & \qquad \bar{v}_4 \vee v_1 \vee v_3 \vee v_5 &= 1 \\ \bar{v}_2 \vee v_1 &= 1 & \bar{v}_6 \vee v_1 \vee v_3 \vee v_5 &= 1 \\ \bar{v}_2 \vee v_3 &= 1 & \bar{v}_7 \vee v_1 &= 1 \\ \bar{v}_3 \vee v_1 &= 1 & \bar{v}_8 \vee v_3 &= 1 \end{aligned}$$

The discriminant of these simultaneous Boolean equations gives us the domain of the objective function. We calculate the value of the objective function for each state of the logic variables and select states with the smallest values (APL program PBP in Appendix G can be used for solving this problem). The discriminant is shown in Fig. 7.16. We can see that there are two minima.

1. $v_1 = v_6 = v_9 = 1$ ($v_i = 0$ otherwise) which means that the prime classes (2, 5, 6), (3, 4) and (1, 2) are selected.

2. $v_5 = v_6 = v_9 = 1$ ($v_i = 0$ otherwise) which means that the prime classes (5, 6), (3, 4) and (1, 2) are selected.

If we denote the prime classes involved in the minimal preserved covers by the same symbols as the logic variables associated with them, we obtain (with a help of the ST-table in Figure 7.13) the minimal forms shown in Figure 7.17. Note that the first minimal form includes four alternatives: $\mathbf{g}(1, v_6) = \mathbf{g}(1, (3, 4)) = 2$ and both $2 \in v_1$ and $2 \in v_9$; thus, either $\mathbf{g}(1, v_6) = v_1$ or $\mathbf{g}(1, v_6) = v_9$. Similarly, $\mathbf{g}(2, v_1) = 2$; thus $\mathbf{g}(2, v_1) = v_1$ or $\mathbf{g}(2, v_1) = v_9$.

Let us note that the Boolean equations can also be solved in their negated OR form producing thus a negation of the domain of the objective function.

$v_{11} v_{10} v_8 v_7 v_6$ \ $v_5 v_4 v_3 v_2 v_1$	0	1	1	2	1	2	2	3	1	2	2	3	2	3	3	4	1	2	2	3	2	3	3	4	2	3	3	4	3	4	4	5
0								3								4								4								5
1		2				3		4		3				4		5	2	3				4		5	3	4				5		6
1								4								5								5								6
2		3				4		5		3				5		6		4				5		6		5				6		7
1								4								5								5								6
2						4		5						5		6						5		6						6		7
2								5								6								6								7
3						5		6						6		7						6		7						7		8
1								4								5								5								6
2		3				4		5		4				5		6	3	4				5		6	4	5				6		7
2								5								6								6								7
3		4				5		6		5				6		7		5				5		7		6				7		8
2								5								6								6								7
3						5		6						6		7						6		7						7		8
3								6								7								7								8
4						6		7						7		8						7		8						8		9
1								4								5								5								6
2		3				4		5		4				5		6	3	4				5		6	4	5				6		7
2								5								6								6								7
3		4				5		6		5				6		7		5				6		7		6				7		8
2								5								6								6								7
3						5		6						6		7						6		7						7		8
3								6								7								7								8
4						6		7						7		8						7		8						8		9
2								5								6	3							6	4							7
3		4				5		6		5				6		7	4	5				6		7	5	6				7		8
3								6								7								7								8
4		5				6		7		6				7		8		6				7		8		7				8		9
3								6								7								7								8
4						6		7						7		8						7		8						8		9
4								7								8								8								9
5						7		8						8		9						8		9						9		10

Figure 7.16 Discriminant for the Boolean equations derived from the simplified table in Figure 7.15.

$z^t \backslash x^t$	y^t			z^{t+1}		
	0	1	2	0	1	2
v_1	0	1	1	v_6	v_1	v_1 or v_9
v_6	1	1	0	v_1	v_1 or v_9	v_9
v_9	0	0	–	v_6	v_1	v_6

$z^t \backslash x^t$	y^t			z^{t+1}		
	0	1	2	0	1	2
v_5	–	1	1	v_5	v_5	v_9
v_6	1	1	0	v_6	v_9	v_9
v_9	0	0	–	v_6	v_5	v_6

Figure 7.17 Minimal forms of the Mealy machine in Figure 7.13a.

This is preferable in most cases since the whole procedure of the solution of the equation can be accomplished in a single chart due to the additivity of the OR function (Chapter 4). Equations of our example would have the form

$$\bar{v}_2\bar{v}_3\bar{v}_6\bar{v}_7\bar{v}_{10} = 0$$
$$\bar{v}_2\bar{v}_6\bar{v}_8\bar{v}_{11} = 0$$
$$\bar{v}_1\bar{v}_2\bar{v}_3\bar{v}_5\bar{v}_7\bar{v}_8 = 0$$
$$\bar{v}_1\bar{v}_3\bar{v}_4\bar{v}_5 = 0$$

for covering constraints and the form

$$v_1\bar{v}_2\bar{v}_6 = 0 \qquad v_4\bar{v}_1\bar{v}_3\bar{v}_5 = 0$$
$$v_2\bar{v}_1 = 0 \qquad v_6\bar{v}_1\bar{v}_3\bar{v}_5 = 0$$
$$v_2\bar{v}_3 = 0 \qquad v_7\bar{v}_1 = 0$$
$$v_3\bar{v}_1 = 0 \qquad v_8\bar{v}_3 = 0$$

for closure constraints. Remember that these are not simultaneous equations but OR equations.

Let us consider now a more general problem of state minimization in which not only the internal states but also the input states (stimuli) are minimized. This is meaningful in all cases where a certain specific number of input states is not assigned.

The state-input minimization employs all concepts introduced for the state minimization as well as modification of some of them for input states. For instance, the concept of compatible internal states can be modified to the concept of compatible input states.

DEFINITION 7.15 Input state $\mathbf{x}_i$ of machine M_1 and input state $\mathbf{x}_j$ of machine M_2 are said to be *compatible*, written as $\mathbf{x}_i \simeq \mathbf{x}_j$, if and only if for every internal state $\mathbf{z}_k$

$$\mathbf{f}_1(\mathbf{x}_i, \mathbf{z}_k) = \mathbf{f}_2(\mathbf{x}_j, \mathbf{z}_k)$$

wherever both $\mathbf{f}_1$ and $\mathbf{f}_2$ are defined, and

$$\mathbf{g}_1(\mathbf{x}_i, \mathbf{z}_k) \simeq \mathbf{g}_2(\mathbf{x}_j, \mathbf{z}_k)$$

wherever both $\mathbf{g}_1$ and $\mathbf{g}_2$ are defined. M_1 and M_2 may refer to the same machine.

On the basis of the compatibility of input states, other concepts, similar to those used for internal state minimization, can be introduced for the input state minimization, namely, the concept of *compatible input sets*, *maximal compatible input sets*, *final input class*, *prime compatible input classes*, and *implied class-sets*. Furthermore, procedures similar to those for determining the final class and the prime classes of internal states can be applied for a determination of the input final class and the input prime classes. The reader himself should be able to establish the above mentioned modified concepts and to modify the procedures for input states.

It can be proven that the prime classes of internal states and input states are the only potential candidates for a minimal form when both the internal and input states are minimized. Thus, a determination of all these prime classes represents again the first step in the minimization procedure. However, the definition of prime classes in this generalized context includes joint closure requirements imposed by pairs of internal state and input state compatible classes. This will be illustrated later.

After all prime classes of internal and input states are determined, the second step of the minimization procedure consists in selecting a subset of the prime classes for which a given objective function reaches its minimum subject to the following constraints:

1. Each internal state must be covered by at least one prime class of internal states.
2. Each input state must be covered by at least one prime class of input states.
3. All selected internal state prime classes must be closed with respect to the next state function.
4. All selected input state prime classes must be closed with respect to the next state function.
5. All selected couples of one prime class of internal states and one prime class of input states must be closed with respect to the next state function and must be consistent with respect to the output function.

The constraints can be expressed in a form of the covering and closure table. It has five sections, one for each of the above specified types of constraints. The table may be simplified by various simplification rules and, finally, described by a set of Boolean equations. The discriminant of the Boolean equations represents the domain of the used objective function. States within the domain for which the objective function reaches its minimum are then selected. They represent directly the minimal preserved covers.

The objective function may be defined, in the simplest case, as the sum of all the logic variables involved or, if reasonable, different weights may be assigned to variables associated with internal states and input states respectively. Of course, other, more sophisticated definitions of the objective function may be used without changing the basic procedure of machine minimization.

As an example, consider the machine whose ST-table is given in Figure 7.18. Suppose that we have already determined the final class {(1, 2, 3), (2, 3, 4), (4, 5, 6)} for the internal states and the final class {(*a*, *b*, *c*), (*d*, *e*)} for the input states. To determine prime classes, we generate all compatible classes of both internal and input states such that each of them has two or more states. Next we determine implied class sets for the compatible classes of internal states and input states separately as well as jointly. It is convenient to display the results in the form shown for our example in Figure 7.19.

Let the compatible classes of internal states and input states be called, respectively, the *row classes* and the *column classes*. Let $R_i = (\mathbf{z}_{i_1}, \mathbf{z}_{i_2}, \ldots, \mathbf{z}_{i_m})$ be a row class and let $C_j = (\mathbf{c}_{j_1}, \mathbf{c}_{j_2}, \ldots, \mathbf{c}_{j_n})$ be a column class. If R_i is selected together with C_j, then the following conditions must be satisfied in the original ST-table:

1. $\mathbf{f}(\mathbf{c}_a, \mathbf{z}_b)$ for all $\mathbf{c}_a \in C_j$ and $\mathbf{z}_b \in R_i$ must be equal whenever specified.
2. $\mathbf{g}(\mathbf{c}_a, \mathbf{z}_b)$ for all $\mathbf{c}_a \in C_j$ and $\mathbf{z}_b \in R_i$ must constitute a row class, say R_k.
3. The row class R_k must be included in any preserved cover which contains R_i and C_j.

If the first or the second condition is not satisfied for a particular pair R_i, C_j, then this pair cannot be selected to any preserved cover. In such a case we cross out the respective entry in the joint class sets section of the table in Figure 7.19. If the two conditions are satisfied, the entry contains either the required row class R_k, provided that R_k is not a subset of any compatible class of the class set of R_i or C_j, or is left blank otherwise. All entries except those with small crosses of the table in Figure 7.19 are determined by these rules. This is an initial form of the table which must be updated on the basis of the

	y^t					z^{t+1}				
z^t \ x^t	*a*	*b*	*c*	*d*	*e*	*a*	*b*	*c*	*d*	*e*
1	–	–	–	1	–	2	4	4	1	–
2	–	0	0	–	0	–	5	5	2	2
3	1	–	–	–	–	2	–	4	–	1
4	–	–	–	0	–	4	–	6	–	–
5	1	1	1	0	0	6	–	–	1	3
6	1	1	1	0	0	5	–	–	–	–

Figure 7.18 Mealy machine used as an example of the state-input minimization.

Input states → / compatible classes		(a, b, c)	(a, b)	(a, c)	(b, c)	(d, e)
Input states → / implied class sets		(2,4) (4,6)	(2,4)	(2,4) (4,6)	∅	(1,3)
(1, 2, 3)	(4, 5)	X	X	X		X
(1, 2)	(4, 5)	X	X	X		X
(1, 3)	∅					
(2, 3)	(1, 2), (4, 5)	X	X	X		×
(2, 3, 4)	(1, 2), (4, 5, 6)	X	X	X		×
(2, 4)	(5, 6)	(4, 5, 6)	(4, 5)	(4, 5, 6)		
(3, 4)	(2, 4), (4, 6)	X		X		
(4, 5, 6)	∅					
(4, 5)	(4, 6)					
(4, 6)	(4, 5)	(4, 5, 6)		(4, 5, 6)		
(5, 6)	∅					
Int. states compatible classes	Int. states implied class sets	Joint implied class sets				

Figure 7.19 A table of compatible classes with class-sets for the state-input minimization.

following rule: If a pair R_i, C_j cannot be selected (the corresponding entry in the table is crossed out), then any other couple R_k, C_j such that R_i is implied by the couple or contained in the implied class set of R_k cannot be selected either. In Figure 7.19, the updated entries are denoted by small crosses. For instance, the pair (2, 3), (d, e) cannot be selected because the pair (1, 2), (d, e) cannot be selected and the implied class set of (2, 3) contains (1, 2). If (2, 3), (d, e) were selected, then the pair (1, 2) (d, e) would have to be selected too. But it cannot be selected. Hence, (2, 3), (d, e) cannot be selected either. This change in the table of class sets may exclude some other pairs of row and column classes. In our example, the pairs (2, 3), (d, e) and (2, 3, 4), (d, e) are the only pairs excluded from the initial table.

After the table of class sets is updated, we have to select prime row and column classes. To define these prime classes, let us define the concepts of row class and column class exclusion first.

DEFINITION 7.16 A row class R_i is *excluded* by a row class R_j if:

1. $R_i \subset R_j$
2. $I(R_i) \supseteq I(R_j)$
3. For every column class C_k, either both pairs (R_i, C_k) and (R_j, C_k) are excluded from selection, or $I(R_i, C_k) \supseteq I(R_j, C_k)$, or R_j is contained in $I(C_k)$.

DEFINITION 7.17 A column class C_i is excluded by a column class C_j if:

1. $C_i \subset C_j$
2. $I(C_i) = I(C_j)$

3. For every row class R_k, either both pairs (R_k, C_i) and (R_k, C_j) are excluded from selection or $I(R_k, C_i) = I(R_k, C_j)$.

DEFINITION 7.18 A row class (or a column class) is a *prime class* if it is not excluded by any other row class (or column class, respectively).

Applying the definition of prime classes to the table of class sets in Figure 7.19, we obtain the table of prime classes which is shown in Figure 7.20, where a logic variable is assigned to each prime class. Now, we can express all constraints in the form of the covering and closure table. For our example, it is shown in Figure 7.21.

Using the rules of simplification summarized in Exercise 7.4-4, the covering and closure table in Figure 7.21 can be simplified as follows:

1. Row class r_6 must be selected ($r_6 = 1$) since states 5 and 6 are covered only by this class. Since r_6 is fixed, this row may be eliminated together with columns 4, 5, 6, A, C, D, F, H, R, S (Rule 1).
2. Columns 3 and J may be deleted (Rule 3 with respect to column 1).
3. Column b may be deleted (Rule 3 with respect to column c).
4. Row r_3 must not be selected ($r_3 = 0$). It may be deleted from the table together with columns B, N, P, Q (Rule 4 with respect to row r_4).
5. Row r_7 must not be selected ($r_7 = 0$). It may be deleted from the table (Rule 4 with respect to row r_4).
6. Row r_5 must not be selected ($r_5 = 0$). It may be deleted from the table together with columns E and T (Rule 5).
7. Row c_2 must not be selected ($c_2 = 0$). It may be deleted from the table together with columns I and L (Rule 4 with respect to row c_1).

		Logic variables →	c_1	c_2	c_3	c_4	c_5	c_6	c_7
		Column prime classes →	(a,b,c)	(a,b)	(b,c)	(d,e)	(a)	(d)	(e)
		Column implied class sets →	(2,4) (4,6)	(2,4)	∅	(1,3)	∅	∅	∅
r_1	(1,2,3)	(4,5)	X	X		X			
r_2	(1,3)	∅							
r_3	(2,3,4)	(1,2), (4,5,6)	X	X		X			
r_4	(2,4)	(5,6)	(4,5,6)	(4,5)					
r_5	(3,4)	(2,4), (4,6)	X						
r_6	(4,5,6)	∅							
r_7	(2)	∅							
Logic variable	Row prime classes	Row implied class sets	Joint implied class sets						

Figure 7.20 Prime classes table derived from the table in Figure 7.19.

	Row covering						Column covering					Row closure						Column closure				Joint closure								
	1	2	3	4	5	6	*a*	*b*	*c*	*d*	*e*	*A*	*B*	*C*	*D*	*E*	*F*	*G*	*H*	*I*	*J*	*K*	*L*	*M*	*N*	*P*	*Q*	*R*	*S*	*T*
r_1	×	×	×									○	×								×	○	○	○						
r_2	×		×																		×									
r_3		×	×	×									○	○		×		×		×					○	○	○			
r_4		×		×											○	×		×		×								○	○	
r_5			×	×												○	○													○
r_6				×	×	×						×		×	×		×		×									×	×	
r_7		×																												
c_1							×	×	×									○	○			○			○			○		○
c_2							×	×												○			○			○			○	
c_3								×	×																					
c_4										×	×										○			○			○			
c_5							×																							
c_6										×																				
c_7											×																			

Figure 7.21 Covering and closure table for the prime classes table in Figure 7.20.

Thus, some of the logic variables involved have been fixed during the simplification procedure ($r_6 = 1$, $r_3 = r_5 = r_7 = c_2 = 0$) and a simplified covering and closure table produced. This new version of the table is shown in Figure 7.22. Boolean equations for the remaining logic variables can be written directly from the table:

$$r_1 \vee r_2 = 1$$
$$r_1 \vee r_4 = 1$$
$$c_1 \vee c_5 = 1$$
$$c_1 \vee c_3 = 1$$
$$c_4 \vee c_6 = 1$$
$$c_4 \vee c_7 = 1$$
$$\bar{c}_1 \vee r_4 = 1$$
$$\bar{r}_1 \vee \bar{c}_1 = 1$$
$$\bar{r}_1 \vee \bar{c}_4 = 1$$

	1	2	*a*	*c*	*d*	*e*	*G*	*K*	*M*
r_1	×	×						○	○
r_2	×								
r_4		×					×		
c_1			×	×			○	○	
c_3				×					
c_4					×	×			○
c_5			×						
c_6					×				
c_7						×			

Figure 7.22 Covering and closure table obtained from the table in Figure 7.21 by simplification rules.

When the map method is used for solution of these equations, the negated OR form is preferable:

$$\bar{r}_1\bar{r}_2 = 0$$
$$\bar{r}_1\bar{r}_4 = 0$$
$$\bar{c}_1\bar{c}_5 = 0$$
$$\bar{c}_1\bar{c}_3 = 0$$
$$\bar{c}_4\bar{c}_6 = 0$$
$$\bar{c}_4\bar{c}_7 = 0$$
$$c_1\bar{r}_4 = 0$$
$$r_1c_1 = 0$$
$$r_1c_4 = 0$$

The discriminant is shown in Figure 7.23. We can see that there exists only one minimum of the objective function

$$\sum_{i=1}^{7} r_i + \sum_{j=1}^{7} c_j$$

subject to all the covering and closure constraints expressed by the Boolean equations. The minimum value is 5 and is reached for $r_2 = r_4 = r_6 = c_1 = c_4 = 1$ and $r_1 = r_3 = r_5 = r_7 = c_2 = c_3 = c_5 = c_6 = c_7 = 0$. Thus, the minimal preserved cover contains row and columns prime classes:

(1, 3)	(*a*, *b*, *c*)
(2, 4)	(*d*, *e*)
(4, 5, 6)	

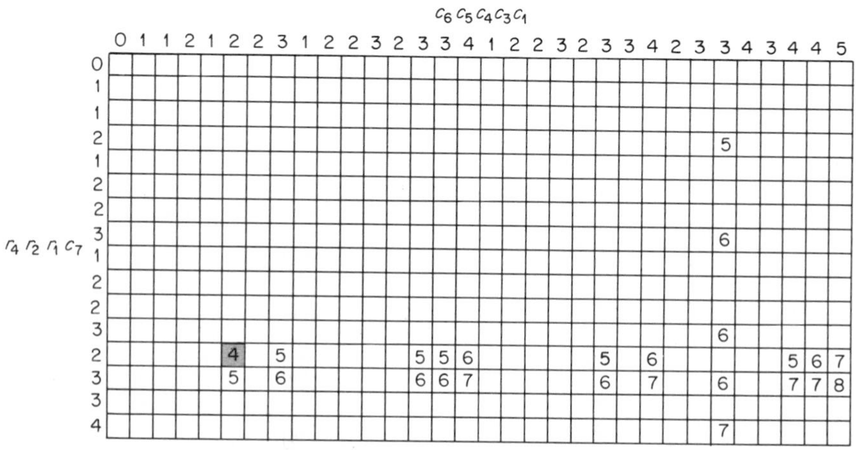

Figure 7.23 Discriminant of the Boolean equations describing the table in Figure 7.22.

The minimal Mealy machine is shown in Figure 7.24, where the symbols of logic variables are used to represent the corresponding prime classes.

The concept of state and machine equivalence, which was introduced for completely specified machines, is not applicable for incompletely specified machines because internal states of the latter differ in applicable input sequences. For the purpose of state minimization, we used the concept of state compatibility which defines a reflexive and symmetric relation on the set of internal states. Another useful relation defined on this set is that of state or *machine inclusion*.

DEFINITION 7.19 State $\mathbf{z}_i$ of machine M_1 is said to be included in state $\mathbf{z}_j$ of machine M_2, written as $\mathbf{z}_i \leq \mathbf{z}_j$, if and only if $A_i \subseteq A_j$ and $\mathbf{L}_1(\mathbf{I}, \mathbf{z}_i) = \mathbf{L}_2(\mathbf{I}, \mathbf{z}_j)$ for all $\mathbf{I} \in A_i$. M_1 and M_2 may refer to the same machine.

It can be readily verified that this relation of state inclusion is reflexive and transitive, but not symmetric. Thus, it represents a partial ordering. The state inclusion represents a basis for the following definition of machine inclusion.

DEFINITION 7.20 Machine M_1 is said to be *included* in machine M_2, written as $M_1 \leq M_2$, if and only if for each state $\mathbf{z}_i$ of M_1 there exists a state $\mathbf{z}_j$ of M_2 such that $\mathbf{z}_i \leq \mathbf{z}_j$.

The machine inclusion is, similarly as the state inclusion, a partial ordering. Note that, if $M_1 \leq M_2$, then M_2 can substitute for M_1 in any system without effecting the performance of the system. In other words, if only input sequences from A_1 are used, then M_1 and M_2 have identical behavior. For this reason, the relation $M_1 \leq M_2$ is sometimes called a *quasi-equivalence* of incompletely specified machines. It enables us to define precisely the minimal form for incompletely specified machines.

DEFINITION 7.21 A Mealy machine M_m is said to be a *minimal form of a Mealy machine M* if

1. $M \leq M_m$.
2. For every Mealy machine M_i, such that $M \leq M_i$, the number of internal states (or the sum of the numbers of internal and input states) of M_i is

	y^t		z^{t+1}	
z^t \ x^t	c_1	c_4	c_1	c_4
r_2	1	1	r_4	r_2
r_4	0	0	r_6	r_4
r_6	1	0	r_6	r_2

Figure 7.24 The final result of the state-input minimization of the Mealy machine specified in Figure 7.18.

greater than or equal to the number of internal states (or the sum of the number of internal and input states) of M_m.

We saw in this section that an incompletely specified Mealy machine may have several minimal forms which are not isomorphic.

Exercises to Section 7.4

*7.4-1 Perform the state minimization for the Mealy machines shown in Figure 7.25.

*7.4-2 Perform the state-input minimization for the Mealy machines shown in Figure 7.26.

7.4-3 Prove that for completely specified machines the set of all maximal compatible classes forms a partition of the states of the machine.

7.4-4 Prove the following rules which can be used for a simplification of the covering and closure table:
Rule 1. If a column c_i has only blank entries except for one cross in row r_j, this row must be selected ($r_j = 1$). Row r_j may be then deleted from the table together with all columns in which it contains crosses.
Rule 2. If a column c_i has only blank entries except for one dot in row r_j, this row must not be selected ($r_j = 0$). Row r_j may be deleted from the table together with all columns in which it contains dots.
Rule 3. If a column c_i has all the crosses and dots of a column c_j, c_i may be deleted from the table.
Rule 4. If a row r_i has all the crosses of a row r_j and for every column c_k in which r_i has a dot, there exists a column c_l in which r_j has a dot, such that, disregarding the entries in rows r_i and r_j, c_k has all the crosses and the dots of c_l, then row r_i must not be selected ($r_i = 0$). Row r_i may be deleted from the table together with all the columns in which it contains dots.
Rule 5. If a row r_i contains only dots and blank entries, it must not be selected ($r_i = 0$). Row r_i may, therefore, be deleted from the table together with all columns in which it contains dots.

7.4-5 Prove that if $c_k = c_{k+1}$, then $c_k = c$.

7.5 ABSTRACT SYNTHESIS: A DISCUSSION OF THE PROBLEM

The *abstract synthesis* of a deterministic sequential switching circuit is the first stage of its synthesis during which the ST-structure is constructed from a description of the required performance of the switching circuit. The description has no standard form and differs from one case to the next. Thus, the abstract synthesis may also be considered as the transformation of different forms of specification into a single form—the ST-structure. The latter con-

	y^t		z^{t+1}	
z^t \ x^t	0	1	0	1
1	0	0	2	6
2	1	1	3	1
3	–	0	–	4
4	–	0	–	5
5	1	–	3	–
6	1	–	7	–
7	–	0	–	8
8	–	1	–	–

(a)

	y^t			z^{t+1}		
z^t \ x^t	*a*	*b*	*c*	*a*	*b*	*c*
1	0	0	1	3	2	3
2	–	0	0	–	1	3
3	0	–	–	1	–	3

(b)

	y^t		z^{t+1}	
z^t \ x^t	0	1	0	1
1	0	0	2	–
2	0	–	5	4
3	1	0	–	1
4	1	0	3	5
5	–	0	2	4

(c)

	y^t			z^{t+1}		
z^t \ x^t	0	1	2	0	1	2
1	–	0	0	–	6	–
2	0	0	–	2	3	2
3	0	1	1	5	1	–
4	0	0	1	2	4	3
5	1	0	–	6	4	–
6	0	–	–	1	–	–

(d)

	y^t		z^{t+1}	
z^t \ x^t	0	1	0	1
1	–	0	2	3
2	1	–	4	5
3	–	0	–	1
4	1	–	2	–
5	–	0	3	–

(e)

	y^t			z^{t+1}		
z^t \ x^t	0	1	2	0	1	2
1	0	1	–	3	5	–
2	0	–	–	3	5	–
3	–	0	–	2	3	1
4	0	–	–	2	3	5
5	–	0	–	–	5	1

(f)

Figure 7.25 Mealy machines involved in Exercise 7.4-1.

stitutes then a standard basis for the second stage of the synthesis, which is usually referred to as the *structure synthesis*.

Although the abstract synthesis depends on the used paradigm of deterministic sequential switching circuits, the resulting ST-structure obtained for one paradigm can always be converted by an algorithmic procedure to a corresponding ST-structure based on another paradigm. Thus, the solution of the abstract synthesis is not unique due to the multiplicity of applicable paradigms. However, it is not unique even when only a single paradigm is applied, as we saw in Section 7.3 and 7.4. Nevertheless, the abstract synthesis is considered to be brought to its end when at least one ST-structure satisfying

	y^t			z^{t+1}		
z^t \ x^t	*a*	*b*	*c*	*a*	*b*	*c*
1	0	0	–	2	1	–
2	–	–	1	1	2	3
3	1	–	1	1	3	–

(a)

	y^t			z^{t+1}		
z^t \ x^t	*a*	*b*	*c*	*a*	*b*	*c*
1	0	0	–	2	3	–
2	1	–	0	1	–	2
3	–	0	–	3	1	–

(b)

	y^t				z^{t+1}			
z^t \ x^t	*a*	*b*	*c*	*d*	*a*	*b*	*c*	*d*
1	–	–	1	–	–	–	5	–
2	0	1	0	–	3	1	2	–
3	0	1	–	0	3	4	–	1
4	–	1	–	–	–	5	2	–
5	0	–	–	0	2	–	3	2

(c)

	y^t							z^{t+1}						
z^t \ x^t	*a*	*b*	*c*	*d*	*e*	*f*	*g*	*a*	*b*	*c*	*d*	*e*	*f*	*g*
1	0	1	–	–	–	1	–	1	3	5	–	5	5	–
2	0	1	1	–	–	–	0	1	3	5	–	–	–	8
3	0	–	–	–	0	–	–	–	4	–	1	1	–	5
4	–	–	–	–	–	–	1	1	4	5	–	1	4	5
5	0	–	0	1	0	–	–	5	–	3	4	1	5	–
6	1	0	1	0	–	–	1	5	4	3	1	1	4	5
7	0	–	1	1	1	0	–	1	2	–	6	7	8	–
8	–	1	–	1	–	0	0	–	2	–	4	–	8	7

(d)

Figure 7.26 Mealy machines involved in Exercise 7.4-2.

the described performance is determined, no matter which paradigm is involved. Once a particular form of the ST-structure is determined, other desirable forms can be derived from it by various algorithmic procedures. An example of such a modification of a given ST-structure is the state (or state-input) minimization.

The Mealy machine is used here as a paradigm for the abstract synthesis. Transformations to forms representing other paradigms, which sometimes may be preferable for the structure synthesis, are described in Chapter 8.

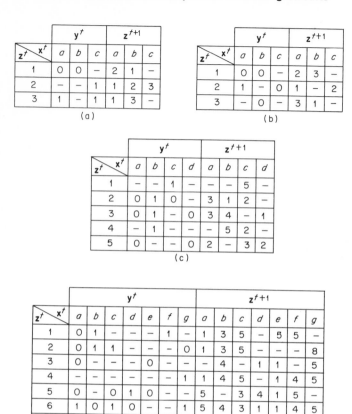

Although it may seem easy at the first sight, the abstract synthesis is one of the most peculiar and difficult problems included in the methodology of switching circuits. The major difficulty consists in the diversity of forms in which the desired performance of the switching circuit could be specified. This includes diverse forms of verbal description, various pulse diagrams, and lists of prescribed input-output sequences. It also happens frequently that a description, though correct, is not complete. In such cases, the procedure of abstract synthesis should involve identification of all properties which are not included in the description of the performance of the system but are essential for implementing correctly the abstract synthesis.

In some cases, the described performance cannot be implemented by a deterministic sequential system with a finite number of states. The procedure of abstract synthesis must involve also these cases, identifying them as insolvable.

At present, the abstract synthesis is solved in most cases by intuition based on the experience of the designer. It is usually a sort of "trial and error" procedure. Since this approach to abstract synthesis cannot be formalized, let us illustrate it by some examples.

EXAMPLE 7.1 Determine a ST-structure of the serial binary adder. It is a switching circuit having two input variables x_1, x_2 and one output variable y (see Figure 7.27a). When a sequence of binary digits representing a binary number comes to one of the inputs in time (in order from the least significant digit) and another similar sequence comes to the other input, the output is supposed to provide the sum of the two binary numbers (again in the order

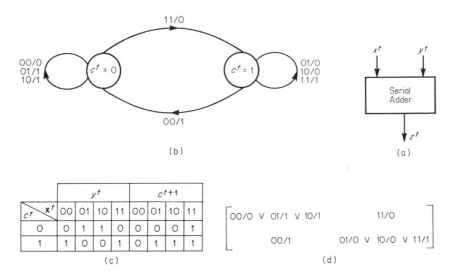

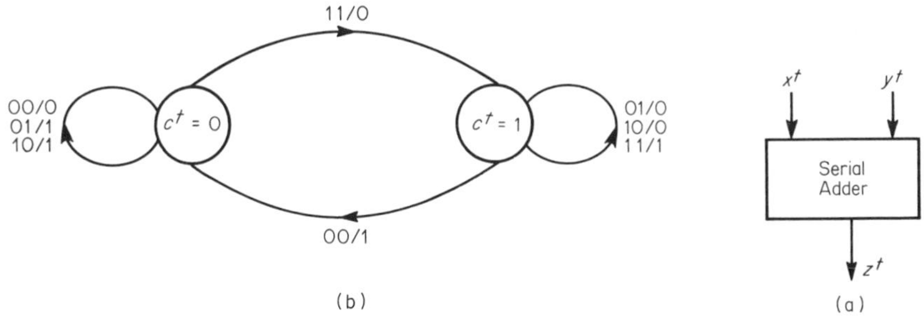

c^t \ x^t	y^t: 00	01	10	11	c^{t+1}: 00	01	10	11
0	0	1	1	0	0	0	0	1
1	1	0	0	1	0	1	1	1

(c)

$$\begin{bmatrix} 00/0 \vee 01/1 \vee 10/1 & 11/0 \\ 00/1 & 01/0 \vee 10/0 \vee 11/1 \end{bmatrix}$$

(d)

Figure 7.27 ST-structure of the serial binary adder.

from the least significant digit to the most significant digit in time). From the algorithm for binary addition, the designer knows that the sum digit y^t at a particular time is given by the formula

$$y^t = (x_1^t + x_2^t + c^t) \text{ (modulo 2)}$$

where c^t is the carry applicable at time t (generated at time $t - 1$). He also knows that the carry c^{t+1} applicable at time $t + 1$ is generated at time t and its value is given by the formula

$$c^{t+1} = \frac{x_1^t + x_2^t + c^t - y^t}{2}$$

It is easy to find that if the values of the carry c^t are considered as internal states, then the above equations represent, respectively, the output function and the transition function of a Mealy machine. Since the carry has only two distinct values, the Mealy machine has two internal states. The ST-diagram, table, or matrix, which are shown in Figure 7.27, follow directly from the above equations.

EXAMPLE 7.2 Determine a ST-structure of a switching circuit with one input and one output variable which exhibits the following performance: "It ascertains whether the sequence 0, 1 at the input is immediately followed by either the sequence 1, 1, 1 or 1, 1, 0 at the input. Only if this happens does the output variable assume the value 1 for the duration of the last of the three stimuli. After the output variable assumes the value 1, the circuit again waits for the arrival of the sequence 0, 1." An intuitive procedure may consist of the following steps:

1. Let $\mathbf{z}_1$ represent the internal state in which the circuit waits for the arrival of the sequence 0, 1. If the first stimulus is 1, then the system still waits for the sequence 0, 1 and, therefore, may stay in the same state. If the first stimulus is zero, the system waits only for 1 now. This is a different condition which must be identified by a different internal state, say $\mathbf{z}_2$. In both cases, the output variable answers the value 0. The determined portion of the ST-diagram is shown in Figure 7.28a.

2. State $\mathbf{z}_1$ has no more transitions oriented from it. If 0 is accepted in the state $\mathbf{z}_2$, the system again waits for 1 and, therefore, may stay in the same state. If 1 is accepted, the system waits now for either the sequence 1, 1, 1 or 1, 1, 0. This new condition must be represented by a new internal state, say $\mathbf{z}_3$. The value of the output variable is 0 in both cases. The determined portion of the ST-diagram can be seen in Figure 7.28b.

3. States $\mathbf{z}_1$ and $\mathbf{z}_2$ have no more transitions oriented from them. If 0 is accepted in the state $\mathbf{z}_3$, the system reaches a condition identical to that represented by state $\mathbf{z}_2$. Thus, the transition may terminate in $\mathbf{z}_2$. If 1 is accepted,

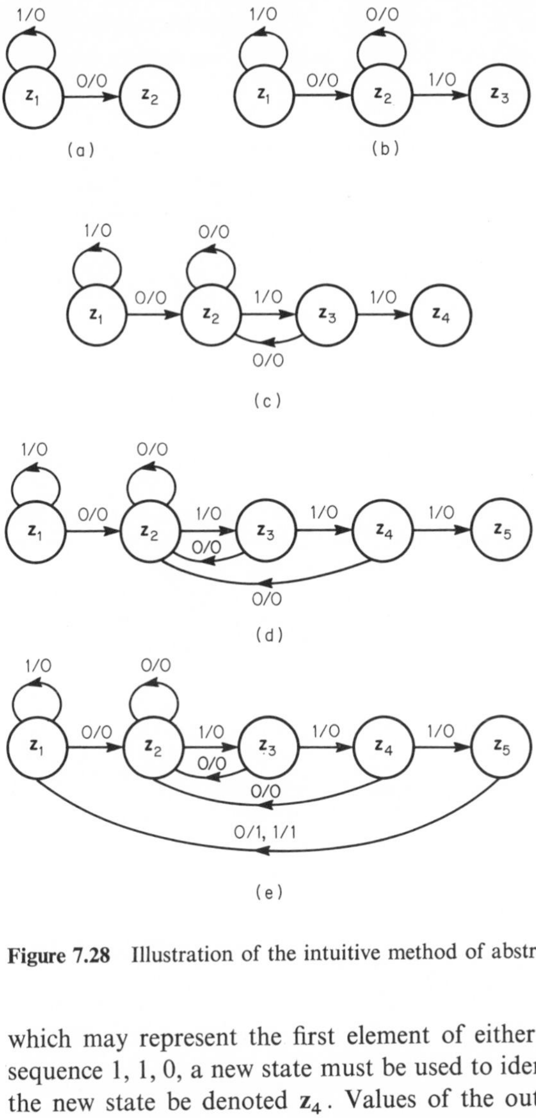

Figure 7.28 Illustration of the intuitive method of abstract synthesis (Example 7.2).

which may represent the first element of either the sequence 1, 1, 1 or the sequence 1, 1, 0, a new state must be used to identify this new condition. Let the new state be denoted $\mathbf{z}_4$. Values of the output variable are again 0 in both of the transitions. This stage of the abstract synthesis is illustrated in Figure 7.28c.

4. Using the same reasoning for state $\mathbf{z}_4$, as that used in Step 3 for state $\mathbf{z}_3$, we obtain the partial ST-diagram shown in Figure 7.28d.

5. Now, regardless of the value of the input variable, the output variable assumes the value 1 and the system waits again for the arrival of the sequence 0, 1. This situation is represented by the state $\mathbf{z}_1$. Thus, the state $\mathbf{z}_5$ passes to

the state $\mathbf{z}_1$. Now, transitions from each state are specified for both stimuli which concludes the abstract synthesis. The final ST-diagram is shown in Figure 7.28e.

The *intuitive method*, which was illustrated by the previous example, is, in many cases, the most powerful method of solving the abstract synthesis. Its power depends, however, considerably on the experience of the person involved. The method cannot be sufficiently formalized so that it cannot be performed by computers. This is a serious disadvantage which stimulated a development of other, more formalized, methods. One of them is presented in the next section of this chapter.

Exercises to Section 7.5

*7.5-1 Find if the ST-diagram in Figure 7.28e represents the minimal form of the respective Mealy machine.

7.5-2 Determine a ST-structure of a switching circuit with one input and one output variable. The output variable is equal to 1 when an input sequence 0, 1, 0, 1, is detected; otherwise the output variable is equal to 0.

*7.5-3 Determine a ST-structure of a switching circuit with one input variable x and two output variables y_1, y_2. Let s^t denote the state identifier corresponding to values of the output variables at time t and let the behavior of the system be given by the equation

$$s^t = \sum_{i=0}^{t} x^i \text{ (modulo 4)}$$

(The described circuit is called the counter modulo 4.)

*7.5-4 Solve Exercise 7.5-3 in case that the circuit has two input variables x_1, x_2 and its behavior is described by the equation

$$s^t = \sum_{i=0}^{t} (x_1^i - x_2^i) \text{ (modulo 4)}$$

(The described circuit is called the reversible or bi-directional counter modulo 4.)

7.5-5 Determine a *ST*-structure of a switching circuit with two input variables x_1, x_2 and one output variable y which meets the following requirements: (i) x_1 and x_2 can never be both equal to 1. (ii) $y^t = \bar{x}_2^t$ if and only if x_1 has been equal to 1 exactly twice since the last time when x_2 was equal to 1. Note: This is an incompletely specified circuit.

7.6 REGULAR EXPRESSIONS

Let A_R denote the set of all input sequences applicable in a reference (initial) state $\mathbf{z}_R$ of a Mealy machine M. This set can be partitioned into equivalence classes of input sequences by either of the following equivalence relations.

DEFINITION 7.22 Two input sequences $\mathbf{I}_j, \mathbf{I}_k \in A_R$ of a Mealy machine M are *state-equivalent* with respect to state $\mathbf{z}_R$, written as $\mathbf{I}_j \overset{s}{=} \mathbf{I}_k$, if and only if

$$\mathbf{G}(\mathbf{I}_j, \mathbf{z}_R) = \mathbf{G}(\mathbf{I}_k, \mathbf{z}_R)$$

If $\mathbf{G}(\mathbf{I}_j, \mathbf{z}_R) = \mathbf{z}_i$, then the *state-equivalence class* containing $\mathbf{I}_j$ will be denoted $Z_{R,i}$.

DEFINITION 7.23 Two input sequences $\mathbf{I}_j, \mathbf{I}_k \in A_R$ of a Mealy machine M are *output-equivalent* with respect to state $\mathbf{z}_R$, written as $\mathbf{I}_j \overset{o}{=} \mathbf{I}_k$, if and only if

$$\mathbf{L}(\mathbf{I}_j, \mathbf{z}_R) = \mathbf{L}(\mathbf{I}_k, \mathbf{z}_R)$$

If $\mathbf{L}(\mathbf{I}_j, \mathbf{z}_R) = \mathbf{y}_i$, then the *output-equivalence class* containing $\mathbf{I}_j$ will be denoted $Y_{R,i}$.

It is easy to verify that both of the relations are reflexive, symmetric and transitive, being thus proper equivalence relations. It follows directly from Definition 7.22 that states $\mathbf{z}_i$ are in a one-to-one correspondence with state-equivalence classes $Z_{R,i}$. Similarly, it follows from Definition 7.23 that responses $\mathbf{y}_i$ are in a one-to-one correspondence with output-equivalence classes $Y_{R,i}$.

Regular expressions are an algebraic tool by which equivalence classes $Z_{R,i}$ or $Y_{R,i}$ can be represented in a compact form. As such, they can be employed in a formalization of the abstract synthesis of sequential machines. Regular expressions may contain symbols $\mathbf{x}_i \in X$ of stimuli, A_R denoting the set union of *all input sequences* acceptable in the reference state $\mathbf{z}_R$ (the universal set in this case), $\varnothing$ representing the *empty* (*null*) *set* and λ corresponding to the *input sequence of zero length* (*null string*), which specifies the reference state $\mathbf{z}_R$. For any subset a of A_R, $\bar{a}$ denotes the complement of a with respect to the universal set A_R. For a pair a, b of subsets of A_R, $a \cup b$, $a \cap b$, and ab denote, respectively, the *union*, *intersection*, and *concatenation* of a and b. The meaning of concatenation can be expressed as "a is followed by b." Concatenation aa, aaa, etc., will be abbreviated as a^2, a^3, etc. Then, $aa^{k-1} = a^k (k \geq 1)$ and $a^0 = \lambda$. For any subset a of A_R, the *star operation* is defined as

$$a^* = \bigcup_{k=0}^{\infty} a^k \tag{7.24}$$

The meaning of this operation can be expressed as "any number of occurences of a, including none." *Parentheses*, if used in a regular expression, denote an order of execution of the operations participating in the expression. To simplify the notation, let us introduce the *convention* that operations are applied in the following order: star operations, negations, concatenations, intersections, unions. If a different order is required, it will be denoted by parentheses. Now, we are in a position to define regular expressions formally.

DEFINITION 7.24 Let X, $\varnothing$ or λ have the meaning as introduced above. Then

1. A single element $\mathbf{x}_i \in X$ (the symbol of a stimulus) or $\varnothing$ or λ are regular expressions.
2. If E is a regular expression, then so is its complement $\bar{E}$ and its star operation E^*.
3. If E_i and E_j are any regular expressions, then so is their union $E_i \cup E_j$, their intersection $E_i \cap E_j$, and their concatenation $E_i E_j$.
4. Only those expressions that can be obtained by a finite number of applications of (i), (ii), and (iii) are regular expressions.

An example of a regular expression is A_R0110. It represents all input sequences ending in 0110. Another example is 100^*1 which represents all sequences which begin with 1 followed by at least one 0 and end with 1. Regular expression $A_R0110 \cup 100^*1$ represents the union of both of the above mentioned sets of input sequences. If two regular expressions describe the same set of input sequences, we consider them equal and may put the equality sign between them. Some equalities of regular expressions, which are useful when manipulating the latter, are listed in Table 7.1. They follow directly

TABLE 7.1. Some Equalities Between Regular Expressions

$\lambda a = a\lambda = a$	$a(b \cap c) = ab \cap ac$
$\varnothing a = a\varnothing = \varnothing$	$(a \cup b)c = ac \cup bc$
$\lambda^* = \varnothing^* = \lambda$	$(a \cap b)c = ac \cap bc$
$\varnothing \cup a = a \cup \varnothing = a$	$a(ba)^* = (ab)^*a$
$a \cup b = b \cup a$	$(ab)^* = (a^*b^*)^*$
$a \cap b = b \cap a$	$(a \cup b)^* = (a^* \cup b^*)^*$
$\overline{a \cup b} = \bar{a} \cap \bar{b}$	$(a \cap b)^* = (a^* \cap b^*)^*$
$\overline{a \cap b} = \bar{a} \cup \bar{b}$	$a^*a^* = a^*$
$a \cup a = a$	$\lambda \cup aa^* = a^*$
$a \cap a = a$	$(a^*b^*) = \lambda \cup (a \cup b)^*b$
$(ab)c = a(bc)$	$a^* = (\lambda \cup a)\lambda^*$
$a(b \cup c) = ab \cup ac$	

from the definitions of the operations involved. Equal regular expressions will be said to be of the same type.

Before we start to investigate the abstract synthesis based on regular expressions, let us show a determination of the state-equivalence classes $Z_{R,i}$ and the output equivalence classes $Y_{R,i}$ from a given ST-structure.

Let us recall that $Z_{R,i}$ denotes all input sequences taking the machine from the reference state $\mathbf{z}_R$ and leaving it in state $\mathbf{z}_i$. Let $a_{j,k}$ denote the set union of all stimuli that cause a direct transition from state $\mathbf{z}_j$ to state $\mathbf{z}_k$. Then, the ST-structure can be described by the following set of regular-expression equations:

$$Z_{R,k} = \bigcup_{j=1}^{r} Z_{R,j}\, a_{j,k} \quad \text{for all } k \neq R$$

$$Z_{R,R} = \bigcup_{j=1}^{r} Z_{R,j} a_{j,R} \cup \lambda \tag{7.25}$$

We want to solve these equations for $Z_{R,k}$ $(k = 1, 2, \ldots, r)$. Notice that each of the equations demonstrates the form

$$Z_{R,k} = Z_{R,k} A_k \cup B_k \tag{7.26}$$

where A_k and B_k are certain regular expressions that do not contain $Z_{R,k}$. We can easily verify that

$$Z_{R,k} = B_k A_k^* \tag{7.27}$$

is a solution of (7.26). Indeed, when we substitute $B_k A_k^*$ for $Z_{R,k}$ on the right-hand side of (7.26), we obtain

$$B_k A_k^* A_k \cup B_k = B_k(\lambda \cup A_k \cup A_k^2 \cup \cdots)$$
$$A_k \cup B_k = (B_k A_k \cup B_k A_k^2 \cup \cdots) \cup B_k = B_k A_k^*.$$

EXAMPLE 7.3 Find the state-equivalence classes $Z_{1,1}$, $Z_{1,2}$, and $Z_{1,3}$ for the ST-structure specified by the diagram in Figure 7.29. Equations (7.25) for this ST-structure have the form:

$$Z_{1,1} = Z_{1,1}1 \cup Z_{1,2}0 \cup Z_{1,3}0 \cup \lambda$$
$$Z_{1,2} = Z_{1,1}0$$
$$Z_{1,3} = Z_{1,2}1 \cup Z_{1,3}1$$

Applying the form (7.27) to each of these equations, we obtain

$$Z_{1,1} = (Z_{1,2}0 \cup Z_{1,3}0 \cup \lambda)1^* \tag{a}$$
$$Z_{1,2} = Z_{1,1}0 \tag{b}$$
$$Z_{1,3} = Z_{1,2}11^* \tag{c}$$

We can substitute now for $Z_{1,2}$ into (c) from (b). This gives us

$$Z_{1,3} = Z_{1,1}011^* \tag{d}$$

After substituting for $Z_{1,2}$ and $Z_{1,3}$ into (a) from (c) and (d), respectively, we obtain

$$\begin{aligned} Z_{1,1} &= (Z_{1,1}0^2 \cup Z_{1,1}011^*0 \cup \lambda)1^* \\ &= Z_{1,1}(0^2 \cup 011^*0)1^* \cup 1^* \end{aligned} \tag{e}$$

An application of the form (7.27) to (e) gives us a regular expression for $Z_{1,1}$:

$$Z_{1,1} = 1^*((0^2 \cup 011^*0)1^*)^* \tag{f}$$

When substituting for $Z_{1,1}$ from (f) into (b), we obtain

$$Z_{1,2} = 1^*((0^2 \cup 011^*0)1^*)0 \tag{g}$$

We can substitute now for $Z_{1,2}$ from (g) into (c) and obtain

$$Z_{1,3} = 1^*((0^2 \cup 011^*0)1^*)^*011^* \tag{h}$$

It becomes rather trivial to find that the output-equivalence classes $Y_{R,i}$ of a Mealy machine are related to the state-equivalence classes by the formula

$$Y_{R,i} = \bigcup Z_{R,j}\mathbf{x} \tag{7.28}$$

where the union is taken for all stimuli $\mathbf{x}$ and all internal states $\mathbf{z}_j$ such that $\mathbf{f}(\mathbf{x}, \mathbf{z}_j) = \mathbf{y}_i$.

EXAMPLE 7.4 Determine $Y_{1,0}$ and $Y_{1,1}$ of the Mealy machine shown in Figure 7.29. Using formula (7.28), we obtain

$$Y_{1,0} = Z_{1,1}0 \cup Z_{1,2}0 \cup Z_{1,2}1 \cup Z_{1,3}1$$

and

$$Y_{1,1} = Z_{1,1}1 \cup Z_{1,3}0$$

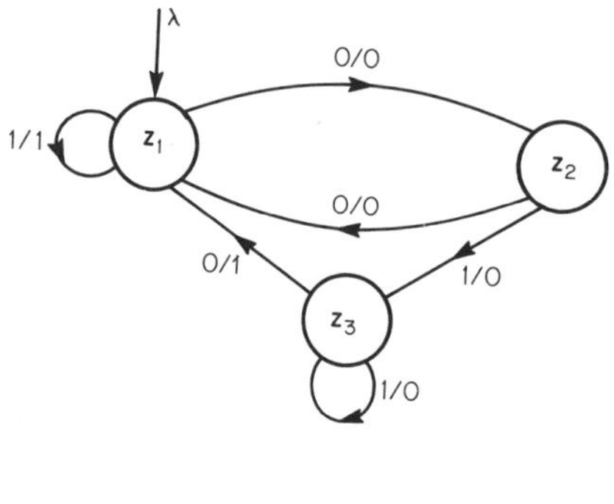

Figure 7.29 Illustration to Examples 7.3. and 7.4.

Substitution for $Z_{1,1}$ $Z_{1,2}$, $Z_{1,3}$ from (f), (g), (h) in Example 7.3 will yield the regular expressions for $Y_{1,0}$ and $Y_{1,1}$. Since the machine has only two responses, clearly, it should be

$$Y_{1,0} = \overline{Y}_{1,1}$$

Indeed,
$$\begin{aligned}\overline{Y}_{1,1} &= \overline{Z_{1,1}1 \cup Z_{1,3}0} = \overline{Z_{1,1}1} \cap \overline{Z_{1,3}0} \\ &= (Z_{1,1}0 \cup Z_{1,2}0 \cup Z_{1,2}1 \cup Z_{1,3}0 \cup Z_{1,3}1) \cap \\ &\quad (Z_{1,1}0 \cup Z_{1,1}1 \cup Z_{1,2}0 \cup Z_{1,2}1 \cup Z_{1,3}1) \\ &= Z_{1,1}0 \cup Z_{1,2}0 \cup Z_{1,2}1 \cup Z_{1,3}1 = Y_{1,0}\end{aligned}$$

Now we are in a position to speak about the role of regular expressions in the solution of the abstract synthesis of finite-state machines. Assume that the behavior, which essentially is a description of output-equivalence classes $Y_{R,i}$ for all responses $\mathbf{y}_i$, can be expressed in the form of regular expressions. If this assumption holds true, then the abstract synthesis can be solved by the procedure which will be described in this section. If, however, the assumption cannot be satisfied, then regular expressions are of no use to the abstract synthesis. The applicability of regular expressions to abstract synthesis depends, primarily, on the form in which the behavior is specified, on the ability of the person solving the abstract synthesis to convert this form into a set of regular expressions and his ability to manipulate the latter. In some cases the conversion is rather trivial. For example, let the behavior be described by the following proposition: "The output is equal to one if and only if the machine is stimulated by two zeroes followed by three ones or by a single one followed by at least one zero after which exactly two ones follow or by three time repeated sequence of a single zero followed by a single one; otherwise the output is equal to zero." This proposition, which describes actually the output-equivalence class $Y_{R,1}$, can be immediately converted to the regular expression

$$Y_{R,1} = 0^2 1^3 \cup 100^* 1^2 \cup (01)^3$$

There are only two output-equivalence classes in this case, namely, $Y_{R,1}$ and $Y_{R,0} = \overline{Y}_{R,1}$.

If the number of output-equivalence classes is greater than one, the corresponding machine has either several two-valued output variables or a single many-valued variable. Let us note, however, that the latter case is possible only if $Y_{R,i} \cap Y_{R,j} = \varnothing$ for all specified responses $\mathbf{y}_i \neq \mathbf{y}_j$.

The abstract synthesis based on regular expressions is, essentially, a procedure by which the output and next state functions are determined from regular expressions of output-equivalence classes $Y_{R,i}$. The concept of a derivative of a regular expression, which is introduced by the following definition, is heavily involved in this procedure.

DEFINITION 7.25 Let E and $\mathbf{I}$ be, respectively, a set of input sequences and a finite input sequence. Then, the *derivative* $D_{\mathbf{I}}E$ of E with respect to $\mathbf{I}$ is the set of all input sequences I_k such that $\mathbf{I}I_k \in E$. Formally,

$$D_{\mathbf{I}}E = \{I_k : \mathbf{I}I_k \in E\}$$

For example, if $E = \{11010, 1111, 010, 1001\}$, then $D_0 E = \{10\}$, $D_1 E = \{1010, 111, 001\}$, $D_{00} E = \varnothing$, $D_{01} E = \{0\}$, $D_{10} E = \{01\}$, $D_{11} E = \{010, 11\}$, $D_{110} E = \{10\}$, etc.

For various problems associated with regular expressions, we need to identify if a regular expression E contains λ. For this purpose we introduce the function

$$\delta(E) = \begin{cases} \lambda & \text{if } \lambda \in E \\ \varnothing & \text{if } \lambda \notin E \end{cases}$$

Some properties of this function are summarized in Table 7.2.

TABLE 7.2. Some Basic Properties of the Function $\delta(E)$

$\delta(x) = \varnothing$ for any $x \in X$
$\delta(\lambda) = \lambda$
$\delta(\varnothing) = \varnothing$
$\delta(E^*) = \lambda$
$\delta(E_1 E_2) = \delta(E_1) \cup \delta(E_2)$
$\delta(E_1 \cup E_2) = \delta(E_1) \cup \delta(E_2)$
$\delta(E_1 \cap E_2) = \delta(E_1) \cap \delta(E_2)$
$\delta(\bar{E}) = \begin{cases} \lambda & \text{if } \delta(E) = \varnothing \\ \varnothing & \text{if } \delta(E) = \lambda \end{cases}$

It follows from the definitions of the operations of regular expressions, the derivative, and the function $\delta(E)$ that the derivative of a regular expression E with respect to a single stimulus $\mathbf{x} \in X$ can be found by repeatedly applying the equalities listed in Table 7.3.

TABLE 7.3. Basic Equalities Concerning Derivatives of Regular Expressions

$D_{\mathbf{x}}\mathbf{x} = \lambda$
$D_{\mathbf{x}}a = \varnothing$ for $a = \lambda$ or $a = \varnothing$ or $a \neq \mathbf{x}$
$D_{\mathbf{x}}(E^*) = (D_{\mathbf{x}}E)E^*$
$D_{\mathbf{x}}(E_1E_2) = (D_{\mathbf{x}}E_1)E_2 \cup \delta(E_1)D_{\mathbf{x}}E_2$
$D_{\mathbf{x}}(E_1 \cup E_2) = D_{\mathbf{x}}E_1 \cup D_{\mathbf{x}}E_2$
$D_{\mathbf{x}}(E_1 \cap E_2) = D_{\mathbf{x}}E_1 \cap D_{\mathbf{x}}E_2$
$D_{\mathbf{x}}(\bar{E}) = \overline{D_{\mathbf{x}}E}$

Using Definition 7.25, the derivative of a regular expression E with respect to a finite input sequence $\mathbf{I} = \mathbf{a}_1\mathbf{a}_2 \cdots \mathbf{a}_m$ can be found recursively as follows:

$$D_{\mathbf{a}_1\mathbf{a}_2}E = D_{\mathbf{a}_2}(D_{\mathbf{a}_1}E)$$
$$D_{\mathbf{a}_1\mathbf{a}_2\mathbf{a}_3}E = D_{\mathbf{a}_3}(D_{\mathbf{a}_1\mathbf{a}_2}E)$$
$$\vdots \qquad \vdots$$
$$D_{\mathbf{I}}E = D_{\mathbf{a}_m}(D_{\mathbf{a}_1\mathbf{a}_2 \cdots \mathbf{a}_{m-1}}E)$$

In the case that $I = \lambda$, we have $D_\lambda E = E$.

EXAMPLE 7.5 Let $E = 1(00^* \cup 1)^*1$. Calculate all derivatives of E with respect to input sequences of length one and two.

Applying some of the equalities in Table 7.3, we obtain:

$$D_0E = D_{00}E = D_{01}E = \varnothing$$
$$D_1E = (00^* \cup 1)^*1$$

$$\begin{aligned} D_{10}E &= D_0(D_1E) = D_0(00^* \cup 1)^*1 \\ &= D_0(00^* \cup 1)(00^* \cup 1)^*1 \\ &= (0^* \cup \varnothing)(00^* \cup 1)^*1 \\ &= 0^*(00^* \cup 1)^*1 \end{aligned}$$

$$\begin{aligned} D_{11}E &= D_1(D_1E) = D_1(00^* \cup 1)^*1 \\ &= D_1(00^* \cup 1)(00^* \cup 1)^*1 \\ &= (D_1(00^*) \cup D_1(1))(00^* \cup 1)^*1 \\ &= (\varnothing \cup \lambda)(00^* \cup 1)^*1 \\ &= (00^* \cup 1)^*1 \end{aligned}$$

It has been seen that the formation of $D_{\mathbf{I}}E$ involves only a finite number of regular expression operations. Thus, if E is a regular expression, then $D_{\mathbf{I}}E$ is a regular expression too. Some properties of $D_{\mathbf{I}}E$, which are important for the abstract synthesis, are the subject of the following theorems.

THEOREM 7.6 An input sequence **I** is represented by a regular expression E if and only if λ is contained in $D_{\mathbf{I}}E$.

PROOF It follows from Definition 7.25 that if $\lambda \in D_{\mathbf{I}}E$, then $\mathbf{I}\lambda = I \in E$ and, conversely, if $\mathbf{I} \in E$, then $\mathbf{I}\lambda \in E$ and $\lambda \in D_{\mathbf{I}}E$. ∎

THEOREM 7.7 Every regular expression E has a finite number $N(E)$ of different derivatives.

PROOF First we can easily determine all derivatives of $\varnothing$, λ, and a single stimulus $\mathbf{x}_i \in X$. We obtain $D_{\mathbf{I}}\varnothing = \varnothing$ for every input sequence **I**, $D_\lambda \lambda = \lambda$ and $D_{\mathbf{I}}\lambda = \varnothing$ for every input sequence **I**, $D_\lambda \mathbf{x}_i = \mathbf{x}_i$, $D_{\mathbf{x}_i}\mathbf{x}_i = \lambda$, and $D_{\mathbf{I}}\mathbf{x}_i = \varnothing$ for every input sequence $\mathbf{I} \neq \lambda$ and $\mathbf{I} \neq \mathbf{x}_i$. Thus, $N(\varnothing) = 1$, $N(\lambda) = 2$ and $N(\mathbf{x}_i) = 3$.

Assume now that E_1 and E_2 are regular expressions that have $N(E_1)$ and $N(E_2)$ types of derivatives, respectively. Clearly, if $E = E_1 \cup E_2$, then $D_{\mathbf{I}}E = D_{\mathbf{I}}E_1 \cup D_{\mathbf{I}}E_2$ so that $N(E) \leq N(E_1) \cdot N(E_2)$. Similarly, if $E = E_1 \cap E_2$, then $D_{\mathbf{I}}E = D_{\mathbf{I}}E_1 \cap D_{\mathbf{I}}E_2$ so that $N(E) \leq N(E_1) \cdot N(E_2)$. In cases of $\bar{E}$ we have $D_{\mathbf{I}}\bar{E} = \overline{D_{\mathbf{I}}E}$ so that $N(\bar{E}) = N(E)$.

To complete the proof we have to show that also E_1E_2 and E_1^* have a finite number of types of derivatives. Let $E = E_1E_2$ and $\mathbf{I} = \mathbf{a}_1\mathbf{a}_2 \cdots \mathbf{a}_m$. Then $D_{\mathbf{a}_1}E = (D_{\mathbf{a}_1}E_1)E_2 \cup \delta(E_1)D_{\mathbf{a}_1}E_2$. Similarly, for input sequences of length two, we have $D_{\mathbf{a}_1\mathbf{a}_2}E = (D_{\mathbf{a}_1\mathbf{a}_2}E_1)E_2 \cup \delta(D_{\mathbf{a}_1}E_1)D_{\mathbf{a}_2}E_2 \cup \delta(E_1)\, D_{\mathbf{a}_1\mathbf{a}_2}E_2$. Generally, for input sequences of length m, we obtain

$$D_{\mathbf{a}_1 \cdots \mathbf{a}_m}E = (D_{\mathbf{a}_1 \cdots \mathbf{a}_m}E_1)E_2 \cup \delta(D_{\mathbf{a}_1 \cdots \mathbf{a}_{m-1}}E_1)D_{\mathbf{a}_m}E_2 \cup \\ \delta(D_{\mathbf{a}_1}E_1)D_{\mathbf{a}_2 \cdots \mathbf{a}_m}E_2 \cup \delta(E_1)D_{\mathbf{a}_1 \cdots \mathbf{a}_m}E_2$$

Thus, $D_{\mathbf{I}}(E_1E_2)$ is a union of $(D_{\mathbf{I}}E_1)E_2$ and at most m derivatives of E_2. Derivatives of E_2 cannot be formed in more than $2^{N(E_2)}$ ways (this is an upper bound and we do not care whether it can be achieved or not). Hence, there are $N(E_1E_2) \leq N(E_1) \cdot 2^{N(E_2)}$.

It remains now to find an upper bound for $N(E^*)$. We have

$$\begin{aligned} D_{\mathbf{a}_1}(E^*) &= (D_{\mathbf{a}_1}E)E^* \\ D_{\mathbf{a}_1\mathbf{a}_2}(E^*) &= (D_{\mathbf{a}_1\mathbf{a}_2}E)E^* \cup \delta(D_{\mathbf{a}_1}E)D_{\mathbf{a}_2}(E^*) \\ &= (D_{\mathbf{a}_1\mathbf{a}_2}E)E^* \cup \delta(D_{\mathbf{a}_1}E)(D_{\mathbf{a}_2}E)E^* \end{aligned}$$

If we proceeded to determine higher order derivatives of E^* we would find that $D_{\mathbf{a}_1 \cdots \mathbf{a}_m}(E^*)$ is a union of at most m derivatives. If E has a finite number $N(E)$ of different derivatives, then there cannot be more than $2^{N(E)}$ of forming these derivatives. Hence, $H(E^*) \leq 2^{N(E)}$. Since any regular expression contains only a finite number of the above mentioned operations applied to $\varnothing$, λ, or $\mathbf{x} \in X$ and each of the operations has a finite number of derivatives, every regular expression has a finite number of derivatives too. ∎

Suppose now that a Mealy machine is in state $\mathbf{z}_j$ described by a set of output-equivalence classes $\{Y_{j,i} : i = 1, 2, \ldots, q\}$ and an input sequence $\mathbf{I}$ is applied which takes the machine to state z_k described by another set of output-equivalence classes $\{Y_{k,i} := 1, 2, \ldots, q\}$. Then, by Definition 7.25, the equation

$$D_{\mathbf{I}} Y_{j,i} = Y_{k,i} \tag{7.29}$$

must be satisfied for every $i = 1, 2, \ldots, q$.

Equation (7.29) is the clue for determining the desired ST-structure from a given set of regular expressions representing the output-equivalence classes $Y_{R,i}$ of a reference state $\mathbf{z}_R$. To make the following explanation easier to comprehend, let us assume, first, that there are only two responses, 0 and 1, and that the reference state is that denoted as $\mathbf{z}_1$. Thus, we assume that a single regular expression representing $Y_{1,1}$ will be used to describe the behavior of the designed system. Clearly, $\mathbf{L}(\mathbf{I}, \mathbf{z}_1) = 0$ for all input sequences which are not included in $Y_{1,1}$. Later, a generalization for more responses will be shown.

To determine the ST-structure from a regular expression representing $Y_{1,1}$, we form all different derivatives of $Y_{1,1}$, and associate one internal state with each of their derivatives. The derivative $D_\lambda Y_{1,1} = Y_{1,1}$ is associated with the reference state $\mathbf{z}_1$. It follows from Theorem 7.7 that the number of different derivatives of $Y_{1,1}$ is finite; hence, the number of associated internal states is finite too.

When two derivatives are compared in case of the Mealy machine (but not in case of the Moore machine), the presence of λ in a derivative can be ignored. This is due to the fact that none of the output or transition functions of the Mealy machine is defined for λ (note that the output function of the Moore machine is defined for λ).

After all different derivatives of $Y_{1,1}$ are determined and internal states assigned to them, we can apply equation (7.29) to determine the transition function.

Let two different derivatives of $Y_{1,1}$ be represented by regular expressions R_j, R_k and associated with internal states $\mathbf{z}_j$, $\mathbf{z}_k$, respectively. Let

$$R_k = D_{\mathbf{x}} R_j = D_{\mathbf{x}}(D_{\mathbf{I}} Y_{1,1}) = D_{\mathbf{Ix}} Y_{1,1}$$

where $\mathbf{I}$ is an input sequence and $\mathbf{x}$ is a single stimulus. Then, clearly, $\mathbf{z}_k = \mathbf{g}(\mathbf{x}, \mathbf{z}_j)$.

Suppose again that the derivatives $D_{\mathbf{I}} Y_{1,1}$ and $D_{\mathbf{Ix}} Y_{1,1}$ represent internal states $\mathbf{z}_j$ and $\mathbf{z}_k$, respectively. Then, there is a transition from $\mathbf{z}_j$ to $\mathbf{z}_k$ and, according to Theorem 7.6, the response associated with this transition is 1 if $D_{\mathbf{Ix}} Y_{1,1}$ contains λ; otherwise the response is 0.

EXAMPLE 7.6 Determine the ST-structure of a Mealy machine whose response is 1 whenever the input sequence is 01 or a single 1 followed by at least one 0 is accepted at the input.

Let $\mathbf{z}_1$ be the reference state. Then, $Y_{1,1} = 01 \cup 100^*$ and the procedure of determining the ST-structure is summarized in Table 7.4. The final ST-diagram is shown in Figure 7.30.

TABLE 7.4. Determination of the ST-Structure Representing $Y_{1,1} = 01 \cup 100^*$ and $Y_{1,0} = \bar{Y}_{1,1}$ (Example 7.6)

Derivatives	States	Transition Function	Output Function
$D_\lambda Y_{1,1} = Y_{1,1}$	$\mathbf{z}_1$	—	—
$D_0 Y_{1,1} = 1$	$\mathbf{z}_2$	$g(0, \mathbf{z}_1) = \mathbf{z}_2$	$f(0, \mathbf{z}_1) = 0$
$D_1 Y_{1,1} = 00^*$	$\mathbf{z}_3$	$g(1, \mathbf{z}_1) = \mathbf{z}_3$	$f(1, \mathbf{z}_1) = 0$
$D_{00} Y_{1,1} = \varnothing$	$\mathbf{z}_4$	$g(0, \mathbf{z}_2) = \mathbf{z}_4$	$f(0, \mathbf{z}_2) = 0$
$D_{01} Y_{1,1} = \lambda$	$\mathbf{z}_5$	$g(1, \mathbf{z}_2) = \mathbf{z}_5$	$f(1, \mathbf{z}_2) = 1$
$D_{10} Y_{1,1} = 0^*$	$\mathbf{z}_6$	$g(0, \mathbf{z}_3) = \mathbf{z}_6$	$f(0, \mathbf{z}_3) = 1$
$D_{11} Y_{1,1} = \varnothing$		$g(1, \mathbf{z}_3) = \mathbf{z}_4$	$f(1, \mathbf{z}_3) = 0$
$D_{000} Y_{1,1} = \varnothing$		$g(0, \mathbf{z}_4) = \mathbf{z}_4$	$f(0, \mathbf{z}_4) = 0$
$D_{001} Y_{1,1} = \varnothing$		$g(1, \mathbf{z}_4) = \mathbf{z}_4$	$f(1, \mathbf{z}_4) = 0$
$D_{010} Y_{1,1} = \varnothing$		$g(0, \mathbf{z}_5) = \mathbf{z}_4$	$f(0, \mathbf{z}_5) = 0$
$D_{011} Y_{1,1} = \varnothing$		$g(1, \mathbf{z}_5) = \mathbf{z}_4$	$f(1, \mathbf{z}_5) = 0$
$D_{100} Y_{1,1} = 0^*$		$g(0, \mathbf{z}_6) = \mathbf{z}_6$	$f(0, \mathbf{z}_6) = 1$
$D_{101} Y_{1,1} = \varnothing$		$g(1, \mathbf{z}_6) = \mathbf{z}_4$	$f(1, \mathbf{z}_6) = 0$
$D_{110} Y_{1,1} = \varnothing$		—	—
$D_{111} Y_{1,1} = \varnothing$		—	—

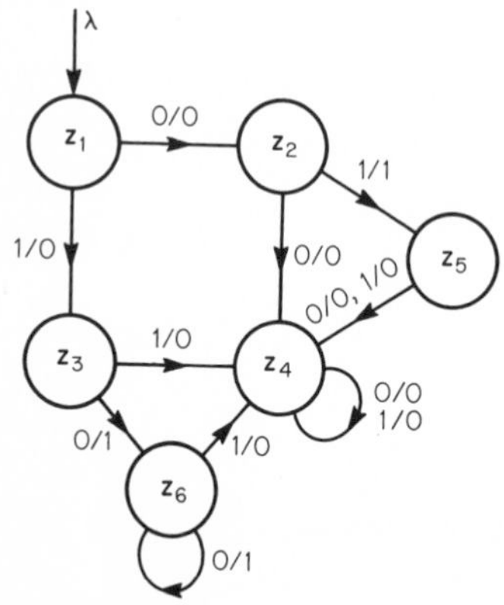

Figure 7.30 Result of the abstract synthesis performed in Table 7.4 (Example 7.6).

Let us consider now a sequential switching circuit with more than one output. Clearly, its behavior can be represented by a set of regular expressions, one for each two-valued output variable. For convenience, the set of regular expressions can be viewed as a vector $Y_{R,1} = (Y_1, Y_2, \ldots, Y_q)_{R,1}$, where, Y_i is a regular expression representing the output-equivalence class of all input sequences which, when applied at a reference state $\mathbf{z}_R$, make the ith output variable equal to 1.

DEFINITION 7.26 Two vectors of regular expressions are equal if and only if they have the same number of components and all pairs of corresponding components are equal regular expressions.

DEFINITION 7.27 The derivative of a vector of regular expressions $\mathbf{R} = (R_1, R_2, \ldots, R_n)$ with respect to an input sequence $\mathbf{I}$ is denoted by $D_{\mathbf{I}}\mathbf{R}$ and defined by $D_{\mathbf{I}}\mathbf{R} = (D_{\mathbf{I}}R_1, D_{\mathbf{I}}R_2, \ldots, D_{\mathbf{I}}R_{\mathbf{I}})$.

Definitions 7.26 and 7.27 give us a clue for a generalization of the abstract synthesis, based on regular expressions, for multiple output switching circuits. We determine all distinct derivatives of $Y_{R,1}$ and assign an internal state to each of them. Then, we identify the transition and output functions. The transition function is identified by exactly the same rule as for a single output. Obviously, the rule is applied to vectors rather than single regular expressions in the multiple output case. A significant difference is that we obtain a set of output functions, one for each component of the respective vectors $D_{\mathbf{I}}Y_{R,1}$.

EXAMPLE 7.7 Let the behavior of a sequential switching circuit with two output variables be described by $Y_{1,1} = (0^*1, 1(10)^*)$.

The whole procedure is summarized in Table 7.5 and the final ST-diagram is in Figure 7.31.

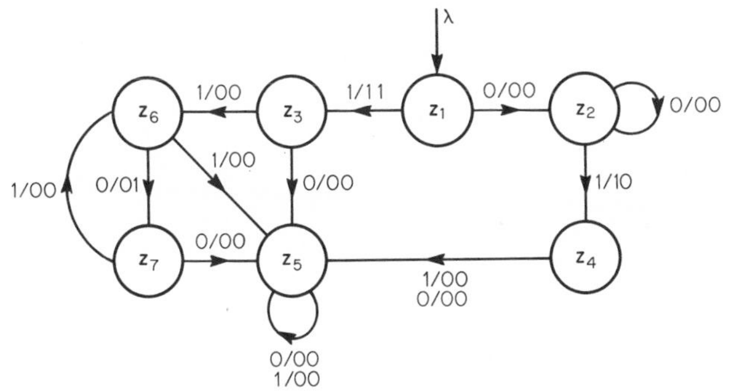

Figure 7.31 Results of the abstract synthesis performed in Table 7.5 (Example 7.7).

TABLE 7.5. Determination of the ST-Structure Representing $Y_{1,1} = (0^*1, 1(10)^*) = (A, B)$ (Example 7.7)

Derivatives	States	Transition Function	Output Functions
$D_\lambda Y_{1,1} = (A, B)$	$\mathbf{z}_1$	—	—
$D_0 Y_{1,1} = (A, \varnothing)$	$\mathbf{z}_2$	$\mathbf{g}(0, \mathbf{z}_1) = \mathbf{z}_2$	$\mathbf{f}(0, \mathbf{z}_1) = (0, 0)$
$D_1 Y_{1,1} = (\lambda, (10)^*)$	$\mathbf{z}_3$	$\mathbf{g}(1, \mathbf{z}_1) = \mathbf{z}_3$	$\mathbf{f}(1, \mathbf{z}_1) = (1, 1)$
$D_{00} Y_{1,1} = (A, \varnothing)$		$\mathbf{g}(0, \mathbf{z}_2) = \mathbf{z}_2$	$\mathbf{f}(0, \mathbf{z}_2) = (0, 0)$
$D_{01} Y_{1,1} = (\lambda, \varnothing)$	$\mathbf{z}_4$	$\mathbf{g}(1, \mathbf{z}_2) = \mathbf{z}_4$	$\mathbf{f}(1, \mathbf{z}_2) = (1, 0)$
$D_{10} Y_{1,1} = (\varnothing, \varnothing)$	$\mathbf{z}_5$	$\mathbf{g}(0, \mathbf{z}_3) = \mathbf{z}_5$	$\mathbf{f}(0, \mathbf{z}_3) = (0, 0)$
$D_{11} Y_{1,1} = (\varnothing, 0(10)^*)$	$\mathbf{z}_6$	$\mathbf{g}(1, \mathbf{z}_3) = \mathbf{z}_6$	$\mathbf{f}(1, \mathbf{z}_3) = (0, 0)$
$D_{000} Y_{1,1} = (A, \varnothing)$		—	—
$D_{001} Y_{1,1} = (\lambda, \varnothing)$		—	—
$D_{010} T_{1,1} = (\varnothing, \varnothing)$		$\mathbf{g}(0, \mathbf{z}_4) = \mathbf{z}_5$	$\mathbf{f}(0, \mathbf{z}_4) = (0, 0)$
$D_{011} Y_{1,1} = (\phi, \phi)$		$\mathbf{g}(1, \mathbf{z}_4) = \mathbf{z}_5$	$\mathbf{f}(1, \mathbf{z}_4) = (0, 0)$
$D_{100} Y_{1,1} = (\phi, \phi)$		$\mathbf{g}(0, \mathbf{z}_5) = \mathbf{z}_5$	$\mathbf{f}(0, \mathbf{z}_5) = (0, 0)$
$D_{101} Y_{1,1} = (\phi, \phi)$		$\mathbf{g}(1, \mathbf{z}_5) = \mathbf{z}_5$	$\mathbf{f}(1, \mathbf{z}_5) = (0, 0)$
$D_{110} Y_{1,1} = (\phi, (10)^*)$	$\mathbf{z}_7$	$\mathbf{g}(0, \mathbf{z}_6) = \mathbf{z}_7$	$\mathbf{f}(0, \mathbf{z}_6) = (0, 1)$
$D_{111} Y_{1,1} = (\phi, \phi)$		$\mathbf{g}(1, \mathbf{z}_6) = \mathbf{z}_5$	$\mathbf{f}(1, \mathbf{z}_6) = (0, 0)$
$D_{1100} Y_{1,1} = (\phi, \phi)$		$\mathbf{g}(0, \mathbf{z}_7) = \mathbf{z}_5$	$\mathbf{f}(0, \mathbf{z}_7) = (0, 0)$
$D_{1101} Y_{1,1} = (\phi, 0(10)^*)$		$\mathbf{g}(1, \mathbf{z}_7) = \mathbf{z}_6$	$\mathbf{f}(1, \mathbf{z}_7) = (0, 0)$

Exercises to Section 7.6

*7.6-1 Determine regular expressions of the equivalence classes $Z_{1,1}$ and $Z_{1,2}$ of the Mealy machine whose ST-diagram is in Figure 7.32.

7.6-2 Using regular expressions, determine the ST-structure of a Mealy machine whose behavior is described by the following proposition: "The output is equal to 1 if the input sequence contains two consecutive zeroes but does not end in 01."

7.6-3 Minimize the number of internal states of the ST-structures in Figures 7.30 and 7.31.

*7.6-4 Modify formula (7.24) for the Moore machine.

7.6-5 Modify the abstract synthesis based on regular expressions for the Moore machine and solve Exercise 7.6-2 in terms of the Moore machine.

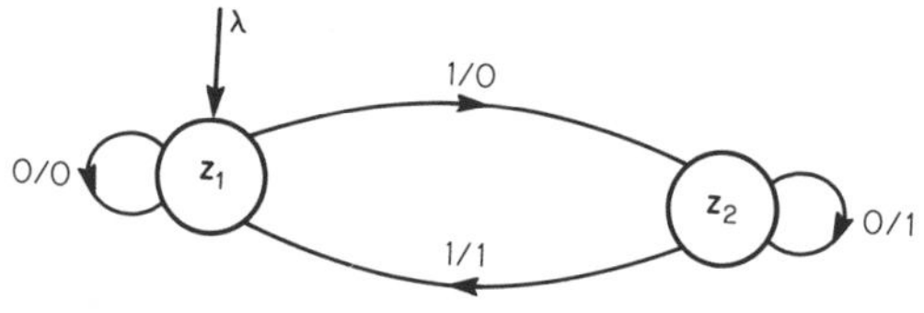

Figure 7.32 Illustration to Exercises 7.6-1.

7.6-6 Transform the Mealy machine obtained in Exercise 7.6-2 to the corresponding Moore machine and compare the latter with the result of Exercise 7.6-5.

7.6-7 Determine the ST-structure of a sequential switching circuit with two outputs whose behavior is described by $Y_{1,1} = ((0 \cup 10^*1)^*10^*1, (0 \cup 1)^*01)$.

Comprehensive Exercises to Chapter 7

7.1 A machine $M = (X, Y, Z, \mathbf{f}, \mathbf{g})$ is said to be strongly connected if for any pair of states $\mathbf{z}_i, \mathbf{z}_j \in Z$ there exists at least one input sequence $\mathbf{I}$ such that $\mathbf{G}(\mathbf{I}, \mathbf{z}_i) = \mathbf{z}_j$. Assume now that machines M_1 and M_2 are strongly connected. Prove that if there exists a state of machine M_1 which is equivalent to a state of machine M_2, then machine M_1 is equivalent to machine M_2.

7.2 Derive an algorithm to ascertain if there exists an input sequence $\mathbf{I}$ that will take a machine from a given state $\mathbf{z}_i$ to state $\mathbf{z}_j$. Show an application of this algorithm to determine whether or not a given machine is strongly connected.

7.3 The following algorithm for state-minimization of incompletely specified machines is suggested:

(a) Complete the given ST-table in all possible ways.

(b) Perform the state minimization for each completely specified machine formed in (a).

(c) From machines obtained under (b), select the machine with the least number of states.

Show that the algorithm is false.

*7.4 Determine the ST-structure of a switching circuit with one input x and one output y. Let N_0 be the number of synchronizing pulses which have arrived while $x = 0$ and let N_1 be the number of synchronizing pulses which have arrived while $x = 1$. Let $y = 0$ except when $1 < (2N_0 - 3N_1)(\text{mod } 5) < 4$.

7.5 A finite-state machine with no inputs is called autonomous machine.

(a) Show that the output sequence of an autonomous machine is periodic.

(b) Determine the maximum length of this period.

7.6 Find whether or not each of the following statements describes a behavior realizable by a finite-state machine:

(a) The output is one if the input sequence contains k ones and $k+1$ zeroes ($k = 0, 1, 2, \ldots$); otherwise the output is zero.

(b) The output is one if the input sequence contains more ones than zeroes; otherwise it is zero.

(c) The machine reverses sequences of finite length m in the sense that $y^{km+i} = x^{km-i}$, where $i = t$ (modulo m) and $k = \lfloor (t/m)$; x^t and y^t denote, respectively, the input value and the output value at time t.

(d) The output at time t is equal to one ($y^t = 1$) for $t = 2k + 1$ ($k \geq 0$) if and only if the input sequence contains k ones followed by a single zero followed by k ones; otherwise the output is zero.

(e) The output is one for all input sequences in which every zero is immediately preceded by at least k ones and is immediately followed by exactly k ones, where k is a specified integer; otherwise the output is zero.

7.7 Let M be a strongly connected completely specified Mealy machine. Prove that the Moore representation of M is also strongly connected completely specified machine.

7.8 Prove the following equalities between regular expressions:
(a) $(01 \cup 10 \cup 0)^* = (10 \cup 0^*01)^* 0^*$.
(b) $(1010)^*(\lambda^* \cup \lambda(1010)^*) \cup 10 = (1010)^* \cup 10$.
(c) $0(0 \cup 1)^* \cup \lambda \cup (1 \cup 0)^* 00(0 \cup 1)^* = ((1^*0)^*01^*)^*$.
(d) $1(0 \cup 21)^* = (10^*2)^*10^*$.
(e) $(0 \cup 12)^* 1 = 0^*1(20^*1)^*$.

7.9 Give a concise word description of the sets of sequences described by the following expressions:
*(a) 0^*11^*0.
(b) $(1 \cup 11)^*(\lambda \cup 0)(\lambda \cup 0)$.
(c) $1(0 \cup 1)^*101$.
(d) $(00 \cup (11)^*0)^*10$.
(e) $(1 \cup (0 \cup 10)^*)^*$.
(f) $(1 \cup 00^*1) \cup (1 \cup 00^*1)(0 \cup 10^*1)^*$.
(g) $(10)^*(01)^*(00 \cup 11)^*$.

7.10 The reverse A^r of a set A of sequences is the set that consists of the reverses of the sequences in A. For instance, if $1101 \in A$, then $1011 \in A^r$. Prove that if A can be described by a regular expression, then A^r can be described by a regular expression too.

7.11 Let $A = (00)^*(0 \cup 10^*)^* \cup 10^*(01^*10^*)^*$. Determine A^r as defined in Exercise 7.10.

Reference Notations to Chapter 7

The concept of the finite-state machine was introduced and investigated initially by Huffman [25], Moore [35], and Mealy [33]. Their work was extended in several other papers [1, 2, 10, 15, 40]. The literature devoted to the problem of the state minimization is rather extensive. Paul and Unger [37] and Ginsburg [15, 16] started an exploration of the state minimization for incompletely specified machines. The method described in Section 7.4 is due to Grasseli and Luccio [17, 18]. They are also authors of the state-input minimization [19, 20, 30]. References 3, 4, 11, 28, 34, and 38 represent a sample of other approaches to the state minimization of incompletely specified machines.

A general method for the abstract synthesis of finite-state machines was proposed by Tal and Gusev [22, 41, 42] and further explored by Jerome [26]. Other methods for the abstract synthesis were presented by Gill [14], Gray and Harrison [21], and Fuchs [12].

Regular expressions and their meaning for the finite-state machines were first shown by Kleene [29]. Various properties of regular expressions and their applications have been thoroughly investigated by Brzozowski and several others [5, 6, 7, 8, 9, 13, 27, 31, 32 36]. A good survey of regular expressions was prepared by Havel [23, 24]. An axiomatic theory of regular expressions was elaborated by Salomaa [39].

References to Chapter 7

1. **Aufenkamp, D. D.,** and **F. E. Hohn,** "Analysis of Sequential Machines," *TC* **EC-6** (4) 276–285 (Dec. 1957).
2. **Aufenkamp, D. D.,** "Analysis of Sequential Machines," *TC* **EC-7** (4), 299–306 (Dec. 1958).
3. **Beatty, J. C.,** and **R. E. Miller,** "Some Theorems for Incompletely Specified Sequential Machines with Applications to State Minimization," *CRSA* 124–136 (Sept. 1962).
4. **Bouchet, A.,** "An Algebraic Method for Minimizing the Number of States in an Incomplete Sequential Machine," *TC* **C-17** (8) 795–798 (Aug. 1968).
5. **Brzozowski, J. A.,** "A Survey of Regular Expressions and Their Applications," *TC* **EC-11** (3) 324–335 (June 1962).
6. **Brzozowski, J. A.,** and **J. F. Poage,** "On the Construction of Sequential Machines from Regular Expressions," *TC* **EC-12** (4) 402–403 (Aug. 1963).
7. **Brzozowski, J. A.,** and **E. J. McCluskey,** "Signal Flow Graph Techniques for Sequential Circuit State Diagrams," *TC* **EC-12** (2) 67–76 (April 1963).
8. **Brzozowski, J. A.,** "Derivatives of Regular Expressions," **JACM 11** (4) 481–494 (Oct. 1964).
9. **Brzozowski, J. A.,** "Regular Expressions from Sequential Circuits," *TC* **EC-13** (6) 741–744 (Dec. 1964).
10. **Copi, I., C. C. Elgot,** and **J. B. Wright,** "Realization of Events by Logical Nets," *JACM* **5** (2) 181–196 (April 1958).
11. **Elgot, C. C.,** and **J. D. Rutledge,** "Machine Properties Preserved under State Minimization, *CRSA*, 62–70 (Sept. 1962).
12. **Fuchs, J.,** "Determination of the Sequential Machine State Diagram," M.Sc. Thesis, UCLA, 1969.
13. **Ghiron, H.,** "Rules to Manipulate Regular Expressions of Finite Automata," *TC* **EC-11** (4) 574–575 (Aug. 1962).
14. **Gill, A.,** "Realization of Input-Output Relations by Sequential Machines," *JACM*, **5** (1) 33–42 (Jan. 1966).
15. **Ginsburg, S.,** "On the Reduction of Superfluous States in a Sequential Machine," *JACM* **6** (2) 259–282 (April 1959).
16. **Ginsburg, S.,** "Synthesis of Minimal-State Machines," *TC* **EC-8** (4) 441–449 (Dec. 1959).
17. **Grasselli, A.,** and **F. Luccio,** "A Method for Minimizing the Number of Internal States in Incompletely Specified Sequential Networks," *TC* **EC-14** (3) 350–359 (June 1965).
18. **Grasselli, A.,** "Minimal Closed Partitions for Incompletely Specified Flow Tables," *TC* **EC-15** (2) 245–249 (April 1966).
19. **Grasselli, A.,** and **F. Luccio,** "A Method for the Combined Row-Column Reduction of Flow Tables," *CRSA*, 136–147 (Oct. 1966).
20. **Grasselli, A.,** and **F. Luccio,** *Some Covering Problems in Switching Theory*, in G. Biorci, ed., Network and Switching Theory, Academic Press, New York, 1968, pp. 536–551.

21. **Gray, J. N.,** and **M. A. Harrison,** "The Theory of Sequential Relations," *IC* **9** (5) 435-468 (Oct. 1966).
22. **Gusev, L. A.,** and **A. A. Tal,** "The Possibilities of Constructing Algorithms for the Abstract Synthesis of Sequential Machines Using the Questionnaire Language," *ARC* **26** (3) 507–514, March 1965.
23. **Havel, I. M.,** "Regular Expressions over Generalized Alphabet and Design of Logical Nets," *KYB* **4** (6) 516–537 (1968).
24. **Havel, I. M.,** "The Theory of Regular Events," *KYB* **5** (5 and 6), 400–419 520–544 (1969).
25. **Huffman, D. A.,** "The Design of Sequential Switching Circuits," *JFI* **257** (2 and 4), 161–190 257–303 (March, and April 1954); reprinted in *SM*.
26. **Jerome, E. J.,** "A Mathematical Approach to the Abstract Synthesis of Sequential Discrete Systems," M.Sc. Thesis, McGill University, Montreal, Canada, Dec. 1970.
27. **Johnson, M. D.,** and **R. B. Lackey,** "Sequential Machine Synthesis Using Regular Expressions," *CD* **7** (9) 44–47 (Sept. 1968).
28. **Kella, J.,** "State Minimization of Incompletely Specified Sequential Machines," *TC* **C-19** (4) 342–348 (April 1970).
29. **Kleene, S. C.,** "Representation of Events in Nerve Sets and Finite Automata," *AS*, 3–41.
30. **Luccio, F.,** "Reduction of the Number of Columns in Flow Table Minimization," *TC* **EC-15** (5) 803–805 (Oct. 1966).
31. **McNaughton, R.,** and **H. Yamanda,** "Regular Expressions and State Graphs for Automata," *TC* **EC-9** (1) 39–49 (March 1960); reprinted in *SM*.
32. **McNaughton, R.,** "An Introduction to Regular Expressions," *AAT*, 35–54.
33. **Mealy, G. H.,** "A Method for Synthesizing Sequential Circuits," *BSTJ* **34** (5) 1045–1079 (Sept. 1955).
34. **Meisel, W. S.,** "A Note on Internal State Minimization in Incompletely Specified Sequential Networks," *TC* **EC-16** (4) 508–509 (Aug. 1967).
35. **Moore, E. F.,** "Gedanken-Experiments on Sequential Machines," *AS*, 129–153.
36. **Ott, G.,** and **N. Feinstein,** "Design of Sequential Machines from their Regular Expressions," *JACM* **8** (4), 585–600 (Oct. 1961).
37. **Paull, M. C.,** and **S. H. Unger,** "Minimizing the Number of States in Incompletely Specified Sequential Switching Functions," *TC* **EC-8** (3) 356–367 (Sept. 1959).
38. **Prather, R. E.,** "Minimal Solutions of Paull-Unger Problems," Mathematical Systems Theory, **3**, No. 1, 76–85 (March 1969).
39. **Salomaa, A.,** "Two Complete Axiom Systems for the Algebra of Regular Events," *JACM* **13** (1) 158–169 (Jan. 1966).
40. **Seshu, S.,** et al., "Transition Matrices of Sequential Machines," *IRE Trans. on Circuit Theory* **CT-6** (1) 5–12 (1959).
41. **Tal, A. A.,** "Questionnaire Language and the Abstract Synthesis of Minimal Sequential Machines," *ARC* **25** (6) 846–859 (June 1964).
42. **Tal, A. A.,** "The Abstract Synthesis of Sequential Machines from the Answers to Questions of the First Kind in the Questionnaire Language," *ARC* **26** (4) 675–680, April 1965.

8

SYNCHRONOUS SEQUENTIAL SWITCHING CIRCUITS: STRUCTURE SYNTHESIS

8.1 DISCUSSION OF THE PROBLEM

Present responses of sequential switching circuits do not depend, in contradistinction to memoryless circuits, uniquely on present stimuli. Nevertheless, they are determined uniquely by some sequences of present and past stimuli. The required sequences of stimuli must thus be stored in the circuit. There is no need to store them solely in the direct form. They may be stored in the form of output variables, each of which is related to certain sequences of stimuli. They may be also stored in the form of special internal variables.

A general paradigm of deterministic switching circuits is shown in Figure 8.1. Two blocks are distinguished: (1) *Functional generator* which is a memoryless portion of the sequential circuits, i.e., it contains only memoryless elements. (2) *Memory* which contains all memory elements included in the sequential circuit.

All possible couplings between the functional generator, the memory, and the environment are considered in the *general paradigm* illustrated by Figure 8.1. Various special models of sequential switching circuits are obtained when some of these couplings are exluded. For instance, the Mealy machine model does not consider any coupling between the memory and the environment.

The *structure synthesis* of sequential switching circuits is the following problem: The switching circuit is given by its ST-structure representing the Mealy machine (or the Moore machine). Find a UC-structure which:

(a) Implements the given ST-structure.
(b) Contains only elements from a specified set of elements (this set must contain elements representing a complete set of logic functions and at least one memory element).
(c) Satisfies certain constraints concerning its form (number of levels, fan-in, fan-out, etc.).

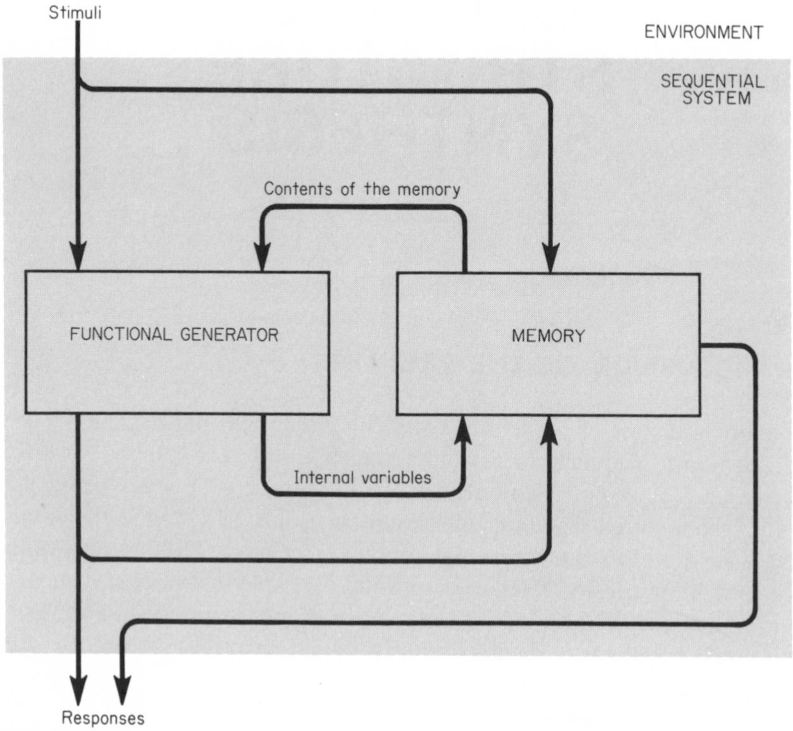

Figure 8.1 General paradigm of sequential switching circuits.

(d) Produces the minimum or a value reasonably near to the minimum of a specified objective function defined on a set of UC-structures (the objective function represents an approximation of our "criteria of goodness" such as the lowest cost, the highest reliability, etc.).

Two basic steps are involved in the structure synthesis of sequential switching circuits: (1) A transformation of the given ST-structure to a UC-structure of the memory and a behavior of the functional generator. (2) Synthesis of the functional generator.

The second step represents the synthesis of a memoryless switching circuit. This is treated in Chapter 6. Thus, it remains to discuss the first step. As a rule, it includes the following partial problems:

(i) The *selection of a specific paradigm* (model) of sequential switching circuits.
(ii) The *coding of internal variables*, which is usually referred to as the *state assignment problem.*
(iii) The *determination of all logic functions represented by the functional generator.*

Certain selection of a subset of the couplings shown in Figure 8.1 together with a UC-structure of the memory represents a specific model of sequential circuits. For a given ST-structure, generally, only some of the specific models allow the objective function to reach its absolute minimum. Consequently, an improper selection of the model makes it impossible to reach the absolute minimum of the objective function. Similarly, an improper coding of internal variables, even though applied to a proper model makes it impossible to reach the absolute minimum of the objective function. If both the selected model and the code of internal variables are proper, then it is the matter of proper synthesis of the functional generator to reach the absolute minimum of the objective function. Thus, there are three levels of optimization involved in the synthesis of sequential switching circuits as illustrated in Figure 8.2.

8.2 THE CONCEPT OF SYNCHRONIZATION

Synchronous sequential switching circuits, which are the subject of this chapter, are characterized by the following constraints concerning their stimuli and internal states:

1. Values of all input variables (both external and internal) of the functional generator are defined only during certain time intervals identified by a sequence of synchronizing pulses generated by an independent source.
2. Values of the input variables of the functional generator do not change when they are defined.

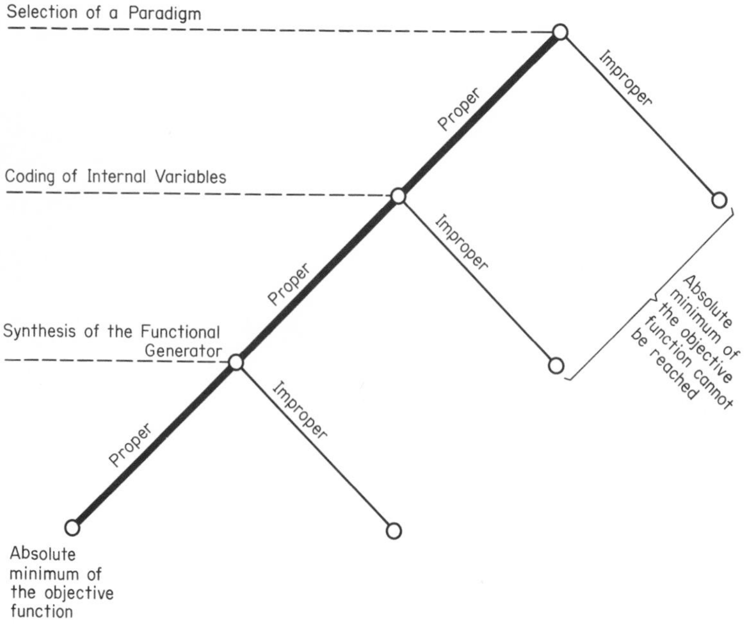

Figure 8.2 Three levels of optimization in synthesis of sequential circuits.

The first constraint can be implemented by connecting an AND gate stimulated by the source of synchronizing pulses to each input of the functional generator. This is shown in Figure 8.3 for the Mealy model. The source of synchronizing pulses, which is usually called a *clock*, is denoted by *C*. Although it is not a necessity, the clock usually generates a simple periodic sequence with two time intervals, Δt_0 and Δt_1, per period. During the intervals Δt_0 and Δt_1, the clock pulses have magnitudes which represent, respectively, the logic values 0 and 1.

The second of the above constraints imposes certain requirements on the environment. Namely, the values of the external inputs must not change during the intervals Δt_1. In addition, values of Δt_0 and Δt_1 must be properly selected. Let Δt_m and Δt_M denote, respectively, the minimal and maximal delay in the memory. Similarly let Δt_f and Δt_F denote, respectively, the minimal and the maximal delay in the functional generator. Then, clearly, the inequalities

$$\begin{aligned} \Delta t_0 &> \Delta t_M + \Delta t_F \\ \Delta t_1 &< \Delta t_m + \Delta t_f \end{aligned} \tag{8.1}$$

must be satisfied to satisfy the second constraint of synchronous circuits. In addition, it is required that

$$\Delta t_1 > \Delta t_F + \Delta t_S \tag{8.2}$$

where Δt_S denotes the maximal time to switch the used memory elements.

Thus, in synchronous circuits, both external and internal input variables of the functional generator are defined and do not change during time intervals $t_i (i = 1, 2, 3, \ldots)$, where

$$i \cdot \Delta t_0 + (i-1)\Delta t_1 \leq t_i < i(\Delta t_0 + \Delta t_1) \tag{8.3}$$

These time intervals are usually called *sampling times.* The value of variable v_j at sampling time t_i will be denoted v_j^i. Thus, $v_j(t_i) = v_j^i$ by convention. For obvious reasons, letter t will be frequently used for the identification of sampling times.

Exercises to Section 8.2

8.2-1 Discuss the possibility of synchronizing sequential switching circuits by connecting the AND gates stimulated by the clock to outputs of the functional generator rather than to its inputs.

8.2-2 Discuss possible synchronizing schemes for the general paradigm of sequential switching circuits shown in Figure 8.1.

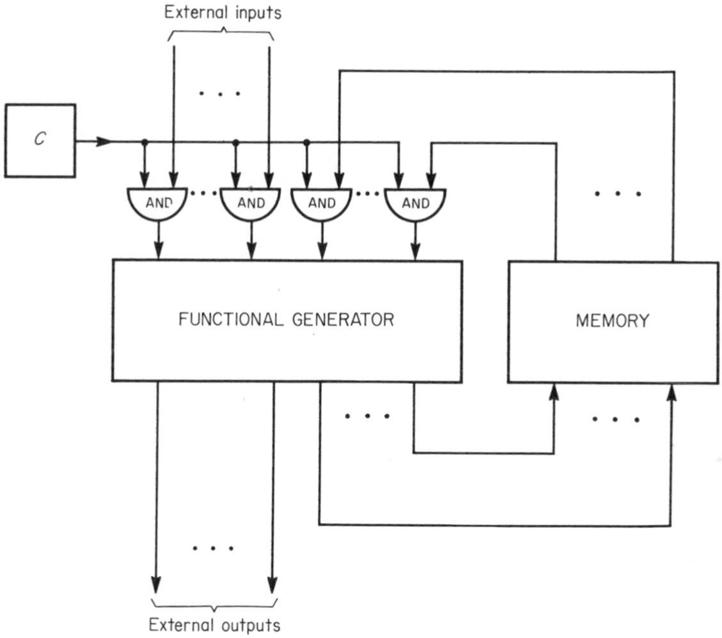

Figure 8.3 A possible way of synchronizing sequential switching circuits.

8.3 MEMORY ELEMENTS

In the sense of the explanation given in Chapter 5, we consider as a *memory element* every logic element in the operator of which it is impossible—in the given context—to neglect the time delay or from whose operator it is impossible to eliminate some internal variables.

It is our convention that memory elements used in a sequential switching circuit constitute the block called "memory"; the block called "functional generator" does not contain any memory elements.

The simplest memory element, which is called DELAY, is described by the operator

$$w^{t+1} = v^t \tag{8.4}$$

where v and w are, respectively, the input and output variable, and t is the identifier of sampling times. This is the only memory element that does not change the stored values at various sampling times t. Each value is stored for exactly one sampling time.

All memory elements other than DELAY transform the stored values. Let $v_i (i = 1, 2, \ldots, p)$ and $w_j (j = 1, 2, \ldots, q)$ be, respectively, the input and output variables of a memory element. Then, its behavior can be described, generally, by a set of logic functions

$$w_j^{t+1} = f_j(v_1^t, v_2^t, \ldots, v_p^t) \tag{8.5}$$

where $j = 1, 2, \ldots, q$. As long as we consider the memory element as a self-contained system, we can regard $v_1, v_2, \ldots, v_p$ as independent variables. However, when the memory element is connected to the functional generator, as shown in Figure 8.4, these variables become functions

$$v_k^t = g_k(x_1^t, x_2^t, \ldots, x_n^t, w_1^t, w_2^t, \ldots, w_q^t),\ k = 1, 2, \ldots, p \tag{8.6}$$

of both the external and internal input variables of the latter. Although these functions do not follow directly from the given *ST*-structure, the latter yields the functions

$$w_j^{t+1} = h_j(x_1^t, x_2^t, \ldots, x_n^t, w_1^t, w_2^t, \ldots, w_q^t) \tag{8.7}$$

where $j = 1, 2, \ldots, q$. Functions (8.6) can be obtained by solving simultaneous equations (8.5) and (8.7)

EXAMPLE 8.1 Let us consider a memory element with two input variables v_1, v_2 and one output variable w, the behavior of which is described by the following table:

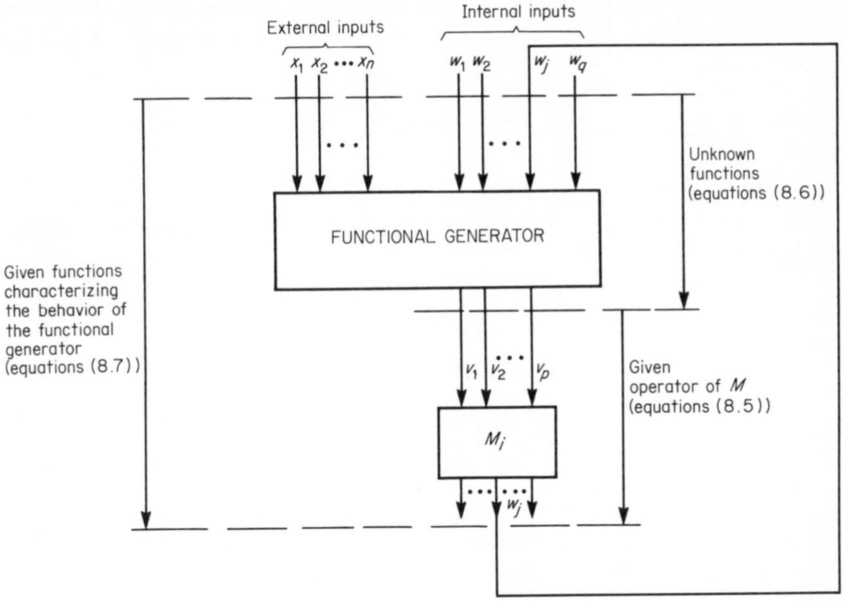

Figure 8.4 Illustration of the effect of chosen memory elements on the functional generator.

v_1^t	v_2^t	w^t	w^{t+1}
0	0	0	0
0	0	1	1
0	1	0	0
0	1	1	0
1	0	0	1
1	0	1	1

Input state $v_1^t = v_2^t = 1$ is forbidden. The operator of this element, which is usually called set-reset FLIP-FLOP (or SR FLIP-FLOP), is represented by the equations

$$w^{t+1} = v_1^t \vee \bar{v}_2^t w^t$$
$$v_1^t v_2^t = 0 \qquad \text{(a)}$$

Variable w^{t+1} is a function (specified in the given ST-structure) of input variables of the functional generator. It can be written as

$$w^{t+1} = w^t H_1 \vee \bar{w}^t H_2 \qquad \text{(b)}$$

where H_1 and H_2 are certain functions of all input variables of the functional generator except the variable w^t. When solving simultaneous equations (a) and (b), we obtain

$$v_1^t = H_2\bar{w}^t$$
$$v_2^t = \bar{H}_1 w^t$$

There are four meaningful memory elements with one input v and one output w:

$$\text{DELAY}: w^{t+1} = v^t$$
$$\text{INVERTED DELAY}: w^{t+1} = \bar{v}^t$$
$$\text{TRIGGER (COUNTER MODULO 2)}: w^{t+1} = \bar{v}^t w^t \vee v^t \bar{w}^t$$
$$\text{INVERTED TRIGGER}: w^{t+1} = \bar{v}^t \bar{w}^t \vee v^t w^t$$

If not specified otherwise, DELAYS, which are the only pure memory elements, will always be considered as building blocks of the memory. A transformation to other memory elements can be accomplished by the procedure explained in this section. Occasionally, TRIGGERS and SR FLIP-FLOPS will be used to illustrate various aspects of synthesis of sequential circuits. In block diagrams, memory elements will be represented by triangles labeled by abbreviations of the individual types of elements. Symbols of some memory elements are shown in Figure 8.5.

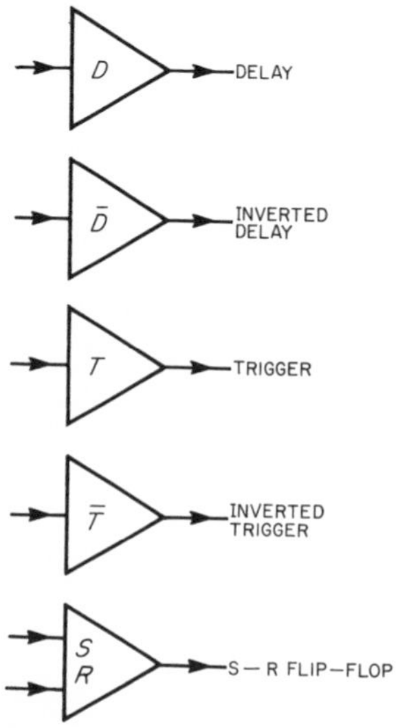

Figure 8.5 Used symbols of some memory elements.

Exercises to Section 8.3

8.3-1 Solve Example 8.1 for:
- (a) INVERTED DELAY.
- *(b) TRIGGER.
- *(c) The so-called JK FLIP-FLOP, which is identical with the SR FLIP-FLOP except that $w^{t+1} = \bar{w}^t$ for $v_1^t = v_2^t = 1$.
- (d) A memory element whose operator is specified by the equations
  $w^{t+1} = \bar{v}_1^t(\bar{v}_2^t(\bar{v}_3^t w^t \vee v_3^t \bar{w}^t) \vee v_2^t \bar{v}_3^t)$
  $v_1^t(v_2^t \vee v_3^t) \vee v_2^t v_3^t = 0.$

8.3-2 Implement JK FLIP-FLOP by AND, OR, and NOT elements, and
- (a) SR FLIP-FLOP.
- (b) TRIGGER.

8.3-3 Implement SR FLIP-FLOP by NAND elements and
- (a) DELAYS.
- (b) INVERTED TRIGGERS.

8.4 MODELS OF SEQUENTIAL SWITCHING CIRCUITS

The general paradigm of sequential switching circuits, which will be called a *sequential discrete system*, is roughly illustrated by the block diagram in Figure 8.1. This paradigm yields various particular models which differ one from another in couplings between the functional generator, the memory, and the environment, and in a span (depth) of the memory. All these aspects are included in the following definition.

DEFINITION 8.1 A *sequential discrete system with a memory span k* is a quintuple

$$S = (X, Y, Z, \mathbf{f}, \mathbf{g}) \tag{8.8}$$

where X is a finite set of input states (stimuli), Y is a finite set of output states (responses), Z is a finite set of internal states, $\mathbf{f}$ is an output function

$$X^{a+1} \times Y^b \times Z^c \to Y$$

$\mathbf{g}$ is a transition function

$$X^{a+1} \times Y^b \times Z^c \to Z$$

$k = \max(a, b, c)$, and a, b, c are positive integers including zero.

Let

$$\mathbf{x}^t = (x_1^t, x_2^t, \ldots, x_p^t)$$
$$\mathbf{y}^t = (y_1^t, y_2^t, \ldots, y_q^t)$$
$$\mathbf{z}^t = (z_1^t, z_2^t, \ldots, z_r^t)$$

be the input state, the output state, and the internal state at time t, respectively. Then,

$$y_j^{t-\delta_j} = f_j(\mathbf{x}^{t-\delta_j}, \mathbf{x}^{t-\delta_j-\alpha}, \mathbf{y}^{t-\delta_j-\beta}, \mathbf{z}^{t-\delta_j-\gamma})$$

and

$$\mathbf{z}^t = \mathbf{g}(\mathbf{x}^t, \mathbf{x}^{t-\alpha}, \mathbf{y}^{t-\beta}, \mathbf{z}^{t-\gamma})$$

for all $\alpha \in A \subset K$, all $\beta \in B \subset K$, all $\gamma \in C \subset K$, and $\delta_j = 0$ or $\delta_j \in K$ ($j = 1, 2, \ldots, q$), where $K = \{1, 2, \ldots, k\}$ and $\max(\alpha, \beta, \gamma) = k$. Clearly, $a = |A|$, $b = |B|$, $c = |C|$. Note that $\delta_j = 0$ if y_j is an output variable of the functional generator and $\delta_j > 0$ if y_j is a variable stored in the memory.

The following classes of sequential discrete systems (models) may be distinguished:

1. *kth order* ($k \geq 1$) *finite-state models*: $A = \varnothing$, $B = \varnothing$, $C \neq \varnothing$ (Fig. 8.6a).
2. *Finite-memory models*: $C = \varnothing$ and $A \neq \varnothing$ or $B \neq \varnothing$ (Figure 8.6b).
3. *kth order* ($k \geq 1$) *combined models*: $C \neq \varnothing$ and $A \neq \varnothing$ or $B \neq \varnothing$ (Fig. 8.6c).

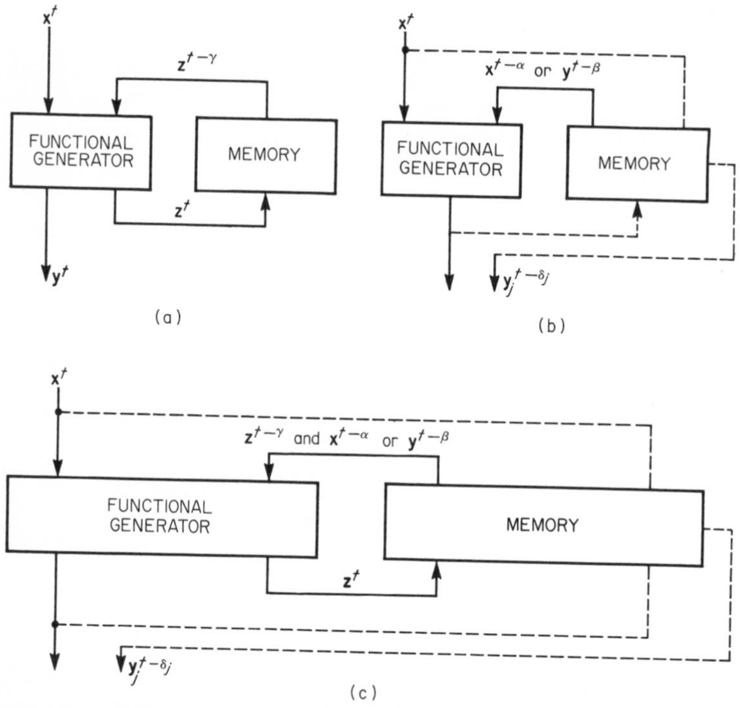

Figure 8.6 Block diagrams of basic classes of models of sequential discrete systems.

Each of the above specified models is applicable for a certain class of behaviors. A question arises: Are all of the models applicable for the same class of behaviors? An answer is given by the following four theorems.

THEOREM 8.1 The class of behaviors realizable by the first-order finite-state model ($k = 1$) is identical with the class of behaviors realizable by the kth order finite-state model, where k is any finite integer greater than 1.

PROOF Any behavior representable by the first order finite-state model ($k = 1$) can be realized by a degenerate kth order finite-state model whose variables $z_i^{t-\gamma}$ ($i = 1, 2, \ldots, r$; $\gamma = 2, 3, \ldots, k$) are all dummy variables.

For any given kth order finite-state realization, let us consider $C = K$, where $\mathbf{z}^{t-\gamma}$ may be dummy for some $\gamma \neq k$. Let the internal state

$$\mathbf{Z}^{t-1} = (z_1^{t-1}, z_2^{t-1}, \ldots, z_k^{t-1})$$

of a first-order finite-state model be defined as the ordered k-tuple $(\mathbf{z}^{t-1}, \mathbf{z}^{t-2}, \ldots, \mathbf{z}^{t-k})$. Then $\mathbf{Z}^t = (z_1^t, z_2^t, \ldots, z_k^t) = (\mathbf{z}^t, \mathbf{z}^{t-1}, \ldots, \mathbf{z}^{t-k+1})$ and the transition function of the first-order finite-state model

$$\mathbf{Z}^t = \mathbf{G}(\mathbf{x}^t, \mathbf{Z}^{t-1})$$

can be written as

$$\begin{aligned}
\mathbf{z}_1^t &= \mathbf{g}(\mathbf{x}^t, \mathbf{Z}^{t-1}) \\
\mathbf{z}_2^t &= \mathbf{z}_1^{t-1} \\
\mathbf{z}_3^t &= \mathbf{z}_2^{t-1} \\
\vdots\ & \quad \vdots \\
\mathbf{z}_k^t &= \mathbf{z}_{k-1}^{t-1}
\end{aligned}$$

so that the function g is identical with the transition function of the kth order finite-state realization.

If $\delta = \max \delta_j = 0$, then, similarly, the output function

$$\mathbf{y}^t = \mathbf{f}(\mathbf{x}^t, \mathbf{Z}^{t-1})$$

can be put identical with the output function of the kth order finite-state model. If $\delta > 0$, we have to define

$$\mathbf{Z}^{t-1} = (\mathbf{x}^{t-1}, \mathbf{x}^{t-2}, \ldots, \mathbf{x}^{t-\delta}, \mathbf{z}^{t-1}, \mathbf{z}^{t-2}, \ldots, \mathbf{z}^{t-\delta-k})$$

Then,

$$y_j^{t-\delta_j} = f_j(\mathbf{x}^t, \mathbf{Z}^{t-1}),$$

where some components of $\mathbf{Z}^{t-1}$ or $\mathbf{x}^t$ may be dummy variables, can be made identical with the output functions of the kth order finite-state model. ∎

THEOREM 8.2 The class of behaviors realizable by the kth order combined model ($k \geq 1$) is identical with the class of behaviors realizable by the first order finite-state model.

PROOF Any behavior representable by the first order finite-state model can be realized by a degenerate kth order combined model whose variables $x_i^{t-\delta_j-\alpha}$ $(i = 1, 2, \ldots, p;\ \alpha \in A)$, $y_j^{t-\delta_j-\beta}(j = 1, 2, \ldots, q;\ \beta \in \mathrm{B})$, $z_l^{t-\delta_j-\gamma}(l = 1, 2, \ldots, r;\ \gamma \in C;\ \gamma \neq 1$ are all dummy variables.

For any given kth order combined realization, let us consider $A = B = C = K$, where $\mathbf{x}^{t-\alpha}$, $\mathbf{y}^{t-\beta}$, $\mathbf{z}^{t-\gamma}$ may be dummy for some α, β, γ. Let the internal state $\mathbf{Z}^{t-1}$ of a first order finite-state model be defined as the ordered $(k^3 - 1)$-tuple $(\mathbf{x}^{t-2}, \mathbf{x}^{t-3}, \ldots, \mathbf{x}^{t-k}; \mathbf{y}^{t-1}, \mathbf{y}^{t-2}, \ldots, \mathbf{y}^{t-k}; \mathbf{z}^{t-1}, \mathbf{z}^{t-2}, \ldots, \mathbf{z}^{t-k})$. Then $\mathbf{Z}^t = (\mathbf{x}^{t-1}, \mathbf{x}^{t-2}, \ldots, \mathbf{x}^{t-k+1}; \mathbf{y}^t, \mathbf{y}^{t-1}, \ldots, \mathbf{y}^{t-k+1}; \mathbf{z}^t, \mathbf{z}^{t-1}, \ldots, \mathbf{z}^{t-k+1})$ and the transition function of the first order finite-state model

$$\mathbf{Z}^t = \mathbf{G}(\mathbf{x}^t, \mathbf{Z}^{t-1})$$

can be written as a set of functions among which the function

$$\mathbf{z}^t = \mathbf{g}(\mathbf{x}^t, \mathbf{Z}^{t-1})$$

is included. This function can be put identical with the transition function of the combined model. Similarly, if $\delta = \max \delta_j = 0$, then the output function

$$\mathbf{y}^t = \mathbf{f}(\mathbf{x}^t, \mathbf{Z}^{t-1})$$

can be set identical with the output function of the combined model. If $\delta > 0$, we have to define

$$\mathbf{Z}^{t-1} = (\mathbf{x}^{t-1}, \mathbf{x}^{t-2}, \ldots, \mathbf{x}^{t-k-\delta}, \mathbf{y}^{t-1}, \mathbf{y}^{t-2}, \ldots, \mathbf{y}^{t-k-\delta}, \mathbf{z}^{t-1}, \mathbf{z}^{t-2}, \ldots, \mathbf{z}^{t-k-\delta})$$

Then,

$$y_j^{t-\delta_j} = f_j(\mathbf{x}^t, \mathbf{Z}^{t-1}),$$

where some components of $\mathbf{Z}^{t-1}$ or $\mathbf{x}^t$ may be dummy variables, can be set identical with the output function of the combined model. ▌

THEOREM 8.3 The class of behaviors realizable by the k_1th order combined model ($k_1 \geq 1$) is identical with the class of behaviors realizable by the k_2th order finite-state model ($k_2 \geq 1$).

PROOF Follows directly from Theorems 8.1 and 8.2. ▌

THEOREM 8.4 The class of behaviors realizable by the finite-memory model is a proper subset of the class of behaviors realizable by the first order finite-state model.

PROOF Let the internal state $\mathbf{Z}^{t-1}$ of the first order finite-state model be defined as the ordered $2k$-tuple

$$(\mathbf{x}^{t-\delta-1}, \mathbf{x}^{t-\delta-2}, \ldots, \mathbf{x}^{t-\delta-k}, \mathbf{y}^{t-\delta-1}, \mathbf{y}^{t-\delta-2}, \ldots, \mathbf{y}^{t-\delta-k})$$

where $\delta = \max \delta_j$. Then,

$$\mathbf{Z}^t = \mathbf{G}(\mathbf{x}^t, \mathbf{Z}^{t-1})$$

and

$$y_j^{t-\delta_j} = f_j(\mathbf{x}^t, \mathbf{Z}^{t-1}),$$

where some components of $\mathbf{Z}^{t-1}$ or $\mathbf{x}^t$ may be dummy variables. Thus, every behavior realizable by the finite-memory model is realizable by the first-order finite-state model.

Now, let us consider the ST-structure shown in Figure 8.7. Its implementation by the first order finite-state model is represented by the equations

$$y^t = x^t \bar{z}^{t-1}$$
$$z^t = \bar{x}^t z^{t-1} \vee x^t \bar{z}^{t-1}$$

We will show that the ST-structure cannot be implemented by the finite-memory model. Let state 0 be the initial state. Then the response is 1 if the stimulus is 1 and if the system has accepted an even number of zeros each of which can be followed by any number of ones. Thus, it must be somehow identified whether the number of accepted zeros is even or odd. In the finite-memory model, no internal variables are allowed. Hence, the complete input/output "history" must be stored by the system. No finite memory is sufficient to satisfy this requirement. ∎

Exercises to Section 8.4

*8.4-1 Let a behavior of the functional generator of the finite-memory model be represented by the equation

$$y^t = y^{t-1}\bar{x}^t \vee \bar{y}^{t-1}x^t$$

where x is the input variable and y is the output variable. Using the scheme of the proof of Theorem 8.4, find the corresponding first-order finite-state model.

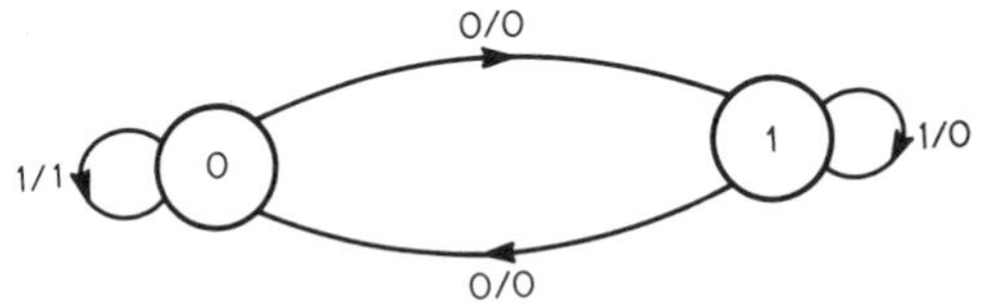

Figure 8.7 ST-structure which cannot be implemented by the finite-memory model.

8.4-2 Let a behavior of the functional generator of the second order finite-state model be represented by the equations

$$z^t = z^{t-2}\bar{z}^{t-1} \vee x^t z^{t-1}$$
$$y^t = \bar{z}^{t-2} \vee z^{t-1} \vee \bar{x}^t$$

Using the scheme of the proof of Theorem 8.1, find the corresponding first-order finite-state model.

8.4-3 Determine the ST-structures of the systems specified in the previous two exercises.

8.4-4 Find some ST-structures which cannot be implemented by the finite-memory model.

8.5 FINITE-STATE MODELS

A given ST-structure directly represents the output function f and the transition function g of the *first-order finite-state model.* To implement these functions by a switching circuit, internal states have to be coded by some logic variables first. A proper code must be selected to make it possible to reach the absolute minimum of the objective function.

In this section, we are going to investigate various properties of kth order finite-state models ($k \geq 1$) regardless of the effect of the code used for the internal states. The problem of determining a proper code for the internal states, which is usually referred to as the state assignment problem, is discussed in Section 8.10 for all models.

Let us consider the first order finite state model with p input variables $x_1, x_2, \ldots, x_p$, q output variables $y_1, y_2, \ldots, y_q$, and r internal variables $z_1, z_2, \ldots, z_r$. Clearly, its memory consists of r DELAYS (or other memory elements) in parallel, one for each internal variable. The truth table describing the behavior of its functional generator consists of four portions that are arranged from left to right in the following order: (1) Internal variables at time $t-1$. (2) Input variables at time t. (3) Output variables at time t. (4) Internal variables at time t. In the first two portions there are values of $r+p$ input variables of the functional generator. The last two portions contain values of $q+r$ output variables of the functional generator.

We begin the construction of the truth table from an arbitrarily chosen state. Values of the internal variables assigned to this state are written into the first portion of the truth table. Then, we consider a transition from this state. Values of input and output variables associated with this transition are recorded, respectively, into the second and the third portion of the same row in the truth table. Values of the internal variables assigned to the state which we reach by this transition are recorded into the same row and the last portion of the truth table. Thus, a transition from one state to another is represented by a single row in the truth table. When we repeat the above described recording procedure for every transition included in the given ST-structure, we obtain the behavior of the functional generator. Although we must be sure that all transitions have been considered, it makes no difference in which order they have been recorded.

EXAMPLE 8.2 Using the first order finite-state model and elements AND, OR, NOT, and DELAY, design a switching circuit which implements the ST-structure shown in Figure 8.8.

Since the given ST-structure contains four internal states, we need at least two internal variables to distinguish them. Let the pair (z_2, z_1) of variables represent the internal state and let $\mathbf{s}_0 = (0, 0)$, $\mathbf{s}_1 = (0, 1)$, $\mathbf{s}_2 = (1, 0)$, $\mathbf{s}_3 = (1, 1)$. Then, the memory contains two delays in parallel and the behavior of the functional generator is represented by the following truth table:

z_2^{t-1}	z_1^{t-1}	x^t	y^t	z_2^t	z_1^t	Transitions from state
0	0	0	1	1	1	$\mathbf{s}_0 = (0, 0)$
0	0	1	0	1	1	
0	1	0	1	0	0	$\mathbf{s}_1 = (0, 1)$
0	1	1	1	1	1	
1	0	0	0	0	1	$\mathbf{s}_2 = (1, 0)$
1	0	1	1	0	1	
1	1	0	0	1	0	$\mathbf{s}_3 = (1, 1)$
1	1	1	0	0	1	

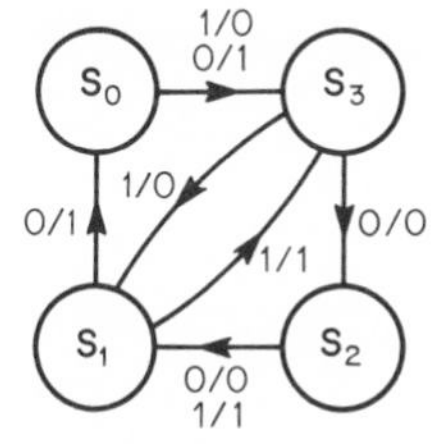

Figure 8.8 ST-structure used in Examples 8.2 and 8.3.

It remains now to implement this truth table. Standard Boolean minimization of the functions involved yields the expressions:

$$y^t = \bar{z}_2^{t-1}\bar{x}^t \vee z_2^{t-1}\bar{z}_1^{t-1}x^t \vee \bar{z}_2^{t-1}z_1^{t-1}$$
$$z_2^t = \bar{z}_2^{t-1}x^t \vee \bar{z}_2^{t-1}\bar{z}_1^{t-1} \vee z_2^{t-1}z_1^{t-1}\bar{x}^t$$
$$z_1^t = \bar{z}_1^{t-1} \vee x^t$$

The UC-structure of the sequential switching circuit based on these expressions is shown in Figure 8.9. We cannot claim that this is the simplest solution obtainable within the first-order finite-state model since we completely ignored the state assignment problem.

Let us now consider *kth order finite-state models*, where $k > 1$. If $\mathbf{z}^t$ denotes the internal state produced by the functional generator at time t, then $\mathbf{Z}^{t-1} = (\mathbf{z}^{t-k}, \mathbf{z}^{t-k+1}, \ldots, \mathbf{z}^{t-2}, \mathbf{z}^{t-1})$ is stored in the memory at time t. This changes to $\mathbf{Z}^t = (\mathbf{z}^{t-k+1}, \mathbf{z}^{t-k+2}, \ldots, \mathbf{z}^{t-1}, \mathbf{z}^t)$ at time $t+1$. Clearly, the memory contains a chain of k delays connected in series for each variable included in $\mathbf{z}^t$.

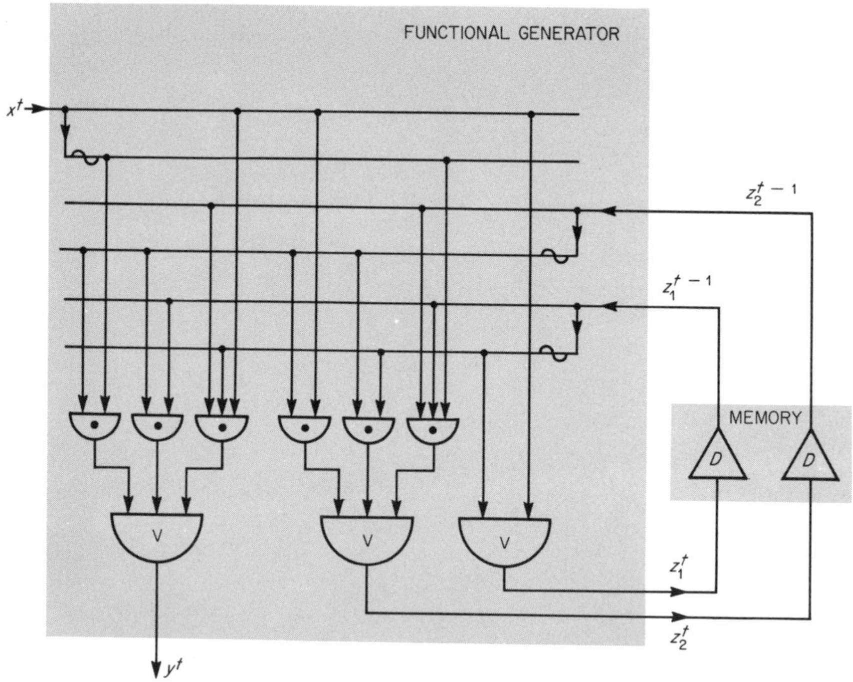

Figure 8.9 A first-order finite-state model implementation of the ST-structure in Figure 8.8.

To determine the behavior of the functional generator for the kth order finite-state model, we have to compile from the given ST-structure all possible sequences of internal states of length k. Each of these sequences can be considered as an internal state of the kth order finite state model and we can determine its output function

$$\mathbf{y}^t = \mathbf{f}(\mathbf{x}^t, \mathbf{Z}^{t-1})$$

and its transition function

$$\mathbf{Z}^t = \mathbf{g}(\mathbf{x}^t, \mathbf{Z}^{t-1})$$

After both of these functions are determined, we can partition internal states of the kth order model into equivalence classes. This identifies equalities of the original states (states of the first order model) from the viewpoint of the kth order model. Clearly, equal states can be coded by the same values of internal variables. This may show that a smaller number of internal variables than that required by the first order model is sufficient.

EXAMPLE 8.3 Using the second order finite-state model and elements AND, OR, NOT, DELAY, design a switching circuit which implements the ST-structure shown in Figure 8.8.

First, we have to determine the set of all possible sequences of two consecutive states. We obtain $\{(\mathbf{s}_0, \mathbf{s}_3), (\mathbf{s}_1, \mathbf{s}_0), (\mathbf{s}_1, \mathbf{s}_3), (\mathbf{s}_2, \mathbf{s}_1), (\mathbf{s}_3, \mathbf{s}_1), (\mathbf{s}_3, \mathbf{s}_2)\}$. Now we construct the following table, which represents the output and transition functions of the second order finite-state model:

	y^t		$\mathbf{Z}^t$	
$\mathbf{Z}^{t-1}$	$x^t = 0$	$x^t = 1$	$x^t = 0$	$x^t = 1$
$(\mathbf{s}_0, \mathbf{s}_3)$	0	0	$(\mathbf{s}_3, \mathbf{s}_2)$	$(\mathbf{s}_3, \mathbf{s}_1)$
$(\mathbf{s}_1, \mathbf{s}_0)$	1	0	$(\mathbf{s}_0, \mathbf{s}_3)$	$(\mathbf{s}_0, \mathbf{s}_3)$
$(\mathbf{s}_1, \mathbf{s}_3)$	0	0	$(\mathbf{s}_3, \mathbf{s}_2)$	$(\mathbf{s}_3, \mathbf{s}_1)$
$(\mathbf{s}_2, \mathbf{s}_1)$	1	1	$(\mathbf{s}_1, \mathbf{s}_0)$	$(\mathbf{s}_1, \mathbf{s}_3)$
$(\mathbf{s}_3, \mathbf{s}_1)$	1	1	$(\mathbf{s}_1, \mathbf{s}_0)$	$(\mathbf{s}_1, \mathbf{s}_3)$
$(\mathbf{s}_3, \mathbf{s}_2)$	0	1	$(\mathbf{s}_2, \mathbf{s}_1)$	$(\mathbf{s}_2, \mathbf{s}_1)$

Both y^t and $\mathbf{Z}^t$ are identical for $(\mathbf{s}_0, \mathbf{s}_3)$ and $(\mathbf{s}_1, \mathbf{s}_3)$. Hence, $(\mathbf{s}_0, \mathbf{s}_3) = (\mathbf{s}_1, \mathbf{s}_3)$. Similarly, $(\mathbf{s}_2, \mathbf{s}_1) = (\mathbf{s}_3, \mathbf{s}_1)$. It follows from these two equalities that $\mathbf{s}_0 = \mathbf{s}_1$ and $\mathbf{s}_2 = \mathbf{s}_3$. One logic variable z suffices to distinguish one of these pairs of states from the other. Let $z = 0$ for $(\mathbf{s}_0, \mathbf{s}_1)$ and $z = 1$ for $(\mathbf{s}_2, \mathbf{s}_3)$. Then we obtain the following table of output and transition functions:

$\mathbf{Z}^{t-1} = (\mathbf{z}^{t-2}, \mathbf{z}^{t-1})$	y^t $x^t = 0$	$x^t = 1$	$\mathbf{Z}^t = (\mathbf{z}^{t-1}, \mathbf{z}^t)$ $x^t = 0$	$x^t = 1$
(0, 1)	0	0	(1, 1)	(1, 0)
(0, 0)	1	0	(0, 1)	(0, 1)
(1, 0)	1	1	(0, 0)	(0, 1)
(1, 1)	0	1	(1, 0)	(1, 0)

Now, it is trivial to determine the truth table representing the behavior of the functional generator (note that z^{t-1}, which is the first component of $\mathbf{Z}^t$, is supplied from the memory and need not be thus produced by the functional generator):

z^{t-2}	z^{t-1}	x^t	y^t	z^t
0	0	0	1	1
0	0	1	0	1
0	1	0	0	1
0	1	1	0	0
1	0	0	1	0
1	0	1	1	1
1	1	0	0	0
1	1	1	1	0

The truth table yields unique minimal Boolean forms for the two functions:

$$y^t = \bar{z}^{t-1}\bar{x}^t \vee z^{t-2}x^t$$
$$z^t = \bar{z}^{t-1}x^t \vee \bar{z}^{t-2}\bar{x}^t$$

The final UC-structure is shown in Figure 8.10.

Exercises to Section 8.5

8.5-1 Using the first-order finite-state model and NAND and DELAY elements, implement the ST-structures given in:
(a) Figure 8.7.
(b) Figure 8.12.
(c) Figure 7.5.
(d) Figure 7.17.
(e) Figure 7.24.

8.5-2 Discuss each of the ST-structures in Exercise 8.5-1 from the standpoint of kth order finite-state models ($k > 1$) and make a design for those models which seem to lead to less complex implementations than that of the corresponding first order finite-state model.

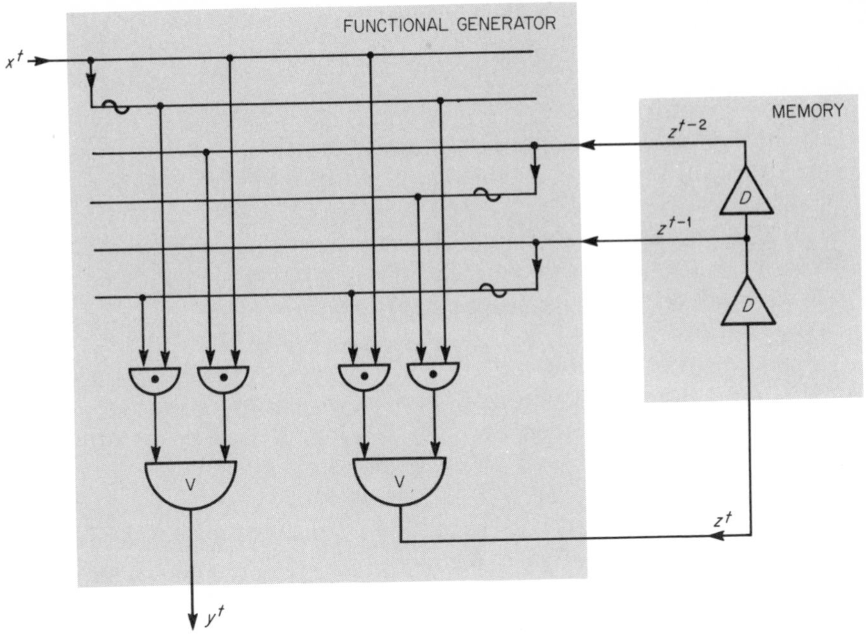

Figure 8.10 A second-order finite-state model implementation of the ST-structure in Figure 8.8.

8.6 FINITE-MEMORY AND COMBINED MODELS

The *finite-memory model*, if it exists for a given ST-structure, may be considered as a degenerate combined model in which the number of internal variables is zero. Both of these models are characterized by storing sequences of stimuli and/or responses of certain length. If states of the given ST-structure are not distinguished by these sequences, some internal variables must be introduced to distinguish them.

To make our unified exposition of finite-memory and combined models easier, we shall introduce some new concepts first.

DEFINITION 8.2 Let a minimal ST-structure of the Mealy type be given. Let $E_k^{(i)}$ denote the set of all sequences containing k consecutive pairs of stimuli and responses that terminate at state i, regardless of the initial state to which they are applied, and let $E_k = \bigcup_i E_k^{(i)}$. Then $e_{k,j} \in E_k$ will be called the *element of uncertainty* $u(e_{k,j})$ if and only if it is included in $u(e_{k,j})$ different sets $E_k^{(i)}$.

DEFINITION 8.3 Let $e_{k,j}$, $u(e_{k,j})$, $E_k^{(i)}$, and E_k have the same meaning as introduced in Definition 8.2. Then, the maximum value of $u(e_{k,j})$ for all $e_{k,j} \in E_k$ will be called the *uncertainty of the given ST-structure with respect to memory span* k; it will be denoted by U_k.

Now, some theorems will be presented which are essential for developing a unified structure synthesis for finite-memory and combined models.

THEOREM 8.5 Let n and U_k be, respectively, the number of internal states in a minimal ST-structure and the uncertainty of the minimal ST-structure with respect to memory span k. Then,

$$1 \leq U_k \leq n. \tag{8.9}$$

PROOF Each element $e_{k,j} \in E_k$ must be included in at least one of the sets $E_k^{(i)}$. Hence $1 \leq U_k$. There are n sets $E_k^{(i)}$. If an element $e_{k,j}$ is included in each of these sets, then $U_k = n$ otherwise, $U_k < n$. ▮

THEOREM 8.6 Every deterministic ST-structure can be realized by the kth order combined model containing

$$n_k = \ulcorner \log_2 U_k \tag{8.10}$$

internal logic variables, where $\ulcorner \log_2 U_k$ represents the smallest integer greater than or equal to $\log_2 U_k$.

PROOF Let $k = 1$. If $U_1 = 1$, every pair of a stimulus and a response included in the ST-structure identifies uniquely an internal state and, thus, no internal variables are needed. Since $\log_2 1 = 0$, the theorem holds.

If $U_1 > 1$, at least one pair of a stimulus and a response identifies U_1 different (distinguishable) internal states. This pair is denoted by $\mathbf{s}/\mathbf{r}$ in Figure 8.11a. Let us suppose now that the stimulus following the pair $\mathbf{s}/\mathbf{r}$ is $\mathbf{s}_1$. In the most unfavorable case, the responses to $\mathbf{s}_1$ may be different for different states, as shown in Figure 8.11b. To determine uniquely the responses $\mathbf{r}_1, \mathbf{r}_2, \ldots, \mathbf{r}_{U_1}$, the pairs $\mathbf{s}/\mathbf{r}$ must be distinguished. This can be done by U_1 states of internal variables. If the states of internal variables are denoted as $\mathbf{I}_1, \mathbf{I}_2, \ldots, \mathbf{I}_{U_1}$, the responses $\mathbf{r}_1, \mathbf{r}_2, \ldots, \mathbf{r}_{U_1}$ are determined as shown in Figure 8.11c, i.e.,

$$\mathbf{s}/\mathbf{r}\ \mathbf{I}_1 \mathbf{s}_1/\mathbf{r}_1$$
$$\mathbf{s}/\mathbf{r}\ \mathbf{I}_2\ \mathbf{s}_1/\mathbf{r}_2$$
$$\vdots$$
$$\mathbf{s}/\mathbf{r}\ \mathbf{I}_{U_1}\ \mathbf{s}_1/\mathbf{r}_{U_1}$$

It is quite obvious that U_1 states of internal variables can be realized by n_1 logic variables, where n_1 is given by equation (8.10). Any other pair $\mathbf{s}'/\mathbf{r}'$ that

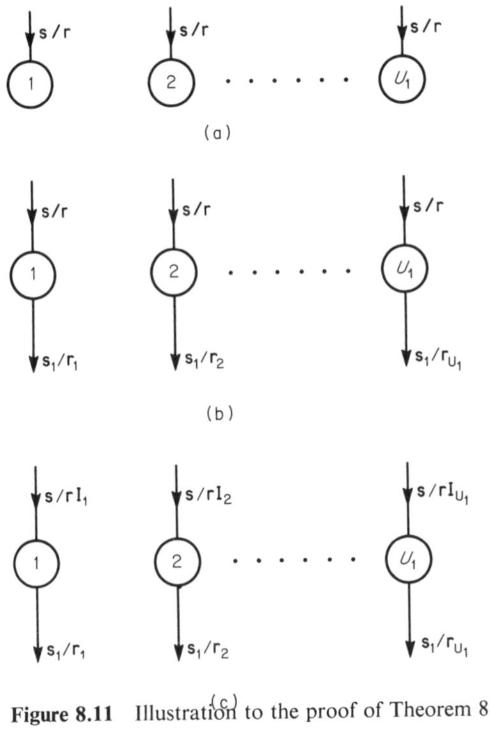

Figure 8.11 Illustration to the proof of Theorem 8.6.

terminates in U_1 states or in a smaller number of states can be, evidently, distinguished by the states of the same internal variables (states $\mathbf{I}_1, \mathbf{I}_2, \ldots, \mathbf{I}_{U_1}$).

Let $k > 1$. Let **s**/**r** in Figure 8.11a be a sequence containing k consecutive pairs of stimuli and responses. Then the proof is analogous to the proof for $k = 1$. ▮

It follows from the proof of Theorem 8.6 that $\ulcorner \log_2 U_k$ represents the maximum number of internal variables required for the kth order combined model ($k \geq 1$) in case of behavior with no input restrictions. If there are some input restrictions, some of the pairs **s**/**r** associated with U_1 different states may be distinguished by the next stimuli so that the number of required internal variables could be even smaller than $\ulcorner \log_2 U_k$.

It also follows from the proof of Theorem 8.6 that in case of $k > 1$ the number of required internal variables could be smaller than $\ulcorner \log_2 U_k$ even if there are no input restrictions (elements of E_k are distinguished by k states of internal variables).

Using Theorem 8.6, we suggest the following procedure of structure synthesis of sequential discrete systems in case that the first order combined model (or the finite-memory model in case that $U_1 = 1$) is used:

1. Elements of $E_1^{(i)}$ are compiled from the given ST-structure for all states i. Example: elements of $E_1^{(i)}(i = 1, 2, \ldots, 5)$ for the ST-structure shown in the form of a diagram in Figure 8.12a are summarized in Figure 8.13a.

2. The uncertainty U_1 and the required number of internal variables n_1 are calculated. If $U_1 = 1$ (and $n_1 = 0$), go to step 5. Example: calculation of U_1 is shown in Figure 8.13a; $n_1 = \lceil \log_2 2 = 1$.

3. Elements of E_1 that terminate in more than one state have to be distinguished by states of internal variables. Example: dots in every column of Figure 8.13a are distinguished by symbols referring to states of internal variables (see Figure 8.13b). It is only required that $a_1 \neq a_2$, $b_1 \neq b_2$, $c_1 \neq c_2$, $d_1 \neq d_2$. Since only one two-valued internal variable is used, we may

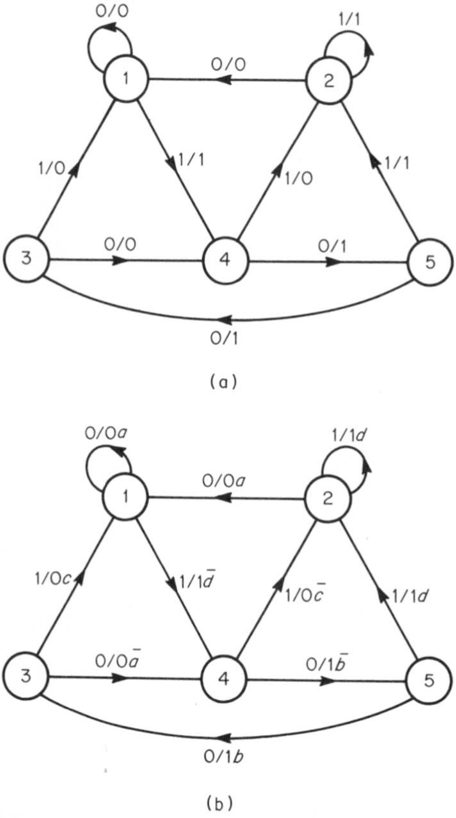

Figure 8.12 Illustration of the procedure of synthesis for the first order combined model.

States	Elements of E_1 (external situations) $e_{11}=0/0$	$e_{12}=0/1$	$e_{13}=1/0$	$e_{14}=1/1$	
1	•		•		$E_1^{(1)}$
2			•	•	$E_1^{(2)}$
3		•			$E_1^{(3)}$
4	•			•	$E_1^{(4)}$
5		•			$E_1^{(5)}$
$u(e_{1j})$	2	2	2	2	$U_1 = \max_j u(e_{1j}) = 2$

(a)

States	Elements of E_1 (external situations) 0/0	0/1	1/0	1/1
1	a_1		c_1	
2			c_2	d_1
3		b_1		
4	a_2			d_2
5		b_2		

(b)

States	Elements of E_1 (external situations) 0/0	0/1	1/0	1/1
1	a		c	
2			$\bar{c}$	d
3		b		
4	$\bar{a}$			$\bar{d}$
5		$\bar{b}$		

(c)

Figure 8.13 Calculation of the uncertainty U_1 for the ST-structure in Figure 8.12a and the procedure of its reduction.

use $a_1 = a$, $a_2 = \bar{a}$, ..., $d_1 = d$, $d_2 = \bar{d}$ (See Figure 8.13c). Let us note that a, b, c, and d are values of a single internal variable.

4. The state-transition structure is supplemented by the states of internal variables. An example is shown in Figure 8.12b.

5. The set E_2 compiled from the supplemented state-transition structure represents the behavior (excitation table) of the functional generator. Example: the behavior of the functional generator corresponding to the state-transition structure in Figure 8.12b is in Table 8.1.

6. Synthesis of the functional generator is completed for the given behavior. This includes a solution of the state assignment problem and the problem of a memoryless switching circuit design. Example: a possible assignment of a, b, c, and d is represented by Table 8.2. When only NAND gates and DELAY elements are used, multiple-output minimization leads to the UC-structure shown in Figure 8.14. It contains three DELAYS and twelve NAND gates.

When the same problem is solved by the first order finite-state model, five internal states must be distinguished so that at least three internal variables are needed. Solving properly the state assignment problem, we obtain the

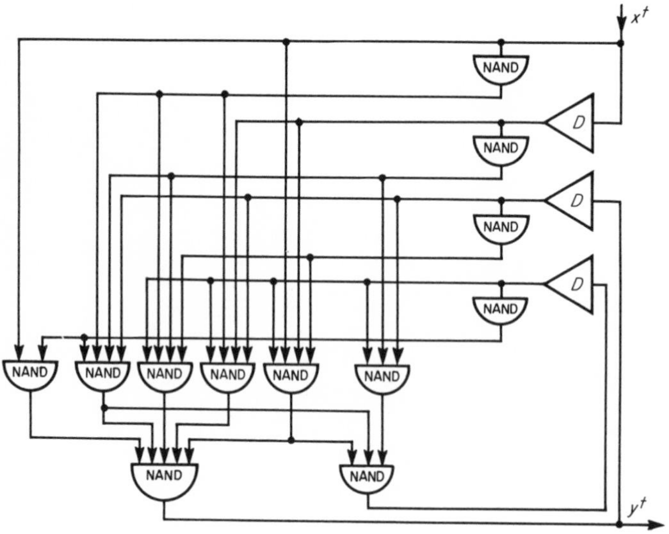

Figure 8.14 A first-order combined model implementation of the ST-structure in Figure 8.12a.

TABLE 8.1.

	x^{t-1}	y^{t-1}	z^{t-1}	x^t	y^t	z^t
Toward state 1	0	0	a	0	0	a
	1	0	c	0	0	a
	1	1	d	0	0	a
	0	1	b	1	0	c
	1	0	$\bar{c}$	0	0	a
Toward state 2	1	1	d	1	1	d
	1	0	$\bar{c}$	1	1	d
	0	1	$\bar{b}$	1	1	d
	1	1	$\bar{d}$	1	0	$\bar{c}$
	0	0	$\bar{a}$	1	0	$\bar{c}$
Toward state 3	0	1	$\bar{b}$	0	1	b
Toward state 4	0	0	a	1	1	$\bar{d}$
	1	0	c	1	1	$\bar{d}$
	0	1	b	0	0	$\bar{a}$
Toward state 5	0	0	$\bar{a}$	0	1	$\bar{b}$
	1	1	$\bar{d}$	0	1	$\bar{b}$

TABLE 8.2.

x^{t-1}	y^{t-1}	z^{t-1}	x^t	y^t	z^t
0	0	0	0	0	0
1	0	1	0	0	0
1	1	0	0	0	0
0	1	1	1	0	1
1	0	0	0	0	0
1	1	0	1	1	0
1	0	0	1	1	0
0	1	0	1	1	0
1	1	1	1	0	0
0	0	1	1	0	0
0	1	0	0	1	1
0	0	0	1	1	1
1	0	1	1	1	1
0	1	1	0	0	1
0	0	1	0	1	0
1	1	1	0	1	0

coding of internal states which is shown in Table 8.3. Applying this code to the ST-structure in Figure 8.12a, we obtain Table 8.4, an excitation table of the functional generator. Multiple-output minimization of the functions involved leads to the UC-structure shown in Figure 8.15. It contains three DELAYS and fifteen NAND gates. Getting the same number of DELAYS we save three gates when we use the combined model in this case.

Let us now consider the kth order combined model. First, the kth order combined model ($k > 1$) has a practical meaning only if $U_k > 1$ and

$$\ulcorner \log_2 U_k < \ulcorner \log_2 U_{k-1} \tag{8.11}$$

Indeed, if $U_k = 1$, there is no need to use internal variables; the finite-memory model is applicable. If (8.11) is not satisfied, $(k-1)$th order combined model is preferable because it requires a smaller number of memory

TABLE 8.3

Internal State	Code z_1	z_2	z_3
1	0	0	0
2	1	1	1
3	0	0	1
4	0	1	0
5	0	1	1

TABLE 8.4.

z_1^{t-1}	z_2^{t-1}	z_3^{t-1}	x^t	z_1^t	z_2^t	z_3^t	y^t
0	0	0	0	0	0	0	0
0	0	0	1	0	1	0	1
1	1	1	0	0	0	0	0
1	1	1	1	1	1	1	1
0	0	1	0	0	1	0	0
0	0	1	1	0	0	0	0
0	1	0	0	0	1	1	1
0	1	0	1	1	1	1	0
0	1	1	0	0	0	1	1
0	1	1	1	1	1	1	1

elements and leads to a realization of functions of a smaller number of variables.

The procedure of synthesis of kth order combined model (including the finite-memory model, if applicable) is a simple generalization of the procedure for the first-order combined model. It can be summarized as follows:

1. Sets $E_k^{(i)}$ are compiled from the given state-transition structure.
2. Uncertainty U_k is determined; if $U_k = 1$, go to step 6.
3 Number of internal variables is calculated by equation (8.10).
4. Uncertainty is removed from ST-structure, i.e., sets $E_k^{(i)}$ are made pairwise disjoint by states of internal variables.

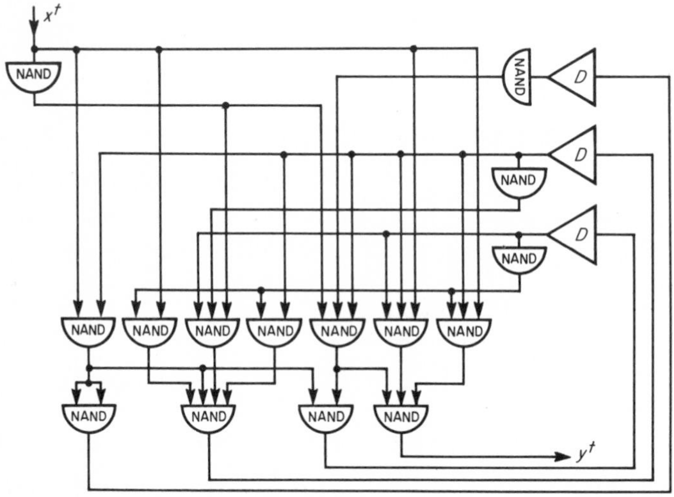

Figure 8.15 A first-order finite-state model implementation of the ST-structure in Figure 8.12a.

5. State assignment problem is solved.

6. Set E_{k+1} is compiled (this set already represents the behavior of the functional generator).

7. Functional generator is designed.

8. Inputs and outputs of the functional generator are properly connected to the memory.

EXAMPLE 8.4 Using the kth order combined model for an appropriate k, design a switching circuit implementing the ST-structure shown in Figure 8.16a.

First, we compile sets $E_1^{(i)}$ from the ST-structure for all states and determine the uncertainty U_1. This is shown in Figure 8.17a. We obtain $U_1 = 3$ and, by formula (8.10), $n_1 = 2$.

Although we can use the first order combined model with two internal variables, this design is left to the reader as an exercise. We will show the design for a kth order combined model, where $k > 1$.

Now, we compile sets $E_2^{(i)}$ from the ST-structure for all states and determine the uncertainty U_2. This is shown in Figure 8.17b. We obtain $U_2 = 2$ and $n_2 = 1$. Thus, the second order combined model requires a smaller number of internal variables than the first-order combined model and, therefore, it is reasonable to consider it.

Next, we have to reduce the uncertainty to 1 by introducing values of a single variable to outputs of individual transitions. Let these values be denoted by distinct letters as shown in Figure 8.16b. These values must distinguish all input/output sequences of length 2 that terminate in more than one state. This leads to the following requirements concerning the values of the introduced internal variable:

For (0/0, 0/0): (a, a) must differ from (h, f).
For (0/0, 1/1): (h, i) must differ from (g, c) and (f, c).
For (1/0, 0/0): (e, f) must differ from (j, h).
For (1/0, 1/1): (j, d) must differ from (e, i).
For 1/1, 0/0): (d, a) must differ from (i, g).
For (1/1, 1/0): (i, j) must differ from (d, e).

Clearly, the above requirements can be described by the following set of simultaneous Boolean equations:

$$a\bar{h} \vee \bar{a}h \vee a\bar{f} \vee \bar{a}f = 1$$
$$(h\bar{g} \vee \bar{h}g \vee i\bar{c} \vee \bar{i}c)(h\bar{f} \vee \bar{h}f \vee i\bar{c} \vee \bar{i}c) = 1$$
$$e\bar{j} \vee \bar{e}j \vee f\bar{h} \vee \bar{f}h = 1$$
$$j\bar{e} \vee \bar{j}e \vee d\bar{i} \vee \bar{d}i = 1$$
$$d\bar{i} \vee \bar{d}i \vee a\bar{g} \vee \bar{a}g = 1$$
$$i\bar{d} \vee \bar{i}d \vee j\bar{e} \vee \bar{j}e = 1$$

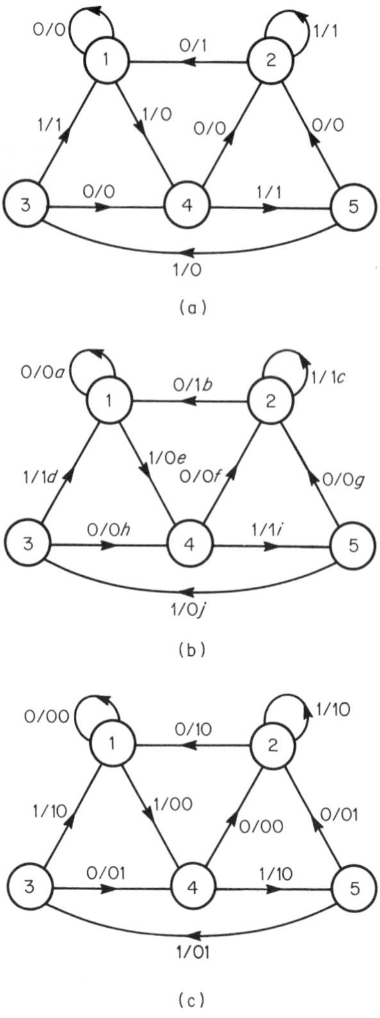

Figure 8.16 Illustration for Example 8.4 (synthesis based on the second-order combined model).

The discriminant of these equations contains all possible collections of values of the internal variable. One possible collection of these values is shown in Figure 8.16c. The state assignment problem which consists of selecting a single collection of values from a set of possible collections with regard to a certain objective criteria is not solved here.

When we compile all possible input/output sequences of length 3 (elements of the set E_3), from the ST-structure in Figure 8.16c, we obtain directly a truth table representing the behavior of the functional generator. Then, it remains to design the functional generator.

States	$e_{1,1}$ 0/0	$e_{1,2}$ 0/1	$e_{1,3}$ 1/0	$e_{1,4}$ 1/1	
	Elements of E_1				
1	•	•		•	$E_1^{(1)}$
2	•			•	$E_1^{(2)}$
3			•		$E_1^{(3)}$
4	•		•		$E_1^{(4)}$
5				•	$E_1^{(5)}$
$u(e_{1,j})$	3	1	2	3	$U_1 = \max_j u(e_{1,j}) = 3$

(a)

States	$e_{2,1}$ 0/0,0/0	$e_{2,2}$ 0/0,0/1	$e_{2,3}$ 0/0,1/0	$e_{2,4}$ 0/0,1/1	$e_{2,5}$ 0/1,0/0	$e_{2,6}$ 0/1,1/0	$e_{2,7}$ 1/0,0/0	$e_{2,8}$ 1/0,1/1	$e_{2,9}$ 1/1,0/0	$e_{2,10}$ 1/1,0/1	$e_{2,11}$ 1/1,1/0	$e_{2,12}$ 1/1,1/1	
	Elements of E_2												
1	•	•			•			•	•	•			$E_2^{(1)}$
2	•			•			•		•			•	$E_2^{(2)}$
3											•		$E_2^{(3)}$
4			•			•	•				•		$E_2^{(4)}$
5				•				•					$E_2^{(5)}$
$u(e_{2,j})$	2	1	1	2	1	1	2	2	2	1	2	1	

$U_2 = \max_j u(e_{2,j}) = 2$

(b)

Figure 8.17 Determination of U_1 and U_2 for the ST-structure in Figure 8.16a.

The reader should complete the synthesis of the second-order combined model and compare it with the designs based on the first-order combined model and the first-order finite-state model.

The reader should be able at this stage to construct Boolean equations representing constraints concerning values of internal variables for any values of k, and n_k, and any given ST-structure.

We have seen that if $U_k = 1$ for a given ST-structure, the kth order combined model degenerates to the finite-memory model. No internal variable is needed in this case. The behavior of the functional generator is represented by the set of all input/output sequences of length $k + 1$ (set E_{k+1}) compiled from the original ST-structure.

If there exists a finite k such that $U_k = 1$ and $U_{k-1} > 1$ for a given ST-structure, then k will be called the *intrinsic memory span* of the system and will be denoted by μ; if no such k exists, then $\mu = \infty$. Clearly, if $\mu = \infty$, the finite memory model is not applicable; otherwise, it is applicable.

A question arises at this point: Is there any upper bound for the memory span k beyond which the finite-memory model is not applicable? To answer the question, it will be helpful to introduce the concept of the ST-structure of pairs of states.

DEFINITION 8.4 Let Z denote the set of all distinguishable states of a ST-structure and let W denote the set of all unordered pairs of states taken from Z. Then, the *ST-structure of pairs of states* (or *PST-structure*) is a function $W \to W$ such that $(\mathbf{z}_i, \mathbf{z}_j) \in W$ is mapped into $(\mathbf{z}_k, \mathbf{z}_l) \in W$ if and only if

$$\mathbf{g}(\mathbf{x}, \mathbf{z}_i) = \mathbf{z}_k$$
$$\mathbf{g}(\mathbf{x}, \mathbf{z}_j) = \mathbf{z}_l$$
$$\mathbf{f}(\mathbf{x}, \mathbf{z}_i) = \mathbf{f}(\mathbf{x}, \mathbf{z}_j) = \mathbf{y}$$

for at least one input/output pair $\mathbf{x}/\mathbf{y}$.

When transforming a given ST-structure to the corresponding PST-structure, it is useful to determine all predecessors for each state and all input/output pairs, first. This can be arranged in the form of the table shown in Figure 8.18. We compile now all pairs of states which are mapped into two different states for the same input/output pair. These pairs of states with their successors represent the PST-structure which can be put similarly as the ST-structure, into the form of a table, a diagram, or a matrix.

EXAMPLE 8.5 Determine the PST-structure for the ST-structure shown in Figure 8.19a.

First, we determine all predecessors for each state and for all input/output pairs and arrange them in the table shown in Figure 8.19b. Then, using Definition 8.4, we can determine directly the PST-structure and put it into, for example, one of the forms that are shown in Figure 8.20. Note that the ith row represents the same pair $\mathbf{p}_i$ of states as the ith column in the PST-matrix. Using a simplified notation, the (i, j) entry of the matrix is equal to 1 if $\mathbf{p}_j$ is a successor of $\mathbf{p}_i$; otherwise we make it equal to 0. The respective input/output pairs can be used instead of ones in the PST-matrix but there is usually no need for doing that.

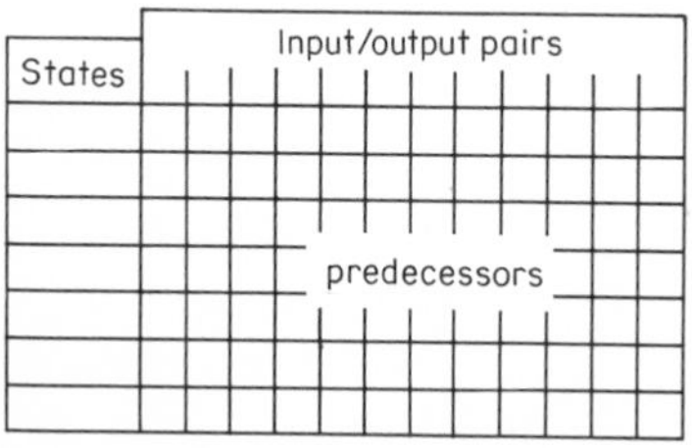

Figure 8.18 A form of the table of predecessors for individual input/output pairs.

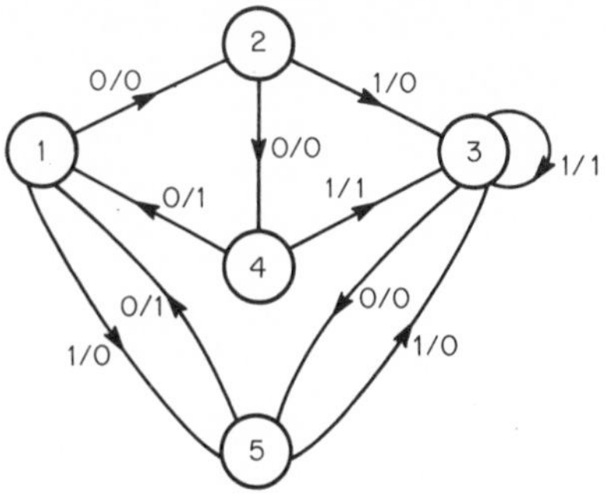

(a)

	0/0	0/1	1/0	1/1
1		4,5		
2	1			
3			2,5	3,4
4	2			
5	3		1	

(b)

Figure 8.19 Illustration to Example 8.5 and 8.6 (synthesis based on the finite memory model).

THEOREM 8.7 The finite-memory model is applicable for a given ST-structure (μ is finite) if and only if the corresponding PST-diagram D does not contain any loop.

PROOF Assume that D contains a loop. Then, by repeating the sequence of stimuli associated with transitions in the loop we can construct an arbitrarily long input/output sequence whose uncertainty is equal to two.

To prove sufficiency, let us assume that D contains no loop. If the finite-memory model is not applicable, there must exist an arbitrarily long path in D. But this is impossible without a loop in D since the number of nodes in D is $n(n-1)/2$, where n is the number of states in the given ST-structure. ▌

Pairs of states	Successors of the pairs of states			
	0/0	0/1	1/0	1/1
(1,2)	(2,4)		(3,5)	
(1,3)	(2,5)			
(1,4)				
(1,5)			(3,5)	
(2,3)	(4,5)			
(2,4)				
(2,5)				
(3,4)				
(3,5)				
(4,5)				

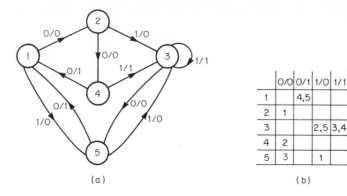

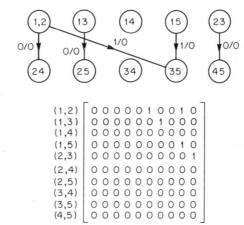

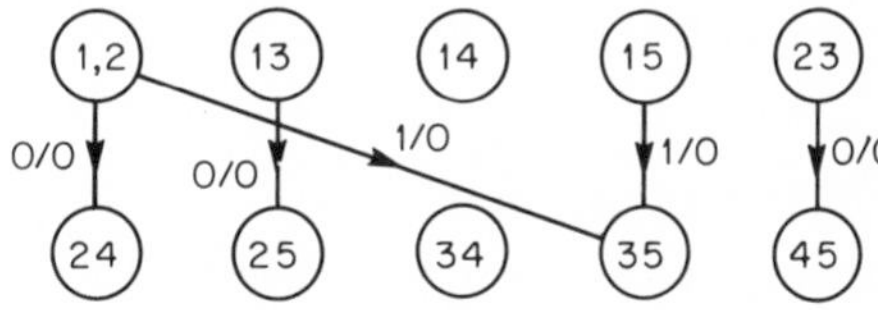

$$
\begin{array}{c}
(1,2)\\ (1,3)\\ (1,4)\\ (1,5)\\ (2,3)\\ (2,4)\\ (2,5)\\ (3,4)\\ (3,5)\\ (4,5)
\end{array}
\begin{bmatrix}
0&0&0&0&0&1&0&0&1&0\\
0&0&0&0&0&0&1&0&0&0\\
0&0&0&0&0&0&0&0&0&0\\
0&0&0&0&0&0&0&0&1&0\\
0&0&0&0&0&0&0&0&0&1\\
0&0&0&0&0&0&0&0&0&0\\
0&0&0&0&0&0&0&0&0&0\\
0&0&0&0&0&0&0&0&0&0\\
0&0&0&0&0&0&0&0&0&0\\
0&0&0&0&0&0&0&0&0&0
\end{bmatrix}
$$

Figure 8.20 Various forms of the PST-structure of the ST-structure in Figure 8.19a.

THEOREM 8.8 Let D be PST-diagram for a given ST-structure and let D contain no loops. If the length of the longest path in D is l, then $\mu = l + 1$.

PROOF Left to reader as an exercise. ∎

THEOREM 8.9 If the intrinsic memory span μ of a sequential system is finite, then

$$\mu \leq n(n-1)/2$$

PROOF From Theorem 8.7, μ is finite if the PST-diagram contains no loops. The PST-diagram contains $n(n-1)/2$ nodes. The longest possible path in such a diagram without loops cannot exceed the length of $l = n(n-1)/2 - 1$. Hence, either $n = \infty$ or, by Theorem 8.8, $\mu = l + 1 \leq n(n-1)/2$. ∎

Theorem 8.9 provides us with the answer to our question concerning the upper bound of the memory span k for which the finite-memory model is applicable: If $U_k > 1$ for $k = n(n-1)/2$, then the finite-memory model cannot be applied.

Loops and the longest paths in a given PST-diagram can be easily identified by inspection. For computer processing, PST-matrices are usually preferable. The following algorithm, which is easy to derive, is suggested.

ALGORITHM 8.1 Given a PST-matrix, find the intrinsic memory span μ of the system: (1) Let $k = 1$. (2) Delete all rows having zeros in all columns. (3) Delete all columns with the same labels as the deleted rows in (2). (4) If there are rows having zeros in all columns, increase k by 1 and go to (2). (5) If the matrix has completely disappeared, then $\mu = k$; otherwise $\mu = \infty$.

Another algorithm, which is based on the ST-structure can be used equally well.

ALGORITHM 8.2 Given a ST-structure containing pair wise distinguishable states $\mathbf{z}_1, \mathbf{z}_2, \ldots, \mathbf{z}_n$, find the intrinsic memory span μ of the system: (1) Let $k = 1$. (2) Compile the sets $E_k^{(1)}, E_k^{(2)}, \ldots, E_k^{(n)}$. (3) (*a*) If $E_k^{(i)} \cap E_k^j \neq \varnothing$ for some i and $j \neq i$, go to (4). (*b*) If $E_k^{(i)} \cap E_k^{(j)} = \varnothing$ for all i and $j \neq i$, then $\mu = k$. (4) (*a*) If $k < n(n-1)/2$, increase k by 1 and go to (2). (b) If $k = n(n-1)/2$, then $\mu = \infty$.

Flow charts of both of these algorithms are shown in Figure 8.21.

EXAMPLE 8.6 Using the finite-memory model, implement, if possible, the ST-structure specified in Figure 8.19a.

From the corresponding PST-diagram, which is in Figure 8.20 (Example 8.5), it is trivial to find that $\mu = 2$ (the longest path in the diagram has the length 1). Hence, the behavior of the functional generator is represented by

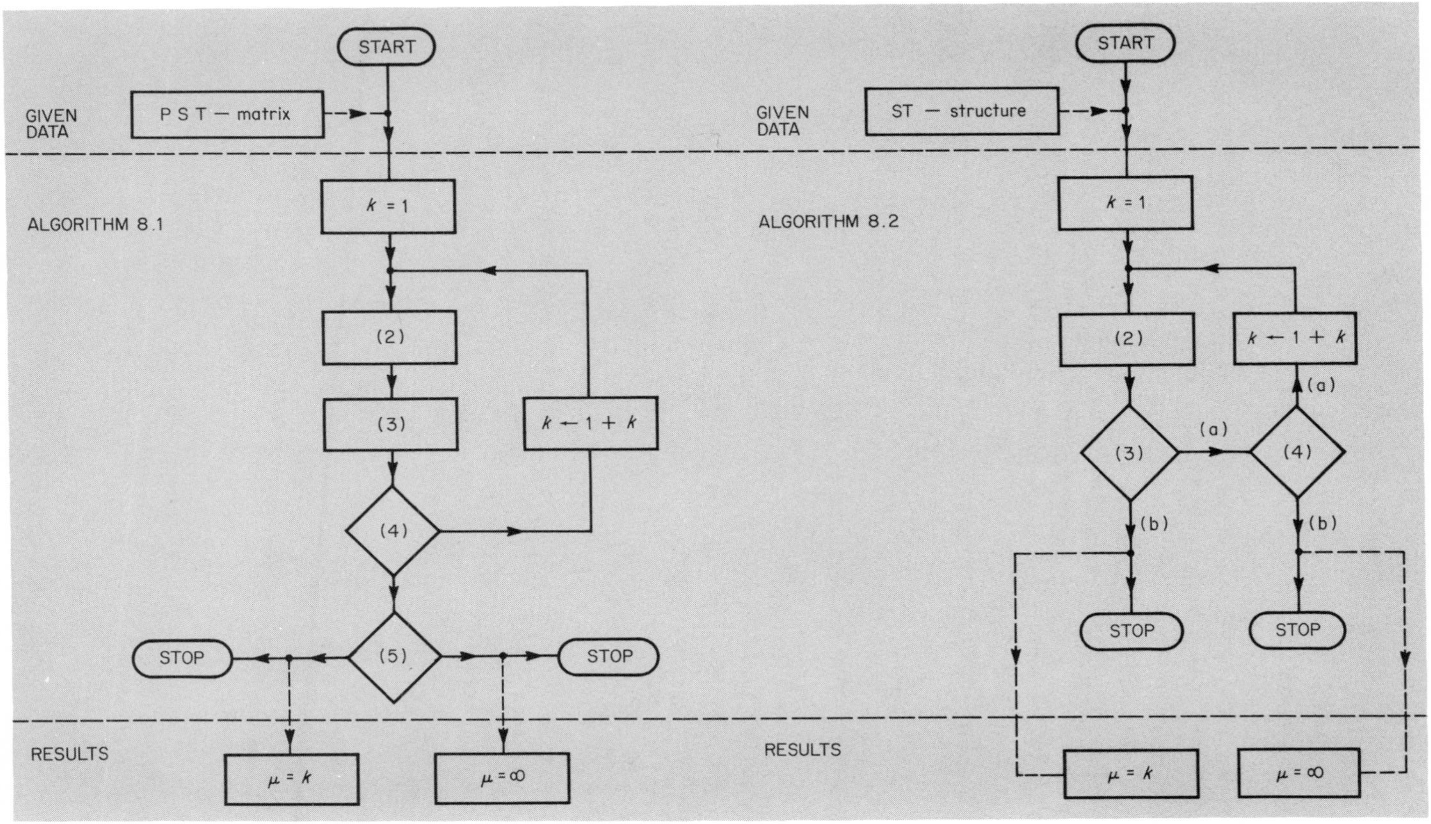

Figure 8.21 Flow charts of two algorithms for calculating the intrinsic memory span of a system.

all possible input/output sequences of length 3. This gives us the following table, where x is the input variable and y is the output variable:

x^{t-2}	y^{t-2}	x^{t-1}	y^{t-1}	x^t	y^t	x^{t-2}	y^{t-2}	x^{t-1}	y^{t-1}	x^t	y^t
0	0	0	0	0	1	1	0	0	0	0	1
0	0	0	0	1	1	1	0	0	0	1	0
0	0	0	1	0	0	1	0	0	1	0	0
0	0	0	1	1	0	1	0	0	1	1	0
0	0	1	0	0	0	1	0	1	0	0	0
0	0	1	0	1	1	1	0	1	0	1	1
0	0	1	1	0	0	1	0	1	1	0	0
0	0	1	1	1	1	1	0	1	1	1	1
0	1	0	0	0	0	1	1	0	0	0	1
0	1	0	0	1	0	1	1	0	0	1	0
0	1	1	0	0	1	1	1	1	1	0	0
0	1	1	0	1	0	1	1	1	1	1	1

A two-level AND-OR implementation is shown in Figure 8.22.

EXAMPLE 8.7 Using the finite-memory model, implement, if possible, the ST-structure in Figure 8.16a.

First, we determine the corresponding PST-diagram (Fig. 8.23). Since this diagram contains a loop (from 15 to 34 and back), $\mu = \infty$ so that the finite-memory model is not applicable (Theorem 8.7).

Exercises to Section 8.6

8.6-1 Using the first order combined model, implement the ST-structure given in:
- (a) Figure 8.16a.
- (b) Figure 8.19a.
- (c) Figure 8.8.
- (d) Figure 7.5.
- (e) Figure 7.17.
- (f) Figure 7.24.

8.6-2 Using the finite-memory model, implement, if possible, the ST-structure in:
- (a) Figure 8.7.
- (b) Figure 8.8.
- (c) Figure 8.12a.
- (d) Figure 7.5.
- (e) Figure 7.17.
- (f) Figure 7.24.

8.6-3 Using such a kth order combined model that $U_k = 2$, implement the ST-structures suggested in two previous exercises.

8.6-4 Prove Theorem 8.8.

8.6-5 Prove that $U_j \leq U_k$ for $j > k$.

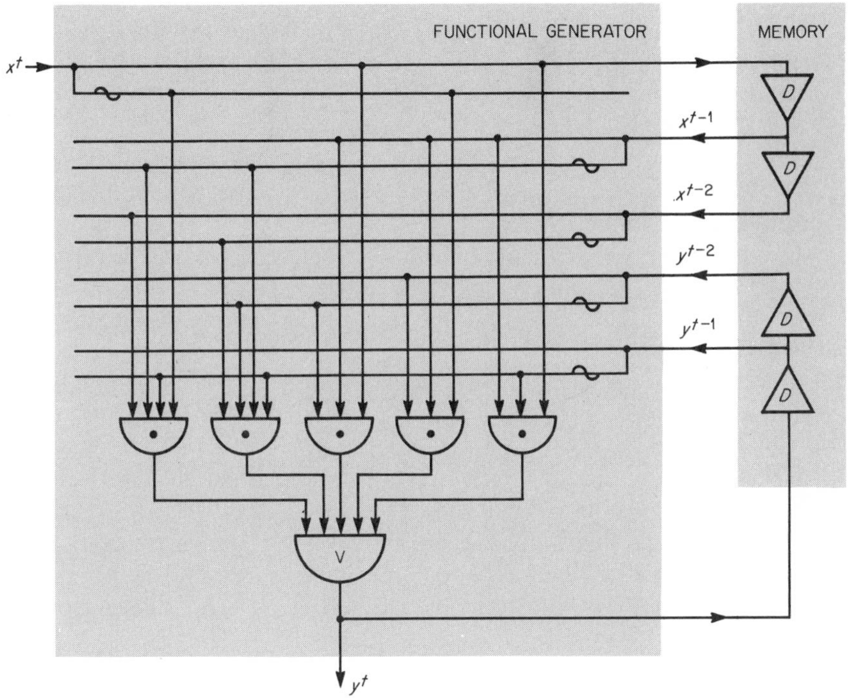

Figure 8.22 A finite-memory model implementation of the ST-structure in Figure 8.19a ($\mu = 2$).

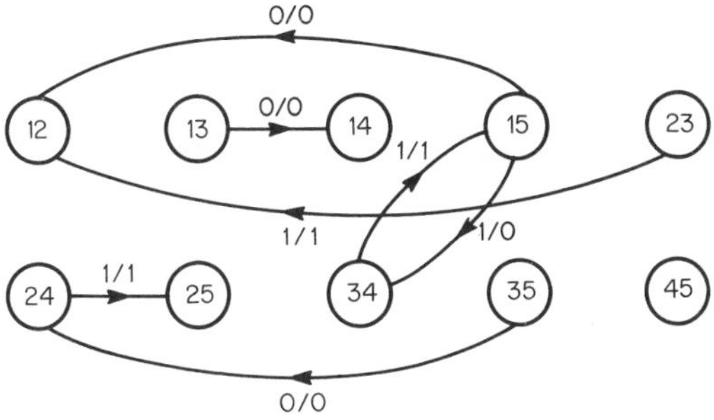

Figure 8.23 PST-diagram of the ST-structure in Figure 8.16a.

8.7 SVOBODA'S METHODICAL APPROACH

Suppose that the performance of sequential switching circuit is given in the form of a matrix $[v_i^t]$ whose entry v_i^t represents the value of the ith logic variable (variable v_i) at the discrete instant of time $t(i = 1, 2, \ldots, m; t = 1, 2, \ldots)$. This matrix will be called the *activity matrix.*

The activity matrix may contain values of input and output variables of the switching circuit as well as values of some internal variables introduced by the designer for various reasons.

The subscript i, which uniquely identifies a logic variable, will be called the *variable identifier*. The superscript t will be referred to as the *time identifier.* The variable identifier determines the rows of the activity matrix, the time identifier its columns.

An example of an activity matrix is presented in Figure 8.24. Variables v_1 and v_2 are regarded as the input variables, variables v_3 and v_4 as the output variables.

When setting up an activity matrix we must also specify the boundary conditions, i.e., determine what states of the variables involved precede the activity matrix and follow upon it. For instance, the boundary conditions of the activity matrix in Figure 8.24 are represented by zeros for all variables.

For a fixed pair (i, t) of integers, the *sampling variable* V_i^T is defined by the equation

$$V_i^T(t) = v_i(t + T) \tag{8.12}$$

which is satisfied for all t in the activity matrix. We see that the sampling variable is a function of time. The pair (i, T) will be called the *sampling pair.*

Clearly, if $T = 0$, then $V_i^T(t)$ represents the value of variable v_i at time t. If $T < 0$, then $V_i^T(t)$ represents a past value of variable v_i with respect to time t. If $T > 0$, then $V_i^T(t)$ represents a future value of variable v_i with respect to time t.

A distinct set of sampling pairs is called a *mask*. Examples of masks are indicated in Figure 8.24. We can consider the mask as a sort of cut-out through which a portion of the activity matrix can be seen.

$t =$			1	2	3	4	5	6	7	8	9	10	11	12	13	14	15	16	17	18	19	20	21	22	23	24	25	26	27	28	29	30	31	32		
$v_1(t) =$	0	0	0	1	1	0	0	0	1	0	1	1	0	0	1	1	0	0	1	1	1	1	1	0	1	0	0	0	0	0	0	1	1	1	0	0
$v_2(t) =$	0	0	0	1	0	0	1	0	1	1	1	0	1	1	0	1	1	0	0	1	1	1	0	1	0	0	0	0	1	1	0	1	1	1	0	0
$v_3(t) =$	0	0	0	1	0	1	0	1	1	0	0	1	0	1	0	0	1	1	1	1	0	1	0	1	0	1	0	0	0	1	1	1	0	1	0	0
$v_4(t) =$	0	0	0	0	0	1	1	0	1	0	1	1	0	0	0	1	1	0	1	0	0	1	0	1	0	1	1	0	0	0	0	1	0	1	0	0
						M_1				M_2				M_3				M_4				M_5						M_6				M				

Figure 8.24 An activity matrix used in Example 8.8.

Every mask represents a set of sampling variables V_i^T. If the mask is placed on an activity matrix $[v_i^t]$ in position t, it uncovers the values $v_i(t + T)$ for all sampling pairs (i, T) that constitute the mask. We say that these values constitute a *sample of the activity* for the given mask and a reference time t.

It follows from equations (8.12) that the sampling variable V_i^0 represents directly the variable v_i no matter whether it is an input or output variable. If v_i is an input variable, then, clearly, $T \leq 0$ for all sampling variables with the subscript i (no future value of v_i is available in the system at any time t). If v_i is an output variable, then it is taken either from the functional generator so that $\delta_i = 0$ in terms of Definition 8.1 or it is taken from the memory so that $\delta_i > 0$. In either case, $T \leq \delta_i$ for all sampling variables with the subscript i. If v_i is not an external variable, there is no restriction concerning values of T in sampling variables V_i^T.

Sampling variables which have the same subscript i and differ only by their superscript T are separated from each other in the switching circuit solely by DELAYS (Figure 8.25). Hence, every mask represents a particular structure of the memory. An example of a mask and the corresponding memory structure is shown in Figure 8.26.

Let T_I and T_i denote, respectively, the maximum and the minimum of T for all sampling variables of a mask which are identified by a particular subscript i. Then, variable v_i requires $T_I - T_i$ DELAYS. The value $D = \max(T_I - T_i)$ for all i is called the *depth of the mask*. If no internal variables are introduced in the activity matrix, then

$$D = \mu. \tag{8.13}$$

It is easy to find that the sampling variables $V_i^{T_I}$ for all i except those which identify external input variables are output variables of the functional generator. We shall call them *dependent variables*. All other sampling variables are potential input variables of the functional generator.

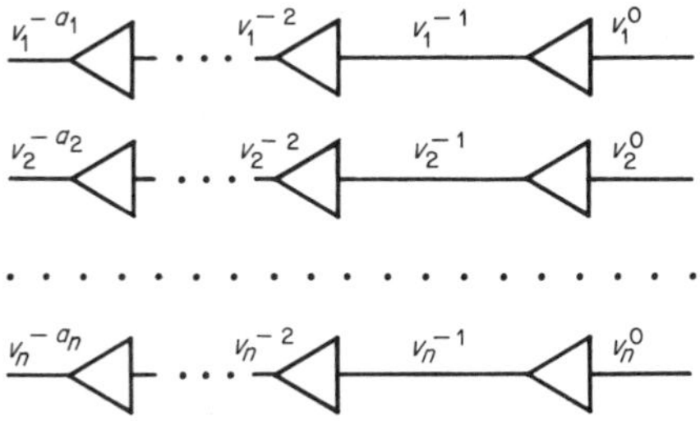

Figure 8.25 Meaning of sampling variables in switching circuits.

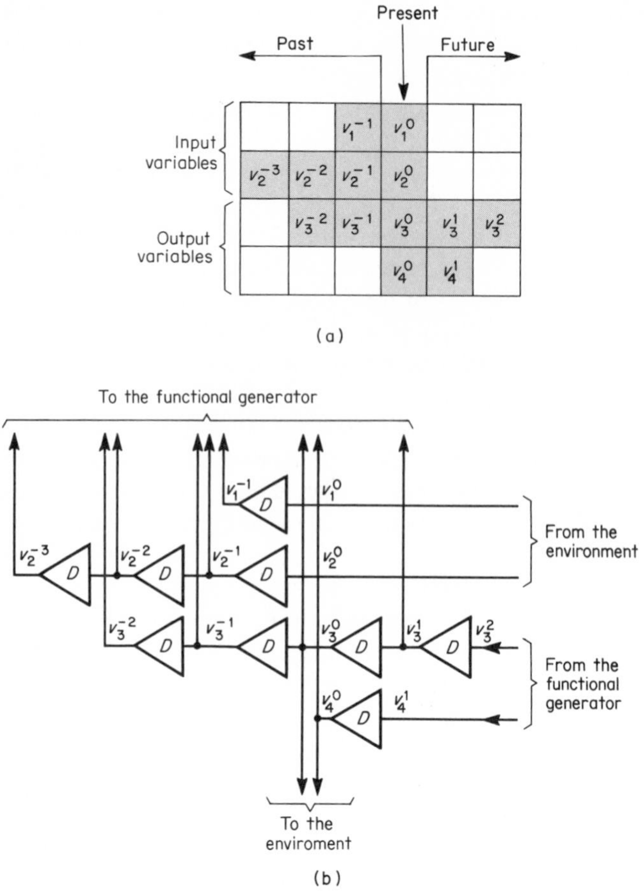

Figure 8.26 A mask and the corresponding memory structure.

The time invariant relation between the sampling variables represented by a particular mask can be obtained by going with the mask through all its possible positions in the direction of time identifier ($t = 1, 2, 3, \ldots$) and writing out the sample of the activity for each position. This relation represents the behavior of the functional generator while the mask determines the structure of the memory.

It is convenient to record samples of the activity directly into an appropriate Marquand chart in which states of the input variables and states of the output variables of the functional generator (the dependent variables) identifies, respectively, columns and rows of the chart. Then, the filled out chart has the same form as the discriminant used when solving logic equations. It is quite clear, therefore, that the chart represents an acceptable solution only if it contains at most one dot in each column so that the output variables of the functional generator are determined at every time uniquely by values of its input variables at the same time t.

If there are columns which contain more than one dot, there is an ambiguity in the determination of values of the dependent variables which is not acceptable for deterministic systems. To eliminate this ambiguity we can modify the mask either without or with some additional (internal) variables.

If we choose a mask too small or unsuitably shaped for the given activity matrix, we cannot arrive at a solution corresponding to a deterministic switching circuit. Conversely, if the mask is too large, i.e., if it contains redundant sampling variables, we arrive at a solution but the procedure involved is unnecessarily toilsome. We therefore try to choose a mask which contains the least number of sampling variables necessary to obtain a deterministic solution for the given activity matrix.

EXAMPLE 8.8 Using AND, OR, NOT, and DELAY elements, implement the activity matrix shown in Figure 8.24, regarding v_1 and v_2 as the input variables, v_3 and v_4 as the output variables.

Let us first show what the relations between the sampling variables would look like for some of the masks presented in Figure 8.24. The masks are redrawn in Figure 8.27 and carry the designation of the sampling variables, the dependent sampling variables being denoted by the shaded areas.

The chart representations of the relations between the sampling variables are shown in Figure 8.28 for the individual masks. We see that mask M_4 is the only satisfactory of all of the masks presented. It contains functions whose chart representations are shown in Figure 8.29. Minimal *sp*-forms are

$$V_4^0 = \overline{V}_3^{-1} V_1^{-1} \vee V_3^{-1} \overline{V}_2^{-1} \overline{V}_1^{-1} \vee \overline{V}_3^{-1} V_2^{-1} V_1^0$$

$$V_3^0 = V_2^{-1} \overline{V}_1^{-1} \overline{V}_1^0 \vee V_3^{-1} \overline{V}_2^{-1} V_1^0 \vee \overline{V}_3^{-1} V_2^{-1} V_1^{-1} \vee \overline{V}_2^{-1} \overline{V}_1^{-1} V_1^0 \vee \overline{V}_3^{-1} V_1^{-1} \overline{V}_1^0$$

and the final scheme of the switching circuit is in Figure 8.30.

Now, let us briefly summarize the entire procedure explained in this section:

1. The activity matrix is given or must be set up (including the boundary conditions) according to the given specification. As already mentioned, the method is of practical importance especially in cases where the activity matrix is explicitly given. That is to say, the activity matrix may be rather difficult to construct for problems specified in a different manner.

2. We choose a suitable mask according to the type of problem to be solved.

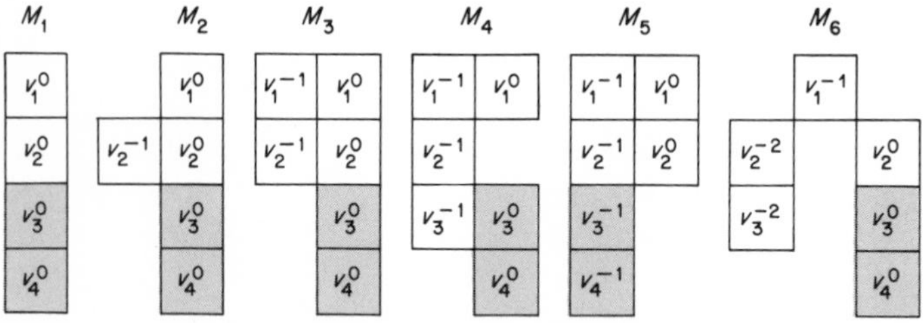

Figure 8.27 Masks that are used in Example 8.8.

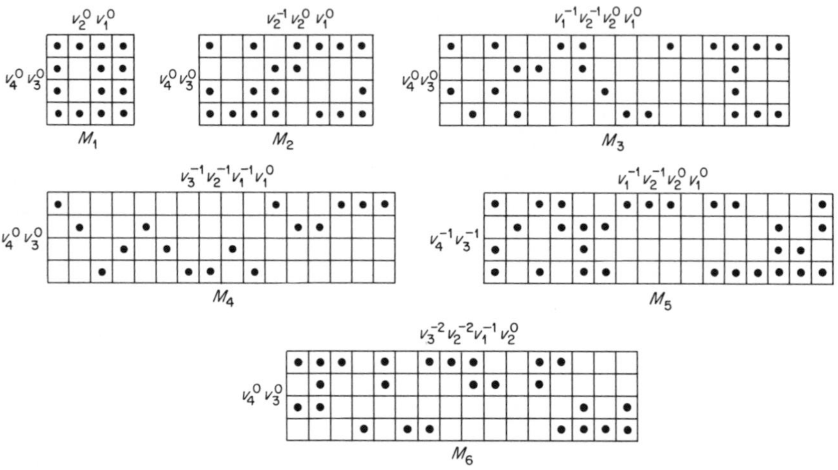

Figure 8.28 Relations between sampling variables in the activity matrix of Figure 8.24 for the masks shown in Figure 8.27.

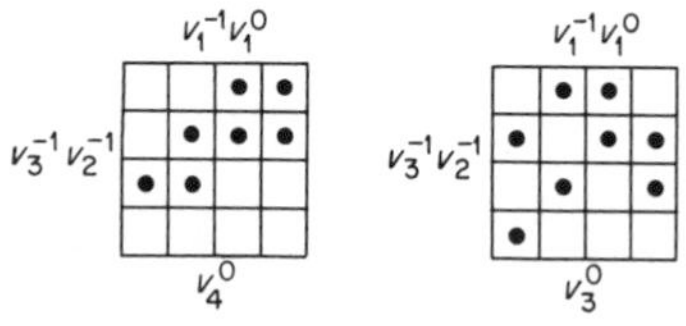

Figure 8.29 Functions represented by mask M_4 when applied to the activity matrix in Figure 8.24.

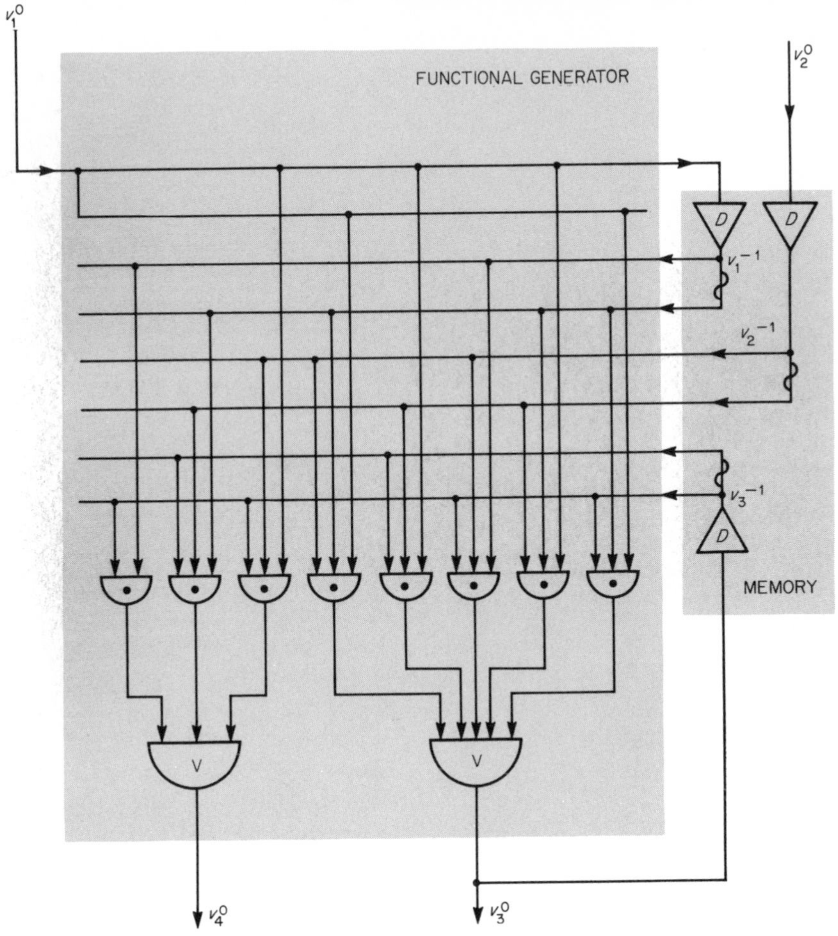

Figure 8.30 An implementation of the activity matrix in Figure 8.24 for mask M_4 in Figure 8.27 (Example 8.8).

3. We draw up a chart, assigning the dependent sampling variables to its rows and the other sampling variables to its columns.

4. We determine the relation between the sampling variables for the chosen chart by sliding the mask (e.g., cut out of paper) over the activity matrix in the direction of the time index and marking in the prepared chart the state of the sampling variables for each position. Should two marked states appear in some column of the chart, we must interrupt the procedure and choose a different mask. The procedure is concluded when we arrive at a relation in which the dependent variables are functions of the other sampling variables (i.e., when there is one marked row in each column of the chart at most).

5. We design properly the functional generator based on the functions representing the dependent variables.

6. We determine, directly from the pattern of the used mask the structure of the memory and then connect the memory properly with the functional generator.

It is interesting to compare various models of the sequential discrete system from the standpoint of the mask. Some typical cases are shown in Figure 8.31.

Exercises to Section 8.7

8.7-1 Determine the mask representing the circuit shown in:
(a) Figure 8.9.
(b) Figure 8.10.
(c) Figure 8.14.
(d) Figure 8.15.
(e) Figure 8.22.

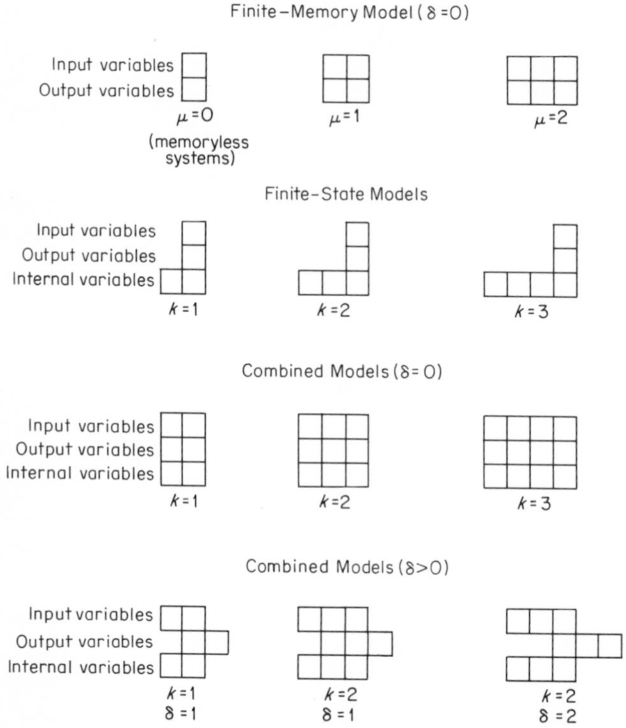

Figure 8.31 Masks for various models of the sequential discrete system.

8.7-2 Using NAND and DELAY elements, implement the following periodical activity matrices, where x is an input variable and y is an output variable:

(a)

$t=$	1	2	3	4	5	6	7	8	9	10	11	12	13	14	15	16
$x=$	0	1	0	1	1	0	1	1	1	0	1	1	1	1	0	0
$y=$	0	1	1	1	0	1	1	0	1	1	1	0	1	0	1	1.

(b)

$t=$	1	2	3	4	5	6	7	8	9	10	11	12	13	14	15	16
$y=$	0	0	0	0	1	1	1	1	0	0	1	0	1	1	0	1.

(c)

$t=$	1	2	3	4	5	6	7	8	9
$y=$	0	0	0	1	1	1	0	1	1.

(d)

$t=$	1	2	3	4	5	6	7	8	9	10	11	12	13	14	15	16	17	18	19	20	21	22	23	24
$x_1=$	0	1	0	1	0	1	0	1	0	1	0	1	0	1	0	1	0	1	0	1	0	1	0	1
$x_2=$	0	0	0	1	0	1	1	0	0	0	0	0	0	1	0	1	1	1	1	0	1	0	1	1
$y_1=$	0	1	1	1	0	1	1	0	1	1	0	0	1	1	0	0	1	1	0	0	1	1	1	1
$y_2=$	0	0	0	0	0	0	0	0	0	1	1	1	1	1	1	1	1	1	0	1	1	1	0	0.

8.8 A ROUGH EVALUATION OF MODELS

It is an advantage if we can compare different models without completing synthesis for all of them. The comparison may show that some of the models are worse than the others for a given problem so that we are allowed to exclude them from our further considerations. This may simplify considerably the whole procedure of synthesis.

Using simple formulas, we can determine from the given ST-structure the following data for every model: (1) number of delays $\# D$; (2) number of input variables of the functional generator $\# I$; (3) number of output variables of the functional generator $\# O$.

In the formulas that follow, p, q, n_k, and N mean, respectively, the number of input variables of the system, the number of output variables of the system, the number of internal variables in case of kth order model, and the number of nonequivalent internal states. We obtain the following formulas by simple considerations:

1. *The kth order finite-state models* ($k = 1, 2, \ldots$):

$$n_1 = \ulcorner \log_2 N$$

$$\ulcorner \frac{1}{k} \log_2 N \leq n_k \leq n_{k-1} (k \geq 2) \tag{8.14}$$

$$\# D = k \cdot n_k$$
$$\# I = p + k \cdot n_k$$
$$\# O = q + n_k$$

2. *The kth order combined models* ($k = 1, 2, \ldots$) or the *finite-memory model* (if $U_k = 1$):

$$n_1 = \ulcorner \log_2 U_1$$

$$\ulcorner \frac{1}{k} \log_2 U_k \leq n_k \leq \ulcorner \log_2 U_k \leq n_{k-1} (k \geq 2)$$

$$\#D = k(p + q + n_k) \tag{8.15}$$

$$\#I = (k+1)p + k(q + n_k)$$

$$\#O = q + n_k$$

Some additional parameters could be calculated, e.g., the number of used states (or the number of "don't care" states) of input variables of the functional generator, the number of possible dummy variables in case of the combined model, etc. All these parameters can be used for an evaluation of the considered models.

Let symbols S_k and C_k denote, respectively, the kth order finite-state models and the kth order combined models, the latter including also the finite-memory model as a special case. Suppose that the models are evaluated in the order $S_1, S_2, \ldots, C_1, C_2, \ldots$.

Model S_1 is always applicable. To evaluate it, we calculate n_1 first and, then, its other parameters.

It is not guaranteed that model S_k with $n_k \leq n_{k-1}$ is applicable for $k \geq 2$. Hence, its applicability must be verified by procedures explained in Section 8.5. First, we consider the smallest possible value of n_k, i.e.,

$$n_k = \min n_k = \ulcorner \frac{1}{k} \log_2 N \tag{8.16}$$

If S_k is applicable for this value of n_k, then we calculate the other parameters of S_k for this value and proceed to calculation of the parameters for S_{k+1}, unless $n_k = 1$ or $k = k_{max}$, where k_{max} is an upper bound for k chosen by the designer. If S_k is not applicable for n_k given by equation (8.16), then we verify its applicability for $n_k = \min n_k + 1$, $n_k = \min n_k + 2, \ldots, n_k = n_{k-1} - 1$.

A flow chart of the procedure by which parameters of all applicable models S_k such that $k \leq k_{max}$ and $n_k < n_{k-1}$ are calculated is in Figure 8.32.

It follows from Theorem 8.6 that every model C_k is applicable. However, it is not reasonable to use model C_k if $n_k = n_{k-1}$.

We start with model C_1 for which U_1 and n_1 are determined first and, then, we calculate the other parameters. Next we proceed to C_2, then to C_3, etc. The procedure terminates when $n_k = 0$ for $k \leq N(N-1)/2$ or $n_k = 1$ for

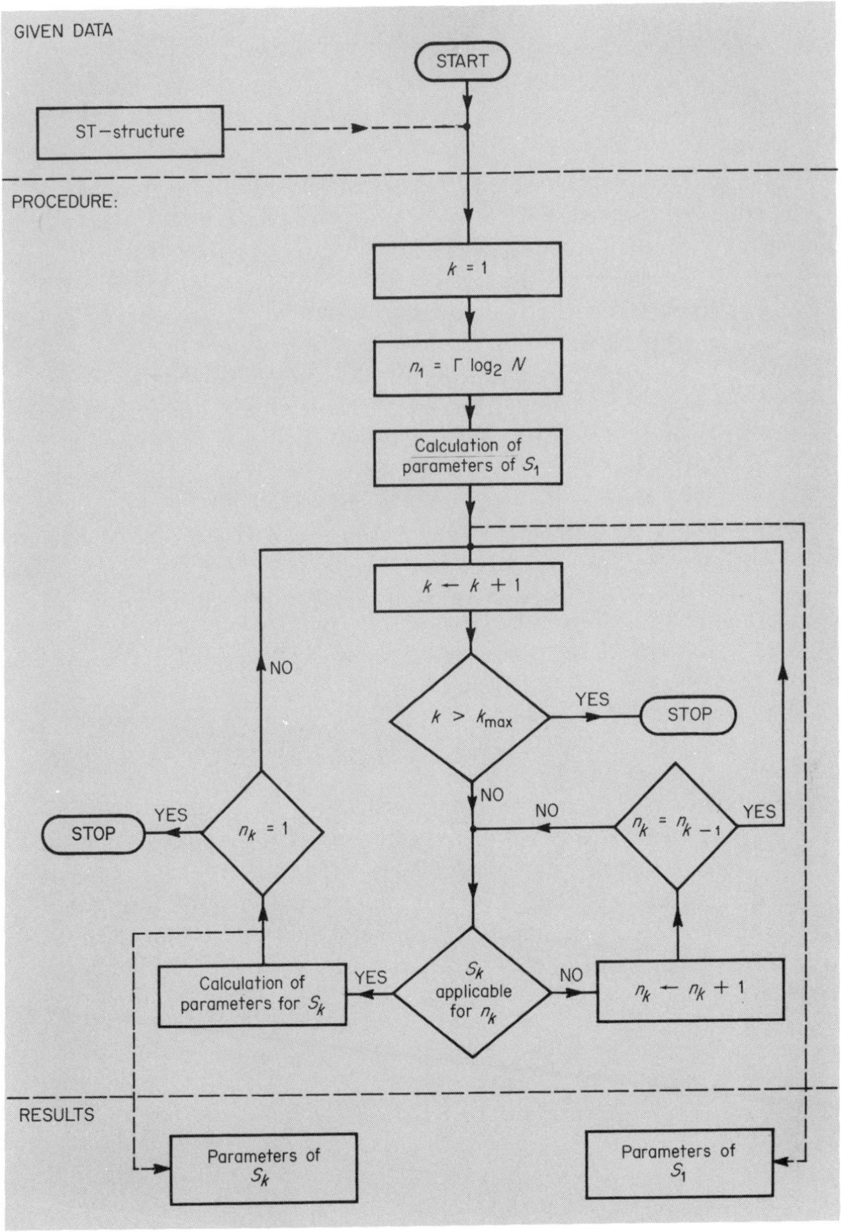

Figure 8.32 A flow chart describing the calculation of parameters for all applicable finite-state models such that $k \leq k_{\max}$.

$k \geq N(N-1)/2$ or $k = k_{max}$, where k_{max} is an upper bound of k chosen by the designer. For each k, we start with the smallest possible value of n_k, i.e.,

$$n_k = \min n_k = \ulcorner \frac{1}{k} \log_2 U_k \qquad (8.17)$$

If C_k is applicable for this value of n_k, then we calculate the other parameters of C_k and either we proceed to C_{k+1} or the procedure terminates. If C_k is not applicable for n_k given by equation (8.17), then we verify its applicability for $n_k = \min n_k + 1$, $n_k = \min n_k + 2, \ldots, n_k = n_{k-1} - 1$.

The procedure by which we calculate parameters for all models C_k such that $k \leq k_{max}$ and $n_k \leq n_{k-1}$ is shown in Figure 8.33.

After we calculate the parameters for all reasonable finite-state and combined models we can use these parameters to compare individual models on the basis of certain objective criteria. For instance, when trying to design a circuit with the smallest number of elements, the comparison can be performed as follows: M_1 is not considered as a favorable model if and only if there exists a model M_2 with some of the parameters $\#D$, $\#I$, $\#O$ smaller but with none of them greater then those for M_1.

A more sophisticated comparison can be based on values of the function

$$F = \#D \cdot W_D + \#I \cdot W_I + \#O \cdot W_O$$

where W_D, W_I, W_O are coefficients (weights) which express statistically the influence of $\#D$, $\#I$, $\#O$ on the value of the used objective function (complexity, cost).

Still another possibility exists. A simple evaluation can be done by a computer and a more sophisticated evaluation left up to the designer. Using his experience and complete knowledge of all circumstances concerning a particular design (the possibility of sharing the memory with other systems, simplicity of diagnostic experiments, etc.), he should be able to decrease considerably the number of offered models and classify only a few of them as the favorable ones.

Exercises to Section 8.8

8.8-1 Prove formulas (8.14) and (8.15).

*8.8-2 Using various parameters of ST-structures, find simple formulas for the number of don't care states at the input of the functional generator of
  (a) the first-order finite-state model!
  (b) the first-order combined model.

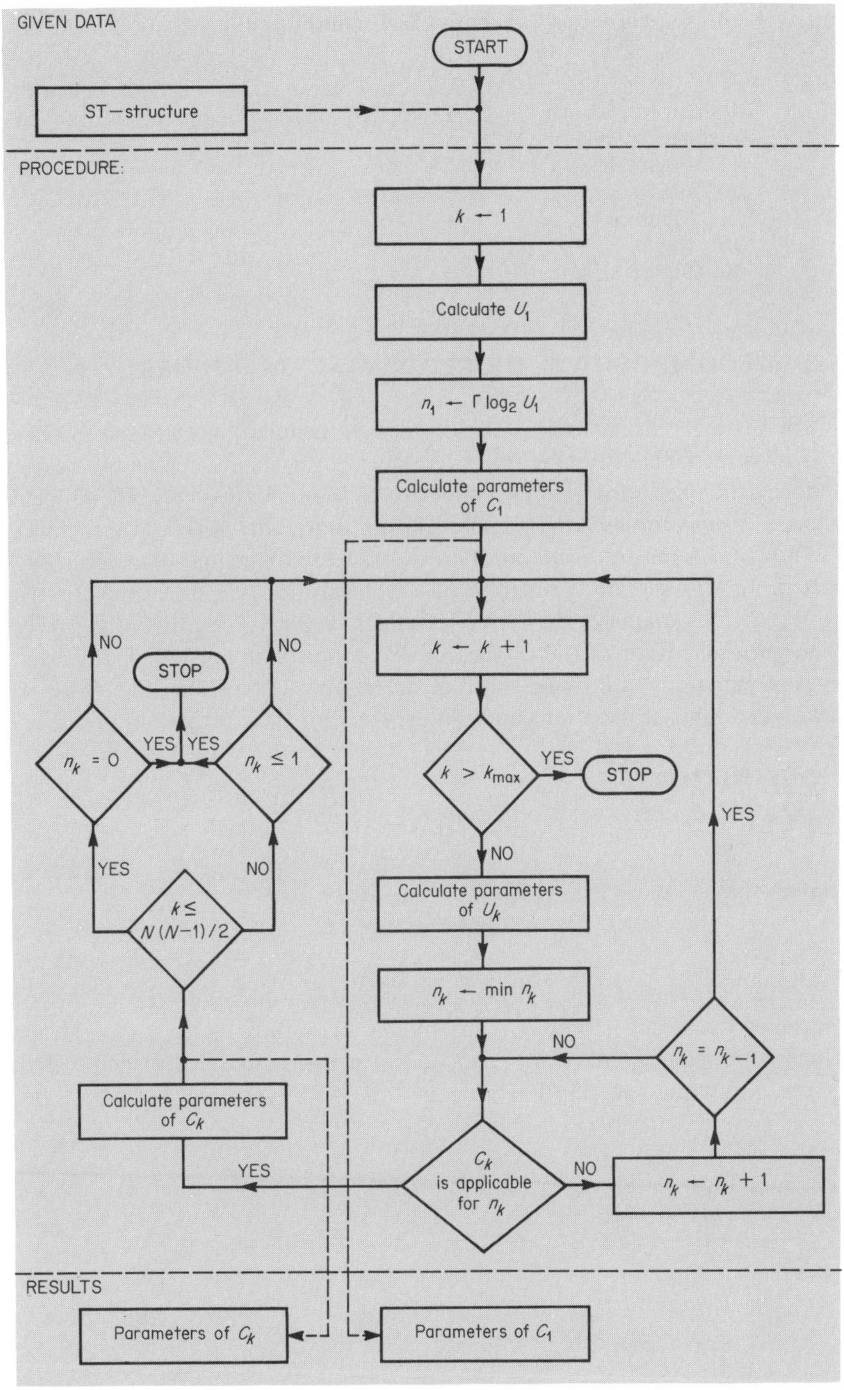

Figure 8.33 A flow chart describing the calculation of parameters for all combined models such that $k \leq k_{\max}$ and $n_k \leq n_{k-1}$.

8.8-3 Let $k_{max} = 3$. Using the flow charts in Figures 8.32 and 8.33, calculate the parameters $\#D$, $\#I$, and $\#O$ for all reasonable models of the ST-structure specified by:
(a) Figure 8.7.
(b) Figure 8.8.
*(c) Figure 8.12.
(d) Figure 8.19.
(e) Figure 8.34.

8.9 DECOMPOSITION OF SEQUENTIAL MACHINES

For various reasons associated with design, installation and maintenance, it is often desirable to implement a complex finite state machine by a set of simpler machines appropriately connected together. In this context, we speak about a *decomposition* of the given machine into partial (composite) machines.

Various schemes of decomposition of sequential machines have been suggested. In this section, principles of decomposition are described for two basic schemes, namely, the *serial* (*cascade*) *decomposition* and the *parallel decomposition.* Both of these types of decompositions are based on certain kind of partitions of internal states of the composite machine. Therefore, we define this kind of partitions and outline some of their properties first.

DEFINITION 8.5 A partition π on the set of states Z of the machine $M = (X, Y, Z, \mathbf{f}, \mathbf{g})$ is said to be *closed* if and only if

$$\mathbf{z}_i = \mathbf{z}_j(\pi)$$

implies that

$$\mathbf{g}(\mathbf{x}, \mathbf{z}_i) = \mathbf{g}(\mathbf{x}, \mathbf{z}_j)(\pi)$$

for all $\mathbf{x} \in X$.

Let us note that if π is a closed partition of the internal states of a given machine, then each stimulus maps blocks of π into blocks of π. Hence, blocks of π can be considered as states of a new, simplified machine, if we ignore the output function $\mathbf{f}$.

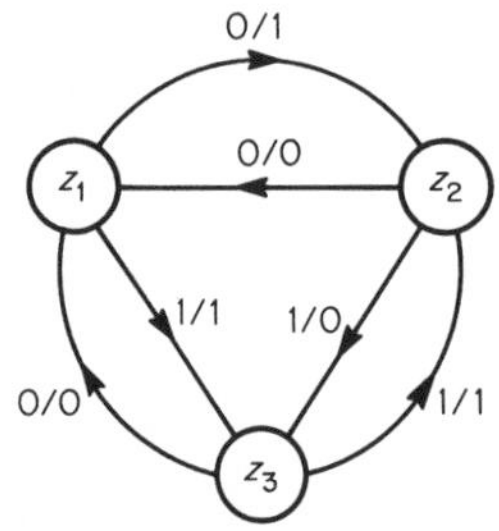

Figure 8.34 Illustration to Exercise 8.8-3e.

EXAMPLE 8.9 Consider the Mealy machine specified in Figure 8.35a. It can be easily verified that $\pi_1 = \{\overline{1, 6}; \overline{2, 5}; \overline{3, 4}\}$ and $\pi_2 = \{\overline{1, 2, 3}; \overline{4, 5, 6}\}$ are closed partitions of the internal states of this machine. If we ignore the output function of the machine, then we obtain for each of these partitions a simplified transition function as shown in Figure 8.35b. Partition $\pi_3 = \{\overline{1, 2}; \overline{3, 4}; \overline{5, 6}\}$ is not closed since block $\overline{1, 2}$ is mapped into block $\overline{4, 6}$ which is not included in the π_3. Similarly, $\overline{3, 4}$ is mapped into $\overline{2, 5}$ and $\overline{5, 6}$ is mapped into $\overline{1, 3}$ but neither $\overline{2, 5}$ nor $\overline{1, 3}$ is included in π_3.

It can be easily shown that if π_1 and π_2 are closed partitions, then $\pi_1 \cdot \pi_2$ and $\pi_1 + \pi_2$ are also closed partitions. Hence, the set C_M of all closed partitions of the internal states of a finite-state machine M forms a lattice under the standard ordering of partitions. It is a sublattice of the lattice of all possible partitions of the states.

The set C_M forms a basis for the investigation of possible decompositions of machine M. It can be determined in two steps: (*1*) For every pair of internal states z_i and z_j, determine the smallest closed partition $\pi_{i,j}$, provided that it exists, which contains z_i and z_j in one block. (*2*) Find all possible sums of the partitions $\pi_{i,j}$ obtained by (*1*).

EXAMPLE 8.10 Determine all closed partitions for the machine whose transition function is specified by the following table:

	a	**b**	**c**
1	6	3	2
2	5	4	1
3	2	5	4
4	1	6	3
5	4	1	6
6	3	2	5

First, we consider states 1, 2. We find that pairs 3, 4 and 5, 6 are implied. Pair 3, 4 implies pairs 1, 2 and 5, 6; pair 5, 6 implies pairs 1, 2 and 3, 4. Hence $\pi_{1,2} = \{\overline{1, 2}; \overline{3, 4}; \overline{5, 6}\}$.

Next we consider states 1 and 3. They imply pairs 2, 4 (stimulus **c**), 2, 6 (stimulus **a**) and 3, 5 (stimulus **b**). Pair 2, 4 implies pairs 1,5 and 4, 6 in addition to pair 1, 3 which has been already identified. Pair 2, 6 does not imply any new pair which has not been identified yet and so does not pair 3, 5. Hence, $\pi_{1,3} = \{\overline{1, 3, 5}; \overline{2, 4, 6}\}$.

z^t \ x^t	z^{t+1} 0	z^{t+1} 1	y^t 0	y^t 1
1	4	3	0	0
2	6	3	0	0
3	5	2	0	1
4	2	5	0	0
5	1	4	1	0
6	3	4	1	1

Given Machine
(a)

	0	1
a=1,6	3,4	3,4
b=2,5	1,6	3,4
c=3,4	2,5	2,5

$\pi_1 = \pi$

	0	1
A=1,2,3	4,5,6	1,2,3
B=4,5,6	1,2,3	4,5,6

$\pi_2 = \tau$

Closed Partitions
(b)

z_1^t \ x_1^t	z_1^{t+1} 0	z_1^{t+1} 1
a	c	c
b	a	c
c	b	b

M_1

z_2^t \ x_2^t	z_2^{t+1} 0,a	0,b	0,c	1,a	1,b	1,c	y_2^t 0,a	0,b	0,c	1,a	1,b	1,c
A	B	B	B	A	A	A	0	0	0	0	0	1
B	A	A	A	B	B	B	1	1	0	1	0	0

M_2

Serial Decomposition
(c)

M_1 =

z_1^t \ x^t	z_1^{t+1} 0	z_1^{t+1} 1	y_1^t 0	y_1^t 1
a	c	c	0,c	1,c
b	a	c	0,a	1,c
c	b	b	0,b	1,b

M_2 =

z_2^t \ x^t	z_2^{t+1} 0	z_2^{t+1} 1	y_2^t 0	y_2^t 1
A	B	A	0,B	1,A
B	A	B	0,A	1,B

x^t	z_1^{t+1}	z_2^{t+1}	y^t
0	a	A	0
0	a	B	1
0	b	A	0
0	b	B	1
0	c	A	0
0	c	B	0
1	a	A	0
1	a	B	1
1	b	A	0
1	b	B	0
1	c	A	1
1	c	B	0

Paralell Decomposition
(d)

Figure 8.35 Simplification of a given Mealy machine based on closed partitions of internal states (Example 8.9).

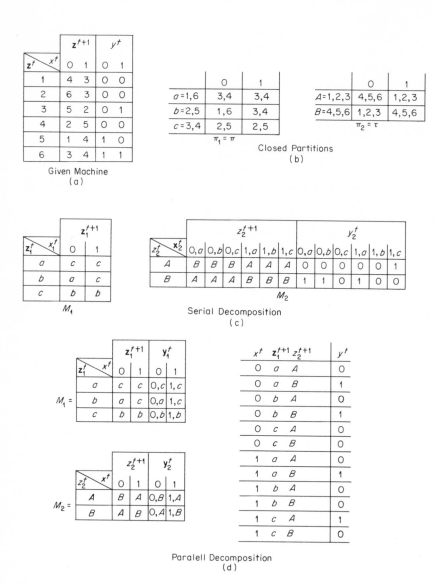

Similarly, we determine partitions $\pi_{i,j}$ for all other pairs of states. We find that $\pi_{1,2}$ and $\pi_{1,3}$ are the only nontrivial closed partitions in this case and their sum forms the trivial partition including all states in one block. The lattice of all closed partitions for this example is shown in Figure 8.36, where $\pi(0)$ and $\pi(I)$ denote, respectively, the trivial partitions $\{\bar{1}; \bar{2}; \bar{3}; \bar{4}; \bar{5}; \bar{6}\}$ and $\{\overline{1, 2, 3, 4, 5, 6}\}$. We see that $\pi(0)$ is identical with the original set of states.

Using the concept of the closed partition of the internal states, we can investigate now some properties of the serial (cascade) composition of two finite-state machines.

DEFINITION 8.6 *The serial composition of two Mealy machines* $M_1 = (X_1, Y_1, Z_1, \mathbf{f}_1, \mathbf{g}_1)$ and $M_2 = (X_2, Y_2, Z_2, \mathbf{f}_2, \mathbf{g}_2)$ such that $Y_1 = X_2$ is the machine $M = (X_1, Y_2, Z_1 \times Z_2, \mathbf{f}, \mathbf{g})$, where

$$\mathbf{g}(\mathbf{x}, (\mathbf{z}_1, \mathbf{z}_2)) = (\mathbf{g}_1(\mathbf{x}, \mathbf{z}_1), \mathbf{g}_2(\mathbf{f}_1(\mathbf{x}, \mathbf{z}_1), \mathbf{z}_2))$$

and

$$\mathbf{f}(\mathbf{x}, (\mathbf{z}_1, \mathbf{z}_2)) = \mathbf{f}_2(\mathbf{f}_1(\mathbf{x}, \mathbf{z}_1), \mathbf{z}_2)$$

provided that $\mathbf{z}_1 \in Z_1$ and $\mathbf{z}_2 \in Z_2$.

In our further discussion, the notion of nontrivial decomposition of a finite-state machine will always mean that each of the component machines has fewer internal states than the composite machine.

THEOREM 8.10 The Mealy machine $M = (X, Y, Z, \mathbf{f}, \mathbf{g})$ has a nontrivial serial (cascade) decomposition if and only if there exists a nontrivial closed partition of the state set Z of M.

PROOF Assume that the machine is implemented by a serial composition of two Mealy machines

$$M_1 = (X_1, Y_1, Z_1, \mathbf{f}_1, \mathbf{g}_1) \quad \text{and} \quad M_2 = (X_2, Y_2, Z_2, \mathbf{f}_2, \mathbf{g}_2).$$

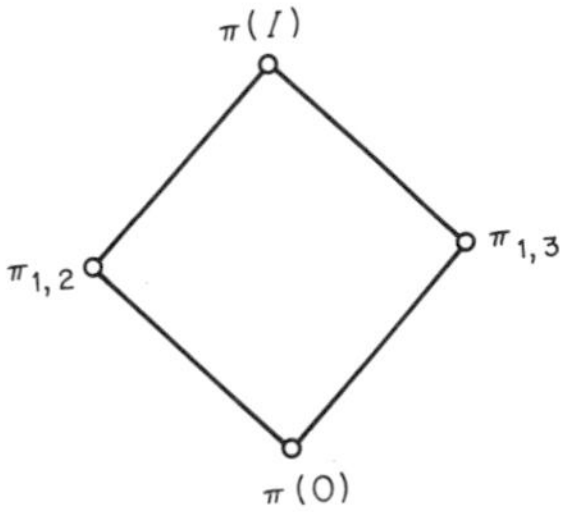

Figure 8.36 Lattice of all closed partitions of internal states for the machine specified in Example 8.10.

Then, $Y_1 = X_2$ and there exists a one-to-one mapping

$$\boldsymbol{\alpha}: Z \to Z_1 \times Z_2$$

The mapping $\boldsymbol{\alpha}$ induces a partition π of Z defined as follows:

$$\mathbf{a} = \mathbf{b}(\pi) \text{ if and only if } \mathbf{a}_1 = \mathbf{b}_1$$

where $\mathbf{a}, \mathbf{b} \in Z$, $\mathbf{a}_1, \mathbf{b}_1 \in Z_1$, $\boldsymbol{\alpha}(\mathbf{a}) = (\mathbf{a}_1, \mathbf{a}_2)$, and $\boldsymbol{\alpha}(\mathbf{b}) = (\mathbf{b}_1, \mathbf{b}_2)$.

It follows from the definition of serial composition that

$$\boldsymbol{\alpha}(\mathbf{g}(\mathbf{x}, \mathbf{a})) = (\mathbf{g}_1(\mathbf{x}, \mathbf{a}_1), \mathbf{g}_2(\mathbf{f}_1(\mathbf{x}, \mathbf{a}_1), \mathbf{a}_2))$$
$$\boldsymbol{\alpha}(\mathbf{g}(\mathbf{x}, \mathbf{b})) = (\mathbf{g}_1(\mathbf{x}, \mathbf{b}_1), \mathbf{g}_2(\mathbf{f}_1(\mathbf{x}, \mathbf{b}_1), \mathbf{b}_2))$$

If $\mathbf{a} = \mathbf{b}(\pi)$, then $\mathbf{a}_1 = \mathbf{b}_1$ and, thus, $\mathbf{g}_1(\mathbf{x}, \mathbf{a}_1) = \mathbf{g}_1(\mathbf{x}, \mathbf{b}_1)$. This means that $\mathbf{g}(\mathbf{x}, \mathbf{a}) = \mathbf{g}(\mathbf{x}, \mathbf{b})(\pi)$ so that π is a closed partition.

$\pi \neq \pi(0)$ because M_1 is required to have fewer states than M and $\pi \neq \pi(I)$ because M_2 is required to have fewer states than M. This completes the proof that the conditions required by the theorem are necessary.

To prove that the conditions are sufficient, let us assume that a nontrivial closed partition π of Z exists. Let π have k blocks and let the largest block contain m elements. Since π is not a trivial partition, $k < r$ and $m < r$, where r is the number of internal states of M. Let τ be a partition of Z which has m blocks and for which

$$\pi \cdot \tau = Z$$

Partition τ may but need not be closed.

Construction of τ: Label the states of each block of π by integers 1, 2, . . ., $n_i(n_i \leq m)$ and place all states with the same label in one block of τ.

Now, we shall consider two Mealy machines M_1 and M_2 which identify, respectively, blocks of π and τ. Since $\pi \cdot \tau = Z$, pairs of internal states of M_1 and M_2 (blocks of π and τ, respectively) are uniquely mapped onto the set Z of states of M. It remains to give such a specification of M_1 and M_2 that their serial composition will represent M. Let

$$M_1 = (X, X \times \pi, \pi, \mathbf{f}_1 = e, \mathbf{g}_\pi)$$

where e is the identity map (at any time, the response of M_1 is equal to the instantaneous stimulus and internal state of M_1), and $\mathbf{g}_\pi$ is the transition function derived from the transition function of M for the partition π. Let the input of M_2 be connected to the output of M_1 (serial composition) so that

$$M_2 = (X \times \pi, Y, \tau, \mathbf{f}_2 = \mathbf{f}, \mathbf{g}_2)$$

where $\mathbf{g}_2$ is a function $(X \times \pi) \times \tau \to \tau$ which is consistent with the function $\mathbf{g}$ with regard to $\boldsymbol{\alpha}$. ▌

The serial composition of a given machine M by two machines M_1 and M_2 as suggested in the proof of Theorem 8.10 is illustrated by the block diagram in Figure 8.37.

EXAMPLE 8.11 Determine a serial decomposition of the Mealy machine specified in Figure 8.35a.

Following the proof of Theorem 8.10, let $\pi = \pi_1 = \{\overline{1, 6}; \overline{2, 5}; \overline{3, 4}\}$ (Figure 8.35b). Then, $\tau = \{\overline{1, 2, 3}; \overline{4, 5, 6}\}$ and, using the symbols introduced in Figure 8.37, we obtain the specification of M_1 and M_2 as shown in Figure 8.35c. Let the first order finite-state model be used for both M_1 and M_2 and let states $\mathbf{z}_1^t$ and $\mathbf{z}_2^t$ be coded by logic variables v_1, v_2, v_3 as follows:

z_1^t	v_2^t	v_1^t
a	0	0
b	0	1
c	1	0

z_2^t	v_3^t
A	0
B	1

Then the behaviors of the functional generators of M_1 and M_2 are given by the following tables:

M_1

v_2^t	v_1^t	x^t	v_2^{t+1}	v_1^{t+1}
0	0	0	1	0
0	0	1	1	0
0	1	0	0	0
0	1	1	1	0
1	1	0	0	1
1	1	0	0	1

M_2

v_2^{t+1}	v_1^{t+1}	x^t	v_3^t	v_3^{t+1}	y^t
0	0	0	0	1	0
0	1	0	0	1	0
1	0	0	0	1	0
0	0	1	0	0	0
0	1	1	0	0	0
1	0	1	0	0	1
0	0	0	1	0	1
0	1	0	1	0	1
1	0	0	1	0	0
0	0	1	1	1	1
0	1	1	1	1	0
1	0	1	1	1	0

Minimal implementation based on AND, OR, NOT, and DELAY elements is shown in Figure 8.38.

Next we shall define the parallel composition and investigate some of its properties.

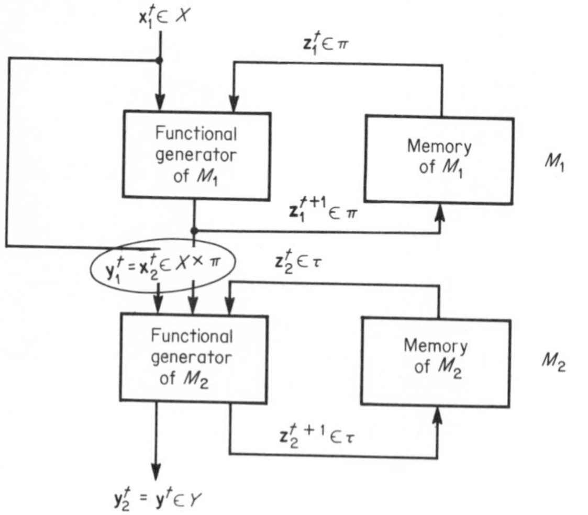

Figure 8.37 Block diagram of the serial composition of two Mealy machines.

DEFINITION 8.7 The *parallel composition of two Mealy machines* $M_1 = (X_1, Y_1, Z_1, \mathbf{f}_1, \mathbf{g}_1)$ and $M_2 = (X_2, Y_2, Z_2, \mathbf{f}_2, \mathbf{g}_2)$ is the machine $M = (X_1 \times X_2, Y_1 \times Y_2, Z_1 \times Z_2, \mathbf{f}, \mathbf{g})$, where

$$\mathbf{f}((\mathbf{x}_1, \mathbf{x}_2), (\mathbf{z}_1, \mathbf{z}_2)) = (\mathbf{f}_1(\mathbf{x}_1, \mathbf{z}_1), \mathbf{f}_2(\mathbf{x}_2, \mathbf{z}_2))$$

and

$$\mathbf{g}((\mathbf{x}_1, \mathbf{x}_2), (\mathbf{z}_1, \mathbf{z}_2)) = (\mathbf{g}_1(\mathbf{x}_1, \mathbf{z}_1), \mathbf{g}_2(\mathbf{x}_2, \mathbf{z}_2))$$

provided that $\mathbf{x}_1 \in X_1$, $\mathbf{x}_2 \in X_2$, $\mathbf{z}_1 \in Z_1$, $\mathbf{z}_2 \in Z_2$

THEOREM 8.11 The Mealy machine $M = (X, Y, Z, \mathbf{f}, \mathbf{g})$ has a nontrivial parallel decomposition if and only if there exist two nontrivial closed partitions, say π_1 and π_2, of Z such that $\pi_1 \cdot \pi_2 = Z$.

PROOF Assume that the machine M is implemented by a parallel composition of two machines M_1 and M_2. Then, there exists a one-to-one mapping

$$\boldsymbol{\alpha}: Z \to Z_1 \times Z_2$$

The mapping $\boldsymbol{\alpha}$ can be used to define two partitions, π_1 and π_2, as follows:

$$\mathbf{a} = \mathbf{b}(\pi_1) \text{ if and only if } \mathbf{a}_1 = \mathbf{b}_1$$

and

$$\mathbf{a} = \mathbf{b}(\pi_2) \text{ if and only if } \mathbf{a}_2 = \mathbf{b}_2$$

where

$$\boldsymbol{\alpha}(\mathbf{a}) = (\mathbf{a}_1, \mathbf{a}_2), \boldsymbol{\alpha}(\mathbf{b}) = (\mathbf{b}_1, \mathbf{b}_2), \mathbf{a}, \mathbf{b} \in Z, \mathbf{a}_1, \mathbf{b}_1 \in Z_1, \mathbf{a}_2, \mathbf{b}_2 \in Z_2$$

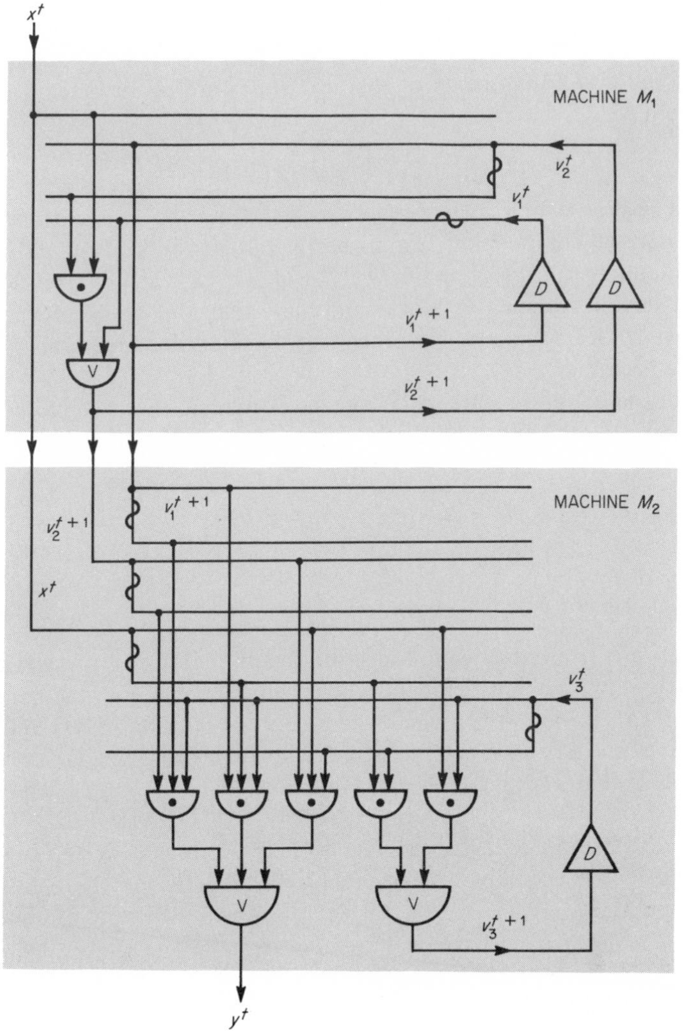

Figure 8.38 A serial decomposition of the Mealy machine specified in Figure 8.35a (Example 8.11).

Then, $\mathbf{a} = \mathbf{b}(\pi_1 \cdot \pi_2)$ implies $\mathbf{a} = \mathbf{b}$ and, thus, $\pi_1 \cdot \pi_2 = Z$.

If $\mathbf{a} = \mathbf{b}(\pi_1)$, then $\boldsymbol{\alpha}(\mathbf{a}) = (\mathbf{a}_1, \mathbf{a}_2)$ and $\boldsymbol{\alpha}(\mathbf{b}) = (\mathbf{a}_1, \mathbf{b}_2)$. But then

$$\boldsymbol{\alpha}(\mathbf{g}(\mathbf{x}, \mathbf{a})) = (\mathbf{g}_1(\mathbf{x}, \mathbf{a}_1), \mathbf{g}_2(\mathbf{x}, \mathbf{a}_2))$$
$$\boldsymbol{\alpha}(\mathbf{g}(\mathbf{x}, \mathbf{b})) = (\mathbf{g}_1(\mathbf{x}, \mathbf{a}_1), \mathbf{g}_2(\mathbf{x}, \mathbf{b}_2))$$

and, therefore, the first components of the next states are again indentical under $\boldsymbol{\alpha}$ and we have

$$\mathbf{g}(\mathbf{x}, \mathbf{a}) = \mathbf{g}(\mathbf{x}, \mathbf{b})(\pi_1)$$

The same could be shown for π_2. Hence, both π_1 and π_2 are closed partitions. Since M is a nontrivial composition, it must be $|Z_1| < |Z|$ and $|Z_2| < |Z|$. To satisfy the mapping $\boldsymbol{\alpha}$, it must be $|Z_1 \times Z_2| = |Z_1| \cdot |Z_2| \geq |Z|$. Hence, both π_1 and π_2 have less than $|Z|$ blocks and more than one block so that they are nontrivial. This completes the proof that the conditions required by the theorem are necessary.

To prove that the conditions are sufficient, let us assume that two closed partitions, π_1 and π_2, of Z exist and that $\pi_1 \cdot \pi_2 = Z$.
Let

$$M_1 = (X, Y_1 = X \times \pi_1, \pi_1, \mathbf{f}_1 = e, \mathbf{g}_{\pi_1})$$

and

$$M_2 = (X, Y_2 = X \times \pi_2, \pi_2, \mathbf{f}_2 = e, \mathbf{g}_{\pi_2})$$

where $\mathbf{g}_{\pi_1}$ and $\mathbf{g}_{\pi_2}$ are, respectively, the transition function derived from $\mathbf{g}$ for the partition π_1 and π_2, and e is the identity map. Since $\pi_1 \cdot \pi_2 = Z$, the pair $(\mathbf{y}_1, \mathbf{y}_2) \in Y_1 \times Y_2 = X \times \pi_1 \times \pi_2$, which represents the outputs of M_1 and M_2, identifies a unique pair in $X \times Z$. Hence, a function φ exists which maps $Y_1 \times Y_2$ onto Y so that M_1 and M_2 with φ implement M. ▌

The parallel composition of a given machine M by two machines M_1 and M_2 as suggested in the proof of Theorem 8.11 is illustrated by the block diagram in Figure 8.39.

EXAMPLE 8.12 Determine a parallel decomposition of the Mealy machine specified in Figure 8.35a.

Following the proof of Theorem 8.11, we can take $\pi_1 = \{\overline{1, 6}; \overline{2, 5}; \overline{3, 4}\}$ and $\pi_2 = \{\overline{1, 2, 3}; \overline{4, 5, 6}\}$ because $\pi_1 \cdot \pi_2 = \{\bar{1}; \bar{2}; \bar{3}; \bar{4}; \bar{5}; \bar{6}\}$. Using the symbols introduced in Figure 8.39, we obtain the specification of M_1 and M_2 with the mapping φ as shown in Figure 8.35d. When the first order finite-state model and the same coding of internal states of M_1 and M_2 is used as in Example 8.11, we obtain the final design shown in Figure 8.40.

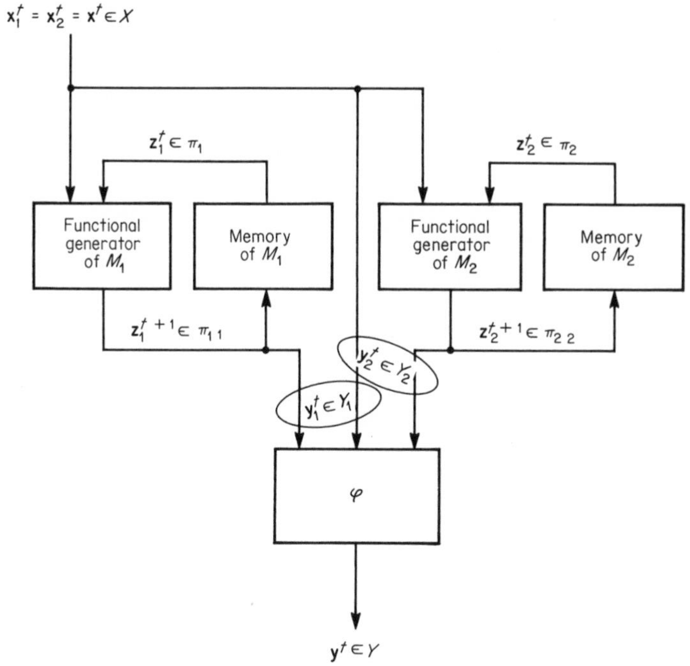

Figure 8.39 Block diagram of the parallel composition of two Mealy machines.

Observe that the parallel decomposition in Figure 8.40 is almost identical with the serial decomposition in Figure 8.38 in this case. More specifically, the only difference consists in variable y^t which depends on v_3^t in case of the serial decomposition but depends on v_3^{t+1} in case of the parallel decomposition. This strong similarity between the serial decomposition and the parallel decomposition is due to the fact that equal partitions were used in both the kinds of decomposition. In the general case, different pairs of partitions may be used. For some pairs of partitions, there exists the serial decomposition but not the parallel decomposition.

Exercises to Section 8.9

8.9-1 Determine the lattice of all closed partitions of internal states of the machines specified in

*(a) Figure 8.41.
(b) Figure 8.19.
(c) Figure 8.16a.
(d) Figure 8.8.

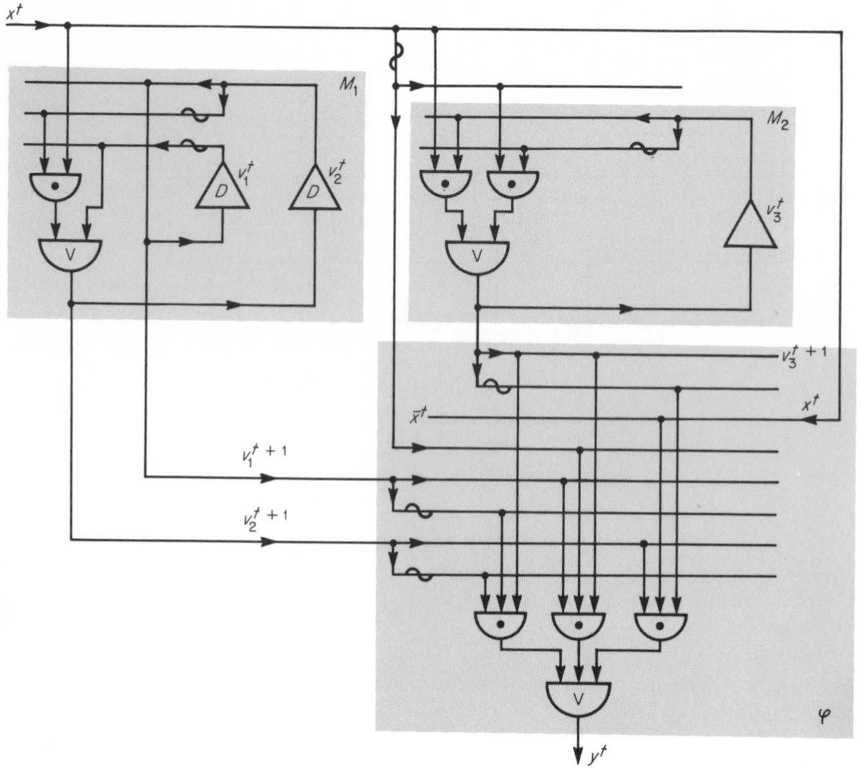

Figure 8.40 A parallel decomposition of the Mealy machine specified in Figure 8.35a (Example 8.12).

8.9-2 Design at least one nontrivial serial decomposition, provided that it exists, for each of the machines specified in Exercises 8.9-1.
Use NAND and DELAY elements and
 (a) the first-order finite-state model.
 (b) the first-order combined model.

8.9-3 Design at least one nontrivial parallel decomposition, provided that it exists, for each of the machines specified in Exercise 8.9-1. Use
 (a) the first-order finite-state model.
 (b) the first-order combined model.

*8.9-4 Suppose that a machines has k nontrivial closed partitions. How many nontrivial serial and parallel decompositions exist for this machine?

8.9-5 Prove that if π_1 and π_2 are closed partitions of internal states of a finite-state machine, then
 (a) $\pi_1 \cdot \pi_2$ is a closed partition.
 (b) $\pi_1 + \pi_2$ is a closed partition.

z^t \ x^t	z^{t+1}		y^t	
	0	1	0	1
1	4	3	0	0
2	6	3	1	1
3	5	2	0	1
4	2	5	1	0
5	1	4	1	1
6	3	4	0	0

(a)

z^t \ x^t	z^{t+1}		y^t	
	0	1	0	1
1	6	4	1	1
2	5	3	1	1
3	2	6	0	0
4	1	5	0	0
5	4	2	0	1
6	3	1	1	0

(b)

z^t \ x^t	z^{t+1}		y^t	
	0	1	0	1
1	2	3	0	0
2	3	4	0	0
3	4	3	0	1
4	5	2	1	0
5	6	1	1	1

(c)

z^t \ x^t	z^{t+1}		y^t	
	0	1	0	1
1	6	3	1	0
2	5	2	0	1
3	4	3	0	0
4	6	3	1	1
5	7	2	0	0
6	1	6	1	1
7	5	7	1	0

(d)

z^t \ x^t	z^{t+1}					y^t				
	a	*b*	*c*	*d*	*e*	*a*	*b*	*c*	*d*	*e*
1	2	1	5	8	3	0	0	0	1	1
2	1	2	6	7	4	0	0	1	0	1
3	4	3	6	6	1	0	1	0	0	1
4	3	4	5	5	2	1	0	0	0	1
5	5	6	3	4	7	1	0	0	1	0
6	6	5	4	3	8	1	0	1	0	0
7	7	8	4	2	5	1	1	0	0	0
8	8	7	3	1	6	0	1	1	0	0

(e)

Figure 8.41 Illustration to some exercises to Section 8.9.

8.9-6 Prove that the set of all closed partitions of internal states of a finite-state machine forms a lattice.

8.9-7 Generalize Theorem 8.11 for more than two partial machines connected in parallel.

8.10 STATE ASSIGNMENT

The *state assignment* for a finite-state machine is the coding of the abstract symbols describing the ST-structure of the machine into values of some logic variables representing these symbols in a switching circuit which implements the machine. When we want to find a state assignment for a given machine that produces, under specified circumstances, the sequential circuit that best satisfies some criteria of goodness, a problem arises which is referred to as the *state-assignment problem.*

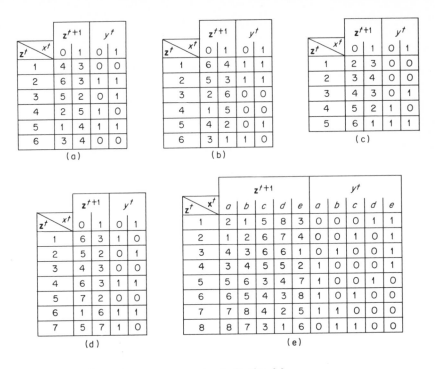

Although the state-assignment problem may concern states of input, output, and internal variables, it is usually assumed that states of both input and output variables are fixed by the environment with which the input and output terminals communicate. As a rule, it is the matter of various macro-level design considerations to select an appropriate code for input and output states. At the level of the synthesis of switching circuits (or the micro-level design), the state-assignment problem is thus limited to selecting values of internal variables only.

The state-assignment problem depends on the model used. For instance, if the finite-memory model is used, the state-assignment problem does not exist at all. The combined model makes the state assignment applicable only to some of the variables stored in the memory; the other variables in the memory represent past values of the input variables and past or future values of the output variables and are thus fixed. Similarly, the kth order finite-state model ($k > 1$) reduces the state assignment problem to a subset of the variables associated with the memory when compared with the first-order finite-state model. Observe also that the state assignment for the finite-state models has a different meaning than the state asignment for the combined models. The former consists in coding internal states while the latter is associated with coding of input-output sequences.

There is a considerable difference between the state assignment problem for synchronous and asynchronous switching circuits. In case of the synchronous circuits, the primary objective in the selection of a state assignment is, usually, that the total cost of the circuit reaches its minimum for a given set of elements and some specified circuit constraints. Clearly, the available elements and the circuit constraints have a great influence upon the state assignment problem. What may be a good assignment for one set of elements and constraints may be very poor for another set.

Thus, the method for solving the state assignment problem depends on the objective criteria, elements used, and required constraints. The number of possibilities grows rapidly with the number of different objective criteria, sets of elements, and sets of constraints and it becomes practically impossible to investigate all of them. For this reason, the state assignment problem for synchronous circuits has been mostly studied under the assumption that the functional generator is a standard two stage circuit (AND-OR, NAND, NOR, etc.), the memory is built by DELAY elements, and the objective function is proportional to the total cost of the circuit.

In case of the asynchronous switching circuits, the state assignment must satisfy certain additional properties which ensure the proper functioning of the circuit. This aspect is discussed in Chapter 9.

To get an idea about the growth of the number of different state assignments with the number of internal states, let us consider the first-order finite-state model. Assume that the functional generator is implemented by a standard two stage circuit and the memory contains only DELAY elements. Assume further that the number of nonequivalent internal states is N and the number of logic variables representing internal states is $n = \lceil \log_2 N$. Then, the state assignment involves the assignment of 2^n possible states of the logic variables to N states of the respective finite-state machine ($N \leq 2^n$). This can be done in $2^n!/(2^n - N)!N!$ ways, not including reordering. Each selected subset of N states of the logic variables can be reordered in $N!$ ways. Hence, the total number of ways of assigning 2^n states of variables to N states of the machine is equal to $2^n!N!/(2^n - N)!N! = 2^n!/(2^n - N)!$. No permutation of the variables involved changes the structure of the switching circuit except for relabelling internal inputs and outputs of the functional generator. There are $n!$ permutations of n variables. Hence, the number A of different state assignments is given by the formula

$$A = \frac{2^n!}{(2^n - N)!n!}$$

Values of A for $2 \leq N \leq 9$ are given in Table 8.5. We see that an exhaustive evaluation of all state assignments is impractical for more than two internal states and impossible for more than four internal states of the finite-state machine.

TABLE 8.5. The Number A of Different State Assignments for N Internal States and n Internal Variables

N	$n = \lceil \log_2 N$	A
2	1	2
3	2	12
4	2	12
5	3	1,120
6	3	3,360
7	3	6,720
8	3	6,720
9	4	172,972,800

Various approaches to the state assignment problem, which try to avoid the evaluation of all possible state assignments, have been suggested in the literature. A simple approach consists in an application of the following two rules:

Rule 1. 2^k of internal states ($k = 1, 2, \ldots, n-1$) that are mapped for the same stimulus into the same next state should be made a k-dimensional cell.

Rule 2. If an internal state is mapped into two next states for adjacent stimuli, then these two next states should be made adjacent.

The rules, which are illustrated in Figure 8.42, represent simple heuristic guidelines to the state-assignment problem rather than an algorithmic procedure. Although they do not guarantee the best solution of the state assignment problem, their application, if possible, produces fairly good state assignments.

If Rule 1 can be applied, then, clearly, 2^k elements of the transition function (rows in its truth table) can be implemented by a single AND gate with 2^k inputs. If Rule 2 is applied, then two states are represented by a single AND gate with $n - 1$ inputs for all functions associated with the next state variables except one.

The application of Rule 1 is frequently contradictory to the application of Rule 2. In such a case, Rule 1 takes precedence over Rule 2 because it provides a more significant simplification of the respective minimal Boolean forms representing variables of the next state.

EXAMPLE 8.13 Using Rule 1 and Rule 2, determine a state assignment for the first-order finite-state model representing the Mealy machine specified in Figure 8.43a and compare it with another state assignment in which the application of the rules is eliminated as much as possible.

Rule 1 can be applied if a state is used 2^k times as an entry in a single column of the next state portion of the ST-table. In our example, state 3 appears twice in the first column representing $\mathbf{z}^{t+1}$, namely, as the next state of states 1 and 2. Thus, states 1 and 2 should be adjacent (should represent a 1-cell) by Rule 1. No other application of Rule 1 exists in this example.

Rule 2 can be applied if there are two different entries in a row and two columns representing adjacent stimuli in the next state portion of the ST-table. In our example, this is satisfied in the first row. Since the entries (next states of state 1) are 3 and 2, states 3 and 2 should be adjacent by Rule 2. Similarly, we find from the second row and the fourth row that states 3 and 4 should be adjacent. No other application of Rule 2 exists in this example.

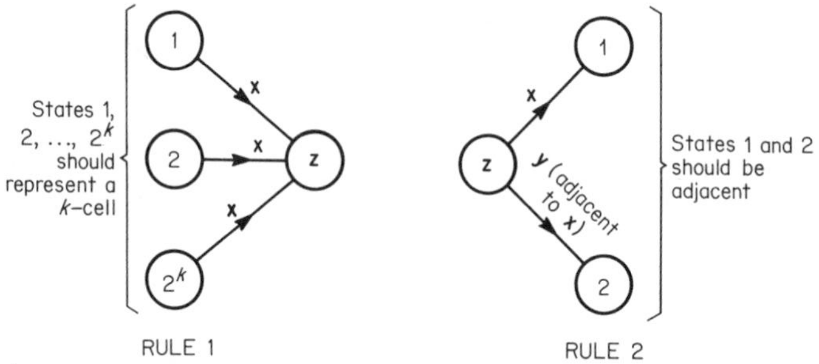

Figure 8.42 Illustration of the two heuristic rules applied in solving the state assignment problem.

z^t \ x^t	z^{t+1}: 0	z^{t+1}: 1	y^t: 0	y^t: 1
1	3	2	0	0
2	3	4	0	1
3	1	1	1	0
4	4	3	1	1

(a)

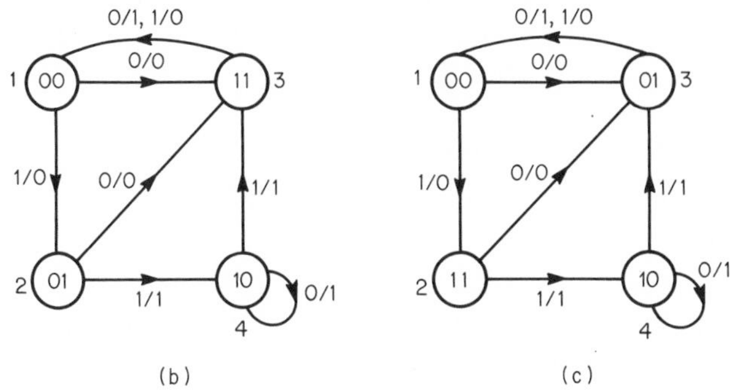

Figure 8.43 Comparison of two different state assignments for the same Mealy machine (Example 8.13).

Thus, having applied the rules, we got the information that pairs of states (1, 2), (2, 3), and (3, 4) should be adjacent. These three requirements are not contradictory and can all be satisfied, e.g., by the following state assignment:

Internal States	State Variables z_2	z_1
1	0	0
2	0	1
3	1	1
4	1	0

This state assignment is represented by the ST-diagram in Figure 8.43b, where states are written in the form $z_2 z_1$. We obtain the following minimal Boolean forms for the next state variables and the output variable:

$$z_2^{t+1} = z_2^t \bar{z}_1^t \vee \bar{z}_2^t z_1^t$$
$$z_1^{t+1} = \bar{z}_1^t x^t \vee \bar{z}_2^t z_1^t \bar{x}^t$$
$$y^t = z_2^t \bar{z}_1^t \vee z_2^t \bar{x}^t \vee \bar{z}_2^t z_1^t x^t$$

This is a unique set of minimal Boolean forms which can be implemented by six AND gates and three OR gates of not more than three inputs each (total number of inputs to the gates is twenty-one), and three NOT elements.

Now let us consider another state assignment which does not satisfy Rule 1 (states 1 and 2 are not adjacent) and satisfies only partially Rule 2 (either states 2, 3 or states 3, 4 are not adjacent), e.g.,

Internal States	State Variables z_2	z_1
1	0	0
2	1	1
3	0	1
4	1	0

This state assignment is represented by the *ST*-diagram in Figure 8.43c and produces the following minimal Boolean forms:

$$z_2^{t+1} = z_2^t \bar{z}_1^t \bar{x}^t \vee \bar{z}_2^t \bar{z}_1^t x^t \vee z_2^t z_1^t x^t$$
$$z_1^{t+1} = \bar{z}_2^t \bar{z}_1^t \vee \bar{z}_1^t x^t \vee z_2^t z_1^t \bar{x}^t$$
$$y^t = z_2^t \bar{z}_1^t \vee z_2^t x^t \vee \bar{z}_2^t z_1^t \bar{x}^t$$

This is again a unique set of minimal Boolean forms. It can be implemented by nine AND gates, three OR gates (total number of inputs to the gates is thirty-two), and three NOT elements. Thus, an application of Rule 1 and Rule 2 helped us to save three AND gates and eleven pins (diodes).

Let us introduce now a *general Boolean representation* of all possible state assignments of a given Mealy machine. Let z_{ij} denote the value of internal variable z_i in state j and let

$$z_i^{\bar{z}_{ij}} = \begin{cases} \bar{z}_i \text{ if } z_{ij} = 0 \\ z_i \text{ if } z_{ij} = 1 \end{cases}$$

Then, each output variable can be expressed by a Boolean form containing the input variables of the machine and the variables $z_i^{\bar{z}_{ij}}$. Similarly, each next state internal variable z_i' can be expressed by a Boolean form

$$z_i' = \bigvee_{j=1}^{N} z_{ij} f_j$$

where f_j are *sp* Boolean expressions containing the input variables of the machine and the variables $z_k^{\bar{z}_{lm}}$ $(k, l = 1, 2, \ldots, n; m = 1, 2, \ldots, N)$.

EXAMPLE 8.14 Determine a general Boolean representation of all possible state assignments of the machine specified in Figure 8.43a.

Let us use the following symbols for the values of the internal state variables in individual states:

Internal States	State Variables z_2	z_1
1	$z_{21} = a$	$z_{11} = b$
2	$z_{22} = c$	$z_{12} = d$
3	$z_{23} = e$	$z_{13} = f$
4	$z_{24} = g$	$z_{14} = h$

Then, the truth table of the functional generator has the following general form:

x	z_2	z_1	y	z_2'	z_1'
0	$z_{21} = a$	$z_{11} = b$	0	$z_{23} = e$	$z_{13} = f$
1	$z_{21} = a$	$z_{11} = b$	0	$z_{22} = c$	$z_{12} = d$
0	$z_{22} = c$	$z_{12} = d$	0	$z_{23} = e$	$z_{13} = f$
1	$z_{22} = c$	$z_{12} = d$	1	$z_{24} = g$	$z_{14} = h$
0	$z_{23} = e$	$z_{13} = f$	1	$z_{21} = a$	$z_{11} = b$
1	$z_{23} = e$	$z_{13} = f$	0	$z_{21} = a$	$z_{11} = b$
0	$z_{24} = g$	$z_{14} = h$	1	$z_{24} = g$	$z_{14} = h$
1	$z_{24} = g$	$z_{14} = h$	1	$z_{23} = e$	$z_{13} = f$

The general Boolean representation of state assignments can be easily determined from the table and has the form

$$y = xz_2^c z_1^d \vee \bar{x}z_2^e z_1^f \vee z_2^g z_1^h$$

$$z_i' = z_{i1}z_2^e z_1^f \vee z_{i2}xz_2^a z_1^b \vee z_{i3}(\bar{x}z_2^a z_1^b \vee \bar{x}z_2^c z_1^d \vee xz_2^g z_1^h) \vee z_{i4}(xz_2^c z_1^d \vee \bar{x}z_2^g z_1^h)$$

where $i = 1, 2$.

After the general Boolean representation of state assignments for a given machine is determined, we can take advantage of the displayed forms to select values of the z_{ij}'s in order to simplify the forms as much as possible. Clearly, the values cannot be selected arbitrarily but the Boolean equations

$$\bigvee_{i=1}^{m} (z_{ij}\bar{z}_{ik} \vee \bar{z}_{ij} z_{ik}) = 1$$

for all pairs $j \neq k$ must be satisfied to ensure that each internal state will be coded uniquely.

Although they do not guarantee reaching the simplest forms, the following two rules are helpful to obtain a fairly good state assignment:

Rule I. Those z_{ij}'s which have the most complex factors in the general Boolean representation of state assignments should be set equal to 0.

Rule II. Such values should be assigned to the z_{ij}'s for which the nonzero terms in the general Boolean representation can be most effectively combined.

EXAMPLE 8.15 Using Rule I and Rule II, determine a state assignment for the machine specified in Figure 8.43a.

The general Boolean representation of state assignments for this machine was determined in Example 8.14. When consulting this representation we find that $z_{i3}(i = 1, 2)$ has the most complex factor. Hence, $z_{13} = f = 0$ and $z_{23} = e = 0$ by Rule I and we obtain

$$y = xz_2^c z_1^d \vee \bar{x}\bar{z}_2 \bar{z}_1 \vee z_2^g z_1^h$$
$$z_i' = z_{i1} \bar{z}_2 \bar{z}_1 \vee z_{i2} x z_2^a z_1^b \vee z_{i4}(xz_2^c z_1^d \vee \bar{x} z_2^g z_1^h)$$

Applying again Rule I, we put $z_{24} = g = 0$ because z_{i4} has the most complex factor. To distinguish state 4 from state 3, we must put $z_{14} = b = 1$. We obtain

$$y = xz_2^c z_1^d \vee \bar{x}\bar{z}_2 \bar{z}_1 \vee \bar{z}_2 z_1$$
$$z_2' = z_{21} \bar{z}_2 \bar{z}_1 \vee z_{22} x z_2^a z_1^b$$
$$z_1' = z_{11} \bar{z}_2 \bar{z}_1 \vee z_{12} x z_2^a z_1^b \vee xz_2^c z_1^d \vee \bar{x}\bar{z}_2 z_1$$

It remains to determine values of pairs $(z_{12}, z_{11}) = (a, b)$ and $(z_{21}, z_{22}) = (c, d)$. These values can be (1, 0) for one of the pairs and (1, 1) for the other; other pairs of values have been already assigned. This means, essentially, that either $z_{11} = b = 0$ or $z_{12} = d = 0$. Applying Rule II, we choose $z_{12} = d = 0$ so that $z_{11} = b = 1$, $z_{21} = d = 1$, and $z_{22} = c = 1$. The final state assignment is

Internal States	State Variables z_2	z_1
1	1	1
2	1	0
3	0	0
4	0	1

It produces the following unique minimal *sp* Boolean forms:

$$y = xz_2\bar{z}_1 \vee \bar{x}\bar{z}_2 \vee \bar{z}_2 z_1$$

$$z_2' = \bar{z}_2\bar{z}_1 \vee xz_2 z_1$$

$$z_1' = \bar{x}\bar{z}_2 \vee x\bar{z}_1$$

We see that we obtained a state assignment which, though different from the state assignment determined in Example 8.13 by the previous method, leads to a circuit with the same complexity (six AND gates, three OR gates, twenty-one pins).

The method of the general Boolean representation of state assignments is not an algorithmic procedure and does not guarantee that we obtain a state assignment representing the simplest two-stage switching circuit. Nevertheless, it guides the designer to determine a fairly good state assignment. Its advantage is that it can be used for any model (*k*th order finite state, combined).

Now, it will be shown that the state assignment problem may be viewed as a *partitioning problem.* Clearly, each internal state variable z_i induces a partition π_i with two blocks on the set of internal states of a given machine. States with $z_i = 0$ identify one block of π_i, states with $z_i = 1$ identify the other block. If the state assignment is proper, then the product of all partitions is equal to the set of internal states (zero partition).

Let π and ρ be two partitions of the Z of internal states of a finite-state machine such that $\pi \cdot \rho = Z$. Let π and ρ contain, respectively, $|\pi|$ and $|\rho|$ blocks. Then, $p = \ulcorner\log_2|\pi|$ of logic variables are needed to distinguish blocks of π and $r = \ulcorner\log_2|\rho|$ logic variables are needed to distinguish blocks of ρ. These two sets of variables may be mutually independent or, at least, one of them may be independent of the other. Such an independence has a favorable influence upon the complexity of that portion of the switching circuit which implements the transition function.

We shall now show that the concept of closed partitions, which was introduced in Section 8.9 and is employed in various types of machine decomposition, is intrinsically related to the above-mentioned independence among variables representing internal states.

THEOREM 8.12 Let M be a Mealy machine with n internal state variables $z_1, z_2, \ldots, z_n$. Then, some of the next state variables, say $z_1', z_2', \ldots, z_m'$ $(m < n)$, depends only on the input variables and variables $z_1, z_2, \ldots, z_m$ (do not depend on the remaining $n - m$ internal state variables) if and only if there exists a closed partition π of the internal states of M such that internal states $\mathbf{z}_i$ and $\mathbf{z}_j$ are in the same block of π if and only if values of variables $z_1, z_2, \ldots, z_m$ assigned to $\mathbf{z}_i$ and $\mathbf{z}_j$ are equal.

PROOF Assume that a closed partition of the internal states of M exists and all states in each block of π are assigned the same values of the variables $z_1, z_2, \ldots, z_m$. Then, the present block of π and the present input state are sufficient to determine the next block of π. This means that values of $z_1', z_2', \ldots, z_m'$ are uniquely determined by values of $z_1, z_2, \ldots, z_m$ and values of input variables.

To prove the converse, let a partition π of internal state of M be formed such that all states which are assigned the same values of $z_1, z_2, \ldots, z_m$ are in the same block of π. Let states $\mathbf{z}_i$ and $\mathbf{z}_j$ belong to the same block of π. Then $\mathbf{z}_i$ and $\mathbf{z}_j$ are represented by the same values of variables $z_1, z_2, \ldots, z_m$. Let the corresponding next state variables $z_1', z_2', \ldots, z_m'$, be dependent only on the variables $z_1, z_2, \ldots, z_m$ and the input variables of M. Then, the successors of $\mathbf{z}_i$ and $\mathbf{z}_j$ for every state of input variables are assigned to the same values of $z_1', z_2', \ldots, z_m'$. Hence, the successors belong to the same block of π so that π is a closed partition. ▌

Observe that the independence between subsets of internal state variables (in the sense of Theorem 8.12) reflects certain decomposition of the machine and vice versa. The decomposition theory, some aspects of which are exposed in Section 8.9, finds thus an important application in the solution of the state assignment problem.

Let π and ρ be two closed partitions of the set of internal states of a finite-state machine M such that their product is the set of internal states. Let blocks of π and ρ be coded, respectively, by internal variables $z_1, z_2, \ldots, z_m$ and $z_{m+1}, z_{m+2}, \ldots, z_n$, where

$$m = \ulcorner \log_2 |\pi|$$

and

$$n - m = \ulcorner \log_2 |\rho|$$

Then, it follows directly from Theorem 8.12 that the next state variables $z_1', z_2', \ldots, z_m'$ and $z_{m+1}', z_{m+2}', \ldots, z_n'$ are independent of variables $z_{m+1}, z_{m+2}, \ldots, z_n$ and $z_1, z_2, \ldots, z_m$ respectively. This property can be easily generalized for three or more closed partitions whose product is the set of internal states.

Note that the application of closed partitions does not specify completely the state assignment for the given machine. It only prescribes certain constraints for the variables z_{ij} involved in the general Boolean representation of state assignments for a given machine. This reduces the number of acceptable state assignments. Other means must be used to specify further which one of the remaining assignments is actually used.

EXAMPLE 8.16 Employing, if possible, an independence among internal state variables, determine a state assignment for the first-order finite-state model of the Mealy machine specified in Figure 8.44.

It is easy to find closed partitions $\pi_1 = \{\overline{1, 3}; \overline{2, 4}\}$ and $\pi_2 = \{\overline{1, 4}; \overline{2, 3}\}$ for the machine. Since $|\pi_1| = |\pi_2| = 2$ and $\pi_1 \cdot \pi_2 = \{\bar{1}; \bar{2}; \bar{3}; \bar{4}\}$, blocks of each of the paritions can be coded by a single variable and these two variables are mutually independent.

Let z_1 and z_2 represent blocks of π_1 and π_2 respectively and let the following symbols be used for the values of variables z_1, z_2 in individual states:

Internal States	State Variables z_2	z_1
1	$z_{21} = a$	$z_{11} = b$
2	$z_{22} = c$	$z_{12} = d$
3	$z_{23} = e$	$z_{13} = f$
4	$z_{24} = g$	$z_{14} = h$

Then, $z_{11} = z_{13} \neq z_{12} = z_{14}$ due to π_1; similarly $z_{21} = z_{24} \neq z_{22} = z_{23}$ due to π_2. The table of values of z_1 and z_2 can be simplified as follows:

Internal States	State Variables z_2	z_1
1	a	b
2	$\bar{a}$	$\bar{b}$
3	$\bar{a}$	b
4	a	$\bar{b}$

z \ x	z' 0	z' 1	y 0	y 1
1	2	4	1	1
2	1	3	1	0
3	4	2	0	1
4	3	1	0	0

Figure 8.44 Illustration to Example 8.16.

The truth table of the functional generator has the general form

x	z_2	z_1	y	z_2'	z_1'
0	a	b	1	$\bar{a}$	$\bar{b}$
1	a	b	1	a	$\bar{b}$
0	$\bar{a}$	$\bar{b}$	1	a	b
1	$\bar{a}$	$\bar{b}$	0	$\bar{a}$	b
0	$\bar{a}$	b	0	a	$\bar{b}$
1	$\bar{a}$	b	1	$\bar{a}$	$\bar{b}$
0	a	$\bar{b}$	0	$\bar{a}$	b
1	a	$\bar{b}$	0	a	b

from which we determine the general Boolean representation of state assignments as follows:

$$
\begin{aligned}
y &= z_2^a z_1^b \vee \bar{x} z_2^{\bar{a}} z_1^{\bar{b}} \vee x z_2^{\bar{a}} z_1^b \\
z_2' &= \bar{a}(\bar{x} z_2^a \vee x z_2^{\bar{a}}) \vee a(\bar{x} z_2^{\bar{a}} \vee x z_2^a) \\
z_1' &= \bar{b} z_1^b \vee b z_1^{\bar{b}}
\end{aligned}
$$

We see that z_2' and z_1' are independent of z_1 and z_2, respectively, as originally intended. We also see that the selection of values for a and b does not have any influence on the final complexity of the Boolean forms, i.e., all of the four possible assignments are equally good. Let us choose $a = b = 0$. Then the functional generator will implement the following Boolean forms:

$$
\begin{aligned}
y &= \bar{z}_2 \bar{z}_1 \vee \bar{x} z_2 z_1 \vee x z_2 \bar{z}_1 \\
z_2' &= \bar{x} \bar{z}_2 \vee x z_2 \\
z_1' &= \bar{z}_1
\end{aligned}
$$

So far we have investigated the possibility of reducing the dependence between variables representing internal states. To show that the freedom which exists in assigning values of internal variables to blocks of the used partitions can frequently be employed to reduce the dependence of the output variables on the internal variables, we shall define a new type of partition first.

DEFINITION 8.8 A partition π of the internal states of a finite-state machine M is said to be *output-consistent* if all internal states contained in a single block of π generate the same values of all output variables for every input state.

Using a slight modification of the proof of Theorem 8.12, we could show that the existence of an output-consistent partition π of the internal states of a finite-state machine guarantees that there exists a state assignment for M such that each of the output variables depends at most on the input variables and the internal variables by which the blocks of π are coded.

EXAMPLE 8.17 The machine specified in Figure 8.45 has a nontrivial output-consistent partition $\pi_1 = \{\overline{1, 5}; \overline{2}; \overline{3, 4}; \overline{6}\}$. It has also a closed partition $\pi_2 = \{\overline{1, 2, 3}; \overline{4, 5, 6}\}$. Since $\pi_1 \cdot \pi_2 = \{\overline{1}; \overline{2}; \overline{3}; \overline{4}; \overline{5}; \overline{6}\}$, the partitions can be adopted for a state assignment with reduced dependencies among the variables involved. π_1 can be coded by two variables, say z_1 and z_2, and only one variable, say z_3, is needed to code the blocks of π_2. Let us use the following coding of blocks of π_1 and π_2:

Blocks of π_1	State Variables	
	z_2	z_1
1, 5	0	0
2	0	1
3, 4	1	0
6	1	1

Blocks of π_2	State Variables z_3
1, 2, 3	0
4, 5, 6	1

Then, we obtain the state assignment

Internal State	State Variables		
	z_3	z_2	z_1
1	0	0	0
2	0	0	1
3	0	1	0
4	1	1	0
5	1	0	0
6	1	1	1

	z'		y	
z \ x	0	1	0	1
1	2	6	0	0
2	3	4	0	1
3	1	5	1	1
4	6	3	1	1
5	4	3	0	0
6	4	1	1	0

Figure 8.45 Mealy machine with a nontrivial output-consistent partition (Example 8.17).

and the following minimal *sp* Boolean forms describing the functional generator:

$$y = z_2\bar{z}_1 \vee \bar{x}z_2 \vee x\bar{z}_2 z_1$$

$$z_3' = \bar{x}z_3 \vee x\bar{z}_3$$

$$z_2' = z_3\bar{z}_1 \vee \bar{z}_2 z_1 \vee x\bar{z}_2 \vee \bar{x}z_3$$

$$z_1' = \bar{z}_3\bar{z}_2\bar{z}_1 \vee \bar{x}z_3 z_2\bar{z}_1$$

Note that these results agree with our expectations: y is independent of z_3 and z_3 is independent of z_1 and z_2.

The requirement of reducing the dependencies of the output variables (the application of an output-consistent partition) frequently conflicts with the requirement of reducing the dependencies of the internal variables (the application of a closed partition). No practical method to resolve such a conflict is available at the present time.

We have seen that closed partitions are quite helpful in solving the state-assignment problem. However, only a small fraction of finite-state machines possess nontrivial closed partitions. For this reason, the application of closed partitions is considerably limited.

Closed partitions, if they exist, enable us to reduce dependence among the variables involved to a self-dependence among individual subsets of these variables. A more general type of reduced dependence, which employs the concepts of a partition pair, will be discussed next.

DEFINITION 8.9 A partition pair (π, ρ) on the set Z of the internal states of a finite-state machine M is an *ordered pair of partitions* of Z such that

$$\mathbf{z}_i = \mathbf{z}_j(\pi)$$

implies

$$\mathbf{g}(\mathbf{x}, \mathbf{z}_i) = \mathbf{g}(\mathbf{x}, \mathbf{z}_j)(\rho)$$

for all stimuli $\mathbf{x}$.

We see that (π, ρ) is a partition pair of M if and only if the blocks of π are mapped into the blocks of ρ by the transition function of M. We also see that a partition π is closed if and only if (π, π) is a partition pair.

The application of partition pairs in the state assignment problem is summarized in the following theorem.

THEOREM 8.13 Let M be a finite-state machine whose internal states are represented by the variables $z_1, z_2, \ldots, z_n$ and let $A \subseteq \{1, 2, \ldots, k\}$ and

$B \subseteq \{1, 2, \ldots, k\}$. Let π_A and π_B denote, respectively, partitions whose blocks are coded by variables $\{z_i: i \in A\}$ and $\{z_j: j \in B\}$. Then, the next state variables $\{z'_j: j \in B\}$ depend only on the input variables and the internal variables $\{z_i: i \in A\}$ if and only if (π_A, π_B) is a partition pair.

PROOF Left to the reader as an exercise. ▮

EXAMPLE 8.18 The machine specified in Figure 8.46 has neither a closed partition nor an output-consistent partition. However, it has a partition pair (π_A, π_B), where $\pi_A = \{\overline{1, 2}; \overline{3, 4}\}$ and $\pi_B = \{\overline{1, 4}; \overline{2, 3}\}$. Let $A = \{1\}$ and $B = \{2\}$, i.e., blocks of π_A and π_B are coded, respectively, by variables z_1 and z_2. Let us use the following coding:

π_B	z_2
1, 4	0
2, 3	1

π_A	z_1
1, 2	0
3, 4	1

Then, we get the state assignment

Internal state	State variables z_2	z_1
1	0	0
2	1	0
3	1	1
4	0	1

and the following minimal *sp* Boolean forms representing the functional generator:

$$y = \bar{x}z_2 \vee z_2 z_1 \vee x\bar{z}_2\bar{z}_1$$

$$z'_2 = \bar{x}\bar{z}_1 \vee x z_1$$

$$z'_1 = \bar{x}z_2 \vee \bar{x}z_1 \vee z_2 z_1 \vee x\bar{z}_2\bar{z}_1$$

Note that, in agreement with out intention, z'_2 is independent of z_2.

z \ x	z′: 0	z′: 1	y: 0	y: 1
1	2	4	0	1
2	3	1	1	0
3	4	3	1	1
4	4	2	0	0

Figure 8.46 Mealy machine with a partition pair (Example 8.18).

Exercises to Section 8.10

*8.10-1 For each of the machines specified in Figure 8.47, determine an economical state assignment based on reduced dependency of the next state variables and/or the output variables on the present state variables.

8.10-2 Prove that if π and ρ are two output-consistent partitions of the set of internal states of a finite-state machine, then $\pi + \rho$ and $\pi \cdot \rho$ are also output-consistent partitions.

8.10-3 Prove that if a closed partition π of the set of internal states of a finite-state machine M is also an output-consistent partition, then M can be reduced to an equivalent machine which has $|\pi|$ internal states.

z \ x	z' 0	z' 1	y 0	y 1
1	2	2	0	1
2	3	1	0	1
3	2	4	1	0
4	3	3	0	1

(a)

z \ x	z' 0	z' 1	y 0	y 1
1	3	5	0	1
2	4	5	0	0
3	2	5	0	0
4	2	5	1	0
5	1	3	1	0

(b)

z \ x	z' a	z' b	z' c	z' d	y a	y b	y c	y d
1	1	4	2	2	0	1	1	0
2	3	5	3	2	1	1	1	0
3	1	6	3	1	1	1	1	0
4	1	5	4	1	0	1	1	0
5	2	5	6	2	0	0	1	0
6	1	4	6	3	0	0	0	1

(c)

z \ x	z' 0	z' 1	y 0	y 1
1	7	2	0	0
2	5	6	0	1
3	2	6	0	0
4	4	1	0	0
5	3	8	0	0
6	2	8	0	0
7	4	3	0	0
8	3	5	1	0

(d)

z \ x	z' a	z' b	z' c	z' d	y a	y b	y c	y d
1	2	3	1	1	A	B	C	C
2	2	4	3	2	A	B	B	C
3	4	3	3	1	A	B	B	C
4	4	4	1	2	C	B	C	A

(e)

z \ x	z' 0	z' 1	y 0	y 1
1	4	2	1	1
2	4	2	0	1
3	1	3	0	0
4	3	1	1	0

(f)

Figure 8.47 Illustration for Exercise 8.10-1.

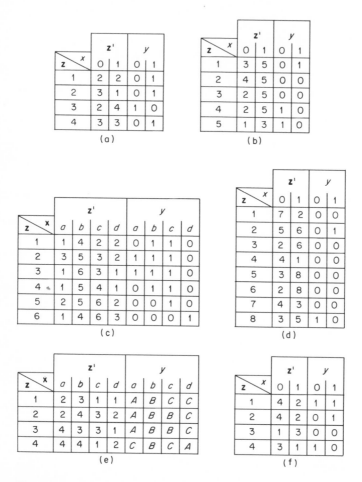

8.10-4 Prove that if no closed partition exists which is also an output-consistent partition of the set of internal states of a finite-state machine M, then M is represented by its minimal form.

8.10-5 Prove Theorem 8.13.

8.10-6 Prove that if (π_1, ρ_1) and (π_2, ρ_2) are partition pairs, then
(a) $(\pi_1 \cdot \pi_2, \rho_1 \cdot \rho_2)$ is a partition pair.
(b) $(\pi_1 + \pi_2, \rho_1 + \rho_2)$ is a partition pair.

8.10-7 Suggest a procedure for determining all possible partition pairs for a given machine.

8.10-8 If (π_1, ρ_1) and (π_2, ρ_2) are partition pairs, we define the partial ordering of partition pairs as $(\pi_1, \rho_1) \leq (\pi_2, \rho_2)$ if and only if $\pi_1 \leq \pi_2$ and $\rho_1 \leq \rho_2$. Prove that the set of partition pairs for a given machine forms a lattice with respect to the partial ordering of partition pairs.

Comprehensive Exercises to Chapter 8

8.1 Using NAND and DELAY elements, design the serial binary adder on the basis of
(a) the first-order finite-state model.
(b) the first-order combined model (or the finite-memory model, if possible).

8.2 Modify the solution of Exercise 8.1 for the SR FLIP-FLOPS.

8.3 Use the cascade method to modify the solution of Exercise 8.2.

*8.4 Determine an activity matrix which contains all information concerning the performance of the serial binary adder.

8.5 Using the Svoboda method (Section 8.7) and the activity matrix determined in Exercise 8.4, design the serial binary adder which is built by:
(a) WOS and DELAY elements.
(b) WOS and TRIGGER elements.
(c) WOS and SR FLIP-FLOP elements.
(d) NOR and INVERTED DELAY elements.
(e) a cascade circuit and DELAY elements.

8.6 Determine an activity matrix which contains all information concerning the performance of the SR FLIP-FLOP; use this activity matrix to design a switching circuit which implements the SR FLIP-FLOP by:
*(a) NAND and DELAY elements.
(b) WOS and DELAY elements.
(c) WOS and TRIGGER elements.

8.7 Discuss the relationship between the problem of model selection, the state-assignment problem, and the decomposition theory for sequential switching circuits.

8.8 Design a serial adder which adds simultaneously three binary numbers and is built by:
(a) NAND and DELAY elements.
(b) NAND and SR FLIP-FLOP elements.
(c) WOS and DELAY elements.

*8.9 The counter modulo m is a sequential switching circuit with one input and appropriate number of outputs whose response identifies uniquely the modulo m of the total number of ones which have been received at the input, i.e.,

$$s^t = \sum_{i=0}^{t} x^i \qquad \text{(modulo } m\text{)}$$

where x^i is the value of the input variable at time i and s^t denotes the state identifier of the output variables at time t. Find a formula for the required number of internal variables of the counter modulo m for:
(a) the first-order finite-state model.
(b) the first-order combined model.

8.10 Using NAND and DELAY elements, design the counter modulo 4 for the:
(a) first-order finite-state model.
(b) first-order combined model.
(c) kth order finite-state model ($k > 1$), if reasonable.

8.11 Using NAND and DELAY elements and the most favorable model, design the counter modulo 10.

8.12 The reverse (bi-directional) counter modulo m with one input is defined by

$$s^t = \sum_{i=0}^{t} (x^i - \bar{x}^i) \qquad \text{(modulo } m\text{)}$$

where x^i and s^t have the same meaning as in Exercise 8.9. Using NAND gates and SR FLIP-FLOPS, design the reverse counter modulo 10 for the most favorable model.

8.13 The reverse counter modulo m with two inputs is defined by

$$s^t = \sum_{i=0}^{t} (x_1^i - x_2^i) \qquad \text{(modulo } m\text{)}$$

where x_1^i, x_2^i denote values of input variables x_1, x_2 at time i, respectively, and s^t is the state identifier of the output variables at time t. Using NAND and DELAY elements, design the reverse counter modulo 10 with two inputs for the most favorable model.

8.14 Using NAND and DELAY elements and the most favorable model, design a switching circuit with one input and appropriate number of outputs which implements the following function:
(a) $s^t = (x^t - x^{t-1})$(modulo 2).
(b) $s^t = x^t + t$ (modulo 3).
(c) $s^t = x^t - t$ (modulo 3).
(d) $s^t = tx^t + 2t$ (modulo 3).
(e) $s^t = (t(x^t - x^{t-1}) + (x + t)$(modulo 3)) (modulo 4).

Reference Notations to Chapter 8

Models of sequential switching circuits, which are discussed in Section 8.4, were first suggested by Klir [21, 22]. Almost all literature devoted to the structure theory of sequential circuits is based on the first-order finite-state model (Mealy or Moore machines). Liu [26] explored kth order finite-state models for $k > 1$. Various properties of the finite-memory model have been investigated by Gill [10, 11], Vairavan [35, 36, 37], and others [17, 27, 29]. A synthesis procedure for the combined models was developed and a simple evaluation of models suggested by Klir and Marin [23]. Kellerman's cost formulas [19, 20] promise further refinement of the model evaluation. The multimodel theory initiated in Reference 23 was explored in new directions by Jerome [16]. A study by McCluskey [28] is somehow related to the combined model. The approach outlined in Section 8.7 was proposed by Svoboda [32, 33].

The study of machine decomposition was initiated by Hartmanis in 1960 [13] and was further developed by others [9, 25, 40, 41, 42]. A systematic presentation of the decomposition theory is given Reference 15; its more formal presentation is shown in Reference 1. Bakerdjian [4] investigates the decomposition in terms of the Svoboda conceptual framework (Section 8.7.).

The state-assignment problem was initiated by Hartmanis and Stearns [14, 31] in the context of the decomposition theory. It was further explored in this direction by several authors [5, 6, 18, 24]. Other approaches to the state-assignment problem were presented elsewhere [2, 3, 7, 8, 30, 34, 38]. A formula for the number of distinct state assignments was derived by Weiner and Smith [39]. The most comprehensive study of the state-assignment problem was done by Haring [12].

References to Chapter 8

1. **Arbib, M. A.,** ed., *Algebraic Theory of Machines, Languages and Semigroups*, Academic Press, New York, 1968.
2. **Armstrong, D. B.,** "A Programmed Algorithm for Assigning Internal Codes to Sequential Machines," *TC* **EC-11** (4) 466–472 (Aug. 1962).
3. **Armstrong, D. B.,** "On the Efficient Assignment of Internal Codes to Sequential Machines," *TC* **EC-11** (5) 611–622 (Oct. 1962).
4. **Bakerdjian, V.,** "The Cascade Decomposition of Finite-Memory Synchronous Sequential Machines," M.Sc. Thesis, McGill University, Montreal, March 1971.
5. **Curtis, H. A.,** "Multiple Reduction of Variable Dependency of Sequential Machines," *JACM* **9** (3) 324–344 (July 1962).
6. **Curtis, H. A.,** "Use of Decomposition Theory in the Solution of the State Assignment Problem for Sequential Machines," *JACM* **10** (3) 386–412 (July 1963).
7. **Davidow, W. H.,** "A State Assignment Technique for Synchronous Sequential Networks," Stanford University Electronics Lab., Rept. 1901–2, July 1961.
8. **Dollota, T. A.,** and **E. J. McCluskey,** "The Coding of Internal States of Sequential Circuits," *TC* **EC-13,** (5) 549–562 (Oct. 1964).
9. **Gill, A.,** "Cascaded Finite-State Machines," *TC* **EC-10** (3) 366–370 (Sept. 1961).
10. **Gill, A.,** *Introduction to the Theory of Finite-State Machines*, McGraw-Hill, New York, 1962.

11. **Gill, A.,** "On the Bound of the Memory of a Sequential Machine," *TC* **EC-14** (3) 464–466 (June 1965).
12. **Haring, D. R.,** *Sequential-Circuits Synthesis: State Assignment Aspects*, MIT Press, Cambridge, Mass.; 1966.
13. **Hartmanis, J.,** "Symbolic Analysis of a Decomposition of Information Processing Machines," *IC* **3** (2) 154–178 (June 1960).
14. **Hartmanis, J.,** "On the State Assignment Problem for Sequential Machines I," *TC* **EC-10** (2) 157–165 (June 1961).
15. **Hartmanis, J.,** and **R. E. Stearns,** *Algebraic Structure Theory of Sequential Machines*, Prentice-Hall, Englewood Cliffs, N. J., 1966.
16. **Jerome, E. J.,** "A Mathematical Approach to the Abstract Synthesis of Sequential Discrete Systems," M.Sc. Thesis, McGill University, Montreal, Dec. 1970.
17. **Kambayashi, Y.,** et al, "On Finite-Memory Machines," *TC* **C-19** (3) 254–258 (March 1970).
18. **Karp, R. M.,** Some Techniques of State Assignment for Synchronous Sequential Machines," *TC* **13** (5) 507–518 (Oct. 1964).
19. **Kellerman, E.,** "A Formula for Logical Network Cost," *TC* **C-17** (9) 881–884 (Sept. 1968).
20. **Kellerman, E.,** "Logical Network Cost," Term Paper, School of Advanced Technology, State University of New York at Binghamton; Jan. 1971.
21. **Klir, G. J.,** "A Note on the Basic Block Diagrams of Finite Automata," *TC* **EC-16** (2) 223–224 (April 1967).
22. **Klir, G. J.,** *An Approach to General Systems Theory*, Van Nostrand Reinhold, New York, 1969.
23. **Klir, G. J.,** and **M. A. Marin,** "A Multimodel and Computer Oriented Methodology for Synthesis of Sequential Discrete Systems," *TS* **SSC-6** (1), 40–48 (Jan. 1970).
24. **Kohavi, Z.,** "Secondary State Assignment for Sequential Machines," *TC* **EC-13** (3) 193–203 (June 1964).
25. **Krohn, K. B.,** and **L. J. Rhodes,** "Algebraic Theory of Machines," in J. Fox, ed., *Mathematical Theory of Automata*, Polytechnic Press, Brooklyn, N.Y., 1963.
26. **Liu, C. L.,** "*k*th Order Finite Automaton," *TC* **EC-12** (5), 470–475 (Oct. 1963).
27. **Masey, J. L.,** "Note on Finite-Memory Sequential Machines," *TC* **EC-15** (4) 658–659 (August 1966).
28. **McCluskey, E. J.,** "Reduction of Feedback Loops in Sequential Circuits and Carry Leads in Iterative Networks," *IC* **6** (2) 99–118 (June 1963).
29. **Newborn, M. M.,** "Maximal Memory Binary Input-Binary Output Finite-Memory Sequential Machines," *TC* **C-17** (1) 67–71 (Jan. 1968).
30. **Schneider, M. I.,** "State Assignment Algorithm for Clocked Sequential Machines," MIT Lincoln Lab., Techn. Rept. 270, May 1962.
31. **Stearns, R. E.,** and **J. Hartmanis,** "On the State Assignment Problem for Sequential Machines II," *TC* **EC-10** (4) 593–603 (Dec. 1961).
32. **Svoboda, A.,** "Synthesis of Logical Systems of Given Activity," *TC* **EC-12** (6) 904–910 (Dec. 1963).
33. **Svoboda, A.,** "Behavior Classification in Digital Systems," *IPM* **10,** 25–42 (1964).
34. **Torng, H. C.,** "An Algorithm for Finding Secondary Assignment of Synchronous Sequential Circuits," *TC* **C-17** (5) 461–469 (May 1968).
35. **Vairavan, K.,** "On the Memory of Finite-State Machines," Ph.D. Dissertation, E.E. Department, University of Notre Dame, April 1968.

36. **Vairavan, K.,** "On the Lower Bound to the Memory of Finite State Machines," *TC* **C-18** (9) 856–861 (Sept. 1969).
37. **Vairavan, K.,** "Input-Output Relations of Finite Memory Systems," *IC* **16** (1) 52–65 (March 1970).
38. **Weiner, P.,** and **E. J. Smith,** "Optimization of Reduced Dependencies for Synchronous Sequential Machines," *TC* **EC-16** (6) 835–847 (Dec. 1967).
39. **Weiner, P.,** and **E. J. Smith,** "On the Number of Distinct State Assignments for Synchronous Sequential Machines," *TC* **EC-16** (2), 220–221 (April 1967).
40. **Yoeli, M.,** "The Cascade Decomposition of Sequential Machines," *TC* **EC-10** (4) 587–592 (Dec. 1961).
41. **Yoeli, M.,** "Cascade-Parallel Decomposition of Sequential Machines," *TC* **EC-12** (3) 322–324 (June 1963).
42. **Zeiger, H. P.,** "Loop-Free Synthesis of Finite-State Machines," Ph.D. Dissertation, E.E. Department, M.I.T., Sept. 1964.

9

ASYNCHRONOUS DETERMINISTIC SEQUENTIAL SWITCHING CIRCUITS: STRUCTURE SYNTHESIS

9.1 PECULIARITIES OF ASYNCHRONOUS CIRCUITS

Asynchronous switching circuits are characterized by the absence of an independent source of synchronizing pulses. As a consequence, discrete times of an asynchronous switching circuit are not defined externally and independently of the circuit. On the contrary, they are fully defined by the circuit itself. More specifically, they are defined by those time intervals during which the values of all the variables associated with the circuit are defined. Two consecutive discrete times are distinguished by a time period during which at least one variable changes its value and is not thus defined (the real value of the corresponding physical quantity belongs to neither of the two predefined disjoint subsets of values). This period between two consecutive discrete times will be called the *transient period.*

Although discrete times are represented by highly uneven periods of time in asynchronous switching circuits, they can be identified by integers just as they are identified in case of synchronous circuits. Thus, there is no difference between synchronous and asynchronous circuits as far as their formal representation is concerned. Both of them can be represented by the Mealy machine or the Moore machine whose output and transition functions can take the form of a table, a diagram, a matrix, or any other convenient form. However, the interpretation of a particular representation for asynchronous circuits is considerably different from its interpretation for synchronous circuits. This is due to the differences in the meaning of discrete time intervals.

In synchronous switching circuits, discrete time intervals (sampling times) do not depend on the *ST*-structure of the circuit because they are defined by a source which is independent of the circuit. At every sampling time, a pair $(\mathbf{x}_a, \mathbf{z}_b)$ of stimulus $\mathbf{x}_a$ and internal state $\mathbf{z}_b$ is mapped into a pair $(\mathbf{y}_c, \mathbf{z}_d)$ of response $\mathbf{y}_c$ and next internal state $\mathbf{z}_d$. There is no essential difference between the case $\mathbf{z}_b = \mathbf{z}_d$ (the circuit does not change its internal state $\mathbf{z}_b$ for $\mathbf{x}_a$) and the case $\mathbf{z}_b \neq \mathbf{z}_d$ (the circuit changes its internal state $\mathbf{z}_b$ for $\mathbf{x}_a$).

In asynchronous circuits, the difference between $\mathbf{z}_b = \mathbf{z}_d$ and $\mathbf{z}_b \neq \mathbf{z}_d$ for the above-specified transition is essential. If $\mathbf{z}_b = \mathbf{z}_d$ at time t, then discrete time $t + 1$ is associated with a new stimulus, say $\mathbf{x}'_a$. On the real time scale, the two time intervals identified as t and $t + 1$ are separated by a transient period associated with the change from $\mathbf{x}_a$ to $\mathbf{x}'_a$. During this period, at least one of the input variables is not defined.

If $\mathbf{z}_b \neq \mathbf{z}_d$ at time t, the situation is quite different. The circuit changes its internal state and this change is associated with a transient period and new discrete time $t + 1$ without any change of the stimulus $\mathbf{x}_a$. The new internal state may again be required to change. Its change generates another transient period and discrete time $t + 2$, etc.

Thus, several state transitions can be made by an asynchronous switching circuit for a single stimulus. The first transition in the sequence is initiated by a change in the stimulus, the other are autonomous. In special cases, a periodical sequence of transitions may be repeated indefinitely.

Real time intervals of transient periods and discrete times of asynchronous switching circuits depend on response times of the used elements, primarily the memory elements, and their interconnection. It is not practically possible to make physical elements with exactly equal response times even if they are of the same type. Differences in response times may produce ambiguities in the transition. Since this is inadmissible, we must ensure proper transitions by some other means when synthesizing asynchronous switching circuits. Three methods can be used for this purpose:

1. The state assignment is constructed so that not more than one internal variable changes its value in each transition.

2. All possible orders of changes of internal variables (due to differences in response times) are considered when several internal variables change their values in a single transition and we ensure that all orders of changes lead finally to the same internal state.

3. Memory elements with significantly different response times are employed and these time relations are utilized in the synthesis.

The most frequently used method is the first one. Sometimes, however, it is difficult to find a suitable state assignment which satisfies the condition quoted. In such a case we invoke the aid of the second method, i.e., we admit the simultaneous change of values of several internal variables in some transitions. If values of m variables are required to change in the course of some particular transition, then we must consider the remaining $2^m - 2$ states of these variables and cause them to terminate in a single state. We thereby unnecessarily lose $2^m - 2$ states. For instance, if a pair $(\mathbf{z}_2, \mathbf{z}_1)$ of internal variables is required to change from (0, 0) to (1, 1) for a particular stimulus $\mathbf{x}$, then we must consider that either $\mathbf{z}_1$ changes first, or $\mathbf{z}_2$ changes first or both of them change simultaneously. Thus we must consider three possible transitions from (0, 0) for the stimulus x: $(0, 0) \to (0, 1)$, $(0, 0) \to (1, 0)$, $(0, 0) \to (1, 1)$. In the first two cases we must complete the original transition $(0, 0) \to (1, 1)$ by prescribing two autonomous transitions: $(0, 1) \to (1, 1)$ and $(1, 0) \to (1, 1)$. This makes a loss of two states, (0, 1) and (1, 0).

The third method is used solely for simple circuits since it leads to a considerably more difficult synthesis. Moreover, this method results in slower circuits.

Various aspects of the state-assignment problem for asynchronous switching circuits are discussed in Section 9.3.

Another important difference between synchronous and asynchronous switching circuits concerns the so-called *hazards*. These are temporary erroneous values of output or next state variables which exist only in asynchronous circuits. They are caused by processes taking place in the transient periods.

Since hazards produce only temporary erroneous values, they often do not have a serious effect on the performance of the circuit. However, if they produce a malfunction of the circuit, we have to eliminate them. This gives rise to new problems, which are meaningful only for asynchronous circuits, e.g., the problem of identification of hazards, the problem of synthesis of hazard-free circuits, etc. Some of these problems are discussed in Section 9.4.

A desirable feature of asynchronous circuits is that they can take full advantage of the speed of used elements because they do not have to wait for the arrival of clock pulses to effect individual transitions. Consequently, asynchronous circuits are usually faster than corresponding synchronous circuits. However, the absence of clock pulses introduces the problem of insuring that the circuit behaves as required independent of variations in signal propagation delays of individual elements and delays associated with connections between elements of the circuit (so-called stray delays). To solve this problem, more elements are usually needed in an asynchronous circuit when compared with the synchronous implementation of the same behavior. Hence, asynchronous switching circuits are, as a rule, faster but more expensive and more difficult to design.

9.2 BASIC CONCEPTS

As a background for a discussion of the state-assignment problem for asynchronous switching circuits (Section 9.3) and various problems concerning hazards (Section 9.4), definitions of some basic concepts meaningful only for asynchronous circuits are presented in this section. In the definitions that follow, we assume that the asynchronous interpretation of a Mealy machine $M = (X, Y, Z, \mathbf{f}, \mathbf{g})$ is given, where the symbols have the same meaning as for synchronous circuits (Chapter 7 and 8). Symbols $\mathbf{x}_i \in X$, $\mathbf{y}_j \in Y$, $\mathbf{z}_k \in Z$ will denote stimuli (input states), responses (output states), and internal states, respectively. The pair $(\mathbf{x}_i, \mathbf{z}_k)$ of the input state and the internal state will be called the *total state* or, simply, the state of the circuit.

DEFINITION 9.1 Let $\mathbf{g}(\mathbf{x}, \mathbf{z}_i) = \mathbf{z}_j$. Then, if $\mathbf{z}_j = \mathbf{z}_i$, the total state $(\mathbf{x}, \mathbf{z}_i)$ is *stable*; otherwise (if $\mathbf{z}_j \neq \mathbf{z}_i$), the state $(\mathbf{x}, \mathbf{z}_i)$ is *unstable*.

Observe, that for a particular stimulus x, the asynchronous circuit autonomously changes its internal state until it reaches a stable state. Then, it stays in the stable state until a new stimulus representing an unstable state is reached. Observe also that the classification of states to stable states and unstable states is meaningless for synchronous switching circuits.

As an example, a Mealy machine with two input variables x_1 and x_2, two output variables y_1 and y_2, and two internal variables z_1 and z_2 is specified in the three basic forms in Figure 9.1. Next state variables are denoted z_1' and z_2'. In the case of the asynchronous interpretation of the machine, we find that it has seven stable states which are encircled in the table form. In the diagram, each stable state is associated with an arc that connects a node with itself (a self-loop). In the matrix, stable states are identified by the elements on the main diagonal (they are underlined in Figure 9.1c).

DEFINITION 9.2 A *fundamental mode asynchronous Mealy machine* is a Mealy machine $M = (X, Y, Z, \mathbf{f}, \mathbf{g})$ which operates under the following rules: (i) The machine has at least one stable state for each stimulus. (ii) The

z_2z_1 \ x_2x_1	$z_2'z_1'$ 00	01	10	11	y_2y_1 00	01	10	11
00	10	00	00	11	–	01	00	–
01	11	11	01	11	–	–	01	–
10	10	00	10	11	01	–	10	–
11	00	11	01	11	–	10	–	11

(a)

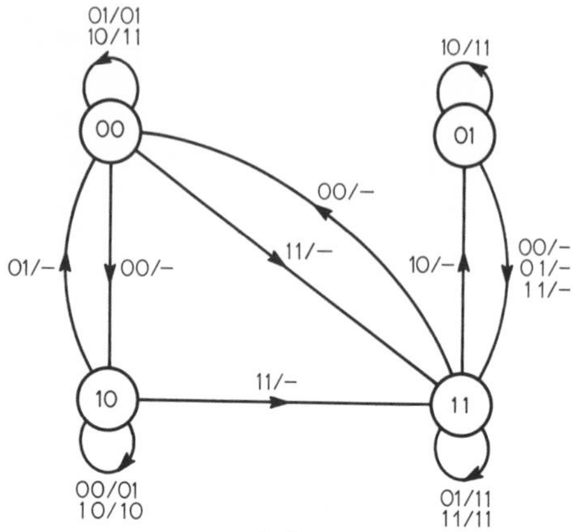

(b)

$$\begin{bmatrix} \underline{01/01 \cup 10/11}, & \varnothing, & 00/-, & 11/- \\ \varnothing, & \underline{10/11}, & \varnothing, & 00/- \cup 01/- \cup 11/- \\ 01/-, & \varnothing, & \underline{00/01 \cup 10/10}, & 11/- \\ \varnothing, & 10/-, & \varnothing, & \underline{01/11 \cup 11/11} \end{bmatrix}$$

(c)

Figure 9.1 An example of the fundamental mode asynchronous Mealy machine.

stimulus (input state) cannot change unless the machine is in a stable state. (iii) The response is of interest only when the machine is in a stable state.

The machine specified in Figure 9.1 satisfies Rules (i) and (iii) of Definition 9.2. It cannot be verified from the machine specification whether or not Rule (ii) is satisfied. To satisfy the rule, the environment of the circuit must be properly designed or the circuit may generate a feedback signal (usually called the *ready signal*) which allows a change of the stimulus only if the circuit is in its stable state. The latter approach is used in circuits which are usually referred to as *Muller circuits* or *speed independent circuits*. They are applicable only when the input source is under the control of the circuit or when a buffer memory is placed between the input source and the circuit.

The fundamental mode asynchronous switching circuits (machines whose input, output and internal states are binary coded) represent by far the most important class of asynchronous switching circuits. For that reason, we shall limit our further exposition in Chapter 9 to this class. Moreover, we shall discuss only the first-order finite-state model.

Since the response of the fundamental mode asynchronous machine is usually of no interest for unstable states, the transition function of a given machine can be modified, without changing the external performance of the machine, by prescribing each unstable state to pass directly to the stable state for the given stimulus. This leads to the following definition of the normal asynchronous Mealy machine.

DEFINITION 9.3 A fundamental mode asynchronous Mealy machine with the property that each unstable state is required to pass directly to a stable state for a particular stimulus is called the *normal asynchronous Mealy machine*.

Obviously, every fundamental mode but not normal asynchronous Mealy machine can be easily modified to its normal counterpart which has the same behavior. For instance, the machine in Figure 9.1 is not normal due to the transition function for stimulus 00. We can make it normal simply by prescribing that each internal state will pass directly to the stable internal state 10 for this stimulus. The table form of this normal machine is shown in Figure 9.2.

x_2x_1 / z_2z_1	$z_2'z_1'$				y_2y_1			
	00	01	10	11	00	01	10	11
00	10	00	00	11	–	01	00	–
01	10	11	01	11	–	–	01	–
10	10	00	10	11	01	–	10	–
11	10	11	01	11	–	10	–	11

Figure 9.2 The normal counterpart of the Mealy machine specified in Figure 9.1.

DEFINITION 9.4 The situation in which, for a particular stimus x, the next internal state differs from the present internal state in values of two or more variables is called a *race condition*. If a race condition exists and there is a possibility that variations in signal propagation delays of the used elements may cause the circuit to reach a stable state different from the one intended, the race condition is called *critical*; otherwise, it is called *non-critical.*

Assume that the asynchronous switching circuit specified in Figure 9.1 is in internal state 00 and accepts stimulus 11. This situation creates a race condition because the internal state is required to change from 00 to 11. However, this is not a critical race condition because the order in which the internal variables change their values does not prevent the circuit from reaching state 11.

Assume now that the circuit is in state 01 and accepts stimulus 00. First, the circuit passes from internal state 01 to internal state 11. This transition does not create a race condition since the value of only one internal variable is changed. The new state is not stable. The circuit passes now from internal state 11 to internal state 00. This requires a change of values of two variables and creates thus a race condition. If both values were changed simultaneously, 11 would pass to 00 and 00 would pass to 10 and, then, the circuit would stay in this stable state until the stimulus changes to 01 or 11. If z_1 changed its value first, the circuit would pass from internal state 11 directly to its stable state. This is allowed since the responses are prescribed only for stable states in this example. However, if z_2 changed its value first, the circuit would pass from 11 to 01 and, then, back to 11. This pair of transitions would be repeated until the stimulus was changed. Thus, the circuit would oscillate between internal states 01 and 11 and the stable state for stimulus 00 would not be reached at all. Hence, the race condition is critical in this case.

Exercises to Section 9.2

9.2-1 Show that neither the number of internal states nor the number of input states can be reduced in case of the Mealy machine specified in Figure 9.1.

9.2-2 Identify all race conditions for the asynchronous switching circuit specified in Figure 9.1 and find which of them are critical.

9.2-3 Give an example of the fundamental mode asynchronous Mealy machine, and find all of its critical race conditions.

9.2-4 Using NAND gates, design for the machine specified in Figure 9.1 a circuit which generates the ready signal used in the Muller circuits.

9.3 STATE ASSIGNMENT

Higher speed of operation is the main reason for which asynchronous switching circuits are frequently preferred to synchronous circuits. In synchronous circuits, the speed of operation does not depend on the circuit design. It is fixed by the frequency of the clock pulses which, of course, must be compatible with some physical properties of the used elements. In asynchronous switching circuits, the speed does depend on the design. For a particular set of element types (modules), a key factor influencing the speed of asynchronous sequential circuits is the maximum number of transition times required to pass from one stable state to another.

To preserve the major advantage of asynchronous switching circuit, the high speed of operation, it is usually required that: (a) The normal asynchronous Mealy machine is used, and (b) all internal state variables that are to undergo a change in a transition are simultaneously excited, and the outcomes are independent of the order in which the variables actually change (noncritical race condition).

Let a basic unit of time, Δt, be defined as the average time required to change the value of an internal state variable. Then, if the two above-specified requirements are satisfied, each transition between two stable states takes place in approximately time Δt. Any state assignment which allows this mode of operation and prevents critical race conditions will be called the *minimum transition time state assignment.*

The problem of finding a minimum transition time state assignment for a given normal asynchronous Mealy machine will be viewed as a partitioning problem. A set of two-block partitions of the set Z of internal states of a given machine will be involved, each of which is induced by a single internal state variable. They will be referred to as *z-variable partitions.* Besides regular partitions of Z, special partitions whose blocks do not contain all elements of Z, will be considered. These special partitions will be called *incomplete.* For instance, $\{\overline{1, 2}; \overline{3, 4}; \overline{5, 6, 7}\}$ is a complete partition of the set $Z = \{1, 2, 3, 4, 5, 6, 7\}$ while $\{\overline{1, 3, 5}; \overline{4, 6}\}$ is an incomplete partition of Z because it does not contain elements 2, 7 of Z.

The method for solving the above specified kind of the state assignment problem is based on the following theorem.

THEOREM 9.1 A state assignment for a normal asynchronous Mealy machine $M = (X, Y, Z, \mathbf{f}, \mathbf{g})$ has all the properties of the minimum transition time state assignment if and only if for every autonomous transition $(\mathbf{x}, \mathbf{z}_a) \to (\mathbf{x}, \mathbf{z}_b)$, where $\mathbf{x} \in X$ and $\mathbf{z}_a, \mathbf{z}_b \in Z$,

(a) if $(\mathbf{x}, \mathbf{z}_c) \rightarrow (\mathbf{x}, \mathbf{z}_d)$ is another autonomous transition associated with the same stimulus $\mathbf{x}$ ($\mathbf{z}_c, \mathbf{z}_d \in Z$), then at least one z-variable partition contains the pair $(\mathbf{z}_a, \mathbf{z}_b)$ in one block and the pair $(\mathbf{z}_c, \mathbf{z}_d)$ in another block,

(b) if $(\mathbf{x}, \mathbf{z}_k)$ is a stable state ($\mathbf{z}_k \in Z$), then at least one z-variable partition contains the pair $(\mathbf{z}_a, \mathbf{z}_b)$ in one block and the internal state $\mathbf{z}_k$ in another block,

(c) for $i \neq j$, internal states $\mathbf{z}_i$ and $\mathbf{z}_j$ are in separate blocks of at least one z-variable partition.

PROOF (a) Since each z-partition may contain at most two blocks, it is sufficient to consider all partitions with two blocks or less of the four internal states $\mathbf{z}_a, \mathbf{z}_b, \mathbf{z}_c, \mathbf{z}_d$. There are eight of them:

$$\pi_1 = \{\overline{\mathbf{z}_a, \mathbf{z}_b, \mathbf{z}_c, \mathbf{z}_d}\} \qquad \pi_5 = \{\overline{\mathbf{z}_a}; \overline{\mathbf{z}_b, \mathbf{z}_c, \mathbf{z}_d}\}$$

$$\pi_2 = \{\overline{\mathbf{z}_a, \mathbf{z}_b}; \overline{\mathbf{z}_c, \mathbf{z}_d}\} \qquad \pi_6 = \{\overline{\mathbf{z}_b}; \overline{\mathbf{z}_a, \mathbf{z}_c, \mathbf{z}_d}\}$$

$$\pi_3 = \{\overline{\mathbf{z}_a, \mathbf{z}_c}; \overline{\mathbf{z}_b, \mathbf{z}_d}\} \qquad \pi_7 = \{\overline{\mathbf{z}_c}; \overline{\mathbf{z}_a, \mathbf{z}_b, \mathbf{z}_d}\}$$

$$\pi_4 = \{\overline{\mathbf{z}_a, \mathbf{z}_d}; \overline{\mathbf{z}_b, \mathbf{z}_c}\} \qquad \pi_8 = \{\overline{\mathbf{z}_d}; \overline{\mathbf{z}_a, \mathbf{z}_b, \mathbf{z}_c}\}$$

The only partition that satisfies the requirements (a) of the theorem is π_2. An assignment will be made for the four states involved such that each of the partitions except π_2 is induced by an internal state variable. Let variable z_i induce partition π_i. Then, the following state assignment may result:

	z_1	z_3	z_4	z_5	z_6	z_7	z_8
$\mathbf{z}_a$	0	0	0	0	1	1	1
$\mathbf{z}_b$	0	1	1	1	0	1	1
$\mathbf{z}_c$	0	0	1	1	1	0	1
$\mathbf{z}_d$	0	1	0	1	1	1	0

Note that the Hamming distance between the assigned code words does not change by any permutation and/or negation of the internal state variables so it is immaterial in which order the partitions appear and which assignment of 0 and 1 to blocks of each individual partition is used.

To show that the above specified state assignment produces critical race conditions, let us consider the transition from $\mathbf{z}_a$ to $\mathbf{z}_b$. We must take into account that in the course of the transition, if imperfect elements are used, the circuit can momentarily assume any of the internal states 0----11, where either 0 or 1 can be substituted for each of the dashes. Next, let us consider the transition from $\mathbf{z}_c$ to $\mathbf{z}_d$. We find that the circuit can momentarily assume any of the internal states 0--11-- in the course of this transition. The two transitions share any of the internal states 0--1111 and introduce thus critical race conditions under the considered state assignment.

Assume now that the state assignment is expanded to include the internal variable z_2 which induces the partition π_2. We obtain:

	z_1	z_2	z_3	z_4	z_5	z_6	z_7	z_8
$\mathbf{z}_a$	0	0	0	0	0	1	1	1
$\mathbf{z}_b$	0	0	1	1	1	0	1	1
$\mathbf{z}_c$	0	1	0	1	1	1	0	1
$\mathbf{z}_d$	0	1	1	0	1	1	1	0

For this state assignment, the intermediate state involved in the transition from $\mathbf{z}_a$ to $\mathbf{z}_b$ are 00----11 and those involved in the transition from $\mathbf{z}_c$ to $\mathbf{z}_d$ are 01--11--. We see that no states are shared by both of the transitions now.

To prove that the condition is sufficient, we recall that the minimum transition time state assignment is defined for the normal machine (each state is either a stable state or passes directly to a stable state) and requires that all variables which change in a transition are excited simultaneously. Consider a state assignment which includes variable z_2 as defined above. In the transition from $\mathbf{z}_a$ to $\mathbf{z}_b$, $\mathbf{z}_2 = 0$ at the beginning and, since all variables that are to undergo a change in the course of a transition are excited simultaneously, $z_2 = 0$ at the whole period of the transition as well as at its end. Hence, there is no chance of $z_2 = 1$ in the whole course of the transition from $\mathbf{z}_a$ to $\mathbf{z}_b$ and, consequently, no states entered by this transition are shared with the states entered by the transition from $\mathbf{z}_c$ to $\mathbf{z}_d$.

(b) This proof is very similar to the proof of (a) and is left to the reader as an exercise.

(c) It can be easily demonstrated that a state assignment satisfying conditions (a) and (b) of the theorem does not necessarily distinguish all internal states from one another. Hence, the requirement (c) must be included. ▌

Before we use Theorem 9.1 to formulate a method for solving the state assignment problem, we shall define some new concepts.

DEFINITION 9.5 Partition π_i is *smaller than or equal* to partition π_j, written as $\pi_i \leq \pi_j$, where π_i or π_j may be incomplete, if and only if all elements included in blocks of π_i are also included in blocks of π_j and each block of π_i appears in a single block of π_j.

DEFINITION 9.6 Let a normal asynchronous Mealy machine M be given. If, for a particular stimulus $\mathbf{x}$ of M, there exist two transitions $(\mathbf{x}, \mathbf{z}_a) \rightarrow (\mathbf{x}, \mathbf{z}_b)$ and $(\mathbf{x}, \mathbf{z}_c) \rightarrow (\mathbf{x}, \mathbf{z}_d)$ or there exists a transition $(\mathbf{x}, \mathbf{z}_a) \rightarrow (\mathbf{x}, \mathbf{z}_b)$ and a stable state $(\mathbf{x}, \mathbf{z}_k)$, where $\mathbf{z}_a, \mathbf{z}_b, \mathbf{z}_c, \mathbf{z}_d, \mathbf{z}_k$ are internal states of M, then the two-block partitions $\{\overline{\mathbf{z}_a, \mathbf{z}_b}; \overline{\mathbf{z}_c, \mathbf{z}_d}\}$ and $\{\overline{\mathbf{z}_a, \mathbf{z}_b}; \overline{\mathbf{z}_k}\}$ are called *essential partitions* of M.

For instance, the partition $\{\overline{1, 2}; \overline{3, 6}\}$ is essential for the normal asynchronous Mealy machine whose transition function is specified in Figure 9.3a. It is essential because $2 \to 1$ and $3 \to 6$ are two different transitions associated with the same stimulus 00. The list of all essential partitions for this machine is as follows:

$$\pi_a = \{\overline{1, 2}; \overline{3, 6}\} \qquad \pi_f = \{\overline{1, 4}; \overline{2, 3}\}$$
$$\pi_b = \{\overline{1, 5}; \overline{3, 6}\} \qquad \pi_g = \{\overline{1, 4}; \overline{3, 5}\}$$
$$\pi_c = \{\overline{1, 3}; \overline{4, 5}\} \qquad \pi_h = \{\overline{1, 3}; \overline{2, 4}\}$$
$$\pi_d = \{\overline{1, 3}; \overline{2, 6}\} \qquad \pi_i = \{\overline{1, 3}; \overline{5, 6}\}$$
$$\pi_e = \{\overline{2, 6}; \overline{4, 5}\} \qquad \pi_j = \{\overline{2, 4}; \overline{5, 6}\}$$

z \ x	z' 00	01	10	11
1	1	3	4	3
2	1	6	3	2
3	6	3	3	3
4	–	4	4	2
5	1	4	3	5
6	6	6	–	5

(a)

Partition identifier	Internal states 1	2	3	4	5	6
a	0	0	1	–	–	1
b	0	–	1	–	0	1
c	0	–	0	1	1	–
d	0	1	0	–	–	1
e	–	0	–	1	1	0
f	0	1	1	0	–	–
g	0	–	1	0	1	–
h	0	1	0	1	–	–
i	0	–	0	–	1	1
j	–	0	–	0	1	1

(b)

Maximal compatible class	Internal states 1	2	3	4	5	6	Internal variables
a, g, j	0	0	1	0	1	1	z_1
b, ē, f	0	1	1	0	0	1	z_2
c, d, h, i	0	1	0	1	1	1	z_3

(c)

Figure 9.3 Illustration of the state-assignment problem for asynchronous switching circuits (Example 9.1).

DEFINITION 9.7 The two-block partitions $\tau_1, \tau_2, \ldots, \tau_n$ induced by the internal state variables $z_1, z_2, \cdots, z_n$ in a minimum transition time state assignment are called the *τ-partitions of the assignment.*

It follows from Theorem 9.1 that a state assignment for a normal asynchronous Mealy machine M has all the properties of the minimum transition time state assignment if and only if each essential partition of M is smaller than or equal to some partition τ_i. Let the essential partitions be denoted $\tau_k(k = 1, 2, \ldots)$. Then the state assignment problem consists in determining the smallest possible number of partitions $\tau_i(i = 1, 2, \ldots)$ for a given set of the essential partitions π_k so that for each π_k there exists a partition τ_i such that $\pi_k \leq \tau_i$.

A trivial minimum transition time state assignment, which does not guarantee that the smallest possible number of internal state variables is used, can be obtained by setting $\pi_k = \tau_i$. The number of variables is equal to the number of essential partitions in this case. For the machine specified in Figure 9.3a, this trivial state assignment is shown in Figure 9.3b. Each row represents an essential partition and at the same time a τ-partition induced by an internal state variable. Internal states of the first block of each partition are coded with zeroes, states included in the second block are coded with ones. Internal states which are not included in a partition are denoted by dashes in the corresponding row and have the meaning of don't cares for the internal state variables.

The trivial state assignment usually contains too many internal state variables. A procedure of their reduction will be described after we introduce some concepts convenient for the formulation of this procedure.

DEFINITION 9.8 Two internal state variables of the state assignment in which essential partitions and internal state variables are in a one-to-one correspondence (formerly referred to as the trivial state assignment) are *compatible* if they are equal wherever both are specified.

For instance, variables which represent partitions a and b in Figure 9.3b are compatible. Similarly, variables representing partitions c and d are compatible. Variables b and c are not compatible because they have different values for state 3.

DEFINITION 9.9 The *union of two compatible variables*, say z_i and z_j, is a variable whose values are equal to both z_i and z_j wherever either is specified and is don't care everywhere else.

For instance, the union of variables representing partitions a and b in Figure 9.3b is the variable $001-01$.

DEFINITION 9.10 Let a subset of variables $z_1, z_2, \ldots, z_m$ be a set of pair-wise compatible variables. Then, this subset is called a *compatible class of variables*. The union of variables of a compatible class is called a compatible class variable. A compatible class may be enlarged by adding a variable if and only if this variable is compatible with every variable in the compatible class. A compatible class to which none of the variables involved can be added is called a *maximal compatible class* within the given set of variables.

For instance, variables representing partitions c, d, h, i in Figure 9.3b form a compatible class because they are pair-wise compatible. Furthermore, they form a maximal compatible class since none of the other variables is compatible with all of the four variables in the class. The compatible class variable, which is the union of all variables in the class, has values 010111.

Note that although variables representing partitions e and f in Figure 9.3b are not compatible, they become compatible when one of the variables is negated.

Now, we describe a procedure for the construction of a minimum state transition time state assignment with the minimum number of internal state variables. The procedure was developed by J. H. Tracey and, therefore, we will refer to it as the *Tracey procedure*. Given a normal asynchronous Mealy machine $M = (X, Y, Z, \mathbf{f}, \mathbf{g})$, the procedure consists of the following steps:

Step 1. We determine all essential partitions of Z and introduce for each of them an internal state variable which induces the same partition of Z. This gives us a trivial minimum transition time state assignment.

Step 2. We determine all maximal compatible classes of the variables involved in the trivial state assignment determined in Step 1. Both assertions and negations of the variables may be used.

Step 3. We select the minimum number of maximal compatible classes such that each essential partition is smaller than or equal to at least one partition induced by a maximal compatible class variable. Clearly, this is a standard covering problem. It can be solved by assigning a logic variable, say c_i, to each maximal compatible class ($c_i = 1$ if the class is chosen and $c_i = 0$ otherwise) and minimizing the pseudo-Boolean function

$$\sum_i c_i$$

subject to the covering constraints expressed by appropriate Boolean equations.

Step 4. We determine the compatible class variable for each of the maximal compatible classes selected in Step 3. The collection of these compatible class variables represents a minimum transition time state assignment with the minimum number of internal state variables.

EXAMPLE 9.1 Determine a minimum transition time state assignment with the minimum number of internal state variables for the machine whose transition function is specified in Figure 9.3a.

Step 1. The trivial minimum transition time state assignment, which is shown in Figure 9.3b has been already determined.

Step 2. From the table in Figure 9.3b, we determine all pairs of compatible variables. For each pair z_i and z_j of variables we must check both z_i with z_j and z_i with $\bar{z}_j$ as possible compatible pairs. We obtain the following compatible pairs, where the variables are identified by identifiers of the corresponding essential partitions (recall that the internal state variables are in a one-to-one correspondence with the essential partitions in the trivial state assignment):

$(a, b)(a, g)(a, j)$	(f, g)
$(b, \bar{e})(b, f)$	(g, j)
$(c, d)(c, e)(c, h)(c, i)$	$(h, i)(h, \bar{j})$
$(d, e)(d, h)(d, i)$	(i, j)
$(\bar{e}, f)$	

From this complete list of compatible pairs, we determine the following maximal compatible classes C:

Maximal Compatible Class	*Variables Involved*
C_1	a, b
C_2	a, g, j
C_3	$b, \bar{e}, f$
C_4	c, d, h, i
C_5	c, e
C_6	d, e
C_7	f, g
C_8	$h, \bar{j}$
C_9	i, j

Step 3. We assign a logic variable c_i to each of the maximal compatible classes $C_i (i = 1, 2 \ldots, 9)$, obtained in Step 2, such that $c_i = 1$ if C_i is selected and $c_i = 0$ otherwise. Then we find minima of the pseudo-Boolean function

$$\sum_{i=1}^{9} c_i$$

subject to constraints expressed by the following set of simultaneous Boolean equations:

$c_1 \vee c_2 = 1$	(a)	$c_3 \vee c_7 = 1$	(f)
$c_1 \vee c_3 = 1$	(b)	$c_2 \vee c_7 = 1$	(g)
$c_1 \vee c_5 = 1$	(c)	$c_4 \vee c_8 = 1$	(h)
$c_4 \vee c_6 = 1$	(d)	$c_4 \vee c_9 = 1$	(i)
$c_3 \vee c_5 \vee c_6 = 1$	(e)	$c_2 \vee c_8 \vee c_9 = 1$	(j)

The letters in the parentheses identify the essential partitions which impose the individual constraints. When solving the equations*, we find that there is only one minimum of the objective function which is reached for $c_i = 1$ for $i = 2, 3, 4$ and $c_i = 0$ otherwise. Thus, the maximal compatible classes (a, g, j), $(b, \bar{e}, f)$, (c, d, h, i) are selected.

Step 4. Using the table in Figure 9.3b, we determine the compatible class variable for each of the three selected maximal compatible classes as shown in Figure 9.3c.

The above described procedure provides us with all possible minimum transition time state assignments with the minimum number of internal state variables. If there are more than one of them, we try to predict which of them will yield a simple implementation. Although the problem is similar to the state assignment problem for synchronous sequential circuits, we observe two differences:

1. The number of state assignments that have to be considered is, as a rule, considerably smaller in case of asynchronous circuits because we consider only minimum transition time state assignments with the minimum number of variables.

2. Certain properties of normal asynchronous machines can be employed to evaluate the given state assignments as will be shown later in this section.

Let $\mathbf{x}_1, \mathbf{x}_2, \ldots, \mathbf{x}_p$ be input states (stimuli), $z_1, z_2, \ldots, z_n$ be internal state variables, and $z'_1, z'_2, \ldots, z'_n$ be next state internal variables of a given normal asynchronous Mealy machine. Then, we can write the transition functions in the general form

$$z'_i = f_{i1}(z_1, z_2, \ldots, z_n)\mathbf{x}_1 \vee f_{i2}(z_1, z_2, \ldots, z_n)\mathbf{x}_2 \vee \cdots \vee f_{ip}(z_1, z_2, \ldots, z_n)\mathbf{x}_p \tag{9.1}$$

where $i = 1, 2, \ldots, n$. A method will be described here by which such a state assignment is selected which tends to minimize Boolean forms of the functions $f_{ij}(z_1, z_2, \ldots, z_n)$, $i = 1, 2, \ldots, n$, $j = 1, 2, \ldots, p$.

* APL program PBP, which is given in Appendix G, can be used.

First, we need to define some new concepts and, then, two theorems will be proven which represent the essence of the method.

DEFINITION 9.11 Given a normal asynchronous Mealy machine M, the set containing a stable internal state and all unstable internal states which pass autonomously to this stable state for a particular stimulus is called a *stable set* of M.

For instance, {1, 2, 5} is a stable set of the machine in Figure 9.3a because state 1 is stable for stimulus 00 and states 2 and 5 are the only unstable states which pass to state 1 for this stimulus.

DEFINITION 9.12 A *stable partition* σ_i of a given normal asynchronous Mealy machine M is a *collection of all stable sets* of M for a particular stimulus $\mathbf{x}_i$.

For the machine in Figure 9.3, $\{\overline{1, 2, 5}; \overline{3, 6}\}$ is the stable partition for stimulus 00. Similarly, $\{\overline{1, 3}; \overline{2, 6}; \overline{4, 5}\}$ is the stable partition for stimulus 01.

THEOREM 9.2 If all the stable states of a stable partition σ_j (the stable partition for stimulus $\mathbf{x}_j$) of a normal asynchronous Mealy machine M are in the same block of a τ-partition, say partition τ_i, then the function f_{ij} in (9.1) is either 0 or 1.

PROOF Let S denote the set of all stable internal states for the stimulus $\mathbf{x}_j$ (all stable states of the stable partition σ_j). Then, if the normal asynchronous Mealy machine is considered, the transition function for $\mathbf{x}_j$ is specified by $\mathbf{g}(\mathbf{x}_j, \mathbf{z}_k) = \mathbf{z}_a$, where $\mathbf{z}_a \in S$ for every $\mathbf{z}_k \in Z$. If the partition τ_i exists which has all states of S in a single block, then for the corresponding internal state variable z_i there is either $z_i = 0$ or $z_i = 1$ for all states of S. Hence, either $z_i' = 0$ or $z_i' = 1$ for $\mathbf{x}_j$. ▮

THEOREM 9.3 The function f_{ij} in (9.1) is equal to z_i if $\sigma_j \leq \tau_i$) (σ_j, τ_i have the same meaning as in Theorem 9.2).

PROOF Left to the reader as an exercise. ▮

Let A_a denote the total number of constant functions $f_{ij}(f_{ij} = 0$ or $f_{ij} = 1)$ in (9.1) for a state assignment a made for a given normal asynchronous Mealy machine and let B_a denote the total number of functions $f_{ij} = z_i$ for the same assignment made for the same machine. These numbers A_a and B_a

can be easily calculated on the basis of the properties specified in Theorems 9.2 and 9.3 for all considered state assignments. Then they can be employed for a rough comparison of the state assignments involved. We select the state assignment a represented by the greatest sum $A_a + B_a$ or, more generally, the greatest sum $c \cdot A_a + B_a$, where c is an appropriate coefficient ($c > 1$ seems reasonable).

EXAMPLE 9.2 The transition function of a normal asynchronous Mealy machine is specified in Figure 9.4. In the same figure, two minimal transition time state assignments with the minimum number of variables are shown. Find which of them is preferable.

To determine A_a and A_b, we identify stable states for each $\mathbf{x}_i$ first:

$$\mathbf{x}_1: 1, 3, 6$$
$$\mathbf{x}_2: 3, 4, 5, 6$$
$$\mathbf{x}_3: 2, 4, 5$$

Now we compare these sets of stable states with τ-partitions induced by variables of both the state assignment a and b. We find that none of the τ-partitions associated with the state assignment b contain any of the above specified sets of stable states in a single block. For the state assignment a, we find that 1, 3, 6 are included in the first block of τ_3 and 2, 4, 5 are in the second block of τ_3. Hence, $A_a = 2$ and $A_b = 0$ by Theorem 9.2.

To calculate B_a and B_b, we determine stable partitions $\sigma_1, \sigma_2, \sigma_3$ for

$\mathbf{z}$ \ $\mathbf{x}_i$	$\mathbf{z}'$			State assignment					
				a			*b*		
	$\mathbf{x}_1$	$\mathbf{x}_2$	$\mathbf{x}_3$	z_3	z_2	z_1	z_3	z_2	z_1
1	1	6	4	0	0	0	0	0	0
2	3	4	2	1	0	1	1	0	1
3	3	3	2	0	0	1	1	1	1
4	1	4	4	1	0	0	1	0	0
5	6	5	5	1	1	0	0	1	1
6	6	6	5	0	1	0	0	1	0

Figure 9.4 Illustration for Example 9.2.

individual stimuli $\mathbf{x}_1, \mathbf{x}_2, \mathbf{x}_3$ of the machine and τ-partitions for both state assignments:

$$\sigma_1 = \{\overline{1, 4}; \overline{2, 3}; \overline{5, 6}\}$$
$$\sigma_2 = \{\overline{1, 6}; \overline{2, 4}; \bar{3}; \bar{5}\}$$
$$\sigma_3 = \{\overline{1, 4}; \overline{2, 3}; \overline{5, 6}\}$$

$$\left.\begin{aligned} \tau_1 &= \{\overline{1, 4, 5, 6}; \overline{2, 3}\} \\ \tau_2 &= \{\overline{1, 2, 3, 4}; \overline{5, 6}\} \\ \tau_3 &= \{\overline{1, 3, 6}; \overline{2, 4, 5}\} \end{aligned}\right\} \text{For the state assignment } a$$

$$\left.\begin{aligned} \tau_1 &= \{\overline{1, 4, 6}; \overline{2, 3, 5}\} \\ \tau_2 &= \{\overline{1, 2, 4}; \overline{3, 5, 6}\} \\ \tau_3 &= \{\overline{1, 5, 6}; \overline{2, 3, 4}\} \end{aligned}\right\} \text{For the state assignment } b$$

The τ-partitions that satisfy the condition of Theorem 9.3 for the state assignment a are $\sigma_1 \le \tau_1, \sigma_1 \le \tau_2, \sigma_2 \le \tau_3, \sigma_3 \le \tau_1, \sigma_3 \le \tau_2$. For the state assignment b we find that only $\alpha_2 \le \tau_3$. Hence, $B_a = 5$ and $B_b = 1$ by Theorem 9.3.

Since for the state assignments a and b we have, respectively, $A_a + B_a = 7$ and $A_b + B_b = 1$, we consider the state assignment scheme a as preferable.

Next we describe a general state assignment scheme such that only one variable is required to change in every transition. Such an assignment may be desirable for various reasons some of which become clear to the reader after reading Section 9.4. The assignment scheme which will be described is based on the assumption that each internal state of the given machine is coded by several states of some logic variables or, in other words, each state is split into several equivalent states.

Let the given machine have r internal states which will be denoted $\mathbf{z}_0, \mathbf{z}_1, \ldots, \mathbf{z}_{r-1}$. Assume that state $\mathbf{z}_i$ is assigned to set $A_i (i = 0, 1, \ldots, r-1)$ of states of n logic variables $z_1, z_2, \ldots, z_n$. Then, the sets A_i must be pairwise disjoint and each pair A_i, A_j must contain a pair of adjacent states of variables $z_1, z_2, \ldots, z_n$.

We take $n = 2^p - 1$, where $p = \ulcorner \log_2 r$, and calculate the following values:

$$\begin{aligned} a_1 &= (z_1 + z_3 + z_5 + z_7 + \cdots + z_n) \quad \text{(modulo 2)} \\ a_2 &= (z_2 + z_3 + z_6 + z_7 + \cdots + z_n) \quad \text{(modulo 2)} \\ a_3 &= (z_4 + z_5 + z_6 + z_7 + \cdots + z_n) \quad \text{(modulo 2)} \\ &\vdots \\ a_p &= (z_{(n+1)/2} + \cdots + z_{n-1} + z_n) \quad \text{(modulo 2)} \end{aligned} \tag{9.2}$$

Clearly, the p-tuples $(a_p, a_{p-1}, \ldots, a_1)$ may be considered as binary numbers which range over the values 0 through $2^p - 1$. Let the set A_i contain all states of the variables $z_1, z_2, \ldots, z_n$ for which the binary number $(a_p, a_{p-1}, \ldots, a_1)$ has a value of $i (i = 0, 1, \ldots, r - 1)$. Then, each set A_i contains 2^{n-p} states of variables $z_1, z_2, \ldots, z_n$ and $A_i \cap A_j = \varnothing$ for each pair $i \neq j$. Careful investigation of properties of (9.2), which is left to the reader as an exercise, shows that for each pair $A_i \neq A_j$ there exists $v \in A_i$ and $w \in A_j$ such that v and w are adjacent.

For instance, if $r = 4$, then $p = 2$, $n = 3$, and the congruences (9.2) have the form

$$a_1 = (z_1 + z_3) \quad \text{(modulo 2)}$$
$$a_2 = (z_2 + z_3) \quad \text{(modulo 2)}$$

The assignment of states of variables z_1, z_2, z_3 to states $\mathbf{z}_0, \mathbf{z}_1, \mathbf{z}_2, \mathbf{z}_3$ is shown in Figure 9.5a, where states z_i are identified by their subscripts i. We can easily check that the adjacency requirement is satisfied.

As a more complex example, let $r = 8$ so that $p = 3$, $n = 7$ and the congruences (9.2) assume the form

$$a_1 = (z_1 + z_3 + z_5 + z_7) \quad \text{(modulo 2)}$$
$$a_2 = (z_2 + z_3 + z_6 + z_7) \quad \text{(modulo 2)}$$
$$a_3 = (z_4 + z_5 + z_6 + z_7) \quad \text{(modulo 2)}$$

The assignment has the form shown in Figure 9.5b.

Exercises to Section 9.3

9.3-1 Prove Part (b) of Theorem 9.1.

9.3-2 Solve the Boolean equations obtained in Example 9.1 and find the minimum of the respective objective function.

$z_2 z_1$ (columns); z_3 (rows)

0	1	2	3
3	2	1	0

(a)

$z_4 z_3 z_2 z_1$ (columns); $z_7 z_6 z_5$ (rows)

0	1	2	3	3	2	1	0	4	5	6	7	7	6	5	4
5	4	7	6	6	7	4	5	1	0	3	2	2	3	0	1
6	7	4	5	5	4	7	6	2	3	0	1	1	0	3	2
3	2	1	0	0	1	2	3	7	6	5	4	4	5	6	7
7	6	5	4	4	5	6	7	3	2	1	0	0	1	2	3
2	3	0	1	1	0	3	2	6	7	4	5	5	4	7	6
1	0	3	2	2	3	0	1	5	4	7	6	6	7	4	5
4	5	6	7	7	6	5	4	0	1	2	3	3	2	1	0

(b)

Figure 9.5 State assignments based on the congruences (9.2).

9.3-3 Prove Theorem 9.3.

9.3-4 Find whether or not the state assignments in Figure 9.4 are the only minimum transition time state assignments with the minimum number of variables for the given machine.

*9.3-5 Determine all minimum transition time state assignments with minimum number of variables for each of the normal asynchronous Mealy machines whose transition functions are specified in Figure 9.6. If there are more than one of these assignments for a particular machine, use Theorems 9.2 and 9.3 to evaluate them (calculate for each the sum $A + B_a$).

z \ x	z' 0	z' 1
1	3	2
2	3	**2**
3	**3**	**3**
4	**6**	6
5	–	6
6	6	**6**

(a)

z \ x	z' *a*	z' *b*	z' *c*	z' *d*
1	**1**	4	2	**1**
2	1	4	**2**	**2**
3	**3**	**3**	4	**3**
4	**3**	**4**	**4**	1

(b)

z \ x	z' 0	z' 1
1	**1**	3
2	3	6
3	**3**	**3**
4	8	5
5	8	**5**
6	7	**6**
7	**7**	3
8	**8**	5

(c)

z \ x	z' *a*	z' *b*	z' *c*	z' *d*
1	**1**	2	**1**	**1**
2	3	**2**	1	**2**
3	**3**	4	1	1
4	1	**4**	1	**4**

(d)

z \ x	z' *a*	z' *b*	z' *c*	z' *d*
1	**1**	2	**1**	**1**
2	**2**	**2**	4	**2**
3	4	**3**	1	**3**
4	**4**	3	**4**	**4**

(e)

Figure 9.6 Transition functions of normal asynchronous Mealy machines (Exercise 9.3-5).

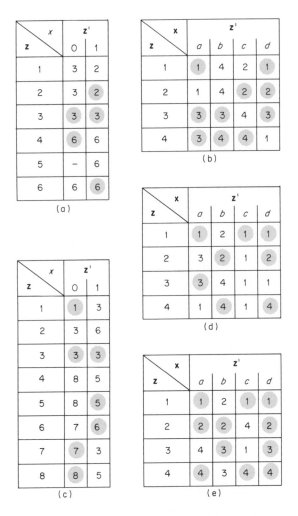

9.4 HAZARDS

Hazards are spurious output signals of short duration which may occur within transient periods associated with changes of input states of memoryless switching circuits or memoryless portions of sequential circuits (functional generators). Although they may cause malfunctions in strictly memoryless switching circuits if their output signals are not properly controlled by clock pulses, their effect on the behavior of asynchronous sequential circuits is far more serious. For this reason, a discussion of hazards is included in this chapter. Clearly, there are no hazards in synchronous switching circuits.

Several types of hazards are introduced by the following definitions.

DEFINITION 9.13 Let $\mathbf{x}_i$ and $\mathbf{x}_j$ be two adjacent input states for both of which a particular output variable is equal to 1. If this output variable can momentarily become equal to 0 in the transient period associated with the transition from $\mathbf{x}_i$ to $\mathbf{x}_j$, then this condition is referred to as the *static 1 hazard*.

An example of a circuit with the static 1 hazard is shown in Figure 9.7. Suppose that the input variables x_3, x_2, x_1 change their values from 1, 1, 1 to 1, 1, 0. For both of these input states, the output of the circuit is 1. If the input state is 1, 1, 1, the outputs of the AND gates G_1 and G_2 are, respectively, 1 and 0. If the input is next changed to 1, 1, 0, then outputs of both the AND gates change their values. The NOT element has some delay so that the output G_1 changes to 0 a little earlier than the output of G_2 changes to 1. Thus, there is a short period of time during which outputs of both G_1 and G_2 are zero so that $f = 0$.

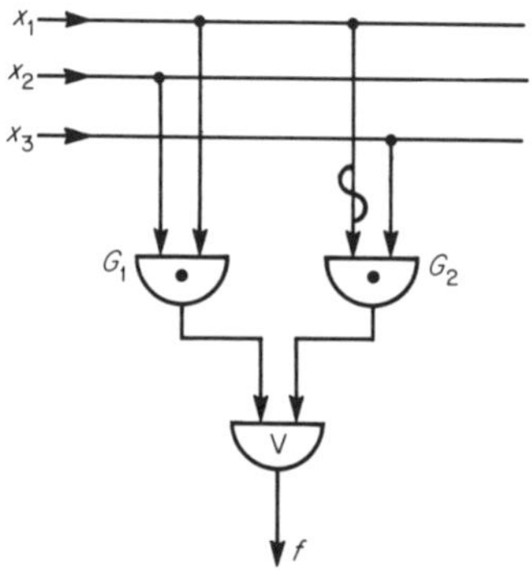

Figure 9.7 Circuit with the static 1 hazard.

DEFINITION 9.14 Let $\mathbf{x}_i$ and $\mathbf{x}_j$ be two adjacent input states for both of which a particular output variable is equal to 0. If this output variable can momentarily become equal to 1 in the transient period associated with the transition from $\mathbf{x}_i$ to $\mathbf{x}_j$, then this condition is referred to as the *static 0 hazard.*

As an example of a circuit with the static 0 hazard, a circuit representing negation $\bar{f}$ of the function f implemented by the circuit illustrated in Figure 9.7 may be used.

DEFINITION 9.15 Let $\mathbf{x}_i$ and $\mathbf{x}_j$ be two adjacent input states such that a particular output variable is equal to 0 for $\mathbf{x}_i$ and is equal to 1 for $\mathbf{x}_j$. If this output variable can momentarily become 0 after it becomes 1 within the transient period associated with the transition from $\mathbf{x}_i$ to $\mathbf{x}_j$ or if it can momentarily become 1 after it becomes 0 with the transient period associated with the transition from $\mathbf{x}_j$ to $\mathbf{x}_i$, then either of these conditions is called the *dynamic hazard.*

For instance, the circuit shown in Figure 9.8 has a dynamic hazard associated with the change from $x_1 = 0$ to $x_1 = 1$ provided that $x_2 = x_3 = x_4 = 1$ and $x_5 = 0$. We see that z_1 and z_3 change from 1 to 0 while z_2 changes from

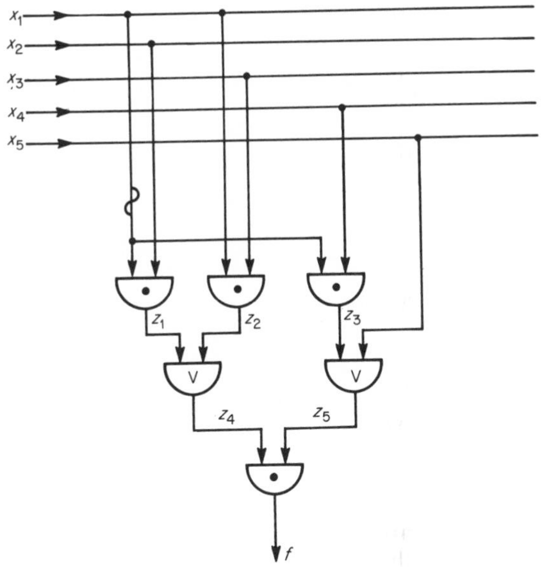

Figure 9.8 Circuit with the dynamic hazard.

0 to 1. If z_1 changes before z_2, then z_4 becomes momentarily equal to 0 (the static 1 hazard). If z_5 changes to 0 only after z_4 has gone through the sequence 1, 0, 1, the output variable takes on the sequence of values 1, 0, 1, 0. This situtation represents a dynamic hazard.

Note that all hazards were defined under the assumption that only one input variable changes its value. Hazards occur for other changes of input states too. However, while hazards based on single variable changes can always be prevented by simple means, hazards based on multiple variable changes are far more difficult to eliminate. The latter are not pursued in this section; they are the subject of Section 9.5.

DEFINITION 9.16 A switching circuit which has neither a static hazard nor a dynamic hazard is called *hazard-free.*

To establish a necessary and sufficient condition for hazard-free switching circuits based on *sp* Boolean forms, we must introduce some new concepts first.

DEFINITION 9.17 A *1-set* of a switching circuit with respect to a particular output variable of the circuit is any minimum subset of input variables, each variable taken either as assertion or as negation, such that when each variable in the set is equal to 1, the respective output variable is equal to 1 independent of values of the other variables.

DEFINITION 9.18 An input state $(\dot{x}_1, \dot{x}_2, \ldots, \dot{x}_p)$ is said to be *covered by a 1-set A* if the two following conditions hold:

(a) If $x_i \in A$, then $\dot{x}_i = 1$
(b) If $\bar{x}_i \in A$, then $\dot{x}_i = 0$

Clearly, if a 1-set does not contain k out of p input variables (either asserted or negated), then it covers 2^k input states. In case of circuits representing *sp* Boolean forms, each 1-set corresponds to variables that are inputs of a single AND gate.

1-sets of the circuit in Figure 9.7 are $\{x_1, x_2\}$ and $\{\bar{x}_1, x_3\}$. The first one covers states for which $x_1 = x_2 = 1$ and either $x_3 = 0$ or $x_3 = 1$. The second one covers states for which $x_1 = 0$, $x_3 = 1$, and either $x_2 = 0$ or $x_2 = 1$. 1-sets of the circuit in Figure 9.8 are $\{\bar{x}_1, x_2, x_4\}$, $\{\bar{x}_1, x_2, x_5\}$, $\{x_1, x_3, x_5\}$.

Next, an important property of hazard-free switching circuits based on *sp* Boolean form will be established by the theorem that follows.

THEOREM 9.4 A switching circuit that implements directly a *sp* Boolean form is hazard-free if and only if each pair of adjacent input states for which the output is equal to 1 is covered by at least one 1-set of the circuit.

PROOF Let $f(x_1, x_2, \ldots, x_p)$ be the function which is implemented by the circuit under consideration. Let $\dot{\mathbf{x}} = (\dot{x}_p, \dot{x}_{p-1}, \ldots, \dot{x}_1)$ and $\tilde{\mathbf{x}} = (\tilde{x}_p, \tilde{x}_{p-1}, \ldots, \tilde{x}_1)$ be a pair of adjacent input states such that $f(\dot{\mathbf{x}}) = f(\tilde{\mathbf{x}}) = 1$. If both $\dot{\mathbf{x}}$ and $\tilde{\mathbf{x}}$ are not covered by a 1-set of the circuit, then no single AND gate has its output equal to 1 for both $\dot{\mathbf{x}}$ and $\tilde{\mathbf{x}}$. Consider the transition from $\dot{\mathbf{x}}$ to $\tilde{\mathbf{x}}$. If the AND gates whose outputs are equal to 1 for $\dot{\mathbf{x}}$ change their outputs to 0 before any AND gate whose output is equal to 1 for $\tilde{\mathbf{x}}$ changes its output to 1, then the OR gate produces a short erroneous output value of 0. Thus, the circuit has a static 1 hazard for the transition from $\dot{\mathbf{x}}$ to $\tilde{\mathbf{x}}$.

Conversely, if each pair $(\dot{\mathbf{x}}, \tilde{\mathbf{x}})$ with the above specified properties is covered by a 1-set, then the output of a single AND gate represents both $\dot{\mathbf{x}}$ and $\tilde{\mathbf{x}}$ and its output is kept 1 during the transition from $\dot{\mathbf{x}}$ to $\tilde{\mathbf{x}}$. Furthermore, no output of any AND gate in the circuit can ever become equal to 1 for any transition between adjacent input states for which $f(x_1, x_2, \ldots, x_n) = 0$. Thus the circuit has no static hazards. Finally, consider a transition between adjacent input states which changes the values of the output of the circuit. If the output changes from 1 to 0, this change can happen only when the OR gate has all its inputs equal to 0. Since no AND gate tends to change its output to 1 in this kind of transition, no dynamic hazard occurs. If the output of the circuit is 0, then outputs of all AND gates are equal to 0. Hence, if the output of the circuit changes from 0 to 1, no AND gate tends to change its output to 0 so that no dynamic hazard occurs. ∎

It follows directly from Theorem 9.4 that if we cover states occupied by a given logic function by such a collection of prime implicants that each pair of adjacent and occupied states is covered by at least one prime implicant, then the corresponding circuit is hazard-free. This suggests the following procedure of synthesis of hazard-free two stage AND-OR switching circuits:

(1) Determine all prime implicants of the given function (or functions).
(2) Compile all pairs of adjacent states which are occupied by the function (or at least by one of the functions).
(3) Cover all the pairs determined in (2) by the minimum number of prime implicants or, more generally, cover all the pairs by such a set of prime implicants for which a given objective function reaches its minimum.

EXAMPLE 9.3 Design a hazard-free two stage AND-OR switching circuit implementing logic function f which is specified in Figure 9.9a.

Following the above described procedure, we determine all prime implicants of the function and all pairs of adjacent states which are occupied by the function and arrange them into the covering table as shown in Figure 9.9b.

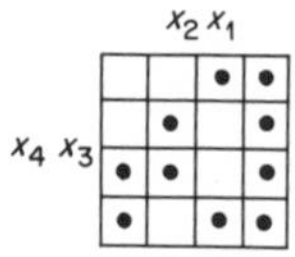

(a)

	2–3	3–7	3–11	5–7	7–15	8–9	8–12	9–11	11–15	12–14	14–15
$\bar{x}_4\, \bar{x}_3\, x_2$	•										
$\bar{x}_4\, x_3\, x_1$				•							
$x_4\, \bar{x}_3\, x_1$								•			
$x_4\, \bar{x}_3\, \bar{x}_2$						•					
$x_4\, x_3\, x_2$											•
$x_4\, x_3\, \bar{x}_1$										•	
$x_4\, \bar{x}_2\, \bar{x}_1$							•				
$x_2\, x_1$		•	•		•				•		

(b)

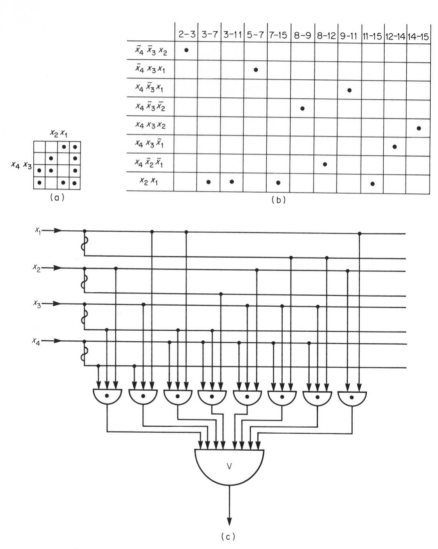

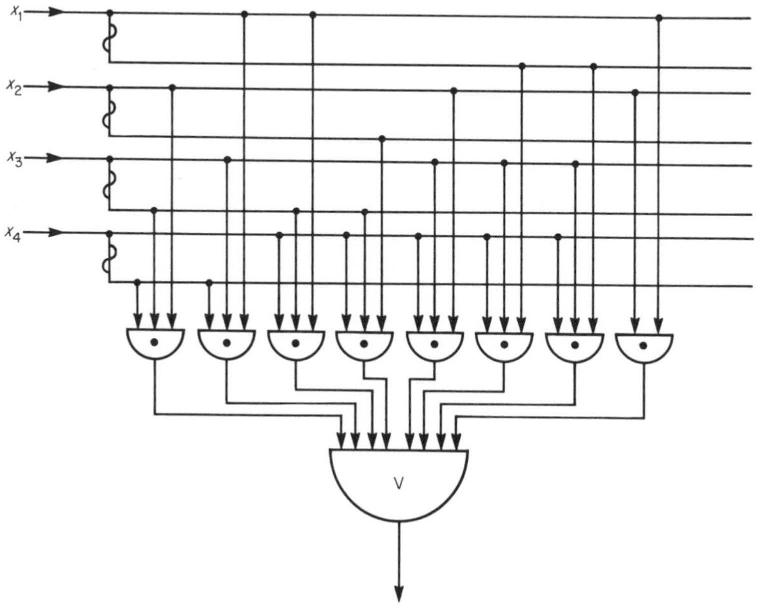

(c)

Figure 9.9 Illustration of the hazard-free two stage AND-OR circuit synthesis (Example 9.3.)

The covering procedure is trivial in this case: We must take all the prime implicants to cover all the pairs in the table. The hazard-free circuit implementing the given function is illustrated in Figure 9.9c.

It is easy to find directly from the chart in Figure 9.9a that there is only one minimal *sp* Boolean form for the function:

$$f = \bar{x}_4 \bar{x}_3 x_2 \vee \bar{x}_4 x_3 x_1 \vee x_4 \bar{x}_3 \bar{x}_2 \vee x_4 x_3 \bar{x}_1 \vee x_2 x_1$$

Thus, three prime implicants (three AND gates) must be added to the minimal form in this case to make the circuit hazard free.

It should be remembered that our procedure for the synthesis of hazard-free circuits is based on the assumption that the input states of the circuit are properly coded so that at most one input variable may change at a time.

Let us consider now asynchronous sequential switching circuits. So far it has been assumed that a DELAY element is placed in each feedback loop associated with an internal state variable. It has been also assumed that the signal delay represented by these elements is sufficiently long to allow the functional generator to reach a steady state before the present internal state changes.

Delays inserted in feedback loops of asynchronous sequential switching circuits considerably slow down their performance. This is undesirable and, therefore, we try to eliminate these delays wherever possible. It turns out that either they can be eliminated completely or no more than one DELAY element is required.

If a memoryless switching circuit with a hazard is used as the functional generator of an asynchronous sequential circuit, the resulting sequential circuit can enter an incorrect internal state due to the hazard. To ensure the correct performance of the circuit, its functional generator must be hazard-free. Unfortunately, this is only a necessary condition. Indeed, if a change of the input state effects one feedback loop before another, and no DELAY elements are used in the feedback loops, the circuit may reach another stable state than the one required. This is called the *essential hazard* and is illustrated by the following example.

EXAMPLE 9.4 Consider the normal asynchronous Mealy machine a portion of whose transition function is specified in Figure 9.10a. An implementation of this portion of the transition function is shown in Figure 9.10b.

Assume that the DELAY elements in the circuit in Figure 9.10b are omitted. Assume further that the total state (x, z_2, z_1) is $(0, 0, 0)$ and that the input changes from 0 to 1. The circuit should pass to the stable state $(1, 0, 1)$ under these conditions. However, if the delay associated with the NOT element is greater than the delay of the OR gate, then z_1 becomes 1 before $\bar{x}$ is changed

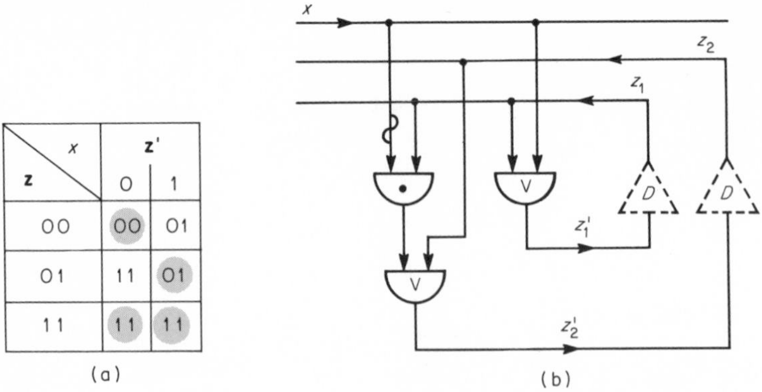

Figure 9.10 Asynchronous sequential circuit with an essential hazard (Example 9.4).

from 1 to 0 and, thus, z_2 becomes 1, i.e., the circuit passes to the erroneous stable state (1, 1, 1). This is an example of the essential hazard. Clearly, it can be prevented by using a DELAY element in the feedback loop for z_1.

Next we proceed to a formalization of the concept of the essential hazard and show its meaning for the implementation of asynchronous circuits without DELAY elements.

DEFINITION 9.19 A total state $\mathbf{z}$ and an input variable x of a normal asynchronous Mealy machine M represents an *essential hazard* if and only if, when M is initially in state $\mathbf{z}$, three consecutive changes in x bring the machine to a state other than the one arrived at after the first change in x.

THEOREM 9.5 If a normal asynchronous Mealy machine does not contain any essential hazards, it can be implemented by an asynchronous switching circuit which does not contain any DELAY elements.

PROOF Since the definition of the essential hazard refers to changes of a single input variable, we need to consider changes between two adjacent input states, say $\mathbf{x}_i$ and $\mathbf{x}_j$. Let $\mathbf{z}_a$ be a stable state for $\mathbf{x}_i$ and let $\mathbf{g}(\mathbf{x}_j, \mathbf{z}_a) \doteq \mathbf{z}_b$. Suppose that the transition from $\mathbf{z}_a$ to $\mathbf{z}_b$ is accomplished by a change of a single variable, say $\mathbf{z}_k$. All cases which can appear for a machine when starting in the stable state $\mathbf{z}_a$ and changing the input state from $\mathbf{x}_i$ to $\mathbf{x}_j$ are shown in the table form in Figure 9.11. Cases (e) and (f) contain an essential hazard for the internal state $\mathbf{z}_a$ and the input variable in which $\mathbf{x}_i$ and $\mathbf{x}_j$ differ one from the other. It suffices thus to consider only cases (a) through (d) to prove the theorem.

(a) No internal state variable changes and, therefore, no erroneous stable state can be reached for any state assignment.

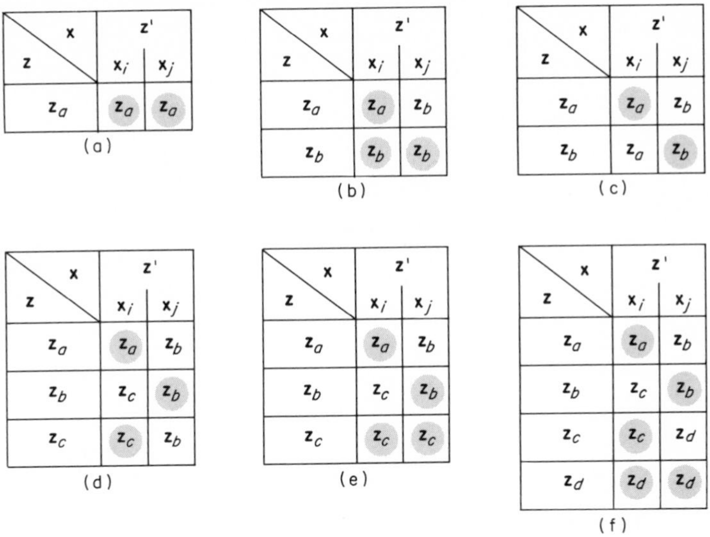

Figure 9.11 Basic patterns for a change of a single input variable (proof of Theorem 9.5).

(b) Since $\mathbf{g}(\mathbf{z}_b, \mathbf{x}_i) = \mathbf{z}_b$, no other variable than z_k will tend to change regardless whether the input change $\mathbf{x}_i$ to $\mathbf{x}_j$ has been sensed or not.

(c) Since the transition is associated with a change of a single variable z_k and $\mathbf{g}(\mathbf{z}_b, \mathbf{x}_i) = \mathbf{z}_a$, no other variable than z_k tends to change.

(d) The transition $\mathbf{g}(\mathbf{x}_i, \mathbf{z}_b) = \mathbf{z}_c$ may cause difficulties. After z_k (associated with the transition from $\mathbf{z}_a$ to $\mathbf{z}_b$) changes, some other variables may not have sensed the input change from $\mathbf{x}_i$ to $\mathbf{x}_j$ yet. If the variable which changes in the transition from $\mathbf{z}_b$ to $\mathbf{z}_c$, say variable z_l, is among them, then z_l may change too. A variable may exist, say z_m $(m \neq k, m \neq l)$, which senses changes of z_k and z_l in an order. If the change of z_l is sensed before the change of z_k, then the machine may pass to a state $\mathbf{z}_d$ which is different from states $\mathbf{z}_a, \mathbf{z}_b, \mathbf{z}_c$. In addition, state $\mathbf{z}_d$ may be stable for the input state $\mathbf{x}_j$. This would cause an erroneous performance. To avoid this error, we add a new internal state $\mathbf{z}_t$ (called a *trap state*) to the machine and define $\mathbf{g}(\mathbf{x}_i, \mathbf{z}_t) = \mathbf{z}_t$ and $\mathbf{g}(\mathbf{x}_j, \mathbf{z}_t) = \mathbf{z}_b$. If state $\mathbf{z}_t$ is coded as $\mathbf{z}_a$ with negated variable z_l, then if the change of z_l is sensed before the change of z_k, no variable z_m exists. After the input change and the change of z_k are finally sensed, the machine will pass to the stable state $\mathbf{z}_b$.

To complete the proof, we have to demonstrate that such a state assignment exists which requires only one variable to change in any transition and enables us to introduce as many trap states as necessary.

Let the given machine have r internal states denoted $\mathbf{z}_0, \mathbf{z}_1, \ldots, \mathbf{z}_{r-1}$ and let $A_0, A_1, \ldots, A_{2^r-1}$ be those sets of states of $2^r - 1$ logic variables, say $z_1, z_2, \ldots, z_{2^r-1}$, which would be assigned to a machine with 2^r internal states by the general state assignment scheme based on equations (9.2). Let A_{2^i} be assigned to state $\mathbf{z}_i$ of the machine and let A_t, where $t = 2^a + 2^b + 2^c$, be assigned to the trap state $\mathbf{z}_t$ which is added to the states $\mathbf{z}_a, \mathbf{z}_b, \mathbf{z}_c$ in case of the transition function in Figure 9.11d. Clearly, $t \leq 2^r - 1$ and $t \neq 2^e$ for any integer e so that A_e exists and is not assigned to any of the states $\mathbf{z}_0$ through $\mathbf{z}_{r-1}$. Furthermore, if $(\mathbf{z}_\alpha, \mathbf{z}_\beta, \mathbf{z}_\gamma)$ is a triple of internal states of the given machine which is different from the triple $(\mathbf{z}_a, \mathbf{z}_b, \mathbf{z}_c)$ but is represented by the pattern of transition function as shown in Figure 9.11d, the $\tau = 2^\alpha + 2^\beta + 2^\gamma \neq t$. This guarantees that A_e is not assigned to two different trap states.

It remains to show that A_t contains that state of variables $z_0, z_1, \ldots, z_{2^r-1}$ which differs from their state assigned to $\mathbf{z}_a$ only in variable z_l. Let $(\dot{z}_0, \ldots, \dot{z}_k, \ldots, \dot{z}_l, \ldots, \dot{z}_{2^r-1})$ be assigned to $\mathbf{z}_a$. Then, $(\dot{z}_0, \ldots, \bar{\dot{z}}_k, \ldots, \dot{z}_l, \ldots, \dot{z}_{2^r-1})$ is assigned to $\mathbf{z}_b$ and $(\dot{z}_0, \ldots, \bar{\dot{z}}_k, \ldots, \bar{\dot{z}}_l, \ldots, \dot{z}_{2^r-1})$ is assigned to $\mathbf{z}_c$. When we take the sum of these three vectors modulo two, we obtain $(\dot{z}_0, \ldots, \dot{z}_k, \ldots, \bar{\dot{z}}_l, \ldots, \dot{z}_{2^r-1})$. This vector is to be assigned to the trap state. Since, by equation (9.2), $a_{i+1} = 1$ and $a_j = 0$ $(j \neq i+1)$ for each state $\mathbf{z}_i$ $(i = a, b, c)$, the modulo two sum of the above specified three vectors in terms of values of a_{i+1} $(i = a, b, c)$ is $2^a + 2^b + 2^c = t$. ∎

To illustrate the proof of Theorem 9.5, the normal asynchronous Mealy machine will be considered whose transition function is specified in Figure 9.12a. The machine has no essential hazards and contains four transition

z \ x	a	b	c	d
z_0	z_1	z_1 ←1—	z_0	—3→ z_3
z_1	z_1	z_1	z_2 ←4—	z_1
z_2	z_1	z_1	z_2	z_3
z_3	z_1	z_3 —2→	z_2	z_3

(a)

Assigned Sets	z \ x	a	b	c	d
A_1	z_0	z_1	z_1	z_0	z_3
A_2	z_1	z_1	z_1	z_2	z_1
A_4	z_2	z_1	z_1	z_2	z_3
A_8	z_3	z_1	z_3	z_2	z_3
A_7	z_4	—	z_1	z_4	—
A_{14}	z_5	—	z_5	z_2	z_5
A_{13}	z_6	—	—	z_6	z_3

(b)

Figure 9.12 An example illustrating the proof of Theorem 9.5.

patterns of the type shown in Figure 9.11d. These four transition patterns are indicated by the arrows labeled 1, 2, 3, 4 and are associated with the trios $(\mathbf{z}_0, \mathbf{z}_1, \mathbf{z}_2)$, $(\mathbf{z}_1, \mathbf{z}_2, \mathbf{z}_3)$, $(\mathbf{z}_0, \mathbf{z}_2, \mathbf{z}_3)$, $(\mathbf{z}_1, \mathbf{z}_2, \mathbf{z}_3)$ of internal states respectively (trios with these properties are referred to as *d-trios*). Since the same trio is involved in 2 and 4, a single trap state can be used for both of these transition patterns. Hence, only three trap states need be added. Let $\mathbf{z}_4$, $\mathbf{z}_5$, $\mathbf{z}_6$ be used, respectively, for 1, 2 (and 4), 3. Then, following the procedure described in the proof of Theorem 9.5, we obtain the machine in Figure 9.12b. The sets A_i are determined by four equations of the type (9.2) which contain 15 variables z_1 through z_{15}.

Exercises to Section 9.4

9.4-1 Suggest a procedure of determining 1-sets for a given logic function directly in its chart representation.

9.4-2 State and prove a theorem analogous to Theorem 9.4 which establishes a necessary and sufficient condition for hazard-free switching circuits based on *ps* Boolean forms.
Hint: Define 0-sets and use the principle of duality in Boolean algebra.

9.4-3 Design a hazard-free switching circuit based on the *sp* Boolean form for each of the following logic functions of n variables:
- (a) $f(s) = 0$ for $s = 7, 10, 11, 13, 14, 15$ and $f(s) = 1$ for $s = 0, 1, 2, 3, 4, 5, 8, 9$, $n = 4$.
- *(b) $f(s) = 1$ for $s = 3, 4, 6, 7$, $n = 3$.
- (c) $f(s) = 1$ for $s = 0, 3, 4, 6, 7, 8, 12, 13, 14, 15$, $n = 4$.
- (d) $f(s) = 1$ for $s = 0, 1, 5, 12$ and $f(s) = 0$ for $s = 2, 4, 6, 7, 8, 9, 13, 14$, $n = 4$.
- (e) $f = x_1x_2 \vee x_1x_3 \vee x_2x_3$.
- (f) $f = x_1 \not\equiv (x_1x_2 \vee (x_3 \not\equiv x_4))$.

9.4-4 Repeat Exercise 9.4-3 for the *ps* Boolean forms.

9.4-5 For each of the following normal asynchronous Mealy machines whose transition functions are specified by their ST-matrices, introduce necessary trap states and determine such a state assignment which makes an implementation without feedback delays possible.

(a)
$$\begin{array}{c} \mathbf{z}_0 \\ \mathbf{z}_1 \\ \mathbf{z}_2 \\ \mathbf{z}_3 \end{array} \begin{bmatrix} \underline{00 \cup 10} & 01 & 11 & \varnothing \\ 10 & \underline{00 \cup 01} & 11 & \varnothing \\ \varnothing & 01 & \underline{10 \cup 11} & 00 \\ \varnothing & \varnothing & 10 \cup 11 & \underline{00 \cup 01} \end{bmatrix}$$

(b)
$$\begin{array}{c} \mathbf{z}_0 \\ \mathbf{z}_1 \\ \mathbf{z}_2 \\ \mathbf{z}_3 \\ \mathbf{z}_4 \\ \mathbf{z}_5 \end{array} \begin{bmatrix} \underline{11} & \varnothing & \varnothing & 00 & 01 & 11 \\ \varnothing & \underline{00} & 10 & 01 & \varnothing & \varnothing \\ \varnothing & 00 & \underline{10 \cup 11} & \varnothing & \varnothing & \varnothing \\ \varnothing & \varnothing & \varnothing & \underline{00 \cup 01} & \varnothing & 01 \\ 11 & \varnothing & 10 & \varnothing & \underline{00 \cup 01} & \varnothing \\ \varnothing & \varnothing & 11 & 00 & \varnothing & \underline{01 \cup 00} \end{bmatrix}$$

9.5 MULTIPLE-INPUT CHANGES

So far, we have required that only one input variable may change in any transition. This is a basic requirement which is almost always applied in the theory of asynchronous sequential switching circuits. It represents a severe restriction of the applicability of asynchronous sequential circuits or limits considerably the speed at which these circuits operate. Several methods have been developed to bypass this restriction. A method which is based on a suitable encoding of the input states is described in this section.

Suppose that an asynchronous sequential switching circuit S is specified by its ST-structure which requires changes of more than one input variable in some transitions. Suppose further that S is implemented by a modified asynchronous circuit S_m which is connected in series with an encoder, usually called a *source box*, as shown in Figure 9.13. The ST-structure of S_m is derived from S as follows:

1. Decode the state of input variables of S so that the ith input variable of S_m be equal to 1 and all other input variables of S_m be equal to 0 for any state $\mathbf{x}_i$.
2. Add a new state of input variables consisting of all zeroes which makes each of the internal states stable. This new state, which is usually referred to as the *spacer*, serves as an auxiliary state of the input variables that appears between any two consecutive input states. Output variables remain unchanged from their previous values for the spacer.

Input states of S_m are generated by the source box on the basis of the input states of S. If the state changes from $\mathbf{x}_i$ to $\mathbf{x}_j$ at the input of the source box, then its output changes from the state with the ith variable equal to 1 to the spacer and, then, to the state with the jth variable equal to 1.

Note that S_m has the same behavior as S and, if the source box operates in the manner described above, S_m does not contain any essential hazards

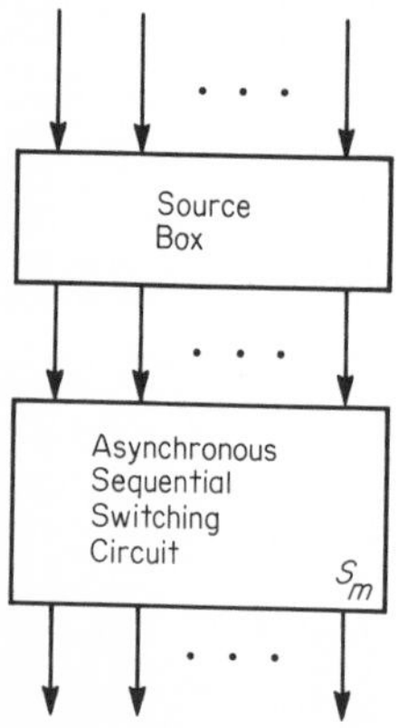

Figure 9.13 Connection of the source box with the modified circuit S_m.

or d-trios. No delays are thus required in the feedback loops of S_m and any state assignment which prevents critical race conditions is acceptable.

The source box is a sequential asynchronous switching circuit whose general ST-structure is shown in Figure 9.14. If the original circuit S has p stimuli, say $\mathbf{x}_1, \mathbf{x}_2, \ldots, \mathbf{x}_p$, then the source box has $p + 1$ internal states (one internal state for each stimulus of S and the spacer). Let the internal states of the source box be represented by the p-tuples $(v_p, v_{p-1}, \ldots, v_1)$. Then the spacer is represented by $v_p = v_{p-1} = \cdots = v_1 = 0$ and the internal state assigned to stimulus $\mathbf{x}_i$ is represented by $v_i = 1$ and $v_j = 0$ for all $j \neq i$. The internal variables of the source box are identical with its output variables, i.e., their states are stimuli to the modified circuit S_m.

EXAMPLE 9.5 Using INVERTED DELAYS, AND, OR, and NOT elements, implement the ST-structure S specified in Figure 9.15a

Since the given ST-structure includes simultaneous changes of two input variables (01 to 10 and vice versa), we design the source box and the modified circuit S_m whose ST-structures are specified in Figures 9.15 b and c, respectively. The state assignment shown in Figure 9.15c can be used to prevent critical race conditions. The following functions can be derived directly from the ST-structures:

Modified Circuit S_m

z_2	z_1	x_3	x_2	x_1	y_2	y_1	z_2'	z_1'	y_2'	y_1'
0	0	0	0	0	0	0	0	0	0	0
0	0	0	0	0	1	1	0	0	1	1
0	0	0	0	1	–	–	0	0	0	0
0	0	0	1	0	–	–	0	1	1	0
0	0	1	0	0	–	–	0	0	1	1
0	1	0	0	0	–	–	0	1	1	0
0	1	0	0	1	–	–	1	1	1	1
0	1	0	1	0	–	–	0	1	1	0
0	1	1	0	0	–	–	0	0	1	1
1	1	0	0	0	–	–	1	1	1	1
1	1	0	0	1	–	–	1	1	1	1
1	1	0	1	0	–	–	1	0	0	1
1	1	1	0	0	–	–	1	0	1	0
1	0	0	0	0	0	1	1	0	0	1
1	0	0	0	0	1	0	1	0	1	0
1	0	0	0	1	–	–	0	0	0	0
1	0	0	1	0	–	–	1	0	0	1
1	0	1	0	0	–	–	1	0	1	0

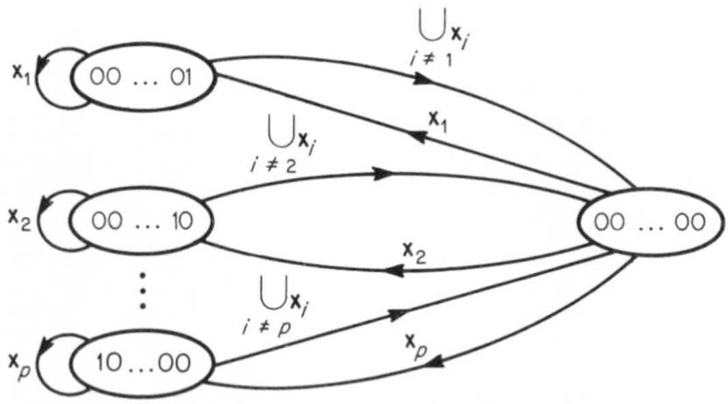

Figure 9.14 General ST-diagram of the source box.

S

z \ x_2x_1	z′ 00	z′ 01	z′ 10	y_2y_1 00	y_2y_1 01	y_2y_1 10
1	1	2	1	00	10	11
2	3	2	1	11	10	11
3	3	4	4	11	01	10
4	1	4	4	00	01	10

(a)

Source Box

$v_3v_2v_1$ \ x_2x_1	$v_3'v_2'v_1'$ 00	$v_3'v_2'v_1'$ 01	$v_3'v_2'v_1'$ 10
000	001	010	100
001	001	000	000
010	000	010	000
100	000	000	100

(b)

S_m

z_2z_1	z \ $v_3v_2v_1$	z′ 000	z′ 001	z′ 010	z′ 100	y_2y_1 000	y_2y_1 001	y_2y_1 010	y_2y_1 100
00	1	1	1	2	1	As before	00	10	11
01	2	2	3	2	1	As before	11	10	11
11	3	3	3	4	4	As before	11	01	10
10	4	4	1	4	4	As before	00	01	10

↑ Spacer (under column z′ 000)

(c)

Figure 9.15 Transformation of circuit *S* to the modified circuit S_m and the source box (Example 9.5).

Source Box

v_3	v_2	v_1	x_2	x_1	v_3'	v_2'	v_1'
0	0	0	0	0	0	0	1
0	0	0	0	1	0	1	0
0	0	0	1	0	1	0	0
0	0	1	0	0	0	0	1
0	0	1	0	1	0	0	0
0	0	1	1	0	0	0	0
0	1	0	0	0	0	0	0
0	1	0	0	1	0	1	0
0	1	0	1	0	0	0	0
1	0	0	0	0	0	0	0
1	0	0	0	1	0	0	0
1	0	0	1	0	1	0	0

Minimal Boolean forms for the source box are

$$v_1' = \bar{v}_3 \bar{v}_2 \bar{x}_2 \bar{x}_1$$
$$v_2' = \bar{v}_3 \bar{v}_1 x_1$$
$$v_3' = \bar{v}_2 \bar{v}_1 x_2$$

and the corresponding implementation is illustrated in Figure 9.16. Minimal hazard free Boolean forms for the modified circuit S_m are

$$z_1' = z_1 \bar{v}_3' \bar{v}_1' \vee \bar{z}_2 v_2' \vee \bar{z}_2 z_1 \bar{v}_3'$$
$$z_2' = z_2 \bar{v}_1' \vee z_1 v_1' \vee z_2 z_1$$
$$y_1' = z_2 z_1 \bar{v}_3' \vee z_2 v_1' \vee z_2 v_2' \vee \bar{z}_2 v_3' \vee z_1 v_1' \vee \bar{z}_2 \bar{z}_1 \bar{v}_2' \bar{v}_1' y_1$$
$$y_2' = \bar{z}_2 \bar{z}_1 \vee z_1 \bar{v}_2' \vee \bar{z}_2 v_2' \vee v_3' \vee \bar{v}_2' \bar{v}_1' y_2$$

and the implementation is illustrated in Figure 9.17. Observe that no DELAY elements are needed in the feedback loops.

A synchronous circuit implementing S (Figure 9.15a) and based on the same state assignment as the asynchronous circuit is shown in Figure 9.18 to get an appreciation of the simplicity of synchronous sequential circuits when compared with asynchronous circuits.

Note that the source box produces the spacer for approximately the delay time Δt_M of the memory elements used (INVERTED DELAYS in Figure 9.16). If all changes at the input of the source box are completed for a single transition within time Δt_I of the first change and Δt_B is the maximum stray delay associated with the source box, then the condition

$$\Delta t_M > \Delta t_I + \Delta t_B$$

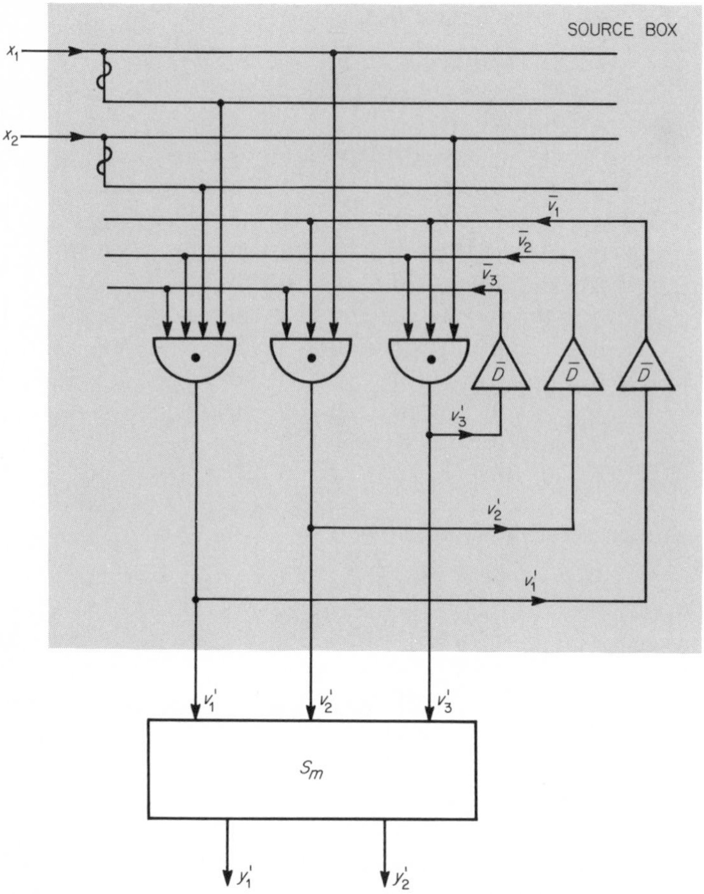

Figure 9.16 Source box for the circuit specified in Figure 9.15a (Example 9.5).

must be satisfied to ensure proper functioning of the source box. If the maximum time required by the modified circuit to reach a stable state is Δt_S, then also

$$\Delta t_M > \Delta t_S$$

must be satisfied.

Exercises to Section 9.5

9.5-1 Show that the minimum time which is permitted between two consecutive changes at the input of the source box is equal to $2(t_M + t_B)$.

9.5-2 Using SR FLIP FLOPS instead of INVERTED DELAYS, implement the ST-structure specified in Example 9.5.

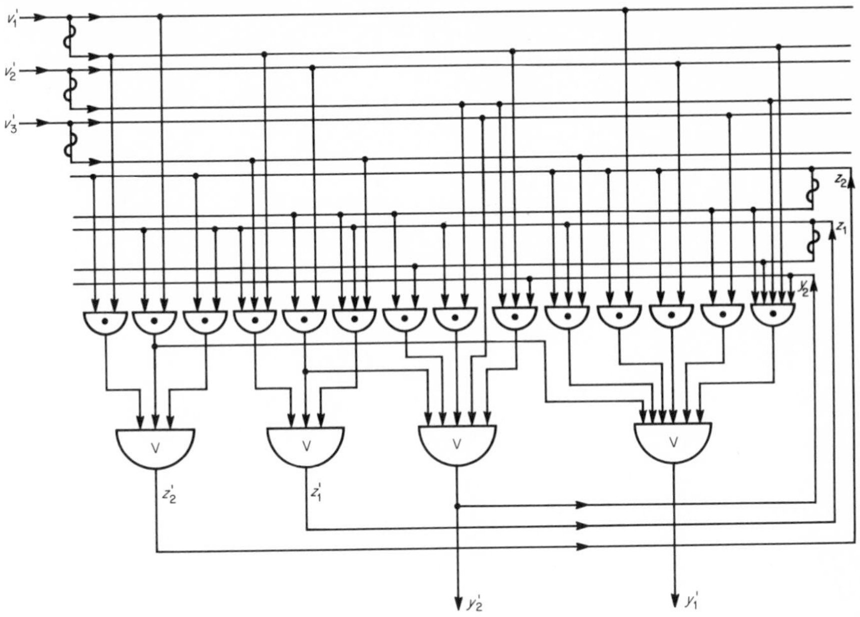

Figure 9.17 Modified circuit S_m for the given circuit S specified in Figure 9.15a (Example 9.5).

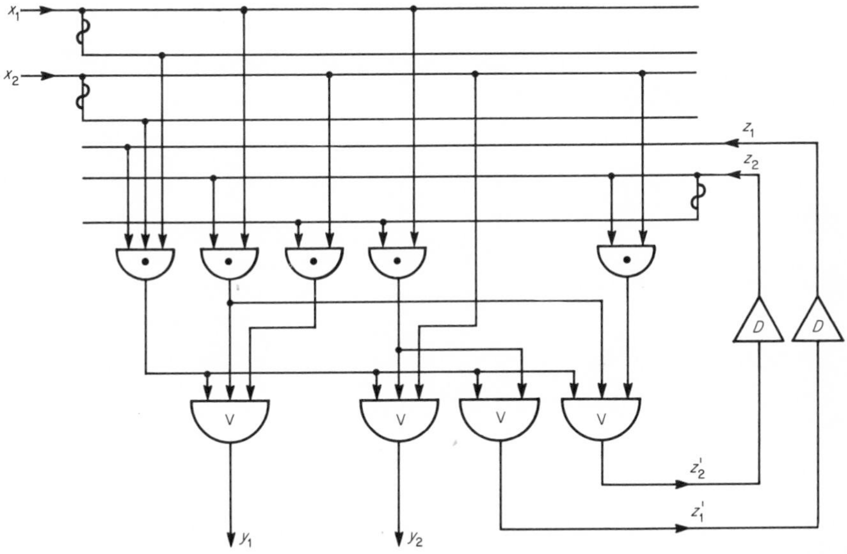

Figure 9.18 Synchronous circuit implementing S (Figure 9.15a).

9.5-3 Using the source box, implement each of the ST-structures specified in Figure 9.19 by
(a) DELAY and NAND elements.
(b) RS FLIP FLOPS and the cascade element.
(c) JK FLIP-FLOPS and NOR elements.

9.6 SUMMARY AND EXAMPLES

The synthesis of asynchronous sequential switching circuits involves, besides all the problems that are associated with the synthesis of synchronous sequential circuits, some additional problems due to race conditions and hazards.

The first step in the synthesis is a determination of the ST-structure for a given description of the behavior of the circuit. This step, which is referred to as the *abstract synthesis*, is the same for both synchronous and asynchro-nous switching circuits. As a rule, either the Mealy machine or the Moore

z \ x_2x_1	z' 00	z' 01	z' 11	y 00	y 01	y 11
1	2	3	1	–	–	1
2	2	3	2	0	–	0
3	4	3	3	–	1	0
4	4	3	2	1	–	–

(a)

z \ x_2x_1	z' 00	z' 10	z' 11	y_2y_1 00	y_2y_1 10	y_2y_1 11
1	1	2	3	00	–	–
2	1	2	3	–	01	–
3	1	2	3	–	–	11

(b)

z \ $x_3x_2x_1$	z' 000	z' 011	z' 100	z' 110	z' 111	y_2y_1 000	y_2y_1 011	y_2y_1 100	y_2y_1 110	y_2y_1 111
1	3	1	1	4	1	–	00	01	–	10
2	2	2	5	4	3	00	11	–	–	–
3	3	1	3	4	3	01	–	10	–	00
4	3	1	3	4	3	–	–	–	11	–
5	2	2	5	5	5	–	–	11	–	01
6	6	1	5	5	5	11	–	–	00	–

(c)

Figure 9.19 Illustration for Exercise 9.5-3.

machine is used as a standard model for performing the abstract synthesis.

The second step consists, usually, in the *minimization of internal states* of the machine involved. Neither this step differs for synchronous and asynchronous circuits. All procedures described in Chapter 7 are thus applicable for asynchronous circuits equally well as they are applicable for synchronous circuits.

The third step, which is not always performed, is a *decomposition* of the given machine into two or more component machines, each of which is then designed separately. It turns out that the parallel decomposition does not give rise to any additional problems for asynchronous switching circuits, i.e., Theorem 8.11 remains valid for asynchronous circuits. On the other hand, Theorem 8.10, which specifies necessary and sufficient conditions for the serial decomposition, is not valid for asynchronous circuits because it is based on the assumption that the component machines operate concurrently. This assumption is not satisfied for asynchronous machines so that the decomposition must be such that the total performance of the serial connection is independent of the order in which the component machines change their states or an appropriate order must be fixed by inserting suitable DELAYS.

The fourth step should be the *evaluation of individual models* for the given machine (or component machines) and the selection of one of them. However, no study has been done in this direction as yet and only little can be adopted from the principles of model evaluation for synchronous switching circuits as described in Section 8.8.

The fifth step is represented by the *state assignment problem*. For asynchronous circuits, this problem has many facets. The chosen state assignment must prevent critical race conditions and should minimize the transition time, the number of internal variables, and the complexity of the functional generator of the circuit.

The sixth step consists in the *investigation of hazards* and in introducing such means by which the given circuit is prevented from those hazards that may produce malfunctions. This includes the introduction of necessary trap states or DELAY elements, a design of the source box and an appropriate modification of the given ST-structure, etc. Recall that this step may influence the state assignment problem in some cases.

The seventh step consists in the *determination of the behavior of the functional generator* and its design. As a rule, the functional generator is required to be hazard-free.

As the eighth step in the synthesis of asynchronous sequential switching circuits, a careful *analysis of the final design* should be made. All essential distributions of stray delays (element and line delays) must be considered in this analysis.

The whole procedure of the synthesis of asynchronous switching circuit will be now illustrated by several examples.

EXAMPLE 9.6 Design an *asynchronous binary counter modulo* 4 (Exercise 8.9) employing AND, OR, NOT, and DELAY elements. States (y_2, y_1) of the output variables should represent, in the binary form, directly the number of the one signals accepted by the input and taken modulo 4.

As the output states may acquire one of four possible states for the same stimulus, the counter must have at least four internal states. Since, however, every change in the stimulus must be distinguished from the others, the counter requires eight internal states in the end. It is easy to determine the ST-structure of the counter, which is shown in Figure 9.20. Since the pairs of responses corresponding to individual internal states are pair-wise different, the ST-structure is already in its minimal form.

To distinguish eight internal states, at least three variables must be used and to prevent critical race conditions, any pair of states i $(i = 1, 2, \ldots, 8)$ and $(i + 1)$ (modulo 8) should differ in the value of exactly one variable. A state assignment with this property is usually referred to as the *Gray code*.

The problem of determining a Gray code for n variables and r states can be solved by finding a closed path in the diagram representing n-dimensional unit cube which passes through r nodes, each of them only once. Such a path, which is called the *Hamiltonian cycle* in graph theory, can be easily determined in the appropriate diagram by inspection. In our example, we use the three-dimensional unit cube as shown in Figure 9.21a. States representing the nodes are denoted by their state identifiers. In Figures 9.21 b and c, two Hamiltonian cycles are shown which give us directly the state assignment I and II as specified in Figure 9.20b. We can find by a simple consideration that all other state assignments can be derived from either I or II by a permutation and/or a negation of the variables. Calculation of parameters $A_{\text{I}}, A_{\text{II}}, B_{\text{I}}, B_{\text{II}}$, as described in Section 9.3, does not show any difference in these two assignments (all of the parameters are equal to zero) so that they may be considered as equally good.

Let us use the state assignment I. Then, we obtain the following Boolean forms which represent the simplest hazard-free two stage design of the functional generator of the counter:

$$z_3' = z_3 z_1 \vee z_3 x \vee z_2 \bar{z}_1 \bar{x} \; (\vee \underline{z_3 z_2})$$

$$z_2' = \underline{z_2 \bar{z}_1} \vee \underline{z_2 x} \vee \bar{z}_3 z_1 \bar{x} \; (\vee \bar{z}_3 z_2)$$

$$z_1' = z_1 \bar{x} \vee z_3 z_2 x \vee \bar{z}_3 \bar{z}_2 x \; (\vee z_3 z_2 z_1 \vee \bar{z}_3 \bar{z}_2 z_1)$$

$$y_2 = \underline{z_2 x} \vee \underline{z_2 \bar{z}_1} \vee z_3 z_1 \bar{x} \; (\vee \underline{z_3 z_2})$$

y_1 is the same as z_1'

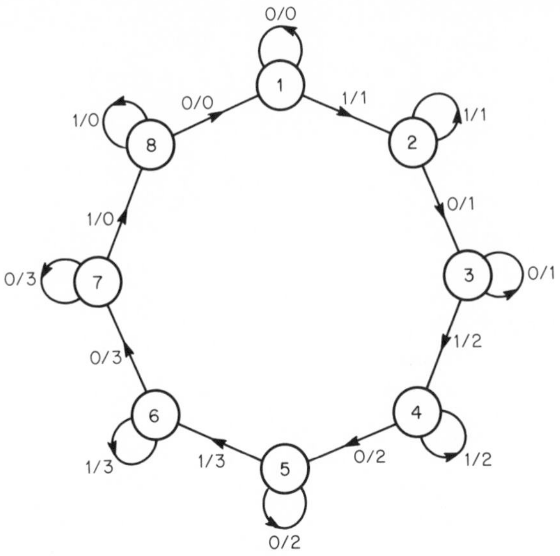

(a)

z \ x	z' 0	z' 1	$y_2 y_1$ 0	$y_2 y_1$ 1	State Assignment I	State Assignment II
1	1	2	00	01	000	000
2	3	2	01	01	001	001
3	3	4	01	10	011	011
4	5	4	10	10	010	111
5	5	6	10	11	110	101
6	7	6	11	11	111	100
7	7	8	11	00	101	110
8	1	8	00	00	100	010

(b)

Figure 9.20 ST-structure of the asynchronous binary counter modulo 4 (Example 9.6).

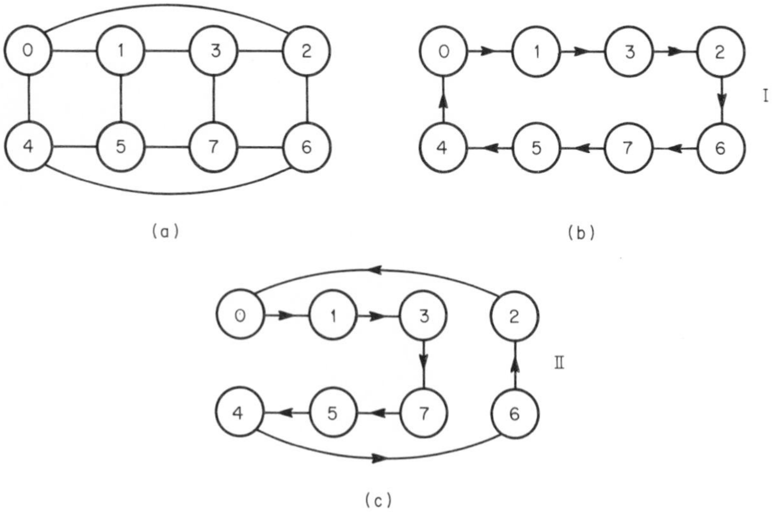

Figure 9.21 A construction of the Gray code for three variables and eight states.

The terms which were added to prevent hazards in the functional generator are enclosed in parentheses. The underlined terms are shared by two functions.

EXAMPLE 9.7 Design an *asynchronous binary reverse* (*bi-directional, two-way*) *counter modulo* 4 (Exercise 8.13). States (y_2, y_1) of the output variables are to represent, in the binary form, directly the difference $(N_1 - N_2)$ (modulo 4) of the number N_1 of the one signals accepted by the first input (variable x_1) and the number N_2 of the one signals accepted by the second input (variable x_2). Only one of the input variables can be equal to one at any time.

The ST-structure of the asynchronous reverse counter, which is shown in Figure 9.22, is an extension of the ST-structure of the one-directional counter (Figure 9.20). Since the triads of responses corresponding to individual internal states are pair-wise different, the ST-structure is already in its minimal form.

To distinguish eight internal states, at least three variables must be used. Critical race conditions can be prevented by using such a state assignment under which a single variable changes in each transition. A convenient procedure for determining this kind of state assignment for any asynchronous reverse counter will be now described.

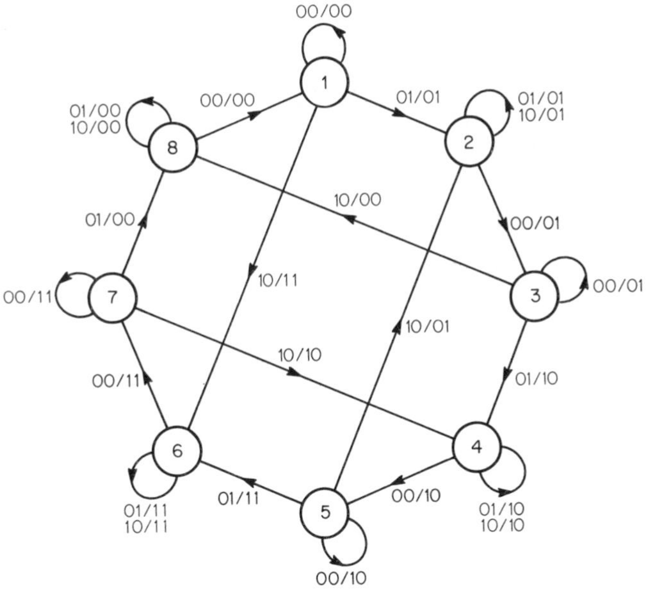

(a)

z \ x_2x_1	z' 00	01	10	y_2y_1 00	01	10	State Assignment (C_j)
1	1	2	6	00	01	11	000
2	3	2	2	01	01	01	001
3	3	4	8	01	10	00	101
4	5	4	4	10	10	10	111
5	5	6	2	10	11	01	011
6	7	6	6	11	11	11	010
7	7	8	4	11	00	10	110
8	1	8	8	00	00	00	100

(b)

Figure 9.22 ST-structure of the reverse asynchronous binary counter modulo 4 (Example 9.7).

Let us consider an asynchronous reverse counter modulo m, i.e., the counter which has $r = 2m$ internal states and $n = \ulcorner \log_2 r$ internal state variables. Then, the state assignment procedure consists of the following steps:

1. We construct a Gray code formed of a sequence (A_i) of $r/2 = m$ binary numbers $(i = 1, 2, \ldots, m)$ with $n - 1$ digits (places) each. In our case $(m = 4,\ r = 8,\ n = 3)$, we are concerned with a sequence of four two digit binary numbers, e.g.,

$$A_1 = 00,\ A_2 = 01,\ A_3 = 11,\ A_4 = 10$$

Recall that the Gray code for $n - 1$ variables can be constructed most conveniently by means of a diagram of the $(n - 1)$-dimensional unit cube.

2. We duplicate all the elements in the sequence (A_i) in such a manner as to obtain a sequence (B_j) for the elements of which we have

$$\begin{aligned} B_1 &= A_1 \\ B_2 = B_3 &= A_2 \\ B_4 = B_5 &= A_3 \\ \vdots \quad & \quad \vdots \\ B_{r-2} = B_{r-1} &= A_m \\ B_r &= A_1 \end{aligned}$$

For our case, we obtain the sequence

$$(B_j) = (00, 01, 01, 11, 11, 10, 10, 00)$$

3. For the numbers of sequence (B_j), we introduce the nth binary order digit in such a manner that its values will regularly alternate by pairs of numbers B_j. This will give us a sequence (C_j), which can be used directly as a race-free state assignment for the asynchronous reverse counter. In our case, we obtain the sequence (C_j) that is shown in Figure 9.22b. Since there is only one Gray code for two variables, provided that we ignore permutations and/or negations of variables, the state assignment in Figure 9.22b is unique. The completion of the design is left to the reader as an exercise (Exercises 9.6-5, 9.6-6).

EXAMPLE 9.8 Design an asynchronous switching circuit capable of indicating the location of an object which can move in both directions in front of two photoelectric cells that are sufficiently far away so that the object cannot obscure both of them simultaneously. The voltage generated by each photocell, which is transmitted via an amplifier to the switching circuit as shown in Figure 9.23, depends on whether the photocell is obscured by the

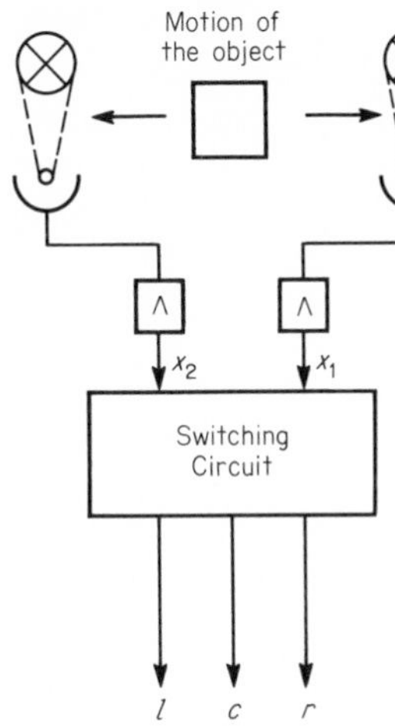

Figure 9.23 The arrangement described in Example 9.8.

object or not. The object cannot change the direction of its move if either of the photocells is obscured. The switching circuit has two input variables x_1, x_2 (each corresponding to one photocell) and three output variables, say l, c, r, each of which is assigned to a particular position of the object (left, center, right). If the left-hand side photocell is obscured, the object is considered to be on the left hand side. Similarly, if the right-hand side photocell is obscured, the object is considered to be on the right-hand side. Let us assume a logic description of the physical quantities involved such that an input variable is equal to one if the corresponding photocell is obscured and an output variable is equal to one if the object is in the position represented by this variable; otherwise the variables are equal to zero.

It is easy to determine the ST-diagram shown in Figure 9.24a. Loop A represents the move of the object from the left-hand side to the center and back, loop B follows its move from the center to the right-hand side. Using Theorem 7.4, we can easily find from the ST-table in Figure 9.24b that the ST-structure contains only one pair of compatible states, namely, $\mathbf{z}_2$ and $\mathbf{z}_6$. The two states can be represented by a single state $\mathbf{z}_{2,6}$ as shown in Figure 9.25.

Next, we must solve the state assignment problem. Using the Tracey procedure, we obtain the following trivial minimum transition time state assignment (pair 2, 6 of compatible states is identified as state 2):

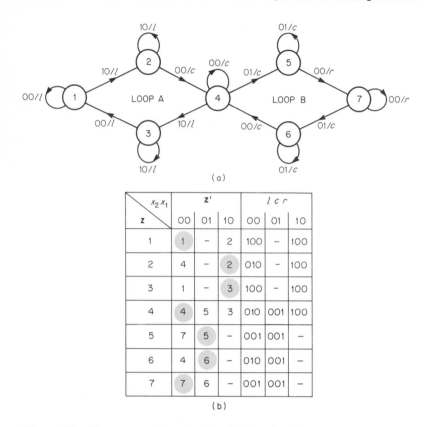

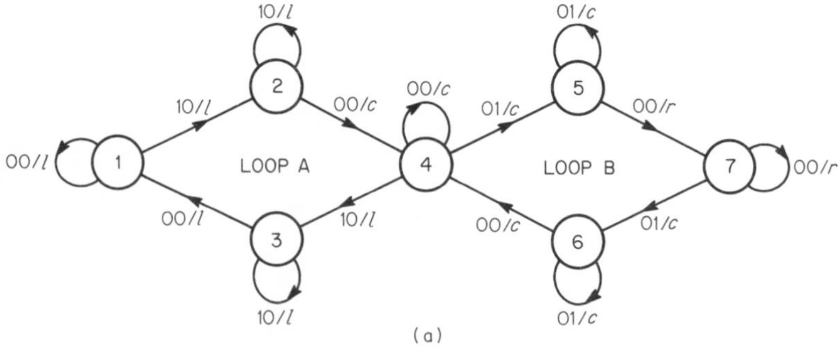

(a)

z \ x_2x_1	z' 00	01	10	$l\,c\,r$ 00	01	10
1	1	–	2	100	–	100
2	4	–	2	010	–	100
3	1	–	3	100	–	100
4	4	5	3	010	001	100
5	7	5	–	001	001	–
6	4	6	–	010	001	–
7	7	6	–	001	001	–

(b)

Figure 9.24 ST-structure of the switching circuit designed in Example 9.8.

	Essential partitions	Internal States 1	2	3	4	5	7
a	$\{\overline{1, 3}; \overline{2, 4}\}$	0	1	0	1	–	–
b	$\{\overline{1, 3}; \overline{5, 7}\}$	0	–	0	–	1	1
c	$\{\overline{2, 4}; \overline{5, 7}\}$	–	0	–	0	1	1
d	$\{\overline{2, 7}; \overline{4, 5}\}$	–	0	–	1	1	0
e	$\{\overline{1, 2}; \overline{3, 4}\}$	0	0	1	1	–	–

Maximal compatible classes are (a, b), $(a, \bar{c})$, (b, c), (d, e). By inspection, we can easily find three minimal coverings of the essential partitions by the compatible classes:

Covering I: (a, b), $(a, \bar{c})$, (d, e)
Covering II: (a, b), (b, c), (d, e)
Covering III: $(a, \bar{c})$, (b, c), (d, e)

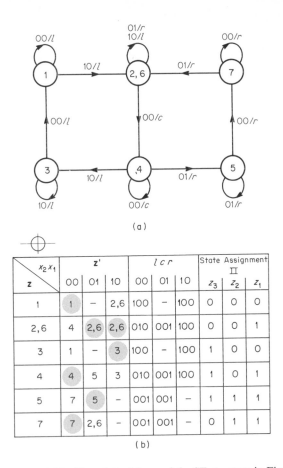

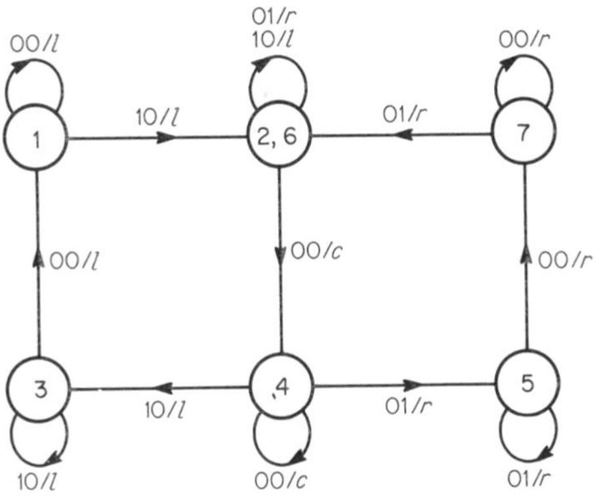

(a)

x_2x_1 / z	z' 00	z' 01	z' 10	lcr 00	lcr 01	lcr 10	State Assignment II z_3	z_2	z_1
1	1	–	2,6	100	–	100	0	0	0
2,6	4	2,6	2,6	010	001	100	0	0	1
3	1	–	3	100	–	100	1	0	0
4	4	5	3	010	001	100	1	0	1
5	7	5	–	001	001	–	1	1	1
7	7	2,6	–	001	001	–	0	1	1

(b)

Figure 9.25 The minimal form of the ST-structure in Figure 9.24.

The corresponding state assignments can be readily determined from the coverings and the trivial state assignment based directly on the essential partitions. We obtain:

States	I	II	III
1	0 0 0	0 0 0	0 0 0
2	0 1 1	0 0 1	0 1 0
3	1 0 0	1 0 0	1 0 0
4	1 1 1	1 0 1	1 1 0
5	1 0 1	1 1 1	1 0 1
6	0 0 1	0 1 1	0 0 1

A calculation of the numbers A_a and B_a (based on Theorems 9.2. and 9.3, respectively) for the individual state assignments which is not shown here, gives us the following results:

$$A_\mathrm{I} = 1, B_\mathrm{I} = 5$$
$$A_\mathrm{II} = 2, B_\mathrm{II} = 6$$
$$A_\mathrm{III} = 1, B_\mathrm{III} = 5$$

State assignment II has both of the numbers larger than the other two state assignments and, therefore, is selected (Figure 9.25b).

Next we determine the logic functions describing the behavior of the functional generator and their minimal hazard-free Boolean forms (Figure 9.26).

The final implementation is illustrated in Figure 9.27. There is no need in this case to try to increase the operation speed of this circuit by eliminating DELAY elements in the feedbacks and modifying, if necessary, appropriately the whole design.

Exercises to Section 9.6

9.6-1   Determine a parallel decomposition of the counter modulo 4 whose ST-structure is specified in Figure 9.20.

9.6-2   Using AND, OR, NOT, and DELAY elements, design the counter modulo 4 as specified in Figure 9.20 for the state assignment II and compare the result with the design for the state assignment I.

$$z_3' = \bar{x}_2 \bar{x}_1 \bar{z}_2 z_1 \lor x_1 z_3 \lor x_2 z_3 \lor (z_3 \bar{z}_2 z_1)$$

$$z_2' = x_1 z_3 \lor \bar{x}_1 z_2 \lor (z_3 \bar{z}_2)$$

$$z_1' = \bar{x}_2 z_1 \lor x_2 \bar{z}_3 \lor (\bar{z}_3 z_1)$$

$$l = x_2 \lor \bar{z}_1$$

$$c = \bar{x}_2 \bar{x}_1 \bar{z}_2 z_1$$

$$r = z_2 \lor x_1$$

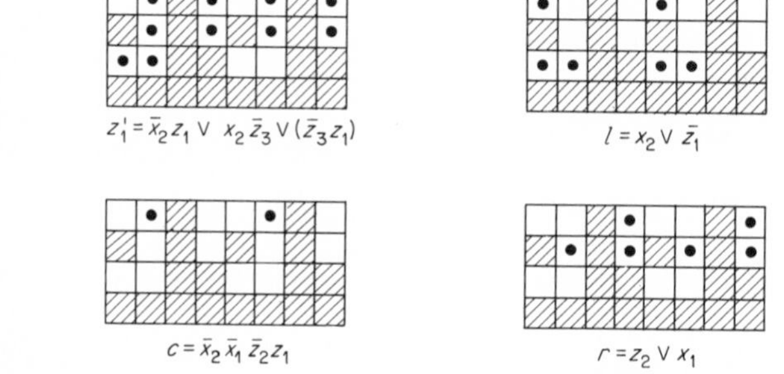

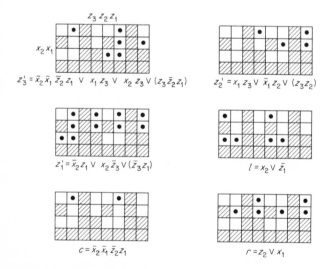

Figure 9.26   Logic functions and Boolean hazard-free minimal forms for the ST-structure in Figure 9.25 (Example 9.8).

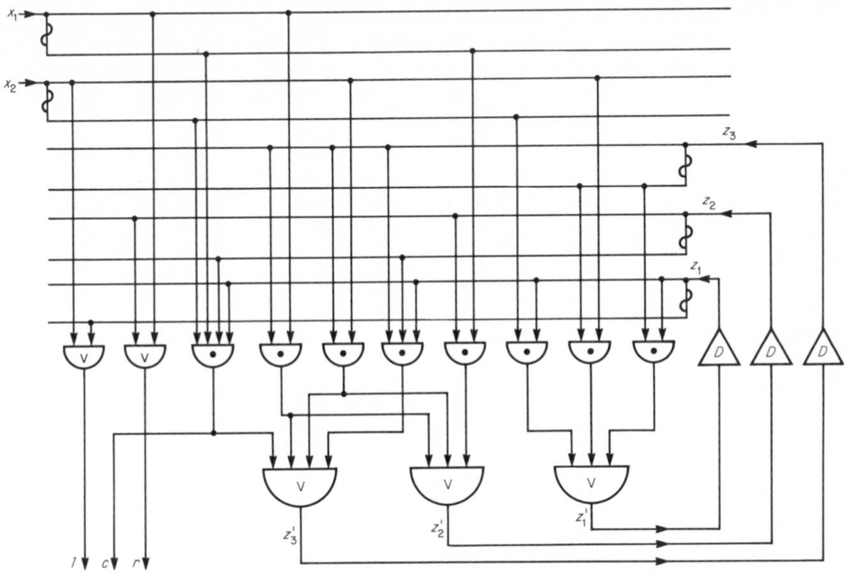

Figure 9.27 The switching circuit which identifies the position of a movable object.

9.6-3 Discuss the possibility of implementing the counter specified in Figure 9.20 without DELAY elements in the feedback loops.

9.6-4 Prove that the procedure for determining a free-race state assignment for the asynchronous reverse counter, as described in Example 9.7, produces a correct state assignment for any given integer $m \geq 2$.

9.6-5 Design a hazard-free functional generator of the asynchronous reverse counter modulo 4 whose ST-structure and free-race state assignment are given in Figure 9.22b. Use AND, OR, and NOT elements and assume that the memory is built by:
(a) DELAY elements.
(b) SR FLIP-FLOPS.

9.6-6 Discuss the possibility of implementing the asynchronous reverse counter without any memory elements in the feedback loops.

9.6-7 Design an asynchronous reverse counter modulo 10 built up by relays.

9.6-8 Design a relay asynchronous switching circuit which identifies the position of a movable object as described in Example 9.8.

Comprehensive Exercises to Chapter 9

9.1 Manually operated contacts are still indispensable devices even in the most advanced computer systems. They enable the operator to control the system. The contact bounce, which is very difficult to eliminate, may produce serious malfunctions if not properly treated. Design a bounce corrector described as follows: Assume that a transfer contact changes its position from x_1 to x_2 at time t_1 as illustrated in Figure 9.28. After making its initial connection at x_2, it bounces several times, producing thus a sequence of short pulses before settling permanently at x_2. The bound corrector must follow the change from x_1 to x_2 but must not allow the erroneous sequence of pulses to appear at its output y as shown in the diagram. A similar prevention must be ensured for the change from x_2 to x_1 (time t_2 in the diagram).

9.2 Design a relay combination lock, which is controlled by two manually operated make contacts x_1 and x_2 and which opens (output variable $y_1 = 1$) when the following sequence is accepted

$$\begin{aligned} t &= 0\ 1\ 2\ 3\ 4\ 5\ 6 \\ x_1(t) &= 0\ 0\ 0\ 0\ 1\ 1\ 0 \\ x_2(t) &= 0\ 1\ 0\ 1\ 1\ 0\ 0 \end{aligned}$$

If any other sequence is accepted then output variable $y_2 = 1$ which causes an alarm. In all other cases $y_1 = y_2 = 0$. The circuit is reset to the original state by a special contact x_3 ($x_3 = 1$ makes the reset, $x_3 = 0$ does not effect the circuit at all).

9.3 Design an asynchronous switching circuit by which you can select by each activation of a push-button one pulse from a sequence of clock pulses as illustrated in Figure 9.29. The whole clock pulse must be selected regardless of the time when the push button is activated.

9.4 Design an asynchronous switching circuit with two input variables x_1, x_2 and one output variable y whose behavior is described as follows:

(i) Only one of the input variables can change at any time.

(ii) $y = 1$ if and only if $x_1 = x_2 = 1$ and variable x_2 has changed last.

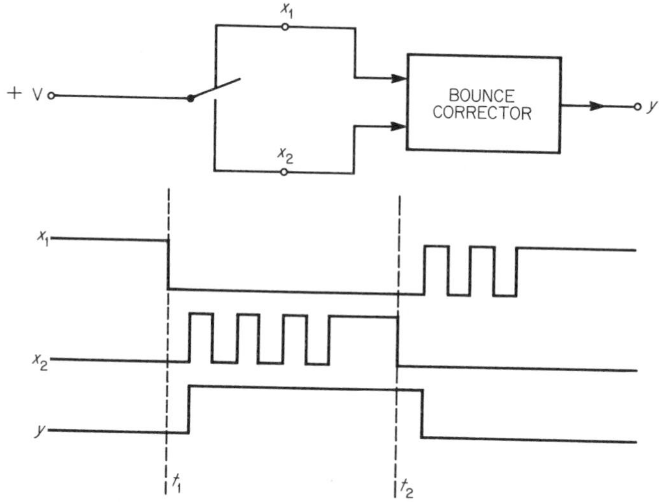

Figure 9.28 Illustration of the meaning of the bounce corrector (Exercise 9.1).

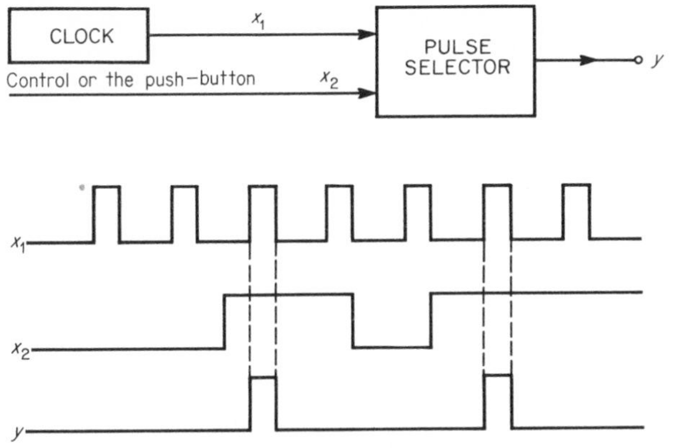

Figure 9.29 Illustration of the meaning of the pulse selector (Exercise 9.3).

Reference Notations to Chapter 9

The foundations of the theory of asynchronous sequential switching circuits was laid by Huffman [12]. Various aspects of the theory has been further developed by many authors. References 1, 2, 4, 5, 7, 10, 11, 16, 17, 18, 20, 29, and 34 represent only a sample from a rather extensive literature in this area. The most comprehensive presentation of the many facets of asynchronous switching circuits was prepared by Unger [32]. This book also contains an excellent bibliography with comments. Russian contributions are well covered in References 19 and 35.

The method described in Section 9.3 for solving the minimum transition time state assignment was developed by Tracey and Maki [21, 31]. The material in Section 9.5 is adopted from Reference 6.

A precise study of various kinds of hazards in switching circuits was started by McCluskey [23] and developed by many others. References 3, 13, 22, 24, 33, and 36 are some of the papers devoted to this subject. A study of decomposition of asynchronous switching circuits started only recently [14, 30]. The speed independent circuits were introduced and developed by Muller and Bartky [26, 27, 28]; their theory is also described in References 9 and 25. Gray codes, which are used in the design of counters, were investigated by Gilbert [8]. The hazard-free state assignment for reverse counters, which is specified in Example 9.7, was suggested by Klir and Seidl [15].

References to Chapter 9

1. **Armstrong, D. B.** et al., "Realization of Asynchronous Sequential Circuits Without Inserted Delay Elements," *TC* **C-17** (2) 129–134 (Feb. 1968).

2. **Armstrong, D. B., A. D. Friedman,** and **P. R. Menon,** "Design of Asynchronous Circuits Assuming Unbounded Gate Delay," *TC* **C-18** (12) 1110–1120 (Dec. 1969).
3. **Eichelberger, E. B.,** "Hazard Detection in Combinational and Sequential Switching Circuits," *IBM J.* **9** (2) 90–99 (March 1965).
4. **Friedman, A. D.,** "Feedback in Synchronous and Asynchronous Sequential Switching Circuits," Ph.D. Dissertation, E. E. Department, Columbus University, 1965.
5. **Friedman, A. D.,** "Feedback in Asynchronous Sequential Circuits," *TC* **EC-15** (5) 740–749 (Oct. 1966).
6. **Friedman, A. D.,** and **P. R. Menon,** "Synthesis of Asynchronous Sequential Circuits with Multiple-Input Changes," *TC* **C-17** (6) 559–566 (June 1968).
7. **Friedman, A. D., R. L. Graham,** and **J. D. Ullman,** "Universal Single Transition Time Asynchronous State Assignment," *TC* **C-18** (6) 541–547 (June 1969).
8. **Gilbert, E. N.,** "Gray Codes and Paths on the *n*-Cube," *BSTJ* **37** (3) 815-826 (May 1958).
9. **Hall, A. D.,** Jr., "Introducing to the Theory of Speed Independent Asynchronous Switching Circuits," Report No. 50, Digital Systems Lab., Princeton University, July 1966.
10. **Harrison, M. A.,** "On Equivalence of State Assignments," *TC* **C-17** (1) 55–57 (Jan. 1968).
11. **Hlavička, J.,** "Essential Hazard Correction Without the Use of Delay Elements," *TC* **C-19** (3) 232–238 (March 1970).
12. **Huffman, D. A.,** "The Synthesis of Sequential Switching Circuits," *JFI*, **257** (3 and 4) 161–190, 275–303 (March, April 1954), reprinted in *SM*.
13. **Huffman, D. A.,** "Design of Hazard-Free Switching Circuits," *JACM* **4,** (1) 47–62 (Jan. 1957).
14. **Kinney, L. L.,** "Decomposition of Asynchronous Sequential Switching Circuits," *TC* **C-19** (6) 515–529 (June 1970).
15. **Klir, G. J.,** and **L. K. Seidl,** "The Codes for the Coincidence Relay Chains" (in Russian), *IPM*, No. 7, 21–35 (1960).
16. **Klir, G. J.,** and **J. Hlavička** "Logical Design of Sequential Asynchronous Switching Circuits," *IPM* No. 11, 135–165 (1965).
17. **Langdon, G. G.,** "Analysis of Asynchronous Circuits Under Different Delay Assumptions," *TC* **C-17** (12) 1131–1143 (Dec. 1968).
18. **Langdon, G. G.,** "Delay-Free Asynchronous Circuits with Constrained Line Delays, *TC* **C-18** (2) 175–181 (Feb. 1969).
19. **Lazarev, V. G.,** and **Pil, E. I.,** *Synthesis of Asynchronous Finite Automata* (in Russian), Izd. "Nauka," Moscow, 1964.
20. **Liu, C. N.,** "A State Variable Assignment Method for Asynchronous Sequential Switching Circuits," *J. ACM* **10** (2) 209–216 (April 1963).
21. **Maki, G. K.,** and **J. H. Tracey,** "State Assignment Selection in Asynchronous Sequential Circuits," *TC* **C-19** (7) 641–644 (July 1970).
22. **McGhee, R. B.,** "Some Aids to the Detection of Hazards in Combinational Switching Circuits," *TC* **C-18** (6) 561–565 (June 1969).
23. **McCluskey, E. J.,** "Transients in Combinational Logic Circuits," in R. H. Wilcox, ed., "Redundancy Techniques for Computing Systems," Spartan Books, Washington, D. C., 1962, 9–46.
24. **Meisel, W. S.,** and **S. S. Kashef,** "Hazards in Asynchronous Sequential Circuits," *TC* **C-18** (8) 752–759 (Aug. 1969).

25. **Miller, R. E.,** *Switching Theory II: Sequential Circuits and Machines*, John Wiley, New York, 1965.
26. **Muller, D. E.,** and **W. S. Bartky,** "A Theory of Asynchronous Circuits I," Report No. 75, Digital Computer Lab., University of Illinois, November 1956.
27. **Muller, D. E.,** and **W. S. Bartky,** "A Theory of Asynchronous Circuits II," Report No. 78, Digital Computer Lab., University of Illinois; March 1957.
28. **Muller, D. E.,** and **W. S. Bartky,** "A Theory of Asynchronous Circuits," *PHU*, 204-243.
29. **Singh, S,** "Asynchronous Sequential Circuits with Feedback," *TC* **C-18** (5) 440–450 (May 1969).
30. **Tan, C. J., P. R. Menon,** and **A. D. Friedman,** "Structural Simplification and Decomposition of Asynchronous Sequential Circuits," *TC* **C-18** (9) 830–838 (Sept. 1969).
31. **Tracey, J. H.,** "Internal State Assignment for Asynchronous Sequential Machines," *TC* **EC-15** (4) 551–560 (Aug. 1966).
32. **Unger, S. H.,** *Asynchronous Sequential Switching Circuits*, John Wiley, New York, 1969.
33. **Unger, S. H.,** "A Row Assignment for Delay-Free Realizations of Flow Tables Without Essential Hazards," *TC* **C-17** (2) 146–151 (Feb. 1968).
34. **Unger, S. H.,** "Asynchronous Sequential Switching Circuits with Unrestricted Input Changes," *CRSA* **11,** 114–121 (Oct. 1970).
35. **Yakubaitis, E. A.,** *Asynchronous Logic Automata* (in Russian), Zinatne, Riga, Latvia, 1966.
36. **Yoeli, M.,** and **S. Rino,** "Application of Ternary Algebra to the Study of Static Hazards," *JACM* **11** (1) 84–97 (June 1964).

10

PROBABILISTIC SWITCHING CIRCUITS

10.1 INTRODUCTION

An ambiguity in the determination of responses is a basic feature of *probabilistic switching circuits* (sometimes called *stochastic switching circuits* in the literature). This means that their output variables are not functions of input and/or internal variables. More specifically, the relation between states of input and/or internal variables and states of output variables is one-to-many. Different responses assigned to the same stimulus and/or internal state are associated with certain probabilities. Clearly, if each of these probabilities is either 0 or 1, then the circuit becomes a deterministic one.

Probabilistic switching circuits contain elements called *random sources* in addition to combinational and/or memory elements. Every random source is associated with a number of mutually independent logic variables whose values are generated at discrete times with certain probabilities. Thus, random sources have no inputs; they have one or more outputs.

Although random sources of any type can be used as building blocks of probabilistic circuits, only the simplest type will be considered here which generates values 0 and 1 of a single logic variable with equal probabilities 0.5. This type is called *Las Vegas source*, *Monte Carlo source*, or *white noise source* in the literature. The name "*Las Vegas Source*" will be adopted here.

10.2 MEMORYLESS CIRCUITS

Let $x_1, x_2, \ldots, x_p$ be input variables of a probabilistic memoryless switching circuit and let $y_1, y_2, \ldots, y_q$ be output variables of this circuit. Then there are 2^p possible stimuli and 2^q possible responses of the circuit. The stimuli and responses will be denoted by $\mathbf{X}_i$ and $\mathbf{Y}_j$ respectively, where i is the identifier of input states ($i = 0, 1, \ldots, 2^p - 1$) and j is the identifier of output states ($j = 0, 1, \ldots, 2^q - 1$). For every logic variable v involved in a probabilistic switching circuit, symbol $v(\mathscr{P})$ will mean that $v = 1$ with probability $\mathscr{P}$. When we say that a variable v is produced with a probability of $\mathscr{P}$, this will always mean that $v = 1$ with probability $\mathscr{P}$; it will also mean that $v = 0$ with probability $1 - \mathscr{P}$.

Using these symbols, the behavior of a probabilistic memoryless switching circuit can be defined as a set of pairs $(\mathbf{X}_i, \mathbf{Y}_j)$ with a conditional probability $\mathscr{P}(\mathbf{Y}_j | \mathbf{X}_i)$ assigned to each pair of this set. Clearly,

$$0 \leq \mathscr{P}(\mathbf{Y}_j | \mathbf{X}_i) \leq 1$$

and

$$\sum_j \mathscr{P}(\mathbf{Y}_j | \mathbf{X}_i) = 1$$

for all responses and a particular stimulus.

Let $\mathscr{P}(y_k | \mathbf{X}_i)$ denote the probability that the output variable y_k is equal to one if stimulus $\mathbf{X}_i$ occurs and let $\mathscr{P}(\bar{y}_k | \mathbf{X}_i)$ denote the probability that $y_k = 0$ under the same condition at the input. Then, clearly,

$$\mathscr{P}(y_k | \mathbf{X}_i) + \mathscr{P}(\bar{y}_k | \mathbf{X}_i) = 1$$

Now let $y_k = 1$ in responses $\mathbf{Y}_{j_k}$ and $y_k = 0$ in all responses $\mathbf{Y}_j \neq \mathbf{Y}_{j_k}$. Then,

$$\mathscr{P}(y_k | \mathbf{X}_i) = \sum_{j_k} \mathscr{P}(\mathbf{Y}_{j_k} | \mathbf{X}_i) \tag{10.1}$$

If

$$0 < \mathscr{P}(y_k | \mathbf{X}_i) < 1$$

for K_i stimuli $\mathbf{X}_i$, then variable y_k represents 2^{K_i} different functions f_l. Suppose that each stimulus $\mathbf{X}_i$ is applied exactly once in a sequence of stimuli. Then, by a simple consideration, the probability $\mathscr{P}(y_k, f_l)$ that variable y_k will represent function f_l within this sequence of stimuli can be calculated by the formula

$$\mathscr{P}(y_k, f_l) = \prod_i \mathscr{P}(\tilde{y}_k | \mathbf{X}_i) \tag{10.2}$$

where $\tilde{y}_k = \bar{y}_k$ if f_l prescribes $y_k = 0$ for $\mathbf{X}_i$ and $\tilde{y}_k = y_k$ if f_l prescribes $y_k = 1$ for $\mathbf{X}_i$. Since y_k represents exactly one function for a particular sequence of stimuli in which each stimulus is applied exactly once, we obtain

$$\sum_l \mathscr{P}(y_k, f_l) = 1$$

EXAMPLE 10.1 Consider a memoryless probabilistic switching circuit with two inputs and two outputs whose behavior is specified in the left portion of the table in Figure 10.1a. Using Formula (10.1), we can calculate probabilities $\mathscr{P}(y_1 \mid \mathbf{X}_i)$ and $\mathscr{P}(y_2 \mid \mathbf{X}_i)$ for all stimuli $\mathbf{X}_0$, $\mathbf{X}_1$, $\mathbf{X}_2$, $\mathbf{X}_3$. Results of this calculation are shown in the right portion of the table.

In Figure 10.1b, the same behavior is given in chart form, where blank squares and dots represent, respectively, zero and one probabilities. Values of other probabilities are written in the respective squares of the chart. Note that input variables determine columns and output variables specify rows. Figure 10.1c shows probabilities of the output variables. They can be determined directly from the chart representation of the behavior.

Variable y_1 represents four functions: $x_2 < x_1$ with probability 0.1875, $\bar{x}_2$

i	$\mathbf{X}_i$ $x_2\ x_1$	$\mathbf{Y}_j$ $y_2\ y_1$	j	$\mathscr{P}(\mathbf{Y}_j \mid \mathbf{X}_i)$	$\mathscr{P}(y_2 \mid \mathbf{X}_i)$	$\mathscr{P}(y_1 \mid \mathbf{X}_i)$
0	0 0	0 0	0	.125	.375	.75
0	0 0	0 1	1	.5		
0	0 0	1 0	2	.125		
0	0 0	1 1	3	.25		
1	0 1	0 0	0	.0	1.0	1.0
1	0 1	0 1	1	.0		
1	0 1	1 0	2	.0		
1	0 1	1 1	3	1.0		
2	1 0	0 0	0	1.0	.0	.0
2	1 0	0 1	1	.0		
2	1 0	1 0	2	.0		
2	1 0	1 1	3	.0		
3	1 1	0 0	0	.25	.0	.75
3	1 1	0 1	1	.75		
3	1 1	1 0	2	.0		
3	1 1	1 1	3	.0		

(a)

$x_2 x_1$ (columns), $y_2 y_1$ (rows)

	$i=0$	1	2	3
$j=0$	.125		•	.25
1	.5			.75
2	.125			
3	.25	•		

(b)

x_1 (columns), x_2 (rows)

	.375	•

$\mathscr{P}(y_2 \mid \mathbf{X}_i)$

	.75	•
		.75

$\mathscr{P}(y_1 \mid \mathbf{X}_i)$

(c)

Figure 10.1 Behavior of a probabilistic memoryless switching circuit (Example 10.1).

and x_1 with probability 0.1875, and $x_2 \leq x_1$ with probability 0.5625. For instance, the probability of $\bar{x}_2$ is calculated, using formula (10.2), as follows:

$$\mathscr{P}(y_1, \bar{x}_2) = \mathscr{P}(y_1 | \mathbf{X}_0) \times \mathscr{P}(y_1 | \mathbf{X}_1) \times \mathscr{P}(\bar{y}_1 | \mathbf{X}_2) \times \mathscr{P}(\bar{y}_1 | \mathbf{X}_3)$$
$$= 0.75 \times 1 \times 1 \times 0.25 = 0.1875$$

Variable y_2 represents two functions: $x_2 < x_1$ with probability 0.625, and $\bar{x}_2$ with probability 0.375.

The *paradigm of probabilistic memoryless switching circuits* which is shown in Figure 10.2 will be used in our further exposition. It contains two partial blocks called *random generator* and *functional generator*.

Random generator contains a set of Las Vegas sources $LV_1, LV_2, \ldots, LV_r$ which generate random logic variables $\lambda_1, \lambda_2, \ldots, \lambda_r$. Both values of each of these variables appear with equal probabilities 0.5. Variables $\lambda_1, \lambda_2, \ldots, \lambda_r$ enter a block in the random generator which is called a *probability transformer*. This block produces another set of variables $\rho_1, \rho_2, \ldots, \rho_m$ with probabilities $\mathscr{P}_1, \mathscr{P}_2, \ldots, \mathscr{P}_m$ respectively. These variables enter the functional generator together with the external input variables $x_1, x_2, \ldots, x_p$ and participate thus in the determination of values of the output variables $y_1, y_2, \ldots, y_q$.

Suppose that the random generator contains r Las Vegas sources. Then these sources generate 2^r states with equal probabilities 2^{-r}. Each state is identified by a single minterm. Hence, the probability that a minterm is equal

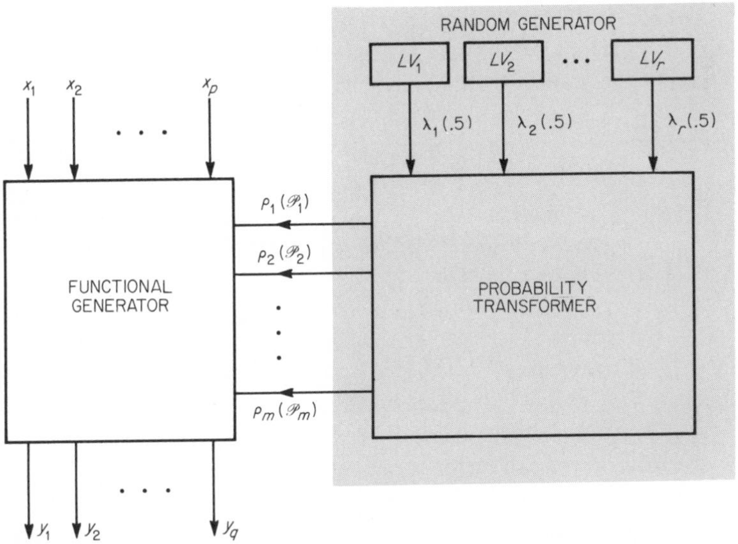

Figure 10.2 A paradigm of probabilistic memoryless switching circuits.

to one is the same as the probability that the corresponding state appears, namely, it is equal to 2^{-r}. It follows immediately that an OR operation of K minterms is equal to one with the probability of $K \cdot 2^{-r}$. Clearly, the OR operation of all minterms ($K = 2^r$) is equal to one with the probability of one, i.e., this OR operation is always equal to one.

Consider now a Boolean p-term containing a out of r variables ($a < r$). Since the term represents an OR operation of 2^{r-a} minterms, the probability that it is equal to one is given by the product $2^{r-a} \times 2^{-r} = 2^{-a}$. For instance, the term $\lambda_3 \bar{\lambda}_1$ of four variables $\lambda_1, \lambda_2, \lambda_3, \lambda_4$ can be written as $\lambda_3 \bar{\lambda}_1(\bar{\lambda}_4 \bar{\lambda}_2 \vee \bar{\lambda}_4 \lambda_2 \vee \lambda_4 \bar{\lambda}_2 \vee \lambda_4 \lambda_2)$ and considered as an OR operation of four minterms. Hence, the probability that $\lambda_3 \bar{\lambda}_1$ is equal to one can be calculated as $4 \times 2^{-4} = 2^{-2}$.

We saw that the probability of an OR operation of two minterms of the variables $\lambda_1, \lambda_2, \ldots, \lambda_r$ is twice as large as the probability of a single minterm. Unfortunately, this rule cannot be generalized for an OR operation of two p-terms which are not minterms. The trouble consists in the fact that some states (minterms) may be represented by both of the terms. These would be counted twice if we simply added the probabilities of the terms. A modified rule for calculation of the probability of an OR operation of two p-terms follows directly from this property: Add the probabilities of the terms and subtract the sum of probabilities of all states (minterms) that are represented by both of the terms. For instance, the terms $\lambda_2 \bar{\lambda}_1$ and $\bar{\lambda}_4 \lambda_3 \lambda_2$ (defined for four variables $\lambda_1, \lambda_2, \lambda_3, \lambda_4$) are equal to one with probabilities 2^{-2} and 2^{-3}, respectively. The terms contain one common minterm, $\bar{\lambda}_4 \lambda_3 \lambda_2 \bar{\lambda}_1$, whose probability is 2^{-4}. Hence, the probability that the OR operation $\lambda_2 \bar{\lambda}_1 \vee \bar{\lambda}_4 \lambda_3 \lambda_2$ is one is equal to the sum of the term probabilities minus the minterm probability, i.e., $2^{-2} + 2^{-3} - 2^{-4} = 5 \times 2^{-4}$.

Let us calculate now the probability $\mathscr{P}_w$ with which the output logic variable $w = f(v_1, v_2, \ldots, v_n)$, where f is a given logic function, is equal to one provided that the input variables $v_1, v_2, \ldots, v_n$ are equal to one with probabilities $\mathscr{P}_1, \mathscr{P}_2, \ldots, \mathscr{P}_n$, respectively. Assume that the input variables are mutually independent. Then, $\mathscr{P}_w$ can be calculated as the sum of probabilities of all minterms of $v_1, v_2, \ldots, v_n$ for which $w = 1$. The probability of a minterm $\tilde{v}_n \tilde{v}_{n-1} \cdots \tilde{v}_1$ is the product $\tilde{\mathscr{P}}_1 \tilde{\mathscr{P}}_2 \cdots \tilde{\mathscr{P}}_n$, where $\tilde{\mathscr{P}}_i = \mathscr{P}_i$ if $\tilde{v}_i = v_i$ and $\tilde{\mathscr{P}}_i = 1 - \mathscr{P}_i$ if $\tilde{v}_i = \bar{v}_i$. For instance, $w = \bar{v}_2 v_1 \vee v_2 \bar{v}_1 \vee v_2 v_1$ is the minterm form of the OR function. If $\mathscr{P}_i$ and $\mathscr{P}_2$ are probabilities of v_1 and v_2, respectively, then $\mathscr{P}_w = (1 - \mathscr{P}_2)\mathscr{P}_1 + \mathscr{P}_2(1 - \mathscr{P}_1) + \mathscr{P}_2 \mathscr{P}_1 = \mathscr{P}_1 + \mathscr{P}_2 - \mathscr{P}_1 \mathscr{P}_2$. Formulas for probabilities of output variables for some logic functions are given in Table 10.1.

Using the paradigm shown in Figure 10.2, we are now able to discuss synthesis of probabilistic memoryless switching circuits. Suppose that the behavior is given in the chart form whose example is in Figure 10.1b. This form

TABLE 10.1. Probability Transformations Made by Some Logic Functions

Logic Function	Minimal *sp* Boolean Form	Probability P_w of the output variable w
NEGATION	$w = \bar{v}_i$	$\mathscr{P}_w = 1 - \mathscr{P}_i$
$v_1 \downarrow v_2$	$w = \bar{v}_1 \bar{v}_2$	$\mathscr{P}_w = (1 - \mathscr{P}_1)(1 - \mathscr{P}_2)$
$v_1 < v_2$	$w = \bar{v}_1 v_2$	$\mathscr{P}_w = (1 - \mathscr{P}_1)\mathscr{P}_2$
$v_1 \not\equiv v_2$	$w = \bar{v}_1 v_2 \vee v_1 \bar{v}_2$	$\mathscr{P}_w = \mathscr{P}_1 + \mathscr{P}_2 - 2\mathscr{P}_1\mathscr{P}_2$
$v_1 \mid v_2$	$w = \bar{v}_1 \vee \bar{v}_2$	$\mathscr{P}_w = 1 - \mathscr{P}_1\mathscr{P}_2$
$v_1 v_2$	$w = v_1 v_2$	$\mathscr{P}_w = \mathscr{P}_1\mathscr{P}_2$
$v_1 \equiv v_2$	$w = \bar{v}_1 \bar{v}_2 \vee v_1 v_2$	$\mathscr{P}_w = 1 - \mathscr{P}_1 - \mathscr{P}_2 + 2\mathscr{P}_1\mathscr{P}_2$
$v_1 \leq v_2$	$w = \bar{v}_1 \vee v_2$	$\mathscr{P}_w = 1 - \mathscr{P}_1 + \mathscr{P}_1\mathscr{P}_2$
$v_1 \vee v_2$	$w = v_1 \vee v_2$	$\mathscr{P}_w = \mathscr{P}_1 + \mathscr{P}_2 - \mathscr{P}_1\mathscr{P}_2$
$v_1 v_2 v_3$	$w = v_1 v_2 v_3$	$\mathscr{P}_w = \mathscr{P}_1\mathscr{P}_2\mathscr{P}_3$
$v_1 \vee v_2 \vee v_3$	$w = v_1 \vee v_2 \vee v_3$	$\mathscr{P}_w = 1 - (1 - \mathscr{P}_1)(1 - \mathscr{P}_2)(1 - \mathscr{P}_3)$
NOR(v_1, v_2, v_3)	$w = \bar{v}_1 \bar{v}_2 \bar{v}_3$	$\mathscr{P}_w = (1 - \mathscr{P}_1)(1 - \mathscr{P}_2)(1 - \mathscr{P}_3)$
NAND(v_1, v_2, v_3)	$w = \bar{v}_1 \vee \bar{v}_2 \vee \bar{v}_3$	$\mathscr{P}_w = 1 - \mathscr{P}_1\mathscr{P}_2\mathscr{P}_3$
M_3	$w = v_1 v_2 \vee v_1 v_3 \vee v_2 v_3$	$\mathscr{P}_w = \mathscr{P}_1\mathscr{P}_2 + \mathscr{P}_1\mathscr{P}_3 + \mathscr{P}_2\mathscr{P}_3 - 2\mathscr{P}_1\mathscr{P}_2\mathscr{P}_3$

can be easily converted into a set of charts, one for each output variable, the example of which is in Figure 10.1c. Each of these charts represents the logic space of input variables. For each state of the input variables, a probability is specified in the chart under which the respective output variable is equal to one. According to our previous convention, zero and one probabilities are indicated in the charts by blank squares and dots, respectively.

The problem of synthesis consists of two partial problems: Synthesis of the random generator and synthesis of the functional generator. The purpose of the random generator is to generate variables $\rho_1, \rho_2, \ldots, \rho_m$ with probabilities $\mathscr{P}_1, \mathscr{P}_2, \ldots, \mathscr{P}_m$, written as $\rho_1(\mathscr{P}_1)$, $\rho_2(\mathscr{P}_2), \ldots, \rho_m(\mathscr{P}_m)$. These are probabilities that are required by individual output variables. For instance, output variables specified in Figure 10.1c require that variables $\rho_1(0.375)$ and $\rho_2(0.75)$ be produced by the random generator.

It should be stressed that proper relationship among variables $\rho_1, \rho_2, \ldots, \rho_m$ must be made if the given behavior requires so. For instance, if $x_1 = x_2 = 0$, then $y_2 = y_1 = 1$ is required to appear with a probability of 0.25 in case of the example specified in Figure 10.1. Variables $\rho_1(0.375)$ and $\rho_2(0.75)$ determine output variables y_2 and y_1, respectively. If ρ_1 and ρ_2 were independent then the probability of $\rho_1\rho_2$ (or $\rho_1 = \rho_2 = 1$) would be $0.375 \times 0.75 = 0.28125$; this value is different from the value 0.25 required for the probability $y_1 y_2$ (or $y_1 = y_2 = 1$). Thus, the probability of $\rho_1\rho_2$ must be fixed to 0.25 by the random generator.

We have already observed that the smallest value of probability which can be produced by the random generator is equal to 2^{-r}, where r is the number of Las Vegas sources used. This is the probability that a minterm of variables $\lambda_1, \lambda_2, \ldots, \lambda_r$ is equal to one. By taking an OR operation of an appropriate number of these minterms, we can produce a variable with probability $K \cdot 2^{-r}(K = 1, 2, \ldots, 2^r)$. Thus, the required probabilities of variables $\rho_1, \rho_2, \ldots, \rho_m$ must be integer multiples of 2^{-r}. It is advantageous, therefore, if the probabilities $\mathscr{P}(\mathbf{Y}_j | \mathbf{X}_i)$, which specify the behavior of the system, are given in the binary number representation. Suppose that the smallest non-zero probability $\mathscr{P}(\mathbf{Y}_j | \mathbf{X}_i)$ is equal to 2^{-r}, where r is an integer. Then, if binary number representation is used, other probabilities involved in the behavior are integer multiples of 2^{-r}. The value of r directly specifies the number of required Las Vegas sources in the random generator.

If probabilities $\mathscr{P}(\mathbf{Y}_j | \mathbf{X}_i)$ are given in the decimal number representation, we must convert them to the binary form first. If an exact conversion is not feasible, a reasonable approximation has to be used. Clearly, the more Las Vegas sources are used, the better approximation can be achieved.

EXAMPLE 10.2 To design a random generator for the behavior discussed in Example 10.1 (Figure 10.1), we convert the probabilities $\mathscr{P}(\mathbf{Y}_j | \mathbf{X}_i)$ to binary form first. We find that the smallest non-zero probability is 0.125 and its binary form $0.001 = 2^{-3}$. We also find that all other probabilities involved are integer multiples of 2^{-3}. Hence, the random generator must contain 3 Las Vegas sources. These sources produce variables $\lambda_1, \lambda_2, \lambda_3$. Eight variables $v_0, v_1, \ldots, v_7$ representing minterms of $\lambda_1, \lambda_2, \lambda_3$ can be derived as shown in Figure 10.3. Each of these eight variables is produced with the probability of $0.125 = 2^{-3}$. They are not independent since only one of them is equal to one at any time, i.e., $v_i v_j = 0$ for every pair $i \neq j$. Hence, the probability of $v_i \vee v_j$, where $i \neq j$, is $2 \times 0.125 = 0.25$. Generally, the probability of an OR operation of k of these variables is equal to $k \times 0.125$. Thus, we can generate variables ρ_1 and ρ_2 by taking OR operations of appropriate numbers of the minterm variables.

Variable ρ_1 is to be generated with probability $0.175 = 6 \times 0.125$. This can be accomplished simply by taking an OR operation of six of the minterm variables. It does not matter which of these variables are taken but it must be six of them. However, a selection of the minterm variables may greatly effect the complexity of the random generator.

Variable ρ_2 is to be generated with probability $0.375 = 3 \times 0.125$. Thus, an OR operation of three minterm variables must be taken. Two of these three variables must be those which have been used in the determination of variable ρ_1 because the probability of $\rho_1 \rho_2$ is required to be $0.25 = 2 \times 0.125$.

The basic scheme of the random generator is shown in Figure 10.3. Clearly, the scheme can be considerably simplified. First, $v_4 \vee v_5 \vee v_6 \vee v_7 = \lambda_3$ and $v_2 \vee v_3 \vee v_6 \vee v_7 = \lambda_2$ so that $\rho_1 = \lambda_3 \vee \lambda_2$. Second, $v_2 \vee v_3 = \bar{\lambda}_3 \lambda_2$ and $v_1 \vee$

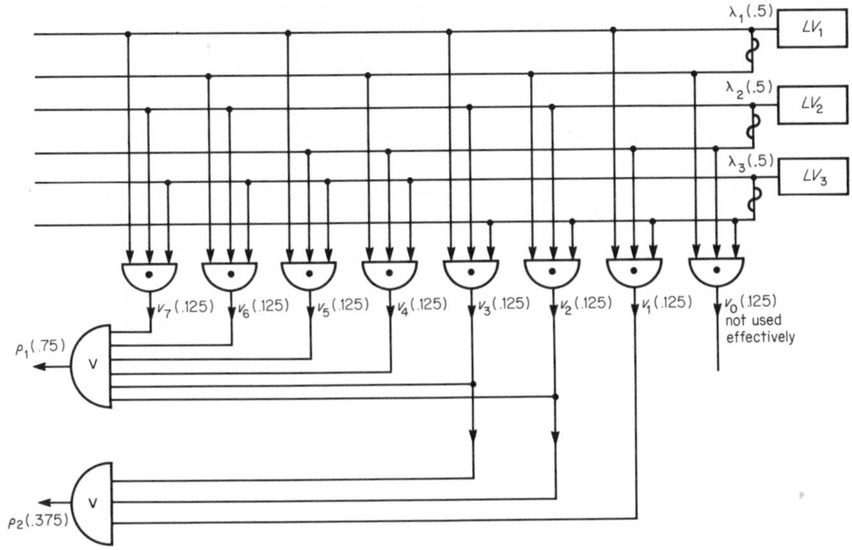

Figure 10.3 Basic scheme of a random generator (Example 10.2).

$v_3 = \bar{\lambda}_3 \lambda_1$ so that $\rho_2 = \bar{\lambda}_3(\lambda_2 \vee \lambda_1)$. The simplified scheme of the random generator is shown in Figure 10.4.

After the random generator is properly designed (including required dependencies among its output variables), the synthesis of the functional generator is very simple. First, we replace numbers indicating probabilities in charts for the output variables by corresponding variables ρ_i ($i = 1, 2, \ldots, m$). The rest of the procedure consists in an implementation of the output variables $y_1, y_2, \ldots, y_q$. They are considered as functions of the input variables $x_1, x_2, \ldots, x_p$ and the random variables $\rho_1, \rho_2, \ldots, \rho_m$.

EXAMPLE 10.3 To design the functional generator for the behavior specified in Figure 10.1, we replace (referring to Example 10.2 and Figure 10.4) the charts in Figure 10.1c by the charts in Figure 10.5a. Assume that we want to use a two level AND-OR circuit with the smallest number of elements. We obtain the minimal Boolean *sp* forms included in Figure 10.5a. The logic diagram shown in Figure 10.5b follows directly from the Boolean forms.

Let us point out that, generally, several variables with the same probability may be required at the output of the random generator. Although they are generated with equal probabilities, each of them has a different relationship to

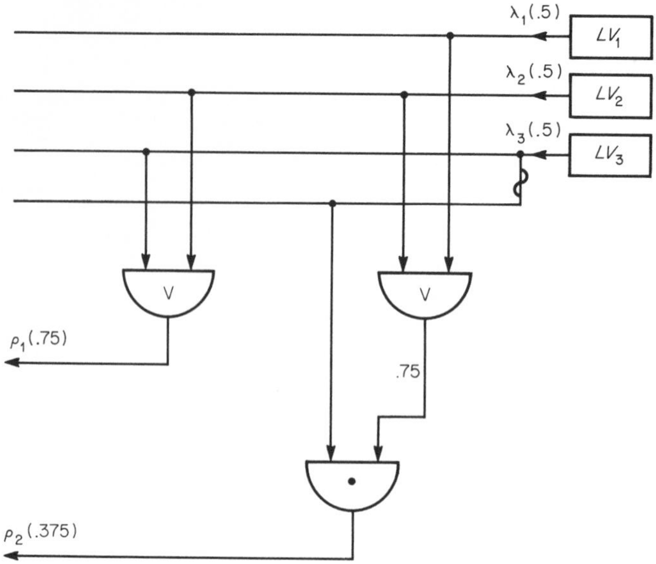

Figure 10.4 A simplified scheme of the random generator in Figure 10.3 (Example 10.2).

other variables produced by the random generator. The conditions under which a single variable produced by the random generator can be employed to control two or more output variables of the circuit are summarized in the following two theorems whose proofs are left to the reader as an exercise.

THEOREM 10.1 Let Y denote the set of all output variables of a probabilistic switching circuit and all possible logic products of two or more of these variables and let $y \in Y$. Then, output variables y_k and y_l of this circuit can be controlled for a particular stimulus $\mathbf{X}_i$ by a single variable, say ρ_a, produced by the random generator if and only if probabilities $\mathscr{P}(y|\mathbf{X}_i)$ for all y containing either y_k or y_l or both are equal.

THEOREM 10.2 Let Y and y have the same meaning as in Theorem 10.1. Let π denote a permutation of output variables of a probabilistic switching circuit and let $\pi(y)$ denote the element assigned to y under the permutation π. Then, the variables $\rho_1, \rho_2, \ldots, \rho_q$ produced by the random generator (some of them may be identical in the sense of Theorem 10.1) can be employed to control output variables $y_1, y_2, \ldots, y_q$ for two different stimuli $\mathbf{X}_i$ and $\mathbf{X}_j$ if and only if there exists a permutation π of the output variables such that

$$\mathscr{P}(y|\mathbf{X}_i) = \mathscr{P}(\pi(y)|\mathbf{X}_j)$$

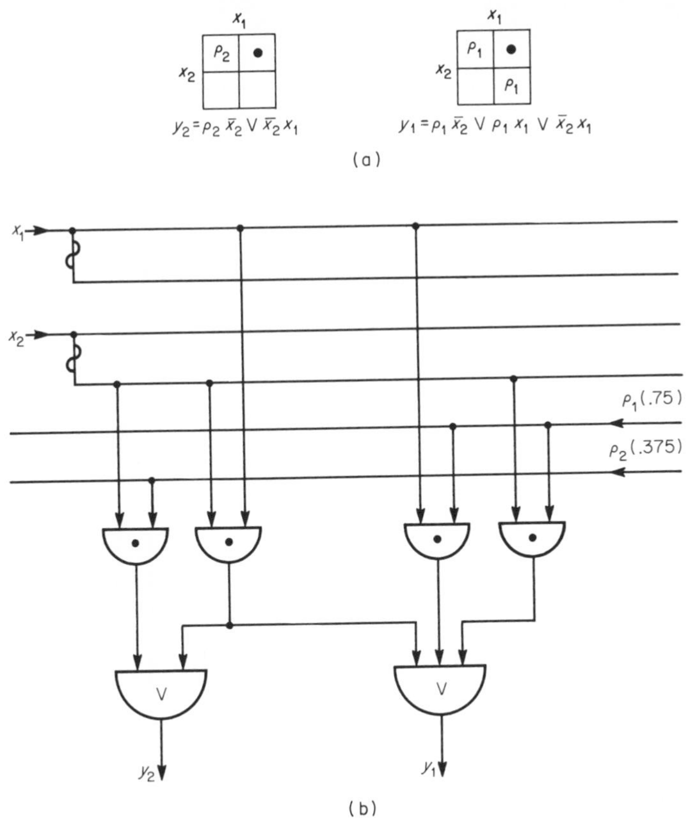

Figure 10.5 Functional generator (Example 10.3).

for all $y \in Y$. If this condition is satisfied and $\rho_1, \rho_2, \ldots, \rho_q$ control, respectively, $y_1, y_2, \ldots, y_q$ for $\mathbf{X}_i$, then $\rho_1, \rho_2, \ldots, \rho_q$ may control, respectively, $\pi(y_1), \pi(y_2), \ldots, \pi(y_q)$ for $\mathbf{X}_j$.

EXAMPLE 10.4 Design a switching circuit implementing the behavior specified in the chart form in Figure 10.6a.

First, we determine the chart forms for the output variables (Figure 10.6b). Then, we investigate the relationship among the output variables y_1, y_2, y_3 for those states $\mathbf{X}_i$ of the input variables which do not produce unique responses. We can do that by calculating probabilities $\mathscr{P}(y_1 y_2 | \mathbf{X}_i)$, $\mathscr{P}(y_1 y_3 | \mathbf{X}_i)$, $\mathscr{P}(y_2 y_3 | \mathbf{X}_i)$ with which two of the output variables are simultaneously equal to one for a particular stimulus $\mathbf{X}_i$, and the probability $\mathscr{P}(y_1 y_2 y_3 | \mathbf{X}_i)$ with which all of the output variables are equal to one. Values of all these probabilities are summarized in Figure 10.6c.

$x_3 x_2 x_1$ (columns); $y_3 y_2 y_1$ (rows)

	.25			.0625	.25	.0625	
	.125		.75	.125	.125	.125	
	.125			.125	.125	.25	.75
				.125		.1875	
•				.25		.125	
	.375			.1875		.125	
				.125	.375	.125	
	.125	•	.25	.0625	.125	.0625	.25

(a)

$x_2 x_1$ (columns); x_3 (rows)

y_3:

•	.5	•	.25
.625	.25	.4375	.25

y_2:

	.25	•	.25
.4375	.375	.625	•

y_1:

	.625	•	•
.5	.5	.5	.25

(b)

X_i	$x_3 x_2 x_1$	$\mathscr{P}(y_3 \mid X_i)$	$\mathscr{P}(y_2 \mid X_i)$	$\mathscr{P}(y_1 \mid X_i)$	$\mathscr{P}(y_3 y_2 \mid X_i)$	$\mathscr{P}(y_3 y_1 \mid X_i)$	$\mathscr{P}(y_2 y_1 \mid X_i)$	$\mathscr{P}(y_3 y_2 y_1 \mid X_i)$
X_1	0 0 1	.5	.25	.625	.125	.5	.125	.125
X_3	0 1 1	.25	.25	•	.25	.25	.25	.25
X_4	1 0 0	.625	.4375	.5	.1875	.25	.1875	.0625
X_5	1 0 1	.5	.625	.25	.5	.125	.125	.125
X_6	1 1 0	.4375	.625	.5	.1875	.1875	.25	.0625
X_7	1 1 1	.25	•	.25	.25	.25	.25	.25

(c)

Figure 10.6 Illustration for Example 10.4.

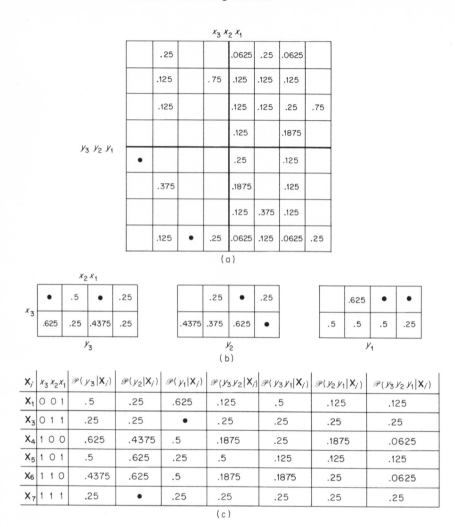

Using Theorem 10.1, we can easily find that variables y_3 and y_2 can be controlled by the same variable of random generator for stimulus $\mathbf{X}_3$. Similarly, y_3 and y_1 can be controlled by the same variable for $\mathbf{X}_7$. Using Theorem 10.2, we find that all probabilities for $\mathbf{X}_3$ and $\mathbf{X}_7$ can be represented by a single variable of the random generator, say variable ρ_1. Using again Theorem 10.2, we find that variables y_3, y_2, y_1 for $\mathbf{X}_1$ can be controlled by the same triads of random variables, say ρ_2, ρ_3, ρ_4, as variables y_3, y_1, y_2 for $\mathbf{X}_5$, respectively. Similarly, variables y_3, y_2, y_1 for $\mathbf{X}_4$ can be controlled by the same random variables, say ρ_5, ρ_6, ρ_7, as variables y_2, y_3, y_1 for $\mathbf{X}_6$, respectively.

Now, we are able to start designing the random generator. The smallest probability involved is $0.0625 = 2^{-4}$. All other probabilities are integer multiples of this value so that the random generator needs four Las Vegas sources. They will produce four random variables $\lambda_1, \lambda_2, \lambda_3, \lambda_4$ from which sixteen minterms can be derived, each of them with the probability of 0.0625. It is useful to add symbols of the introduced random variables to the respective probability values in Figure 10.6c. We obtain the situation shown in Figure 10.7a. If we calculate, for each probability specified in this figure, the number of corresponding minterms of variables $\lambda_1, \lambda_2, \lambda_3, \lambda_4$, we obtain the table in Figure 10.7b. This table gives us directly all constraints concerning coverings of the minterms by individual output variables of the random generator. The covering for each stimulus $\mathbf{X}_i$ can be solved independently of the coverings for other stimuli.

The covering of ρ_1 is very simple. We can take any quadruple of the minterms; e.g., the last four of them so that $\rho_1 = \lambda_4 \lambda_3$. A possible covering for the other variables is shown in Figure 10.8 together with minimal Boolean forms. We can see that $\rho_4 = \rho_5$ under this covering.

$\mathbf{X}_i$	$\mathcal{P}(y_3 \mid \mathbf{X}_i)$	$\mathcal{P}(y_2 \mid \mathbf{X}_i)$	$\mathcal{P}(y_1 \mid \mathbf{X}_i)$	$\mathcal{P}(y_3 y_2 \mid \mathbf{X}_i)$	$\mathcal{P}(y_3 y_1 \mid \mathbf{X}_i)$	$\mathcal{P}(y_2 y_1 \mid \mathbf{X}_i)$	$\mathcal{P}(y_3 y_2 y_1 \mid \mathbf{X}_i)$
$\mathbf{X}_1$	ρ_2 (.5)	ρ_3 (.25)	ρ_4 (.625)	$\rho_2\rho_3$ (.125)	$\rho_2\rho_4$ (.5)	$\rho_3\rho_4$ (.125)	$\rho_2\rho_3\rho_4$ (.125)
$\mathbf{X}_3$	ρ_1 (.25)	ρ_1 (.25)	•	ρ_1 (.25)	ρ_1 (.25)	ρ_1 (.25)	ρ_1 (.25)
$\mathbf{X}_4$	ρ_5 (.625)	ρ_6 (.4375)	ρ_7 (.5)	$\rho_5\rho_6$ (.1875)	$\rho_5\rho_7$ (.25)	$\rho_6\rho_7$ (.1875)	$\rho_5\rho_6\rho_7$ (.0625)
$\mathbf{X}_5$	ρ_2 (.5)	ρ_4 (.625)	ρ_3 (.25)	$\rho_2\rho_4$ (5)	$\rho_2\rho_3$ (.125)	$\rho_3\rho_4$ (,125)	$\rho_2\rho_3\rho_4$ (.125)
$\mathbf{X}_6$	ρ_6 (.4375)	ρ_5 (.625)	ρ_7 (.5)	$\rho_5\rho_6$ (.1875)	$\rho_6\rho_7$ (.1875)	$\rho_5\rho_7$ (.25)	$\rho_5\rho_6\rho_7$ (.0625)
$\mathbf{X}_7$	ρ_1 (.25)	•	ρ_1 (.25)	ρ_1 (.25)	ρ_1 (.25)	ρ_1 (.25)	ρ_1 (.25)

(a)

$\mathbf{X}_i$	$\mathcal{P}(y_3 \mid \mathbf{X}_i)$	$\mathcal{P}(y_2 \mid \mathbf{X}_i)$	$\mathcal{P}(y_1 \mid \mathbf{X}_i)$	$\mathcal{P}(y_3 y_2 \mid \mathbf{X}_i)$	$\mathcal{P}(y_3 y_1 \mid \mathbf{X}_i)$	$\mathcal{P}(y_2 y_1 \mid \mathbf{X}_i)$	$\mathcal{P}(y_3 y_2 y_1 \mid \mathbf{X}_i)$
$\mathbf{X}_1$	ρ_2, 8	ρ_3, 4	ρ_4, 10	$\rho_2\rho_3$, 2	$\rho_2\rho_4$, 8	$\rho_3\rho_4$, 2	$\rho_2\rho_3\rho_4$, 2
$\mathbf{X}_3$	ρ_1, 4	ρ_1, 4	•	ρ_1, 4	ρ_1, 4	ρ_1, 4	ρ_1, 4
$\mathbf{X}_4$	ρ_5, 10	ρ_6, 7	ρ_7, 8	$\rho_5\rho_6$, 3	$\rho_5\rho_7$, 4	$\rho_6\rho_7$, 3	$\rho_5\rho_6\rho_7$, 1
$\mathbf{X}_5$	ρ_2, 8	ρ_4, 10	ρ_3, 4	$\rho_2\rho_4$, 8	$\rho_2\rho_3$, 2	$\rho_3\rho_4$, 2	$\rho_2\rho_3\rho_4$, 2
$\mathbf{X}_6$	ρ_6, 7	ρ_5, 10	ρ_7, 8	$\rho_5\rho_6$, 3	$\rho_6\rho_7$, 3	$\rho_5\rho_7$, 4	$\rho_5\rho_6\rho_7$, 1
$\mathbf{X}_7$	ρ_1, 4	•	ρ_1, 4	ρ_1, 4	ρ_1, 4	ρ_1, 4	ρ_1, 4

(b)

Figure 10.7 Illustration for Example 10.4.

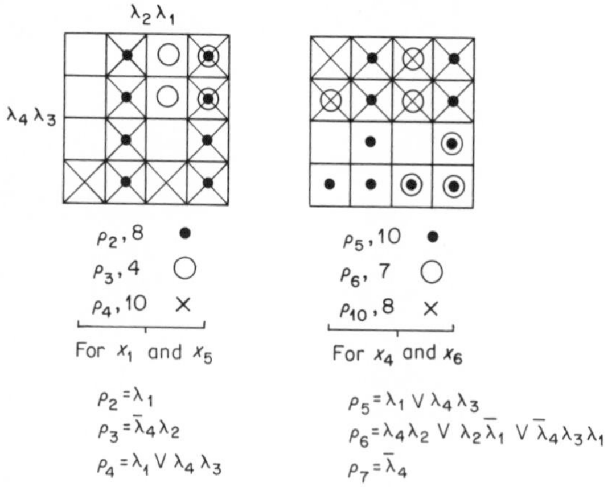

Figure 10.8 Illustration for Example 10.4.

Now, we can substitute symbols of the introduced random variables for the respective probability values in Figure 10.6c. We obtain the functions shown in Figure 10.9. It remains to implement these functions. We obtain, for instance,

$$y_3 = \bar{x}_3\bar{x}_1 \vee \rho_1 x_2 x_1 \vee \rho_2 \bar{x}_2 x_1 \vee \rho_5 \bar{x}_2 \bar{x}_1 \vee \rho_6 x_2 \bar{x}_1$$

$$y_2 = \bar{x}_3 x_2 \bar{x}_1 \vee x_3 x_2 x_1 \vee \rho_1 x_2 x_1 \vee \rho_3 \bar{x}_3 \bar{x}_2 x_1 \vee \rho_4 x_3 x_1 \vee \rho_5 x_3 x_2 \vee \rho_6 x_3 \bar{x}_2 \bar{x}_1$$

$$y_1 = \bar{x}_3 x_2 \vee \rho_1 x_2 x_1 \vee \rho_3 x_3 \bar{x}_2 x_1 \vee \rho_4 \bar{x}_3 x_1 \vee \rho_7 x_3 \bar{x}_1$$

Exercises to Section 10.2

10.2-1 Draw a logic diagram for the circuit designed in Example 10.4.

10.2-2 Draw a random generator with three output variables $\rho_1(.5)$, $\rho_2(.375)$, $\rho_3(.125)$ such that
- (a) only one of these variables is equal to one at any given time.
- (b) $\rho_1\rho_2(0.375)$, $\rho_1\rho_3(0.125)$, $\rho_2\rho_3(0.0)$.
- (c) $\rho_1\rho_2(0.25)$, $\rho_1\rho_3(0.0625)$, $\rho_2\rho_3(0.125)$, $\rho_1\rho_2\rho_3(0.0)$.

10.2-3 Design a probabilistic memoryless switching circuit with one input and one output such that $\mathscr{P}(0|0) = 0.75$ and $\mathscr{P}(1|0) = 0.5$.

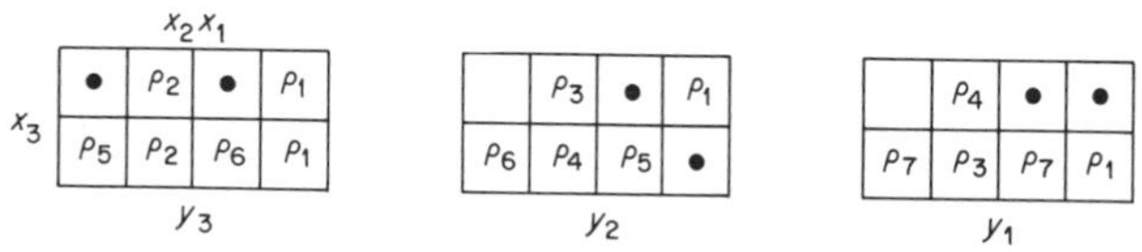

Figure 10.9 Illustration for Example 10.4.

10.2-4 Design probabilistic memoryless switching circuits with two inputs and two outputs whose behaviors are given by the following table:

i	$\mathbf{X}_i$ x_2 x_1	$\mathbf{Y}_j$ y_2 y_1	j	$\mathscr{P}_1(\mathbf{Y}_j\|\mathbf{X}_i)$	$\mathscr{P}_2(\mathbf{Y}_j\|\mathbf{X}_i)$	$\mathscr{P}_3(\mathbf{Y}_j\|\mathbf{X}_i)$
0	0 0	0 0	0	0.25	0.125	1.0
0	0 0	0 1	1	0.25	0.25	0.0
0	0 0	1 0	2	0.25	0.25	0.0
0	0 0	1 1	3	0.25	0.375	0.0
1	0 1	0 0	0	0.5	0.125	0.25
1	0 1	0 1	1	0.0	0.125	0.25
1	0 1	1 0	2	0.0	0.125	0.5
1	0 1	1 1	3	0.5	0.625	0.0
2	1 0	0 0	0	0.75	0.25	0.25
2	1 0	0 1	1	0.0	0.25	0.25
2	1 0	1 0	2	0.25	0.375	0.25
2	1 0	1 1	3	0.0	0.125	0.25
3	1 1	0 0	0	0.25	0.625	1.0
3	1 1	0 1	1	0.25	0.125	0.0
3	1 1	1 0	2	0.0	0.125	0.0
3	1 1	1 1	3	0.5	0.125	0.0

10.2-5 Prove Theorems 10.1 and 10.2.

10.3 SEQUENTIAL CIRCUITS

In this section we want to introduce the notion of *probabilistic sequential switching circuits* and to show that the principles of structure synthesis of probabilistic memoryless circuits, presented in Section 10.2, can be easily adapted for structure synthesis of probabilistic sequential circuits.

A *general paradigm of probabilistic sequential switching circuits* is shown in Figure 10.10. It contains three basic blocks, the *functional generator*, the *memory*, and the *random generator*. These blocks have been described previously. All possible couplings are considered in this general paradigm. When some of the couplings are excluded, we obtain various models. The finite-state model, which contains only the couplings shown in Figure 10.11, will be considered in our further exposition. This model offers three modifications which differ one from another by properties of the output variables of the functional generator. They are referred to as the *Shannon machine*, the *Mealy probabilistic machine*, and the *Moore probabilistic machine*.

DEFINITION 10.1 The *Shannon machine* is a quadruple

$$M = (X, Y, Z, P)$$

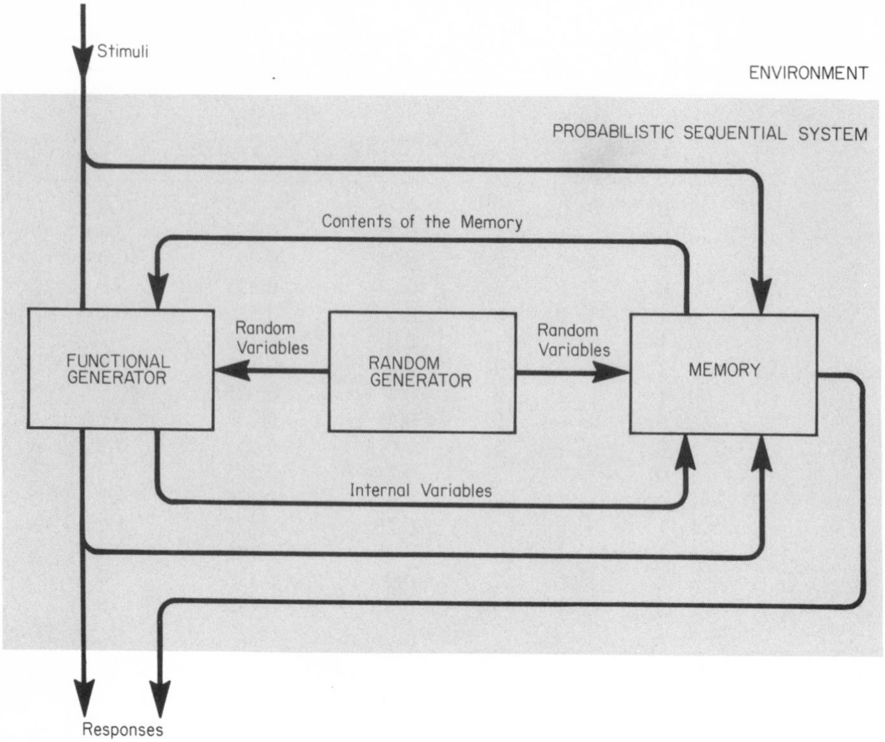

Figure 10.10 General paradigm of probabilistic sequential switching circuits.

where X is a finite set of stimuli; Y is a finite set of responses, Z is a finite set of internal states, and P is a set of conditional probabilities

$$\{\mathscr{P}(\mathbf{y}^t, \mathbf{z}^{t+1} \,|\, \mathbf{x}^t, \mathbf{z}^t)\colon \mathbf{x}^t \in X;\ \mathbf{y}^t \in Y;\ \mathbf{z}^t, \mathbf{z}^{t+1} \in Z\}$$

Symbols $\mathbf{x}^t$, $\mathbf{y}^t$, $\mathbf{z}^t$, $\mathbf{z}^{t+1}$ denote, respectively, the stimulus, the response, the present internal state, and the next internal state.

It is convenient to arrange the conditional probabilities introduced in the previous definition in the chart form. Let rows and columns of the chart be identified by pairs $(\mathbf{y}^t, \mathbf{z}^{t+1})$ and $(\mathbf{x}^t, \mathbf{z}^t)$ respectively. An example of the chart representation of probabilities for a Shannon machine with one input variable, one output variable, and two internal states (one internal variable) is shown in Figure 10.12a. Notice that the sum of all probabilities in each column of the chart is equal to one. Formally,

$$\sum_{(\mathbf{y}^t, \mathbf{z}^{t+1})} \mathscr{P}(\mathbf{y}^t, \mathbf{z}^{t+1} \,|\, \mathbf{x}^t, \mathbf{z}^t) = 1$$

for a particular pair $(\mathbf{x}^t, \mathbf{z}^t)$.

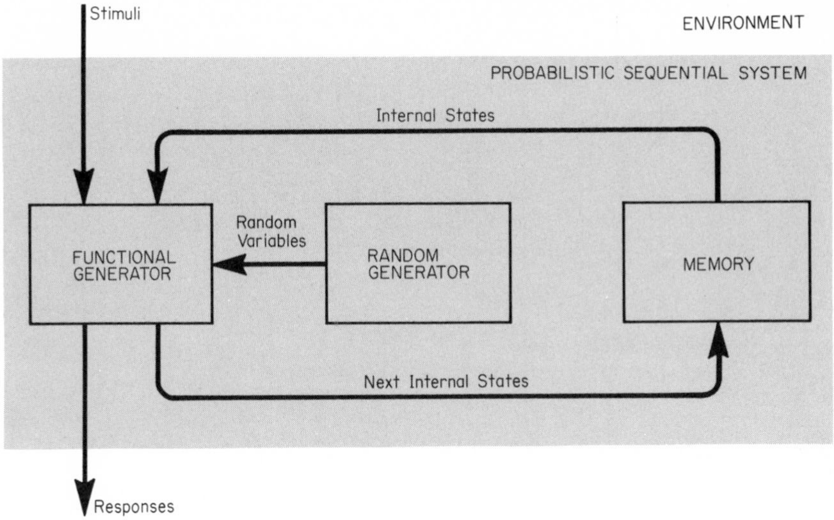

Figure 10.11 Finite-state model of probabilistic sequential switching circuits.

DEFINITION 10.2 The *Mealy probabilistic machine* is a quintuple

$$M = (X, Y, Z, P_y, P_z)$$

where X, Y, Z have the same meaning as in Definition 10.1 and

$$P_y = \{\mathscr{P}(\mathbf{y}^t \mid \mathbf{x}^t, \mathbf{z}^t): \mathbf{x}^t \in X, \mathbf{y}^t \in Y, \mathbf{z}^t \in Z\}$$
$$P_z = \{\mathscr{P}(\mathbf{z}^{t+1} \mid \mathbf{x}^t, \mathbf{z}^t): \mathbf{x}^t \in X, \mathbf{z}^t \in Z, \mathbf{z}^{t+1} \in Z\}$$

are sets of conditional probabilities such that

$$\mathscr{P}(\mathbf{y}^t, \mathbf{z}^{t+1} \mid \mathbf{x}^t, \mathbf{z}^t) = \mathscr{P}(\mathbf{y}^t \mid \mathbf{x}^t, \mathbf{z}^t) \cdot \mathscr{P}(\mathbf{z}^{t+1} \mid \mathbf{x}^t, \mathbf{z}^t)$$

While the Shannon machine allows a statistical dependence between $\mathbf{y}^t$ and $\mathbf{z}^{t+1}$, the Mealy probabilistic machine requires that $\mathbf{y}^t$ and $\mathbf{z}^{t+1}$ are statistically independent. Each of the sets P_y and P_z can be arranged in the chart form. An example is shown in Figure 10.12b. The sum of probabilities in each column of either of the charts must be equal to one or, formally,

$$\sum_{\mathbf{y}^t} \mathscr{P}(\mathbf{y}^t \mid \mathbf{x}^t, \mathbf{z}^t) = 1$$

and

$$\sum_{\mathbf{z}^{t+1}} \mathscr{P}(\mathbf{z}^{t+1} \mid \mathbf{x}^t, \mathbf{z}^t) = 1$$

for a particular pair $(\mathbf{x}^t, \mathbf{z}^t)$.

$\mathcal{P}(y^t, z^{t+1} \mid x^t, z^t)$		(x^t, z^t) 0,0	0,1	1,0	1,1
(y^t, z^{t+1})	0,0	.125		.375	
	0,1	.5	•	.625	.125
	1,0	.25			.125
	1,1	.125			.75

(a)

$\mathcal{P}(y^t \mid x^t, z^t)$		(x^t, z^t) 0,0	0,1	1,0	1,1
y^t	0	.875		.625	.25
	1	.125	•	.375	.75

$\mathcal{P}(z^{t+1} \mid x^t, z^t)$		(x^t, z^t) 0,0	0,1	1,0	1,1
z^{t+1}	0	.5		•	.75
	1	.5	•		.25

(b)

$\mathcal{P}(y^t \mid z^t)$		z^t 0	1
y^t	0	.25	.625
	1	.75	.375

$\mathcal{P}(z^{t+1} \mid x^t, z^t)$		(x^t, z^t) 0,0	0,1	1,0	1,1
z^{t+1}	0	.75	•	.875	
	1	.25		.125	•

(c)

Figure 10.12 Examples of three modifications of the finite-state model of probabilistic switching circuits.

DEFINITION 10.3 The *Moore probabilistic machine* is a quintuple

$$M = (X, Y, Z, P_y, P_z)$$

where X, Y, Z, P_z have the same meaning as in Definition 10.2 and

$$P_y = \{\mathcal{P}(\mathbf{y}^t \mid \mathbf{z}^t) : \mathbf{y}^t \in Y, \mathbf{z}^t \in Z\}$$

is a set of conditional probabilities such that

$$\mathcal{P}(\mathbf{y}^t, \mathbf{z}^{t+1} \mid \mathbf{x}^t, \mathbf{z}^t) = \mathcal{P}(\mathbf{y}^t \mid \mathbf{z}^t) \cdot \mathcal{P}(\mathbf{z}^{t+1} \mid \mathbf{x}^t, \mathbf{z}^t)$$

Although it is not necessary, $\mathbf{y}^t$ is frequently taken to be a function of $\mathbf{z}^t$, i.e., the probabilities $\mathcal{P}(\mathbf{y}^t \mid \mathbf{z}^t)$ are equal either to zero or one.

An example of the Moore probabilistic machine is in Figure 10.12c.

While the Shannon machine does not impose any constraints on the functional generator and the random generator, both the Mealy and Moore probabilistic machines require, in the general case, two independent random generators as shown in Figure 10.13.

The chart representation of the conditional probabilities involved in individual probabilistic machines describes the behavior of the functional generator of the probabilistic sequential circuit. Since the functional generator is a memoryless block, its design can be done by the same method that is described in Section 10.2.

EXAMPLE 10.5 Design the probabilistic switching circuit which is specified in terms of the Shannon machine in Figure 10.12a.

First, we determine probabilities of the output variables of the functional generator and their logic products for individual pairs (x^t, z^t). This is shown in Figure 10.14a. The smallest probability is $0.125 = 2^{-3}$ and all other probabilities are integer multiples of it. Thus, the random generator needs three Las Vegas sources. Neither Theorem 10.1 nor Theorem 10.2 are applicable so

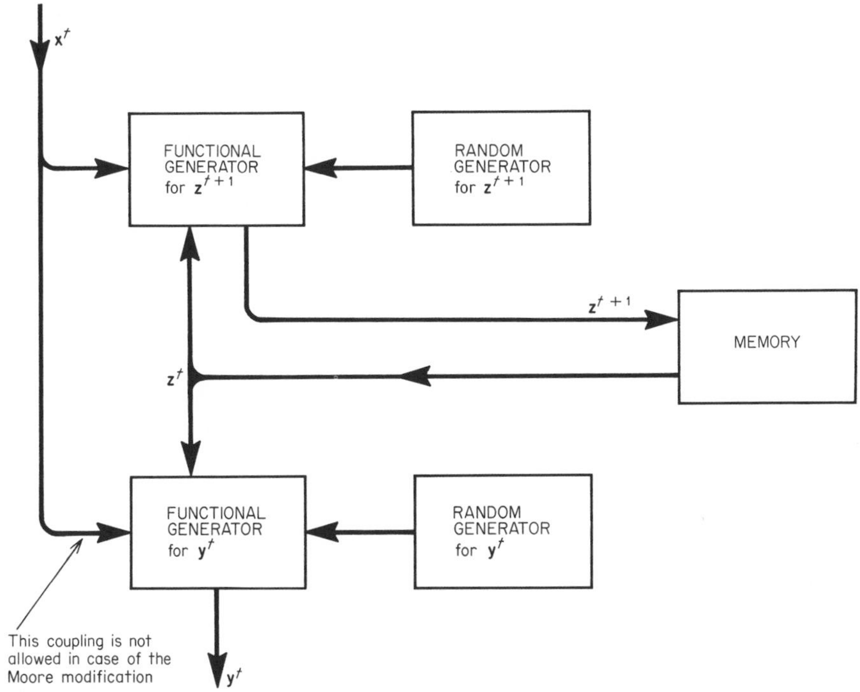

Figure 10.13 The Mealy probabilistic machine

X_i	x^t, z^t	$\mathscr{P}(y^t \mid X_i)$	$\mathscr{P}(z^{t+1} \mid X_i)$	$\mathscr{P}(y^t, z^{t+1} \mid X_i)$
X_0	0,0	.375	.625	.125
X_1	0,1		•	
X_2	1,0		.625	
X_3	1,1	.875	.875	.75

(a)

X_i	x^t, z^t	$\mathscr{P}(y^t \mid X_i)$	$\mathscr{P}(z^{t+1} \mid X_i)$	$\mathscr{P}(y^t, z^{t+1} \mid X_i)$
X_0	0,0	ρ_1, 3	ρ_2, 3	$\rho_1\rho_2$, 1
X_1	0,1		•	
X_2	1,0		ρ_3, 5	
X_3	1,1	ρ_4, 7	ρ_5, 7	$\rho_4\rho_5$, 6

(b)

z^t (columns), x^t (rows):

ρ_1	
	ρ_4

ρ_2	•
ρ_3	ρ_5

(c)

Figure 10.14 Illustration for Example 10.5.

that we obtain the assignment of the output variables of the random generator to the given probabilities as shown in Figure 10.14b. Chart forms of the functions y^t and z^{t+1} follow then immediately (Figure 10.14c). Minimal *sp* Boolean forms of these functions are:

$$y^t = \rho_1 \bar{x}^t z^t \vee \rho_4 x^t z^t$$
$$z^{t+1} = \bar{x}^t z^t \vee \rho_2 \bar{x}^t \vee \rho_3 x^t \bar{z}^t \vee \rho_5 x^t z^t$$

It remains now to design the random generator. A possible covering of states produced by the Las Vegas sources is shown in Figure 10.15. It can be described by the Boolean forms

$$\rho_1 = \lambda_3 \lambda_2 \vee \lambda_3 \lambda_1$$
$$\rho_2 = \lambda_2 \lambda_1 \vee \bar{\lambda}_3 \lambda_2$$
$$\rho_3 = \lambda_3 \vee \lambda_2 \lambda_1$$
$$\rho_4 = \lambda_3 \vee \lambda_2 \vee \lambda_1$$
$$\rho_5 = \bar{\lambda}_3 \vee \lambda_2 \vee \lambda_1$$

The final logic diagram incorporating the functional generator, the random generator, and the memory (represented by a single DELAY) is shown in Figure 10.16.

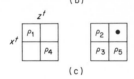

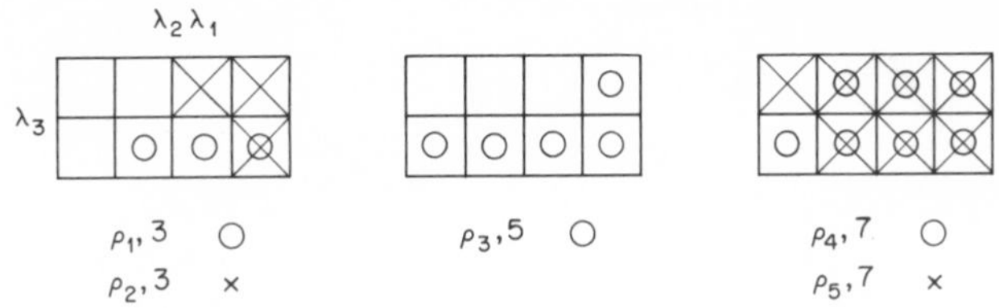

Figure 10.15 Illustration for Example 10.5.

Exercises to Section 10.3

10.3-1 Modify the design in Figure 10.16 in case that instead of DELAY
(a) SR FLIP-FLOP is used.
(b) TRIGGER is used.

10.3-2 Implement the Mealy probabilistic machine and the Moore probabilistic machine which are specified in Figures 10.12b and c, respectively, by a synchronous circuit using
(a) NAND and DELAY elements.
(b) NOR and SR FLIP-FLOPS.

10.4 SVOBODA'S APPROACH

We recall that in the *Svoboda approach*, which is described in Section 8.7, we seek a mask for which dependent sampling variables are functions of other sampling variables. We also recall that we record samples of activity in a chart whose rows are identified by states of the dependent sampling variables. If there is no column in the chart in which two or more samples are recorded, then the mask is proper in the sense that the relationships between the sampling variables involved represents a deterministic switching circuit.

The approach can be generalized for probabilistic switching circuits. Assume that $\mathbf{Y}_j$ and $\mathbf{X}_i$ denote, respectively, states of the dependent sampling variables and states of all other sampling variables within a particular mask. Then, a probabilistic switching circuit is specified by a set of conditional probabilities $\mathscr{P}(\mathbf{Y}_j|\mathbf{X}_i)$ associated with individual samples of activity $(\mathbf{X}_i, \mathbf{Y}_j)$. Clearly, this specification does not differ from the specification of memoryless switching circuits except for the meaning of the sampling variables with respect to the memory. Hence, after the memory is designed for the mask, the synthesis of the functional generator and the random generator is carried out in the same way as the synthesis of memoryless probabilistic switching circuits.

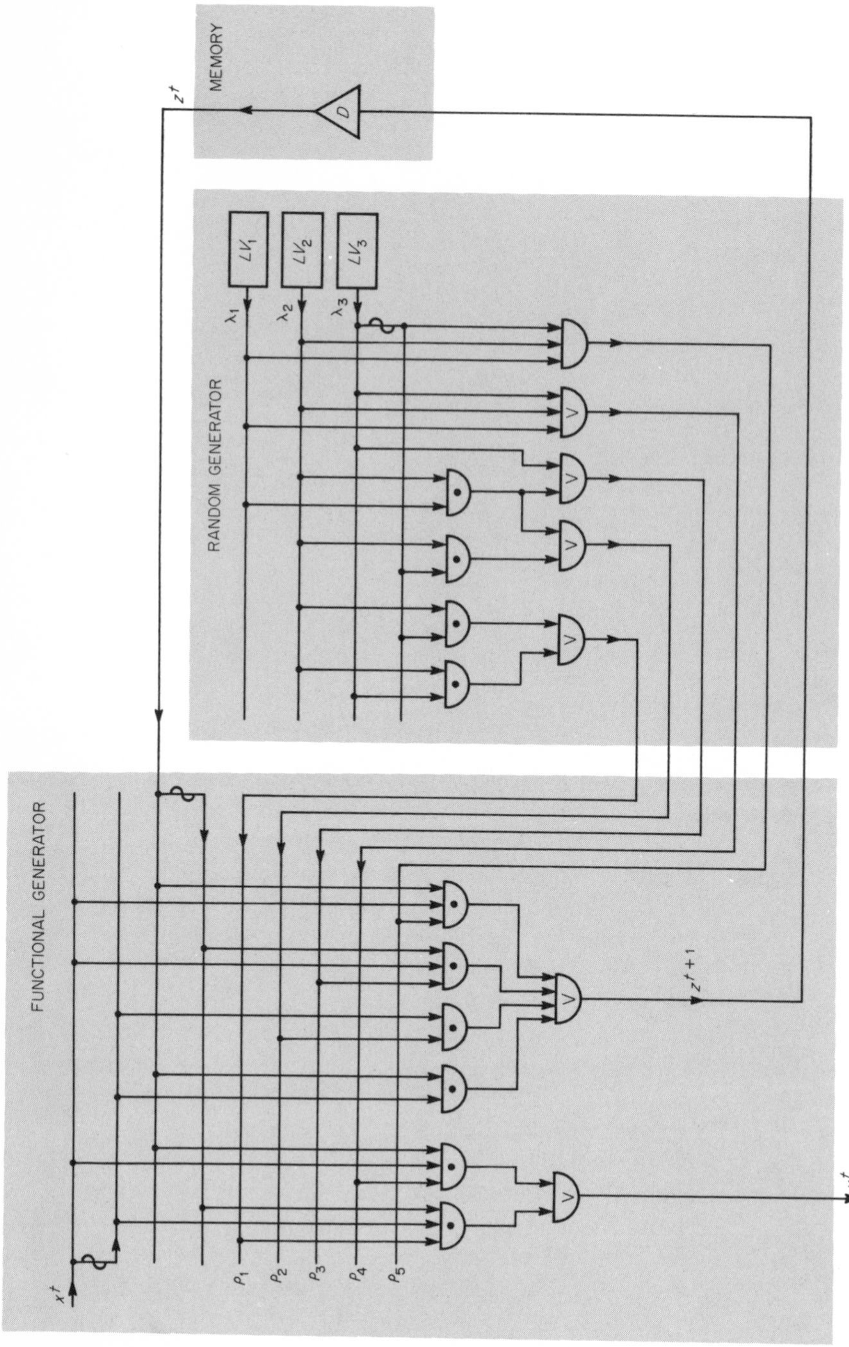

Figure 10.16 An implementation of the Shannon machine specified in Figure 10.12a.

EXAMPLE 10.6 Using the mask shown in Figure 10.17a, design a probabilistic switching circuit to implement the probability distribution of samples of activity which is given in Figure 10.17b.

Two Las Vegas sources suffice to generate the required probabilities. Both Theorems 10.1 and 10.2 can be applied to reduce the number of generated random variables. We obtain the specification of dependent sampling variables V_3^0 and V_2^0 that is shown in Figure 10.17c. Although five random variables are needed, we can make some of them identical by a proper covering of states of the Las Vegas sources. A possible covering is:

$$\rho_1 = \rho_2 = \rho_4 = \lambda_1 \lambda_2$$
$$\rho_3 = \lambda_1$$
$$\rho_5 = \bar{\lambda}_1 \vee \bar{\lambda}_2$$

The final scheme of the circuit is in Figure 10.18.

Exercises to Section 10.4

10.4-1 Design a probabilistic switching circuit implementing the set of samples of activity specified in Figure 8.28 for mask M_1. Assume that the samples in columns 0, 2, 3 appear with the same probability of 0.25, and the samples in column 1 appear with the same probability of 0.5.

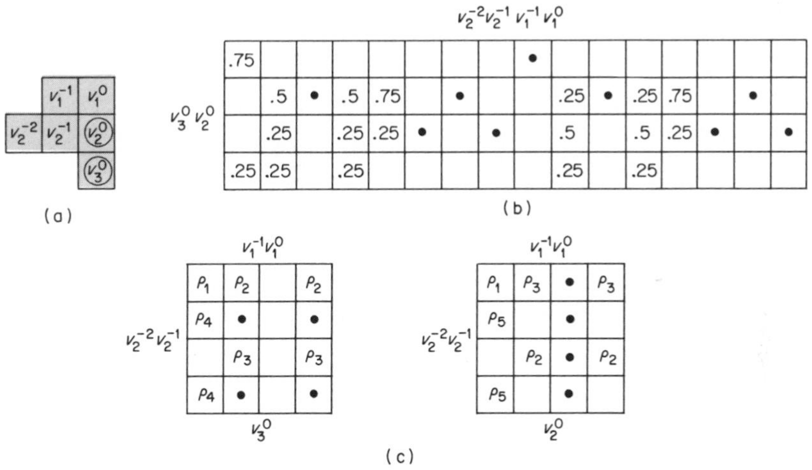

Figure 10.17 Illustration for Example 10.6.

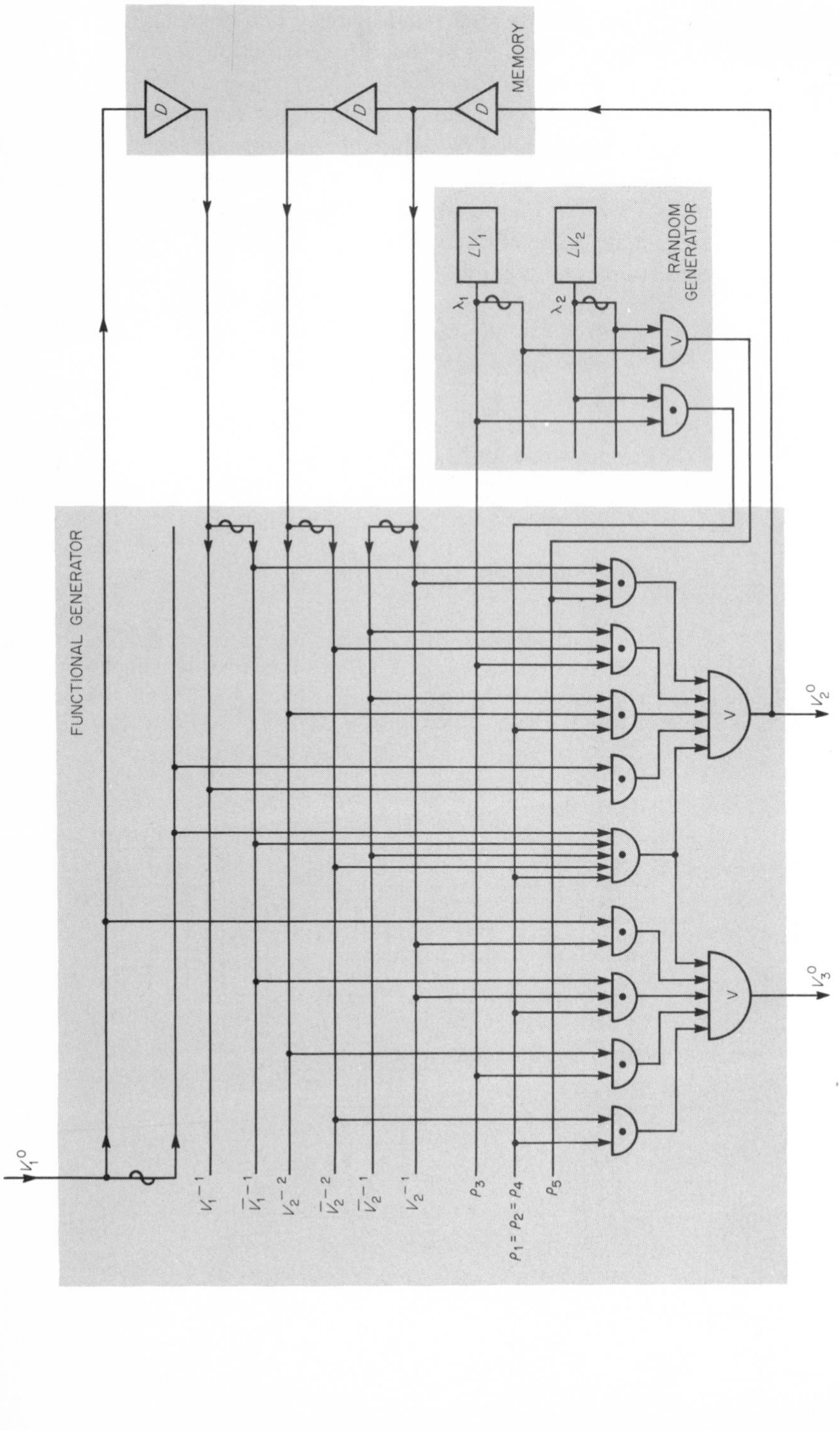

Figure 10.18 A probabilistic sequential switching circuit based on the Svoboda approach (Example 10.6).

10.4-2 Design random generators specified by the following probabilities of samples:

(a)

$\mathscr{P}$	$V^{-1}V^{-2}$ 00	01	10	11
$V^0 = 0$	0.25	0.5	0.0	0.75
1	0.75	0.5	1.0	0.25

(b)

$\mathscr{P}$	$V^{-1}V^{-2}$ 00	01	10	11
$V^0 = 0$	0.125	0.5	0.25	0.625
1	0.875	0.5	0.75	0.375

(c)

$\mathscr{P}$	$V^{-2}V^{-3}$ 00	01	10	11
$V^0V^{-1} = 00$	0.25	0.5	0.375	0.125
01	0.25	0.0	0.375	0.125
10	0.25	0.0	0.125	0.125
11	0.25	0.5	0.125	0.625

10.5 STATE MINIMIZATION

In this section we show that the number of internal states needed to describe a probabilistic machine can be minimized. We also present a survey of basic results relevant to the state minimization of probabilistic machines. Essentially, this is a generalization of the state minimization of deterministic finite-state machines as described in Chapter 7.

Let a Shannon machine $M = (X, Y, Z, P)$ be given, where

$$X = \{\mathbf{x}_1, \mathbf{x}_2, \ldots, \mathbf{x}_p\},\ Y = \{\mathbf{y}_1, \mathbf{y}_2, \ldots, \mathbf{y}_q\}, Z = \{\mathbf{z}_1, \mathbf{z}_2, \ldots, \mathbf{z}_r\},$$

and let the conditional probabilities $\mathscr{P}(\mathbf{y}^t, \mathbf{z}^{t+1} | \mathbf{x}^t, \mathbf{z}^t)$ be arranged in the matrix form $\mathbf{P}$ shown in Figure 10.19. Observe that the matrix, which has $p \cdot r$ rows and $q \cdot r$ columns, is the transpose (rows and columns are interchanged) of the chart form used in Section 10.3 (compare, for instance, Figure 10.19 with Figure 10.12a).

To be able to calculate the conditional probability $\mathscr{P}(\mathbf{v}|\mathbf{u})$, where $\mathbf{u}$ is a sequence of stimuli and $\mathbf{v}$ is a sequence of responses produced by $\mathbf{u}$, we partition matrix $\mathbf{P}$ into $p \cdot q$ square matrices $\mathbf{P}_{i,j}$ as indicated in Figure 10.19. Matrix $\mathbf{P}_{i,j}$ contains probabilities $\mathscr{P}(\mathbf{y}_j, \mathbf{z}^{t+1} | \mathbf{x}_i, \mathbf{z}^t)$ for a particular input output pair $(\mathbf{x}_i, \mathbf{y}_j)$ and all $\mathbf{z}^t, \mathbf{z}^{t+1} \in Z$.

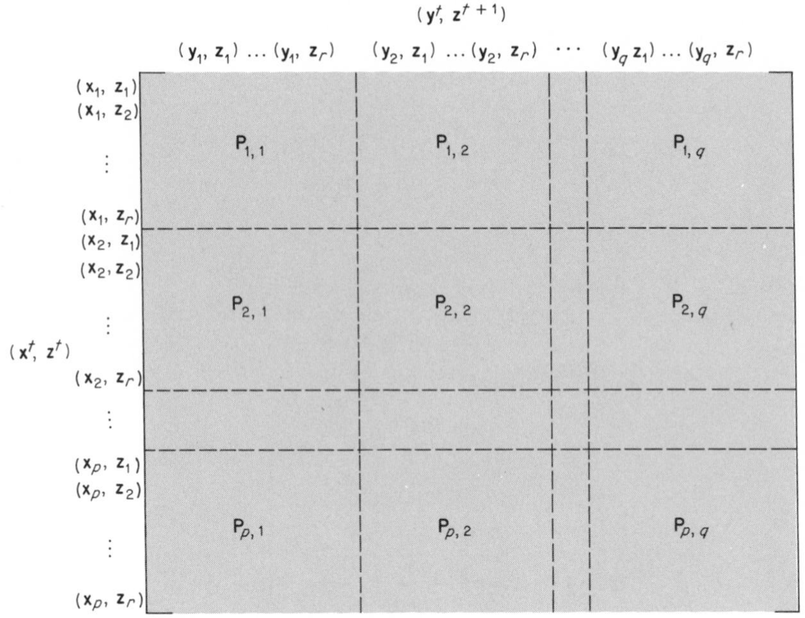

Figure 10.19 Matrix arrangement of probabilities $\mathscr{P}(\mathbf{y}^t, \mathbf{z}^{t+1} | \mathbf{x}^t, \mathbf{z}^t)$ and the meaning of matrices $\mathbf{P}_{i,j}$.

Let the row vector

$$\mathbf{\Phi} = [\phi_1, \phi_2, \ldots, \phi_r]$$

represent an *initial-state probability distribution* on the state set Z and let $\mathbf{c}_r$ denote the column vector containing r rows whose elements are all ones. Then, it is easy to verify that the kth component of the row vector

$$\mathbf{\Phi}(\mathbf{y}_j | \mathbf{x}_i) = \mathbf{\Phi} \times \mathbf{P}_{i,j}$$

represents the probability that state $\mathbf{z}_k$ will be reached when $\mathbf{y}_j$ is observed as a response to $\mathbf{x}_i$. The sum of all elements of $\mathbf{\Phi}(\mathbf{y}_j | \mathbf{x}_i)$, which can be expressed as

$$\mathbf{\Phi}(\mathbf{y}_j | \mathbf{x}_i) \times \mathbf{c}_r$$

represents the probability $\mathscr{P}_{\mathbf{\Phi}}(\mathbf{y}_j | \mathbf{x}_i)$ that, for a given $\mathbf{\Phi}$, the response will be $\mathbf{y}_j$ provided that the stimulus is $\mathbf{x}_i$. Hence,

$$\mathscr{P}_{\mathbf{\Phi}}(\mathbf{y}_j | \mathbf{x}_i) = \mathbf{\Phi} \times \mathbf{P}_{i,j} \times \mathbf{c}_r .$$

Extending the last formula, we can easily derive that the probability $\mathscr{P}_{\mathbf{\Phi}}(\mathbf{v}_j|\mathbf{u}_i)$ of output sequence $\mathbf{v}_j = \mathbf{y}_{j_1}\mathbf{y}_{j_2}\cdots\mathbf{y}_{j_k}$ for input sequence $\mathbf{u} = \mathbf{x}_{i_1}, \mathbf{x}_{i_2}, \ldots, \mathbf{x}_{i_k}$ is given by the formula

$$\mathscr{P}_{\mathbf{\Phi}}(\mathbf{v}_j|\mathbf{u}_i) = \mathbf{\Phi} \times \mathbf{P}_{i_1, j_2} \times \mathbf{P}_{i_2, j_2} \times \cdots \times \mathbf{P}_{i_k, j_k} \times \mathbf{c}_r$$

Let $\mathscr{P}_{\mathbf{\Phi}}(\mathbf{v}_j|\mathbf{u}_i) = \mathbf{\Phi} \times \mathbf{h}(\mathbf{v}_j|\mathbf{u}_i)$. Then $\mathbf{h}(\mathbf{v}_j|\mathbf{u}_i)$ is a column vector containing r rows; its lth row represents the conditional probability of output sequence $\mathbf{v}_j$ given that the input sequence is $\mathbf{u}_i$ and that the machine is initially in state $\mathbf{z}_l$. The column vectors $\mathbf{h}(\mathbf{v}_j|\mathbf{u}_i)$ are of a prime importance to the state minimization of probabilistic machines as will be shown later. Let $\mathscr{P}_{\mathbf{\Phi}}(\mathbf{v}_j|\mathbf{u}_i) = \mathbf{g}_{\mathbf{\Phi}}(\mathbf{v}_j|\mathbf{u}_i) \times \mathbf{c}_r$. Then $\mathbf{g}_{\mathbf{\Phi}}(\mathbf{v}_j|\mathbf{u}_i)$ is a r-component row vector whose lth component expresses the probability of terminating in state $\mathbf{z}_l$ and generating output sequence $\mathbf{v}_j$ given that the initial-state probability distribution is $\mathbf{\Phi}$ and the input sequence is $\mathbf{u}_i$.

Formulas for various other probabilities associated with input and/or output sequences can be derived similarly as shown above. For instance, the state probability distribution $\mathbf{\Phi}(\mathbf{u}_i)$ after input sequence $\mathbf{u}_i$ is given by the formula

$$\mathbf{\Phi}(\mathbf{u}_i) = \mathbf{\Phi} \times \mathbf{P}_{i_1} \times \mathbf{P}_{i_2} \times \cdots \times \mathbf{P}_{i_k},$$

where

$$\mathbf{P}_{i_a} = \sum_{j=1}^{q} \mathbf{P}_{i_a, j}$$

Clearly, the sum of components of $\mathbf{\Phi}(\mathbf{u}_i)$ is equal to one for every $\mathbf{u}_i$.

EXAMPLE 10.7 Given a Shannon machine with $X = \{0, 1\}$, $Y = \{0, 1\}$, $Z = \{z_1, z_2\}$ and

$$\mathbf{P}_{0,0} = \begin{bmatrix} .3 & .1 \\ .6 & .2 \end{bmatrix} \qquad \mathbf{P}_{0,1} = \begin{bmatrix} .2 & .4 \\ .1 & .1 \end{bmatrix}$$

$$\mathbf{P}_{1,0} = \begin{bmatrix} .5 & .2 \\ .7 & .1 \end{bmatrix} \qquad \mathbf{P}_{1,1} = \begin{bmatrix} .0 & .3 \\ .2 & .0 \end{bmatrix}$$

calculate $\mathscr{P}_{\mathbf{\Phi}}(010|110)$, $\mathbf{g}_{\mathbf{\Phi}}(010|110)$, and $\mathbf{\Phi}(010)$.

Suppose the initial-state probability distribution is $\mathbf{\Phi} = [\phi_1, \phi_2]$. Then

$$\mathscr{P}_{\mathbf{\Phi}}(010|110) = [\phi_1, \phi_2] \times \mathbf{P}_{01} \times \mathbf{P}_{11} \times \mathbf{P}_{00} \times \begin{bmatrix} 1 \\ 1 \end{bmatrix}$$

$$= [\phi_1, \phi_2] \times \begin{bmatrix} .06 & .02 \\ .024 & .008 \end{bmatrix} \times \begin{bmatrix} 1 \\ 1 \end{bmatrix}$$

$$= [\phi_1, \phi_2] \times \begin{bmatrix} .08 \\ .032 \end{bmatrix} = .08\,\phi_1 + .032\,\phi_2$$

and

$$\mathbf{g}_{\Phi}(010|110) = [\phi_1, \phi_2] \times \begin{bmatrix} .06 & .02 \\ .024 & .008 \end{bmatrix}$$

$$= [.06\,\phi_1 + .024\,\phi_2,\ .02\,\phi_1 + .008\,\phi_2]$$

Observe that

$$\mathbf{h}(010|110) = \begin{bmatrix} .08 \\ .032 \end{bmatrix}$$

Now, we have

$$\mathbf{P}_0 = \mathbf{P}_{0,0} + \mathbf{P}_{0,1} = \begin{bmatrix} .5 & .5 \\ .7 & .3 \end{bmatrix}$$

$$\mathbf{P}_1 = \mathbf{P}_{1,0} + \mathbf{P}_{1,1} = \begin{bmatrix} .5 & .5 \\ .9 & .1 \end{bmatrix}$$

and

$$\mathbf{\Phi}(010) = \mathbf{\Phi} \times \mathbf{P}_0 \times \mathbf{P}_1 \times \mathbf{P}_0 = [\phi_1, \phi_2] \times \begin{bmatrix} .56 & .44 \\ .576 & .424 \end{bmatrix}$$

$$= [.56\,\phi_1 + .576\,\phi_2,\ .44\,\phi_1 + .424\,\phi_2]$$

Suppose there are three special initial-state probability distributions: $\phi_1 = [1, 0]$, $\phi_2 = [0, 1]$, $\phi_3 = [.5, .5]$. We obtain

$$\mathbf{P}_1(010|110) = .08,\ \mathbf{P}_2(010|110) = .032,\ \mathbf{P}_3(010|110) = .056$$

$$g_1(010|110) = [.06, 0.2],\ g_2(010|110) = [.024, .008],\ g_3(010|110) = [.042, .014]$$

$$\phi_1(010) = [.56, .44],\ \phi_2(10) = [.576, .424],\ \phi_3(10) = [.568, .432]$$

Now we are in a position to introduce the concept of state equivalence for probabilistic machines. It will be considered as a special case of a more general concept of equivalence between two initial-state probability distributions. The following definitions are direct generalizations of those in Section 7.3 for deterministic machines. The reader should compare these two sets of definitions.

DEFINITION 10.4 Let $M_1 = (X, Y, Z_1, P_1)$ and $M_2 = (X, Y, Z_2, P_2)$ be probabilistic machines specified in the Shannon form and let $\mathbf{\Phi}$ and $\mathbf{\Psi}$ be certain initial-state probability distributions given for M_1 and M_2, respectively. We say that $\mathbf{\Phi}$ and $\mathbf{\Psi}$ are *equivalent*, written as $\mathbf{\Phi} \simeq \mathbf{\Psi}$, if and only if

$$\mathscr{P}_{\mathbf{\Phi}}(\mathbf{v} \mid \mathbf{u}) = \mathscr{P}_{\mathbf{\Psi}}(\mathbf{v} \mid \mathbf{u}) \tag{10.3}$$

for all possible input-output sequences $(\mathbf{u}, \mathbf{v})$. Similarly, we say that $\mathbf{\Phi}$ and $\mathbf{\Psi}$ are *k-equivalent*, written as $\mathbf{\Phi} \overset{k}{\simeq} \mathbf{\Psi}$, if and only if (10.3) is satisfied for all possible input-output sequences of length k. If (10.3) is not satisfied for at least one input-output sequence (or at least one input-output sequence of length k), then the distributions $\mathbf{\Phi}$ and $\mathbf{\Psi}$ are said to be *distinguishable* and we write $\mathbf{\Phi} \not\simeq \mathbf{\Psi}$ (or k-distinguishable, respectively, which is written as $\mathbf{\Phi} \overset{k}{\not\simeq} \mathbf{\Psi}$). M_1 and M_2 may refer to the same machine.

DEFINITION 10.5 Let M_1, M_2, $\mathbf{\Phi}$ and $\mathbf{\Psi}$ have the same meaning as in Definition 10.4 and let $Z_1 = \{{}^1\mathbf{z}_1, {}^1\mathbf{z}_2, \ldots, {}^1\mathbf{z}_m\}$, $Z_2 = \{{}^2\mathbf{z}_1, {}^2\mathbf{z}_2, \ldots, {}^2\mathbf{z}_n\}$, $\mathbf{\Phi} = [\phi_1, \phi_2, \ldots, \phi_m]$ and $\mathbf{\Psi} = [\psi_1, \psi_2, \ldots, \psi_n]$. Assume that one component of vector $\mathbf{\Phi}$ and one component of vector $\mathbf{\Psi}$, say ϕ_a and ψ_b, are equal to one and all other components of both ϕ and ψ are equal to zero (we say that probability distributions with this property are degenerate). Assume further that $\mathbf{\Phi} \simeq \mathbf{\Psi}$ (or $\mathbf{\Phi} \overset{k}{\simeq} \mathbf{\Psi}$) in the sense of Definition 10.4. Then state ${}^1\mathbf{z}_a$ of machine M_1 and state ${}^2\mathbf{z}_b$ of machine M_2 are said to be *equivalent* (or *k-equivalent*) and we write ${}^1\mathbf{z}_a = {}^2\mathbf{z}_b$ (or ${}^1\mathbf{z}_a \overset{k}{=} {}^2\mathbf{z}_b$). If $\mathbf{\Phi} \not\simeq \mathbf{\Psi}$ (or $\mathbf{\Phi} \overset{k}{\not\simeq} \mathbf{\Psi}$), we say that states ${}^1\mathbf{z}_a$ and ${}^2\mathbf{z}_b$ are *distinguishable* (or *k-distinguishable*) and we write ${}^1\mathbf{z}_a \neq {}^2\mathbf{z}_b$ (or ${}^1\mathbf{z}_a \overset{k}{\neq} {}^2\mathbf{z}_b$). M_1 and M_2 may refer to the same machine.

DEFINITION 10.6 Two probabilistic machines M_1 and M_2 are *distribution-equivalent*, written as $M_1 = M_2$, if and only if for each initial-state probability distribution $\mathbf{\Phi}$ of M_1 there exists an initial-state probability distribution $\mathbf{\Psi}$ of M_2 such that $\mathbf{\Phi} \simeq \mathbf{\Psi}$ and, conversely, for each $\mathbf{\Psi}$ of M_2 there exists $\mathbf{\Phi}$ of M_1 such that $\mathbf{\Psi} \simeq \mathbf{\Phi}$.

DEFINITION 10.7 Two probabilistic machines M_1 and M_2 are *state equivalent*, written as $M_1 \overset{z}{=} M_2$, if and only if for each state ${}^1\mathbf{z}_i$ of M_1 (a degenerate distribution $\mathbf{\Phi}$ of M_1 concentrated on state ${}^1\mathbf{z}_i$) there exists a state ${}^2\mathbf{z}_j$ of M_2 (a degenerate distribution $\mathbf{\Psi}$ of M_2 concentrated on state ${}^2\mathbf{z}_j$) such that ${}^1\mathbf{z}_i = {}^2\mathbf{z}_j$ and, conversely, for each state ${}^2\mathbf{z}_j$ of M_2 there exists a state ${}^1\mathbf{z}_i$ of M_1 such that ${}^2\mathbf{z}_j = {}^1\mathbf{z}_i$.

Note that state-equivalent machines are distribution-equivalent. The converse, however, does not hold in general.

DEFINITION 10.8 A probabilistic machine M_m with m states is said to be a *minimal-state representation* of a given probabilistic machine M_n with n states if $M_m = M_n$ and no machine M_r with r states exists such that $M_r = M_m$ and $r < m$.

Let us note that when applying Definition 10.8 it is sufficient to require that to each state of the given machine M_n (to each degenerate initial-state probability distribution of M_n) there exists an equivalent distribution of M_m and, conversely, to each state of M_m there exists an equivalent distribution of M_m.

Some theorems will be proven now on the basis of which a procedure of state minimization will be developed.

THEOREM 10.3 Let $(\mathbf{u}, \mathbf{v})$ and $(\mathbf{x}, \mathbf{y})$ be, respectively, an input-output sequence and an input-output pair. Then

$$\mathscr{P}_{\mathbf{\Phi}}(\mathbf{v}|\mathbf{u}) = \sum_{\mathbf{y}} \mathscr{P}_{\mathbf{\Phi}}(\mathbf{vy}|\mathbf{ux})$$

PROOF $\sum_{y} \mathscr{P}_{\mathbf{\Phi}}(\mathbf{vy}|\mathbf{ux}) = \mathbf{g}_{\mathbf{\Phi}}(\mathbf{v}|\mathbf{u}) \sum_{\mathbf{y}} \mathscr{P}(\mathbf{y}|\mathbf{x}) \times \mathbf{c}_r$

$$= \mathbf{g}_{\mathbf{\Phi}}(\mathbf{v}|\mathbf{u}) \times \mathbf{P}_x \times \mathbf{c}_r = \mathbf{g}_{\mathbf{\Phi}}(\mathbf{v}|\mathbf{u}) \times \mathbf{c}_r = \mathscr{P}_{\mathbf{\Phi}}(\mathbf{v}|\mathbf{u}) \quad \blacksquare$$

With the aid of Theorem 10.3 we conclude that k-distinguishibility of distributions or states implies $(k + 1)$-distinguishibility of distributions or states respectively. Similarly, $(k + 1)$-equivalence of distributions or states implies k-equivalence of distributions or states respectively. According to these properties, if E_k denotes the k-equivalence partition of the state set of a probabilistic machine, then, for each k, E_{k+1} is a refinement of E_k though not necessarily a proper refinement of E_k.

The next theorem, due to Carlyle, is of a prime importance to the state minimization of probabilistic machines. Although the state equivalence requires equal probabilities for all input-output sequences (including thus sequences of infinite length), the Carlyle theorem states that only finite input-output sequences of length $l \leq r - 1$, where r is the number of internal states of the given machine, have to be investigated. The Carlyle theorem guarantees thus an existence of a finite procedure of state minimization.

THEOREM 10.4 Let M be a probabilistic machine in the Shannon form with r internal states and let $\mathbf{\Phi}$ and $\mathbf{\Psi}$ be any two initial-state probability distributions on the states of M. Then $\mathbf{\Phi} \overset{r-1}{\simeq} \mathbf{\Psi}$ is a sufficient condition for $\mathbf{\Phi} \simeq \mathbf{\Psi}$.

PROOF From Definition 10.4, $\mathbf{\Phi}$ and $\mathbf{\Psi}$ are k-equivalent if and only if

$$\mathscr{P}_{\mathbf{\Phi}}(\mathbf{v}|\mathbf{u}) - \mathscr{P}\mathbf{\Psi}(\mathbf{v}|\mathbf{u}) = (\mathbf{\Phi} - \mathbf{\Psi}) \times \mathbf{h}(\mathbf{v}|\mathbf{u})$$

vanishes on the linear space (a subspace of r-dimensional space) represented by the column vectors $\mathbf{h}(\mathbf{v}|\mathbf{u})$ for all input-output sequences of length k. Let L_k denote this linear space. We can prove the theorem by demonstrating that for $k \geq r - 1$ all spaces L_k are identical. First,

$$\mathbf{h}(\mathbf{v}|\mathbf{u}) = \sum_{\mathbf{y}} \mathbf{h}(\mathbf{vy}|\mathbf{ux})$$

follows directly from Theorem 10.3 so that L_k is a subspace (not necessarily a proper one) of L_{k+1} for every k. Next, we can show that if $L_k = L_{k+1}$, then $L_{k+1} = L_{k+2}$. Indeed, L_{k+2} is represented by column vectors $\mathbf{h}(\mathbf{yv}|\mathbf{xu})$ for all input-output pairs $(\mathbf{x}, \mathbf{y})$ and for all input-output sequences $(\mathbf{u}, \mathbf{v})$ of length $k + 1$ so that

$$\mathbf{h}(\mathbf{yv}|\mathbf{xu}) = \mathbf{P}_{x,y} \times \mathbf{h}(\mathbf{v}|\mathbf{u})$$

where $\mathbf{h}(\mathbf{v}|\mathbf{u})$ is a vector from L_{k+1}. Since $L_{k+1} = L_k$, as assumed, $\mathbf{h}(\mathbf{v}|\mathbf{u})$ is a vector from L_k or a linear combination of vectors from L_k. Hence, $\mathbf{h}(\mathbf{yv}|\mathbf{xu})$ is a linear combination of vectors in L_{k+1}. Suppose now that there exists an integer J such that $L_k = L_J$ for all $k \geq J$ and $L_k \neq L_J$ for $k < J$. Then the dimension of L_{k+1} must be greater than the dimension L_k for all $k < J$, i.e.,

$$\begin{aligned}
&\dim(L_1) \geq \dim(L_0) + 1 \\
&\dim(L_2) \geq \dim(L_1) + 1 \geq \dim(L_0) + 2 \\
&\quad\vdots \qquad\qquad\qquad\qquad \vdots \\
&\dim(L_J) \geq \dim(L_{J-1}) + 1 \geq \cdots \geq \dim(L_0) + J
\end{aligned}$$

where $\dim(L_k)$ denotes the dimension of the linear space L_k. Since $\mathbf{H}(\mathbf{v}|\mathbf{u})$ are r-component vectors, only r of them can be linearly independent. Hence, J exists and

$$\dim(L_J) \leq r$$

Combining both inequalities for $\dim(L_J)$, we obtain

$$\dim(L_0) + J \leq \dim(L_J) \leq r$$

Since L_0 is represented by a single vector (vector c_r) for any machine, $\dim(L_0) = 1$. Hence, $J \leq r - 1$. ▮

In our further discussion, $\mathbf{H}$ will denote a matrix whose columns are taken from the set of vectors $\mathbf{h}(\mathbf{v}|\mathbf{u})$ for all input-output sequences of a given machine and form a basis for the linear space L_J introduced in the proof of Theorem 10.4. Similarly, $\mathbf{H}_k$ will denote a matrix whose columns form a basis for the linear space $L_k (k < J)$. Clearly, neither $\mathbf{H}$ nor $\mathbf{H}_k$ are, generally, unique.

EXAMPLE 10.8 Determine a matrix $\mathbf{H}$ for the probabilistic machine given in Example 10.7.

The machine has two states so that, by Theorem 10.4, it is sufficient to determine vectors $\mathbf{h}(\mathbf{v}|\mathbf{u})$ for input-output sequences of length less than two. We obtain

$$\mathbf{h}(\lambda|\lambda) = \begin{bmatrix} 1 \\ 1 \end{bmatrix}$$

$$\mathbf{h}(0|0) = \mathbf{P}_{0,0} \times \mathbf{c}_r = \begin{bmatrix} .3 & .1 \\ .6 & .2 \end{bmatrix} \times \begin{bmatrix} 1 \\ 1 \end{bmatrix} = \begin{bmatrix} .4 \\ .8 \end{bmatrix}$$

$$\mathbf{h}(1|0) = \mathbf{P}_{0,1} \times \mathbf{c}_r = \begin{bmatrix} .2 & .4 \\ .1 & .1 \end{bmatrix} \times \begin{bmatrix} 1 \\ 1 \end{bmatrix} = \begin{bmatrix} .6 \\ .2 \end{bmatrix}$$

$$\mathbf{h}(0|1) = \mathbf{P}_{1,0} \times \mathbf{c}_r = \begin{bmatrix} .5 & .2 \\ .7 & .1 \end{bmatrix} \times \begin{bmatrix} 1 \\ 1 \end{bmatrix} = \begin{bmatrix} .7 \\ .8 \end{bmatrix}$$

$$\mathbf{h}(1|1) = \mathbf{P}_{1,1} \times \mathbf{c}_r = \begin{bmatrix} .0 & .3 \\ .2 & .0 \end{bmatrix} \times \begin{bmatrix} 1 \\ 1 \end{bmatrix} = \begin{bmatrix} .3 \\ .2 \end{bmatrix}$$

Clearly, only two of the five vectors can be linearly independent. For instance,

$$\mathbf{h}(1|0) = \mathbf{h}(\lambda|\lambda) - \mathbf{h}(0|0)$$
$$\mathbf{h}(0|1) = .6\mathbf{h}(\lambda|\lambda) + .25\mathbf{h}(0|0)$$
$$\mathbf{h}(1|1) = .4\mathbf{h}(\lambda|\lambda) - .25\mathbf{h}(0|0)$$

Thus,

$$\mathbf{H} = [\mathbf{h}(\lambda|\lambda), \mathbf{h}(0|0)] = \begin{bmatrix} 1 & .4 \\ 1 & .8 \end{bmatrix}$$

is one possible form of matrix $\mathbf{H}$.

We can now state a theorem which characterizes equivalences and equivalence classes in terms of linear spaces and matrices $\mathbf{H}$ or $\mathbf{H}_k$ as introduced above. The theorem follows directly from Definitions 10.4 and 10.5 and from basic properties of matrices so that no proof is given.

THEOREM 10.5 Let M be a probabilistic machine with r states, let $\mathbf{H}$ be a matrix whose columns form a basis of the linear space of all vectors $\mathbf{h}(\mathbf{v}|\mathbf{u})$ of M, and let $\mathbf{H}_k$ be a matrix whose columns form a basis of the linear space of vectors $\mathbf{h}(\mathbf{v}|\mathbf{u})$ for all input-output sequences of length k. Then:

1. States $\mathbf{z}_i$ and $\mathbf{z}_j$ of M are equivalent (or k-equivalent) if and only if rows i and j of $\mathbf{H}$ (or $\mathbf{H}_k$) are identical.

2. Initial-state probability distributions $\mathbf{\Phi}$ and $\mathbf{\Psi}$ on the state of M are equivalent (or k-equivalent) with regard to M if and only if $\mathbf{\Phi} \times \mathbf{H} = \mathbf{\Psi} \times \mathbf{H}$

(or $\mathbf{\Phi} \times \mathbf{H}_k = \mathbf{\Psi} \times \mathbf{H}_k$). Clearly, the same version of $\mathbf{H}$ (or $\mathbf{H}_k$) must be used on both the left-hand side and the right-hand side of the equation.

3. Let $C(\mathbf{\Phi})$(or $C_k(\mathbf{\Phi})$) denote the equivalence class (or k-equivalence class) of initial-state probability distributions of M which contains $\mathbf{\Phi}$ and let $\mathbf{\Psi} = [\psi_1, \psi_2, \ldots, \psi_r]$.
Then

$$C(\mathbf{\Phi}) = \{\mathbf{\Psi} : \mathbf{\Phi} \times \mathbf{H} = \mathbf{\Psi} \times \mathbf{H} \text{ and } \psi_i \geq 0 \text{ for all } i\}$$

and

$$C_k(\mathbf{\Phi}) = \{\mathbf{\Psi} : \mathbf{\Phi} \times \mathbf{H} = \mathbf{\Psi} \times \mathbf{H}_k \text{ and } \psi_i \geq 0 \text{ for all } i\}$$

Suppose now that the matrix $\mathbf{H}$ of a probabilistic machine M with r states contains c columns, where $c \leq r$. Since $\mathbf{H}$ contains r rows, $r - c$ of these rows are linearly dependent upon the other c rows of $\mathbf{H}$. This means that there is a possibility to eliminate up to $r - c$ states from the state set of M although, due to other requirements, it is not guaranteed that we can really eliminate them. To clarify this point, let us assume first that two rows of $\mathbf{H}$, say rows a and b ($a < b$), are equal. This means, obviously, that either state $\mathbf{z}_a$ or state $\mathbf{z}_b$ is redundant and can be eliminated from the state set of M. When one of these states is eliminated, a new machine, say M', is created. Machines M and M' are equivalent in the sense of Definition 10.7.

Assume that state $\mathbf{z}_b$ is eliminated. Then both the matrices $\mathbf{P}_{i,j}$ and the initial-state probability distribution $\mathbf{\Phi}$ of M must be approximately modified to obtain corresponding matrices $\mathbf{P}'_{i,j}$ and distribution $\mathbf{\Phi}'$ of the new machine M'. We can derive the following modification rules:

1. For each matrix $\mathbf{P}_{i,j}$ of M, $\mathbf{P}'_{i,j}$ of M' is formed from $\mathbf{P}_{i,j}$ by deleting row b and column b and replacing column a with the sum of columns a and b.

2. For the given $\mathbf{\Phi}$ of M, $\mathbf{\Phi}'$ of M' is formed from $\mathbf{\Phi}$ by deleting component (column) b and replacing component a with the sum of components a and b.

Let us assume now that there is no pair of equal rows in $\mathbf{H}$. Assume further that the bth row h_b of $\mathbf{H}$ is linearly dependent upon the first c rows h_i ($i = 1, 2, \ldots, c$) of $\mathbf{H}$, i.e.,

$$h_b = \sum_{i=1}^{c} \alpha_i h_i \tag{10.4}$$

Then, it can be shown by a sizable construction, which is omitted here, that $\mathbf{z}_b$ can be eliminated if and only if (10.4) is satisfied, all coefficients α_i are non-negative ($\alpha_i \geq 0$), and

$$\sum_{i=1}^{c} \alpha_i = 1$$

If all these conditions are satisfied, then, according to nomenclature used in the literature, h_b is called the convex combination of $h_1, h_2, \ldots, h_c$.

Now, let h_b be a convex combination of $h_1, h_2, \ldots, h_c$ and let state $\mathbf{z}_b$ be eliminated. Then the following modifications must be made:

1. For each matrix $\mathbf{P}_{i,j}$ of M, matrix $\mathbf{P}'_{i,j}$ of the reduced machine M' is formed from $\mathbf{P}_{i,j}$ by deleting row b and column b and replacing each column i ($i = 1, 2, \ldots, c$) with the sum of column i and column b multiplied by α_i.
2. For the given $\mathbf{\Phi}$ of M, $\mathbf{\Phi}'$ of M' is formed by deleting component b of $\mathbf{\Phi}$ and replacing each component i of $\mathbf{\Phi}(i = 1, 2, \ldots, c)$ with the sum of component i and component b multiplied by α_i, i.e.,

$$\phi'_i = \phi_i + \alpha_i \phi_b \qquad \text{for all} \quad i = 1, 2, \ldots, c$$

EXAMPLE 10.9 Consider a Shannon probabilistic machine with one input logic variable, one output logic variable, three internal states and the following probability matrices $\mathbf{P}_{i,j}$:

$$\mathbf{P}_{0,0} = \begin{bmatrix} .2 & .2 & .2 \\ .2 & .2 & .2 \\ .2 & .2 & .2 \end{bmatrix} \qquad \mathbf{P}_{0,1} = \begin{bmatrix} .4 & .0 & .0 \\ .1 & .3 & 0 \\ .2 & .1 & .1 \end{bmatrix}$$

$$\mathbf{P}_{1,0} = \begin{bmatrix} 1 & .0 & .0 \\ .0 & .0 & .6 \\ .6 & .1 & .1 \end{bmatrix} \qquad \mathbf{P}_{1,1} = \begin{bmatrix} .0 & .0 & .0 \\ .4 & .0 & .0 \\ .2 & .0 & .0 \end{bmatrix}$$

To minimize internal states of the machine, we calculate column vectors $\mathbf{h}(\mathbf{v}|\mathbf{u})$ for all input-output sequences of length smaller than three first (Theorem 10.4). This means that we have to calculate four vectors for sequences of length one and sixteen vectors for sequences of length two. If we include the vector $\mathbf{h}(\lambda|\lambda)$, the total number of involved vectors $\mathbf{h}(\mathbf{v}|\mathbf{u})$ is twenty one. It follows from the calculation, which is not shown here*, that only two vectors form the basis for the linear space L_2 in this particular case. We may take any pair of linearly independent column vectors $\mathbf{h}(\mathbf{v}|\mathbf{u})$ to form the matrix $\mathbf{H}$. For instance,

$$\mathbf{H} = [\mathbf{h}(\lambda|\lambda), \mathbf{h}(1\ \ 1)] = \begin{bmatrix} 1 & .0 \\ 1 & .4 \\ 1 & .2 \end{bmatrix}$$

A state of the given machine can be eliminated if and only if the corresponding row in $\mathbf{H}$ is a convex combination of the other two rows. When we try the first row, we find that

$$h_1 = -h_2 + 2h_3$$

*) Computer program PSTMIN given in Appendix G can be used.

This is not a convex combination since $\alpha_1 = -1 < 0$. Hence, the first state cannot be eliminated. We obtain the same result for the second state since

$$h_2 = -h_1 + 2h_3$$

The third row, for which

$$h_3 = .5h_1 + .5h_2$$

is a convex combination of the first two rows and, consequently, the third state can be eliminated. To execute this state reduction, we add the third column of each matrix $\mathbf{P}_{i,j}$ multiplied by 0.5 to both its first and second columns ($\alpha_1 = \alpha_2 = .5$ in this case) and, then, eliminate both its third row and third column. We obtain matrices

$$\mathbf{P}'_{0,0} = \begin{bmatrix} .3 & .3 \\ .3 & .3 \end{bmatrix} \qquad \mathbf{P}'_{0,1} = \begin{bmatrix} .4 & 0 \\ .1 & .3 \end{bmatrix}$$

$$\mathbf{P}'_{1,0} = \begin{bmatrix} 1 & 0 \\ .3 & .3 \end{bmatrix} \qquad \mathbf{P}'_{1,1} = \begin{bmatrix} 0 & 0 \\ .4 & 0 \end{bmatrix}$$

of the reduced machine, which represents a minimal form of the given machine. Obviously, any given initial distribution $\mathbf{\Phi} = [\phi_1, \phi_2, \phi_3]$ must be appropriately modified too. We obtain $\mathbf{\Phi}' = [\phi_1 + .5\phi_3, \phi_2 + .5\phi_3]$.

Let us mention, without a proof, two simple rules by which some of the rows which are not convex combinations of other rows (so-called *extremal rows*) can be easily identified in the **H** matrix:

1. If there is only one minimal element in a column of the **H** matrix, the row corresponding to this element is an extremal row.

2. If there is only one maximal element in a column of the **H** matrix, the row corresponding to this element is an extremal row.

For instance, the second column of the **H** matrix determined in Example 10.9 contains a single minimal element (.0) in the first row, and a single maximal element (.4) in the second row. Hence, the first two rows are extremal which means that the first two states of the given machine cannot be eliminated.

Exercises to Section 10.5

10.5-1 Minimize the Shannon machine which has one input logic variable and one output logic variable and whose probability matrices $\mathbf{P}_{i,j}$ have the form:

*(a) $$\mathbf{P}_{0,0} = \begin{bmatrix} .1 & .3 & .2 \\ .0 & .1 & .2 \\ .05 & .2 & .2 \end{bmatrix} \qquad \mathbf{P}_{0,1} = \begin{bmatrix} .1 & .1 & .2 \\ .2 & .1 & .4 \\ .15 & .1 & .3 \end{bmatrix}$$

$$\mathbf{P}_{1,0} = \begin{bmatrix} .2 & .1 & .2 \\ .2 & .0 & .4 \\ .2 & .05 & .3 \end{bmatrix} \qquad \mathbf{P}_{1,1} = \begin{bmatrix} .0 & .1 & .4 \\ .2 & .0 & .2 \\ .1 & .05 & .3 \end{bmatrix}$$

(b) $\mathbf{P}_{0,0} = \begin{bmatrix} .0 & .2 & .4 \\ .1 & .0 & .4 \\ .0 & .0 & .2 \end{bmatrix}$ $\mathbf{P}_{0,1} = \begin{bmatrix} .0 & .0 & .4 \\ .0 & .1 & .4 \\ .0 & .0 & .8 \end{bmatrix}$ $\mathbf{P}_{0} = \begin{bmatrix} 0 & .2 & .8 \\ .1 & .1 & .8 \\ 0 & 0 & 1. \end{bmatrix}$

$\mathbf{P}_{1,0} = \begin{bmatrix} .0 & .1 & .2 \\ .0 & .3 & .2 \\ .0 & .0 & .4 \end{bmatrix}$ $\mathbf{P}_{1,1} = \begin{bmatrix} .1 & .0 & .6 \\ .0 & .1 & .4 \\ .0 & .0 & .6 \end{bmatrix}$ $\mathbf{P}_{1} = \begin{bmatrix} .1 & .1 & .8 \\ 0 & .4 & .6 \\ 0 & 0 & 1. \end{bmatrix}$

10.5-2 Let $\mathbf{\Phi} = [.3, .6, .1]$. Calculate:

*(a) $\mathbf{\Phi}(1101)$ and $\mathbf{\Phi}(0101)$ for the machine specified in Exercise 10.5-1a.

(b) $\mathbf{\Phi}(0011)$ and $\mathbf{\Phi}(0010)$ for the machine specified in Exercise 10.5-1b.

*(c) $\mathscr{P}_{\mathbf{\Phi}}(11\,|\,00)$ and $\mathscr{P}_{\mathbf{\Phi}}(01\,|\,10)$ for the machine specified in Exercise 10.5-1a.

(d) $\mathscr{P}_{\mathbf{\Phi}}(101\,|\,011)$ and $\mathscr{P}_{\mathbf{\Phi}}(010\,|\,000)$ for the machine specified in Exercise 10.5-1b.

10.5-3 Show that, in the general case, there is more than one minimal form for a given Shannon machine.

Comprehensive Exercises to Chapter 10

10.1 Let m_i and M_i be, respectively, minterms and maxterms of n statistically independent logic variables. Let values of each of these variables be generated with equal probabilities of 0.5. Show that

$$\sum_{i=0}^{2^n-1} \mathscr{P}(m_i) = 1 \quad \text{and} \quad \sum_{i=0}^{2^n-1} \mathscr{P}(M_i) = n - 1$$

where $\mathscr{P}(m_i)$ and $\mathscr{P}(M_i)$ are the probabilities that m_i or M_i is equal to one, respectively.

10.2 Consider logic variables generated by n independent Las Vegas sources. Calculate:

*(a) The probability that the OR operation of all these variables is equal to one.

*(b) The probability that an AND operation of k maxterms is equal to one.

(c) The probability that the OR operation of all Boolean p-terms that contain v variables ($v \leq n$) is equal to one.

(d) The probability that the output variable of the logic function F_3^2 is equal to one.

10.3 Discuss a possiblity of using a single Las Vegas source for generating 2^r states of variables $\lambda_1, \lambda_2, \ldots, \lambda_r$, each with the probability of 2^{-r}.

Reference Notations to Chapter 10

Structure analysis and synthesis of probabilistic memoryless switching circuits was initiated and developed by Warfield [24–27]. Some work in this area has been done by Pospelov [16] and other Russian authors.

A probabilistic sequential machine was suggested (under a different name) by Shannon as early as in 1948 [19]. It was used as a mathematical model of discrete communication channels with noise. The foundation of the theory of probabilistic (stochastic) machines was laid by Carlyle [4, 5] and Rabin [17]. Their work has been extended by Bacon [1, 2], Even [6], Ott [10, 11], Page [12], Paz [13, 14], Salomaa [18], and others [9, 20, 21]. Synthesis of probability transformers was investigated by Gill [7]. Structure theory of probabilistic machines was initiated by Nieh and Carlyle [8].

The books by Booth [3] and Paz [15] are recommended for further information. Russian sources can be found in Reference 16.

Most of the material included in Sections 10.3 and 10.5 is based on Carlyle's work [4, 5]. The material in Section 10.4 is based on the concepts introduced by Svoboda [22, 23].

References to Chapter 10

1. **Bacon, G. C.,** "Minimal-State Stochastic Finite State Systems," *IEEE Trans. on Circuit Theory* **CT-11**, 307–308 (1964).
2. **Bacon, G. C.,** "The Decomposition of Stochastic Automata," *IC* **7** (3) 320–339 (Sept. 1964).
3. **Booth, T. L.,** *Sequential Machines and Automata Theory*, John Wiley, New York, 1967.
4. **Carlyle, J. W.,** "Reduced Forms for Stochastic Sequential Machines," *J. Math. Anal. Appl.* **7** (2) 167–175 (Oct. 1963).
5. **Carlyle, J. W.,** "Stochastic Finite-State System Theory," in L. A. Zadeh and E. Polak, eds., *System Theory*, McGraw-Hill, New York, 1969, pp. 387–423.
6. **Even, S.,** "Comments on the Minimization of Stochastic Machines," *TC* **14** (4) 634–637 (Aug. 1965).
7. **Gill, A.,** "Synthesis of Probability Transformers," *JFI* **274** (1) 1–19 (July 1962).
8. **Nieh, T. T.,** and **J. W. Carlyle,** "On a Measure of Complexity for Stochastic Sequential Machines," *CRSA* **9,** 34–41 (Oct. 1968).
9. **Nieh, T. T.,** "Stochastic Sequential Machines with Prescribed Performance," *IC* **13,** (2) 99–113 (Aug. 1969).
10. **Ott, G. H.,** "Reconsider the State Minimization Problem for Stochastic Finite State Systems," *CRSA* 267–273 (Oct. 1966).
11. **Ott, G.,** "Theory and Applications of Stochastic Sequential Machines," Sperry Rand Research Center Report No. SRRC-RR-66-39, May 1966.
12. **Page, C. V.,** "Equivalences Between Probabilistic and Deterministic Machines," *IC* **9** (5) 469–520 (Oct. 1966).
13. **Paz, A.,** "Some Aspects of Probabilistic Automata," *IC* **9** (1) 26–59 (Feb. 1966).
14. **Paz, A.,** "Minimization Theorems and Techniques for Sequential Stochastic Systems," *IC* **11** (1/2) 155–166 (July/August 1967).
15. **Paz, A.,** *Probabilistic Automata*, Academic Press, New York, 1971.
16. **Pospelov, D. A.,** *Probabilistic Automata* (in Russian), Izd. "Energia," Moscow, 1970.
17. **Rabin, M. O.,** "Probabilistic Automata," *IC* **6** (3) 230–245 (Sept. 1963); reprinted in *SM*.

18. **Salomaa, A.,** "On Events Represented by Probabilistic Automata of Different Types," *Can. J. Math.*, **20,** 242–251 (1968).
19. **Shannon, C. E.,** "The Mathematical Theory of Communication," *BSTJ* **27** (3 and 4) 379–423, 623–656 (July and Oct. 1948).
20. **Souza, C. R.,** and **J. Leake,** "Relationships Among Distinct Models and Notions of Equivalence for Stochastic Finite-State Systems," *TC* **C-18** (7) 633–641 (July 1969).
21. **Starke, P. H.,** "Theory of Stochastic Automata," *KYB* **2** (6) 475–482 (1966).
22. **Svoboda, A.,** "Synthesis of Logical Systems of Given Activity," *TC* **EC-12** (6) 904–910 (Dec. 1963).
23. **Svoboda, A.,** "Behavior Classification in Digital Systems," *IPM*, No. 10, 25–42 (1964).
24. **Warfield, J. N.,** "Switching Circuits as Topological Models in Discrete Probability Theory," *TC* **EC-7** (3) 251–252 (1958).
25. **Warfield, J. N.,** *Principles of Logic Design*, Ginn and Company, New York, 1963.
26. **Warfield, J. N.,** "Synthesis of Switching Circuits to Yield Prescribed Probability Relations," *CRSA* **6,** 303–309 (Oct. 1965).
27. **Warfield, J. N.,** "Switching Networks as Models of Stochastic Processes," *AAT*, 81–123.

PART III

APPENDIXES

APPENDIX A
POLYNOMIAL REPRESENTATION OF NUMBERS

If N_{10} is a positive decimal integer written as

$$a_m a_{m-1} \cdots a_1 a_0$$

where $a_m, a_{m-1}, \ldots, a_0$ are decimal digits ($0 \leq a_i \leq 9$), we interpret it as

$$N_{10} = a_m 10^m + a_{m-1} 10^{m-1} + \cdots + a_1 10^1 + a_0 10^0 \tag{1}$$

This representation of numbers is called the polynomial (positional) representation to the base (radix) 10.

The base 10 may be replaced by any integer $B \geq 2$. The positive integer N_B written as

$$b_n b_{n-1} \cdots b_1 b_0$$

where

$$0 \leq b_i \leq B - 1, \quad (i = 0, 1, \ldots, n)$$

is then interpreted as

$$N_B = b_n B^n + b_{n-1} B^{n-1} + \cdots + b_1 B^1 + b_0 B^0 \tag{2}$$

This is the general polynomial representation to the base B.

Let us now suppose that a positive decimal integer $a_m a_{m-1} \cdots a_1 a_0$ is given and we want to find a corresponding number $b_n b_{n-1} \cdots b_1 b_0$ such that $N_{10} = N_B$.

In view of equation (2), we may write

$$a_m a_{m-1} \cdots a_1 a_0 = b_n B^n + b_{n-1} B^{n-1} + \cdots + b_1 B^1 + b_0 B^0$$

On rearranging the right-hand side we get

$$a_m a_{m-1} \cdots a_1 a_0 = (\cdots((b_n B + b_{n-1})B + b_{n-2})B + \cdots + b_1) B + b_0 \tag{3}$$

From this equation it follows that, by dividing the right-hand side by B, we get the quotient $(\cdots((b_n B + b_{n-1})B + \cdots + b_{n-2})B + \cdots + b_1)$ and the remainder b_0. Dividing the quotient again we get a new quotient $(\cdots(b_n B + b_{n-1})B + \cdots + b_2)$

and the remainder b_1. Repeating this procedure we get the remainders $b_2, b_3, \ldots, b_n$.

What is correct for the right-hand side of (3) must also be correct for the left-hand side. Thus, if we repeatedly divide a decimal number $a_m a_{m-1} \cdots a_1 a_0$ by B until the quotient is less than 1 and if we write down the corresponding remainders from right to left, we get the number $b_n b_{n-1} \cdots b_1 b_0$.

EXAMPLE Let us find the binary representation ($B = 2$) of the decimal number 27.

$$\begin{aligned} 27:2 &= 13 \\ 1 &\to b_0 \\ 13:2 &= \ \ 6 \\ 1 &\to b_1 \\ 6:2 &= \ \ 3 \\ 0 &\to b_2 \\ 3:2 &= \ \ 1 \\ 1 &\to b_3 \\ 1:2 &= \ \ 0 \\ 1 &\to b_4 \end{aligned}$$

The binary representation is $b_4 b_3 b_2 b_1 b_0 = 1\ 1\ 0\ 1\ 1$.

If a number $b_n b_{n-1} \cdots b_1 b_0$ is given in base-B representation (where $B \neq 10$), it can be converted to the decimal representation simply by calculating the sum

$$b_n B^n + b_{n-1} B^{n-1} + \cdots + b_1 B^1 + b_0 B^0$$

EXAMPLE Convert the ternary number ($B = 3$) 21202012 to the decimal representation.

$$\begin{aligned} &2 \cdot 3^7 + 1 \cdot 3^6 + 2 \cdot 3^5 + 0 \cdot 3^4 + 2 \cdot 3^3 + 0 \cdot 3^2 + 1 \cdot 3^1 + 2 \cdot 3^0 \\ &= 4374 + 729 + 486 + 0 + 54 + 0 + 3 + 2 \\ &= 5648 \end{aligned}$$

Suppose now that F_{10} is a positive decimal fraction written as

$$0.\, a_{-1} a_{-2} a_{-3} \cdots a_{-m}$$

where $a_{-1}, a_{-2}, \ldots, a_{-m}$ are decimal digits. We interpret it as

$$F_{10} = a_{-1} 10^{-1} + a_2 10^{-2} + \cdots + a_{-m} 10^{-m}$$

The general polynomial representation with the base B of a positive fraction F_B has the form

$$F_B = b_{-1} B^{-1} + b_{-2} B^{-2} + \cdots + b_{-n} B^{-n} + \cdots$$

where $0 \leq b_i \leq B - 1 (i = -1, -2, \ldots, -n)$. This form can be rewritten as

$$F_B = B^{-1}(b_{-1} + B^{-1}(b_{-2} + B^{-2}(b_{-3} + \cdots + B^{-1} b_{-n})) \cdots)$$

When multiplying F_B by B, we obtain

$$b_{-1} + B^{-1}(b_{-2} + B^{-2}(b_{-3} + \cdots + B^{-1} b_{-n})) \cdots)$$

i.e., the number

$$b_{-1}.\, b_{-2} b_{-3} \cdots b_{-n} \cdots$$

When multiplying the fractional portion of this number by B, we obtain the number

$$b_{-2}.b_{-3}b_{-4}\cdots b_{-n}\cdots$$

When proceeding in the same fashion, we generate a sequence of digits

$$b_{-1}, b_{-2}, b_{-3}, \ldots, b_{-n}, \ldots$$

EXAMPLE Convert the decimal number 0.79 to the ternary representation.

$$\begin{aligned} 0.79 \times 3 &= \underline{2}.37 \qquad (b_{-1}=2)\\ 0.37 \times 3 &= \underline{1}.11 \qquad (b_{-2}=1)\\ 0.11 \times 3 &= \underline{0}.33 \qquad (b_{-3}=0)\\ 0.33 \times 3 &= \underline{0}.99 \qquad (b_{-4}=0)\\ 0.99 \times 3 &= \underline{2}.97 \qquad (b_{-5}=2) \end{aligned}$$

We can continue to get more digits of the ternary representation. Nevertheless, 0.21002 is an approximate ternary representation of the decimal fraction 0.79. Observe that fractions which are represented by a finite number of digits in one polynomial number representation cannot often be represented by a finite number of digits in another representation. For instance, 1/3 cannot be represented by a finite number of digits in the decimal representation although its ternary representation is 0.1.

TABLE A.1. Polynomial Representation of Numbers for Different Bases B

Number in Decimal Notation	Number in Nondecimal Notation							
$B=10$	$B=9$	$B=8$	$B=7$	$B=6$	$B=5$	$B=4$	$B=3$	$B=2$
0	0	0	0	0	0	0	0	0
1	1	1	1	1	1	1	1	1
2	2	2	2	2	2	2	2	10
3	3	3	3	3	3	3	10	11
4	4	4	4	4	4	10	11	100
5	5	5	5	5	10	11	12	101
6	6	6	6	10	11	12	20	110
7	7	7	10	11	12	13	21	111
8	8	10	11	12	13	20	22	1000
9	10	11	12	13	14	21	100	1001
10	11	12	13	14	20	22	101	1010
11	12	13	14	15	21	23	102	1011
12	13	14	15	20	22	30	110	1100
13	14	15	16	21	23	31	111	1101
14	15	16	20	22	24	32	112	1110
15	16	17	21	23	30	33	120	1111
16	17	20	22	24	31	100	121	10000
17	18	21	23	25	32	101	122	10001
18	20	22	24	30	33	102	200	10010
19	21	23	25	31	34	103	201	10011
20	22	24	26	32	40	110	202	10100

TABLE A.2. The Powers of 2

2^n	n	2^{-n}
1	0	1.0
2	1	0.5
4	2	0.25
8	3	0.125
16	4	0.062 5
32	5	0.031 25
64	6	0.015 625
128	7	0.007 812 5
256	8	0.003 906 25
512	9	0.001 953 125
1 024	10	0.000 976 562 5
2 048	11	0.000 488 281 25
4 096	12	0.000 244 140 625
8 192	13	0.000 122 070 312 5
16 384	14	0.000 061 035 156 25
32 768	15	0.000 030 517 578 125
65 536	16	0.000 015 258 789 062 5
131 072	17	0.000 007 629 394 531 25
262 144	18	0.000 003 814 697 265 625
524 288	19	0.000 001 907 348 632 812 5
1 048 576	20	0.000 000 953 674 316 406 25
2 097 152	21	0.000 000 476 837 158 203 125
4 194 304	22	0.000 000 238 418 579 101 562 5
8 388 608	23	0.000 000 119 209 289 550 781 25
16 777 216	24	0.000 000 059 604 644 775 390 625
33 554 432	25	0.000 000 029 802 322 387 695 312 5
67 108 864	26	0.000 000 014 901 161 193 847 656 25
134 217 728	27	0.000 000 007 450 580 596 923 828 125
268 435 456	28	0.000 000 003 725 290 298 461 914 062 5
536 870 912	29	0.000 000 001 862 645 149 230 957 031 25
1 073 741 824	30	0.000 000 000 931 322 574 615 478 515 625
2 147 483 648	31	0.000 000 000 465 661 287 307 739 257 812 5
4 294 967 296	32	0.000 000 000 232 830 643 653 869 628 906 25
8 589 934 592	33	0.000 000 000 116 415 321 826 934 814 453 125
17 179 869 184	34	0.000 000 000 058 207 660 913 467 407 226 562 5
34 359 738 368	35	0.000 000 000 029 103 830 456 733 703 613 281 25
68 719 476 736	36	0.000 000 000 014 551 915 228 366 851 806 640 625
137 438 953 472	37	0.000 000 000 007 275 957 614 183 425 903 320 312 5
274 877 906 944	38	0.000 000 000 003 637 978 807 091 712 951 660 156 25
549 755 813 888	39	0.000 000 000 001 818 989 403 545 856 475 830 078 125
1 099 511 627 776	40	0.000 000 000 000 909 494 701 772 928 237 915 039 062 5
2 199 023 255 552	41	0.000 000 000 000 454 747 350 886 464 118 957 519 531 25
4 398 046 511 104	42	0.000 000 000 000 227 373 675 443 232 059 478 759 765 625
8 796 093 022 208	43	0.000 000 000 000 113 686 837 721 616 029 739 379 882 812 5
17 592 186 044 416	44	0.000 000 000 000 056 843 418 860 808 014 869 689 941 406 25
35 184 372 088 832	45	0.000 000 000 000 028 421 709 430 404 007 434 844 970 703 125
70 368 744 177 664	46	0.000 000 000 000 014 210 854 715 202 003 717 422 485 351 562 5
140 737 488 355 328	47	0.000 000 000 000 007 105 427 357 601 001 858 711 242 675 781 25
281 474 976 710 656	48	0.000 000 000 000 003 552 713 678 800 500 929 355 621 337 890 625
562 949 953 421 312	49	0.000 000 000 000 001 776 356 839 400 250 464 677 810 668 945 312 5
1 125 899 906 842 624	50	0.000 000 000 000 000 888 178 419 700 125 232 338 905 334 472 656 25
2 251 799 813 685 248	51	0.000 000 000 000 000 444 089 209 850 062 616 169 452 667 236 328 125
4 503 599 627 370 496	52	0.000 000 000 000 000 222 044 604 925 031 308 084 726 333 618 164 062 5
9 007 199 254 740 992	53	0.000 000 000 000 000 111 022 302 462 515 654 042 363 166 809 082 031 25
18 014 398 509 481 984	54	0.000 000 000 000 000 055 511 151 231 257 827 021 181 583 404 541 015 625
36 028 797 018 963 968	55	0.000 000 000 000 000 027 755 575 615 628 913 510 590 791 702 270 507 812 5
72 057 594 037 927 936	56	0.000 000 000 000 000 013 877 787 807 814 456 755 295 395 851 135 253 906 25
144 115 188 075 855 872	57	0.000 000 000 000 000 006 938 893 903 907 228 377 647 697 925 567 626 953 125
288 230 376 151 711 744	58	0.000 000 000 000 000 003 469 446 951 953 614 188 823 848 962 783 813 476 562 5
576 460 752 303 423 488	59	0.000 000 000 000 000 001 734 723 475 976 807 094 411 924 481 391 906 738 281 25
1 152 921 504 606 846 976	60	0.000 000 000 000 000 000 867 361 737 988 403 547 205 962 240 695 953 369 140 625
2 305 843 009 213 693 952	61	0.000 000 000 000 000 000 433 680 868 994 201 773 602 981 120 347 976 684 570 312 5
4 611 686 018 427 387 904	62	0.000 000 000 000 000 000 216 840 434 497 100 886 801 490 560 173 988 342 285 156 25
9 223 372 036 854 775 808	63	0.000 000 000 000 000 000 108 420 217 248 550 443 400 745 280 086 994 171 142 578 125
18 446 744 073 709 551 616	64	0.000 000 000 000 000 000 054 210 108 624 275 221 700 372 640 043 497 085 571 289 062 5
36 893 488 147 419 103 232	65	0.000 000 000 000 000 000 027 105 054 312 137 610 850 186 320 021 748 542 785 644 531 25
73 786 976 294 838 206 464	66	0.000 000 000 000 000 000 013 552 527 156 068 805 425 093 160 010 874 271 392 822 265 625
147 573 952 589 676 412 928	67	0.000 000 000 000 000 000 006 776 263 578 034 402 712 546 580 005 437 135 696 411 132 812 5
295 147 905 179 352 825 856	68	0.000 000 000 000 000 000 003 388 131 789 017 201 356 273 290 002 718 567 848 205 566 406 25
590 295 810 358 705 651 712	69	0.000 000 000 000 000 000 001 694 065 894 508 600 678 136 645 001 359 283 924 102 783 203 125
1 180 591 620 717 411 303 424	70	0.000 000 000 000 000 000 000 847 032 947 254 300 339 068 322 500 679 641 962 051 391 601 562 5
2 361 183 241 434 822 606 848	71	0.000 000 000 000 000 000 000 423 516 473 627 150 169 534 161 250 339 820 981 025 695 800 781 25
4 722 366 482 869 645 213 696	72	0.000 000 000 000 000 000 000 211 758 236 813 575 084 767 080 625 169 910 490 512 847 900 390 625

APPENDIX B
A GUIDE TO THE LITERATURE

B.1 TEXTBOOKS

The following list of basic textbooks for switching circuit theory courses is followed by a list of basic topics in this theory and Table B.1 showing the coverage of these topics by individual textbooks. Rows of the table represent the textbooks; its columns are assigned to the topics. If topic *i* is included in textbook *j* and is worth being recommended due to some reasons (quality of presentation, complete, up-to-date, a special approach is used, special information is provided, etc.) then a dot is placed in the square situated in *i*th column and *j*th row; otherwise, the square is left blank.

Basic Textbooks

1. **Caldwell, S. H.,** *Switching Circuits and Logical Design*, John Wiley, New York, 1958.
2. **Curtis, H. A.,** *A New Approach to the Design of Switching Circuits*, Van Nostrand Reinhold, New York, 1962.
3. **Flegg, H. G.,** *Boolean Algebra and Its Applications*, John Wiley, New York, 1964.
4. **Givone, D. D.,** *Introduction to Switching Circuit Theory*, McGraw-Hill, New York, 1970.
5. **Harrison, M. E.,** *Introduction to Switching and Automata Theory*, McGraw-Hill, New York, 1965.
6. **Higonnet, R.,** and **R. Grea,** *Logical Design of Electrical Circuits*, McGraw-Hill, New York, 1958.
7. **Hill, H. J.,** and **G. R. Peterson,** *Introduction to Switching Theory and Logical Design*, John Wiley, New York, 1968.
8. **Hoernes, G. E.,** and **M. F. Heilweil,** *Introduction to Boolean Algebra and Logic Design*, McGraw-Hill, New York, 1964.

9. **Hohn, F. E.,** *Applied Boolean Algebra*, Macmillan, New York, 1964.
10. **Hu, S. T.,** *Mathematical Theory of Switching Circuits and Automata*, Univ. of California Press, Berkeley and Los Angeles, 1968.
11. **Humprey, W. S.,** *Switching Circuits with Computer Applications*, McGraw-Hill, New York, 1958.
12. **Hurley, R. B.,** *Transistor Logic Circuits*, John Wiley, New York, 1961.
13. **Keister, W., A. E. Ritchie,** and **S. H. Washburn,** *The Design of Switching Circuits*, Van Nostrand Reinhold, New York, 1951.
14. **Klir, G. J.,** and **L. K. Seidl,** *Synthesis of Switching Circuits*, Iliffe, London, 1968; Gordon and Breach, New York, 1969.
15. **Kohavi, Z.,** *Switching and Finite Automata Theory*, McGraw-Hill, New York, 1970.
16. **Krieger, M.,** *Basic Switching Circuit Theory*, Macmillan, New York, 1967.
17. **Ledley, R. S.,** *Digital Computer and Control Engineering*, McGraw-Hill, New York, 1960.
18. **Lewin, D.,** *Logical Design of Switching Circuits*, Nelson, London, 1968.
19. **Maley, G. A.,** and **J. Earle,** *The Logic Design of Transistor Digital Computers*, Prentice-Hall, Englewood Cliffs, N.J., 1963.
20. **Marcus, M. P.,** *Switching Circuits for Engineers*, Prentice-Hall, Englewood Cliffs, N.J., 1962 (2nd ed., 1967.)
21. **McCluskey, E. J.,** *Introduction to the Theory of Switching Circuits*, McGraw-Hill, New York, 1965.
22. **Mendelson, E.,** *Boolean Algebra and Switching Circuits*, McGraw-Hill, New York, 1970.
23. **Miller, R. E.,** *Switching Theory*, John Wiley, New York, 1965, Vol. I: *Combinational Circuits*, Vol. II: *Sequential Circuits and Machines.*
24. **Moisil, G. C.,** *The Algebraic Theory of Switching Circuits*, Pergamon Press, New York, 1967.
25. **Mukhopadhyay, A.,** ed., *Recent Developments in Switching Theory*, Academic Press, New York, 1971.
26. **Oberman, R. M. M.,** *Disciplines in Combinational and Sequential Circuits Design*, McGraw-Hill, New York, 1971.
27. **Phister, M.,** *Logical Design of Digital Computers*, John Wiley, New York, 1958.
28. **Prather, R. E.,** *Introduction to Switching Theory: A Mathematical Approach*, Allyn and Bacon, Boston, Mass., 1967.
29. **Roginskii, V. N.,** *The Synthesis of Relay Switching Circuits*, Van Nostrand Reinhold, New York, 1963.
30. Staff of Harvard Computation Laboratory, *Synthesis of Electronic Computing and Control Circuits*, Harvard Univ. Press, Cambridge, Mass., 1951.
31. **Torng, H. C.,** *Logical Design of Switching Systems*, Addison-Wesley, Reading, Mass., 1964.
32. **Tou, J. T.,** ed., *Applied Automata Theory*, Academic Press, New York, 1968.
33. **Unger, S. H.,** *Asynchronous Sequential Switching Circuits*, John Wiley, New York, 1969.
34. **Warfield, J. N.,** *Principles of Logic Design*, Ginn and Company, Boston, Mass., 1963.

35. **Wickes, W. E.,** *Logic Design with Integrated Circuits,* John Wiley, New York, 1968.
36. **Wood, P. E.,** *Switching Theory,* McGraw-Hill, New York, 1968.

Basic Topics

1. Introduction to Boolean Algebra
2. Complete Sets of Logic Functions
3. Minimization of Boolean Expressions
4. Solution of Boolean Equations
5. Decomposition of Logic Functions
6. Classification of Logic Functions
7. Logic Description of Physical Systems
8. Circuits with Branch-Type Elements
9. Circuits with Gate-Type Elements
10. Boolean Matrices and Their Applications
11. Symmetric Functions and Circuits
12. State Minimization
13. Abstract Synthesis of Sequential Circuits
14. State Assignment
15. Decomposition of Sequential Machines
16. Hazards
17. Probabilistic Circuits and/or Machines
18. Threshold Logic
19. Cellular Structures
20. Linear Sequential Circuits
21. Reliable Circuits Using Unreliable Elements
22. Diagnostics and/or Identification Experiments
23. Finite-Memory Model

B.2 PERIODICALS

The most important periodicals in the area of the methodology of Switching circuits are the *IEEE Transactions on Computers* (formerly entitled *IRE Transactions on Electronic Computers*) and the *IEEE Conference Records of Annual Symposia on Switching and Automata Theory* (formerly entitled *Proceedings of the Annual Symposia on Switching Circuit Theory and Logical Design*). Both of them are published by the Institute of Electrical and Electronics Engineers, 345 East 47th Street, New York, N.Y. 10017. The *Transactions* have been published since 1952 (quarterly until 1961, bimonthly since 1962, monthly since 1968). The *Records* have been published annually since 1961.

There are several other periodicals which occasionally publish papers related to the methodology of switching circuits:

Automation and Remote Control, Translated from Russian by Consultants Bureau, a Division of Plenum Publishing Corporation, 227 West 17th Street, New York, N.Y. 10011; since 1940; monthly.

TABLE B.1. The Coverage of Basic Topics by Individual Text books

	1	2	3	4	5	6	7	8	9	10	11	12	13	14	15	16	17	18	19	20	21	22	23
1	●		●				●	●	●		●												
2	●				●		●		●														
3	●		●					●		●	●												
4	●		●				●	●	●			●	●	●		●			●				
5	●		●			●		●	●	●	●	●									●		
6	●						●	●			●												
7	●		●		●		●		●		●	●	●	●				●		●			
8	●		●				●																
9	●						●																
10	●	●	●		●	●					●	●											
11	●					●		●	●	●													
12	●		●				●		●														
13	●							●															
14	●	●	●	●			●	●	●	●	●	●	●	●									
15	●				●		●	●	●		●	●	●	●	●	●		●		●	●	●	●
16	●		●						●			●	●	●									
17	●		●	●	●																		
18	●		●				●	●	●			●		●									
19	●		●									●	●	●		●							
20	●		●				●	●	●		●	●	●	●		●							
21	●		●				●	●	●			●	●	●		●							
22	●		●					●	●														
23	●		●		●		●	●	●	●	●	●	●	●		●							
24	●						●	●															
25		●				●					●								●				
26	●						●	●	●	●													
27	●		●	●			●		●														
28	●		●		●	●			●			●	●	●									
29	●						●	●					●										
30	●		●			●			●														
31							●	●	●			●		●									
32																	●	●	●	●			
33												●		●		●							
34	●		●														●						
35	●		●				●		●														
36	●		●				●	●	●			●		●		●		●			●		●

Bell System Technical Journal, published by the American Telephone and Telegraph Company, 195 Broadway, New York, N.Y. 10007; since 1922; 10 times a year.

Computer Design, published by Computer Design Publishing Corporation, 221 Baker Avenue, Concord, Md. 01742; since 1962; monthly.

Cybernetics: Translated from Russian by the Faraday Press, 84 Fifth Avenue, New York, N.Y. 10011; since 1965; bimonthly.

Electronics and Communication in Japan, translated from Japanese by the Institute of Electrical and Electronics Engineers, 345 East 47th Street, New York, N.Y. 10017; since 1918; monthly.

Engineering Cybernetics: translated from Russian by Scripta Publishing Corporation, 1511 K Street, N.W., Washington, D.C., 20005; since 1963; bimonthly.

IBM Journal of Research and Development, published by IBM Corporation, Armonk, New York 10504; since 1957; bimonthly.

IEEE Transactions on Circuit Theory, published by the Institute of Electrical and Electronics Engineers, 345 East 47th Street, New York, N.Y. 10017; since 1954.

IEEE Transactions on Education, published by the Institute of Electrical and Electronics Engineers, 345 East 47th Street, New York, N.Y. 10017; since 1958; quarterly.

Information and Control, published by Academic Press, 111 Fifth Avenue, New York, N.Y. 10003; since 1957; monthly.

Information Processing Machines, published by Academia, Praha, Czechoslovakia; since 1953; annually.

Journal of Computer and System Science, published by Academic Press, 111 Fifth Avenue, New York, N.Y. 10003; since 1967; bimonthly.

Journal of the Association for Computing Machinery, published by the Association for Computing Machinery, 211 East 43rd Street, New York, N.Y. 10017; since 1954; quarterly.

Journal of the Franklin Institute, published by the Franklin Institute, Philadelphia, Pa. 19103; since 1826; monthly.

Proceedings of the IEEE, published by the Institute of Electrical and Electronics Engineers, 345 East 47th Street, New York, N.Y. 10017; since 1913 (originally *Proceedings of the AIEE*, later *Proceedings of the IRE*); monthly.

B.3 COLLECTIONS OF PAPERS

In addition to the above-specified periodicals, various collections of papers containing important information relevant to the methodology of switching circuits have been published occasionally. Although not complete, the following list contains the most important of them:

1. **Aiken, H.**, ed., *Proc. of an International Symposium on the Theory of Switching* (2 volumes), Harvard Univ. Press, Cambridge, Mass., 1959.
2. **Aiken, H.**, ed., *Switching Theory in Space Technology*, Stanford Univ. Press, Stanford, Calif., 1963.

3. **Arbib, M. A.,** ed., *Algebraic Theory of Machines, Languages, and Semigroups,* Academic Press, New York, 1968.
4. **Biorci, G.,** ed., *Network and Switching Theory,* Academic Press, New York, 1968.
5. **Caianiello, E. R.,** ed., *Automata Theory,* Academic Press, New York, 1966.
6. **Fox, J.,** ed., *Mathematical Theory of Automata,* Polytechnic Press, Brooklyn, N.Y., 1963.
7. **Gavrilov, M. A.,** ed., *Structure Theory of Switching Circuits* (in Russian), Izd. Akademii Nauk U.S.S.R., Moscow, 1963.
8. **Gavrilov, M. A.,** ed., *Theory of Finite and Probabilistic Automata* (in Russian), Izd. "Nauka," Moscow, 1965.
9. **McCluskey,** and **T. C. Bartee,** eds., *A Survey of Switching Circuit Theory,* McGraw-Hill, New York, 1962.
10. **Moore, E. F.,** ed., *Sequential Machines: Selected Papers,* Addison-Wesley, Reading, Mass., 1964.
11. **Mukhopadhyay, A.,** ed., *Recent Developments in Switching Theory,* Academic Press, New York, 1971.
12. **Shannon, C. E.,** and **J. McCarthy,** ed., *Automata Studies,* Princeton Univ. Press, Princeton, N.J., 1956.
13. **Tavanetz, I. V.,** et al., eds., *Applications of Logic in Science and Engineering* (in Russian), Izd. Akademii Nauk U.S.S.R., Moscow, 1960.
14. **Tou, J. T.,** ed., *Applied Automata Theory,* Academic Press, New York, 1968.
15. **Wilcox, R. H.,** and **W. C. Mann,** eds., *Redundancy Techniques for Computing Systems,* Spartan Books, Washington, D.C., 1962.
16. **Yablonskii, S. V.,** ed., *A Collection of Papers on Mathematical Logic and Its Applications to Some Problems of Cybernetics* (in Russian), Trudy Matem. Inst. V. A. Steklova, Vol. 51, Izd. Akademii Nauk U.S.S.R., Moscow, 1958.
17. **Zadeh, L. A.,** and **E. Polak,** eds., *System Theory,* McGraw-Hill, New York, 1969.

B.4 SELECTED REFERENCES

Although the greater part of the methodology of switching circuits is covered in this book, several topics were intentionally left out to keep the size of the book within reasonable limits. To help the reader in his further study, basic references covering each of these topics are selected in the following paragraphs.

THRESHOLD LOGIC Quite intensive research work in threshold logic has been done since 1959 even though commercially competitive threshold elements were developed only in the late 1960s. According to Winder [15], "about 200 papers and at least 15 Ph.D. theses dealing with the subject of threshold logic have been published" in the period from 1959 to 1969. The subject is well covered in five books [2,6,8,11,12] and in two excellent survey articles [14,15], where almost all publications on threshold logic available in 1969 are listed. References 1,3,4,5,7,9, 10,13,16, and 17 cover most of the research done in threshold logic from 1969 to 1971.

1. **Cohen, S.,** and **R. O. Winder,** "Threshold Gate Building Blocks," *TC* **C-18** (9) 816–823 (Sept. 1969).
2. **Dertouzos, M. L.,** *Threshold Logic: A Synthesis Approach,* The MIT Press, Cambridge, Mass., 1965.
3. **Ghosh, S.,** et al., "Multigate Synthesis of General Boolean Functions by Threshold Logic Elements," *TC* **C-18** (5) 451–456 (May 1969).
4. **Hadlock, F. O.,** and **C. L. Coates,** "Realization of Sequential Machines with Threshold Elements," *TC* **C-18** (5) 428–439 (May 1969).
5. **Hruz, B.,** "Unateness Test of a Boolean Function and Two General Synthesis Methods Using Threshold Logic Elements," *TC* **C-18** (2) 122–131 (Feb. 1969).
6. **Hu, S.,** *Threshold Logic,* University of California Press, Berkeley and Los Angeles, 1965.
7. **Hwa, H. R.,** and **C. L. Sheng,** "An Approach for the Realization of Threshold Functions of Order r," *TC* **C-18** (10) 923–939 (Oct. 1969).
8. **Lewis, P. M. II,** and **C. L. Coates,** *Threshold Logic,* John Wiley, New York, 1967.
9. **Lyons, R. E.,** "The Synthesis of Redundant Threshold-Logic Elements," *TC* **C-19** (5) 429–443 (May 1970).
10. **Mow, C.,** and **K. Fu,** "Loop-Free Threshold Element Structures," *TC* **C-18** (3) 257–267 (March 1969).
11. **Muroga, S.,** Threshold Logic and Its Applications, John Wiley, New York, 1971.
12. **Sheng, C. L.,** *Threshold Logic,* Academic Press, New York, 1969.
13. **Slivinski, T. A.,** "An Extension of Threshold Logic," *TC* **C-19** (4) 319–341 (April 1970).
14. **Winder, R. O.,** "Fundamentals of Threshold Logic," *AAT,* 235-318.
15. **Winder, R. O.,** "The Status of Threshold Logic," *RCA Review* **30,** 62–84 (March 1969).
16. **Winder, R. O.,** "Chow Parameters in Threshold Logic," *JACM* **18** (2) 265–289 (April 1971).
17. **Yen, Y. T.,** "Some Theoretical Properties of Multithreshold Realizable Functions" *TC* **C-17,** (11) 1081–1088 (Nov. 1968).

CELLULAR LOGIC A Study of mathematical models of cellular arrays (one-, two-, or three-diemensional) consisting of identical or similar logic elements (cells) is usually referred to as cellular logic. Although the study of cellular networks was initiated in the late 1950s [4], its importance was properly appreciated only later, in the 1960s, with the advent of integrated circuits and, especially, in the context of large-scale integration.

The first comprehensive study of analysis and synthesis of cellular networks was presented by Hennie in his monograph [2], which has become a classic in the area of cellular logic. There are several excellent survey papers [1,3,5,6,7], where the reader can find references to almost all important papers on cellular logic published up to 1971.

1. **Elpas, B.,** "The Theory of Multirail Cascades," *RDST*, 315–367.
2. **Hennie, F. C., III,** *Iterative Arrays of Logical Circuits*, The MIT Press and John Wiley, New York, 1961.
3. **Kautz, W. H.,** "Programmable Cellular Logic," *RDST*, 369–422.
4. **McCluskey, E. J.,** "Iterative Combinational Switching Networks—General Design Considerations," *TC* **EC-7** (4) 285–291 (Dec. 1958).
5. **Minnick, R. C.,** "A Survey of Microcellular Research," *JACM* **14** (2) 203–241 (April 1967).
6. **Minnick, R. C.,** "Cellular Networks," in G. Biorci, ed., *Network and Switching Theory*, Academic Press, New York, 1968, pp. 496–520.
7. **Mukhopadhyay, A.,** and **H. S. Stone,** "Cellular Logic," *RDST*, 255–313.

DIAGNOSTICS Some basic ideas of diagnostics of switching circuits (testing by external experimentation) were expressed by Moore in terms of the gedanken experiments as early as in 1956 [19]. Later, these ideas were published in a more developed form by Gill [7]. The problem of testing complex modules is becoming increasingly difficult with the advent of large-scale integration. Although various testing procedures for memoryless or sequential switching circuits have been elaborated since the late 1950s, the situation is still unsatisfactory and, therefore, the research in diagnostics is given a high priority.

The "state of the art" in diagnostics of switching circuits at the beginning of the 1970s is well described in two books [5,6]. References 1,2,9,11,16,18,20,22,23, and 25 represent a sample of papers devoted to diagnostics of memoryless circuits. References 8,13,15,21, and 24 describe testing procedures for sequential circuits. Both of these classes of circuits are treated in References 3 and 4. Testing of cellular networks is covered in References 10 and 17. The property of diagnosability has been introduced as an additional objective criterion in the synthesis of sequential switching circuits [12,14].

1. **Amar, V.,** and **N. Condulmari,** "Diagnosis of Large Combinational Networks," *TC* **EC-16** (5) 675–680 (Oct. 1967).
2. **Armstrong, D. B.,** "On Finding a Nearly Minimal Set of Fault Detection Tests for Combinational Logic Nets," *TC* **EC-15** (1) 66–73 (Feb. 1966).
3. **Brown, A.,** and **H. W. Young,** Algebraic Logic Network Analysis: Toward an Algebraic Theory of the Analysis and Testing of Digital Networks, IBM Technical Report, TR 00.1974, Poughkeepsie Lab., N.Y., Jan. 1970.
4. **Chang, H. Y.,** "An Algorithm of Selecting an Optimum Set of Diagnosis Tests," *TC* **EC-14** (5) 706–711 (Oct. 1965).
5. **Chang, H. Y.,** et al., *Fault Diagnosis of Digital Systems*, John Wiley, New York, 1970.
6. **Friedman, A. D.,** and **P. R. Menon,** *Fault Detection in Digital Circuits*, Prentice Hall, Englewood Cliffs, N.J., 1971.

7. **Gill, A.,** Introduction to the Theory of Finite-State Machines, McGraw-Hill, New York, 1962.
8. **Gonenc, G.,** "A Method for the Design of Fault Detection Experiments," *TC* **C-19** (6) 551–558 (June 1970).
9. **Hornbuckle, G. D.,** and **R. N. Spann,** "Diagnosis of Single-Gate Failures in Combinational Circuits," *TC* **C-18** (3) 216–220 (March 1969).
10. **Kautz, W. H.,** "Testing for Faults in Cellular Logic Arrays," *CRSA* **8,** 161–174 (Oct. 1967).
11. **Kautz, W. H.,** "Fault Testing and Diagnosis in Combinational Digital Circuits," *TC* **C-17** (4) 352-366 (April 1969).
12. **Kohavi, Z.,** and **P. Lavallee,** "Design of Sequential Machines with Fault-Detection Capabilities," *TC* **EC-16** (4) 473–484 (1967).
13. **Kohavi, I.,** and **Z. Kohavi,** "Variable-Length Distinguishing Sequences and Their Application to the Design of Fault-Detection Experiments," *TC* **C-17** (8) 729-795 (Aug. 1968).
14. **Kohavi, Z.,** *Switching and Finite Automata Theory*, McGraw-Hill, New York, 1970.
15. **Mandelbaum, D.,** "A Measure of Efficiency of Diagnostic Tests Upon Sequential Logic," *TC* **EC-13** (5) 530 (Oct. 1964).
16. **Marinos, P. N.,** "Derivation of Minimal Complete Sets of Test-Input Sequences Using Boolean Difference," *TC* **C-20** (1) 25–32 (Jan. 1971).
17. **Menon, P. R.,** and **A. D. Friedman,** "Fault Detection in Iterative Logic Arrays, *TC* **C-20** (5) 524–535 (May 1971).
18. **Poage, J. F.,** "Derivation of Optimum Tests to Detect Faults in Combinational Circuits," in J. Fox, ed., *Mathematical Theory of Automata*, Polytechnic Press, Brooklyn, N.Y., 1963, pp. 483–528.
19. **Moore, E. F.,** "Gedanken-Experiments," *AS*, 129–153.
20. **Powell, T. J.,** "A Procedure for Selecting Diagnostic Tests," *TC* **C-18** (2) 168–175 (Feb. 1969).
21. **Putzolo, G. R.,** and **J. P. Roth,** "A Heuristic Algorithm for the Testing of Asynchronous Circuits," *TC* **C-20** (6) 639–647 (June 1971).
22. **Roth, J. P.,** et al., "Programmed Algorithms to Compute Tests to Detect and Distinguish Between Failures in Logic Circuits," *TC* **EC-16** (5) 567–580 (Oct. 1967).
23. **Sellers, F. F.,** et al., "Analyzing Errors with the Boolean Difference," *TC* **C-17** (7) 676–683 (July 1968).
24. **Seshu, S.,** and **D. N. Freeman,** "The Diagnosis of Asynchronous Sequential Switching Systems," *TC* **EC-11** (4) 459–465 (Aug. 1962).
25. **Whitney, G. E.,** "Algebraic Fault Analysis for Constrained Combinational Networks," *TC* **C-20** (2) 141–148 (Feb. 1971).

REDUNDANT LOGIC The idea of using redundancy in switching circuits to increase the irreliability (often referred to as redundant logic) was proposed by John von Neumann in a paper presented orally in 1952 and published in 1956 [8]. At the same time, Moore and Shannon presented an important contribution to redundancy principles for relay circuits [5]. Kochen [2] suggested an extension of the

Moore-Shannon work. Other extensions of this basic work and some new approaches to the use of redundancy in switching circuits are summarized in Reference 9; this book also contains an extensive bibliography. Another book, more of a textbook type, was published by Pierce in 1965 [7]. The book contains an excellent bibliography, many historical and bibliographical comments, and problems. References 1,3,4, and 6 represent a sample of more recent contributions.

1. **Klaschka, T. F.,** "Two Contributions to Redundancy Theory," *CRSA* **8,** 175–183 (Oct. 1967).
2. **Kochen, M.,** "Extension of Moore-Shannon Model for Relay Circuits," *IBMJ* 3 (2) 169–186 (April 1959).
3. **Lyons, R. E.,** "The Synthesis of Redundant Threshold-Logic Elements," *TC* **C-19** (5) 429–443 (May 1970).
4. **Mine, H.,** and **Y. Koga,** "Basic Properties and a Construction Method for Fail-Safe Logical System," *TC* **EC-16** (3) 282-289 (June 1967).
5. **Moore, E. F.,** and **C. E. Shannon,** "Reliable Circuits Using Less Reliable Relays," *JFI* **262** (3,4) 191–208, 281–297 (Sept., Oct. 1956).
6. **Pierce, W. H.,** "Interwoven Redundant Logic," *JFI* **277** (1) 55–85 (Jan. 1964).
7. **Pierce, W. H.,** *Failure-Tolerant Computer Design*, Academic Press, New York, 1965.
8. **Von Neumann, J.,** "Probabilistic Logics and the Synthesis of Reliable Organisms From Unreliable Components," *AS*, 43–98.
9. **Wilcox, R. H.,** and **W. C. Mann,** eds., *Redundancy Techniques for Computing Systems*, Spartan Books, Washington, D.C., 1962.

LINEAR SEQUENTIAL CIRCUITS The first systematic study of linear sequential circuits was presented by Huffman in 1956 [5]. In the later 1950s, his work was extended by several authors whose papers were collected by Kautz [6]. A thorough mathematical treatment of the theory of linear sequential circuits was prepared by Gill [2]. The book contains an extensive bibliography and gives a survey of basic applications of linear sequential circuits. Another book, on a higher level of abstraction, was written by Harrison [4]. Among other topics, it contains an excellent study of decompositions of linear sequential machines. Applications of linear sequential circuits to error correction is well described by Peterson [7]. A survey paper by Gill [3], supplemented by an excellent bibliography, and Chapter 8 in Reference 1 are also good presentations of the theory of linear sequential circuits.

1. **Booth, T. L.,** *Sequential Machines and Automata Theory*, John Wiley, New York, 1967.
2. **Gill, A.,** *Linear Sequential Circuits: Analysis, Synthesis, and Applications*, McGraw-Hill, New York, 1966.
3. **Gill, A.,** "Linear Modular Systems," in L. A. Zadeh and E. Polak, eds., *System* Theory, McGraw-Hill, New York, 1969, pp. 179-231.
4. **Harrison, M. A.,** *Lectures on Linear Sequential Machines*, Academic Press, New York, 1969.

5. **Huffman, D. A.,** "The Synthesis of Linear Sequential Coding Networks," in C. Cherry, ed., *Information Theory*, Academic Press, New York, 1956.
6. **Kautz, W. H.,** *Linear Sequential Switching Circuits: Selected Technical Papers*, Holden-Day, San Francisco, 1965.
7. **Peterson, W. W.,** *Error-Correcting Codes*, The MIT Press, Cambridge, Mass., 1961.

CIRCUITS WITH MULTIPLE-VALUED VARIABLES Switching circuits with multiple-valued variables were proposed in the 1950s [3,14,15,19] and have been studied ever since [2,6,12,21,25,29]. Ternary switching circuits (circuits with three-valued variables) have been of particular interest [9,10,11,17,24,27], often within the context of the threshold logic [8,16,18,22]. The problem of completeness of logic functions for multiple-valued logics have been investigated primarily by Russian authors [1,5,26,28,29]. Kloss and Nethiporuk [13] studied the general problem of classification of multiple-valued functions defined on multiple-valued variables. Logic equations in a multiple-valued logic algebra were considered by several authors [7,12,30]. Although switching circuits with multiple-valued variables seem promising for the future, their advantage over regular switching circuits is still questionable at this stage of the development of technology [4,23]. The book by Rescher [20] is an excellent survey of multiple-valued logics with an extensive bibliography. Unfortunately, there is neither a book nor a survey article describing the methodology of switching circuits with multiple-valued variables.

1. **Aizenberg, N. N.,** and **Z. L. Rabinovich,** "Certain Classes of Functionally Complete Operations and Canonical Forms of Representation of Functions of Multi-valued Logic," *Kibernetika* **1** (2) 37–45 (1965).
2. **Allen, C. M.,** and **D. D. Givone,** "Design of Multiple-Valued Logic Systems I: A Computer Oriented Minimization Technique," Digitial Systems Lab., E.E. Dept., SUNY at Buffalo, N.Y., May 1966.
3. **Berlin, R. D.,** "Synthesis of n-Valued Switching Circuits," *TC* **EC-7** (1) 52–56 (March 1958).
4. **Cassee, P. R.,** and **M. J. O. Strutt,** "Is There Any Advantage of Ternary Logic as Compared with Binary?" *TC* **C-19** (6) 559 (June 1970).
5. **Gavrilov, G. P.,** "On the Functional Completeness in Logics with Countable Sets of Values," *PC* **15**, 5–64 (1965).
6. **Givone, D. D.,** and **R. W. Snelsire,** "The Design of Multiple-Valued Logic Systems," Digital Systems Lab., E. E. Dept., SUNY at Buffalo, N.Y., June 1968.
7. **Givone, D. D.,** et al., "A Method of Solution for Mutliple-Valued Logic Expressions," *TC* **C-20** (4) 464–467 (April 1971).
8. **Hanson, W. H.,** "Ternary Threshold Logic," *TC* **EC-12** (3), 191–197 (June 1963).
9. **Hermann, R. L.,** "Selection and Implementation of a Ternary Switching Algebra," *AFIPS* **32**, 283–290 (Spring 1968).
10. **Hurst, S. L.,** "An Extension of Binary Minimization Techniques to Ternary Equations," *Computer Journal* **11** (3) 277–286 (Nov. 1968).

11. **Ivas'kiv, Yu. L.,** "On a Class of Three-Valued Algebras and Its Application to the Synthesis of Ternary Logical Networks of Ternary Elements," *Kibernetika* **1** (2) 46–52 (1965).
12. **Klir, G. J.,** and **L. K. Seidl,** *Synthesis of Switching Circuits*, Iliffe, London, 1968; Gordon and Breach, New York, 1969.
13. **Kloss, B. M.,** and **E. I. Nethiporuk,** "On Classification of Functions of Many-Valued Logic," *PC* **9**, 27–69 (1963).
14. **Lee, C. Y.,** and **W. H. Chen,** "Several Valued Combinational Switching Circuits," *Trans. AIEE* **75** (*Commun. and Electronics*) 278–283 (July 1956).
15. **Lowenschuss, O.,** "Nonbinary Switching Theory," IRE National Convention Record, Pt 4, 305–317, 1958.
16. **Merrill, R.,** "Some Properties of Ternary Threshold Logic," *TC* **EC-13** (5) 632–635 (Oct. 1969).
17. **Merrill, R. D.,** "A Tabular Minimization Procedure for Ternary Switching Functions," *TC* **EC-15** (4) 578–585 (Aug. 1966).
18. **Merrill, R. D.,** "Symmetric Ternary Switching Functions: Their Detection and Realization with Threshold Logic," *TC* **EC-16** (5) 624–637 (Oct. 1967).
19. **Muehldorf, E.,** "Multivalued Switching Algebras and Their Application to Digital Systems," *Proc. NEC* **15**, 467–480 (Oct. 1959).
20. **Rescher, N.,** *Many-Valued Logic*, McGraw-Hill, New York, 1969.
21. **Romankevich, A. M.,** "Minimization Methods for Functions of Many-Valued Logic," *Kibernetika* **1** (4) 38–42 (1965).
22. **Santos, J.,** et al., "Threshold Synthesis of Ternary Digital Systems," *TC* **EC-15** (1) 105–107 (Feb. 1966).
23. **Turecki, A. T.,** "The Ternary Number System for Digital Computers," *CD* **7** (2) 66–71 (Feb. 1968).
24. **Valentinuzzi, M. E.,** "Three-Valued Propositional Calculus of Lukasiewicz and Three-Position Double Switches," *TC* **EC-16** (1) 39–44 (Feb. 1967).
25. **Vranesic, Z. G.,** et al., "A Many-Valued Algebra for Switching Systems," *TC* **C-19** (10) 964-971 (Oct. 1970).
26. **Yablonskii, S. V.,** "Constructions of Functions in K-Valued Logic" (in Russian), *Trudy Matem. Inst. V. A. Steklova*, No. 51, Izd. Ak. Nauk U.S.S.R., Moscow, 6-142, 1958.
27. **Yoeli, M.,** and **G. Rosenfeld,** "Logical Design of Ternary Switching Circuits," *TC* **EC-14** (1) 19–28 (Feb. 1965).
28. **Zakharova, E. Yu.,** "Some Sufficient Conditions of Completeness in P_k," *PC* **16**, 239–246 (1966).
29. **Zakharova, E. Yu.,** "A Criterion of Completeness of Systems of Functions from P_k," *PC* **18**, 5–10 (1967).
30. **Zavisca, E.,** "Synthesis Techniques in Multiple-Valued Logic Systems," Ph.D. Dissertation, E.E. Dept., SUNY at Buffalo, N.Y., Sept. 1970.

OTHER TOPICS The problem of assigning logic elements to modules so that a specified objective function reaches its minimum, often referred to as the problem of modular partitioning, has become important with the advent of medium- and large-scale integration [2,7,11].

Lechner [4] and Menger [6] developed a Fourier-like transform for logic functions and suggested some of its applications to the analysis and synthesis of both memoryless and sequential switching circuits. This novel methodological approach to switching circuits, which is also referred to as the harmonic analysis of switching circuits, is quite promising from the computational point of view. An excellent survey is given in Reference 3.

So-called fuzzy logic, based on the Zadeh idea of the fuzzy set [14], was suggested by Marinos [5]. There is a possibility that the fuzzy logic will be employed as a tool for describing switching circuits. As such it might be superior to ordinary logic when solving certain problems, e.g., those involving transient processes.

A large number of articles on switching circuits, primarily those by Russian authors, is devoted to problems associated with various theoretical estimates. References 1,8,9,10,12, and 13 represent a sample of the literature in this area.

1. **Glagolev, V. C.,** "Some Estimates for Disjunctive Normal Boolean Forms," *PC* **19** 75–94 (1967).
2. **Lawler, E. L.,** et al., "Module Clustering to Minimize Delay in Digital Networks," *TC* **C-18** (1) 47–57 (Jan. 1969).
3. **Lechner, R. L.,** "Harmonic Analysis of Switching Functions," *RDST*, pp. 122–228.
4. **Lechner, R. J.,** "A Transform Approach to Logic Design," *TC* **C-19** (7) 672–640 (July 1970).
5. **Marinos, P. N.,** "Fuzzy Logic and Its Application to Switching Systems," *TC* **C-18** (4) 343–348 (April 1969).
6. **Menger, K. S.,** "A Transform for Logic Networks," *TC* **C-18** (3) 241–250 (March 1969).
7. **Podraza,** et al., "Efficient MSI Partitioning for a Digital Computer," *TC* **C-19** (11) 1020–1028 (Nov. 1970).
8. **Sholomov, L. A.,** "Criteria of Complexity of Boolean Functions," *PC* **17**, 92–127 (1966).
9. **Sholomov, L. A.,** "On Functions Representing the Complexity of Systems of Indefinite Boolean Functions," *PC* **19**, 121–139 (1967).
10. **Smith, D. R.,** "Complexity of Partially Defined Combinational Switching Functions," *TC* **C-20** (2) 204–208 (Feb. 1971).
11. **Stone, H. S.,** "An Algorithm for Modular Partitioning," *JACM* **17** (1) 182–195 (Jan. 1970).
12. **Vasil'ev, Yu., L.,** "On Comparison of Complexity of Irredundant and Minimal Disjunctive Normal Forms," *PC* **10**, 5–61 (1963).
13. **Yablonskii, S. V.,** "On Algorithmic Difficulties of the Synthesis of Minimal Switching Circuits," *PC* **2**, 75–121 (1959).
14. **Zadeh, L. A.,** "Fuzzy Sets," *IC* **8** (3) 338–353 (June 1965).

APPENDIX C
GLOSSARY OF SYMBOLS

$\varnothing$	Empty set
$A, B, \ldots, X, Y, \ldots$	Sets
$a, b, \ldots, x, y, \ldots$	Elements of sets
$x \in X$	x belongs to X
$x \notin X$	x does not belong to X
$\{x_i\}$	Set of elements x_i $(i \in I)$
$\lvert X \rvert$	Number of elements of X
$X = Y$	X and Y are equal sets
$X \neq Y$	X and Y are different sets
$X \subseteq Y$	X is a subset of Y
$X \subset Y$	X is a proper subset of Y
$X \nsubseteq Y$	X is not a subset of Y
$X \not\subset Y$	X is not a proper subset of Y
$X \cup Y$	Union of X and Y
$\bigcup_{i \in I} X_i$	Union of X_i $(i \in I)$
$X \cap Y$	Intersection of X and Y
$\bigcap_{i \in I} X_i$	Intersection of X_i $(i \in I)$
$X \times Y$	Cartesian product of X and Y
$(x_1, x_2, \ldots, x_n)$	Ordered n-tuple
$\times_{i \in I} X_i$	Cartesian product of X_i $(i \in I)$
X^n	Cartesian product $\underbrace{X \times X \times \cdots \times X}_{n \text{ times}}$
$\bar{X}$	Complement with respect to a given universal set
π, ρ, σ, τ	Partitions
$a = b(\pi)$	a and b belong to the same block of partition π
$\pi \cdot \rho$	Product of partitions π and ρ
$\pi + \rho$	Sum of partitions π and ρ
$\pi \leq \rho$	Partition π is smaller than or equal to partition ρ

$f, g, h, \ldots$	Functions
f^{-1}	Inverse of function f
$x \to y$	x is mapped into y
$x \leftrightarrow y$	x is mapped into y and y is mapped into x (one-to-one correspondence)
$x, y, \ldots, X, Y, \ldots$	Logic variables
$\dot{x}, \tilde{x}$	Values of variable x
$\bar{x}$ or x^0	NEGATION of x
x or x^1	ASSERTION of x
$x \cdot y$ or xy	AND function (logic product) of x and y
$\prod_{i \in I} x_i$	Product (logic or arithmetic) of variables x_i $(i \in I)$
$x \vee y$	OR function (disjunction) of x and y
$\bigvee_{i \in I} x_i$	OR function (disjunction) of variables x_i $(i \in I)$
$x \mid y$	NAND function of x and y
$x \downarrow y$	NOR function of x and y
$x \equiv y$	EQUIVALENCE of x and y
$x \not\equiv y$	NONEQUIVALENCE of x and y
$x \leq y$	INEQUALITY of x and y
$x < y$	PROPER INEQUALITY of x and y
$x \circ y$	Binary operation on variables x and y $(x, y \in A)$
M_n	Majority function of n variables
$\overline{M}_n$	Minority function of n variables
F_n^m	Logic function "m out of n"
F_n^{m+}	Logic function "at least m out of n"
F_n^{m-}	Logic function "at most m out of n"
$\{a_i\}(x_1, x_2, \ldots, x_n)$	Symmetric function of variables $x_1, x_2 \ldots, x_n$ $(i \in I)$
$a \times b$	Arithmetic product of a and b
$a + b$	Arithmetic sum of a and b
$\sum_{i \in I} a_i$	Arithmetic sum of a_i $(i \in I)$
$a - b$	Arithmetic difference
∞	Infinity
$n!$	n factorial, i.e., $1 \times 2 \times 3 \times \cdots \times n$
$\binom{n}{r}$	Binomial coefficients, i.e., $n!/(r!(n-r)!)$
$a \bmod r$	a modulo r, i.e., the remainder corresponding to $a:r$ (a, r are integers)
$\ulcorner a$	Ceiling, i.e., the smallest integer larger than or equal to a
$\llcorner a$	Floor, i.e., the largest integer smaller than or equal to a
$\log_b a$	Logarithm with base b of number a
$\mathbf{z}_1, \mathbf{z}_2, \ldots, \mathbf{z}_i, \ldots$	Internal states of a finite-state machine
$\mathbf{z}_R$	Reference (initial) internal state of a finite-state machine
$\mathbf{z}_i = \mathbf{z}_j$	$\mathbf{z}_i$ and $\mathbf{z}_j$ are equivalent
$\mathbf{z}_i \neq \mathbf{z}_j$	$\mathbf{z}_i$ and $\mathbf{z}_j$ are distinguishable
$\mathbf{z}_i \overset{k}{=} \mathbf{z}_j$	$\mathbf{z}_i$ and $\mathbf{z}_j$ are k-equivalent

$\mathbf{z}_i \overset{k}{\neq} \mathbf{z}_j$	$\mathbf{z}_i$ and $\mathbf{z}_j$ are k-distinguishable
$\mathbf{z}_i \simeq \mathbf{z}_j$	$\mathbf{z}_i$ and $\mathbf{z}_j$ are compatible
$\mathbf{z}_i \not\simeq \mathbf{z}_j$	$\mathbf{z}_i$ and $\mathbf{z}_j$ are incompatible
$\mathbf{z}_i \overset{k}{\simeq} \mathbf{z}_j$	$\mathbf{z}_i$ and $\mathbf{z}_j$ are k-compatible
$\mathbf{z}_i \overset{k}{\not\simeq} \mathbf{z}_j$	$\mathbf{z}_i$ and $\mathbf{z}_j$ are k-incompatible
ab	Concatenation of input (or input-output) sequences or regular expressions a and b
a^*	Star operation of regular expression a, i.e., $\bigcup_{i=0}^{\infty} a^i$
λ	Input sequence of length 0
$D_I R$	Derivative of regular expression R with respect to input sequence I
$\delta(R)$	λ if $\lambda \in R$, $\varnothing$ if $\lambda \notin R$
$\mathbf{a}, \mathbf{b}, \ldots, \mathbf{x}, \mathbf{y}, \ldots$	Vectors (one-dimensional arrays)
$\mathbf{A}, \mathbf{B}, \ldots, \mathbf{X}, \mathbf{Y}, \ldots$	Matrices (two-dimensional arrays)
$\lvert\mathbf{X}\rvert$	Determinant of matrix X
$\mathbf{X} = \mathbf{Y}$	Matrix X is equal to matrix Y
$\bar{\mathbf{X}}$	Negation of Boolean matrix X
$\mathbf{X} \vee \mathbf{Y}$	OR operation of Boolean matrices X and Y
$\mathbf{X} \cdot \mathbf{Y}$ or $\mathbf{XY}$	AND operation of X and Y
$\mathbf{X} \times \mathbf{Y}$	Matrix product of X and Y
$\mathbf{X}^n$	Matrix product $\underbrace{X \times X \times \cdots \times X}_{n \text{ times}}$
D	Discriminant of a set of logic or pseudo-logic relations
s	State identifier
$\dot{s}, \tilde{s}$	State identifiers of specific states
$d(\dot{s}, \tilde{s})$	Hamming distance between states $\dot{s}$ and $\tilde{s}$
$\dim(L_i)$	Dimension of linear space L_i
$\mathscr{P}(a \mid b)$	Conditional probability of a provided that b occurs
t	Time (continuous or discrete)
t_0	Time instant
Δt	Interval of time

APPENDIX D
WOS MODULE CATALOG

TABLE D.1. Catalog of Functions Represented by the Wos Module: $y = \bar{x}_1\bar{x}_3 \vee x_2\bar{x}_3 \vee x_1\bar{x}_2x_3$

Functional Identifier	Functions								Input Configuration with Negations			Input Configuration with Zero and One Functions		
	1	2	4	8	16	32	64	128	1	2	3	1	2	3
0	0	0	0	0	0	0	0	0	y_1	$\bar{y}_1$	$\bar{y}_1$	0	0	1
*3	1	1	0	0	0	0	0	0	y_2	y_3	y_3			
*5	1	0	1	0	0	0	0	0	y_1	y_3	y_3			
10	0	1	0	1	0	0	0	0	y_1	$\bar{y}_3$	$\bar{y}_1$	y_1	y_3	1
12	0	0	1	1	0	0	0	0	y_2	$\bar{y}_3$	$\bar{y}_2$	y_2	y_3	1
*15	1	1	1	1	0	0	0	0	y_1	y_1	y_3			
*17	1	0	0	0	1	0	0	0	y_1	y_2	y_2			
30	0	1	1	1	1	0	0	0	$\bar{y}_1$	y_2	y_3			
34	0	1	0	0	0	1	0	0	y_1	$\bar{y}_2$	$\bar{y}_1$	y_1	y_2	1
*45	1	0	1	1	0	1	0	0	y_1	y_2	y_3			
48	0	0	0	0	1	1	0	0	y_2	$\bar{y}_3$	$\bar{y}_3$	y_3	y_2	1
*51	1	1	0	0	1	1	0	0	y_1	y_1	y_2			
54	0	1	1	0	1	1	0	0	$\bar{y}_1$	y_3	y_2			
*57	1	0	0	1	1	1	0	0	y_1	y_3	y_2			
60	0	0	1	1	1	1	0	0	y_2	$\bar{y}_2$	$\bar{y}_3$	1	y_2	y_3
*63	1	1	1	1	1	1	0	0	y_2	y_3	y_2			
68	0	0	1	0	0	0	1	0	y_1	$\bar{y}_2$	$\bar{y}_2$	y_2	y_1	1
*75	1	1	0	1	0	0	1	0	y_2	y_1	y_3			
80	0	0	0	0	1	0	1	0	y_1	$\bar{y}_3$	$\bar{y}_3$	y_3	y_1	1
*85	1	0	1	0	1	0	1	0	y_1	y_1	y_1			
86	0	1	1	0	1	0	1	0	$\bar{y}_2$	y_3	y_1			
*89	1	0	0	1	1	0	1	0	y_2	y_3	y_1			
90	0	1	0	1	1	0	1	0	y_1	$\bar{y}_1$	$\bar{y}_3$	1	y_1	y_3
*95	1	1	1	1	1	0	1	0	y_1	y_3	y_1			

TABLE D.1 *Continued*

Functional Identifier	Functions								Input Configuration with Negations			Input Configuration with Zero and One Functions		
	1	2	4	8	16	32	64	128	1	2	3	1	2	3
*99	1	1	0	0	0	1	1	0	y_3	y_1	y_2			
*101	1	0	1	0	0	1	1	0	y_3	y_2	y_1			
102	0	1	1	0	0	1	1	0	y_1	$\bar{y}_1$	$\bar{y}_2$	1	y_1	y_2
106	0	1	0	1	0	1	1	0	y_2	$\bar{y}_3$	$\bar{y}_1$			
108	0	0	1	1	0	1	1	0	y_1	$\bar{y}_3$	$\bar{y}_2$			
*119	1	1	1	0	1	1	1	0	y_1	y_2	y_1			
120	0	0	0	1	1	1	1	0	y_1	$\bar{y}_2$	$\bar{y}_3$			
135	1	1	1	0	0	0	0	1	y_1	$\bar{y}_2$	y_3			
136	0	0	0	1	0	0	0	1	y_1	y_2	$\bar{y}_1$			
147	1	1	0	0	1	0	0	1	y_1	$\bar{y}_3$	y_2			
149	1	0	1	0	1	0	0	1	y_2	$\bar{y}_3$	y_1			
153	1	0	0	1	1	0	0	1	y_1	$\bar{y}_1$	y_2	y_1	0	y_2
154	0	1	0	1	1	0	0	1	y_3	y_2	$\bar{y}_1$			
156	0	0	1	1	1	0	0	1	y_3	y_1	$\bar{y}_2$			
160	0	0	0	0	0	1	0	1	y_1	y_3	$\bar{y}_1$			
165	1	0	1	0	0	1	0	1	y_1	$\bar{y}_1$	y_3	y_1	0	y_3
166	0	1	1	0	0	1	0	1	y_2	y_3	$\bar{y}_1$			
169	1	0	0	1	0	1	0	1	$\bar{y}_2$	y_3	$\bar{y}_1$			
175	1	1	1	1	0	1	0	1	y_1	$\bar{y}_3$	y_3	y_3	y_1	0
180	0	0	1	0	1	1	0	1	y_2	y_1	$\bar{y}_3$			
187	1	1	0	1	1	1	0	1	y_1	$\bar{y}_2$	y_2	y_2	y_1	0
192	0	0	0	0	0	0	1	1	y_2	y_3	$\bar{y}_2$			
195	1	1	0	0	0	0	1	1	y_2	$\bar{y}_2$	y_3	y_2	0	y_3
198	0	1	1	0	0	0	1	1	y_1	y_3	$\bar{y}_2$			
201	1	0	0	1	0	0	1	1	$\bar{y}_1$	y_3	$\bar{y}_2$			
207	1	1	1	1	0	0	1	1	y_2	$\bar{y}_3$	y_3	y_3	y_2	0
210	0	1	0	0	1	0	1	1	y_1	y_2	$\bar{y}_3$			
221	1	0	1	1	1	0	1	1	y_1	$\bar{y}_2$	y_1	y_1	y_2	0
225	1	0	0	0	0	1	1	1	$\bar{y}_1$	y_2	$\bar{y}_3$			
238	0	1	1	1	0	1	1	1	y_1	y_2	$\bar{y}_2$			
243	1	1	0	0	1	1	1	1	y_2	$\bar{y}_3$	y_2	y_2	y_3	0
245	1	0	1	0	1	1	1	1	y_1	$\bar{y}_3$	y_1	y_1	y_3	0
250	0	1	0	1	1	1	1	1	y_1	y_3	$\bar{y}_3$			
252	0	0	1	1	1	1	1	1	y_2	y_3	$\bar{y}_3$			
255	1	1	1	1	1	1	1	1	y_1	$\bar{y}_1$	y_1	0	0	0

* Only assertions are used.

APPENDIX E

PN-EQUIVALENCE CLASSES OF LOGIC FUNCTIONS

The following table lists representatives of PN-equivalence classes of logic functions for $n \leq 4$. The labels have the following meaning: CT, counter; FI, functional identifier; STATES, states of the input variables for which the output variable is equal to 1 (in case of some functions of four variable, i.e., functions for which CT = 238 through 401, $\overline{\text{STATES}}$ indicates the states for which the output variable is equal to 0 and $\overline{\text{FI}}$ represents the complement of the functional identifier with respect to $2^{16} - 1$, i.e., $\overline{\text{FI}} = 65{,}535 - \text{FI}$); L, number of literals in the minimal *sp* Boolean form; and E, number of functions in the equivalence class.

TABLE E1

CT	FI	STATES	L	E
		Logic functions of one variable ($n = 1$)		
0	0		0	1
1	1	0	1	2
2	3	0, 1	0	1
		Logic functions of two variables ($n = 2$)		
0	0		0	1
1	1	0	2	4
2	3	0, 1	1	4
3	6	1, 2	4	2
4	7	0, 1, 2	2	4
5	15	0, 1, 2, 3	0	1

TABLE E1 *continued*

CT	FI	STATES	L	E
	Logic functions of three variables ($n = 3$)			
0	0		0	1
1	128	7	3	8
2	129	0, 7	6	4
3	144	4, 7	5	12
4	192	6, 7	2	12
5	104	3, 5, 6	9	8
6	152	3, 4, 7	6	24
7	224	5, 6, 7	3	24
8	240	4, 5, 6, 7	1	6
9	232	3, 5, 6, 7	6	8
10	202	1, 3, 6, 7	4	24
11	120	3, 4, 5, 6	7	24
12	105	0, 3, 5, 6	12	2
13	153	0, 3, 4, 7	4	6
14	47	0, 1, 2, 3, 4	4	24
15	103	0, 1, 2, 5, 6	6	24
16	151	0, 1, 2, 4, 7	9	8
17	63	0, 1, 2, 3, 4, 5	2	12
18	111	0, 1, 2, 3, 5, 6	6	12
19	126	1, 2, 3, 4, 5, 6	7	4
20	127	0, 1, 2, 3, 4, 5, 6	3	8
21	255	0, 1, 2, 3, 4, 5, 6, 7	0	1
	Logic functions of four variables ($n = 4$)			
0	0		0	1
1	1	0	4	16
2	3	0, 1	3	32
3	9	0, 3	6	48
4	129	0, 7	7	32
5	32769	0, 15	8	8
6	7	0, 1, 2	4	96
7	67	0, 1, 6	7	192
8	16387	0, 1, 14	8	64
9	41	0, 3, 5	10	64
10	4105	0, 3, 12	10	48
11	8201	0, 3, 13	12	96
12	15	0, 1, 2, 3	2	24
13	23	0, 1, 2, 4	7	64
14	39	0, 1, 2, 5	5	192
15	135	0, 1, 2, 7	8	192
16	4103	0, 1, 2, 12	9	96
17	8199	0, 1, 2, 13	8	192
18	32775	0, 1, 2, 15	10	96
19	195	0, 1, 6, 7	5	48
20	1091	0, 1, 6, 10	11	192

CT	FI	STATES	L	E
21	2115	0, 1, 6, 11	11	192
22	16451	0, 1, 6, 14	8	96
23	32835	0, 1, 6, 15	11	192
24	49155	0, 1, 14, 15	6	16
25	105	0, 3, 5, 6	10	16
26	553	0, 3, 5, 9	14	16
27	1065	0, 3, 5, 10	14	96
28	16425	0, 3, 5, 14	15	64
29	36873	0, 3, 12, 15	8	12
30	24585	0, 3, 13, 14	11	24
31	31	0, 1, 2, 3, 4	5	192
32	4111	0, 1, 2, 3, 12	7	96
33	151	0, 1, 2, 4, 7	10	64
34	279	0, 1, 2, 4, 8	9	16
35	535	0, 1, 2, 4, 9	8	192
36	2071	0, 1, 2, 4, 11	11	192
37	32791	0, 1, 2, 4, 15	13	64
38	103	0, 1, 2, 5, 6	7	192
39	1063	0, 1, 2, 5, 10	6	192
40	2087	0, 1, 2, 5, 11	11	384
41	16423	0, 1, 2, 5, 14	10	384
42	2183	0, 1, 2, 7, 11	11	96
43	4231	0, 1, 2, 7, 12	13	192
44	8327	0, 1, 2, 7, 13	12	384
45	32903	0, 1, 2, 7, 15	9	192
46	12295	0, 1, 2, 12, 13	8	192
47	36871	0, 1, 2, 12, 15	12	96
48	24583	0, 1, 2, 13, 14	10	96
49	40967	0, 1, 2, 13, 15	9	192
50	1219	0, 1, 6, 7, 10	10	192
51	5187	0, 1, 6, 10, 12	12	64
52	9283	0, 1, 6, 10, 13	14	192
53	33859	0, 1, 6, 10, 15	15	192
54	18499	0, 1, 6, 11, 14	11	192
55	617	0, 3, 5, 6, 9	16	64
56	16937	0, 3, 5, 9, 14	19	16
57	5161	0, 3, 5, 10, 12	14	48
58	63	0, 1, 2, 3, 4, 5	3	96
59	159	0, 1, 2, 3, 4, 7	7	96
60	287	0, 1, 2, 3, 4, 8	8	96
61	543	0, 1, 2, 3, 4, 9	8	192
62	2079	0, 1, 2, 3, 4, 11	8	96
63	4127	0, 1, 2, 3, 4, 12	6	192
64	8223	0, 1, 2, 3, 4, 13	9	384
65	32799	0, 1, 2, 3, 4, 15	11	192
66	12303	0, 1, 2, 3, 12, 13	6	96
67	36879	0, 1, 2, 3, 12, 15	10	48

TABLE E1 *continued*

CT	FI	STATES	L	E
68	407	0, 1, 2, 4, 7, 8	15	64
69	663	0, 1, 2, 4, 7, 9	13	192
70	2199	0, 1, 2, 4, 7, 11	14	192
71	32919	0, 1, 2, 4, 7, 15	12	64
72	33047	0, 1, 2, 4, 8, 15	15	16
73	1559	0, 1, 2, 4, 9, 10	10	192
74	2583	0, 1, 2, 4, 9, 11	10	384
75	16919	0, 1, 2, 4, 9, 14	12	192
76	33303	0, 1, 2, 4, 9, 15	11	192
77	10263	0, 1, 2, 4, 11, 13	12	192
78	34839	0, 1, 2, 4, 11, 15	10	192
79	231	0, 1, 2, 5, 6, 7	8	32
80	2151	0, 1, 2, 5, 6, 11	13	192
81	4199	0, 1, 2, 5, 6, 12	10	192
82	8295	0, 1, 2, 5, 6, 13	9	384
83	32871	0, 1, 2, 5, 6, 15	12	192
84	5159	0, 1, 2, 5, 10, 12	13	192
85	9255	0, 1, 2, 5, 10, 13	8	192
86	33831	0, 1, 2, 5, 10, 15	9	192
87	6183	0, 1, 2, 5, 11, 12	11	192
88	18471	0, 1, 2, 5, 11, 14	13	384
89	34855	0, 1, 2, 5, 11, 15	9	384
90	49191	0, 1, 2, 5, 14, 15	10	192
91	6279	0, 1, 2, 7, 11, 12	17	96
92	10375	0, 1, 2, 7, 11, 13	15	192
93	34951	0, 1, 2, 7, 11, 15	9	48
94	12423	0, 1, 2, 7, 12, 13	12	384
95	36999	0, 1, 2, 7, 12, 15	13	192
96	24711	0, 1, 2, 7, 13, 14	14	192
97	41095	0, 1, 2, 7, 13, 15	10	192
98	28679	0, 1, 2, 12, 13, 14	6	48
99	45063	0, 1, 2, 12, 13, 15	10	96
100	57351	0, 1, 2, 13, 14, 15	10	48
101	3267	0, 1, 6, 7, 10, 11	9	32
102	5315	0, 1, 6, 7, 10, 12	10	96
103	9411	0, 1, 6, 7, 10, 13	14	96
104	37955	0, 1, 6, 10, 12, 15	18	64
105	42051	0, 1, 6, 10, 13, 15	14	96
106	1641	0, 3, 5, 6, 9, 10	13	48
107	37929	0, 3, 5, 10, 12, 15	18	8
108	127	0, 1, 2, 3, 4, 5, 6	5	64
109	319	0, 1, 2, 3, 4, 5, 8	7	192
110	1087	0, 1, 2, 3, 4, 5, 10	6	384
111	16447	0, 1, 2, 3, 4, 5, 14	8	192
112	415	0, 1, 2, 3, 4, 7, 8	11	192
113	671	0, 1, 2, 3, 4, 7, 9	11	192

CT	FI	STATES	L	E
114	4255	0, 1, 2, 3, 4, 7, 12	10	192
115	8351	0, 1, 2, 3, 4, 7, 13	12	192
116	4383	0. 1, 2, 3, 4, 8, 12	5	48
117	8479	0, 1, 2, 3, 4, 8, 13	12	192
118	33055	0, 1, 2, 3, 4, 8, 15	13	96
119	4639	0, 1, 2, 3, 4, 9, 12	9	384
120	16927	0, 1, 2, 3, 4, 9, 14	12	384
121	6175	0, 1, 2, 3, 4, 11, 12	9	192
122	10271	0, 1, 2, 3, 4, 11, 13	12	192
123	12319	0, 1, 2, 3, 4, 12, 13	8	384
124	36895	0, 1, 2, 3, 4, 12, 15	10	192
125	24607	0, 1, 2, 3, 4, 13, 14	13	192
126	40991	0, 1, 2, 3, 4, 13, 15	9	384
127	28687	0, 1, 2, 3, 12, 13, 14	8	96
128	2455	0, 1, 2, 4, 7, 8, 11	14	96
129	33175	0, 1, 2, 4, 7, 8, 15	14	64
130	1687	0, 1, 2, 4, 7, 9, 10	14	192
131	2711	0, 1, 2, 4, 7, 9, 11	14	384
132	17047	0, 1, 2, 4, 7, 9, 14	16	192
133	33431	0, 1, 2, 4, 7, 9, 15	11	192
134	10391	0, 1, 2, 4, 7, 11, 13	16	192
135	34967	0, 1, 2, 4, 7, 11, 15	14	192
136	3607	0, 1, 2, 4, 9, 10, 11	11	192
137	5655	0, 1, 2, 4, 9, 10, 12	11	64
138	9751	0, 1, 2, 4, 9, 10, 13	12	384
139	34327	0, 1, 2, 4, 9, 10, 15	14	192
140	10775	0, 1, 2, 4, 9, 11, 13	9	96
141	18967	0, 1, 2, 4, 9, 11, 14	12	384
142	35351	0, 1, 2, 4, 9, 11, 15	11	384
143	49687	0, 1, 2, 4, 9, 14, 15	13	192
144	26647	0, 1, 2, 4, 11, 13, 14	14	64
145	43031	0, 1, 2, 4, 11, 13, 15	13	192
146	2279	0, 1, 2, 5, 6, 7, 11	12	64
147	6247	0, 1, 2, 5, 6, 11, 12	15	192
148	10343	0, 1, 2, 5, 6, 11, 13	14	384
149	34919	0, 1, 2, 5, 6, 11, 15	11	192
150	12391	0, 1, 2, 5, 6, 12, 13	10	192
151	36967	0, 1, 2, 5, 6, 12, 15	12	192
152	24679	0, 1, 2, 5, 6, 13, 14	8	192
153	41063	0, 1, 2, 5, 6, 13, 15	11	384
154	37927	0, 1, 2, 5, 10, 12, 15	16	192
155	25639	0, 1, 2, 5, 10, 13, 14	11	192
156	22567	0, 1, 2, 5, 11, 12, 14	12	384
157	51239	0, 1, 2, 5, 11, 14, 15	10	384
158	14471	0, 1, 2, 7, 11, 12, 13	14	192
159	39047	0, 1, 2, 7, 11, 12, 15	16	48
160	26759	0, 1, 2, 7, 11, 13, 14	15	96

TABLE E1 *continued*

CT	FI	STATES	L	E
161	28807	0, 1, 2, 7, 12, 13, 14	12	96
162	7363	0, 1, 6, 7, 10, 11, 12	12	64
163	5737	0, 3, 5, 6, 9, 10, 12	16	16
164	255	0, 1, 2, 3, 4, 5, 6, 7	7	8
165	383	0, 1, 2, 3, 4, 5, 6, 8	9	64
166	639	0, 1, 2, 3, 4, 5, 6, 9	9	192
167	2175	0, 1, 2, 3, 4, 5, 6, 11	8	192
168	32895	0, 1, 2, 3, 4, 5, 6, 15	8	64
169	831	0, 1, 2, 3, 4, 5, 8, 9	6	32
170	1343	0, 1, 2, 3, 4, 5, 8, 10	6	192
171	2367	0, 1, 2, 3, 4, 5, 8, 11	10	384
172	16703	0, 1, 2, 3, 4, 5, 8, 14	11	192
173	33087	0, 1, 2, 3, 4, 5, 8, 15	12	192
174	3135	0, 1, 2, 3, 4, 5, 10, 11	4	96
175	5183	0, 1, 2, 3, 4, 5, 10, 12	8	192
176	9279	0, 1, 2, 3, 4, 5, 10, 13	8	192
177	17471	0, 1, 2, 3, 4, 5, 10, 14	8	384
178	33855	0, 1, 2, 3, 4, 5, 10, 15	11	384
179	49215	0, 1, 2, 3, 4, 5, 14, 15	7	96
180	2463	0, 1, 2, 3, 4, 7, 8, 11	10	48
181	4511	0, 1, 2, 3, 4, 7, 8, 12	9	192
182	8607	0, 1, 2, 3, 4, 7, 8, 13	16	384
183	33183	0, 1, 2, 3, 4, 7, 8, 15	12	192
184	1695	0, 1, 2, 3, 4, 7, 9, 10	9	48
185	4767	0, 1, 2, 3, 4, 7, 9, 12	12	384
186	8863	0, 1, 2, 3, 4, 7, 9, 13	12	192
187	17055	0, 1, 2, 3, 4, 7, 9, 14	15	192
188	12447	0, 1, 2, 3, 4, 7, 12, 13	11	384
189	37023	0, 1, 2, 3, 4, 7, 12, 15	8	96
190	24735	0, 1, 2, 3, 4, 7, 13, 14	10	96
191	37151	0, 1, 2, 3, 4, 8, 12, 15	10	48
192	24863	0, 1, 2, 3, 4, 8, 13, 14	15	96
193	41247	0, 1, 2, 3, 4, 8, 13, 15	11	192
194	12831	0, 1, 2, 3, 4, 9, 12, 13	11	192
195	21023	0, 1, 2, 3, 4, 9, 12, 14	12	384
196	37407	0, 1, 2, 3, 4, 9, 12, 15	14	384
197	49695	0, 1, 2, 3, 4, 9, 14, 15	12	192
198	14367	0, 1, 2, 3, 4, 11, 12, 13	11	384
199	38943	0, 1, 2, 3, 4, 11, 12, 15	11	96
200	26655	0, 1, 2, 3, 4, 11, 13, 14	16	96
201	28703	0, 1, 2, 3, 4, 12, 13, 14	11	192
202	45087	0, 1, 2, 3, 4, 12, 13, 15	11	384
203	57375	0, 1, 2, 3, 4, 13, 14, 15	8	192
204	61455	0, 1, 2, 3, 12, 13, 14, 15	4	12
205	10647	0, 1, 2, 4, 7, 8, 11, 13	20	64
206	35223	0, 1, 2, 4, 7, 8, 11, 15	15	96

CT	FI	STATES	L	E
207	3735	0, 1, 2, 4, 7, 9, 10, 11	16	192
208	5783	0, 1, 2, 4, 7, 9, 10, 12	16	64
209	9879	0, 1, 2, 4, 7, 9, 10, 13	16	384
210	34455	0, 1, 2, 4, 7, 9, 10, 15	12	192
211	10903	0, 1, 2, 4, 7, 9, 11, 13	13	96
212	19095	0, 1, 2, 4, 7, 9, 11, 14	15	384
213	35479	0, 1, 2, 4, 7, 9, 11, 15	13	384
214	49815	0, 1, 2, 4, 7, 9, 14, 15	14	192
215	26775	0, 1, 2, 4, 7, 11, 13, 14	14	64
216	43159	0, 1, 2, 4, 7, 11, 13, 15	15	96
217	7703	0, 1, 2, 4, 9, 10, 11, 12	12	192
218	11799	0, 1, 2, 4, 9, 10, 11, 13	11	192
219	36375	0, 1, 2, 4, 9, 10, 11, 15	12	96
220	38423	0, 1, 2, 4, 9, 10, 12, 15	16	64
221	26135	0, 1, 2, 4, 9, 10, 13, 14	10	192
222	42519	0, 1, 2, 4, 9, 10, 13, 15	13	384
223	27159	0, 1, 2, 4, 9, 11, 13, 14	15	96
224	51735	0, 1, 2, 4, 9, 11, 14, 15	14	384
225	59415	0, 1, 2, 4, 11, 13, 14, 15	11	32
226	6375	0, 1, 2, 5, 6, 7, 11, 12	11	32
227	14439	0, 1, 2, 5, 6, 11, 12, 13	16	192
228	39015	0, 1, 2, 5, 6, 11, 12, 15	11	192
229	26727	0, 1, 2, 5, 6, 11, 13, 14	14	192
230	43111	0, 1, 2, 5, 6, 11, 13, 15	12	192
231	28775	0, 1, 2, 5, 6, 12, 13, 14	9	48
232	57447	0, 1, 2, 5, 6, 13, 14, 15	10	96
233	58407	0, 1, 2, 5, 10, 13, 14, 15	12	24
234	55335	0, 1, 2, 5, 11, 12, 14, 15	9	96
235	30855	0, 1, 2, 7, 11, 12, 13, 14	12	48
236	15555	0, 1, 6, 7, 10, 11, 12, 13	9	8
237	38505	0, 3, 5, 6, 9, 10, 12, 15	14	2

CT	$\overline{\text{FI}}$	$\overline{\text{STATES}}$	L	E
238	127	0, 1, 2, 3, 4, 5, 6	5	64
239	319	0, 1, 2, 3, 4, 5, 8	8	192
240	1087	0, 1, 2, 3, 4, 5, 10	7	384
241	16447	0, 1, 2, 3, 4, 5, 14	8	192
242	415	0, 1, 2, 3, 4, 7, 8	11	192
243	671	0, 1, 2, 3, 4, 7, 9	11	192
244	4255	0, 1, 2, 3, 4, 7, 12	10	192
245	8351	0, 1, 2, 3, 4, 7, 13	12	192
246	4383	0, 1, 2, 3, 4, 8, 12	4	48
247	8479	0, 1, 2, 3, 4, 8, 13	12	192
248	33055	0, 1, 2, 3, 4, 8, 15	12	96
249	4639	0, 1, 2, 3, 4, 9, 12	8	384

TABLE E1 *continued*

CT	$\overline{\text{FI}}$	$\overline{\text{STATES}}$	L	E
250	16927	0, 1, 2, 3, 4, 9, 14	12	384
251	6175	0, 1, 2, 3, 4, 11, 12	9	192
252	10271	0, 1, 2, 3 4, 11, 13	12	192
253	12319	0, 1, 2, 3, 4, 12, 13	7	384
254	36895	0, 1, 2, 3, 4, 12, 15	10	192
255	24607	0, 1, 2, 3, 4, 13, 14	12	192
256	40991	0, 1, 2, 3, 4, 13, 15	9	384
257	28687	0, 1, 2, 3, 12, 13, 14	7	96
258	2455	0, 1, 2, 4, 7, 8, 11	15	96
259	33175	0, 1, 2, 4, 7, 8, 15	13	64
260	1687	0, 1, 2, 4, 7, 9, 10	14	192
261	2711	0, 1, 2, 4, 7, 9, 11	15	384
262	17047	0, 1, 2, 4, 7, 9, 14	16	192
263	33431	0, 1, 2, 4, 7, 9, 15	11	192
264	10391	0, 1, 2, 4, 7, 11, 13	17	192
265	34967	0, 1, 2, 4, 7, 11, 15	13	192
266	3607	0, 1, 2, 4, 9, 10, 11	11	192
267	5655	0, 1, 2, 4, 9, 10, 12	10	64
268	9751	0, 1, 2, 4, 9, 10, 13	12	384
269	34327	0, 1, 2, 4, 9, 10, 15	14	192
270	10775	0, 1, 2, 4, 9, 11, 13	10	96
271	18967	0, 1, 2, 4, 9, 11, 14	13	384
272	35351	0, 1, 2, 4, 9, 11, 15	11	384
273	49687	0, 1, 2, 4, 9, 14, 15	12	192
274	26647	0, 1, 2, 4, 11, 13, 14	15	64
275	43031	0, 1, 2, 4, 11, 13, 15	13	192
276	2279	0, 1, 2, 5, 6, 7, 11	13	64
277	6247	0, 1, 2, 5, 6, 11, 12	15	192
278	10343	0, 1, 2, 5, 6, 11, 13	14	384
279	34919	0, 1, 2, 5, 6, 11, 15	11	192
280	12391	0, 1, 2, 5, 6, 12, 13	11	192
281	36967	0, 1, 2, 5, 6, 12, 15	12	192
282	24679	0, 1, 2, 5, 6, 13, 14	9	192
283	41063	0, 1, 2, 5, 6, 13, 15	11	384
284	37927	0, 1, 2, 5, 10, 12, 15	17	192
285	25639	0, 1, 2, 5, 10, 13, 14	11	192
286	22567	0, 1, 2, 5, 11, 12, 14	11	384
287	51239	0, 1, 2, 5, 11, 14 15	10	384
288	14471	0, 1, 2, 7, 11, 12, 13	13	192
289	39047	0, 1, 2, 7, 11, 12, 15	15	48
290	26759	0, 1, 2, 7, 11, 13, 14	14	96
291	28807	0, 1, 2, 7, 12, 13, 14	11	96
292	7363	0, 1, 6, 7, 10, 11, 12	11	64
293	5737	0, 3, 5, 6, 9, 10, 12	15	16
294	63	0, 1, 2, 3, 4, 5	4	96
295	159	0, 1, 2, 3, 4, 7	7	96

CT	$\overline{FI}$	$\overline{STATES}$	L	E
296	287	0, 1, 2, 3, 4, 8	7	96
297	543	0, 1, 2, 3, 4, 9	8	192
298	2079	0, 1, 2, 3, 4, 11	8	96
299	4127	0, 1, 2, 3, 4, 12	6	192
300	8223	0, 1, 2, 3, 4, 13	10	384
301	32799	0, 1, 2, 3, 4, 15	11	192
302	12303	0, 1, 2, 3, 12, 13	6	96
303	36879	0, 1, 2, 3, 12, 15	10	48
304	407	0, 1, 2, 4, 7, 8	16	64
305	663	0, 1, 2, 4, 7, 9	13	192
306	2199	0, 1, 2, 4, 7, 11	15	192
307	32919	0, 1, 2, 4, 7, 15	12	64
308	33047	0, 1, 2, 4, 8, 15	14	16
309	1559	0, 1, 2, 4, 9, 10	10	192
310	2583	0, 1, 2, 4, 9, 11	10	384
311	16919	0, 1, 2, 4, 9, 14	13	192
312	33303	0, 1, 2, 4, 9, 15	12	192
313	10263	0, 1, 2, 4, 11, 13	13	192
314	34839	0, 1, 2, 4, 11, 15	11	192
315	231	0, 1, 2, 5, 6, 7	8	32
316	2151	0, 1, 2, 5, 6, 11	13	192
317	4199	0, 1, 2, 5, 6, 12	11	192
318	8295	0, 1, 2, 5, 6, 13	9	384
319	32871	0, 1, 2, 5, 6, 15	13	192
320	5159	0, 1, 2, 5, 10, 12	13	192
321	9255	0, 1, 2, 5, 10, 13	9	192
322	33831	0, 1, 2, 5, 10, 15	10	192
323	6183	0, 1, 2, 5, 11, 12	12	192
324	18471	0, 1, 2, 5, 11, 14	13	384
325	34855	0, 1, 2, 5, 11, 15	10	384
326	49191	0, 1, 2, 5, 14, 15	10	192
327	6279	0, 1, 2, 7, 11, 12	17	96
328	10375	0, 1, 2, 7, 11, 13	15	192
329	34951	0, 1, 2, 7, 11, 15	9	48
330	12423	0, 1, 2, 7, 12, 13	13	384
331	36999	0, 1, 2, 7, 12, 15	13	192
332	24711	0, 1, 2, 7, 13, 14	14	192
333	41095	0, 1, 2, 7, 13, 15	10	192
334	28679	0, 1, 2, 12, 13, 14	7	48
335	45063	0, 1, 2, 12, 13, 15	10	96
336	57351	0, 1, 2, 13, 14, 15	10	48
337	3267	0, 1, 6, 7, 10, 11	10	32
338	5315	0, 1, 6, 7, 10, 12	10	96
339	9411	0, 1, 6, 7, 10, 13	14	96
340	37955	0, 1, 6, 10, 12, 15	18	64
341	42051	0, 1, 6, 10, 13, 15	14	96
342	1641	0, 3, 5, 6, 9, 10	13	48

TABLE E1 *continued*

CT	$\overline{\text{FI}}$	$\overline{\text{STATES}}$	L	E
343	37929	0, 3, 5, 10, 12, 15	19	8
344	31	0, 1, 2, 3, 4	5	192
345	4111	0, 1, 2, 3, 12	7	96
346	151	0, 1, 2, 4, 7	11	64
347	279	0, 1, 2, 4, 8	10	16
348	535	0, 1, 2, 4, 9	9	192
349	2071	0, 1, 2, 4, 11	12	192
350	32791	0, 1, 2, 4, 15	13	64
351	103	0, 1, 2, 5, 6	8	192
352	1063	0, 1, 2, 5, 10	7	192
353	2087	0, 1, 2, 5, 11	11	384
354	16423	0, 1, 2, 5, 14	11	384
355	2183	0, 1, 2, 7, 11	11	96
356	4231	0, 1, 2, 7, 12	13	192
357	8327	0, 1, 2, 7, 13	12	384
358	32903	0, 1, 2, 7, 15	9	192
359	12295	0, 1, 2, 12, 13	9	192
360	36871	0, 1, 2, 12, 15	12	96
361	24583	0, 1, 2, 13, 14	11	96
362	40967	0, 1, 2, 13, 15	9	192
363	1219	0, 1, 6, 7, 10	11	192
364	5187	0, 1, 6, 10, 12	13	64
365	9283	0, 1, 6, 10, 13	15	192
366	33859	0, 1, 6, 10, 15	15	192
367	18499	0, 1, 6, 11, 14	12	192
368	617	0, 3, 5, 6, 9	16	64
369	16937	0, 3, 5, 9, 14	18	16
370	5161	0, 3, 5, 10, 12	14	48
371	15	0, 1, 2, 3	2	24
372	23	0, 1, 2, 4	8	64
373	39	0, 1, 2, 5	6	192
374	135	0, 1, 2, 7	9	192
375	4103	0, 1, 2, 12	10	96
376	8199	0, 1, 2, 13	9	192
377	32775	0, 1, 2, 15	10	96
378	195	0, 1, 6, 7	6	48
379	1091	0, 1, 6, 10	10	192
380	2115	0, 1, 6, 11	12	192
381	16451	0, 1, 6, 14	8	96
382	32835	0, 1, 6, 15	11	192
383	49155	0, 1, 14, 15	7	16
384	105	0, 3, 5, 6	11	16
385	553	0, 3, 5, 9	14	16
386	1065	0, 3, 5, 10	15	96
387	16425	0, 3, 5, 14	14	64
388	36873	0, 3, 12, 15	9	12

CT	$\overline{FI}$	$\overline{STATES}$	L	E
389	24585	0, 3, 13, 14	12	24
390	7	0, 1, 2	5	96
391	67	0, 1, 6	8	192
392	16387	0, 1, 14	9	64
393	41	0, 3, 5	11	64
394	4105	0, 3, 12	11	48
395	8201	0, 3, 13	12	96
396	3	0, 1	3	32
397	9	0, 3	7	48
398	129	0, 7	8	32
399	32769	0, 15	9	8
400	1	0	4	16
401	0		0	1

APPENDIX F

ANSWERS TO SELECTED EXERCISES

CHAPTER 1

Section 1.1

1.1-3 (b) and (c).

1.1-4 $x_t = 1 - x^{t-1}x^{t-2}$ for all of the activities.

CHAPTER 2

Section 2.1

2.1-1 (a) $d(s_1, s_2) = 4$.

2.1-3 (a) $s_m = 20$.

(c) $s_m = 2^b - 1 - a$.

2.1-5 $d(s_1, s_2) = d(s_{r_1}, s_{r_2}) + d(s_{c_1} + s_{c_2})$.

Section 2.2

2.2-2 (a) $w_1 = w_2 = \cdots = w_n = 1, T = n - 1$ (other solutions are possible).

(d) $w_1 = w_2 = \cdots = w_n = 1, T = m - 1$ (other solutions are possible).

2.2-3 All of them except EQUIVALENCE and NONEQUIVALENCE.

2.2-4 NONEQUIVALENCE: $f = 0$ if $x_1 = x_2 = \cdots = x_n = 0$ or $x_1 = x_2 = \cdots = x_n = 1$; otherwise $f = 1$. EXCLUSIVE OR: logic function "one out of n."

2.2-5 (a) $i = 128$.

2.2-6 (b) $f(s) = 1$ for $s = 6$; otherwise $f(s) = 0$.

2.2-7 (c) $f(0) = \sim, f(1) = 1, f(2) = f(3) = 0$.

2.2-9 (a) $M_3 = x_1x_2 + x_1x_3 + x_2x_3 - 2x_1x_2x_3$.

Section 2.3

2.3-1 (c), (f), (g).

2.3-3 (a) $x_1 \not\equiv x_2 = (x_1 | (x_2 | x_2)) | ((x_1 | x_1) | x_2)$.

(c) $x_1 \not\equiv x_2 = (x_1 \bar{x}_2) \vee (\bar{x}_1 x_2)$.

2.3-5 (a) $x_1 \circ x_2$ has the same values for all states where $x_1 \neq x_2$.

(b) Functions ZERO, AND, OR, NAND, NOR, EQUIVALENCE, NONEQUIVALENCE, ONE.

Section 2.4

2.4-1 (a) $x < (y | ((z \downarrow v) \vee w))$

2.4-2 (b) $xy | xx | yy | | |$

Section 2.7

2.7-2

x_3	x_2	x_1	
0	0	0	11111111111111111111111111111111111111111111111111111111
0	0	1	00000000000000000000000000001111111111111111111111111111
0	1	0	00000000000000111111111111110000000000000011111111111111
0	1	1	00000001111111000000011111110000000111111100000001111111
1	0	0	00001110001111000011100001110000111000111100001110001111
1	0	1	00110010010011001101101100110011001001001100110110110001
1	1	0	01010100100101010100100101010101011011010101011011010101
1	1	1	00000000000000000000000000000000000000000000000000000000

2.7-5 $f_2 = x_1 x_2 = \overline{x_1 \leq \bar{x}_2}$

$f_3 = x_1 > x_2 = \overline{x_1 \leq x_2}$

$f_7 = x_1 \not\equiv x_2 = \overline{(x_1 \leq x_2) \leq (\bar{x}_1 \leq \bar{x}_2)}$

$f_8 = x_1 \vee x_2 = \bar{x}_1 \leq x_2$, $f_9 = x_1 \downarrow x_2 = \overline{\bar{x}_1 \leq x_2}$

$f_{10} = x_1 \equiv x_2 = \overline{(x_1 \leq x_2) \leq \overline{(\bar{x}_1 \leq \bar{x}_2)}}$

$f_{15} = x_1 | x_2 = x_1 \leq \bar{x}_2$.

2.7-6 The set G is complete if such functions $F_1, F_2, \ldots, F_m$ can be found that

$$
\begin{aligned}
f_1 &= F_1(g_1, g_2, \ldots, g_q) \\
f_2 &= F_2(g_1, g_2, \ldots, g_q) \\
&\vdots \\
f_m &= F_m(g_1, g_2, \ldots, g_q)
\end{aligned}
$$

2.7-7 (a) $x_1 \not\equiv x_2 = (\bar{x}_1 x_2) \vee (x_1 \bar{x}_2)$

(c) $F_3^2 = (x_1 x_2 \bar{x}_3) \vee (x_1 \bar{x}_2 x_3) \vee (\bar{x}_1 x_2 x_3)$

Section 2.8

2.8-1 (a) All logic expressions in $P_1, P_2, \ldots, P_m$ containing functions from C_1 are replaced by equal logic expressions containing functions from C_2.
(b) If a tabular representation of all functions in C_2 can be derived from $Q_1, Q_2, \ldots, Q_n$, this set of axioms is complete.

Section 2.9

2.9-4 (a) $(x \vee \bar{y} \vee z)(\bar{x} \vee y \vee \bar{z})$.
(c) $(x \vee \bar{y}(z \vee \bar{v}))(\bar{x} \vee v(\bar{y} \vee \bar{z})) \vee \bar{w}$.

2.9-7 From (II): $0 \vee 0 = 0$, $1 \vee 0 = 1$, $0 \cdot 1 = 0$, $1 \cdot 1 = 1$. From (II) and (III): $0 \vee 1 = 1$, $1 \cdot 0 = 0$. From (IV) and previous results: $1 \vee 1 = 1$, $0 \cdot 0 = 0$. From (V) and (II): $\bar{0} = 1$, $\bar{1} = 0$.

Section 2.10

2.10-2 (a)
$$f = \bar{x}_5\bar{x}_4x_3\bar{x}_2\bar{x}_1 \vee \bar{x}_5\bar{x}_4x_3x_2\bar{x}_1 \vee \bar{x}_5x_4\bar{x}_3\bar{x}_2\bar{x}_1 \vee \bar{x}_5x_4\bar{x}_3x_2\bar{x}_1 \vee$$
$$\bar{x}_5x_4\bar{x}_3\bar{x}_2x_1 \vee \bar{x}_5x_4\bar{x}_3x_2x_1 \vee \bar{x}_5x_4x_3\bar{x}_2\bar{x}_1 \vee \bar{x}_5x_4x_3\bar{x}_2x_1 \vee$$
$$\bar{x}_5x_4x_3x_2\bar{x}_1 \vee \bar{x}_5x_4x_3x_2x_1 \vee x_5\bar{x}_4\bar{x}_3\bar{x}_2\bar{x}_1 \vee x_5\bar{x}_4x_3\bar{x}_2\bar{x}_1 \vee$$
$$x_5\bar{x}_4x_3x_2\bar{x}_1 \vee x_5\bar{x}_4x_3x_2x_1 \vee x_5x_4\bar{x}_3\bar{x}_2\bar{x}_1 \vee x_5x_4x_3\bar{x}_2\bar{x}_1 \vee$$
$$x_5x_4x_3x_2\bar{x}_1 \vee x_5x_4x_3x_2x_1$$
$$f = (x_5 \vee x_4 \vee x_3 \vee x_2 \vee x_1)(x_5 \vee x_4 \vee x_3 \vee x_2 \vee \bar{x}_1)(x_5 \vee x_4 \vee x_3 \vee \bar{x}_2 \vee x_1)$$
$$(x_5 \vee x_4 \vee x_3 \vee \bar{x}_2 \vee \bar{x}_1)(x_5 \vee x_4 \vee \bar{x}_3 \vee x_2 \vee \bar{x}_1)(x_5 \vee x_4 \vee \bar{x}_3 \vee \bar{x}_2 \vee \bar{x}_1)$$
$$(\bar{x}_5 \vee x_4 \vee x_3 \vee x_2 \vee \bar{x}_1)(\bar{x}_5 \vee x_4 \vee x_3 \vee \bar{x}_2 \vee x_1)(\bar{x}_5 \vee x_4 \vee x_3 \vee \bar{x}_2 \vee \bar{x}_1)$$
$$(\bar{x}_5 \vee x_4 \vee \bar{x}_3 \vee x_2 \vee \bar{x}_1)(\bar{x}_5 \vee \bar{x}_4 \vee x_3 \vee x_2 \vee \bar{x}_1)(\bar{x}_5 \vee \bar{x}_4 \vee x_3 \vee \bar{x}_2 \vee x_1)$$
$$(\bar{x}_5 \vee \bar{x}_4 \vee x_3 \vee \bar{x}_2 \vee \bar{x}_1)(\bar{x}_5 \vee \bar{x}_4 \vee \bar{x}_3 \vee x_2 \vee \bar{x}_1)$$

Simplest standard form:
$$f = \bar{x}_5x_4 \vee x_3\bar{x}_1 \vee x_5x_3x_2 \vee x_5\bar{x}_2\bar{x}_1$$
$$f = (x_5 \vee x_4 \vee x_3)(x_5 \vee x_4 \vee \bar{x}_1)(\bar{x}_5 \vee x_3 \vee \bar{x}_2)(\bar{x}_5 \vee x_2 \vee \bar{x}_1)$$

Section 2.12

2.12-1 (a) $f_1 \not\equiv f_2 = \bar{y}\bar{x} \vee \bar{v}\bar{z}\bar{x} \vee zyx$
$f_1 \equiv f_2 = \bar{y}x \vee \bar{z}x \vee zy\bar{x} \vee vy\bar{x}$
$f_1 \vee f_2 = \bar{y} \vee x \vee \bar{v}\bar{z}$
$f_1 \cdot f_2 = \bar{y}x \vee \bar{z}x$
$f_1 \leq f_2 = v \vee z \vee x$
$f_1 > f_2 = \bar{v}\bar{z}\bar{x}$
$f_1 | f_2 = \bar{x} \vee zy$
$f_1 \downarrow f_2 = vy\bar{x} \vee zy\bar{x}$.

2.12-2 (a) $(v \vee z \vee y \vee x)(\bar{v} \vee \bar{z} \vee \bar{y} \vee \bar{x})$

Comprehensive Exercises to Chapter 2

2.3 (a) $f_1 = x_1(x_4x_3\bar{x}_2 \vee x_4\bar{x}_3x_2) \vee \bar{x}_1(x_3\bar{x}_2 \vee x_4\bar{x}_3x_2)$
$f_1 = x_2x_4\bar{x}_3 \vee \bar{x}_2(x_3\bar{x}_1 \vee x_4x_3)$
$f_1 = x_3(\bar{x}_2\bar{x}_1 \vee x_4\bar{x}_2) \vee \bar{x}_3x_4x_2$
$f_1 = x_4(x_3\bar{x}_2 \vee \bar{x}_3x_2) \vee \bar{x}_4x_3\bar{x}_2\bar{x}_1$

$f_1 = (\bar{x}_1 \vee x_4 x_3 \bar{x}_2 \vee x_4 \bar{x}_3 x_2)(x_1 \vee x_3 \bar{x}_2 \vee x_4 \bar{x}_3 x_2)$
$f_1 = (\bar{x}_2 \vee x_4 \bar{x}_3)(x_2 \vee x_3 \bar{x}_1 \vee x_4 x_3)$
$f_1 = (\bar{x}_3 \vee \bar{x}_2 \bar{x}_1 \vee x_4 \bar{x}_2)(x_3 \vee x_4 x_2)$
$f_1 = (\bar{x}_4 \vee x_3 \bar{x}_2 \vee \bar{x}_3 x_2)(x_4 \vee x_3 \bar{x}_2 \bar{x}_1)$.

2.5 (a) $f_1 = 0 \cdot \bar{x}_1 \bar{x}_2 \vee 0 \cdot \bar{x}_1 x_2 \vee 1 \cdot x_1 \bar{x}_2 \vee x_3 \cdot x_1 x_2$
$f_1 = 0 \cdot \bar{x}_1 \bar{x}_2 \bar{x}_3 \vee 0 \cdot \bar{x}_1 \bar{x}_2 x_3 \vee 0 \cdot \bar{x}_1 x_2 \bar{x}_3 \vee 0 \cdot \bar{x}_1 x_2 x_3 \vee 1 \cdot x_1 \bar{x}_2 \bar{x}_3 \vee$
$1 \cdot x_1 \bar{x}_2 x_3 \vee x_4 \cdot x_1 x_2 \bar{x}_3 \vee x_4 \cdot x_1 x_2 x_3$.

2.7 $d(\dot{x}, \tilde{x}) = \sum_{i=1}^{n} (\bar{\dot{x}}_i \tilde{x}_i \vee \dot{x}_i \bar{\tilde{x}}_i)$.

2.9 (a) $\dot{x}_i = 0$ if neither x_i nor $\bar{x}_i$ participate in the p-term, $\dot{x}_i = 1$ if x_i is in the p-term, and $\dot{x}_i = 2$ if $\bar{x}_i$ is in the p-term.

2.10 (a) $t = 33, 25, 57, 2, 40$.

2.13 (a) 2^n, (c) $3^n - 1$.

2.14 (b) $\bar{x}_3 \bar{x}_2 x_1 \vee \bar{x}_3 x_2 \bar{x}_1 \vee x_3 \bar{x}_2 \bar{x}_1$.

2.15 (a) 2^{n-m}.

2.18 (b) $M_3 = ((x_2 x_1) \not\equiv (x_3 x_1)) \not\equiv (x_3 x_2)$.

2.19 (a) $f_1 = \bar{x}_3 \bar{x}_2 \bar{x}_1 \vee \bar{x}_3 \bar{x}_2 x_1 \vee \bar{x}_3 x_2 \bar{x}_1 \vee x_3 \bar{x}_2 x_1 \vee x_3 x_2 \bar{x}_1$
$f_1 = (x_3 \vee \bar{x}_2 \vee \bar{x}_1)(\bar{x}_3 \vee x_2 \vee x_1)(\bar{x}_3 \vee \bar{x}_2 \vee \bar{x}_1)$.

2.21 (a) Domain: $s = 0, 1, 2, 3, 6, 7$.

CHAPTER 3

Section 3.2

3.2-4 (a) $f_1 = YX \vee \bar{Y}x \vee \bar{Y}\bar{X}\bar{y}$
$f_1 = (\bar{Y} \vee X)(Y \vee \bar{X} \vee x)(Y \vee \bar{y} \vee x)$.
(e) $f_5 = Xy \vee \bar{Y}x \vee \bar{Y}y$
$f_5 = (y \vee x)(\bar{Y} \vee X)$.

3.2-5 (a)

Critical Vertex	Cell	A Note
H0	H0	Theorem 3.1
E1	E15	
E2	E26	
G3	CG3	
B0	B0246	
F3	BF37	
C7	C67	Theorem 3.2
C4	CD4	
D3	D13	
D5	DH5	
G6	EFGH6	
H7	FH67	

Section 3.3

3.3-3 (a) $\bar{x}_2\bar{x}_1, \bar{x}_3\bar{x}_1, x_4\bar{x}_1, x_4x_3x_2, \bar{x}_4x_3\bar{x}_2, \bar{x}_4\bar{x}_3x_2$.

(b) $x_2\bar{x}_1, \bar{x}_2x_1, x_3, x_4\bar{x}_1, x_4\bar{x}_2, \bar{x}_4x_1, \bar{x}_4x_2$.

(f) $\bar{x}_3x_2, x_4\bar{x}_3x_1, \bar{x}_4x_3\bar{x}_2x_1, x_5\bar{x}_4x_2x_1, x_5\bar{x}_4x_3x_1, x_5\bar{x}_4x_3\bar{x}_2, \bar{x}_5\bar{x}_3x_1, \bar{x}_5\bar{x}_4\bar{x}_2x_1$.

(g) $\bar{x}_4\bar{x}_2, \bar{x}_4\bar{x}_1, \bar{x}_2x_1, \bar{x}_3x_2\bar{x}_1, x_4\bar{x}_3x_2, x_4\bar{x}_3x_1$.

Section 3.4

3.4-3 $f = x_3\bar{x}_1 \vee \bar{x}_3x_1 \vee x_4x_3\bar{x}_2 \vee \bar{x}_4x_2x_1$ (continuation of Exercise 3.3-3c).
$f = \bar{x}_4\bar{x}_3\bar{x}_1 \vee \bar{x}_4\bar{x}_3x_2 \vee \bar{x}_4x_2\bar{x}_1 \vee \bar{x}_3x_2\bar{x}_1 \vee x_3\bar{x}_2x_1 \vee x_4\bar{x}_2x_1 \vee x_4x_3x_1$ (continuation of Exercise 3.3-3d).

Section 3.5

3.5-2 (b) $f_1 = x_1$
$f_2 = \bar{x}_2x_1 \vee x_2\bar{x}_1$
$f_3 = x_3\bar{x}_2\bar{x}_1 \vee \bar{x}_3x_1 \vee \bar{x}_3x_2$
$f_4 = x_4 \vee x_3x_1 \vee x_3x_2$.

(c) $f_1 = x_4\bar{x}_3 \vee \bar{x}_4\bar{x}_3\bar{x}_2 \vee x_4\bar{x}_2x_1 \vee x_4x_2\bar{x}_1$
$f_2 = x_4x_3 \vee \bar{x}_4\bar{x}_3\bar{x}_2 \vee x_4\bar{x}_2x_1 \vee x_4x_2\bar{x}_1$.

Section 3.6

3.6-4 (a) $f = x_3x_2x_1 \vee \bar{x}_1(\bar{x}_3 \vee \bar{x}_2)$.

(b) $f = x_1(x_3 \vee x_2) \vee \bar{x}_4x_3x_2$

(c) $f = \bar{x}_5\bar{x}_3\bar{x}_2 \vee \bar{x}_4(\bar{x}_5 \vee \bar{x}_3)(\bar{x}_2 \vee x_1)$.

(d) $f = \bar{x}_1(\bar{x}_2 \vee x_3)(x_2 \vee \bar{x}_3) \vee x_1(\bar{x}_2 \vee \bar{x}_3)(x_2 \vee x_3)$.

3.6-5 $f = \bar{x}_4x_3x_2 \vee x_1(x_3 \vee x_2)$
$f = x_2x_1 \vee \bar{x}_4x_3(x_2 \vee x_1)$
$f = x_3x_1 \vee \bar{x}_4x_2(x_3 \vee x_1)$
$f = x_2x_1 \vee x_3(x_1 \vee \bar{x}_4x_2)$
$f = x_3x_1 \vee x_2(x_1 \vee \bar{x}_4x_3)$
$f = (x_3 \vee x_2x_1)(x_1 \vee \bar{x}_4x_2)$
$f = (x_2 \vee x_3x_1)(x_1 \vee \bar{x}_4x_3)$
$f = (x_3 \vee x_1)(\bar{x}_4x_2 \vee x_3x_1)$
$f = (x_2 \vee x_1)(\bar{x}_4x_3 \vee x_2x_1)$.

Comprehensive Exercises to Chapter 3

3.5 (a) $(\bar{x}_1|\bar{x}_1)|(x_2|\bar{x}_3)|(\bar{x}_2|x_3)$ or $x_1|(x_2|\bar{x}_3)|(\bar{x}_2|x_3)$.

(b) $(\bar{x}_1|x_2|x_3|\bar{x}_4)|(\bar{x}_1|x_2|x_3|\bar{x}_4)$.

3.6 (a) $x_1|(x_2|(x_3|x_3)) \mid ((x_2|x_2) \mid x_3)$.

(b) $((x_1|x_1)|x_2|x_3(x_4|x_4))|((x_1|x_1)|x_2|x_3|(x_4|x_4))$.

3.8 (a) $(\bar{x}_1 \downarrow x_2 \downarrow x_3) \downarrow (\bar{x}_3 \downarrow \bar{x}_2)$.

(b) $(x_2 \downarrow \bar{x}_3) \downarrow (x_1 \downarrow x_1)$ or $(x_2 \downarrow \bar{x}_3) \downarrow x_1$.

(c) $(x_1 \downarrow x_2) \downarrow (x_1 \downarrow x_2)$.

CHAPTER 4

Section 4.1

4.1-1 $\bar{x}yz \vee xy\bar{z} \vee \bar{x}\bar{y}z \vee x\bar{y}\bar{z} = 1$ or, after simplification, $\bar{x}z \vee x\bar{z} = 1$. The equation is satisfied only if $z = \bar{x}$.

4.1-2 (b) $2^{m \cdot 2^n}$.

(d) The equation has two solutions: $z^1 = y$, $z^2 = y \vee x$.

Section 4.2

4.2-2 No. Only the second form expresses the fact that the equations are expected to hold simultaneously.

4.2-4 (a) No.

(b) No.

Give reasons for the answers!

Section 4.3

4.3-1 Function $\bar{h}(x_1, x_2, \ldots, x_n)$ represents the domain.

4.3-2 (a) ${}^1y_1 = 0$, ${}^1y_2 = x_2 \vee x_1$, ${}^2y_1 = \bar{x}_2\bar{x}_1$, ${}^2y_2 = x_2 \vee x_1$.

(d) There are 16 solutions in the whole domain. One of the solutions is $y_2 = 0, y_1 = 0$.

Section 4.4

4.4-1 All cases except the case when (4.1) contains one equation.

4.4-2 $A_r = \prod_{c=0}^{2^n-1} b_c$, where $b_c = a_c$ if $a_c \neq 0$ and $b_c = 2^m$ if $a_c = 0$.

(A_r is the number of different solutions in the restricted domain.)

4.4-3 (a) $y_1 = 0$ with the domain restricted by the equation $\bar{x}_1\bar{x}_2 = 0$.

Section 4.5

4.5-2 (a) $y_1 = x_1, y_2 = y_3 = 0$.

(b) $y_1 = \bar{x}_1, y_2 = \bar{x}_2$.

4.5-3 2^{p+q}.

Section 4.6

4.6-1 (a) $(\bar{f}_1\bar{g}_1 \vee f_1g_1)(\bar{f}_2\bar{g}_2 \vee f_2g_2)\overline{(\bar{f}_3\bar{g}_3 \vee f_3g_3)} \vee$
$\overline{(\bar{f}_1\bar{g}_1 \vee f_1g_1)(\bar{f}_2\bar{g}_2 \vee f_2g_2)}(\bar{f}_3\bar{g}_3 \vee f_3g_3)$.

(b) $\overline{\bar{f}_1g_1} \cdot \overline{\bar{f}_2g_2} \cdot \overline{\bar{f}_3g_3} \vee \bar{f}_1g_1 \vee \bar{f}_2g_2 \vee \bar{f}_3g_3$.

4.6-4 (a) $y_1 = 0$ with the domain restricted by the equation $\bar{x}_1\bar{x}_2 \vee x_1x_2 = 1$.

Section 4.7

4.7-2 (a) One solution $y_1 = x_1 x_2$ in the whole domain of the logic space (x_1, x_2).
4.7-4 (a) One minimum for $x_5 = x_4 = x_3 = x_2 = x_1 = 1$, where $f = -7$, one maximum for $x_5 = x_4 = x_3 = x_2 = x_1 = 0$, where $f = 0$.
(b) One minimum for $x_5 = x_3 = 0$ and $x_4 = x_2 = x_1 = 1$, where $f = -7$, and one maximum for $x_4 = x_3 = x_1 = 0$, and $x_5 = x_2 = 1$, where $f = 7$.

Section 4.8

4.8-2 (a) 256 decompositions; $y_1 = x_1$, $y_2 = x_2 x_3$.
4.8-4 (a) $f = (x_3 \vee x_2)(x_1 \vee \bar{x}_2)$.
(b) $f = (\bar{x}_3 \vee \bar{x}_2)(x_3 \vee x_2)$.

Comprehensive Exercises to Chapter 4

4.4 $y_1 = \bar{x}_1$, $y_2 = \bar{x}_1 + \bar{x}_2$.

CHAPTER 5

Section 5.2

5.2-1 $y_1(t_0) = \bar{y}_2(t_0) = \bar{z}(t_0)$
$z(t_0 + \Delta t) = z(t_0) \cdot v(t) \vee x_1(t)$.

Section 5.3

5.3-1 Complete set of logic functions is obtained either for *HL* or for *LH*. To construct a general logic function, at least two logic elements must be connected in series. The logic value at the output of the first element is then the opposite of that required at the input of the second element.
5.3-2 $y(t_0 + \Delta t) = e_1(t) \vee x_1(t) \vee x_2(t) \vee \cdots$ for Figure 5.1a and $y(t_0 + \Delta t) = e_1(t) \cdot x_1(t) \cdot x_2(t) \cdots$ for Figure 5.1b provided that all propositions are of *H* type. The student should derive operators for other combinations of the types of propositions.

Section 5.4

5.4-1 It is permissible to apply signals of value 1 simultaneously to both inputs of the relay, i.e., $x_1 = 1 = x_2$; with a bi-stable element this is impossible. The operators are identical for all other states, as can be seen from the relevant chart.

Section 5.5

5.5-1 The operators remain unchanged.

5.5-2 $y = e \cdot (\overline{-e_b})(x_1 \vee x_2 \vee \cdots) \vee e \cdot (-e_b)$.

Section 5.6

5.6-1 *PN*, current operated: $z(t_0 + \Delta t) = z(t_0)\overline{{}^0x(t)} \vee \overline{{}^0x(t)} \cdot {}^1x(t)$

$y(t_0) = x(t) \cdot z(t_0)$

PN, voltage operated: $z(t_0 + \Delta t) = z(t_0)\overline{{}^0x(t)} \vee \overline{{}^0x(t)} \cdot {}^1x(t)$

$y(t_0) = \bar{z}(t_0) \cdot x(t)$

PP, current operated: $z(t_0 + \Delta t) = z(t_0) \vee {}^1x_1(t) \vee {}^1x_2(t)$

$y(t_0) = x(t) \cdot z(t_0)$

PP, voltage operated: $z(t_0 + \Delta t) = z(t_0) \vee {}^1x_1(t) \vee {}^1x_2(t)$.

$y(t_0) = \bar{z}(t_0) \cdot x(t)$

5.6-2 $y = \bar{a}$

$y = 0$

$y = a$

$y = a \vee b$

$y(t_0 + 2\Delta t) = \bar{y}(t_0)b(t) \vee \bar{a}(t)$.

Section 5.7

5.7-1 $z(t_0 + \Delta t) = \bar{x}_1(t)\bar{x}_2(t)$

$y(t_0) = z(t_0)$.

CHAPTER 6

Section 6.1

6.1-6 (b) $y_1 = \bar{x}_1, y_2 = \bar{x}_2 \vee \bar{x}_1$.

6.1-7 $y_1 = \bar{x}_4\bar{x}_2 \vee y_2\bar{x}_1$

$y_2 = x_4\bar{x}_3x_2 \vee y_1\bar{x}_1$.

Section 6.4

6.4-1 (a) NAND[NAND($\bar{x}_2, \bar{x}_4$), NAND($x_3, \bar{x}_4$), NAND($\bar{x}_1, x_2, x_3$)].

(b) NOR[NOR($x_2, \bar{x}_4$), NOR($\bar{x}_2, x_3$), NOR($\bar{x}_1, \bar{x}_4$)].

Section 6.5

6.5-2 13 NAND gates.

6.5-4 $2^{n-1} - 1$.

Section 6.7

6.7-3 (a) $f_A = f_{AB} \vee f_{AC}$, $f_B = f_{AB} \vee f_{BC}$, $f_C = f_{AC} \vee f_{BC}$.

Comprehensive Exercises to Chapter 6

6.10 $F = (x_2 \vee x_1)\overline{x_4 x_3 x_1}$

6.12 (b) No. It preserves both 0 and 1.

CHAPTER 7

Section 7.2

7.2-4 $(qr)^{pr}$.

7.2-6 (a) $F(012012012, 4) = 001001001$.
(c) $G(2222222, 5) = 3$.
(f) $L(12120001212, 5) = 1$.

Section 7.3

7.3-1 The minimal form of M_4 is

z^t \ x^t	y^t 0	y^t 1	z^{t+1} 0	z^{t+1} 1
a	0	1	b	a
3	0	1	a	3
4	0	1	3	a
5	0	1	4	a
b	1	1	5	b

$a = \{1, 2\}$
$b = \{6, 7\}$

7.3-12 $\sum_{i=1}^{n} r_i(r_i - 1)/2$.

7.3-13 (a) Minimum is $m + 1$, maximum is r.

Section 7.4

7.4-1 The minimal form of the machine in Figure 7.25c is:

z^t \ x^t	y^t 0	y^t 1	z^{t+1} 0	z^{t+1} 1
a	0	0	a	b
b	1	0	b	a

$a = \{1, 2, 5\}$
$b = \{3, 4\}$

7.4-2 The minimal form of the machine in Figure 7.26a is:

x^t / z^t	y^t		z^{t+1}	
	a	B	a	B
1	0	0	II	1
II	1	1	1	II

II = {2, 3}
$B = \{b, c\}$

Section 7.5

7.5-1 The ST-diagram represents the minimal form.

7.5-3

x^t / z^t	$y_2^t y_1^t$		z^{t+1}	
	0	1	0	1
0	00	01	0	1
1	01	10	1	2
2	10	11	2	3
3	11	00	3	0

7.5-4

$x_2^t x_1^t$ / z^t	$y_2^t y_1^t$				z^{t+1}			
	00	01	10	11	00	01	10	11
0	00	01	11	00	0	1	3	0
1	01	10	00	01	1	2	0	1
2	10	11	01	10	2	3	1	2
3	11	00	10	11	3	0	2	3

Section 7.6

7.6-1 $Z_{1,1} = (0 \cup 10^*1)^*$
$Z_{1,2} = (0 \cup 10^*1)^*10^*$.

7.6-4 $Y_{R,i} = \bigcup Z_{R,j}$, where the union is taken for all internal states z_j such that $f(z_j) = y_i$.

Comprehensive Exercises to Chapter 7

7.4

z^t \ x^t	y^t		z^{t+1}	
	0	1	0	1
0	1	1	2	2
1	1	1	3	3
2	0	0	4	4
3	0	0	0	0
4	0	0	1	1

7.9 (a) At least one 1 which may be preceded by an arbitrary number of 0's and is always followed by a single 0.

CHAPTER 8

Section 8.3

8.3-1 (b) $v^t = w^t\bar{H}_1 \vee \bar{w}^t H_2$.
(c) $v_1^t = H_2, v_2^t = \bar{H}_1$.

Section 8.4

8.4-1 $Z^t = Z^{t-1}\bar{x}^t \vee \bar{Z}^{t-1}x^t$
$y^t = Z^{t-1}\bar{x}^t \vee \bar{Z}^{t-1}x^t (=Z^t)$.

Section 8.8

8.8-2 (a) $2^{n_1+p} - T$, where T is the total number of transitions.
(b) $2^{n_1+p+q} - S$, where S is the total number of different input-output sequences of length 2.

8.8-3 (c) First order finite-state model: $\#D = 3$, $\#I = 4$, $\#O = 4$. Second order finite-state model: $\#D = 4$, $\#I = 5$, $\#O = 3$. Third order finite-state model: $\#D = 3$, $\#I = 4$, $\#O = 2$ (its applicability must be tested). First order combined model: $\#D = 3$, $\#I = 4$, $\#O = 2$. Third order combined model (finite-memory model): $\#D = 6$, $\#I = 7$, $\#O = 1$.

Section 8.9

8.9-1 (a) For the machine in Figure 8.41e, there exist four nontrivial closed partitions of internal states:

$$\pi_1 = \{\overline{1, 2}; \overline{3, 4}; \overline{5, 6}; \overline{7, 8}\}$$
$$\pi_2 = \{\overline{1, 2, 3, 4}; \overline{5, 6, 7, 8}\}$$
$$\pi_3 = \{\overline{1, 4}; \overline{2, 3}; \overline{5, 8}; \overline{6, 7}\}$$
$$\pi_4 = \{\overline{1, 2, 7, 8}; \overline{3, 4, 5, 6}\}.$$

8.9-4 k nontrivial serial decompositions; no parallel decomposition is guaranteed.

Section 8.10

8.10-1 For the machine in Figure 8.47a, we get

States	$z_2 z_1$
1	0 0
2	1 1
3	0 1
4	1 0

Comprehensive Exercises to Chapter 8

8.4 Any activity that contains all input-output sequences of length 2 compiled from the ST-structure of the serial binary adder.

8.6 (a) One possible activity matrix:

$t =$	0	1	2	3	4	5	6	7	8	9	10	11	12	13
$v_1^t =$	0	0	0	0	0	1	0	1	1	0	0	1	0	0
$v_2^t =$	0	0	1	1	0	0	1	0	0	0	0	0	0	1
$w^t =$	0	0	0	0	0	1	0	1	1	1	1	1	1	0

Implementation: $w^t = v_1^t \vee w^{t-1}\bar{v}_2^t$.

8.9 $n_1 = \ulcorner \log_2 m$ for both the models.

CHAPTER 9

Section 9.3

9.3-5 For the machine in Figure 9.6e, we have:

States	z_2 z_1
1	0 0
2	0 1
3	1 0
4	1 1

Section 9.4

9.4-3 (b) $f = x_3\bar{x}_1 \vee x_3 x_2 \vee x_2 x_1$

CHAPTER 10

Section 10.5

10.5-1 (a) $P'_{0,0} = \begin{bmatrix} .2 & .4 \\ .1 & .2 \end{bmatrix}$ $P'_{0,1} = \begin{bmatrix} .2 & .2 \\ .4 & .3 \end{bmatrix}$

$P'_{1,0} = \begin{bmatrix} .3 & .2 \\ .4 & .2 \end{bmatrix}$ $P_{1,1} = \begin{bmatrix} .2 & .3 \\ .3 & .1 \end{bmatrix}$

10.5-2 (a) $\Phi(1101) = [.00111, .00579, .9931,]$
$\Phi(0101) = [.00054, .00318, .99628]$.
(c) $\mathscr{P}_\Phi(01 \mid 10) = .281$, $\mathscr{P}(11 \mid 00) = .382$.

Comprehensive Exercises to Chapter 10

10.2 (a) $1 - 0.5^n$.
(b) $(1 - 0.5^n)^k$.

APPENDIX G
SELECTED COMPUTER PROGRAMS

SOLUTION OF LOGIC AND/OR PSEUDO-LOGIC RELATIONS

The following APL program *SBR* determines the discriminant and the functions of dependent variables for a given set of logic and/or pseudo-logic relations. Both the discriminant and the functions are printed in the form of matrices which have the same meaning as the Marquand charts used in Chapter 4. The following symbols are used for defining the functions.

0: Logic value "0".
1: Logic value "1".
D: Don't care state.
R: Don't care state with a restriction.
∅: State that does not belong to the domain of the function.

No subroutine is required by the program. To run the program, variables *IND* (the number of independent variables), *DEP* (the number of dependent variables), and *NEQ* (the number of relations) must be specified.

First, after the program is started by printing *SBR*, the user is instructed to assign rows of matrix *S* (generated by the program) to individual independent variables *X*1, *X*2, . . . and dependent variables *Y*1, *Y*2, ... Then he is asked to specify the relations *D*1, *D*2, . . . and the logic function *D* imposed upon the relations.

```
        ∇SBR[⎕]∇

    ∇ SBR
      S←((IND+DEP),2*IND+DEP)ρ0
      I←0
     AA:I←I+1
      S[I;]←(2*IND+DEP)ρ((2*I-1)ρ0),(2*I-1)ρ1
      →(I≠IND+DEP)/AA
      I←0
     SA:I←I+1
      'ENTER: X';I;'←S[';I;';]'
      0ρ⎕
      →(I≠IND)/SA
      I←0
     SB:I←I+1
      'ENTER: Y';I;'←S[';IND+I;';]'
      0ρ⎕
      →(I≠DEP)/SB
      I←0
     SC:I←I+1
      'ENTER RELATION D';I
      0ρ⎕
      →(I≠NEQ)/SC
      'ENTER THE FUNCTION OF THE RELATIONS'
      D←(2*DEP,IND)ρ⎕
      (4ρ'
      '),'THE DISCRIMINATION FUNCTION D:
      '
      D
      T← 3 3 ρ 4 1 1 0 2 3 0 3 3
      S←5I0
      LGY←1=(DEPρ2)⊤¯1+⍳2*DEP
      FY←(DEP,(2*⌊IND÷2),2*⌈IND÷2)ρ0
      I←0
     L1:I←I+1
      J←0
     L2:J←J+1
      K←0
     L3:K←K+1
      N←+/(~LGY[K;])∧D[;J+(I-1)×2*⌈IND÷2]
      A←+/LGY[K;]∧D[;J+(I-1)×2*⌈IND÷2]
      AUX←((N=0,2*DEP-1),((N>0)∧N<2*DEP-1))/[1] T
      FY[K;I;J]←((A=0,2*DEP-1),((A>0)∧A<2*DEP-1))/,AUX
      →(K<DEP)/L3
      C←3=FY[;I;J]
      →(0=+/C)/L6
      L←0
      S←(2,DEP,3*DEP)ρ0
     LX:L←L+1
      S[;L;]←(2ρ2)⊤(3*DEP)ρ((3*DEP-L)ρ0),((3*DEP-L)ρ1),(3*DEP-L)ρ2
      →(L≠DEP)/LX
      L←1
     L4:L←L+1
      MASK←(∧/[1](1=2⊥S[;;L])/[1] LGY)∧∧/[1](2=2⊥S[;;L])/[1]~LGY
      →((+/[1] MASK=D[;J+(I-1)×2*⌈IND÷2])=2*DEP)/L5
      →(L<3*DEP)/L4
      →L6
     L5:FY[;I;J]←FY[;I;J]-C
     L6:→(J<2*⌈IND÷2)/L2
      →(I<2*⌊IND÷2)/L1
      '
```

```
      THE SOLUTIONS FOR THE DEPENDENT VARIABLES:
      '
[57]  TC←'01DRQ'
[58]  I←0
[59] L7:I←I+1
[60]  J←0
[61] L8:J←J+1
[62]  →(J=⌈(2*⌊IND÷2)÷2)/L9
[63]  ((3×2*⌈IND÷2)⍴ 0 1 0)\TC[1+FY[I;J;]]
[64]  →L9+1
[65] L9:(((3×2*⌈IND÷2)⍴ 0 1 0)\TC[1+FY[I;J;]]),'  =  Y';1+DEP-I
[66]  →(J≠2*⌊IND÷2)/L8
[67]  '

      '
[68]  →(I≠DEP)/L7
```

```
       IND←3
       DEP←3
       NEQ←5
       SBR
ENTER: X1←S[1;]
⎕:
       X1←S[1;]

ENTER: X2←S[2;]
⎕:
       X2←S[2;]

ENTER: X3←S[3;]
⎕:
       X3←S[3;]

ENTER: Y1←S[4;]
⎕:
       Y1←S[4;]

ENTER: Y2←S[5;]
⎕:
       Y2←S[5;]

ENTER: Y3←S[6;]
⎕:
       Y3←S[6;]

ENTER RELATION D1
⎕:
       D1←((X3∧((Y3∧(~Y1))∨(Y2∧(~X1))))∨X2)=((Y3∧(~X3))∨(X1∧(~Y2)))

ENTER RELATION D2
⎕:
       D2←1=((~X3)∧(X2∧X1))∨(X3∧X2∧Y3∧(~Y2))∨((~X3)∧X1∧(~Y3)∧(~Y2))

ENTER RELATION D3
⎕:
       D3←1=(X3∧(~X2)∧(~Y3)∧Y2)∨((~X3)∧(~X2)∧(~X1)∧(~Y3)∧(~Y1))
```

```
ENTER RELATION D4
⎕:
      D4←1=(X3∧Y3∧(~Y2)∧Y1)∨((~Y3)∧Y1∧X3∧(~X2)∧(~X1))

ENTER RELATION D5
⎕:
      D5←1=(~Y3)∧Y2∧(~X3)∧X2

ENTER THE FUNCTION OF THE RELATIONS
⎕:
      D←D1≠D2∨D3∨D4∨D5


THE DISCRIMINATION FUNCTION D:

 0 1 0 0 1 0 0 1
 1 1 0 0 0 0 0 1
 0 1 1 1 1 0 0 0
 1 1 1 1 1 0 0 0
 0 0 1 0 0 1 1 0
 0 0 1 0 0 1 1 0
 0 0 1 0 0 0 0 0
 0 0 1 0 0 1 0 0


THE SOLUTIONS FOR THE DEPENDENT VARIABLES:

 0  0  R  0    =  Y3
 0  1  1  0


 D  D  R  1    =  Y2
 R  R  0  0


 1  D  R  D    =  Y1
 R  R  D  D
```

```
ENTER RELATION D1
⎕:
      D1←Y1=X1∧(~Y3)

ENTER RELATION D2
⎕:
      D2←Y2=X2∧(~Y1)

ENTER RELATION D3
⎕:
      D3←Y3=X3∧(~Y2)

ENTER THE FUNCTION OF THE RELATIONS
⎕:
      D←D1∧D2∧D3
```

```
THE DISCRIMINATION FUNCTION D:

 1  0  0  0  0  0  0  0
 0  1  0  1  0  0  0  0
 0  0  1  0  0  0  1  0
 0  0  0  0  0  0  0  0
 0  0  0  0  1  1  0  0
 0  0  0  0  0  0  0  0
 0  0  0  0  0  0  0  0
 0  0  0  0  0  0  0  0

THE SOLUTIONS FOR THE DEPENDENT VARIABLES:

 0   0   0   0       =   Y3
 1   1   0   ⍉


 0   0   1   0       =   Y2
 0   0   1   ⍉


 0   1   0   1       =   Y1
 0   0   0   ⍉
```

PSEUDO-BOOLEAN PROGRAMMING

The following APL program *PBP* determines values of a given pseudo-Boolean objective function subject to some constraints expressed by a set of logic and/or pseudo-logic relations, and identifies the maxima and minima of the objective function.

No subroutine is required by the program. To run the program, variables *N* (the number of variables) and *M* (the number of relations) must be specified.

First, after the program is started by printing *PBP*, the user is instructed to assign rows of matrix *S* (generated by the program) to individual variables *X*1, *X*2, ... in the natural order. Then, he is asked to specify the relations, the logic function imposed upon the relations, and the pseudo-Boolean objective function.

```
      ∇PBP[⎕]∇

    ∇ PBP
[1]   S←(N,2*N)⍴0
[2]   I←0
[3]  L1:I←I+1
[4]   S[I;]←(2*N)⍴((2*I-1)⍴0),(2*I-1)⍴1
[5]   →(I≠N)/L1
[6]   I←0
[7]   'ENTER: XI←S[I;], I=1,2,...,';N
```

```
[8]   L2:I←I+1
[9]    0⍴⎕
[10]   →(I≠N)/L2
[11]   I←0
[12]  L3:I←I+1
[13]   'ENTER DISCRIMINATION FUNCTION D';I
[14]   0⍴⎕
[15]   →(I≠M)/L3
[16]   'ENTER DISCRIMINATION FUNCTION D'
[17]   D←((2*⌈N÷2),(2*⌊N÷2))⍴⎕
[18]   'ENTER OBJECTIVE FUNCTION H'
[19]   H←((2*⌈N÷2),(2*⌊N÷2))⍴⎕
[20]   H←H×D
[21]   MAX←⌈/⌈/H
[22]   MIN←⌊/⌊/H
[23]   '

       THE DISCRIMINATION FUNCTION:
       '
[24]   D
[25]   '

       THE OBJECTIVE FUNCTION:
       '
[26]   H
[27]   '
       THE MAXIMUM = ';MAX
[28]   '
       THE MINIMUM = ';MIN
     ∇



       N←8
       M←3
       PBP
ENTER: XI←S[I;], I=1,2,....,8
⎕:
       X1←S[1;]

⎕:
       X2←S[2;]

⎕:
       X3←S[3;]

⎕:
       X4←S[4;]

⎕:
       X5←S[5;]

⎕:
       X6←S[6;]

⎕:
       X7←S[7;]

⎕:
       X8←S[8;]

ENTER DISCRIMINATION FUNCTION D1
⎕:
       D1←(X4+X3)>((2×X2)-3×X1)
```

```
ENTER DISCRIMINATION FUNCTION D2
⎕:
      D2←(X1∨(~X8))=(X2∨X3∨X4∨(~X7))

ENTER DISCRIMINATION FUNCTION D3
⎕:
      D3←(X5≠X6)≤(X1≠X8)

ENTER DISCRIMINATION FUNCTION D
⎕:
      D←(D1∧D2)∨(D1∧D3)∨(D2∧D3)
ENTER OBJECTIVE FUNCTION H
⎕:
      H←(2×((~X1)∧X2))+(5×(X5∨X3))-(7×X4)+(X6∨X7∨X8)∧(X1∨(~X3)∨X5)
```

```
THE DISCRIMINATION FUNCTION:

 1 1 1 1 1 1 1 1 1 1 1 1 1 1 1 1
 0 1 0 1 1 1 0 1 1 1 0 1 1 1 0 1
 0 1 0 1 1 1 0 1 1 1 0 1 1 1 0 1
 1 1 1 1 1 1 1 1 1 1 1 1 1 1 1 1
 0 1 1 1 1 1 1 1 1 1 1 1 1 1 1 1
 0 1 0 1 1 1 0 1 1 1 0 1 1 1 0 1
 0 1 0 1 1 1 0 1 1 1 0 1 1 1 0 1
 0 1 1 1 1 1 1 1 1 1 1 1 1 1 1 1
 0 1 0 1 1 1 0 1 1 1 0 1 1 1 0 1
 0 1 0 1 1 1 0 1 1 1 0 1 1 1 0 1
 0 1 0 1 1 1 0 1 1 1 0 1 1 1 0 1
 0 1 0 1 1 1 0 1 1 1 0 1 1 1 0 1
 1 1 0 1 1 1 0 1 1 1 0 1 1 1 0 1
 1 0 0 1 1 1 0 1 1 1 0 1 1 1 0 1
 1 0 0 1 1 1 0 1 1 1 0 1 1 1 0 1
 1 1 0 1 1 1 0 1 1 1 0 1 1 1 0 1
```

```
THE OBJECTIVE FUNCTION:

  0  0  2  0  5  5  7  5 ¯7 ¯7 ¯5 ¯7 ¯2 ¯2  0 ¯2
  0  5  0  5  5  5  0  5 ¯2 ¯2  0 ¯2 ¯2 ¯2  0 ¯2
  0 ¯1  0 ¯1  5  4  0  4 ¯8 ¯8  0 ¯8 ¯2 ¯3  0 ¯3
  4  4  6  4  4  4  6  4 ¯3 ¯3 ¯1 ¯3 ¯3 ¯3 ¯1 ¯3
  0 ¯1  1 ¯1  5  4  7  4 ¯8 ¯8 ¯6 ¯8 ¯2 ¯3  0 ¯3
  0  4  0  4  4  4  0  4 ¯3 ¯3  0 ¯3 ¯3 ¯3  0 ¯3
  0 ¯1  0 ¯1  5  4  0  4 ¯8 ¯8  0 ¯8 ¯2 ¯3  0 ¯3
  0  4  6  4  4  4  6  4 ¯3 ¯3 ¯1 ¯3 ¯3 ¯3 ¯1 ¯3
  0 ¯1  0 ¯1  5  4  0  4 ¯8 ¯8  0 ¯8 ¯2 ¯3  0 ¯3
  0  4  0  4  4  4  0  4 ¯3 ¯3  0 ¯3 ¯3 ¯3  0 ¯3
  0 ¯1  0 ¯1  5  4  0  4 ¯8 ¯8  0 ¯8 ¯2 ¯3  0 ¯3
  0  4  0  4  4  4  0  4 ¯3 ¯3  0 ¯3 ¯3 ¯3  0 ¯3
 ¯1 ¯1  0 ¯1  5  4  0  4 ¯8 ¯8  0 ¯8 ¯2 ¯3  0 ¯3
  4  0  0  4  4  4  0  4 ¯3 ¯3  0 ¯3 ¯3 ¯3  0 ¯3
 ¯1  0  0 ¯1  5  4  0  4 ¯8 ¯8  0 ¯8 ¯2 ¯3  0 ¯3
  4  4  0  4  4  4  0  4 ¯3 ¯3  0 ¯3 ¯3 ¯3  0 ¯3

THE MAXIMUM = 7

THE MINIMUM = ¯8
```

STATE MINIMIZATION OF FINITE-STATE DETERMINISTIC MACHINES

The following APL program *STATEMIN* determines the minimal form of a given completely specified Mealy or Moore machine.*) The program uses, essentially, the technique of state minimization by pair tables, which is described in Section 7.3.

The program requires a subroutine called *DFT* (*D*igital *F*ormat *T*able) which is used only for controlling the format of printed ST-tables.

First, after the program is started by printing *STATEMIN*, the user is asked to specify the type of the machine (Moore or Mealy), the number of its internal and input states, and individual rows of the ST-table, each of which must be entered in the order: Present state, outputs, next states (integers are used for representing input, output, and internal states). The given ST-table is then printed with headings and properly spaced. Finally, the partition of internal states into equivalence classes and the ST-table of the minimum form of the machine are printed.

```
      ∇ STATEMIN;I;N;TYPE;PAIRS1;PAIRS2;PAIRS;ADD;ADD1;COMP;CLASS;RHO;J;
        HEAD1;HEAD2;EX;SP;R;LINE;IT;ALPHA
[1]     'MOORE OR MEALY MACHINE?'
[2]     TYPE←⍞
[3]     'HOW MANY INTERNAL STATES?'
[4]     TABLE←⍳0×N←⎕×I←1
[5]     'HOW MANY INPUT STATES?'
[6]     X←0,⍳⎕-1
[7]     'ENTER THE STATE TABLE ONE ROW AT A TIME.'
[8]     TABLE←TABLE,⎕
[9]     →(N≥I←I+1)/¯1+⌶26
[10]    →(∧/TYPE='MOORE')/MOORE
[11]  MEALY:PAIRS1←PAIRS2←⍳I←0×⍴⍴TABLE←(N,1+2×⍴X)⍴TABLE
[12]    PAIRS2←PAIRS2,ADD←((+/TABLE[I+⍳N-I;1+⍳⍴X]=((N-I),⍴X)⍴TABLE[I+1;1+
        ⍳⍴X])∊⍴X)/I+⍳N-I
[13]    PAIRS1←PAIRS1,(⍴ADD)⍴I+1
[14]    →(N>I←I+1)/¯2+⌶26
[15]    I←1+0×⍴⍴PAIRS←⍉(2,(⍴PAIRS1))⍴PAIRS1,PAIRS2
[16]    PAIRS1←PAIRS←PAIRS,[2]((⍴PAIRS[;⍳2])⍴TABLE[,PAIRS[;⍳2];I+1+⍴X])
[17]    →((⍴X)≥I←I+1)/¯1+⌶26
[18]    →REDUCE
[19]  MOORE:PAIRS1←PAIRS2←⍳I←0×⍴⍴TABLE←(N,2+⍴X)⍴TABLE
[20]    PAIRS2←PAIRS2,ADD←((TABLE[I+⍳N-I;2]=(N-I)⍴TABLE[I+1;2])∊1)/I+⍳N-I
[21]    PAIRS1←PAIRS1,(⍴ADD)⍴I+1
[22]    →(N>I←I+1)/¯2+⌶26
[23]    I←1+0×⍴⍴PAIRS←⍉(2,(⍴PAIRS1))⍴PAIRS1,PAIRS2
```

*) The program was prepared by Mr. Leighton B. Brown at the School of Advanced Technology, SUNY at Binghamton.

```
[24]  PAIRS1←PAIRS←PAIRS,[2]((ρPAIRS[;⍳2])ρTABLE[,PAIRS[;⍳2];2+I])
[25]  →((ρX)≥I←I+1)/¯1+⌶26
[26] REDUCE:J←1
[27]  COMP←⍳0×I←1
[28]  COMP←COMP,+/((+/PAIRS1[I;(2×J)+⍳2]⍳PAIRS1[; 1 2])∊3),PAIRS1[I;1+
      2×J]=PAIRS1[I;2+2×J]
[29]  →((ρPAIRS1[;1])≥I←I+1)/¯1+⌶26
[30]  PAIRS1←COMP/[1] PAIRS1
[31]  →((ρX)≥J←J+1)/¯4+⌶26
[32]  →((ρPAIRS[;1])=ρPAIRS1[;1])/FCLASS
[33]  →REDUCE+0×ρρPAIRS←PAIRS1
[34] FCLASS:CLASS←RHO←⍳0×ρρPAIRS←PAIRS[; 1 2]
[35]  ADD←⍳0×ρADD1←PAIRS[1;]
[36]  ADD1←ADD1,ADD←,(COMP←∨/PAIRS∊ADD1)/[1] PAIRS
[37]  PAIRS←(~COMP)/[1] PAIRS
[38]  →((ρADD)>0)/¯2+⌶26
[39]  I←1
[40]  ADD1←ADD1[I],(ADD1≠ADD1[I])/ADD1
[41]  →((ρADD1)≥I←I+1)/¯1+⌶26
[42]  CLASS←CLASS,ADD1←ADD1[⍋ADD1]
[43]  RHO←RHO,ρADD1
[44]  →((ρPAIRS[;1])>0)/¯9+⌶26
[45]  'IF THE ORIGINAL MACHINE IS:
      '
[46]  →(∧/TYPE='MEALY')/MEALYOUT
[47]  IT←8-N←⌈/((ρX)> 0 13 15 18 22 28)/0,⍳5
[48]  ⎕←HEAD1←((5-N)ρ' '),'Q(T)',((5-N)ρ' '),'Z(T)',((⌊3+
      0.5×IT×¯1+ρX)ρ' '),'Q(T+1)'
[49]  ⎕←HEAD2←((18-3×N)ρ' '),'X(T)=',(7-N)↓EX←(IT,0) DFT X
[50]  ' '
[51]  →TABLEOUT
[52] MEALYOUT:IT←8-N←⌈/((ρX)> 0 7 8 9 11 14)/0,⍳5
[53]  ⎕←HEAD1←((5-N)ρ' '),'Q(T)',((⌊0.5×IT+IT×¯1+ρX)ρ' '),'Z(T)',((⌊(-
      IT×0.5)+IT×ρX)ρ' '),'Q(T+1)'
[54]  ⎕←HEAD2←((10-2×N)ρ' '),'X(T)=',(7-N)↓EX,EX←(IT,0) DFT X
[55]  ' '
[56] TABLEOUT:(IT,0) DFT TABLE
[57]  '


      THE PARTITION IS:
      '
[58]  I←1+COMP←0
[59]  COMP←COMP,+/RHO[⍳I]
[60]  →((ρRHO)≥I←I+1)/¯1+⌶26
[61]  I←1
[62]  ⎕←CLASS[COMP[I]+⍳COMP[I+1]]
[63]  →((ρCOMP)>I←I+1)/¯1+⌶26
[64]  '


      THE EQUIVALENT MINIMUM FORM IS:
      '
[65]  HEAD1
[66]  HEAD2
[67]  ' '
[68]  ALPHA←' ABCDEFGHIJKLMNOPQRSTUVWXYZABCDEFGHIJKLMNOPQRSTUVWXYZ'
[69]  J←I←1
[70]  →(∧/TYPE='MEALY')/MFMEALY
[71]  LINE←(SP←(6-N)ρ' '),ALPHA[1 28 + 26 26 ⊤I-1],(IT,0) DFT TABLE[(R
      ←CLASS[COMP[I+1]]);2]
```

```
[72]  LINE←LINE,SP,ALPHA[1 28 + 26 26 ⊤¯1++/(CLASS⍳TABLE[R;2+J])>COMP]
[73]  →((⍴X)≥J←J+1)/¯1+⌶26
[74]  LINE
[75]  →((⍴RHO)≥I←I+J←1)/¯4+⌶26
[76]  →0
[77] MFMEALY:LINE←(SP←(6-N)⍴' '),ALPHA[1 28 + 26 26 ⊤I-1],(IT,0) DFT
      TABLE[(R←CLASS[COMP[I+1]]);1+⍳⍴X]
[78]  LINE←LINE,SP,ALPHA[1 28 + 26 26 ⊤¯1++/(CLASS⍳TABLE[R;J+1+⍴X])>
      COMP]
[79]  →((⍴X)≥J←J+1)/¯1+⌶26
[80]  LINE
[81]  →((⍴RHO)≥I←I+J←1)/MFMEALY
      ∇
```

```
    ∇ Z←W DFT X;D;E;F;G;H;I;J;K;L;Y
[1]   D←' 0123456789.¯'
[2]   →(∨/W≠⌊W←,W+(H←0)×L←1<⍴⍴X)/DFTERR+0×F←2
[3]   →(3 2 1 <⍴⍴X)/(DFTERR+F←0), 2 3 +⌶26
[4]   →(⍴⍴⍴X←((∨/ 1 2 =⍴W)⌽ 1 2)⍉(1,⍴,X)⍴X)/2+⌶26
[5]   X←(0 1 1 /⍴X)⍴X
[6]   →((∧/(⍴W)≠ 1 2 ,2×E←1⍴⌽⍴X),1≠⍴W)/(DFTERR×F←1),3+⌶26
[7]   I←1+⌈/0,,⌊10⍟|X+1>|X
[8]   W←(2+I+W+(W≠0)+∨/,X<0),W
[9]   →(∨/2>-/[1] W←⍉(E,2)⍴W)/DFTERR+0×F←2
[10]  Z←((K←1⍴⍴X),+/W[1;])⍴' '
[11]  X←⌊0.5+X×10*(⍴X)⍴W[2;]
[12] DFTLP:→(E<H←H+1)/DFTEND
[13]  J←1+⌊10|(|Y←X[;H])∘.÷10*¯1+⌽⍳I←W[1;H]
[14]  J←(,J)×G←,⍉(⌽⍴J)⍴(,⍉(J≠1)∨.∧(⍳I)∘.≤⍳I-F+1),(K×1+F←W[2;H])⍴1
[15]  →(∧/0≤Y)/2+⌶26
[16]  J[1+(⍴J)|¯1+(I-+/(K,I)⍴G)+I×¯1+⍳K]←12×Y<0
[17]  J←(K,I)⍴J
[18]  →(0=F)/3+⌶26
[19]  J←J[;(1⌽⍳G),(G←-/W[;H])+⍳F]
[20]  J[;G]←11
[21]  →DFTLP×⍴⍴⍴Z[;(+/W[1;⍳H-1])+⍳I]←D[1+J]
[22] DFTEND:→L/0
[23]  →0×⍴Z←,Z
[24] DFTERR:'DFT ',(3 6 ⍴' RANK LENGTHDOMAIN')[F+1;],' PROBLEM.'
    ∇
```

```
      STATEMIN
MOORE OR MEALY MACHINE?
MOORE
HOW MANY INTERNAL STATES?
⎕:
      4
HOW MANY INPUT STATES?
⎕:
      2
ENTER THE STATE TABLE ONE ROW AT A TIME.
⎕:
      1 0 3 2
⎕:
      2 1 4 1
⎕:
      3 0 1 4
⎕:
      4 1 2 3
```

```
IF THE ORIGINAL MACHINE IS:

      Q(T)       Z(T)          Q(T+1)
                     X(T)=0          1

        1          0         3         2
        2          1         4         1
        3          0         1         4
        4          1         2         3


THE PARTITION IS:

1  3
2  4


THE EQUIVALENT MINIMUM FORM IS:

      Q(T)       Z(T)          Q(T+1)
                     X(T)=0          1

        A          0         A         B
        B          1         B         A
```

```
       STATEMIN
MOORE OR MEALY MACHINE?
MEALY
HOW MANY INTERNAL STATES?
⎕:
       5
HOW MANY INPUT STATES?
⎕:
       2
ENTER THE STATE TABLE ONE ROW AT A TIME.
⎕:
       1 0 0 2 5
⎕:
       2 0 0 5 4
⎕:
       3 1 0 4 1
⎕:
       4 1 0 3 5
⎕:
       5 0 0 2 4


IF THE ORIGINAL MACHINE IS:

      Q(T)          Z(T)                 Q(T+1)
           X(T)=0          1          0          1

        1          0          0          2          5
        2          0          0          5          4
        3          1          0          4          1
        4          1          0          3          5
        5          0          0          2          4
```

THE PARTITION IS:

1
2 5
3
4

THE EQUIVALENT MINIMUM FORM IS:

Q(T)	Z(T) X(T)=0	1	Q(T+1) 0	1
A	0	0	B	B
B	0	0	B	D
C	1	0	D	A
D	1	0	C	B

IF THE ORIGINAL MACHINE IS:

Q(T)	Z(T) X(T)=0	1	2	Q(T+1) 0	1	2
1	1	0	0	2	2	5
2	0	1	1	1	4	4
3	1	0	0	2	2	5
4	0	1	1	3	2	2
5	1	0	0	6	4	3
6	0	1	1	8	9	6
7	1	0	0	6	2	8
8	1	0	0	4	4	7
9	0	1	1	7	9	7

THE PARTITION IS:

1 3 8
2 4
5 7
6
9

THE EQUIVALENT MINIMUM FORM IS:

Q(T)	Z(T) X(T)=0	1	2	Q(T+1) 0	1	2
A	1	0	0	B	B	C
B	0	1	1	A	B	B
C	1	0	0	D	B	A
D	0	1	1	A	E	D
E	0	1	1	C	E	C

MEMORY SPAN DETERMINATION OF FINITE-STATE MACHINES

The following APL program *MEMORYSPAN* determines the memory span of a given completely specified Mealy or Moore machine.*) The program is based, essentially, on ALGORITHM 8.1 defined in Section 8.6.

Similarly as the previous program, this program requires the subroutine *DFT* which controls the format of the printed ST-tables.

The input is exactly the same as in the program *STATEMIN*. The output consists of the ST-table properly spaced and with headings, and the specification of the memory span.

```
     ∇ MEMORYSPAN;I;N;TYPE;PAIRS1;PAIRS2;PAIRS;ADD;ADD1;COMP;CLASS;RHO;
       J;HEAD1;HEAD2;EX;SP;R;LINE;IT
[1]    'MOORE OR MEALY MACHINE?'
[2]    TYPE←⍞
[3]    'HOW MANY INTERNAL STATES?'
[4]    TABLE←⍳0×N←⎕×I←1
[5]    'HOW MANY INPUT STATES?'
[6]    X←0,⍳⎕-1
[7]    'ENTER THE STATE TABLE ONE ROW AT A TIME.'
[8]    TABLE←TABLE,⎕
[9]    →(N≥I←I+1)/¯1+⍳26
[10]   PAIRS1←PAIRS2←⍳I←0
[11]   PAIRS2←PAIRS2,I+⍳N-I+0×⍴PAIRS1←PAIRS1,(N-I)⍴I+1
[12]   →(N>I←I+1)/¯1+⍳26
[13]   INDEX←+/((J,2)⍴ 100 1)×PAIRS←⍉(2,J←+/⍳N)⍴PAIRS1,PAIRS2
[14]   →(∧/TYPE='MOORE')/MOORE
[15]  MEALY:I←1+0×⍴⍴TABLE←(N,1+2×⍴X)⍴TABLE
[16]   PAIRS←PAIRS,[2]((⍴PAIRS[;⍳2])⍴TABLE[,PAIRS[;⍳2];I+1+⍴X])
[17]   →((⍴X)≥I←I+1)/¯1+⍳26
[18]   →MATRIX
[19]  MOORE:I←1+0×⍴⍴TABLE←(N,2+⍴X)⍴TABLE
[20]   PAIRS←PAIRS,[2]((⍴PAIRS[;⍳2])⍴TABLE[,PAIRS[;⍳2];2+I])
[21]   →((⍴X)≥I←I+1)/¯1+⍳26
[22]  MATRIX:MX←(J,J)⍴0×I←1
[23]   COMP←⍳0×J←2
[24]   COMP←COMP,INDEX⍳+/ 100 1 ×PAIRS1[⍋PAIRS1←PAIRS[I;J+ 1 2]]
[25]   →((⍴PAIRS[1;])>J←J+2)/¯1+⍳26
[26]   MX[I;COMP]←=/[1] TABLE[PAIRS[I; 1 2];2+X×5=+/TYPE='MEALY']
[27]   →((⍴MX[;1])≥I←I+1)/¯4+⍳26
[28]   MX←COMP/[2] MX←(COMP←≠/PAIRS[; 1 2])/[1] MX
[29]   I←¯1
[30]   MX←COMP/[2] MX←(COMP←0≠+/MX)/[1] MX
[31]   I←I+1
[32]   →(0≠+/~COMP)/¯2+⍳26
[33]   'IF THE FINITE STATE MACHINE MODEL IS:
       '
[34]   →(∧/TYPE='MEALY')/MEALYOUT
```

*) The program was prepared by Mr. Leighton B. Brown at the School of Advanced Technology, SUNY at Binghamton.

```
[35]  IT←8-N←⌈/((ρX)> 0 13 15 18 22 28)/0,⍳5
[36]  ⎕←HEAD1←((5-N)ρ' '),'Q(T)',((5-N)ρ' '),'Z(T)',((⌊3+
      0.5×IT×¯1+ρX)ρ' '),'Q(T+1)'
[37]  ⎕←HEAD2←((18-3×N)ρ' '),'X(T)=',(7-N)↓EX←(IT,0) DFT X
[38]  ' '
[39]  →TABLEOUT
[40] MEALYOUT:IT←8-N←⌈/((ρX)> 0 7 8 9 11 14)/0,⍳5
[41]  ⎕←HEAD1←((5-N)ρ' '),'Q(T)',((⌊0.5×IT+IT×¯1+ρX)ρ' '),'Z(T)',((⌊
      IT×0.5)IT×ρX)ρ' '),'Q(T+1)'
[42]  ⎕←HEAD2←((10-2×N)ρ' '),'X(T)=',(7-N)↓EX,EX←(IT,0) DFT X
[43]  ' '
[44] TABLEOUT:(IT,0) DFT TABLE
[45]  →(0 0 =ρMX)/3+⍳26
[46]  '

      THEN THE MEMORY SPAN OF THE CORRESPONDING FINITE
      MEMORY MODEL IS INFINITE.'
[47]  →0
[48]  '

      THEN THE MEMORY SPAN OF THE CORRESPONDING FINITE
      MEMORY MODEL IS ':I;'.'
```

```
      MEMORYSPAN
MOORE OR MEALY MACHINE?
MEALY
HOW MANY INTERNAL STATES?
⎕:
      5
HOW MANY INPUT STATES?
⎕:
      2
ENTER THE STATE TABLE ONE ROW AT A TIME.
⎕:
      1 0 0 2 5
⎕:
      2 0 0 4 3
⎕:
      3 0 1 5 3
⎕:
      4 1 1 1 3
⎕:
      5 1 0 1 3



IF THE FINITE STATE MACHINE MODEL IS:

    Q(T)          Z(T)               Q(T+1)
         X(T)=0          1          0          1

      1         0          0          2          5
      2         0          0          4          3
      3         0          1          5          3
      4         1          1          1          3
      5         1          0          1          3


THEN THE MEMORY SPAN OF THE CORRESPONDING FINITE
MEMORY MODEL IS 2.
```

```
      MEMORYSPAN
MOORE OR MEALY MACHINE?
MEALY
HOW MANY INTERNAL STATES?
⎕:
      9
HOW MANY INPUT STATES?
⎕:
      2
ENTER THE STATE TABLE ONE ROW AT A TIME.
⎕:
      1 0 0 5 4
⎕:
      2 0 1 3 3
⎕:
      3 0 0 7 9
⎕:
      4  1 1 3 3
⎕:
      5 0 0 3 3
⎕:
      6 0 0 2 5
⎕:
      7 0 0 8 6
⎕:
      8 0 0 5 5
⎕:
      9 0 0 1 8

IF THE FINITE STATE MACHINE MODEL IS:

    Q(T)          Z(T)                  Q(T+1)
          X(T)=0        1          0          1

     1        0         0          5          4
     2        0         1          3          3
     3        0         0          7          9
     4        1         1          3          3
     5        0         0          3          3
     6        0         0          2          5
     7        0         0          8          6
     8        0         0          5          5
     9        0         0          1          8

THEN THE MEMORY SPAN OF THE CORRESPONDING FINITE
MEMORY MODEL IS INFINITE.
```

```
      MEMORYSPAN
MOORE OR MEALY MACHINE?
MOORE
HOW MANY INTERNAL STATES?
⎕:
      2
HOW MANY INPUT STATES?
⎕:
      2
```

```
ENTER THE STATE TABLE ONE ROW AT A TIME.
⎕:
      1 0 1 2
⎕:
      2 1 2 1




IF THE FINITE STATE MACHINE MODEL IS:

      Q(T)      Z(T)        Q(T+1)
                       X(T)=0       1

       1         0          1       2
       2         1          2       1

THEN THE MEMORY SPAN OF THE CORRESPONDING FINITE
MEMORY MODEL IS 1.
```

STATE MINIMIZATION OF PROBABILISTIC MACHINES

The following APL program *PSTMIN* determines a minimal form of a given probabilistic machine.*) It is based, essentially, on the procedure described in Section 10.5. The following subroutines are used:

CONVERT: A calculation of the *H* matrix.

PERMUTE: A calculation of all sequences of a specified length made up of symbols from the input or output alphabet. It is used as a subroutine in *CONVERT*.

RDC (*Red*uction): A transformation of the original machine to the minimal machine, based on the available *H* matrix.

CLEANUP: A reduction of computational errors which may arise during the procedure.

The program requires two arguments: *O* and *M*. *O* stands for the number of output symbols (responses) of the machine, *M* represents the probability matrix as specified in Section 10.5 (Figure 10.19).

Before the program can be run, variable *M* must be defined as a stochastic matrix. Then, the program is executed by printing *O PSTMIN M*, where *O*

*) The program was prepared by Mr. Sirajul Islam, School of Advanced Technology, SUNY at Binghamton.

stands for the number of output symbols. Thus, if the machine has 4 output symbols, we must print 4 *PSTMIN M*. If, by mistake, an improper matrix *M* is defined, the program prints the message "*THE INPUT MATRIX IS NOT A STOCHASTIX MATRIX*". Otherwise, the program prints the original matrix *M*, the *H* matrix, and either a minimal form of the machine or the message "*THE GIVEN MACHINE IS IN A MINIMAL FORM*".

```
        ∇PSTMIN[⎕]∇

     ∇ O PSTMIN M;H;X
[1]    →(~∧/(,M≥0),1=+/M)/L1
[2]    H←O CONVERT M
[3]    'THE H MATRIX IS ';H
[4]    X←M RDC H
[5]    →(∧/(ρX)=ρM)/L2
[6]    'A MINIMAL FORM OF THE MACHINE IS';X
[7]    →0
[8]   L2:'THE GIVEN MACHINE IS IN A MINIMAL FORM'
[9]    →0
[10]  L1:'THE INPUT MATRIX IS NOT A STOCHASTIC MATRIX'
     ∇
```

```
        ∇CLEANUP[⎕]∇

     ∇ Z←CLEANUP A;II;JJ;N;M
[1]    M←(ρA)[1]
[2]    N←(ρA)[2]
[3]   TS1:II←0
[4]   TS1I:→(M<II←II+1)/OWT
[5]    JJ←0
[6]   TS1J:→(N<JJ←JJ+1)/TS1I
[7]    →((|A[II;JJ])>10*¯10)/TS1J
[8]    A[II;JJ]←0
[9]    →TS1J
[10]  OWT:Z←A
     ∇
```

```
        ∇PERMUTE[⎕]∇

     ∇ B←X PERMUTE A;E;I;D
[1]    B←((E←(D←ρA)*X),X)ρI←0
[2]   L1:B[;I+1]←Eρ,⍉((D*X-I+1),D)ρA
[3]    →((X-1)≥I←I+1)/L1
     ∇
```

```
        ∇CONVERT[⎕]∇

     ∇ H←O CONVERT M;K1;K2;K3;K4;H1;B1;B2;B3
[1]    Q←(ρM)[2]÷O
[2]    I←(ρM)[1]÷Q
```

```
[3]    H←H1←(Q,1)⍴K1←1
[4]   L6:B1←K1 PERMUTE ¯1+⍳O
[5]    B2←K1 PERMUTE ¯1+⍳I
[6]    K2←1
[7]   L1:K3←1
[8]   L5:B3←B2[K2;],B1[K3;]
[9]    K4←1
[10]  L2:H1←M[((Q×B3[K4])+⍳Q);(Q×B3[K4+K1])+⍳Q]+.×H1
[11]   →(K1≥K4←K4+1)/L2
[12]   →(=/,H1)/L3
[13]   →(0=~LDPNDNCY⍉H,H1)/L3
[14]   H←H,H1
[15]  L3:H1←H[;,1]
[16]   →((O*K1)≥K3←K3+1)/L5
[17]   →((I*K1)≥K2←K2+1)/L1
[18]   →(Q>K1←K1+1)/L6
     ∇
```

```
                 ∇RDC[⎕]∇

     ∇ X←M RDC H;K1;B5;A;I;A1;A2;B1;J;A5;L;P1;P2;A3;B;B2;K
[1]    K1←0
[2]    B5←⍳0
[3]    A←(⍴H)⍴2
[4]    I←2
[5]   L3:→(1≠+/H[;I]=A1←⌊/H[;I])/L1
[6]    A[A2;]←H[(A2←H[;I]⍳A1);]
[7]   L1:→(1≠+/H[;I]=A1←⌈/H[;I])/L2
[8]    A[A2;]←H[(A2←H[;I]⍳A1);]
[9]   L2:→((⍴H)[2]≥I←I+1)/L3
[10]   B←A[(B1←(~2∊A[;1])/⍳(⍴H)[1]);]
[11]   I←1
[12]  L6:→(I∊B1)/L7
[13]   →(0=B2←LDPNDNCY B,[1] H[I;])/L5
[14]   B5←B5,I
[15]   →L7
[16]  L5:B←B,[1] H[I;]
[17]  L7:→((⍴H)[1]≥I←I+1)/L6
[18]   →(0=⍴B5)/L9
[19]   I←1
[20]  S1:J←B5[I]
[21]   A5←H[J;]⌹XXX←⍉B[(⍳A3);⍳A3←⌊/⍴B]
[22]   →((⍴A5)=+/A5≥1E¯17)/S2
[23]  S3:→((⍴B5)≥I←I+1)/S1
[24]   →D3
[25]  S2:K←1
[26]  S5:K1←K1+1
[27]   M[;K]←M[;K]+A5[1+(⍴A5)|1+K1]×M[;J×⌈K÷Q]
[28]  D5:K←K+1
[29]   →(0=J|K)/D5
[30]   →(K≤(⍴M)[2])/S5
[31]   L←1
[32]   K←J
[33]  D1:M[K;]←2
[34]   L←L+1
[35]   →((⍴M)[1]≥K←L×J)/D1
[36]   L←1
[37]   K←J
```

```
[38] D2:M[;K]←2
[39]  L←L+1
[40]  →((ρM)[2]≥K←L×J)/D2
[41]  →S3
[42] D3:P1←+/M[;1]≥2
[43]  P2←+/M[1;]≥2
[44]  M←((ρM)-P1,P2)ρ(2>,M)/,M
[45] L9:X←CLEANUP M
    ∇
```

```
        THE GIVEN MACHINE IS

   0.10      0.30      0.2      0.10      0.10      0.2
   0.00      0.10      0.2      0.20      0.10      0.4
   0.05      0.20      0.2      0.15      0.10      0.3
   0.20      0.10      0.2      0.00      0.10      0.4
   0.20      0.00      0.4      0.20      0.00      0.2
   0.20      0.05      0.3      0.10      0.05      0.3


      2 PSTMIN M

THE H MATRIX IS
   1              0.6
   1              0.3
   1              0.45
THE MINIMAL MACHINE IS
  0.2             0.4             0.2             0.2
  0.1             0.2             0.4             0.3
  0.3             0.2             0.2             0.3
  0.4             0.2             0.3             0.1


      THE GIVEN MACHINE IS

    0.3       0.1       0.2       0.4
    0.6       0.2       0.1       0.1
    0.5       0.2       0.0       0.3
    0.7       0.1       0.2       0.0


      2 PSTMIN M2

THE H MATRIX IS
   1              0.4
   1              0.8
THE GIVEN MACHINE IS IN A MINIMAL FORM


        THE GIVEN MACHINE IS

   1.00   1.00   1.00   1.00   1.00   1.00   1.00   1.00
   1.00   1.00   1.00   1.00   1.00   1.00   1.00   1.00
   1.00   1.00   1.00   1.00   1.00   1.00   1.00   1.00
   1.00   1.00   1.00   1.00   1.00   1.00   1.00   1.00


      2 PSTMIN   M3
THE INPUT MATRIX IS NOT A STOCHASTIC MATRIX
```

```
        THE GIVEN MACHINE IS

    0.2         0.2         0.2         0.4         0           0
    0.2         0.2         0.2         0.1         0.3         0
    0.2         0.2         0.2         0.2         0.1         0.1
    1           0           0           0           0           0
    0           0           0.6         0.4         0           0
    0.6         0.1         0.1         0.2         0           0

      2 PSTMIN M1

THE H MATRIX IS
    1           [illegible]
    1           0.6
    1           0.8
THE MINIMAL MACHINE IS
    0.3         0.3         0.4         0
    0.3         0.3         0.1         0.3
    1           0           0           0
    0.3         0.3         0.4         0
```

INDEX

Abhyankar, S., 122
Absorption, law of, 66
Absorption of negation, law of, 68
Abstract algebra, 65
Activity, 5
  matrix, 362
Adder, serial binary, 307, 308
Adjacent
  state, 23
  vertex, 90, 91
Adjoint state, 277
Aggressivity of 0 and 1, law of, 68
Algebra
  abstract, 65
  Boolean, 66, 85, 259
  explicit definition, 65
  implicit definition, 65
  logic, 65
    proper, 65
  theory of, 65
Amplifier, pneumatic, 192, 193
Analysis, 13, 199–213, 443
  harmonic, 513
Analyzer, Boolean, 162, 259
AND function, 33, 34
Applicable input sequence, 283, 310
Aranovich, B. I., 259
Architecture, systems, 216
Arnold, R. F., 260
Ashenhurst, R. L., 259
Assertion, 32, 34
Associative
  function, 45
  law, 68
Asynchronous switching circuit, 19, 406, 407, 455
At least *m* out of *n* function, 35
At most *m* out of *n* function, 35
Atomic proposition, 16
Autonomous transition, 407

Bakerdjian, V., 403
Bartky, W. S., 455
Behavior, 5, 8, 10, 11, 199, 321
  logic, 166
  physical, 166
Bell, N. R., 162
Biconditional, 34
Binary functional identifier, 31
Bi-stable logic element, 175, 176
Bit, 16, 17
Black box, 13, 14
  problem, 14
Block diagram, 38
Boda, M., 19
Boole, G., 66, 85, 162
Boolean
  algebra, 66, 85, 259
  analyzer, 162, 259
  expression, 67
    absolute minimal, 116
  form
    minimal, 87, 101
    psp, 118, 120
    sps, 118, 120
  function, 69
  matrix, 207–213, 259, 362
    connection, 209
    normal, 207
    output, 209
    primitive connection, 209
    reduced connection, 210
    symmetric, 207
Booth, T. L., 493
Born, R. C., 260
Bounce corrector, 453, 454
Bound vertex, 89

Branch-type
  element, 195
  switching circuits, 201, 239–249
Break contact, 239
Bridge, 206
  switching circuits, 205, 206
Brzozowski, J. A., 324
Burks, A. W., 259
BUT-NOT function, 34

Carlyle, J. W., 486, 493
Carlyle theorem, 486
Cascade
  decomposition, 374
  element, 225, 230
  method, 225–230, 237
  switching circuit, 225–230, 238
Catalog of WOS module, 517, 518
Cell of a cube, 25, 88
Cellular logic, 507, 508
Chart
  Karnaugh, 26, 27, 29, 84
  logic, 25, 71
  Marquand, 25–30
  operation, 80
  Svoboda, 84
  ternary, 101
Class
  column, 298
    excluded, 299
  compatible, 285
    maximal, 291, 418
    prime, 290, 297
  final, 285, 289, 290, 297
  of variables, 418
  output-equivalence, 311
  state-equivalence, 311
Clock, 330
  pulses, 408
Closed partition, 374–377, 393
Closure
  condition, 285
  law of, 66
Code, 17
  Gray, 444, 448, 455
Collections of papers in switching theory, 505, 506
Column
  class, 298
  class, excluded, 299
  distinguished, 107
Combinational switching circuits, 19
Combined model
  first order, 348
  *k*th order, 336, 370
Commutative
  function, 45
  laws, 66
Compatible
  class, 285
    maximal, 291, 418
    of variables, 418
    prime, 290, 297
  input states, 296
  states, 284, 286
  variables, 417
Compatibility table, 288
Complement, 34
Completely specified machine, 272
Complete set of logic functions, 53, 54, 60–62
  minimal, 53, 62, 63
Complete *z*-variable partition, 413
Composition, 37
Conditional, 34
Conjunction, 34
Connection, extensional, 17
Connection matrix, Boolean, 209
Contact
  break, 239
  make, 239
  network, 239–242
  transfer, 239
  tree, 241–245
Continuous system, 11
Contracted solutions, 139
Convex combination, 489
Cost, total, 157
Counter
  modulo 2, 334
  modulo 4, asynchronous, 444, 445
  modulo 4, reverse, 446–448
Coupling, 8
Cover
  preserved, 285, 289
    minimal, 291, 297
Covering
  condition, 285
  minimal, 106, 155
  problem, 156
Covering and closure table, 293, 300, 301
Criterion, objective, 87, 215, 217

Critical race condition, 412, 413
Cryotrons, 186–188
Cube, unit, 23, 24
Curtis, H. A., 259
Cycle, Hamiltonian, 444

Decomposition, 154–156, 259
  cascade, 374
  of machines, 374, 394, 403, 443, 455
  parallel, 374, 380, 382, 383
  serial, 374, 377–380
Definite logic function, 31
Degenerate logic functions, 49, 50
Delay element, 176, 177, 332, 334, 363, 431, 432, 439
  inverted, 334
DeMorgan laws, 68
Dependent logic variable, 31, 127
Derivative of regular expression, 316
Determinant, 209
Deterministic system, 11
  discrete, 267
Diagnostics, 508, 509
Diode elements, 170–172
Direct transition, 11
Discrete
  system, 11
    deterministic, 267
  time, 406, 407
Discriminant, 130–132, 139, 151, 152
  partial, 132, 133
Disjoint state, 277
Disjunction, 34
Distance, Hamming, 22–24
Distinguishable states, 274, 485
Distinguished column, 107
Distributive law, 68
Domain of logic function, 31
Dominated prime implicant, 107
Don't care state, 35, 71, 104
  conditional, 141
D-trio, 435
Double negation, law of, 68
Dual expression, 69
Duality principle, 69
Dunham, B., 260
Dynamic hazard, 427

Electronic tube elements, 172–174
Element
  branch-type, 195
  cascade, 225, 230
  delay, 176, 177, 332, 334, 363, 431, 432, 439
    inverted, 334
  diode, 170–172
  electronic tube, 172–174
  fluidic, 189, 192, 194, 195
  gate, 195
  magnetic, 183–186
  memory, 19, 332–334, 408
  of a system, 8
  transistor, 179–182
Elementary information, 16
Environment, 3, 8
Equation
  logic, 126–138
  standard, 127, 128
Equivalence, 34
Equivalent
  machines, 274, 485
  states, 274, 485
Error correction, 510
Essential hazard, 431, 432
Excluded middle, law of, 66, 68
Exclusive OR function, 34
Experiment, gedanken, 508
Explicit definition of algebra, 65
Expression
  absolute minimal, 116
  Boolean, 67
  dual, 69
  logic, 37–43
    proper, 38
  regular, 311, 312, 315, 324, 325
Extensional connections, 17
Extremal rows, 491

Falsum, 34
Feed-forward switching circuits, 200
Final class, 285, 289, 290, 297
Finite-memory model, 336, 345, 370, 403
Finite-state model
  first order, 340
  *k*th order, 336, 342, 369, 402
  second order, 343
Flip-flop
  JK, 334
  SR, 333
Fluidic elements, 189, 192, 194, 195
Folding, 247–249

Formal divisor, 102, 103
Formally divisible term, 102, 103
Forslund, D. C., 260
Fourier transform for logic functions, 513
Free vertex, 89
Fuchs, J., 324
Function
  associative, 45
  at least *m* out of *n*, 35
  at most *m* out of *n*, 35
  Boolean, 69
  commutative, 45
  last-output, 273
  linear, 54, 55, 57
  majority, 35
  minority, 35
  monotonic, 56, 57
  m out of *n*, 35
  mutually dual, 54
  NAND, 33, 34
  NEITHER-NOR, 34
  N-equivalent, 52
  next state, 265
  nondegenerate, 49, 50
  NOR, 33, 34
  NOT, 34
  NOT-BOTH, 34
  objective, 87, 88, 215, 298, 429
  ONE, 32, 34, 54, 57
  OR, 33, 34
  output, 265
  output-sequence, 272
  P-equivalent, 51
  Pierce, 34
  PN-equivalent, 52
  primary, 165
  pseudo-logic, 143, 149
  self-dual, 54, 57
  terminal state, 273
  threshold, 35, 175
  transition, 265
  ZERO, 32, 34, 54, 57
Functional
  generator, 265, 327, 329–332, 341, 349, 354, 363, 368, 443, 461, 463, 465, 471, 477
  identifier, binary, 31
  identifier, ternary, 32
Fuzzy
  logic, 513
  set, 513

Gate elements, 195
Gate-type switching circuits, 201
Gavrilov, M. A., 122
Gedanken experiments, 508
General Boolean representation of state assignment, 390–393, 396
Generator
  functional, 265, 327, 329–332, 341, 349, 354, 363, 368, 443, 461, 463, 465, 471, 477
  random, 19, 461, 469, 471, 477, 563–566
Gilbert, E. N., 455
Gill, A., 324, 403, 508, 510
Gimpel, J. F., 122
Ginsburg, S., 324
Givone, D. D., 162
Glossary of symbols, 514–516
Grasseli, A., 324
Gray code, 444, 448, 455
Gray, J. N., 324
Greniewski, H., 19
Grids, Svoboda, 76–80
Group minimization, 113
Gusev, L. A., 324

Hamiltonian cycle, 444
Hammer, P. L., 162
Hamming, R. W., 84
Hamming distance, 22–24
Haring, D. R., 403
Harmonic analysis for switching circuits, 513
Harrison, M. A., 85, 260, 324, 510
Hartmanis, J., 403
Havel, I. M., 325
Hawkins, J. K., 196
Hazard, 408, 426–428, 431, 443, 455
  dynamic, 427
  essential, 431, 432
  static 1, 426
  static 0, 427
Hazard-free switching circuit, 428–431
Hellerman, L., 260
Hennie, F. C., III, 507
Hierarchy of functions, 38
  in Boolean algebra, 68
H-matrix, 487–491
H_k-matrix, 487
Hohn, F. E., 259
Huffman, D. A., 324, 455, 510
Huntington, E. V., 67, 85

Identifier
  functional, binary, 31
  functional, ternary, 32
  state, 22
  term, 101, 102
  time, 362
  variable, 362
Identity, 34
Implicant, 100
  essential prime, 106
  prime, 100, 103, 429–431
Implication, 34
Implicit definition of algebra, 65
Incompatible states, 284
Incompletely specified machine, 272
Incomplete z-variable partition, 413
Indefinite logic function, 31, 35, 71
Independent logic variable, 31, 127
Indirect transition, 11
Inequality, 34
  proper, 34
Information, elementary, 16
Inhibition, 34
Initial-state probability distribution, 482, 485
  distinguishable, 485
  equivalent, 485, 488, 489
  k-distinguishable, 485
  k-equivalent, 485
Initial state, 272
Input, partial, 7
Input logic variable, 31
Input/output
  pair, 269
  sequence, 271
  state assignment, 217
Input sequences
  applicable, 283, 310
  output equivalent, 311
  state equivalent, 311
Input state, compatible, 296
Instantaneous state, 12
Internal state, 11, 265, 407
  minimization, 443
Intrinsic memory span, 355, 358, 359
Inverted trigger, 334
Invertion, 34
Irredundant form, 101
Isomorphic machines, 280, 281

Jerome, E. J., 324, 403
JK flip-flop, 335

Karnaugh chart, 26, 27, 29, 84
Karnaugh, M., 84
Kautz, W. H., 510
Kazakov, V. D., 122
k-compatible
  class, 285
  states, 284
k-distinguishable states, 275, 485
Kempisty, M., 19
k-equivalent states, 275, 485
k-incompatible states, 284
King, W. F., 260
Kleene, S. C., 324
Klir, G. J., 85, 260, 402, 403, 455
Kloss, B. M., 511
Kochen, M., 509
Kudielka, V., 85

Last-output function, 273
Law
  associative, 68
  commutative, 66, 68
  DeMorgan, 68
  distributive, 68
  of absorption, 66
  of absorption of negation, 68
  aggressivity of 0 and 1, 68
  of closure, 66
  of double negation, 68
  of neutrality of 0 and 1, 68
  of the excluded middle, 66, 68
Lawler, E. L., 122, 162
Lechner, R. J., 513
Ledley, R. S., 162
Lee, C. Y., 84
Linear function, 54, 55, 57
Linear programming, 122
Linear sequential switching circuits, 510
Linear space, 487
  dimension of, 487
Liu, C. L., 402
Lo, A. W., 196
Löfgren, L., 195, 196
Logic
  algebra, 65
    proper, 65
  behavior, 166
  chart, 25, 71
    Marquand, 25–30

Karnaugh, 26, 27, 29, 84
cellular, 507, 508
circuit, 16
element, 165
bi-stable, 175, 176
equations, 126–138
expression, 37–43
proper, 38
function
complete set of, 53, 54, 60–62
definite, 31
degenerate, 49, 50
domain of, 31
indefinite, 31, 35, 71
minimal complete set of, 53, 62, 63
function symmetric, 231–238
map, 25
operation, 31
operator, 33
primitive, universal, 252–256, 260
product, 33, 34
relations, 126
space, 23
sum, 34, 72
variable, 21
dependent, 31, 127
independent, 31, 127
input, 31
output, 31
Lowenschuss, O., 260
Luccio, F., 324
Lukasiewicz, J., 46, 84
Lunts, A. G., 259

Machine
completely specified, 272
decomposition, 374, 394, 403, 443, 455
equivalent, 274, 485
inclusion, 303
incompletely specified, 272
isomorphic, 280, 281
Mealy, 265, 266, 319, 402, 407, 409–411, 442
fundamental mode, 409–411
normal asynchronous, 411
probabilistic, 471, 473, 475
Moore, 265, 266, 319, 402, 407, 442
probabilistic, 471, 474, 475
Shannon, 471–473, 475, 478, 481
Macro-design, 215, 216
Magnetic logic elements, 183–186
Maitra, K. K., 162
Majority function, 35
Make contact, 239
Maki, G. K., 455
Map, logic, 25
Marcus, M. P., 260
Marin, M. A., 162, 259, 403
Marinos, P. N., 513
Marquand, A., 84
Marquand chart, 25–30, 72, 90, 364
Mask, 362–368, 477
depth, 363
Matrix
activity, 362
Boolean, 207–213, 259
connection, 209
normal, 207
output, 209
primitive connection, 209
reduced connection, 210
Boolean symmetric, 207
stochastic, 11, 12
Maximal compatible class, 291, 418
Maxterm, 72
Mazer, L., 260
McCluskey, E. J., 122, 403, 455
Mealy, G. H., 324
Mealy machine, 265, 266, 319, 402, 407, 409–411, 442
fundamental mode, 409–411
normal asynchronous, 411
probabilistic, 471, 473, 475
Memory, 265, 327, 332, 368, 471
element, 332–334, 408
structure, 364
Memoryless switching circuits, 19, 154
Memory span, intrinsic, 355, 358, 359
Menger, K. S., 85, 513
Meo, A. R., 260
Merging, 246
Method
cascade, 225–230, 237
Petrick, 111, 112
Quine-McCluskey, 107, 111–113
Svoboda, 88–99
Micro-design, 215, 216
Minimal Boolean form, 87, 101
Minimal covering, 106, 155
Minimal-state representation, 486
Minimization
group, 113

state, 267, 274, 276, 283, 443
state-input, 296
Minimum transition state assignment, 413, 419, 420, 455
Minor, 209
Minority function, 35
Minterm, 72
Model
combined, *k*th order, 336
evaluation, 443
combined, first order, 348
finite-memory, 336, 345, 370, 403
finite-state, first order, 340
finite-state, *k*th order, 336
finite-state, second order, 343
Modified circuit, 436–439, 441
Modular partitioning, 512
Module, 216, 508, 512
NAND, 217
universal, 252, 256, 260
WOS, 216, 217
Molecular proposition, 17, 44
Monotonic function, 56, 57
Moore, E. F., 324, 508, 509
m out of *n* function, 35
Mueller, R. K., 122
Mukhopadhyay, A., 260
Muller circuits, 411
Muller, D. E., 260, 455
Mutually dual functions, 54

Nadler, M., 84
Nakashima, A., 19
NAND
function, 33, 34
module, 216, 508, 512
Negation, 32, 34
NEITHER-NOR function, 34
N-equivalent functions, 52
Nethiporuk, E. I., 511
Network
contact, 239–242
symmetric, 245–250
Neumann, J. von, 509
Neutrality of 0 and 1, law of, 68
Neutral relay, 169
Next state, 265
function, 265
Nieh, T. T., 493
Noncritical race condition, 412
Nondegenerate logic functions, 49, 50
Nonequivalence, 33, 34
NOR function, 33, 34
Normal Boolean matrix, 207
Notation, Polish, 46–48, 84
NOT-BOTH function, 34
NOT function, 34
Number representation, polynomial, 497–499

Object, 3
Objective
criterion, 87, 215, 217
function, 87, 88, 215, 298, 429
pseudo-logic function, 157
Occupied vertex, 88
Oliva, P., 85
ONE function, 32, 34
Operation
chart, 80
logic, 31
star, 311
Operator, 168
logic, 33
OR function, 33, 34
Organization, 7, 8, 12
Output
Boolean matrix, 209
equivalence class, 311
function, 265
partial, 7
Output-consistent partition, 396, 397
Output equivalent input sequences, 311
Output logic variable, 31
Output-sequence function, 272

Pair, input-output, 265
Pairs table, 278, 279
Pankajam, S., 260
Paradigm of
probabilistic memoryless circuits, 461
probabilistic sequential circuits, 471, 472
Parallel decomposition, 374, 380, 382, 383
Parentheses, 37
Partial
discriminant, 132, 133
input, 7
output, 7
stimulus, 8
Partition
closed, 374–377, 393
essential, 416

output-consistent, 396, 397
stable, 421
z-variable, 413
Partition pair, 398, 399
Partitioning, modular, 512
Patt, Y. N., 260
Paul, M. C., 324
Paz, A., 493
P-equivalent logic functions, 51
Perfect induction, proof by, 43
Periodicals in switching theory, 503, 505
Peterson, W. W., 510
Petrick function, 122
Petrick method, 111, 112
Physical behavior, 166
Pierce function, 34
Pierce, W. H., 509
PN-equivalence classes of logic functions of n variables ($n \leq 4$), 519–529
PN-equivalent logic functions, 52
Pneumatic
amplifier, 192, 193
logic elements, 189
relay, 189–191
PNN-equivalence relation, 52
Polarized relay, 166
Polish notation, 46–48, 84
Polynomial representation of numbers, 497–499
Pospelov, D. A., 492
Post, E. L., 85
Povarov, G. N., 85, 233, 260
Prather, R. E., 260
Precedence rule, 38
Preparata, F. P., 260
Preserved cover, 285, 289
minimal, 291, 297
Primary function, 165
Prime compatible class, 290, 297
Prime implicant, 100, 103, 429–431
dominated, 107
essential, 106
ps, 119
table, 106, 115
table, reduced, 107
Primitive, universal logic, 252–256, 260
Primitive connection matrix, Boolean, 209
Principle of duality, 69
Prior, A. N., 84
Probabilistic switching circuit, 19, 458
memoryless, 459–462, 492
sequential, 471
Probabilistic system, 11
Probability transformer, 461, 493
Product, logic, 33, 34
Program, 7, 8, 12
Proof by perfect induction, 43
Proper inequality, 34
Proposition
atomic, 16
molecular, 17, 44
Propositional connectives
extensional, 44
intensional, 44
Pseudo-Boolean relation, 88
Pseudo-logic function, 143, 149, 150, 152, 157
Ps standard form, expanded, 72
Psp Boolean form, 118, 120
Ps-prime implicant, 119
PST-diagram, 357, 358
PST-matrix, 358
PST-structure, 356
P-term, 72

Quantity, 3
ideal value of, 15
input, 7
output, 7
perfect value of, 15, 16
value of, 3
Quasi-equivalence, 303
Quine, W. V., 121
Quine-McCluskey method, 107, 111–113

Rabin, M. O., 493
Race condition, 412
critical, 412, 413
noncritical, 412
Random generator, 19, 461, 463–466, 469, 471, 477
Range of logic function, 31
Ready signal, 411
Reduced connection Boolean matrix, 210
Redundant logic, 509, 510
Reference state, 272
Regular expression, 311, 312, 315, 324, 325
derivative, 316
vector, 321
Relation
logic, 126
PNN-equivalence, 52

pseudo-Boolean, 88
time-invariant, 5, 364
Relay
combination lock, 454
neutral, 169
pneumatic, 189–191
polarized, 166
Reliability level, 158
Rescher, N., 511
Resolution level, 3–5
Response, 7, 265
partial, 8
time, 407
Richards, R. K., 196
Roginskij, V. M., 260
Roth, J. P., 122
Row class, 298
excluded, 299
prime, 300
Rudeanu, S., 162

Salomaa, A., 325, 493
Sample of activity, 363, 477
Sampling pair, 362
Sampling time, 329, 407
Sampling variable, 362–364
Schissler, L. R., 259
Schmitz, G., 260
Schoeffler, J. D., 19, 162
Scidmore, A. K., 260
Seidl, L. K., 260, 455
Self-dual function, 54, 57
Semon, W., 257
Serial decomposition, 374, 377–380
Series-parallel switching circuit, 205, 206
Sequence
input-output, 271
length, 271
Sequential
discrete system, 335, 336
switching circuit, 19, 264, 329
Serial binary adder, 307, 308
Set, fuzzy, 513
Shannon, C. E., 19, 259, 260, 492, 509
Shannon machine, 471–473, 475, 478, 481
Sheffer, H. M., 85
Sheffer stroke, 34
Short, R. A., 259
Signal, 17
Singleton, 277
Slepian, D., 85
Smith, E. J., 403
Source
box, 436–440
Las Vegas, 458, 461, 464
MonteCarlo, 458
white noise, 458
Space
linear, 487
logic, 23
of *p*-terms, 102
Spacer, 436, 437, 439
Speed independent circuits, 411, 455
Spiro, K., 19
Sps Boolean forms, 118, 120
Sp standard form, 72
contracted, 74
SR flip-flop, 333
Stable set of internal states, 421
Stable state, 11, 409
Standard
equation, 127, 128
form, ps, 72
form, sp, 72
pseudo-Boolean form, 149
Star operation, 311
State
adjacent, 23
adjoint, 277
compatible, 284, 286
compatibility, 283
disjoint, 277
distinguishable, 274, 485
don't care, 35, 71, 104
conditional, 141
equivalent, 274, 485, 488
incompatible, 284
initial, 272
input, compatible, 296
internal, 11, 265, 407
minimization, 267, 274, 276, 283
next, 265
reference, 272
stable, 11, 409
total, 409
trap, 433, 434
unstable, 11, 409
State assignment, 385, 407, 422, 423
general Boolean representation of, 390–393, 396
input/output, 217

minimum transition time, 413, 419, 420, 455
problem, 329, 385–388, 398, 403, 408, 409, 443, 449
State-equivalence class, 311
State equivalent input sequences, 311
State identifier, 22
State-input minimization, 296
Static 0 hazard, 427
Static 1 hazard, 426
ST-diagram, 268, 269
Stearns, R. E., 403
S-term, 72
ST-matrix, 268, 269
Stimulus, 7, 265
partial, 8
Stochastic
matrix, 11, 12
switching circuit, 458
Structure, 7, 8
memory, 364
PST, 356
ST, 12, 199, 267, 304–306, 319, 340, 341, 369
UC, 8, 12, 199
UC, proper, 199
ST-structure, 12, 199, 269, 304–306, 319, 340, 341, 369
uncertainty, 346
ST-table, 268, 269
Subcube, 25
Sum, logic, 34, 72
Svoboda, A., 84, 85, 88, 101, 122, 126, 162, 259, 403, 493
Svoboda
approach, 477
chart, 84
grids, 76–80
method, 88–99
Switching circuit, 4, 15, 16
asynchronous, 19, 406, 407, 455
branch-type, 201, 239–249
bridge, 205, 206
cascade, 225–230, 238
combinational, 19
feed-forward, 200
gate-type, 201
hazard-free, 428–431
linear sequential, 510
memoryless, 19, 154
probabilistic, 19, 458
memoryless, 459–462, 492
sequential, 471
sequential, 19, 264, 329
series–parallel, 205, 206
stochastic, 458
symmetric, 231
synchronous, 19, 329, 407
two stage, 220–224
with multiple-valued variables, 511, 512
Symmetric
Boolean matrix, 207
logic function, 231–238, 260
network, 245–250
switching circuit, 231
Synchronous switching circuit, 19, 329, 407
Synthesis, 13, 214–220
abstract, 267, 304–307, 315, 324, 442
of asynchronous sequential circuits, 442, 444
of probabilistic memoryless switching circuits, 462
structure, 305, 328, 348
System, 3, 4, 8
continuous, 11
deterministic, 11
discrete, 267
discrete, 11
probabilistic, 11
Systems architecture, 216

Table
compatibility, 288
covering and closure, 293, 300, 301
pairs, 278, 279
Petrick, 122
prime implicant, 106, 115
truth, 17
Tabular representation of logic functions, 71, 75
Tal, A. A., 324
TANT circuit, 257
Term
identifier, 101, 102
sharing, 116
Terminal-state function, 273
Ternary
chart, 101
functional identifier, 32
Textbooks for switching theory, 501–504
Theorem, Carlyle, 486
Theory, of algebra, 65